Active Solar Systems

Solar Heat Technologies: Fundamentals and Applications
Charles A. Bankston, editor-in-chief

1. *History and Overview of Solar Heat Technologies*
Donald A. Beattie, editor

2. *Solar Resources*
Roland L. Hulstrom, editor

3. *Economic Analysis of Solar Thermal Energy Systems*
Ronald E. West and Frank Kreith, editors

4. *Fundamentals of Building Energy Dynamics*
Bruce Hunn, editor

5. *Solar Collectors, Energy Storage, and Materials*
Francis de Winter, editor

6. *Active Solar Systems*
George Löf, editor

7. *Passive Solar Buildings*
J. Douglas Balcomb, editor

8. *Passive Cooling*
Jeffrey Cook, editor

9. *Solar Building Architecture*
Bruce Anderson, editor

10. *Implementation of Solar Thermal Technology*
Ronal Larson and Ronald E. West, editors

Active Solar Systems

edited by George Löf

The MIT Press
Cambridge, Massachusetts
London, England

This book was set in Times Roman by Asco Trade Typesetting Ltd., Hong Kong, and was printed and bound by in the United States of America.

Library of Congress Cataloging-in-Publication Data

Active solar systems / edited by George Löf.
   p. cm.—(Solar heat technologies; 6)
 Includes bibliographical references and index.
 ISBN 0-262-12167-0
 1. Solar energy. 2. Solar heating. 3. Solar air conditioning.
I. Löf, George O. G. II. Series.
TJ809.95.S68   1988 vol. 6
[TJ810]
697'.78 s—dc20
[621.47]   92-25252
                                 CIP

Contents

Series Foreword

Charles A. Bankston

This series of volumes summarizes research, development, and implementation of solar thermal energy conversion technologies carried out under federal sponsorship during the last eleven years of the National Solar Energy Program. During the period from 1975 to 1986, the U.S. Department of Energy's Office of Solar Heat Technologies spent more than $1.1 billion on research and development, demonstration, and technology support projects, and the National Technical Information Center added more than 30,000 titles on solar heat technologies to its holdings. So much work was done in such a short period of time that little attention could be paid to the orderly review, evaluation, and archival reporting of the significant results.

In response to the concern that the results of the national program might be lost, this documentation project was conceived. It was initiated in 1982 by Frederick H. Morse, Director of the Office of Solar Heat Technologies, Department of Energy, who had served as technical coordinator of the 1972 NSF/NASA study "Solar Energy as a National Resource" that helped start the National Solar Energy Program. The purpose of the project has been to conduct a thorough, objective technical assessment of the findings of the federal program using leading experts from both the public and private sectors, and to document the most significant advances and findings. The resulting volumes are neither handbooks nor textbooks, but benchmark assessments of the state of technology and compendia of important results. There is a historical flavor to many of the chapters, and volume 1 of the series will offer a comprehensive overview of the programs, but the emphasis throughout is on results rather than history.

The goal of the series is to provide both a starting point for the new researcher and a reference tool for the experienced worker. It should also serve the needs of government and private-sector officials who want to see what programs have already been tried and what impact they have had. And it should be a resource for entrepreneurs whose talents lie in translating research results into practical products.

The scope of the series is broad but not universal. It is limited to solar technologies that convert sunlight to heat in order to provide energy for application in the building, industrial, and power sectors. Thus it explicitly excludes photovoltaic and biological energy conversion and such thermally driven processes as wind, hydro, and ocean thermal power. Even with this limitation, though, the series assembles a daunting amount of information. It represents the collective efforts of more than 200 authors and editors. The volumes are logically divided into those dealing with general topics such as the availability, collection, storage, and economic analysis of solar energy and those dealing with applications.

Volume 6 covers active solar sstems of building, including sections on domestic and commercial water heating, residential and commercial space heating, and residential and commercial space cooling. The emphasis in this volume is systems: their design, analysis, performance, and economics. Most of the active solar systems described in this volume are not now commercial products. Many were built and tested as part of the research and development or the demonstration programs carried out by the Department of Energy, Department of Housing and Urban Development, and other federal and state agencies. However, the scope of systems covered is very broad and includes virtually every configuration that has been built and tested to date.

The reader with an interest in active systems for buildings aplications may also want to consult volume 5, which covers the components and materials of active systems, volume 4, which treats the fundamental energy dynamics of buildings, and volume 9, which deals with the architectural integration of energy systems and buildings.

Volumes 5 and 6 together constitute a comprehensive compendium of the results of active solar research development and demonstration programs spanning more than a decade and a half. They cover every aspect of active solar technologies from material properties to system performance and economics, and the treatment of topics ranges from fundamental physics to practical applications data. Anyone with an interest in active solar technology, from the researcher to the decision maker to the end user, will find these two volumes a treasure trove of valuable information.

Preface

The term "active solar systems" is commonly applied to assemblies of equipment that provide heating, cooling, or hot water in dwellings and commercial buildings. Usually excluded is equipment that supplies electricity or products of biological processes. Also excluded are materials and designs that, by permitting the passive supply of solar energy to a building, usually through windows or other transparent and translucent surfaces, can reduce the demand for conventional energy.

As with the other books in this series, the primary purpose of this volume is the documentation of technological programs and developments, supported primarily by the U.S. Department of Energy, which facilitated the practical use of solar energy. Three interrelated technologies are covered—the heating of water for consumptive use, space heating in buildings, and the use of solar energy in equipment for cooling buildings. Similar equipment, such as solar collectors and heat storage units, is used in all three applications. Other components, for example, the subsystems producing cold air or cold water by use of a solar heat supply, are specific to a particular application.

The federal program, initiated in 1973 and first conducted by the National Science Foundation (NSF), next by the Energy Research and Development Administration (ERDA), and then by the Department of Energy (DOE), provided most of the financial support of the research, development, and demonstration of these technologies. Starting with a very limited base, investigators in university laboratories, government research centers, and in industry successfully developed the technology necessary for commercialization of solar energy use in buildings.

Although the federally funded program is the principal subject of this book, earlier developments and some projects sponsored by other organizations are also included. The extent of all these activities is much too large for describing all of them or for detailing any of them in this book. References accompanying each chapter offer readers an opportunity to obtain additional information on the various topics.

Historical Perspective

Prior to the start of U.S. government support of solar energy research in 1973, the state-of-the-art in solar energy for buildings ranged from the use of practical residential solar water heating in warm climates to negligible applications of solar cooling technology; limited research on solar space heating had been done, and about a dozen experimental systems in dwellings had been built and tested.

Solar water heaters had been sold by several manufacturers in California and Florida since the start of the century, so the primary thrust of the NSF, ERDA, and DOE programs on this application was the improvement of existing technology through the development of better designs, construction materials, performance information, and standards, and in widespread demonstrations of systems.

The development of active solar space heating in the United States started in the late 1930s and early 1940s with several experimental systems at universities in Massachusetts and Colorado, sponsored by research foundations, a U.S. government department, and private sources. By 1945, full-size systems providing part of the heat requirements in four houses had been built, tested, and documented. Solar heat was collected in water in two of the buildings; in the other two, air was heated in the solar collectors. Heat was stored in tanks of water, bins of small rocks, or in containers of fused salts.

Funded by private sources, foundation grants, and an industrial sponsor, another half-dozen U.S. buildings were provided with solar heating systems during the next two decades. Although these pioneering efforts provided technical and design information, they did not stimulate significant commercial interest until after 1973, when U.S. government support of development and demonstration programs began. The lack of a substantial technical base and experience led to the funding of design studies and experiments, mainly in universities and technical institutes, followed by the construction and demonstration of many full-scale solar heating systems of various types in residential and commercial buildings.

Solar cooling research was initiated in 1973 simultaneously with development and demonstration of solar water heating and space heating. Absorption air conditioners designed for steam or natural gas supply were modified for solar heat input, systems for their use were designed and tested, and full-scale demonstrations in residential and commercial buildings were made.

The development and application of active solar systems under the National Solar Energy Program are presented in the twenty-two chapters of this book. Most of the material relates to work conducted between 1973 and 1985 when government and industry were most active. Research and commercialization have continued beyond that time, however, and a short postscript relating to this more recent period should be recorded.

Research and development in the area of water heating and space heating, toward system quality improvement, has continued on a reduced scale. Design and operating handbooks have been produced, and standards for solar collectors and water heaters have been promulgated and implemented. System designs have become more uniform and reliable.

The market for solar space heating systems disappeared with the termination of the government subsidy program in 1985, and the sale of solar water heaters retracted mainly to traditional locations in southern California and Florida. System quality and reliability were substantially improved during this period.

Solar cooling research has continued primarily in the direction of desiccant systems regenerated by use of solar heat. Commercialization has been sporadic, with no sustained industrial supply of solar cooling systems.

Significance of National Solar Energy Program in the Development of Active Solar Systems

The federal program in active solar systems had the following beneficial results:

• Provided the basic information for applying solar energy to the heating of water and to the heating and cooling of buildings

• Demonstrated the practicality of using solar energy to supply hot water and to heat and cool buildings

• Created an industry for manufacture, distribution, marketing, and installation of solar systems

• Saved fuels equivalent to more than two million barrels of oil per year

• Provided the technical base for widespread future use of solar energy to reduce U.S. dependency on fossil fuels

Although the present price of fossil fuels places solar heat at an economic disadvantage, there can be no question that the inevitable increase in fuel prices will make the solar technology developed in the National Solar Energy Program of great usefulness.

Environmental concerns, political and strategic uncertainties, and other factors may become even more important than fuel prices in accelerating solar energy use.

Three types of applications—hot water, space heating, and space cooling—and three types of federal programs—research and development (R&D), demonstration, and tax credits (subsidy)—interacted to various degrees. Hot water technology required the least new development, so demonstration and tax credit support were of greatest importance and had major impact on establishing this application in residential and commercial buildings. Space heating required more research and development than did water heating, so federal R&D projects greatly influenced the introduction of this technology. Demonstration was a major program in solar space heating, though often misused for testing unproven and ineffective systems. The tax credit program provided needed support for commercializing the space heating technology found to be successful in the demonstration program. Solar cooling was primarily an R&D activity, although projects in commercial buildings were included in the demonstration program. Even with tax credit support, the high cost of these cooling systems prevented significant commercialization after conclusion of the demonstration program.

Potential of Active Solar Heating and Cooling

The large, dispersed use of energy for supplying heating, cooling, and hot water in buildings, estimated to be about one-fourth of the national energy demand, and the on-site availability of sufficient solar radiation to furnish most of those energy requirements make these applications of great potential importance. Solar water heating is currently the most competitive with conventional energy, primarily because of its year-round use, but water heating requires only a small fraction of total heat use in buildings. Growth of this application is inevitable at rates dependent on combinations of prices, incentives, legislation, and social and political circumstances.

Space heating constitutes the major use of energy in most U.S. buildings. Solar energy can be used to supply almost half the annual requirements for hot water and space heating in residential buildings at a cost per unit heat delivery about the same as that of water heating alone. Actively promoted, this application can be expected to grow at nearly the same rate as solar water heating.

Solar cooling technology is not yet at the practical stage of development, but when it has been adequately developed, its use in combination with solar heating and hot water should become practical in U.S. buildings that need year-round temperature control.

If the cost of cooling equipment suitable for solar operation is comparable to the cost of purchasing and operating conventional cooling systems, the year-round use of solar can then provide a cost saving by sharing solar hardware costs between the two (or three, with hot water) applications. Where little or no space heating is required, solar cooling will probably grow more slowly unless there is year-round demand for cooling, as in commercial buildings with high internal cooling loads.

To summarize, there is good potential for growth of solar water heating in this field, but the fractional reduction in U.S. fuel use will be limited. Solar space heating represents a much larger market and fuel savings potential, but higher consumer investment per system and geographic limitations will require the development of effective marketing strategies. Commercialization of solar cooling will require technological improvements, whereupon combinations with solar heating systems should become viable.

Guide to the Book

Each of the twenty-two chapters in this book has been written by a specialist and reviewed by others with comparable experience. Although the documentation of the data and results of other investigations is the focus of each chapter, interpretations, explanations, reservations, and occasional exceptions are expressed by the authors. Every effort has been made by the editors to present as objectively as possible summaries of the great volume of work covered in this book.

The chapters are grouped into four sections containing approximately equal numbers of topics. The sections are:

• Design, Analysis, and Control Methods
• Solar Water Heating
• Solar Space Heating
• Solar Cooling

In the five chapters of the first section of the book, methods for designing and controlling active solar systems and for predicting their performance are presented. The use of analysis, theoretical models, and simulation methods in system design and the techniques of optimization are shown. Control theory and its practical application to achieve desired system performance are reviewed.

The sections on water heating (five chapters) and space heating (seven chapters) have similar formats, starting with an introduction and summary, followed by chapters containing system descriptions, results of experimental and field test programs, costs, and future prospects. Each introductory chapter provides background and historical information, an overview of the current status of the application, and a brief summary of topics and chapters in the sections.

Solar cooling development, described in five chapters, has followed three technical paths—engine-driven vapor compression, absorption, and desiccant. Substantial differences in scope and emphasis in the research programs as well as in the technologies themselves are recognized in chapters on each of the three processes. These chapters are preceded by an introduction and followed by an overview and evaluation of future prospects.

Each section or group of chapters in this book is presented independently of the others, thereby making practical the separate consideration of each application. As frequently noted, however, combinations of heating with hot water, heating with cooling, and combinations of all three are in use, so the technical and economic interdependence of these applications in the corresponding sections of the book needs to be recognized.

It should also be noted that other volumes in this series, particularly *Solar Collectors, Energy Storage, Materials* and *Economic Analysis of Solar Thermal Energy Systems*, relate closely to topics in this book. The reader is referred to those publications as well as to the literature cited at the end of each chapter in this book.

Acknowledgments

The list of experts who advanced the knowledge and application of solar heating and cooling and whose work is the basis for the information in this book is far too long for full recognition here. Their contributions are cited in the reference sections of each chapter. But the editor wishes to acknowledge here the major contributions of the planners, authors, and reviewers of this work.

To Dr. Frederick Morse, whose leadership of the DOE Solar Heating and Cooling Program made possible the massive accomplishments described in this book, goes the deep appreciation of the editor. Without his vision and dedication, this work would not have been possible.

The multiple responsibilities of the editor in chief, handled so skillfully by Charles Bankston, and assisted by Lynda McGovern-Orr, greatly facilitated the scheduling and tracking of the numerous writing, editing, and reviewing tasks involved in a work of this type. His advice, overview, and encouragement are much appreciated.

To the twenty authors of the chapters in this book, profiled in the contributors section, go the special thanks of the editor. Although each is a specialist in the subject of his or her chapter, extensive effort had to be devoted by every writer to the assembly of a great amount of information from almost countless sources, followed by review, appraisal, and authoritative presentation. The authors are to be congratulated for jobs well done.

The selection of subject matter and its presentation were the responsibilities of the authors, but nearly forty reviewers, also specialists in these subjects, are complimented and thanked for their substantial assistance in assuring the quality of this book. On behalf of the authors, the editor thanks the following experts who reviewed, commented, suggested changes, and materially improved the results of this undertaking: David Anand, Philip Anderson, Charles Bankston, Robert Barber, Frank Biancardi, Wendell Bierman, John Clark, Barry Cohen, William Duff, Hunter Fanney, Charles Hansen, James Hedstrom, James Hill, Alan Hirshberg, K. G. T. Hollands, Richard Jordan, Jan Kreider, Zalman Lavan, Robert LeChevalier, Noam Lior, Arthur McGarity, Richard Merriam, Jeffrey Morehouse, Stanley Mumma, Alwin Newton, Terry Penney, Ari Rabl, Douglas Root, William Sholten, Elmer Streed, Gary Vliet, Mashuri Warren, Ronald West, Byard Wood, Marvin Yarosh, and John Yellott.

Finally, the editor expresses deep appreciation to less visible but vitally important participants. The management services of Jack Roberts and

Oscar Hillig of the ETEC division of Rockwell International and the editorial and logistic contributions of Nancy Reese and Barbara Miller of the National Renewable Energy Laboratory and Melissa Vaughn of the MIT Press are gratefully acknowledged.

I DESIGN, ANALYSIS, AND CONTROL METHODS

1 Overview of Modeling and Simulation

John A. Duffie

1.1 The Nature of Solar Processes

Active systems are assemblies of collectors, storage units, fluid transport devices (pipes, ducts, pumps, blowers, etc.), heat exchangers, loads (the building to be heated, hot water to be supplied, industrial-process energy need to be met, etc.), auxiliary energy sources, air conditioners, heat engines, controls, and other components. An example of a space and water heating system is shown in figure 1.1. The basic question addressed in modeling and simulation is: "How can the performance of these components and systems be represented mathematically (or numerically), so as to lead to understanding of how active systems function and how to predict their output?"

In essence, modeling is the development of the theoretical basis for performance calculation, and simulation is the use of the theory to predict system performance. Solar processes have certain characteristics that increase modeling difficulties:

• They are transient in nature, being driven by time-dependent forcing functions and loads.

• They are driven by forcing functions (weather) that are partially random in nature.

• They are nonlinear in their response to solar radiation.

Because of these characteristics, mathematical solutions to the equations describing the systems are generally impractical and numerical solutions (by computer) are used.

The procedure that has been developed to a high degree of effectiveness includes two major steps. The first is the formulation of component models, i.e., mathematical relationships that describe how components function. These component models may be based on first principles (energy balances, mass balances, rate equations, and equilibrium relationships) at one extreme or empirical curve fits to operating data from specific machines (such as absorption chillers) at the other. In most cases, the models that have been developed have been compared with or based on experimental measurements of component performance to lend confidence that the models adequately represent the performance of the hard-

John A. Duffie

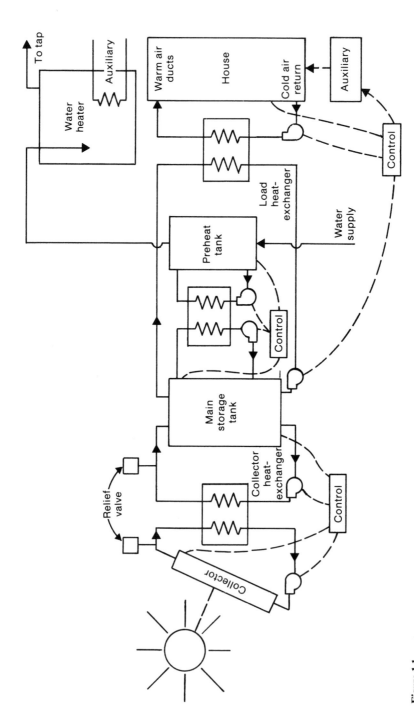

Figure 1.1
A liquid-based space heating system with pumps and controls.

ware. The result of this work are component models in which we have great confidence for a wide variety of equipment. As new component hardware is conceived or developed, new models must be formulated.

The second step is the development of means for simultaneous solution of the component models for all of the components that are included in a particular system. The most widespread approach to the solution hag been to program the component models, i.e., write component subroutines, and then simultaneously solve them with an executive program, as will be described in chapter 2 in the TRNSYS program. The alternative is to algebraically combine the component models into system models for particular system configurations and then solve the system models. The advantages and disadvantages of these two approaches are noted in this report. In either case, the solutions to the models are obtained by using, as forcing functions, hour-by-hour meteorological data (solar radiation, ambient temperature, wind speed, humidity, etc.) and hour-by-hour loads on the system (which may also be driven by weather).

1.2 Numerical and Physical Experiments

Under DOE sponsorship, programs have been developed that are applicable to a very wide range of solar and nonsolar processes. We can now readily do numerical experiments. We "build" systems on paper by assembling the appropriate component subroutines and "operate" them in the computer in whatever climate we wish, as long as we have meteorological data for the location of interest or a climatically similar one.

Two types of information are obtained from these simulations or numerical experiments. First, information on process dynamics is available, for example, the time-temperature history of any part of the system. This history is useful in establishing such matters as criteria for materials of construction and pressures to be expected in liquid systems. Parasitic (operating) power and the time it is used can be established, as can the time of use of auxiliary energy, which may be of critical interest to utilities supplying electricity and gas. Second, and this is often more important, the long-term integrated system performance can be computed, including the amount of solar energy usefully delivered to meet a load and the amount of auxiliary energy needed. These figures are essential for economic analysis of a process.

Simulations constitute a powerful tool for the developer and designer of solar processes . They are not a substitute for physical experiments, as noted below. With some limitations, however, they have some very significant advantages over physical experiments:

• Simulations can be done in two to four orders of magnitude less time than physical experiments . In general, a few hours or days are used to get up a simulation using a program like TRNSYS, and the CPU time to do a year' s simulation is some seconds or minutes on a mainframe computer. Assembling hardware and operating a comparable physical experiment might take one to two years.

• The costs of simulations are two to four orders of magnitude less than the costs of comparable physical experiments.

• Simulations can be detailed to various levels, depending on the effects the user wishes to study. With the programs now available, very simple simulations can be done to show gross effects; with user-written component subroutines, very detailed performance calculations can be done.

• Numerical experiments can be repeated, design variables changed, and the systems operated in the identical weather; thus, changes in the system design or configuration can be evaluated independently of weather. Physical experiments, on the other hand, are not repeatable because of weather variability.

Physical experiments are, of course, essential. They can show design and operating problems that a simulation of thermal performance cannot indicate. They provide necessary experience with physical layouts, assembly, and mechanical design that must be in hand before systems can be manufactured. Numerical experiments can be viewed as the theoretical study to be undertaken as part of the process of planning and conducting a physical experiment, with both types of experiment preceding the development of commercial systems. These steps are illustrated in the discussions of DOE-sponsored system development in parts II through IV of this book.

1.3   Utility of Simulation

Simulations of solar energy systems have proven to be useful for several purposes. First, use of numerical experiments in which key design parameters or system configurations are systematically changed aids in understanding how solar energy systems work. This knowledge then can lead to development of better systems or to improved performance of conventional systems. For example, a recent DOE-sponsored study of control strategy of domestic hot water systems indicates that optimum collector flow rates may be an order of magnitude less than those in use in 1980s commercial systems. The reduced flow rates can lead to improvement in annual system performance of 10% to 20%, provided storage tanks can be designed to achieve very high degrees of stratification and other operating criteria are met. The use of these low flow rates had been suggested earlier, and simulations provided the means for their quantitative evaluation. The next step is to see how the necessary stratification can be maintained. (Physical experiments in the United States and Europe are providing information on maintenance of stratification.)

Second, simulations are powerful tools for research and development of new systems and processes. Before physical experiments are undertaken, many numerical experiments can be done quickly and inexpensively to assist in the selection and design of the hardware for a system physical experiment. The effects of system configuration or design changes on long-term system performance can be assessed for a variety of climates.

Third, simulations can readily evaluate effects of new developments in materials or components on long-term system performance and, thus, on economics. Conversely, if system performance goals are set, simulations can indicate the material properties or component performance levels that must be achieved to reach those goals.

Two examples illustrate this use of simulations. The first is a component design example of selecting collectors with one or two covers. Collectors with one cover have lower optical losses but higher thermal losses than those with two covers, other factors being the same. One-cover collectors also cost less than two-cover ones. The temperatures at which collectors of liquid systems operate vary with time; if, on the average, the collector operates at temperatures not far above ambient, the one-cover collector will probably be best. If temperatures of operation are higher, the two-cover collector may be best. Simulations provide quantitative information on system thermal performance on which to base selections.

A second example concerns the effects of stratification in storage tanks on system performance of solar water heaters. Simulations show that if tanks can be designed to achieve very high degrees of stratification, low flow rates through the collector loop may lead to significant performance improvements. Thus the benefits of successful development of tanks that will stay stratified can be quantitatively estimated.

Fourth, simulations are used in the design of some systems for specific applications. In general, this use is restricted to large, one-of-a-kind applications where the cost of the system is high compared to the cost of doing the simulations, as is the case with industrial process heat systems or hot water systems for institutional buildings. Given a specific design problem, simulations provide a useful means of "trying" systems of various configurations and then sizing the selected system to achieve an optimum mix of solar and other energy sources.

A DOE-sponsored study of an industrial process heat application provides an interesting example of this use of simulations. The application involved a meat packing plant in Iowa. Heat for cleanup water for the plant was to be supplied by a combination of three sources: solar, waste heat from other operations in the plant, and steam. Many simulations of various system configurations and sizes were done. The study indicated that an extensive system of waste heat recovery equipment without solar contribution was the most economical one, and this system was installed in the plant.

Finally, simulations are used in the development of design methods for standard types of systems. Used by people who are not simulation programmers, these design methods involve calculations that can be done by hand, with hand-held calculators or with microcomputers. They are intended to eliminate the need for simulations in the design of systems of common types. This development has been significant enough that it is reviewed in detail in chapter 3. The topic is introduced in section 1.5 of this chapter.

1.4 Simulation Programs

Prior to NSF/ERDA/DOE sponsorship, there were a few studies of simulation of solar processes. For example, the University of Queensland in Australia developed the use of analog computers in simulation of simple processes, a UCLA study dealt with digital computers in simulation of space heating systems, and an Ohio State University study dealt with heating and cooling systems. These projects produced interesting, specific results and pointed the way toward what could be done, but they were severely limited in scope.

Several types of solar process simulation programs have been developed with DOE support. They can be broadly classified into general purpose and special purpose programs.

General purpose programs are modular and include subroutines for each individual component or subassemblies of components. These components and subassemblies are "assembled" into system programs, with interconnections (corresponding to piping and wiring in a physical system) made by appropriate programming techniques. These general purpose modular programs are widely applicable to transient processes. The major program of this type supported by the DOE solar program is TRNSYS, developed at the University of Wisconsin. TRNSYS has evolved through a series of versions, the latest being TRNSYS 13. It is written in standard FORTRAN and is compatible with a very wide range of mainframe computers. Approximately two thousand programs have been distributed in various versions to a wide variety of users (A&E firms, laboratories, industries, universities, utilities, etc.). The key to the usefulness of this program has been the careful documentation that enables the user to understand what is behind the program, how it is to be used, and how the user can prepare and use his own component subroutines.

DOE-2 is a related but far larger general purpose program, designed primarily for building-load calculations. It also includes solar process capabilities. DOE-2 is not readily transportable from one computer to another, but it has been modified to run on several mainframe computers. The user accesses the program on government-owned facilities or on commercial networks such as CYBERNET. Some of the solar process part of DOE-2 was derived from TRNSYS.

Special purpose programs are written for limited or specific system configurations and often involve algebraic combinations of the component models prior to programming. Special purpose programs have the advantage of being less expensive to run but lack the versatility of modular programs. They are generally not documented or transportable and are of use mainly to their authors. Examples of such programs are SERIPH, a special purpose program designed for evaluation of industrial process applications of solar energy, and the programs written for specific system evaluations as part of the SOLCOST design method.

Other DOE-sponsored simulation programs have been written at Los Alamos National Laboratory and Sandia Laboratories. They have been used in-house for many tasks but have not been documented and made available to other organizations.

Simulations programs developed under DOE auspices are reviewed in detail in chapter 2.

1.5   Design

Simulations, while useful for many purposes, are not appropriate for day-to-day design of common solar heating systems that are routinely installed by builders, plumbing and sheet-metal contractors, and other installers. Simulations are expensive relative to the cost of the systems, and the versatility is not needed. Under DOE sponsorship, several simulation-based design methods have been developed and checked with physical measurements. There methods are much simpler to use than simulations. The essential ideas behind the methods are noted in introduction to chapter 3.

The most widely used of the methods, developed at the University of Wisconsin under DOE sponsorship, is the f-chart method. It was developed as a general design method applicable to common, standard kinds of systems, simple enough that a designer could do the thermal performance calculations by hand or with a hand-held calculator, obviating the need for large computing facilities. The pattern of development of f-chart was very similar to that of other engineering developments. Many (numerical) experiments were done, and the results of these experiments were correlated in terms of appropriate dimensionless groups of variables. These variables include the collector-describing parameters as obtained from standard collector tests, the collector area, and information about ducts or piping and heat exchangers. They also include the heating loads of the building and hot water loads that are calculated by standard ASHRAE methods or any other method the designer wishes to use. The designer calculates the dimensionless parameters for each month, applies simple correction factors for storage capacity and several other design parameters if they are not at or near their standard values, and uses the f-chart correlations (graphs or equations) to obtain monthly performance estimates of solar contributions and auxiliary energy required. The monthly values are summed to get annual performance, with the final and significant result in the form of the annual fraction of the total load that is met by solar energy and with the quantity of energy that must be purchased.

The f-chart method has been published in the technical literature, has been the subject of a book and chapters in handbooks, and is widely available for use by designers. It has also been programmed for mainframe computers and microcomputers as well as for hand-held calculators. The FCHART programs that include economic analysis as well as thermal performance estimates have found widespread use by system manufacturers, distributors, and designers. They have also been the basis for development of even simpler design tools.

Quite a different approach was taken in another design method, SOLCOST, developed by Martin Marietta Corporation under DOE sponsorship. In this method, "shortcut" simulations of standard systems are performed for a "good" day and a "poor" day, with the results weighted to estimate monthly system performance. The designer does not do the calculations himself; he submits data on the system of interest to an agency having the program. The agency does the calculations and returns the thermal and economic results to him.

Another powerful set of design methods is based on "utilizability" concepts. Utilizability is the fraction of the total solar radiation in a specified time period that is above a critical radiation level. This concept is illustrated in figure 1.2. The time period can be a month, giving monthly average daily utilizability, $\bar{\phi}$, or it can be a particular hour of the day (e.g., 10–11 A.M.) for a month, giving monthly average hourly utilizability, $\bar{\phi}$. The critical radiation level is the radiation intensity at which radiation absorbed by a collector equals thermal losses from the collector; any radiation below that level cannot be usefully collected.

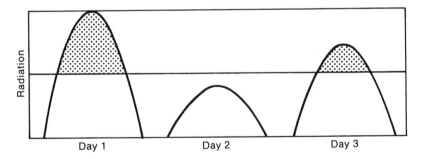

Figure 1.2
Utilizability is the fraction of the total radiation that is above a critical radiation level. The ratio of shaded area to total area under the curves, summed for all data for a month, is $\bar{\phi}$ for that month.

General methods have been developed for calculating ϕ and $\bar{\phi}$ from monthly average radiation (a commonly available statistic) and geometric factors. It is possible to estimate critical radiation levels for many active systems. From the critical radiation level and the monthly radiation, the average useful gain from a system is readily determined.

Utilizability design methods have been applied to many solar systems, including those with seasonal energy storage, heat-pump solar energy systems, and industrial and domestic water heaters. The same ideas have also been applied to design of passive heating and photovoltaic power systems. In each case, the results obtained from utilizability calculations have been compared with results for the same system obtained from simulations and with measurements on physical experiments to assure that the design methods are reliable. Methods for ϕ and $\bar{\phi}$ have been published and are available as computer programs that have been widely used by designers. Many of the design calculations in the latest FCHART programs are done by utilizability methods.

Design methods developed under DOE sponsorship are reviewed in more detail in chapter 3.

1.6   Controls

There are many approaches to the control of active solar energy systems, ranging from on/off control to optimal controls based on various objective functions. The major control variable in the collector loop is fluid-flow rate. The controls on energy distribution systems are generally designed for optimum combinations of solar and auxiliary energy use. Simulation programs have control component subroutines that model anything from simple on/off controls to microprocessor controls with multiple control functions. Simulation techniques are used to evaluate the effects of control functions and strategies on system performance.

An example of DOE-sponsored controls research, applicable to systems where parasitic power is not negligible, concerns the selection of flow rates for solar heating systems that lead to the maximum net energy gain. This study shows that the optimum strategy is not on/off control but rather the control of collector flow rate, such that the correct balance is always maintained to maximize the difference between collected energy and parasitic energy for pumps. These studies were a combination of analysis and simulation, with the results compared with experimental measurements.

Control studies are reviewed in chapter 4.

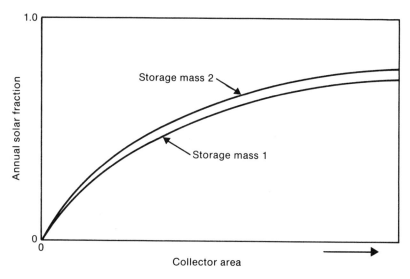

Figure 1.3
Typical solar contributions to meeting a space heating load, for two storage sizes.

1.7   Economic Analyses and Optimization

Thermal performance can be estimated by simulation methods or design methods. The effects of changes in system configuration or changes in design variables on annual system performance are calculated. The principal design variable is most often collector area; an example of the results of these calculations for a space heating operation in a northern U.S. climate is shown in figure 1.3 for two storage volumes. With this thermal performance information, an economic analysis can be done to determine optimum design. The first costs of a system go up as collector area goes up, but the annual output per unit area of collector goes down (i.e., the slope of the F vs. A_c curve decreases as A_c increases), and an economic optimum normally exists at some collector area corresponding to solar fractions between 0 and 0.8 or so.

No universally accepted economic figure of merit or yardstick is used in economic calculations. DOE-sponsored methods generally use life-cycle cost (or savings), with the calculations set up to find the system showing the lowest life-cycle cost (or highest life-cycle savings). Payback times and cash flow analyses are also used by manufacturers and vendors of solar heating systems.

Some of the significant economic factors are: system first cost, maintenance and operating costs, insurance, fuel (auxiliary) energy costs, and interest, inflation, and discount rates. The problem involves comparing solar energy systems characterized by high investments and low operating costs with conventional systems that are characterized by low investments but high future operating costs. The results of economic analyses, and thus of system designs, are particularly sensitive to current and projected fuel prices and the system installed costs.

A life-cycle cost analysis can be complex and time consuming. A method has been developed to reduce such calculations to two parameters, each of which is a combination of many economic factors. These parameters are the ratio of the life-cycle fuel cost savings to the first year's fuel cost savings, and the ratio of the life-cycle expenditures on the solar energy system to the investment in that system. These two parameters are easily calculated, and optimum designs of many systems can readily be determined from them.

The methods for economic analyses developed for DOE are included in design programs such as FCHART and SOLCOST, and an economic subroutine is available as part of the TRNSYS executive program. Tables and charts have been developed to facilitate hand calculations. Work on economic analysis and optimization techniques is reviewed in chapter 5.

1.8   Meteorological Data Requirements

Simulations and design methods require two different kinds of data. Simulation computations are generally done with hourly meteorological data (solar radiation, ambient temperature, and possibly other factors, such as wind speed and humidity). The SOLMET program of the National Oceanic and Atmospheric Administration (NOAA), sponsored by DOE, has provided a data base of 23 years of hourly data for 26 U.S. stations. For each of these stations, a "typical meteorological year" (TMY) has been developed. When used in simulations of most solar processes, TMY represents, in a satisfactory way, 23 years of data.

Most design methods require monthly average solar radiation and monthly average ambient temperature. A radiation data base of monthly averages has been developed as part of SOLMET program for 248 U.S. stations. These data are built into design programs such as FCHART. Continuing work is needed to improve this data base.

Most solar radiation data are measured with pyranometers on a horizontal surface. The information desired is hourly or monthly average radiation on a sloped plane of a collector. Significant progress has been made in developing the necessary means for obtaining the desired quantities from the data available. This procedure has involved such studies as correlations of diffuse fractions with clearness indices, and distribution of diffuse radiation over the sky dome and its effect on radiation on sloped collectors or windows. General methods of estimating utilizability ϕ and $\bar{\phi}$, solar radiation statistics, referred to in section 1.5 above, have also been developed, and these methods are integral parts of design programs.

1.9 Validation

As with any other theory, there need to be comparisons of the results of simulations and design procedures with measurements from physical experiments. DOE's concern for the ability of simulations and design methods to reflect the real world has resulted in validation efforts of several types.

Careful experiments have been used to verify component models. For example, building on the work of several pioneers in the field, collectors have been very carefully studied by organizations such as the National Bureau of Standards and by developers and manufacturers. Many comparisons of experiments and theory have been made. There is reasonable confidence in the collector models of all but a few of the newest collectors. (These collector experiments and models have then been used as the basis of development of standard collector test procedures for commercially produced equipment that will provide the performance parameters needed in design procedures or simulations.) Other component models have been similarly examined.

System experiments also have been compared with system simulations. Very careful experiments sponsored by DOE—at Colorado State University, for example—have been used to validate TRNSYS simulations by comparisons of process dynamics (such as storage tank temperature vs. time) with predictions, and by comparisons of short-term energy balances (over time spans of days or weeks). If the simulations are carefully done, the results are remarkably close for process dynamics and very good for energy balances.

Measurements of seasonal and annual system performance in laboratories and in the field have been used to check the predictions of design methods (which, in turn, are based on simulations). It is necessary to use measured meteorological data and loads as inputs to the design method. (Predicted loads are notably unreliable regardless of what kind of heating system is used to meet those loads.) When the design methods are properly used, with systems for which they are meant to apply and which are well built and operated, the measured and predicted performances agree very well.

There have also been intercomparisons of simulation results obtained with DOE- sponsored programs (those of Los Alamos and TRNSYS) with results of simulation programs developed independently in other International Energy Agency countries. Standard-building heating and industrial-process heat problems were used with a year's meteorological data for U.S. and European locations. After an iteration or two, it was found that results of these rather different codes were essentially identical, indicating that agreement is satisfactory if adequate care is exercised in using the programs.

Thus, a pattern of validation efforts indicates, individually and collectively, that simulations and design methods, when properly used, represent physical realities in a very satisfactory manner.

1.10 Problems and Accomplishments

The use of the tools that have been developed for DOE has not been without problems. These problems can be categorized into two general types. First, design and simulation methods have been misused. All engineering calculation methods have limitations that must be appreciated and observed by the user, and solar process calculations are no exception. It is easy to do an inadequate job of simulating a solar process. The simulator may overlook or misunderstand important features of the system he is modeling. The particular component models he is using may have been simplified in a way appropriate for many purposes but may be inadequate for his immediate problem. Part of the art of simulation is to know the necessary level of detail that should be included. The results of a simulation are only as good as the knowledge and skill of the simulator allows. It has been clearly shown by comparison with careful experiments that simulations, when properly done, can represent reality to a very high degree.

Design methods, like simulations, can include various levels of detail. It is possible, for example, to include such effects as pipe or duct losses, heat exchanger effectiveness, collector flow rate, etc. on collector performance, but users of the methods may unjustifiably ignore these effects. Thus, even though the design methods are quite simple, they may be used improperly. Better education of designers and better documentation can relieve this problem.

The most significant problem with design methods has probably been in their applications to systems for which they were not intended. For example, FCHART has been applied to series solar energy/heat pump heating systems; it was not developed for such systems and will not give realistic design information for these systems. It has been shown that design methods, when properly used in design of systems that are well built, are very good predictors of long-term system performance.

The second problem with design and simulation methods lies not with the methods but with the systems to which they are applied. Many systems have been well designed and well built, and measurements on these systems generally are in good agreement with predictions. However, in the nascent solar industry, some systems have been poorly built and their performance does not measure up to that predicted by the design method. Time, increasing skills of installers, and development of industry standards will improve system performance.

One must also keep in mind that designs are based on average or typical years, that a particular year's weather may be very different from the average, and, thus, that performance for that year may be better or poorer than average. There is no way of avoiding the use of historical meteorological data to predict future performance.

Design methods can be further improved and extended to new kinds of systems. Education of users of the methods will improve the applications that are being made of existing and new methods, and future documentation of the methods should be aimed at making designers more aware of the importance of the range of design factors that can affect system performance.

Under DOE sponsorship, powerful tools have been developed for studying new systems, predicting the thermal performance of systems, and optimizing designs. The same or closely related techniques that have been developed and applied to active systems have also been applied to passive and hybrid systems. These methods have been widely used in the solar-energy and related industries and have had a substantial positive impact on their development.

2 Detailed Simulation Methods and Validation

William A. Beckman

2.1 Introduction

2.1.1 Overview

The development of solar simulation capabilities has greatly assisted in the promotion of practical solar systems. What makes a solar system different from a conventional heating system so that it requires a complex computer program for design? After all, conventional heating systems have been designed for many years without the aid of the computer. Why, then, is solar different?

Although it is technically feasible to design a 100% solar energy system for almost any climate, economic considerations require a backup system for reliability. If a heating system is sufficiently large in midwinter to supply all required energy, then it is much too large in the spring and fall. Even in industrial systems with constant loads all year around, designing a system to cover the few sequences of insufficient sunshine will result in overdesign for the remainder of the year. Since the capital cost of solar equipment is very high, the cost of delivering solar energy to a system is a strong function of the collector area. Overdesign in a conventional system has only very modest penalties since the capital costs are low. Once it is decided that a 100% solar system is not economically sound, then the decision is how much conventional energy and how much solar energy will result in the minimum total cost. The problem would be simple if it were not for the nonlinear nature of solar system performance. Doubling the collector area will not double the useful energy produced. This nonlinear performance is difficult to generalize in an easy-to-use form.

Insight into proper design could be obtained by building systems and measuring fuel consumption. However, full-scale solar systems are expensive to build, and obtaining reliable performance data has proven to be even more costly than building the systems themselves.

Simulation programs enable the designer and researcher to look at a large number of alternatives with a reasonable expenditure of time and money. Such programs are essential tools in the design of large systems but have been replaced by design methods in the sizing of small systems. Chapter 3 describes these shortcut design tools that have been generally developed from large simulation programs. Performing a simulation study of a proposed solar system should not be considered a guarantee for a successful design. In spite of careful analysis including detailed simulation studies, many solar systems have had poor or, in some cases, no performance due to the inexperience of the designer with the problems associated with building reliable solar systems.

This chapter presents a short history of the development of simulation programs and then discusses the fundamental differences between the few simulation programs that are widely used. The component models and load models that are a central part of all of these programs are discussed in some detail. Finally, validation efforts are described.

2.1.2 Early History

The first simulation study of a solar process was done by Sheridan et al. (1967), who used an analog computer to study solar water-heating systems. They investigated short-term dynamics but made no long-term performance predictions due to the difficulties of using real data in analog simulations. While on leave at the University of Wisconsin, Close (1967) used an analog simulator on a digital computer to study the operation of solar water heaters. He used a factorial design method to cover a wide range of operating and design options and was able to provide general guidelines. In both of these studies, weather data were represented by analytic functions, and only a few days of actual operation were simulated due to the high cost of performing annual simulations. These early studies used analog simulation, sometimes even on a digital computer, because it was easy to translate the physical systems into electrical models. However, dramatic reductions in the cost of digital simulation soon overcame the advantages of analog computers, and most modeling has since used digital computers.

Buchberg and Roulet (1968) used a digital computer and a full year of real weather data to model a solar house-heating system. They used a pattern search technique to find the optimum design for the particular house and climate. Löf and Tybout (1973) performed the first general solar system studies that looked at a number of climates, buildings, and collector costs. They presented optimum collector areas and tilts for a wide range of conditions. This was the first "what if" simulation study to look at the potential for widespread use of solar energy.

The simulation studies before 1973 were supported by various foundations, universities, and private sponsors and thus, as a group, lacked coherent research direction. In 1972, the National Science Foundation (NSF) offered a national plan for solar energy utilization. The study by Butz, Beckman, and Duffie (1974) was the first simulation study supported by this predecessor of the Department of Energy solar program. Their work was the first general simulation of a solar heating and cooling system. Although the study was limited to the climate of Albuquerque, New Mexico, it clearly demonstrated the potential advantages of combining heating, domestic hot water, and cooling. The computer code used in this study was not intended to be a tool for general use. However, the NSF was supporting Colorado State University (CSU) to build an experimental solar-heated house, and the CSU researchers needed to look at various design options before the system was built. The University of Wisconsin was asked to use the Butz program to look at options such as collector area and tilt to give guidelines to the designers. Although the proposed CSU system was similar to the system used in the Albuquerque study, considerable effort was required to modify the program to accommodate the design features of the CSU house.

The difficulty of modifying a large computer program to suit particular needs led the University of Wisconsin Solar Energy Laboratory to propose to the NSF the development of a modular solar simulation computer program. The initial version of the code that was to become TRNSYS was finished in 1975. (The original name of the program was TRANSYS for TRANsient SYStems, but most FORTRAN compilers permit only six characters: hence, TRNSYS is pronounced TRANSYS.)

2.1.3 Simulation Programs

The first two simulation programs developed under U.S. government sponsorship (including NSF, ERDA, and DOE) were TRNSYS and the Los Alamos Scientific Laboratory (LASL) programs. The LASL programs have never been fully documented since they were designed as in-house research tools. The TRNSYS program was originally conceived as a general purpose, public domain program and has been continuously updated. Version 5.1 was first widely distributed in 1975, and version 13.1 became available in early 1990 (Klein et al. 1990). The program is written in ANSI standard FORTRAN so that it is compatible with all mainframe computers. The latest version uses FORTRAN 77. A microcomputer version using Microsoft FORTRAN is also available.

The TRNSYS program has a library of about fifty subroutines, some of which were contributed by users outside of the University of Wisconsin. Table 2.1 is a complete list of the TRNSYS library. Most of these subroutines represent particular pieces of equipment, such as collectors, pumps, and tanks. However, some of the library models are combinations of physical systems that form a subassembly. These combined models are provided for many common systems that frequently need to be modeled. They offer advantages to the model builder in reducing the effort required to construct a TRNSYS deck. They can also significantly reduce the computation time. Other components are used for general simulation tasks, such as input and output, plotting, and summarizing the results.

The physical and mathematical formulations of all models are completely described in the manual. The library can be expanded by the user to add specific models or modify the existing models. Instructions are given for writing FORTRAN code for use in TRNSYS to satisfy the user's particular needs. This flexibility has been the primary reason for the continued popularity of the program.

The models in TRNSYS are described by either algebraic or differential equations. The TRNSYS executive program is designed to simultaneously solve the set of equations describing the system. Rather than worry about numerical methods or the details of modeling, the user only connects the various components together, using the TRNSYS language, to form a system model. At this stage of systems analysis, the user acts as a plumber or electrician of sorts, since he or she connects the various "pipes and wires" between components. The user becomes the design engineer when assigning specific sizes to the various components and interpreting the results.

The large, public-domain building thermal analysis program, DOE-2 (York and Tucker 1980), has solar simulation capabilities in a section called CBS (Component Based Simulator). CBS is similar to TRNSYS in that individual components are modeled in separate subroutines, and the user only connects the components together to build a system model. However, all CBS models are algebraic; the differential equations are solved either analytically or numerically within the component model. As in TRNSYS, models of common systems are also available as preassembled subroutines. This practice greatly simplifies the task of the user and reduces the execution time of the program.

Table 2.1
The TRNSYS library

Utility Components	*Building Loads and Structures*
Data reader	Energy/(degree-hour) house
Time-dependent forcing function	Detailed zone (transfer function)
Algebraic operator	Roof and attic
Radiation processor	Overhang and wingwall shading
Quantity integrator	Window
Psychrometrics	Thermal storage wall
Load profile sequencer	Attached sunspace
Collector array shading	Multizone building
Convergence promoter	*Hydronics*
Weather data generator	Pump/fan
Solar Collectors	Flow diverter/mixing valve/tee piece
Linear thermal efficiency data	Pressure relief valve
Detailed performance map	Pipe
Single or biaxial incidence angle modifier	*Controllers*
Theoretical flat-plate	Differential controller with hysteresis
Theoretical CPC	Three-stage room thermostat
Thermal Storage	Microprocessor controller
Stratified liquid storage (finite difference)	*Output*
Algebraic tank (plug flow)	Printer
Rockbed	Plotter
Equipment	Histogram plotter
On/off auxiliary heater	Simulation summarizer
Absorption air conditioner	Economics
Dual-source heat pump	*User-Contributed Components*
Conditioning equipment	PV/thermal collector
Cooling coil	Storage battery
Cooling tower	Regulator/inverter
Chiller	Electrical subsystem
Heat Exchangers	*Combined Subsystems*
Heat exchanger	Liquid collector-storage system
Waste-heat recovery	Air collector-storage system
Utility Subroutines	Domestic hot water system
Data interpolation	Thermosyphon solar water heater
First order differential equations	
View factors	
Matrix inversion	
Least squares curve fitting	
Psychrometric calculations	

The SOLTES program (Fewell and Grandjean 1978) was developed to model large-scale, total energy systems in which a solar system is connected to an electric-power-producing system and "waste heat" is used for some thermal requirement. The collector models are primarily for concentrating collectors, and the storage medium is usually designed for high-temperature operation.

The SOLIPH program was developed by the Solar Energy Research Institute, SERI, now the National Renewable Energy Institute, NREL (Kutscher 1983) to model a wide range of industrial process heat systems. This program models specific systems and thus does not have the flexibility to model systems that were not considered in the original program. However, the authors can modify the code to investigate new systems. Because the program is not general, it does not have a high overhead. Consequently, it is very fast and can be used to look at a large number of design variables for little cost. The code is available from NREL, but the program is not documented. The program was used primarily as a survey tool in the SERI industrial process heat studies.

Many simulation programs that have been developed since 1972 are described in two survey articles (SERI 1980, Merriam and Rancatore 1982). None of the DOE-supported solar simulation programs are in general use today except TRNSYS and the TRNSYS derivative part of DOE-2, CBS. Consequently, many of the comments in the remainder of the chapter will relate to the development and validation of the component models of the TRNSYS program.

2.2   Component Models

2.2.1   Flat-Plate Collector Models

The first serious study of a solar energy system was made by Hottel and Woertz (1942). That classic paper described the fundamental relationships that are still used today to model solar collectors. Their study included the effects of temperature on collector performance, the effects of optical losses, and even anticipated selective surfaces. Little advancement occurred until 1953, when Austin Whillier received his Ph.D. from MIT under the direction of Hoyt Hottel. Their work, presented five years later at the 1958 Arizona Conference on the Use of Solar Energy (Hottel and Whillier 1958), derived the form of the collector equation that is used today. This Hottel-Whillier equation has the form of

$$Q_u = AF_r[I(\tau\alpha) - U_1(T_{in} - T_{amb})].\tag{2.1}$$

Bliss (1959) presented many of the Hottel-Whillier ideas in the first generally available paper on the subject of collector modeling since the Hottel-Woertz paper.

In spite of its simplicity, or perhaps because of it, equation 2.1 has stood the test of time. It expresses the useful output of a solar collector as the product of the maximum output of the collector operating at its most advantageous temperature level, times its effectiveness, F_r. The quantity F_r, generally called the collector heat removal factor, is analogous to the conventional heat exchanger effectiveness. It is the ratio of the actual heat gain to the theoretical maximum gain, given the loss characteristics of the collector. For a collector, the maximum heat gain is obtained when the whole collector is operating at the fluid inlet temperature.

Equation 2.1 does not account for the thermal capacitance of the collector. Generally, thermal capacitance has a negligible effect on the useful gain of the collector as shown by Klein, Duffie, and Beckman (1974) and others. For some collectors, the thermal capacitance is not negligible. Approximate methods for predicting the dynamics of collectors with modest amounts of thermal capacitance are given by Duffie and Beckman (1991). Mathematical models of collectors that contain large amounts of fluid, such as some evacuated tubular collectors, must include both the transit time and the thermal capacitance of the fluid. Mather (1982) has analytically studied this type of collector.

Collector models have been developed by Phillips (1979) to study the two-dimensional conduction of heat in the absorber. For collectors of conventional design and operation, these studies indicate that the one-dimensional heat flow used in the Hottel-Whillier model is valid. However, when the fluid flow rate is very low, as might be the case in thermosyphon systems, the two-dimensional heat flow becomes more important. General guidelines as to when the Hottel-Whillier model breaks down are given by Phillips.

Equation 2.1 has been used to develop collector test standards as discussed in chapter 6 of volume III of this series, *Economic Analysis of Solar Thermal Energy Systems*. If $F_r(\tau\alpha)$ and $F_r U_1$ are independent of operating conditions, then collector test data can be plotted as efficiency versus $(T_{in} - T_{amb})/I$, and the intercept and slope of the resulting straight line are

$F_r(\tau\alpha)$ and F_rU_1, respectively. This model, in which the two parameters are obtained by testing the collector, is probably the most common collector model. If the test data do not fall on a straight line, Dunkle and Cooper (1975) showed that the dependence of the top loss coefficient on the temperature level can be accommodated if U_1 is assumed to be a linear function of temperature. Finally, equation 2.1 does not take into account the variation in collector performance with solar incidence angle. Hottel and Woertz (1942) experimentally verified the Fresnel equations for calculating off-normal transmittance of solar collectors using glass covers. Souka and Safwat (1966) suggested that the ratio of $(\tau\alpha)$ at an incidence angle θ to $(\tau\alpha)$ at normal incidence can be given by

$$K_{\tau\alpha} = 1 + b_0(1/\cos\theta - 1),\tag{2.2}$$

where b_0 is an experimentally determined constant called the incidence angle modifier constant.

The transmittance of diffuse sky and ground-reflected radiation can also be estimated by the incidence angle modifier by defining an equivalent beam angle for the diffuse components of radiation. Brandemuehl and Beckman (1980) followed the suggestion of Hottel and Woertz and presented a graph of the appropriate beam incidence angle for both components of diffuse radiation.

Collector models based on fluid temperatures other than the inlet temperature have been proposed and are in use in some countries outside the United States. The early versions of the ASHRAE collector test standard 95–77 used the average fluid temperature, calculated by averaging the inlet and outlet temperatures, in place of the inlet fluid temperature. Advocates of air collectors have suggested that the outlet fluid temperature would be more appropriate, since, in a system with pebble bed storage, the collectors are usually exposed to a constant inlet temperature. These model variations do not cause any problems or offer any advantages, since it is easy to convert test data from any of the three forms to any other form if the fluid mass flow rate is known (Duffie and Beckman 1991, pp. 316–317).

2.2.2 Collector Model—Evacuated Tubes

The idea for eliminating convection losses by evacuating the collector was first proposed by Speyer (1965). Since then, many different designs have been proposed. Ortabasi and Buehl (1975) describe a flat absorber coupled to a U tube so that the inlet and outlet are at the same end of the evacuated tube. Another liquid-heating collector is described by Mather and Beekley (1976). It uses two concentric glass tubes to form the evacuated space and a third inside tube to direct the fluid to one end of the tube and back. The inside surface of the second glass tube is the absorber surface. Grunes (1976) describes a modification of this collector for use with air as the working fluid. Some recent designs have used heat pipes to reduce the complex manifold problems.

The models developed for evacuated tubular collectors have generally used the Hottel-Whillier form. Problems arise due to nonsymmetric optical properties and, in some designs, the large mass of fluid within the collector. The optical characteristics of an evacuated tubular collector with a tubular absorber are very different from the conventional flat-plate absorber. The collector efficiency can increase as the incidence angle increases, whereas in a flat-plate absorber the efficiency decreases. The optical characteristics will also depend on the azimuth angle of the sun relative to the collector; in a flat-plate collector the optical properties usually vary only with the solar incidence angle as measured from the collector normal. McIntire and Reed (1983) showed how a biaxial incidence angle modifier can be used to describe these nonsymmetric optical properties. Theuinisen and Beckman (1985) developed a model for estimating the transmittance of evacuated tubular collectors with diffuse back reflectors.

Mather (1982) studied the thermal capacitance and transit time effects of evacuated tubular collectors containing large amounts of fluid. He developed a mathematical model to predict the thermal performance of these collectors.

2.2.3 Collector Models—Concentrating

Many different geometries are possible with concentrating collectors. The detailed analyses of common designs have been studied by Winston (1974), Winston and Hinterberger (1975), Welford and Winston (1978, 1981), Rabl (1976a, 1976b), and others. The optical and heat transfer characteristics of concentrating collectors are difficult to generalize for use in simulation programs. However, the Hottel-Whillier model, developed for flat-plate collectors, can be used if it is recognized that the area intercepting solar radiation is different from the heat loss area and that the optical losses must account for radiation that does not reach the absorber surface because it is outside of the collector "acceptance angle." The useful output of all collectors can be expressed in terms of an optical efficiency and a loss coefficient based on the net collector aperture area (instead of the actual loss area). With these definitions, equation 2.1 is also valid for concentrating collectors.

The optical efficiency of a concentrating collector is sometimes modified so that only beam radiation is considered in equation 2.1. For purposes of simulation, either approach is satisfactory. All simulation programs have algorithms to convert total radiation on a horizontal surface to either beam or total radiation on surfaces of any orientation. Of course, if experimental data are available for the beam and diffuse components of solar radiation, they should be used in the simulation. The problems of predicting radiation on surfaces of arbitrary orientation are discussed in chapter 3 of volume II of this series, *Solar Resources*.

The thermal loss coefficients of concentrating collectors are often strong functions of temperature due to the large effects of thermal radiation. If the operating range of the collector is known, then a linear model, linearized near this operating temperature, will usually yield satisfactory results when used in a simulation model. If these temperature effects are deemed too large, then a nonlinear model may be necessary. The simplest nonlinear model consists of tabular values of efficiency at various temperature levels and possibly also at various solar radiation levels. This type of information is generally obtained from collector tests.

The optical efficiency of concentrating collectors is usually a very strong function of the incidence angle of solar radiation. For very high concentration ratios, these effects are so important that two-axis tracking is necessary. If the beam radiation falls outside of a very narrow range, then the optical efficiency approaches zero. For modest concentration ratios, as encountered in parabolic trough collectors, only one-axis tracking is necessary. However, it is necessary to know the optical efficiency over a range of angles, since the beam solar radiation is not always normal to the collector axis. For low concentration ratios, such as those obtained with stationary or periodically moved compound parabolic concentrators (CPCs), it is also necessary to know the optical efficiency as a function of beam radiation incidence angle. Except for two-axis tracking collectors, measurements of the optical efficiency are usually performed in the two primary axes. Interpolation is necessary to determine the optical efficiency at any incidence angle. McIntire and Reed (1983) have proposed a method for interpolation.

2.2.4   Storage—Liquid

The modeling of fully mixed liquid storage tanks is very simple. The energy input from the collector minus the energy supplied to the load and the environment by losses must be equal to the rate of storage of energy in the tank. In equation form, this is expressed as

$$(mC_p)dT/dt = Q_u - Q_{\text{load}} - Q_{\text{loss}}. \tag{2.3}$$

The storage losses can be expressed as the tank conductance area product, UA_t, times the temperature difference between the tank and the tank's environment. Generally, theoretical values of UA_t underestimate tank losses unless connecting pipes and supports are included in the calculation.

Stratified tanks are very difficult to model except in the limit of perfect stratification. Kuhn, von Fuchs, and Zob (1980) present an algorithm for predicting the maximum thermal performance of a storage tank. Their model, which has been incorporated into TRNSYS, is algebraic and does not involve differential equations. Fluid that enters the tank seeks the location where no temperature inversions will occur. This assumption results in the maximum degree of stratification. Wuestling, Duffie, and Klein (1983) have used this model to determine the optimum flow rate in a liquid-based water heating system.

The variable-volume tank permits different inlet and outlet flow rates. The level is allowed to vary between fixed high and low limits. When the low level is reached, the load flow rate is adjusted to maintain the level. When the high limit is reached, the collector flow rate must be reduced to prevent overfilling.

The models available for predicting the complete hydrodynamic conditions within a tank are much too time-consuming to be used in conventional simulations. The two limits of perfect stratification and perfect mixing can give the limits on thermal performance of a system. Consequently, a complete solution to the complex mixing problem within a tank is often not necessary for predicting system performance.

2.2.5   Storage—Pebble Bed

The original Schumann (1929) model of heat transfer in a packed bed is the basis of all recent models of pebble bed storage systems. The basic assumptions are "plug" flow, no axial conduction, constant properties, no mass transfer, and no temperature gradients within the pebbles. These conditions lead to two partial differential equations for the fluid and bed temperatures. The form of these equations requires very small space and time increments for numerical solutions. Hughes, Klein, and Close (1976) and Mumma and Marvin (1976) recommend a numerical technique for numerically solving the Schumann equations. Their results have been checked experimentally by von Fuchs (1981). Kuhn, von Fuchs, and Zob (1980) reviewed the many available analytical and numerical solutions to the Schumann equations.

For many practical designs, heat transfer between air and pebbles is such that the air and pebbles are at essentially the same temperature at any level in the bed. With this assumption, Hughes, Klein, and Close (1976) propose an infinite NTU (Number of Transfer Units) model so that only one differential equation is necessary. For large values of NTU (i.e., greater than 10), the predicted thermal performance of a system using the infinite NTU model is within 95% of the predicted performance using the complete equations. Actually, their model introduces numerical distortion of the thermal wave front that coincidentally approximates the actual situation.

The TRNSYS model uses the infinite NTU approach for solving the Schumann equations and recommends five as the minimum number of nodes. This number represents a compromise between accuracy and simulation cost. If the NTU is less than 10, corresponding to very slow flow, then this model will not give satisfactory results for predicting long-term system performance.

Work at the University of Waterloo has investigated rock bed conduction and flow nonuniformity (Hollands, Sullivan, and Shewen 1984).

2.2.6   Storage—Phase Change

Models of phase-change energy storage systems are similar to pebble bed storage models. Morrison and Abdel-Khalik (1978) developed two coupled partial differential equations for the fluid temperature and phase-change material internal energy. They numerically solved the equations for a variety of geometric and thermodynamic properties and showed that the infinite NTU model, similar to that in a pebble bed storage unit, is a good approximation for most practical systems. Their model assumed complete reversibility and no temperature gradients in the storage material, and had no subcooling or superheating so that the performance estimates are somewhat optimistic. Jurinak and Abdel-Khalik (1979) used the same model and developed a relationship between the thermodynamic properties of the phase-change material and the effective thermal capacitance of a hypothetical pebble bed. The hypothetical pebble bed heat capacity was determined such that when it replaced the phase-change storage unit, the two systems produced the same thermal performance.

2.2.7 Storage—Chemical

Models of chemical storage units have been developed by Offenhartz (1976) and Offenhartz et al. (1980) for use in TRNSYS. However, because the technology is very young, no general models are available. Studies of chemical heat pumps and other reversible chemical reactions for use in solar applications have led to specific models. This technology is in a very early stage of maturity, and models will be generated as necessary to study particular processes.

2.3 Weather Models

The two major forcing functions that drive the models in a solar simulation are the weather and the load (which may or may not be weather-driven). The U.S. National Climatic Center (1978) has a variety of data tapes including SOLMET data from 26 locations from 1952 to 1975 (see table 2.2).

The data from each of the 26 cities have been processed into a typical meteorological year (TMY) to represent long-term weather conditions for

Table 2.2
SOLMET weather data stations

Fort Worth, TX	Ely, NV
Lake Charles, LA	Phoenix, AZ
Columbia, MO	Santa Maria, CA
Apalachicola, FL	Bismarck, ND
Miami, FL	Great Falls, ND
Brownsville, TX	Medford, OR
Charleston, SC	Seattle, WA
Nashville, TN	Fresno, CA
Dodge City, KS	Cape Hatteras, NC
Caribou, ME	Sterling, VA
Madison, WI	Boston, MA
El Paso, TX	New York, NY
Albuquerque, NM	Omaha, NE

each location. The data are from actual months, but months from different years were selected to form a complete composite year. Thus, a year may consist of January 1955, February 1962, etc. The transitions from month to month were smoothed to avoid discontinuities. The process for selecting a particular month was first to calculate the mean, standard deviation, and other statistics for each of the measurements. Weighting factors were assigned to each measurement, and the month that most closely represented the weighted average was selected. This process has been shown to yield excellent agreement between the long-term system performance calculated with ten years of data and the performance calculated with the typical meteorological year. These TMY data have been reformatted for use in TRNSYS and are available from the University of Wisconsin Solar Energy Laboratory.

Some cautions are necessary in using this type of data. Extremes in weather are masked by this process, which may lead to incorrect predictions. For example, Braun (see Duffie and Beckman 1991, pp. 542, modeled a seasonal storage solar system for 21 consecutive years. The average year simulation predicted that no auxiliary energy would be required, whereas only 4 of the 21 consecutive years were shown by simulation not to require auxiliary energy.

Another danger can arise if the TMY data do not represent the actual correlations between variables. For example, the correct average dry bulb temperature and solar radiation may be reproduced on a TMY data set, but the two variables may not occur in the correct sequence. This could result in different predictions for a solar air-conditioning system when using TMY data compared with a 20-year simulation.

Some attempts have been made to estimate long-term performance by considering weather sequences that are much shorter than a year. The goal is to find a few weeks of weather data that yield the same predicted annual performance as would be found from a multiyear simulation. Unfortunately, no general rule for selecting the short sequence has been found. It is possible to select such a sequence from a particular application, but each application requires a new study. A significant effort is required to ensure that the selected short year truly represents the correct long-term average for the important weather variables.

The general conclusion is that if large sums of money are involved in constructing a system or if the system is to be a public demonstration, one should simulate the performance with as much weather data as are available. If only long-term trends or feasibility studies are required, TMY data should be adequate.

2.4 Load Models

Except for domestic hot water, the most common load for a solar system is a building requiring heating. The modeling of building energy requirements is the subject of volume IV of this series and will not be treated here except to indicate how the time variation of loads affects solar system performance. A number of studies have shown that if the solar system storage is sized for approximately one day, then the solar system performance will not depend significantly upon the time distribution of the load if the load is distributed "reasonably" throughout the day. However, the exact nature of this effect is difficult to quantify, and extreme situations will usually produce significant reductions in performance. For example, a system without storage that has a nighttime load will have no solar contribution, whereas a similar system that meets only a small fraction of a daytime load will have excellent solar performance. A good example of a system without storage that has excellent solar performance is a house heating system that is designed to supply a small fraction of the annual load. (Actually, the house itself does provide some energy storage.)

Day-to-day variations in the load, for example, the variations that might be expected in a domestic hot water system with laundry use concentrated on one day, can result in a dramatic reduction in system performance when compared with a more uniform load profile. However, the design of solar systems usually involves comparing two or more alternatives with the same load profile. Even if the load profile is not known with great precision, the choice between two designs involving the same system type generally will be clear based on economic considerations.

2.5 Validation

The validation of a complex computer code is very difficult. Possibly the best validation procedure is code-to-code comparisons. When differences arise, detailed investigations will lead to discussions as to the "correct" way the problem should be solved. This interaction is invaluable in developing useful codes. A requirement is that at least two different codes must be available for comparison.

Presumably, all codes are tested to ensure they satisfy the first and second laws of thermodynamics. Generally, first law violations are easy to discover and are usually the result of programming errors and not conceptual difficulties. Second law violations are harder to discover, but careful analysis of both instantaneous and integrated energy transfers can point out problems. The major differences in simulation codes, as far as the validity of the output is concerned, stem from different assumptions about the heat-transfer rate equations.

The validation of solar system simulation codes was initiated in the late 1970s by the Energy Research and Development Administration in the Systems Simulation and Economic Analysis (SSEA) working group. This group of researchers from the Los Alamos Scientific Laboratory, the Solar Energy Research Institute, the National Bureau of Standards, Sandia Laboratory, the University of Texas, the University of Wisconsin, and Colorado State University compared the predicted output of their programs for specified systems. The systems were selected to be very simple so that the reasons for observed differences could be easily determined. Initially, differences were found in conventions and assumptions among the programs. Some of the differences were easily removed, such as the use of solar time in some programs and local time in others. Some of the differences were not resolved so easily. The breakup of total radiation into its beam and diffuse components was found to be a major cause of differences predicted by the programs. This is still a subject of research by DOE laboratories and contractors.

These early code-to-code comparisons by the researchers directly involved in the development of the programs were essential for the development of reliable algorithms. Without these comparisons, errors would probably have existed in some of the programs for many years.

Other comparisons have been made between various codes by SERI personnel who were not directly involved in the code generation. Wortman, O'Doherty, and Judkoff (1981) compared three passive programs against simple structures that could be modeled analytically. With this technique, the "correct" answer is known. Some errors were discovered in the codes using these comparisons. This method can test only selected, and generally simple, portions of the codes and does not necessarily provide a better validation method than code-to-code comparisons of complex systems. Eden and Morgan (1981) compared the solar portion of DOE-2 (i.e., the Component Based Simulator, CBS) and TRNSYS, and concluded that the agreement between the two programs was remarkable.

The small working-group meetings of the late 1970s led to the establishment of large technical meetings sponsored by DOE (System Simulation and Economic Analysis, 1978 and on and which evolved into the ASME Annual Solar Energy Conference). Through these SSEA meetings, a large body of researchers has been made aware of problems and solutions. These meetings have continued, providing a forum for researchers to discuss advances in modeling techniques.

Following the success of the SSEA code comparisons, the International Energy Agency formed a working group to compare programs from different countries. A similar set of code-to-code comparisons gave confidence that models developed by different research groups with very different modeling philosophies could generate codes that produced similar results. The results of these comparisons are reported by Freeman and Hedstrom (1980).

The comparison of detailed experimental measurements with predictions from a simulation model is a difficult task. The experimental difficulties of keeping recording equipment running over long periods of time mean that only relatively short time spans of uninterrupted data are available. Mitchell, Beckman, and Pawelski (1980) reported on a study of the Colorado State University (CSU) solar house in which three time periods were available with continuous data of up to three weeks in duration. One of the periods was used to "calibrate the model," that is, the quantities that could not be measured were inferred by comparing experimental measurements with predictions. The other two periods were simulated without further adjustments to the model. For this particular experiment, the differences between measured integrated energy quantities and predicted values differed by less than 7%. This agreement is within the accuracy of the measurements.

2.6 Conclusions

The purpose of simulation is to gain a better understanding of a process. When a system is well understood, detailed simulations should not be necessary, at least for preliminary designs. For example, a few years ago even the smallest systems were simulated (or the designer simply guessed at the performance), but today the interactions between the various components are well understood, and design methods, such as f-chart (see chapter 3), have replaced detailed simulations of most domestic hot water

systems and building heating systems. New designs should be simulated
before they are built. Although this procedure does not ensure a proper
design, many potential problems can be resolved in a much less expensive
manner than by building costly experiments. Of course, at some point the
actual hardware must be put together and tested under real conditions.
The results of some of these experiments must then be used to check the
simulation. This information exchange between experiments and simulations makes it possible for future simulation programs to predict actual
performance of even more complex systems.

References

Bliss, R. W. 1959. The derivation of several "plate efficiency factors" useful in the design of
flat-plate solar-heat collectors. *Solar Energy* 3:55.

Brandemuehl, M., and W. A. Beckman. 1980. Transmission of diffuse radiation through CPC
and flat-plate collector glazings. *Solar Energy* 24:511.

Buchberg, H., and J. R. Roulet. 1968. Simulation and optimization of solar collector and
storage for house heating. *Solar Energy* 12:13.

Butz, L. W., W. A. Beckman, and J. A. Duffie. 1974. Simulation of a solar heating and cooling
system. *Solar Energy* 16:129.

Close, D. J. 1967. A design approach for solar processes. *Solar Energy* 11:112.

Duffie, J. A., and W. A. Beckman. 1991. *Solar engineering of thermal processes* 2nd Ed. New
York: Wiley.

Dunkle, R., and P. I. Cooper. 1975. A proposed method for the evaluation of performance
parameters of flat-plate solar collectors. *Proc. of ISES Conf.* Los Angeles, CA.

Eden, A., and M. Morgan. 1981. A comparison of DOE-2 and TRNSYS solar heating
simulation. *Proc. of ASME Annual Solar Energy Conf.* Reno, NV.

Fewell, M. E., and N. R. Grandjean. 1978. *User's manual for computer code SOLTES-1*
(Simulation of Large Thermal Energy Systems). SAND 78-1315 (revised), also *Proc. SSEA*,
1981.

Freeman, T. F., and J. C. Hedstrom. 1980. Validation of solar system simulation codes by the
International Energy Agency. *Proc. of Systems Simulation and Economic Analysis Conf.* San
Diego, CA.

Grunes, H. 1976. Utilization and operational characteristics of evacuated tubular solar collectors using air as the working fluid. M.S. thesis, Univ. of Wisconsin-Madison.

Hollands, K. G. T., H. F. Sullivan, and E. C. Shewen. 1984. Flow uniformity in rock beds.
Solar Energy 32:343.

Hottel, H. C., and B. B. Woertz. 1942. Performance of flat-plate solar collectors. *Trans.
ASME* 64:91.

Hottel, H. C., and A. Whillier. 1958. Evaluation of flat-plate collector performance. *Trans. of
the Conf. on the Use of Solar Energy* 2(1):74.

Hughes, P. J., S. A. Klein, and D. J. Close. 1976. Packed bed thermal storage models for solar
air heating and cooling systems. Transactions of ASME, *J. Heat Transfer* 98:336.

Jurinak, J. J., and S. I. Abdel-Khalik. 1979. Sizing phase change energy storage systems for air-based solar heating systems. *Solar Energy* 22:355.

Klein, S. A., et al. 1990. *TRNSYS—A transient system simulation program, users manual*, 38-13. The University of Wisconsin-Madison Engineering Experiment Station.

Klein, S. A., J. A. Duffie, and W. A. Beckman. 1974. Transient considerations of flat-plate solar collectors. Transactions of ASME, *J. Engin. for Power* 96A:109.

Kuhn, J. K., G. F. von Fuchs, and A. P. Zob. 1980. *Developing and upgrading of solar system thermal energy simulation models*. Final Report. Seattle, WA: Boeing Computer Services.

Kutscher, C. F. 1983. *The development of SOLIPH—A detailed computer model of solar industrial process heat systems*. SERI/TP-253-183. Golden, CO: Solar Energy Research Institute. Also in *Proc. of the ASME Solar Energy Division Fifth Annual Conference*. Orlando, FL, April 1983.

Löf, G. O. G., and R. A. Tybout. 1973. Cost of house heating with solar energy. *Solar Energy* 14:253.

McIntire, J., and K. A. Reed. 1983. Orientational relationships for optically non-symmetric solar collectors. *Solar Energy* 31:405.

Mather, G. R. 1982. Transient response of solar collectors. Transactions of ASME 104:165.

Mather, G. R., and D. F. Beekley. 1976. Performance of an evacuated tubular collector using non-imaging reflectors. *Proc. AS of ISES Conf.* Winnipeg, Canada.

Merriam, R. L., and R. J. Rancatore. 1982. *Evaluation of existing programs for simulation of residential building energy use*. EA-2575 Final Report. Palo Alto, CA: Electric Power Research Institute.

Mitchell, J. W., W. A. Beckman, and M. J. Pawelsky. 1980. Comparisons of measured and simulated performance for CSU house 1. *J. Solar Energy Eng.* 102:192.

Morrison, D. J., and S. I. Abdel-Khalik. 1978. Effects of phase-change energy storage on the performance of air-based and liquid based solar heating systems. *Solar Energy* 20:57.

Mumma, S. D., and W. C. Marvin. 1976. *A method of simulating the performance of a pebble bed thermal energy storage unit*. ASME paper 76-HT-73.

Offenhartz, P. O'D. 1976. Chemical methods of storing thermal energy. *Proc. AS of ISES* 8:48. Winnipeg, Canada. See also SSEA 1980.

Offenhartz, P. O'D., F. C. Brown, R. W. Mar, and R. W. Carling. 1980. A heat pump and thermal storage system for solar heating and cooling based on the reaction of calcium chloride and methanol vapor. Transactions of ASME, *J. Solar Energy Eng.* 102:59.

Ortabasi, U., and W. M. Buehl. 1975. Analysis and performance of an evacuated tubular collector. *Proc. of ISES Conf.* Los Angeles, CA.

Phillips, W. F. 1979. The effects of axial conduction on collector heat removal. *Solar Energy* 23:187.

Rabl, A. 1976a. Comparison of solar concentrators. *Solar Energy* 18:93.

Rabl, A. 1976b. Optical and thermal properties of compound parabolic concentrators. *Solar Energy* 18:497.

Schumann, T. E. W. 1929. Heat transfer: a liquid flowing through a porous bed. *J. Franklin Institute* 208:504.

Sheridan, N. R., K. J. Bullock, and J. A. Duffie. 1967. Study of solar processes by analog computer. *Solar Energy* 11:69.

Solar Energy Research Institute. 1980. *Analysis methods for solar heating and cooling applications*. SERI/SP-35-232R, 2nd ed. Golden, CO: Solar Energy Research Institute.

Souka, A. F., and H. H. Safwat. 1966. Optimum orientations for the double exposure flat-plate collector and its reflectors. *Solar Energy* 10:170.

Speyer, E. 1965. Solar energy collection with evacuated tubes. Transactions of ASME, *J. Eng. for Power* 87:270.

System Simulation and Economic Analysis. *Proc. of SSEA*, 1978 through 1983.

Theuinisen, P. H., and W. A. Beckman. 1985. Optical characteristics of tubular solar collectors. *Solar Energy* 35:311.

U.S. National Climatic Center. 1978. *SOLMET Manual*. Vols. 1, 2. Asheville, NC.

von Fuchs, G. F. 1981. Validation of a rock bed thermal energy storage model. *Proc. of ASME 3d Annual Solar Energy Conf*. Reno, NV.

Welford, W. T., and R. Winston. 1978. *The optics of nonimaging concentrators*. New York: Academic Press.

Welford, W. T., and R. Winston, 1981. Principles of optics applied to solar energy concentrators. Chapter 3 in *Solar Energy Handbook*, ed. J. F. Kreider and F. Kreith. New York: McGraw-Hill.

Winston, R. 1974. Solar concentrators of novel design. *Solar Energy* 16:89.

Winston, R., and H. Hinterberger. 1975. Principles of cylindrical concentrators for solar energy. *Solar Energy* 17:255.

Wortman, D., B. O'Doherty, and R. Judkoff. 1981. The implementation of an analytical verification technique on three building energy analysis codes. *Proc. of ASME 3rd Annual Solar Energy Conf*. Reno, NV.

Wuestling, M. D., S. A. Klein, and J. A. Duffie. 1985. Promising control alternatives for solar heating systems. ASME, *J. Solar Energy Eng*. 107:215.

York, D. A., and E. F. Tucker, eds. 1980. *DOE-2 reference manual (version 2.1)*. LA-7689-M. Los Alamos, NM: Los Alamos Scientific Laboratory.

3 Design Methods for Active Solar Systems

S. A. Klein

3.1 Introduction

The performance of all solar energy systems depends on the weather. In a solar heating system, both the energy collected and energy demanded (i.e., the heating load) are functions of the solar radiation, the ambient temperature, and other meteorological variables. The weather may be viewed as a set of time-dependent forcing functions imposed upon solar energy systems. These forcing functions are neither completely random nor deterministic; they are best described as irregular functions of time both on short (e.g., hourly) and long (e.g., seasonal) time scales.

This irregular behavior of the weather complicates the analyses of solar energy systems. In general, solar energy systems exhibit a nonlinear dependence on the weather, which is further complicated by time lags introduced from thermal capacitance or storage effects. It is usually not possible to accurately analyze these systems by observing their response to short-term or average weather conditions. Because the forcing functions are time-variant for both the short and long term, the analyses of these systems often require an examination of their performance over a long time period. As a result, experiments are very expensive and time-consuming, and it is difficult to vary parameters to see their effect on the system performance.

Computer simulations can provide analyses of solar energy systems. When supplied with meteorological data, mathematical models can be formulated to simulate the transient performance of these systems. The models provide the same information on thermal performance as physical experiments with significantly less time and expense. Simulations can be used directly as a design tool by repeatedly simulating the performance of the system with different values of the design parameters. The use of simulations in this way is often warranted in the design of large or complex systems (e.g., the heating/cooling systems of institutional buildings) or new types of systems (e.g., chemical heat pumps). The general transient system simulation program, TRNSYS (Klein et al. 1973) has proved useful for these applications.

However, the direct use of detailed computer simulations in the routine design of each solar energy system is not satisfactory. Much time, money, and expertise is required to obtain useful results from a simulation. In addition, access to computer facilities and hourly meteorological data for the location of interest are required. The widespread use of solar energy requires simplified design procedures, especially for standard types of systems where the use of simulations cannot be justified. Few design methods for solar energy systems were available before federal sponsorship of solar energy research.

A design method is defined as a means of estimating the long-term average performance of a particular type or class of solar energy system, which requires relatively little calculation effort and uses readily available meteorological data. The results obtained from a design method are not expected to be as accurate or as complete as the information available from a detailed simulation program. Experience has shown, however, that design methods using local climatic data can provide sufficiently accurate estimates of system performance for design and economic evaluation purposes.

The following sections of this chapter review the design methods that have been developed for active solar energy systems. For classification purposes, existing design methods are categorized and discussed as correlation methods, simplified simulations, or utilizability-based methods.

3.2   Correlation Methods

Correlations are empirical or semiempirical relationships of long-term system performance to system parameters and local meteorological data. Correlations are developed by first identifying important design variables or variable groups and then using the results from simulations (or other correlations based on simulations) to develop relationships between these variables and long-term system performance. This approach is analogous to that used in other engineering problems where numerous physical measurements are made and correlated. In solar processes, simulations are used because they can be controlled and repeated, and because they are far less costly than physical experiments.

The correlation design methods that have been developed for active solar systems are reviewed in this section. Note that a separate classification has been made for design methods that make use of the solar radiation utilizability function. These methods, which could also be classified as correlation methods, are reviewed in section 3.4.3.

3.2.1 The F-Chart Method

The f-Chart method (Klein, Beckman, and Duffie 1976, 1977; Beckman, Klein, and Duffie 1977) provides a means of easily determining the thermal performance of active solar space heating systems (using either liquid or air as the working fluid) and solar domestic hot water (SDHW) systems. The f-Chart method is essentially a correlation of the results of hundreds of simulations of solar heating systems. The conditions of the simulations were varied over the ranges anticipated in practical system designs. The resulting correlations give f, the fraction of the monthly heating load (for space heating and hot water) supplied by solar energy as a function of two dimensionless variables involving collector characteristics, heating loads, and local weather. Meteorological data required for the f-Chart method are monthly average daily radiation on the collector and monthly average ambient temperatures.

3.2.1.1 System Configurations

A schematic diagram of the water storage solar heating system treated by the f-Chart method is shown in figure 3.1. This system may use either water or an antifreeze solution as the collector heat-transfer fluid. Energy is stored in the form of sensible heat in a water storage tank. If the collector heat-transfer fluid is not water, a heat exchanger is used between the collectors and the tank. A water-to-air heat exchanger is used to transfer heat from the storage tank to the building. An additional water-to-water heat exchanger is used in systems having a domestic water heating subsystem to transfer energy from the main storage tank to a domestic hot water preheat tank. A conventional furnace and domestic water heater are provided to supply energy when the energy in the storage tanks is depleted. Controllers, relief valves, pumps, and piping make up the remaining equipment.

The standard configuration of a solar air heating system with a pebble bed storage unit is shown in figure 3.2. Other arrangements of fans and dampers can be devised to result in an equivalent flow circuit. Energy required for domestic hot water is provided in some systems by heat exchange with the hot air leaving the collector to a domestic water preheat tank, as in the water storage system. The hot water is further heated, if necessary, by a conventional water heater.

The f-Chart correlation for solar domestic water heating system performance was developed for the two-tank system shown in figure 3.3.

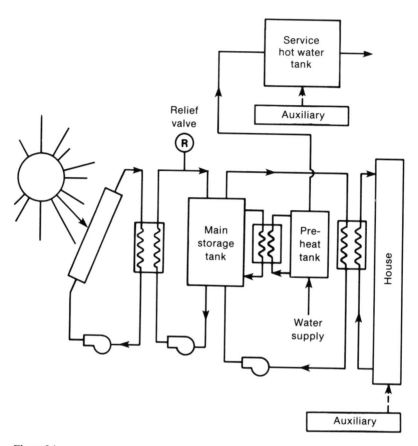

Figure 3.1
Water storage space and domestic water heating system. Source: Beckman, Klein, and
Duffie 1977.

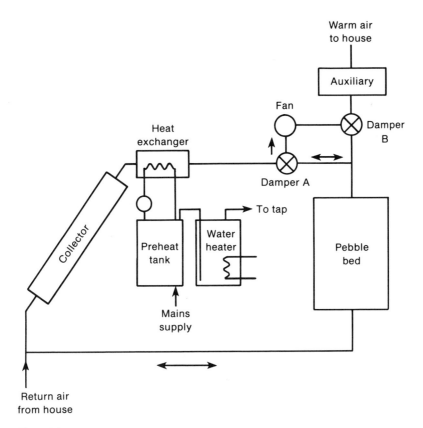

Figure 3.2
Pebble bed storage space and domestic water heating system. Source: Beckman, Klein, and Duffie 1977.

In this system, energy from the collector loop is transferred from the collector fluid to the preheat storage tank directly or via an external heat exchanger. (A direct system in which water from the preheat tank is circulated through the collectors is thermally equivalent to a system having a perfect heat exchanger.) A typical-use profile, developed by Mutch (1974), was assumed in the development of the correlation. Losses from the auxiliary tank were not considered in the correlation.

3.2.1.2   Chart Calculation Method
The f-Chart method is based on identification of the important dimensionless variables of solar heating systems and the use of detailed computer

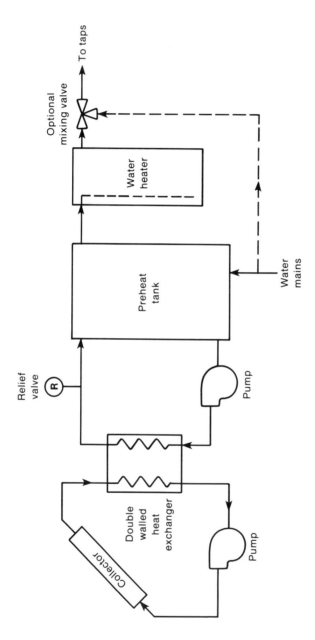

Figure 3.3
Two-tank solar domestic hot water system. Source: Braun, Klein, and Pearson 1983.

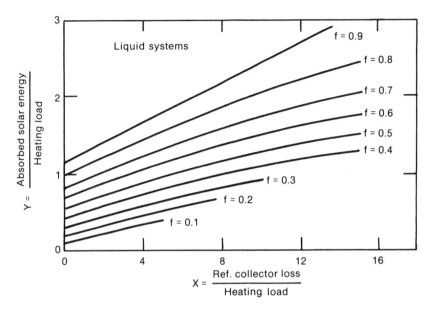

Figure 3.4
F-chart for water storage systems. Source: Beckman, Klein, and Duffie 1977.

simulations to develop correlations between these variables and the long-term system performance. The correlations, in graphical and equation form, are referred to as the "f-Charts."

The two dimensionless variables found to be of primary importance in solar heating systems are X and Y, where

$$X = \frac{\text{reference collector energy loss during a month}}{\text{total heating load during a month}},$$

$$Y = \frac{\text{total energy absorbed on the collector plate during a month}}{\text{total heating load during a month}}.$$

In the f-Chart method, system performance is expressed in terms of f, which is the fraction of the heating load supplied by solar energy during each month. (It is assumed that auxiliary energy supplies the remaining part of the load.) The relationship between f and the dimensionless variables X and Y is shown in figures 3.4 and 3.5 for the water and pebble bed storage systems, respectively.

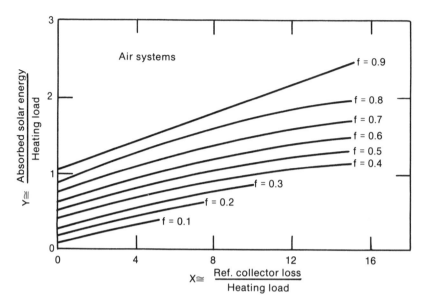

Figure 3.5
F-chart for pebble bed storage systems. Source: Beckman, Klein, and Duffie 1977.

The equations for X and Y can be written as follows:

$$X = F_R U_L \times (F_R'/F_R) \times (T_{\text{ref}} - \overline{T}_a) \times (\Delta t) \times (A/L),$$

$$Y = F_R(\tau\alpha) \times (F_R'/F_R) = (\overline{\tau\alpha})/\tau\alpha) \times \overline{H}_T \times (A/L).$$

$F_R U_L$ and $F_R(\tau\alpha)$ can be obtained theoretically (Duffie and Beckman 1980) or from the slope and intercept of the ASHRAE 93-77 (1977) collector test results. F_R'/F_R is a collector-heat exchanger penalty factor that can be estimated as described by de Winter (1975). T_{ref} is a reference temperature chosen to be 212°F (100°C) and Δt is the number of hours (or seconds, depending on the unit system) in a month. $(\overline{\tau\alpha})/(\tau\alpha)$ is a factor that accounts for the variation in absorbed radiation with the solar incidence angle (Klein 1979). A is the collector area and L is the monthly total load on the solar energy (and auxiliary) system. The meteorological data required are $\overline{T}_a$, the monthly average ambient daytime temperature, and $\overline{H}_T$, the monthly average solar radiation per unit area incident on the collector plane. Radiation data on tilted surfaces are generally not available and must be estimated from horizontal data. One method for converting horizontal data for tilted surfaces is described by Klein and Theilacker (1981).

The f-Chart method proceeds as follows. The dimensionless variables X and Y are calculated for each calendar month. The intersection of the X and Y values in figure 3.4 or 3.5 locates a value of f which is the fraction of the heating load supplied by solar energy for that month. The calculations are repeated for each month. The fraction of the annual heating load supplied by solar energy, F, is found from

$$F = \sum fL / \sum L.$$

Economic calculations ordinarily require only the value of F to describe the thermal performance of the system.

Three system design parameters were held at fixed values in the development of the f-Charts. These are the storage capacity per unit collector area, the collector fluid flow rate per unit collector area, and the relative load heat exchanger size (for water storage systems). Increasing the storage capacity of the pebble bed or water tank from the value assumed in the development of the f-Charts results in a slight increase in solar fraction, since the increased storage capability is offset by a larger area for storage energy losses. The effect of changes in storage capacity are considered by a correction factor to the value of X. Increases in the collector flow rate cause the collector heat removal factor, F_R, to increase, which tends to increase collector efficiency. A countereffect, however, is that increasing the flow rate tends to reduce the degree of thermal stratification in the storage unit. This countereffect is considered in the f-Chart for pebble bed storage systems by correcting the value of X as a function of flow rate. A correction factor for water storage (space and domestic hot water) was not developed because the storage tank was assumed to be fully mixed (unstratified). Wuestling et al. (1983) have shown that the improvement in stratification resulting from reduced collector flow rates can more than offset the reduction in the collector heat removal factor in domestic hot water systems. The size of the load heat exchanger can have a significant effect on the performance of a water storage space heating system. When this heat exchanger is reduced in size, the storage tank temperature must be increased to supply the same amount of heat. The increased storage temperature results in a reduction in collector efficiency. This effect is considered by use of a correction factor that alters the value of Y.

The calculations required to apply the f-Chart method can be done graphically by hand or more simply with the aid of an inexpensive electronic calculator that can evaluate the exponential functions used to curve-fit the f-Charts. Alternatively, programs for personal computers are available to do these calculations.

3.2.1.3 Validation

The f-Chart method has been validated in two ways. First, the performance predictions of f-Chart have been compared with calculations made by TRNSYS for 14 locations in the United States by Beckman, Duffie, and Associates (1976). The standard error of the differences between the TRNSYS simulations and the f-Chart results was about 2.5%. Secondly, f-Chart has been compared with experimental results. Many experimental data on solar heating system performance have become available since the f-Chart method was developed. Fanney and Liu (1980) and Fanney and Klein (1983) at the National Bureau of Standards have comparet the experimental performance of five active solar domestic hot water systems with f-Chart predictions for a two-year period. In these latter comparisons, the f-Chart predictions were found to be 1% to 5% lower than the experimental measurements. Duffie and Mitchell (1983) have compared f-Chart predictions and experimental performance for 20 space heating and 11 solar domestic hot water systems. They conclude that good agreement between experimental and predicted performance is obtained for systems having the configurations assumed in the development of the f-Chart method. Other studies have also shown good agreement between experimental results and f-Chart predictions (Kelly 1982).

3.2.1.4 Limitations and Disadvantages

The f-Chart method is widely used within the solar community. It has been used for both design of new systems and performance analyses of existing systems. Although very useful in the applications for which it was developed, the f-Chart has limitations.

The f-Charts were developed for standard types of solar space and domestic water heating systems. The parameter ranges for which the method is applicable are limited. Using the f-Chart method for systems other than those for which it was originally developed or for design parameters outside the range of the correlations may produce erroneous results. Although the systems and parameter ranges for which the f-Charts are applicable are clearly indicated in the original references, the method has been incorrectly applied to space heating systems with no storage, solar cooling systems, and systems having control strategies that differ from those for which the method was designed.

Performance estimates obtained from the f-Chart method have been shown to be in good agreement with the experimental performance of many carefully monitored systems (see section 3.3.1.3). However, the f-Chart method was not intended to be used for performance prediction and cannot be expected to predict the performance of real systems where there are many factors (such as improper installation) that prevent a system from operating as designed.

To calculate the annual system performance, the f-Chart method requires twelve monthly performance evaluations. Although the f-Chart correlations are relatively simple, they can result in significant computational effort when the method is used repetitively, as it often is in searching for an optimum design. Moreover, the calculation of the monthly heating loads and monthly solar radiation on the collector surface (which are inputs to the f-Chart method) are involved. The calculational effort required by the f-Chart method has led several investigators to seek simpler methods.

The f-Chart method estimates only the thermal performance of a system. Parasitic energy use by the pumps, fans, and controllers is not accounted for. (Parasitic energy use is ordinarily a small fraction, e.g., 5%, of the total energy use in well-designed systems.) The inclusion of parasitic energy along with the required auxiliary energy in the reports of Chaffin and Wessling (1982) on solar domestic hot water explains, in part, why the f-Chart predictions of solar fractions were found to be higher than measured.

The only meteorological data input to the f-Chart method are monthly solar radiation and ambient temperatures. Specific local effects, such as high mountains or typical morning fog, cannot be accounted for. Evidence suggests (Beckman, Duffie and Associates 1976, Mosley and Clark 1981) that the f-Chart method significantly underpredicts system performance in overcast climates, e.g., in Seattle, Washington, during the winter months. The underprediction occurs because the distribution of solar radiation is very different from that built into the f-Charts. The distribution of solar radiation is more directly accounted for by the utilizability methods reviewed in section 3.4.

3.2.2   Correlations Based on the F-Chart Method

Numerous investigators have developed correlations of space and domestic water heating system performance based on results obtained from f-Chart rather than from detailed simulations. Correlations of this kind inherently contain all of the restrictions of the f-Chart method plus additional restrictions, as noted. Their major advantage is that they provide a means of designing solar energy systems with less computational effort than required by the f-Chart method.

Computing capabilities have greatly increased in the last few years. Programs that quickly do the calculations required for the f-Chart method are available for personal computers. The current low cost of computing equipment and the availability of these programs reduce the advantage of the simpler but more restrictive methods described in this section.

3.2.2.1   Ward's Correlation

Ward (1976) used the f-Chart method to calculate the performance of water storage solar space heating systems in 11 locations with three collector types. In these calculations, Ward assumed that: (1) there is no heat exchanger between the collector and storage; (2) domestic water heating is not provided; (3) the collectors face the equator with a slope equal to the latitude plus 15 degrees; and (4) the storage and load heat exchanger sizes are the standard values assumed in the construction of the f-Charts. He then correlated the annual solar fraction, F, to an equation of the form

$$F = a + b \ln(AS_1/L_1),$$

where A is the collector area, S_1 is the total solar radiation per unit area on a horizontal surface in January, L_1 is the January heating load for the building, and a and b are constants dependent on the collector design. A disadvantage of Ward's method, aside from its restricted applicability, is the need to determine the constants a and b for each collector type. In addition, Huck (1976) has found the accuracy of Ward's method to depend somewhat on location.

3.2.2.2   The Relative Areas Method

Barley and Winn (1978) used the f-Chart method to calculate the annual solar fraction for space and domestic water heating systems, which was then correlated to a relationship of the form

$$F = C_1 + C_2 \ln(A/A_0),$$

where A_0 is the collector area that results in an annual solar fraction of 0.5, and C_1 and C_2 are location-dependent constants. A relationship to estimate A_0 was also developed in terms of the collector parameters $F_R'(\tau\alpha)$ and $F_R'U_L$, the house overall energy loss coefficient, and two additional location-dependent parameters. Restrictions of their method, beyond those inherent in the f-Chart method, are: (1) the degree-day model must be used to estimate the space heating load; (2) the collector orientation must be directly towards the equator at the latitude angle for SDHW systems and the latitude plus 15 degrees for space heating systems; and(3) the storage size, collector air flow rate, and load heat exchanger size must be the standard values assumed for f-Chart.

3.2.2.3   The GFL Method

Lameiro and Bendt (1978) developed a correlation for annual solar traction in the form of

$$F = 1 - \mathrm{Exp}(-RA - SA^2),$$

where R and S are quadratic functions of $F_R'U_L/F_R(\tau\alpha)$. The six coefficients for the R and S functions depend on system type (water or pebble-bed storage) and location. The values of these constants are given for 151 locations.

Lameiro and Bendt used the F-CHART 3 [Hughes et al., 1978] program to produce the data upon which their correlation is based. Their correlation is limited to systems that supply both space and domestic water heating. The water heating load is assumed to be 78 gal./day (300 liter/day) heated from 52° to 140°F (11° to 60°C). The correlation is not applicable for SDHW systems. The collectors are assumed to face south at the latitude angle. Correction factors for storage size, load heat exchanger size, and collector air flow rate are not provided.

3.2.2.4   The W-Chart Method

Kreider, Cowing, and Lohner (1982) have developed a correlation of annual solar fraction of the following form:

$$F = C_1 \ln(1 + C_2 A/L),$$

where C_1 and C_2 are location-dependent constants. The data for the correlation were obtained from the f-Chart method. This correlation is quite restrictive in that it assumes that: (1) the collector parameters, $F_R(\tau\alpha)$ and $F_R U_L$ are 0.73 and 1 Btu/ft^2 h °F (5 W/m^2 °C), respectively; (2) the collector-heat exchanger penalty factor is 0.96; (3) the main water temperature is 60°F (15.6°C); (4) the collector faces south at a slope equal to the latitude; and (5) the storage size, collector air flow rate, and load heat exchanger size are the standard values used in the f-Chart method.

3.2.2.5 Solgraf/DHW

Wright (1980) has used f-Chart to develop a series of four graphs that simplify annual predictions of SDHW system performance by hand. The user must obtain values for eight system parameters, just as if he were using the original f-Chart method. Then, performance factors are obtained from each of the graphs and multiplied together to give the annual productivity of the system. Wright indicates that this graphical procedure provides results that are within 5% of those obtained with the F-CHART 3 program.

3.2.3 Extensions to the F-Chart Method

The f-Chart method was developed originally for the three solar heating systems shown in figures 3.1, 3.2, and 3.3. The studies reviewed in this section are aimed at extending the types of system configurations for which the f-Chart method can be applied.

3.2.3.1 Alternative Solar Water Heating Systems

A variety of solar water heating system designs are available that differ from the system configuration originally assumed in the development of the f-Chart correlation. Examples are one-tank systems in which auxiliary energy is supplied at the upper portion of the tank, internal heat exchange systems in which the collector fluid flows through a jacket or coils within the preheat tank, and systems using tempering valves that mix hot water from the solar tank with cold supply water to achieve the desired hot water temperature. In addition, it was unclear how the load distribution affected the performance of a solar water heating system and how tank losses could be accounted for with the f-Chart correlation.

Buckles and Klein (1980) undertook a study to determine the applicability of the f-Chart method for systems having one storage tank, internal heat exchangers, tempering values and water-use patterns that differed from those assumed in the development of the f-Chart method. The findings from this study are as follows.

1. Losses from the auxiliary storage tank can be directly considered in the f-Chart correlation by adding them to the load required for the water heating before applying the f-Chart correlation. These energy losses are assumed to take place from the auxiliary heater set temperature to the environmental temperature through the overall area-conductance product of the tank.

2. The performance of one-tank as well as two-tank systems can be estimated with the f-Chart correlation. The auxiliary losses for a one-tank system are assumed to occur from the portion of the tank above the auxiliary heater thermostat sensor.

3. Tempering valves are an essential part of a solar domestic hot water system for safety reasons but have little effect (less than 2%) on monthly average system performance.

4. Heat exchangers, internal or external, of equivalent effectiveness result in the same thermal performance in a solar water heating system. The effectiveness of internal coils is, however, often lower than that of an external heat exchanger due to free convection heat transfer resistance at the interface between the coil and the water in the tank.

5. Differences in the hourly load use pattern from that assumed in the development of the f-Chart correlation have relatively little effect on monthly average system performance. The worst load use pattern in terms of efficient solar energy utilization is one in which hot water is required during early morning hours before energy collection begins. This load pattern maximizes the tank energy losses and reduces collector efficiency because of high daytime tank temperatures. The most favorable load use pattern is during the late afternoon. Buckles and Klein (1980) found a difference in solar fraction of 7% for a base-case system subjected to these extremes in load use patterns. Day-to-day variations in water use were found to have a significant effect. A series of simulations was run so that the total weekly draw was constant, but there was a wide variation in the day-to-day use. The effect of day-to-day use patterns was found to depend on the storage capacity. The day-to-day draw variation was found to decrease solar fraction for a base-case system by 8% compared with an identical system subjected to the same draw each day.

3.2.3.2 Phase-Change Energy Storage
Jurinak and Abdel-Khalik (1979) used TRNSYS to simulate the long-term performance of air-based solar heating systems. The results of these simu-

lations were then used to devise a correction factor to the f-Chart for pebble-bed storage systems so that it can be applied to systems having phase-change energy storage. The correction factor is in the form of an effective thermal capacitance for phase-change storage units in terms of the melting temperature, latent heat of fusion, and specific heats of the solid and liquid phases of the phase-change material. This effective capacitance is used in place of the thermal capacitance of the pebble bed in the relationship that adjusts the f-Chart X parameter to account for storage capacities differing from the standard value used in the development of the f-Chart.

3.2.3.3 Parallel Solar Heat Pump Systems

Anderson, Mitchell, and Beckman (1980) used simulations to investigate the performance of combined solar and heat pump systems in which the heat pump operates independently, i.e., in parallel with a conventional water or pebble bed storage solar heating system. The simulation results were used to derive a design method for this system configuration. They show that the f-Chart method can be used directly to determine the fraction of the load supplied by solar energy, since the solar system is the primary energy source and operates independently of the heat pump, which serves as a backup energy source. The solar energy system affects the heat pump by causing it to operate less often than a stand-alone heat pump during sunny and/or warm periods in which the solar energy system can supply a major fraction of the total load. The authors show, however, that the effect of the solar energy system on the seasonal performance of the heat pump is small. Thus, a conventional bin method can be used to estimate the performance of the heat pump in these systems with the load reduced by the contribution of the solar system.

3.2.3.4 Active-Passive Interactions

Evans, Klein, and Duffie (1985) used TRNSYS to simulate the hour-by-hour performance of combined active-passive space heating systems. The simulation results show that the presence of a passive system reduces the performance of the active system primarily by reducing the heating load on the active components and by altering the time distribution of the load to more unfavorable times (e.g., at night). The simulation results were then used to develop a general correction factor to be applied to the f-Chart method to account for the effects passive systems have on the performance of active systems.

3.2.4 The Solar-Load Ratio Method

A simple method of estimating the performance of water and pebble bed storage solar space heating systems was developed by Balcomb and Hedstrom (1976) by correlating the results from an (unpublished) hourly simulation program. The method was presented in detailed and simplified forms.

The detailed form involves monthly calculations of system performance similar to those used in the f-Chart method. For each month, the solar-load ratio (SLR) is computed as the ratio of the monthly solar radiation incident on the collector to the monthly building load. Then, the monthly solar fraction, f, is found from the following relationship:

$$f = \begin{cases} 1.06 - 1.366\,\mathrm{EXP}(-0.55\,\mathrm{SLR}) + 0.306\,\mathrm{EXP}(-1.05\,\mathrm{SLR}) \\ \qquad\qquad\qquad\qquad\qquad\qquad \text{for SLR} < 5.66 \\ 1 \text{ for SLR} > 5.66 \end{cases}$$

These calculations are repeated for each month, and an annual solar fraction is obtained as the load-weighted average of the monthly values in the same manner as in the f-Chart method.

The simplified method requires only a single annual calculation. A table of values of LC, the load-collector ratio for annual solar fractions of 25%, 50%, and 75%, has been drawn up from the results of the detailed method for 85 locations. To use this table, it is necessary to calculate the building thermal loss coefficient in units of Btu per degree-day. Then, the collector area needed to achieve an annual solar fraction of 25%, 50%, or 75% is found as the ratio of the building thermal loss coefficient to LC.

The primary disadvantage of the SLR method is that it is rather restrictive. Neither the SLR parameter nor the correlation for monthly solar fraction involves any system parameters other than the collector area. The correlation was developed for "standard" water and pebble bed storage systems with single-glazed, black surface collectors tilted at the latitude plus 10 degrees. The thermal storage was assumed to be 15 Btu/ft^2 °F (85 W/m^2 °C). Other factors, such as the load heat exchanger size, the collector flow rates, and pipe/duct losses, were also assumed for these standard systems.

The simulation program used to develop the SLR correlation was also used to produce the two solar design handbooks for Los Alamos, New

Mexico, and seven cities in the Pacific region (Balcomb et al. 1975, 1976). The handbooks present annual performance results for these locations for the "standard" systems. An important contribution of these handbooks is a series of sensitivity curves that indicate the effect of changes in the system design parameters from the values assumed in the standard systems. For example, sensitivity curves are provided for number of glazings, collector solar absorptance, load heat exchanger size, etc. It is unclear, however, how these sensitivity curves are to be used when two or more design parameters differ from the standard values.

In a later study, Schnurr, Hunn, and Williamson (1981) extended the SLR design method to space and water heating systems for commercial buildings. The DOE-2 simulation program (York and Tucker 1980) was used to investigate the performance of these systems. The simulation results were used to develop new correlations for these systems in the commercial building environment. These correlations are less restrictive than the original SLR correlations. However, the authors indicate that the correlations should only be used in the preliminary design phase.

3.2.5   Cooling Design Charts

Anand, Abarcar, and Allen (1980) and Anand and Kumar (1983) have developed a design method for solar cooling systems in which an absorption chiller is used. Long-term performance estimates for solar cooling systems were obtained from the SHASP (1978) simulation program. In a manner analogous to the development of the f-Charts for space and water heating, f_c, the monthly fraction of the cooling load supplied by solar energy, was correlated with two dimensionless variables, X and Y. X and Y are defined as in the f-Chart method except that T_{ref} is 204.8°F (96.1°C) rather than 212°F (100°C). The load on the solar energy system is the building cooling load divided by the average cost of performance (COP) of the absorption machine.

The disadvantage of this approach is that at the relatively high temperatures needed for solar cooling, the solar collectors often operate at high critical levels, which causes the useful energy gain of the collector to depend on the distribution of solar radiation, as explained in section 3.4. (This, in fact, was the motivation for the development of the $\bar{\phi}$, f-Chart design method, described in section 3.4.3.1, which can also be used for solar cooling applications.) Anand, Abarcar, and Allen (1980) noted a location dependence in their correlation of f_c to X and Y. In their later study, Anand and Kumar (1983) developed separate correlations for four different regions in the United States and for absorption chiller capacities of 2 and 3 tons.

3.3   Simplified Simulation Methods

A detailed simulation program by which long-term system performance at short time (e.g., hourly) intervals can be calculated is a powerful tool that can provide a great deal of information. Used as design tools, however, detailed simulations suffer from the disadvantages noted in the introduction. In general, they are too costly, complex, and inconvenient for routine design of typical systems despite the increasing computing capability of personal computers.

Rather than correlate long-term performance results from simulations, some investigators have chosen to devise simplified simulations that require much less computational effort than the large hour-by-hour simulations programs. Computational effort is lessened primarily by reducing the amount of meteorological data involved (i.e., by using fewer days and/ or taking longer time steps). The different approaches to the development of simplified simulations are reviewed in this section.

3.3.1   SOLCOST

SOLCOST (Connolly et al. 1976) was developed to serve as a design method for nonengineers for solar space heating, water heating, and absorption cooling systems. SOLCOST is a simulation program derived from MITAS (Holmstead 1976), a general thermal network analysis program. To overcome the disadvantages of simulations for design purposes, Connolly et al. incorporated the following simplifications and features into SOLCOST:

1. *Reduced computing time*
The computing time and thus the cost of using a simulation program such as MITAS to do hourly simulations of long-term solar energy system performance is prohibitive. To reduce this cost, SOLCOST uses a single hypothetical day to represent each month.

2. *Reduced expertise requirement*
Simulation programs tend to be complicated, even after they are formulated, because many parameters must be supplied to describe the sys-

tem. To simplify its use, SOLCOST has a data bank that includes default values for many of the parameters in standard systems. Mandatory and optional parameters are identified so that the user needs to provide only minimum information.

3. *Elimination of need for computer facilities*
A valid criticism of simulation programs for use as a design tool is that they generally require large computer facilities. This is true of SOLCOST. A further problem is that SOLCOST must be run on the type of computer for which it was developed. To eliminate these disadvantages, the developers of SOLCOST made it accessible through a design service company that runs SOLCOST for a nominal fee to the user. Input data forms were devised (ERDA 1977) to simplify user input. Thus, the user need not ever use a computer to obtain design information from SOLCOST.

The algorithms used in SOLCOST to calculate solar radiation and system thermal performance are unique and somewhat controversial. SOLCOST is based on a simulation approach in which the following energy balance is performed on the storage component at half-hour intervals:

$$T_{new} = T_{old} + \Delta\tau(q_u - q_{loss} - q_{load})/MC_p, \tag{3.1}$$

where

T_{new} is the fluid storage temperature at the end of the time interval;

T_{old} is the fluid storage temperature at the beginning of the time interval;

q_u is the rate of useful energy delivered to storage from the collector bank;

q_{loss} is the rate of energy loss from storage;

q_{load} is the rate at which energy is removed from storage to supply the load;

$\Delta\tau$ is the time step (0.5 hour);

MC_p is the capacity of the storage component.

This energy balance necessarily assumes that the storage component is at a uniform temperature (i.e., unstratified) at any given moment.

The controversial aspects of the SOLCOST algorithms arise from how the monthly average insolation on the collector surface is calculated, how clear- and cloudy-day collector performance is weighted to account for variability in solar energy, and how a single day is used to represent the monthly system performance.

The insolation model is based on the method described in chapter 59 of the ASHRAE Applications Handbook (1978). The model requires clearness numbers rather than daily horizontal radiation data as input. Average hourly solar fluxes on the collector plane are computed for an average clear day and an average cloudy day for the fifteenth day of each month. According to Winn and Duong (1976), "this model requires rather extensive input data and computation times to generate (the) solar flux incident on a tilted collector." In a later study, Winn et al. (1977) indicate that this model predicts the horizontal radiation fairly accurately but overpredicts the tilted flux in winter months.

The clear- and cloudy-day solar fluxes are used to determine clear- and cloudy-day solar collector efficiencies by use of the familiar Hottel and Whillier collector model (Duffie and Beckman 1980). The rate of useful energy collection, q_u, is then calculated from

$$q_u = (PP\eta_{clear}I_{clear} + (1 - PP)\eta_{cloudy}I_{cloudy})A, \tag{3.2}$$

where η_{clear} and η_{cloudy} are the collector efficiencies for the clear and cloudy conditions, I_{clear} and I_{cloudy} are the clear and cloudy solar fluxes, and PP is the percent possible sunshine for the location divided by 100. Percent possible sunshine data are incorporated into SOLCOST. According to Connolly et al. (1976), "this weighting of the clear and cloudy day collector performance with the percent of possible sunshine accounts for the variability of the incoming solar energy due to random cloudiness conditions." The validity of this approach has not been investigated in the technical literature. The utilizability concepts described in sections 3.4 indicate that this approach will tend to increase the underestimation of collector output as the collector inlet temperature increases.

The energy balance used in SOLCOST for the one-day simulation requires an initial value for the temperature of the solar water tank at the beginning of the day. The choice of this temperature can have a significant effect on the calculated performance for a one-day period. In early versions, default values for the initial (i.e., dawn) storage temperatures were placed in the SOLCOST data bank. The user was free to change these values, but few users would know what values to supply. Validations studies by Winn et al. (1977) showed that the default values resulted in performance estimates that were in good agreement with the f-Chart method for solar fractions between 0.5 and 0.75, but significant departures were observed at higher and lower solar fractions. SOLCOST was later modified (Hull 1980) to internally determine the long-term average dawn storage temperature by repeatedly simulating the one-day performance until the difference between dawn temperatures on two successive days is less than one degree. In practice, three to five repetitions of the one-day performance are required to establish this temperature, which significantly adds to the computing cost.

The loads on the solar system can be entered into SOLCOST in many ways. The user may directly enter UA, the building energy loss coefficient—area product. Alternatively, previous fuel usage records can be entered, and SOLCOST will use these to compute the UA. SOLCOST will also compute a UA consistent with existing building energy use standards if the building dimensions are supplied. Finally, a thermal network for a predefined building can be accessed by SOLCOST to obtain hourly loads including the effects of the thermal capacitance of the structure. Use of this option significantly increases the computing cost. In addition, Winn and Duong point out that to change the building structure from the specific house predefined in SOLCOST requires the MITAS program to be recompiled.

Validations results for SOLCOST are reported by Hull (1980) for three test sites. The agreement of the estimated and measured solar fractions for these three cases is within a few percent. SOLCOST has also been compared with f-Chart by Winn et al. (1977). These comparisons resulted in identification of an error in early versions of SOLCOST.

3.3.2 Other Simulation Design Methods

The development of a simulation program to serve as a design tool is an attractive goal, because simulations potentially offer greater flexibility in the system types and parameters that can be investigated. Considerable research effort has been devoted to the development of simplified simulation techniques for passive solar and building energy analysis programs. Barley (1979) has shown that accurate results may be obtained with daily rather than hourly or shorter time steps in a building energy simulation. Ahrens (1979) describes a technique to select multiday segments of real hourly weather data for each month, which when used in a building energy simulation program, provide results that are in good agreement with full-month simulations. Byrne (1981) describes a method using a quadratic programming technique to reduce the number of simulations needed to locate an optimum design. Although developed for passive solar applications, these techniques can be applied to active solar simulations as well. In addition, these techniques may be directly applicable, since a detailed building energy analysis is necessary to appropriately design active solar equipment.

Several simplifying techniques have been investigated for active systems simulations. Anand and Allen (1976) developed a stochastic model for solar radiation and wet and dry bulb temperatures that they used in place of actual weather data in a simulation study of solar absorption cooling. The advantage of a stochastic weather model is that the weather data can be represented more compactly, saving computer costs for data storage. Also, there is the possibility of needing less simulation time to obtain the statistical information from a stochastic model than with actual weather data. Anand and Kumar (1983) later concluded, however, that the stochastic approach "has the limitation that the phenomenon of persistence, i.e., the tendency of very hot or cold days occurring in succession, is not taken into account."

Duff, Favard, and Den Braven (1981) have developed a simulation program for solar heating, cooling, and hot water systems that uses event-incremented rather than fixed-size time steps. In this approach, cosine functions are used to describe the variation of meteorological parameters such as solar radiation and ambient temperatures over the course of a day. With these functions, the equations describing the system performance can be solved analytically for the period over which they are applicable. Events, such as pumps and auxiliary heaters turning on or off, or domestic hot water use, affect the system equations. Event-incrementing moves the simulation forward in time from one event to the next, which generally results in a much larger time step than that used in fixed-step size simulations. Thus, computing cost is significantly reduced. The authors report good agreement with conventional simulations for the systems they investigated.

Considerable advances have been made in recent years in both reducing computer costs and improving simulation methodology. Despite these advances, simulations remain awkward design tools. None of the simulation-based design methods has enjoyed widespread popularity. Additional research is needed in the development of simplified simulation methods.

3.4   Utilizability Methods

Solar radiation utilizability is defined as the fraction of the solar radiation incident on a surface that exceeds a specified threshold or critical level. Utilizability is a statistic of solar radiation data that is analogous to degree-days, a statistic of ambient temperature data.

Utilizability depends on the distribution of solar radiation. This dependence is illustrated in figure 3.6 in which two 3-day sequences of solar radiation appear. Sequence A consists of three average days, while sequence B consists of a clear, an overcast, and an average day. The sequences were chosen to have the game total solar radiation. However, the daily utilizability for critical level $I_{c,1}$ (i.e., the ratio of the shaded area to the total area under the curves) is greater for sequence B than for sequence A. The effect of radiation distribution becomes more pronounced at higher critical levels. For example, the utilizability corre-

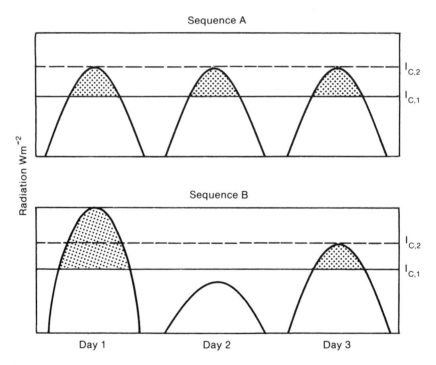

Figure 3.6
Effect of distribution on daily solar radiation utilizability. Source: Klein 1979.

sponding to critical level $I_{c,2}$ is zero for sequence A but greater than zero for sequence B.

The utilizability concept was originally developed to calculate the thermal performance of flat-plate solar collectors. In recent years, the utilizability concept has also been applied to systems with concentrating collectors as well as to passive and photovoltaic systems. For flat-plate collectors and concentrators, the solar radiation of figure 3.6 can be interpreted as the radiation falling on the collector aperture. The area below the critical level represents the absorbed radiation that is necessary to overcome collector losses. The radiation above the critical level is then the "utilizable" portion of the incident radiation.

The utilizability method necessarily assumes that the critical level is constant for the period of analysis. This is not the usual operating condition for most active solar systems. In the more usual case, the critical level varies from hour to hour and from day to day, as conditions such as the storage tank and ambient temperatures change. Variable critical levels can be accounted for by a correction factor that compensates for the error introduced by assuming the critical level to be constant. Correlations for the correction factor in terms of system parameters can then be developed from simulation results. The advantage of this combined utilizability-correlation approach is that it allows the design method to deal with the statistical variability of solar radiation and the system dynamics. The accuracy of utilizability-based design methods is, as a result, usually better than that of the correlation methods described in section 3.2. The development of the utilizability concept is reviewed in the next section. Correlations for the utilizability function are listed in section 3.4.2. Section 3.4.3 reviews the active system design methods that are based on solar radiation utilizability.

3.4.1 Development of the Utilizability Concept

The utilizability concept was originally developed by Whillier (1953) to simplify the calculations needed in the evaluation of flat-plate solar collector performance. The concept of utilizability follows directly from the Hottel-Whillier collector equation (Duffie and Beckman 1980), which relates the rate of useful energy collection, q_u, to the design parameters of the collector and its operating conditions:

$$q_u = A[F_R(\tau\alpha)I_T - F_R U_L(T_i - T_a)]^+. \tag{3.3}$$

The collector parameters, $F_R(\tau\alpha)$ and $F_R U_L$, can be determined from theory or from the ASHRAE 93-77 (1977) collector test. The superscript $^+$ is used to indicate that only positive values of the quantity in brackets are considered. In practice, a controller would be employed to prevent fluid circulation whenever the solar radiation is not sufficient to overcome thermal losses from the collector.

Equation (3.3) can be rearranged into

$$q_u = AF_R(\tau\alpha)[I_T - I_c]^+, \tag{3.4}$$

where I_c is the critical radiation level defined as

$$I_c = \frac{F_R U_L}{F_R(\tau\alpha)}(T_i - T_a). \tag{3.5}$$

In this application, I_c is the radiation level needed to maintain the collector plate at the fluid inlet temperature.

Equation (3.4) can be used to calculate Q_i, the long-term average energy collection for an hourly period, e.g., 9–10 A.M. in January. If I_T and I_c are interpreted as hourly totals rather than rates, and $(\tau\alpha)$ and I_c are considered to be constant for a particular hour and month, then

$$Q_i = AF_R(\tau\alpha) \sum [I_T - I_c]^+/N, \tag{3.6}$$

where N represents a large number of hourly radiation observations sufficient to represent long-term average conditions for the hour and month. On an hourly basis, utilizability is defined as the fraction of the long-term average hourly radiation, $\bar{I}_T$, which is above the critical level and can be written as

$$\phi = \frac{\sum\limits^{N} [I_T - I_c]^+}{\sum\limits^{N} I_T} = \frac{\sum\limits^{N} [I_T - I_c]^+}{N\bar{I}_T}. \tag{3.7}$$

ϕ is the ratio of the shaded area to the sum of the shaded and hatched areas for the radiation sequence in figure 3.7. In terms of ϕ, equation (3.6) becomes

$$Q_i = AF_R(\tau\alpha)\bar{I}_T\phi. \tag{3.8}$$

If sufficient radiation data are available, ϕ can be determined as a function of I_c for each hour, month, and location directly from equation (3.7).

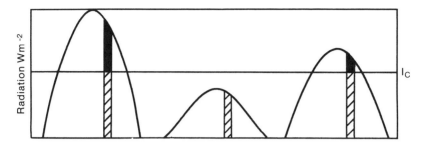

Figure 3.7
Hourly utilizability, the ratio of the shaded to total areas. Source: Klein and Beckman 1984.

To simplify the calculations, Whillier prepared cumulative frequency distributions of hourly solar radiation in dimensionless form. He then determined ϕ by numerical integration and showed that for a given month, location, and collector orientation, plots of ϕ versus I_c/I_T (called ϕ- curves) are independent of time of day. Whillier also showed that, over a long term, solar radiation is ordinarily symmetric about solar noon, so that it is only necessary to determine ϕ for each hourly interval from solar noon rather than for each hour of the day. This simplification would not hold in locations that, for example, typically have morning fog or where there are large mountains to the east or west.

Liu and Jordan (1960, 1963, 1977) generalized Whillier's ϕ-curve method using their statistical analyses of solar radiation data. They introduced the clearness index, $\overline{K}$, the ratio of the monthly average radiation on a horizontal surface to the extraterrestrial (i.e., the maximum possible) radiation. Liu and Jordan showed that the cumulative distribution of daily radiation on a horizontal surface depends primarily on the clearness index. They also showed that the ratio of the average hourly to daily horizontal radiation depends only on the hours from solar noon and the day length. With these results, Liu and Jordan constructed ϕ-curves for clearness indices of 0.3, 0.4, 0.5, 0.6, and 0.7, which were independent of month and location. A generalized ϕ-curve for $\overline{K} = 0.5$ is shown in figure 3.8. Liu and Jordan also developed a method to (approximately) include the effect of collector tilt into the ϕ-curves for collectors facing directly toward the equator by providing curves for a range of values of R_D, the ratio of daily beam radiation on the tilted surface to that on a horizontal surface.

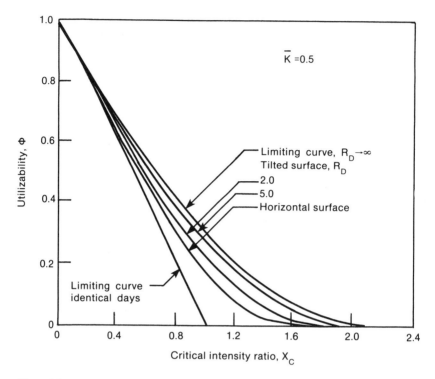

Figure 3.8
Generalized ϕ-curve for $\overline{K} = 0.5$. Source: Duffie and Beckman 1980.

3.4.2 Utilizability Correlations

Utilizability is the ratio of solar energy quantities, and as a result, the time covered must be specified. Whillier (1953) and Liu and Jordan (1963) defined what will be referred to here as hourly utilizability, the long-term average fraction of the solar radiation incident on a surface for a month during an hourly period. Correlations for daily and annual utilizability also have been developed. Daily utilizability, designated $\overline{\phi}$, is defined for the period between sunrise and sunset each month, whereas annual utilizability applies to radiation for the entire year.

3.4.2.1 Hourly Correlations
Use of Liu and Jordan's generalized ϕ-curve method requires significant computational effort. The calculations are difficult to automate because

analytical expressions for the ϕ-curves are not available. In addition, the method used to incorporate the effect of surface orientation was approximate and limited to surfaces facing directly toward the equator.

The defining equation for ϕ can be written in terms of the probability distribution of insolation levels. Liu and Jordan (1960) showed that the probability distribution of daily total radiation for a horizontal surface depends primarily on the clearness index. At the time of their study, the availability of computational facilities and radiation data was rather limited. Later studies with substantially more data by Theilacker (1980) and Bendt, Collares-Pereira, and Rabl (1981) confirmed Liu and Jordan's results with exceptions noted for coastal locations such as Miami, Florida, and Seattle, Washington. Liu and Jordan also suggested that hourly radiation distributions are nearly identical with daily distributions. Theilacker has found hourly distributions to display greater variability than daily distributions for the same clearness index value. However, the hourly and daily distributions are similar enough to allow accurate values of ϕ to be obtained by use of the daily radiation probability distributions. Analytical expressions for these probability distributions have been developed by Cole (1977), Bendt, Collares-Pereira, and Rabl (1981), and Hollands and Huget (1983).

Hollands and Huget (1983) have analytically integrated their curve-fit to Liu and Jordan's radiation distribution function to arrive at an analytical equation for ϕ. This equation is algebraically complicated, but it is suitable for computer implementation and applicable for surfaces of any orientation.

Clark, Klein, and Beckman (1983) directly correlated with clearness indices values of ϕ obtained by numerical integration of many years of hourly radiation data. Like Hollands and Huget's, the Clark correlation is applicable for surfaces of any orientation. An advantage of the Clark correlation over Hollands and Huget's is that it is algebraically simpler. Clark has compared utilizability values obtained by use of his correlation with values obtained from Hollands and Huget's correlation and from numerical integration of long-term (15 to 23 years) hourly radiation data for Madison, Wisconsin; Albuquerque, New Mexico; and Seattle, Washington. The comparisons demonstrate that the Clark and the Hollands and Huget correlations are of similar accuracy. Both compare to the numerical integration results with a root mean square error of about 5%.

3.4.2.2 Daily Correlations

In some applications, the critical level should be considered constant for all hours of the day. In this case, a radiation-weighted daily utilizability, $\bar{\phi}$, can be defined as the monthly-average fraction of the solar radiation above the critical level between sunrise and sunset. $\bar{\phi}$ can be evaluated by use of a radiation-weighted average of hourly values of ϕ for all daylight hours:

$$\bar{\phi} = \frac{\sum\limits^{\text{day}} \bar{I}_T \phi}{\sum\limits^{\text{day}} \bar{I}_T}. \tag{3.9}$$

However, methods for directly evaluating $\bar{\phi}$ with less computation effort have been developed.

If sufficient hourly data were available for every location of interest, $\bar{\phi}$ could be evaluated numerically from

$$\bar{\phi} = \frac{\sum\limits^{\text{days}} \sum\limits^{N} [I_T - I_c]^+}{\sum\limits^{\text{days}} \sum\limits^{N} I_T} = \frac{\sum\limits^{\text{days}} \sum\limits^{N} [I_T - I_c]^+}{N\bar{H}_T}, \tag{3.10}$$

where $\bar{H}_T$ is the monthly-average daily total radiation on a tilted surface that can be calculated from horizontal data, as described, for example, by Klein and Theilacker (1981). Klein (1978) has developed a correlation for $\bar{\phi}$ by curve-fitting values calculated in this manner. Rather than use actual radiation data for one or several locations, Klein generated hourly radiation data from the statistical information given in Liu and Jordan (1960). Hourly radiation data generated in this manner accurately represent long-term average conditions. Values of $\bar{\phi}$ were generated for all 12 months and for a broad range of critical levels, surface orientations, and clearness indices. The resulting correlation is a function of a dimensionless critical level (defined as the ratio of the critical level to the radiation at noon on the average day of the month), a geometry factor (which incorporates the effects of collector orientation, location, and time of the year), and $\bar{K}$, the monthly-average daily clearness index. The correlation is limited to surfaces facing directly toward the equator.

Theilacker (1980) has proposed a minor simplification to Klein's correlation, which also improves its overall accuracy. Theilacker reports a 2.7% standard deviation between his correlation and 8400 $\bar{\phi}$ values obtained by numerical integration of hourly radiation data. Theilacker's correlation is also limited to surfaces facing directly toward the equator. However, he has developed correlations for east- and west-facing vertical surfaces as well as for overhang-shaded surfaces facing the equator.

Collares-Pereira and Rabl (1979) have developed monthly-average daily utilizability correlations for five collector types: flat-plate, compound parabolic concentrator (east-west orientation), east-west tracking, polar tracking, and two-axis tracking. Because radiation for one of these collector types must be within the acceptance angle of the collector as well as above the critical level to be useful, Collares-Pereira and Rabl define utilizability in a manner slightly different from that given by equation (3.10). Their monthly-average daily utilizability, $\bar{\phi}_{CPR}$, is the fraction of the radiation incident on the surface while the collector is operating, that is, above the critical level. The collector operating time may be dictated by thermal or optical considerations.

Two main steps are needed to calculate utilizability with the Collares-Pereira and Rabl method. First, it is necessary to calculate $\bar{H}_{coll}$, the monthly-average radiation incident on the surface during the collector operating period. Simple equations are given for $\bar{H}_{coll}$ for each collector type in terms of the collector "cutoff" time. However, an iterative calculation procedure is needed to determine the cutoff time. The utilizability, $\bar{\phi}_{CPR}$, is then found from a correlation for each collector type involving $\bar{H}_{coll}$, the cutoff time, the critical level, and the monthly-average clearness index. Collares-Pereira and Rabl report an overall accuracy of 5% for useful energy calculated with their procedure. The Collares-Pereira and Rabl, and Klein utilizability definitions are related by

$$\bar{\phi}\bar{H}_T = \bar{\phi}_{CPR}\bar{H}_{coll}. \tag{3.11}$$

Lunde (1980) has proposed two methods for calculating monthly-average daily utilizability for surfaces facing the equator. First, he has prepared tables for a few locations, which give the monthly total radiation above seven critical levels. These tables were constructed by numerically integrating hourly radiation on tilted surfaces for several years. Second, Lunde correlates $\bar{\phi}$ to critical level with a simple exponential relation involving a single factor that he correlates to the monthly clearness index and the incidence angle of the sun on the surface at solar noon.

Evans, Rule, and Wood (1982) have developed an empirical relationship for $\bar{\phi}$ that is in terms of flat-plate collector parameters rather than a critical radiation level. This relationship, designed for collectors facing directly toward the equator, is algebraically very simple and is reported to be of similar accuracy to the equations of Klein (1978). The relation is a quadratic function of $(T_i - T_a)/\overline{K}'$ with coefficients that depend on the collector parameters $F_R U_L$ and $F_R(\tau\alpha)$. K' is a modified monthly clearness index that accounts for the effects of collector tilt. Since utilizability can be used for applications that do not involve flat-plate collectors, Evans, Rule, and Wood have developed an additional correlation for $\overline{\phi}$ as a quadratic function of critical level with coefficients dependent on $\overline{K}'$. The standard error of both of these correlations is less than 4%. The authors show that using the monthly-average daytime ambient temperature in place of the 24-hour average temperature does not result in a significant improvement in the accuracy of these correlations.

3.4.2.3   Annual Correlations

Rabl (1981) has developed correlations for the average yearly total energy above a specified critical level for flat-plate and concentrating solar collectors. His correlations can be used to determine a yearly average utilizability by dividing the total energy above the critical level by the total energy for a critical level of zero. Rabl's correlations are derived from hourly calculations using meteorological data for 26 stations in the United States. The correlations are second-order polynomials in terms of critical level, latitude, and yearly average insolation. Rabl's correlations are limited to applications in which the critical level can be assumed to be a constant value for the entire year. However, Rabl shows that small variations of the critical level about its mean yearly value have only a small effect on the annual energy collected. It may not be possible, however, to accurately determine the mean yearly critical level for some systems, particularly if short-term storage is used.

3.4.3   Utilizability-Based Design Methods

The critical level for solar collectors is defined as the radiation level required to maintain the collector absorber at the fluid inlet temperature. Radiation must be above this level to produce a useful energy gain. A relationship for the critical level in terms of standard flat-plate parameters is given in equation (3.5). An analogous equation can be written for concentrating collectors.

The utilizability concept assumes that the critical level is constant over the period of analysis. The actual critical level in most solar energy systems, particularly those having energy storage, varies from hour to hour and from day to day primarily because of the variation in the collector fluid inlet temperature. It is usually possible, however, to identify lower and/or upper limits on the critical level for a period such as a month. These limits on the critical level can be used to determine the upper and lower limits on system performance.

The maximum energy collection will occur when the collector fluid inlet temperature is always at its minimum. For space heating systems, the minimum temperature is approximately 68°F (20°C), whereas for solar air-conditioning with an absorption chiller, the minimum useful temperature (i.e., the temperature below which the chiller does not operate) may be 176°F (80°C). With a constant critical level, the monthly-average daily energy collection, $\overline{Q}$, can be directly calculated from

$$\overline{Q} = AF_R(\overline{\tau\alpha})\overline{H}_T\overline{\phi}. \tag{3.12}$$

The values of $\overline{Q}$ for the upper and lower critical levels correspond to the lower and upper limits on system performance. The upper limit can be used to determine the thermal or economic feasibility of a proposed system early in the design phase.

3.4.3.1 General Solar Energy System Analysis

When the critical level varies considerably during a month (as it generally does in systems having short-term thermal energy storage), equation (3.12) cannot be used to accurately estimate system performance, since an average critical level and thus an average utilizability cannot be determined. Estimates of the performance of such systems can often be obtained from a design method developed by correlating the results of detailed simulations. One example of a design method is the f-Chart method for active space and domestic water heating systems (see section 3.2.1). A more general design method that can be applied to systems for these and other applications is the $\overline{\phi}$, f-Chart method (Klein and Beckman 1979). With appropriate utilizability correlations, the $\overline{\phi}$, f-Chart method can be applied to determine the performance of systems having concentrating solar collectors.

The $\overline{\phi}$, f-Chart method is composed of two steps. The first step is the estimation of the maximum useful energy collection for the system. This calculation uses equation (3.12) with $\overline{\phi}$ evaluated at a critical level cor-

S. A. Klein

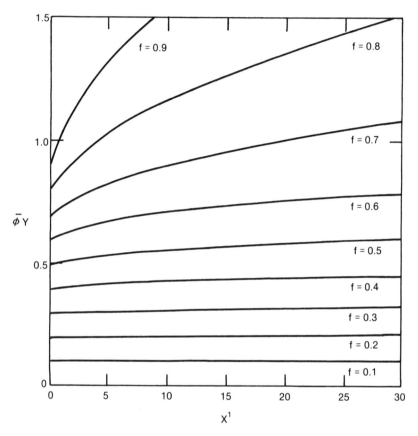

Figure 3.9
$\bar{\phi}$, f-chart for a storage capacity of 17 Btu/ft² °F (350 kJ/m² °C). Source: Klein and Beckman 1979.

responding to the minimum collector fluid inlet temperature. The second step corrects the maximum performance estimate to account for variations in the critical level and for storage energy losses. The correction factor, developed from simulation results, depends primarily on the storage capacity and the fraction of the load supplied by the system. The $\bar{\phi}$, f-Chart is a graphical representation of the correction factor. A $\bar{\phi}$, f-Chart for a storage capacity of 17 Btu/ft² °F (350 kJ/m² °C) is shown in figure 3.9. The ordinate of this plot represents the ratio of the maximum possible energy collection to the load on the system for a monthly period. The

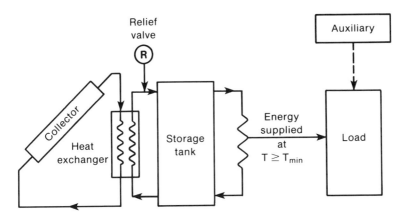

Figure 3.10
Closed-loop solar energy system. Source: Klein and Beckman 1979.

abscissa is the ratio of a reference collector energy loss to the load on the system for the month.

The $\bar{\phi}$, f-Chart method was originally developed for closed-loop liquid systems having the configuration shown in figure 3.10, where energy is supplied above a minimum useful temperature. Braun, Klein, and Pearson (1983) have modified the method to make it applicable to open-loop systems (such as the domestic water heating system shown in figure 3.3) as well.

Compared with the f-Chart method, the $\bar{\phi}$, f-Chart method offers both advantages and disadvantages. The major advantage of the $\bar{\phi}$, f-Chart method is that it can be used for applications other than space and domestic water heating, such as solar air-conditioning or process-heating with either flat-plate or concentrating collectors. A second advantage is that the $\bar{\phi}$, f-Chart method generally agrees more closely with simulation results than the f-Chart method for the space and water heating systems for which f-Chart is applicable, because the variability of solar radiation is incorporated into the utilizability function. The major disadvantage of the $\bar{\phi}$, f-Chart method is that it requires substantially more computational effort than the f-Chart method, which precludes it for use with hand calculations. Storage energy losses and load heat exchanger limitations are built into the f-Charts, whereas the result from the $\bar{\phi}$, f-Chart must be modified for these considerations. The $\bar{\phi}$, f-Chart method, however, requires far less computation than simulations and it can easily be programmed onto a personal computer.

Procedures similar to the $\bar{\phi}$, f-Chart method have been developed by Stickford (1976), Lunde (1979b, 1981), and Chauhan and Goodling (1982). Further modifications to the basic method have been made by Svard, Mitchell, and Beckman (1981) to enable the evaluation of solar-assisted series heat pump systems.

3.4.3.2 Long-Term Storage Systems

An energy balance on the storage tank in a liquid storage system over a one-month period can be written as

$$\Delta U = Q_{in} - Q_{load} - Q_{loss}, \tag{3.13}$$

where

ΔU is the change in internal energy of the liquid storage, which is the product of the mass of storage, its specific heat, and the temperature change during the month;

Q_{in} is the monthly useful energy supplied to the tank from the solar collector system;

Q_{loss} is the energy loss from storage during the month;

Q_{load} is the energy removed from storage to supply the load during the month.

If an appropriate critical level could be determined, Q_{in} would be the product of the number of days in the month, and $\bar{Q}$, the monthly-average daily useful energy collection from equation (3.12). An appropriate critical level is difficult to determine for systems having short-term storage, since the tank temperature variation (and, therefore, the critical level) is strongly coupled to weather conditions for individual days. For long-term storage, however, the tank temperature variation is damped and the critical level is more nearly constant. Braun, Klein, and Mitchell (1981) show that for systems having more than 760 l (200 gal) of water storage per square meter of flat-plate collector area, the assumption of a constant critical level evaluated at the monthly-average tank temperature (determined from the storage tank energy balance) results in performance estimates that compare well with detailed simulations. Lunde (1979a) and Baylin and Sillman

(1980) have also developed similar methods using utilizability for performance estimates of long-term storage systems.

3.4.3.3 Active Charge/Passive Discharge Systems

Swisher (1981) has analyzed the performance of active charge/passive discharge systems in which the energy obtained from a flat-plate collector array is stored in a mass or water wall within the building structure. Energy is passively removed from storage to provide space heat by radiation and convection. Swisher developed models of these systems for use with the TRNSYS program and simulated their performance in 10 U.S. locations for a range of system parameters.

Swisher then developed a design method for these systems based on the utilizability concept. He shows that the monthly-average daily useful energy from these systems is given by equation (3.12). The critical level at which $\bar{\phi}$ is evaluated depends on three dimensionless parameters representing the solar-to-load ratio, the solar-to-storage ratio, and the solar-to-room coupling. Using TRNSYS results, Swisher developed correlations for the critical level for both the water and mass wall storage systems.

3.4.3.4 Collector Operating Time

Mitchell, Theilacker, and Klein (1981) have shown that the utilizability function can provide an estimate of the collector operating time in active systems as well as an estimate of thermal performance. Their reasoning is as follows.

Consider a situation in which the critical radiation level is constant at a value I_c over a monthly period. The average daily useful energy collection is given by equation (3.12). The monthly-average daily collector operating time, $\bar{t}$, is the number of hours during the month in which the solar radiation exceeds the critical level, divided by the number of days in the month. Now, assume that the critical level is raised by a small amount, ΔI_c. The reduction in average daily useful energy collections that would result is given approximately by the product of $F_R(\tau\alpha)A\Delta I_c$ and $\bar{t}$. In the limit as ΔI_c approaches zero, $\bar{t}$ can be found from

$$\bar{t} = \lim_{I_c \to 0} \frac{AF_R(\overline{\tau\alpha})\overline{H}_T\bar{\phi}|_{I_c} - AF_R(\overline{\tau\alpha})\overline{H}_T\bar{\phi}|_{I_c+\Delta I_c}}{AF_R(\overline{\tau\alpha})\Delta I_c} = -\overline{H}_T \frac{d\bar{\phi}}{dI_c}\bigg|_{I_c}. \tag{3.14}$$

The parasitic energy use for a collector pump having an on/off controller is then the product of $\bar{t}$ and the power use when the pump is operating.

3.5   Conclusions

Design methods for active solar systems have been classified as correlations, simplified simulations, or utilizability-based methods. The correlations methods are easy to apply and amenable to hand calculations. They are also easy to understand. For these reasons, correlations methods, such as the f-Chart method, are the most widely used design methods. The limitation of correlation methods is that they are highly empirical and are not reliable for systems that differ in any respect from those for which the method was derived.

Simplified simulation methods are at an early state of development. Though these methods require far less computation than a detailed simulation, the computation they require is substantial, and, as a result, these methods cannot be considered to be hand-calculation methods. To reduce the necessary computation, simplifications, such as the use of an average day for each month, are employed that may introduce significant errors in the calculated results for some system designs. A major advantage of simulation methods is their flexibility. They can be used with minor modifications with a variety of system types. Computing capabilities have greatly increased in recent years with the advent of the microcomputer. As these capabilities continue to increase and as solar designers become familiar with the use of these tools, simplified simulations may become the dominant type of design method.

Utilizability-based methods usually require somewhat more computation effort than correlation methods, but less than simplified simulations. Utilizability-based methods tend to provide more accurate results than correlation methods, since the utilizability factor provides a means of accounting for the dependence of system performance on the location-specific distribution of solar weather. Utilizability methods have an advantage over simulation methods as well, because the utilizability factor is based on long-term average conditions. To obtain the same statistical significance, simulations would require extensive weather data. Although the utilizability concept was originally developed for estimating the performance of flat-plate solar collectors, it has been found to be useful for many other applications, including systems with concentrating collectors, photovoltaic systems, and passive systems. Utilizability is a statistic of solar weather and provides a simple way of determining the effect of radiation distribution on any solar process.

References

Ahrens, E., et al. 1979. A method to abbreviate hourly climate data for computer simulation of annual energy use in buildings. *Proc. 4th Nat. Passive Solar Conf.* Kansas City, MO, pp. 282–86.

Anand, D. K., and R. W. Allen. 1976. Solar powered absorption air conditioning system performance using real and synthetic weather data. *Proc. ASES/ISES Annual Meeting.* Winnipeg, Canada, pp. 208–27.

Anand, D. K., R. W. Allen, and E. O. Bazques. 1977. Solar air conditioning performance using stochastic weather models. *Energy* 1:319–23.

Anand, D. K., R. B. Abarcar, and R. W. Allen. 1980. Long-term solar cooling system performance via a simplified design method. *Proc. System Simulation and Economic Analysis Conf.* San Diego, CA, pp. 285–90.

Anand, D. K., and B. Kumar. 1983. Regional simplified solar cooling design charts. *Proc. ASME Solar Energy Division 5th Annual Meeting.* Albuquerque, NM, pp. 390–97.

Anderson, J. V., J. W. Mitchell, and W. A. Beckman. 1980. A design method for parallel solar heat pump systems. *Solar Energy* 25:155–63.

ASHRAE 1977. *Standard 93–77.* Methods of testing to determine the thermal performance of solar collectors. New York: American Society of Heating, Refrigeration, and Air-Conditioning Engineers.

ASHRAE. 1978. *Applications handbook.* New York: American Society of Heating, Refrigeration, and Air-Conditioning Engineers.

Balcomb, J. D., et al. 1975. *Solar heating handbook for Los Alamos.* Report LA-5967. Los Alamos Scientific Laboratory.

———. 1976. *ERDA's Pacific regional solar heating handbook.* Report LA-6242-MS. Los Alamos Scientific Laboratory.

Balcomb, J. D., and J. C. Hedstrom. 1976. A simplified method for calculating the required solar collector area array size for space heating. *Proc. ASES/ISES Annual Meeting.* Winnipeg, Canada, pp. 280–94.

Barley, C. D. 1979. Passive solar modeling with daily weather data. *Proc. 4th Nat. Passive Solar Conf.* Kansas City, MO, pp. 207–11.

Barley, C. D., and C. B. Winn. 1978. Optimal sizing of solar collectors by the method of relative areas. *Solar Energy* 21:279–89.

Baylin, F., and S. Sillman. 1980. *Systems analysis techniques for annual cycle thermal energy storage solar systems.* SERI RR-721-676. Golden, CO: Solar Energy Research Institute.

Beckman, W. A., J. A. Duffie, and Associates. 1976. *Further verification for the solar heating system design charts (the f-Charts) developed at the University of Wisconsin Solar Energy Laboratory.* Contract #608312 to National Bureau of Standards, Task 2 Report.

Beckman, W. A., S. A. Klein, and J. A. Duffie. 1977. *Solar heating design by the f-Chart method.* New York: Wiley Interscience.

Bendt, P., M. Collares-Pereira, and A. Rabl. 1981. The frequency distribution of daily insolation values. *Solar Energy* 27:1.

Braun, J. E., S. A. Klein, and J. W. Mitchell. 1981. Seasonal storage of energy in solar heating. *Solar Energy* 26:403

Braun, J. E., S. A. Klein, and K. A. Pearson. 1983. An improved design method for solar water heating systems. *Solar Energy* 31:597.

Buckles, W. E., and S. A. Klein. 1980. Analysis of solar domestic hot water heaters. *Solar Energy* 25:417–24.

Byrne, S. J. 1981. Simulation programs as design tools: an optimization technique. *Proc. 6th Nat. Passive Solar Conf.* Portland, OR, pp. 294–97.

Chaffin, D. J., and F. C. Wessling. 1982. Comparisons of F-CHART 4.0 predictions with experiment results. *Proc. ASES/ISES Annual Meeting.* Houston, TX, pp. 399–404.

Chauhan, R. S., and J. S. Goodling. 1982. A simple method to calculate yearly solar fraction of space heating systems. *Proc. ASES/ISES Annual Metting.* Houston, TX, pp. 383–86.

Clark, D. R., S. A. Klein, and W. A. Beckman. 1983. Algorithm for evaluating the hourly radiation utilizability function. *J. Solar Energy Eng.* 105:281.

Cole, R. 1977. Long-term average performance predictions for compound parabolic concentrator solar collectors. *Proc. ASES/ISES Annual Meeting.* Orlando, FL.

Collares-Pereira, M., and A. Rabl. 1979. Derivation of method for predicting long-term average energy delivery of solar collectors, and a simple procedure for predicting long-term average performance of nonconcentrating and of concentrating solar collectors. *Solar Energy* 23:223.

Connolly, M., et al. 1976. Solar heating and cooling computer analysis—a simplified sizing design method for non-thermal specialists. *Proc. ASES/ISES Annual Meeting.* Winnipeg, Canada, pp. 220–34.

de Winter, F. 1975. Heat exchanger penalties in the double loop solar heating systems. *Solar Energy* 17:335–38.

Duff, W. S., G. J. Favard, and K. R. Den Braven. 1981. Development of a day-by-day simulation of solar systems. *Proc. ASME Solar Energy Division 3rd Annual Meeting,* pp. 112–21.

Duffie, J. A., and W. A. Beckman. 1980. *Solar engineering of thermal processes.* New York: Wiley Interscience.

Duffie, J. A., and J. W. Mitchell. 1983. F-chart: predictions and measurements. *J. Solar Energy Eng.* 105:3–9.

Erbs, D. G., S. A. Klein, and J. A. Duffie. 1982. Estimation of the diffuse radiation fraction for hourly, daily, and monthly average global radiation. *Solar Energy* 28:293.

ERDA 1977. *SOLCOST solar hot water handbook and introduction to SOLCOST.* DSE-2531/2. Energy Research and Development Administration.

Evans, B. L., S. A. Klein, and J. A. Duffie. 1985. A design method for active-passive hybrid space heating systems. *Solar Energy* 35:189–98.

Evans, D. L., T. T. Rule, and B. D. Wood. 1982. A new look at long-term collector performance and utilizability. *Solar Energy* 28:13.

Fanney, A. H., and S. T. Liu. 1980. Comparing experimental and computer-predicted performance of solar hot water systems. *ASHRAE Journal* 86(1).

Fanney, A. H., and S. A. Klein. 1983. Performance of solar domestic hot water systems at the National Bureau of Standards—measurements and predictions. *J. Solar Energy Eng.* 105:311–21.

Hollands, K. G. T., and R. G. Huget. 1983. A probability density function for the clearness index with applications. *Solar Energy* 30:195.

Holmstead, G. M. 1976. *Martin Marietta interactive thermal analysis system—user's manual.* Denver, CO: Engineering Department, Martin Marietta Corporation.

Huck, S. E. 1976. Design charts for solar heating systems. M.S. thesis, Dept. of Mechanical Engineering, Colorado State University, Fort Collins, CO.

Hughes, P. J., et al. 1978. *F-CHART user's manual, version 3*. Engineering Experiment Station Report 49–3. University of Wisconsin, Madison, WI.

Hull, D. E. 1980. SOLCOST heat pump algorithms. *Proc. ASES/ISES Annual Meeting*. Phoenix, AZ, pp. 364–68.

Hull, D. E., and R. T. Giellis. 1980. SOLCOST: a solar energy design program. *Proc. System Simulation and Economic Analysis Conf*. San Diego, CA, pp. 297–302.

Jurinak, J. J., and S. I. Abdel-Khalik. 1979. Sizing phase-change energy storage units for air-based solar heating systems. *Solar Energy* 22:355–59.

Kelly, C. J., Jr. 1982. Comparison of field measured performance with F-CHART 3.0 and F-CHART 4.0 predictions. *Proc. ASME Solar Energy Division 4th Annual Meeting*. Albuquerque, NM, pp. 47–58.

Klein, S. A. 1976. A design procedure for solar heating systems. Ph.D. diss., Dept. of Chemical Engineering, University of Wisconsin-Madison.

———. 1978. Calculation of flat-plate collector utilizability. *Solar Energy* 21:393.

———. 1979. Calculation of the monthly-average transmittance-absorptance product. *Solar Energy* 23:547–52.

Klein, S. A., and W. A. Beckman. 1979. A general design method for closed-loop solar energy systems. *Solar Energy* 22:269.

Klein, S. A., and W. A. Beckman. 1989. Review of solar radiation utilizability. J. Solar Energy Eng. 106:393–402.

Klein, S. A., W. A. Beckman, and J. A. Duffie. 1976. A design procedure for solar heating systems. *Solar Energy* 18:113–27.

———. 1977. A design procedure for solar air heating systems. *Solar Energy* 19:509–13.

Klein, S. A., and J. C. Theilacker. 1981. An algorithm for calculating monthly-average radiation on inclined surfaces. *J. Solar Energy Eng.* 103:29–33.

Klein, S. A., et al. 1973. *TRNSYS—a transient system simulation program*. Solar Energy Laboratory, Engineering Experiment Station Report 38.

Kreider, J. F., T. Cowing, and T. Lohner. 1982. Direct calculation of cost-optimal solar DHW gystem size for the United States. *Proc. ASME Solar Energy Division 4th Annual Meeting*. Albuquerque, NM, pp. 115–18.

Lameiro, G. F., and P. Bendt. 1978. *The GFL method for sizing solar energy space and water heating systems*. SERI-30. Golden, CO: Solar Energy Research Institute.

Liu, B. Y. H., and R. C. Jordan. 1960. The interrelationships and characteristic distribution of direct, diffuse, and total radiation. *Solar Energy* 4(3):1–19.

———. 1963. The long-term average performance of flat-plate solar energy collectors. *Solar Energy* 7(53).

———. 1977. Applications of solar energy for heating and cooling of buildings. In *ASHRAE GRP 170*. New York: ASHRAE.

Lunde, P. J. 1979a. Prediction of the performance of solar heating systems utilizing annual storage. *Solar Energy* 22:69.

———. 1979b. Prediction of the performance of solar heating systems over a range of storage capacities. *Solar Energy* 23:115–21.

————. 1980. *Solar thermal engineering.* New York: Wiley.

————. 1981. The base temperature method for the prediction of the performance of solar hot water and space heating systems. *Proc. ASES/ISES Annual Meeting.* Philadelphia, PA, pp. 611–15.

Mitchell, J. C., J. C. Theilacker, and S. A. Klein. 1981. Calculation of monthly average collector operating time and parasitic energy requirements. *Solar Energy* 26:555.

Mosley, J., and G. Clark. 1981. The effects of detailed solar climatology on the accuracy of f-Chart. *Proc. ASES/ISES Annual Meeting.* Philadelphia, PA, pp. 607–8.

Mutch, J. J. 1974. *Residential water heating, fuel consumption, economics, and public policy.* RAND Corp., Dept. R1498, NSF.

Rabl, A. 1981. Yearly average performance of the principle solar collector types. *Solar Energy* 27:215.

Schnurr, N. M., B. D. Hunn, and K. D. Williamson. 1981. The solar load ratio method applied to commercial building active solar system sizing. *Proc. ASME Solar Energy Division 3rd Annual Meeting,* pp. 204–15.

SHASP. 1978. *Solar heating and air conditioning simulation program.* Solar Energy Projects, Mechanical Engineering Department, University of Maryland, College Park, MD.

Stickford, G. H. 1976. An average technique for predicting the performance of a solar energy collector system. *Proc. ASES/ISES Annual Meeting.* Winnipeg, Canada, pp. 295–315.

Svard, C. D., J. W. Mitchell, and W. A. Beckman. 1981. Design procedure and application of solar-assisted series heat pump systems. *J. Solar Energy Eng.* 103:135–43.

Swisher, J. 1981. Active charge/passive discharge solar heating systems: thermal analysis. *Proc. ASES/ISES Annual Meeting.* Philadelphia, PA, pp. 597–601.

Theilacker, J. C. 1980. An investigatiotn of monthly-average utilizability for flat-plate collectors. M.S. thesis, Dept of Mechanical Engineering, University of Wisconsin-Madison.

Ward, J. C. 1976. Minimum cost sizing of solar heating systems. *Proc. ASES/ISES Annual Meeting.* Winnipeg, Canada, pp. 336–46.

Whillier, A. 1953. Solar energy collection and its utilization for house heating. Ph.D. diss., Dept. of Mechanical Engineering, MIT.

Winn, C. B., et al. 1977. *Model validation studies of solar systems phase III.* Contract No. EG-77-C-02-4299. Solar Environmental Engineering Co., Inc., P.O. Box 1919, Fort Collins, CO.

Winn, C. B., and N. Duong. 1976. *Validation of solar house design programs phases I and II.* Final Report. Solar Environmental Engineering Co., Inc., P.O. Box 1919, Fort Collins, CO.

Wright, W. A. 1980. A method for generating graphic solar design tools. *Proc. ASES/ISES Annual Meeting.* Phoenix, AZ, pp. 286–90.

Wuestling, M. D., S. A. Klein, J. A. Duffie, and J. Braun. 1983. Investigation of promising control alternatives for solar water heating systems. *Proc. ASES Annual Meeting.* Minneapolis, MN. See also M. D. Wuestling, M.S. thesis, Solar Energy Laboratory, University of Wisconsin-Madison, 1983.

York, D. A., and E. F. Tucker. 1980. *DOE-2 reference manual, version 2.1.* Report LA-7689-M. Los Alamos Scientific Laboratory.

4 Controls in Active Solar Energy Systems

C. Byron Winn

4.1 Introduction

This chapter discusses the control of active solar heating and cooling systems for buildings. The two principal areas discussed are control of fluid flow in the collector loop and zone temperature control. Specific controllers described include bang-bang, proportional, integral, proportional-integral-differential (PID), adaptive, and optimal controllers. The chapter principally addresses work that has been supported by the U.S. Department of Energy (DOE). First, control theory is reviewed as it applies to zone temperature control and collector loop control. Next, controllers for collecting solar energy are discussed, followed by a discussion of controllers for distribution of energy obtained from solar and off-peak power. Finally, recommendations for future research are presented.

4.2 Review of Control Theory

This section reviews control theory and discusses bang-bang, proportional, integral, derivative, and PID controllers. It concludes by describing design methods for controllers. Both trial-and-error and analytical designs are discussed.

To illustrate various controllers, consider the control of temperature inside a conditioned space. For simplicity, let us use a one-node model for a building for which the governing equation may be written as

$$C_E \frac{dT_e}{dt} = u - UA(T_e - T_a), \tag{4.1}$$

where C_E is the heat capacitance, T_e is the enclosure temperature, t is the time, u is the control variable, UA is the area-conductance product, and T_a is the ambient temperature (symbols used in this chapter are also defined in the nomenclature). It is convenient, though not necessary, to examine the system in the frequency domain instead of the time domain. This is done by taking the Laplace transform of the terms in equation 4.1, which results in

$$C_E s T_e(s) = U(s) - UA[T_e(s) - T_a(s)] + C_E T_e(0), \tag{4.2}$$

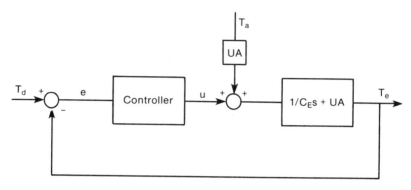

Figure 4.1
Block diagram of a feedback control system.

where s is the Laplace transform variable and $U(s)$ is the Laplace transform of the control function $u(t)$. Solving for the frequency domain representation of the enclosure temperature $T_e(s)$ results in

$$T_e(s) = \frac{U(s)}{(C_E s + UA)} + \frac{U A T_A(s)}{(C_E s + UA)} + \frac{C_E T_e(0)}{(C_E s + UA)}. \tag{4.3}$$

Now consider the case of feedback control, where the enclosure temperature T_e is compared with the desired enclosure temperature T_d. In this case an error signal is used to determine the control. Figure 4.1 shows a block diagram that represents a feedback control system.

The system responds differently for different control actions u. These actions and responses are discussed in the following paragraphs.

4.2.1 Bang-Bang Control

4.2.1.1 Zone Temperature Control
The most commonly used controllers are bang-bang (on/off) controllers. These are the simplest controllers to implement and are also the least expensive. Figure 4.2 shows a block diagram representing a bang-bang (on/off) controller.

A bang-bang controller in a solar space heating system activates the fluid mover in the distribution system whenever the building thermostat calls for heat; that is, whenever the thermostat senses a temperature below the bottom of the dead band, the heating system is activated. The dead band is a region from 1° to 2°C (2° to 4°F) on either side of the set

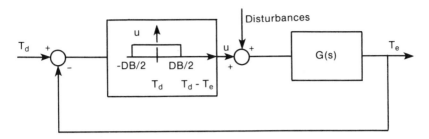

Figure 4.2
Block diagram of a bang-bang controller.

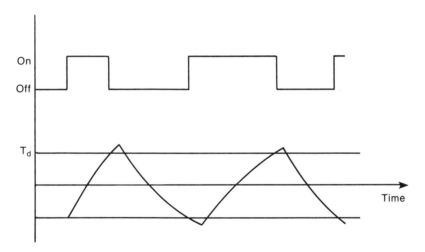

Figure 4.3
Temperature variation using a bang-bang controller.

point temperature. Bang-bang controllers in solar space heating systems normally use a two-stage thermostat. When the temperature in the building drops below the dead band for the first stage, the controller attempts to provide energy from the solar system; if the energy in the solar system is insufficient to meet the demand and the temperature continues to decrease, the controller turns on the auxiliary heating unit when the temperature drops below the bottom of the dead band for the second stage. Figure 4.3 illustrates a typical temperature response for a building using a single-stage bang-bang controller. Note that both cycling and overshoot occur. Overshoot occurs when the building temperature increases beyond the top of the dead band before the controller shuts off the heating and also when the building temperature decreases below the bottom of the dead band before heating is activated. Bang-bang controllers always experience some amount of overshoot. A dead band must be included to prevent the system from chattering about the set point.

The time variation of the temperature may be determined by taking the inverse Laplace transform of equation (4.3). When $u = u_{max}$, and for constant values of ambient temperature, the solution is

$$T_e(t) = \left(\frac{u_{max}}{UA} + T_a\right)(1 - e^{-(UA/C_E)t}) + T_e(0)e^{-(UA/C_E)t}. \tag{4.4}$$

The term UA/C_E is the inverse of the time constant for the building, which is sometimes included in the set of performance specifications. The value for u_{max} must, of course, be large enough to cause T_e to reach the top of the dead band under the most extreme value for T_a.

4.2.1.2   Collector Loop Control

Bang-bang controllers are used to collect solar energy as well as to distribute it for heating or cooling. A bang-bang controller collects solar energy by turning the fluid mover on whenever the temperature measured at the collector exceeds the temperature measured near the lower portion of the storage (or in the living space if solar heat is being used directly) by some preset amount ΔT_{on}. The controller turns the fluid mover off whenever the temperature difference between the collector outlet and the storage or living space drops below another preset amount ΔT_{off}. ΔT_{off} normally ranges from 0.5° to 1°C (1° to 2°F) while ΔT_{on} ranges between four to six times ΔT_{off}. The value for ΔT_{off} is set to ensure that the fluid mover is not operated when the amount of energy being collected would not be worth the cost incurred in collecting that energy. The value for ΔT_{on} is set based on a compromise between collecting energy and experiencing cycling. Cycling characteristically occurs with bang-bang controllers. This is discussed in section 4.3. Numerous studies (Alcone and Herman 1981; Herczfeld et al. 1978; Kahwaji and winn 1985; Kovarik and Lesse 1976; McDonald, Farris, and Melsa 1978; Orbach, Rorres, and Fischl 1981; Schlesinger 1977; Winn 1983a) have been sponsored by DOE in this area, and these are discussed in the following sections.

4.2.1.3 Control System Hardware

Solar system controls consist of power relays that switch electric valves and pumps in the liquid system, or blowers and dampers in the air system, and of auxiliary heating units that respond to temperatures or temperature differences in both systems. Controls for solar systems are basically the same as conventional heating, ventilating, and air conditioning (HVAC) controls; however, solar systems generally have more control functions than typical HVAC systems. Moreover, solar systems have interlocks that prevent undesirable or hazardous sequences of operation.

THERMOSTATS. Solar heating systems generally use a two-stage heat, indoor thermostat; a two-stage heat, one-stage cool type is recommended for solar heating and cooling systems. Variations feature "on," "off," or "automatic" switches from heating to cooling or vice versa to meet the need.

The single-stage component of the thermostat provides cooling of indoor space by using a dead band, a small range in temperature between start and stop signals that is programmed into the controller, which in turn controls the cooling system. The dead band for most thermostats is around 3°C (6°F). The heating operation is a bit more complex. Upon demand for heat, the first stage calls for the solar system to provide heat. If the building heat loss is greater than the solar system can provide, the temperature in the building will continue to drop to stage two, and the auxiliary system will be called upon to provide heat. The auxiliary system can provide sufficient heat for the building by itself, or in combination with the solar system, to raise the temperature in the room to the upper temperature limit of stage one, which stops the heating system. The upper temperature dead band is nominally about 1°C (2°F).

TEMPERATURE SENSORS. There are many types of temperature sensors that can be used in the control subsystem, such as thermocouples, thermistors, silicon transistors, bimetallic elements, and liquid- or vapor-expansion units. Bimetallic elements and liquid- or vapor-expansion units are seldom used because other temperature sensors are more durable and dependable. Thermocouples are frequently used for temperature measurements; however, they are not often used in controls because of their low voltage output (in the millivolt range); without amplification, the voltage is insufficient to be used in controls.

Thermistors and silicon transistors are used in the control subsystem because their voltage outputs (in the 0–10 range) are high enough to serve the control functions. The voltage outputs from thermistors are nonlinear, and calibration circuitry must be provided for the nonlinearity. The voltage outputs from silicon transistors are linear in the normal operating temperature range of solar heating and cooling systems, and provide for simpler circuitry to control the system.

The locations of temperature sensors are not particularly critical, but there are some preferred locations. Sensors measure the temperature of air or liquid as it leaves the collector, in the solar storage tank or rock bed, and in the preheat water tank. The sensor in the conditioned space is the thermostat.

The sensor that measures the fluid temperature at the collector outlet can be located in the manifold, which collects the fluid from the total array of collectors. The sensor should preferably be in contact with the fluid but can be in contact with the pipe, provided there is good thermal contact between the sensor and the pipe. If the sensor is attached to the outside of the outlet pipe, it should be well insulated so that it does not lose heat to the surroundings and register a low temperature. The sensor should also be located near the outlet so that it can register the fluid temperature when the sun is heating the collector but the fluid is not circulating. Sensors in the outlet manifold register the increase in temperature, but the sensor located far from the manifold does not, and useful energy cannot then be collected. Wherever the sensor is located, its characteristics should be checked out when the system is installed.

The sensor in the storage tank should be located near the inside bottom third of the tank. When no fluid is circulating, the temperature at the top of the tank will be slightly higher than at the bottom, but while the fluid is circulating, the fluid in the tank is usually well mixed and the temperature will be uniform, unless the circulation rates are very low or unless anti-mixing (stratification) devices are used.

The sensor in a service water preheat tank should be located near the top one-third of the tank. If it is near the bottom, the temperature at the top could be several degrees higher. The reason for this is that when a household uses hot water, cold water enters the preheat tank near the bottom. Although the preheat tank is thermally mixed when the pump is started, frequent cycling can result from the sensor registering locally cold water temperature. Cycling for water heating in an air system is not particularly harmful because only one pump for the preheat cycle is involved; however, in a hydronic system, which uses two pumps, frequent cycling can waste electric energy. In both air and liquid systems, more heat than necessary would be lost from the pipes and heat exchangers because of frequent cycling.

The sensor in the pebble bed should be located at the bottom (or cold) end of storage. When heat is being stored, the bottom (or outlet) end of storage will determine if storage is "full."

CONTROL PANELS.   Usually a central control panel is convenient for consolidating the circuits and relays that provide the control functions. The panel can house the relays and provide for some adjustment of the temperature limits. It is best to acquire a control panel from the solar equipment manufacturer as a prewired unit to serve the system. A prewired control panel only needs to be connected to appropriate temperature sensors, motor, auxiliary, and valve and damper controls. The manufacturer will provide the necessary hook-up instructions. The power for the control panel will usually be household 115-V single-phase AC line power.

4.2.1.4   Typical Control Subsystems

AIR SYSTEMS.   Figure 4.4 is a sketch of a typical configuration for an air system. The temperature sensors are indicated by T_{ci}, T_{co}, T_s, and T_E. T_{ci} measures the collector inlet temperature T_i, T_{co} measures the collector outlet temperature T_0, T_S measures the storage temperature T_s, and T_E measures the enclosure temperature T_e. BD-1 and BD-2 represent backdraft dampers, D-1 and D-2 represent manual dampers, and MD-1 and MD-2 represent motorized dampers. The controller operates the blowers and motorized dampers to collect and distribute heat for the house. Figure 4.5 shows a flow chart of the control strategy.

The controller first decides whether the zone needs heat by comparing the actual temperature with the desired temperature. If the temperature measured by the temperature sensor in the zone indicates that the zone does not need heat, then the controller determines whether or not heat can be stored by comparing the collector inlet temperature T_i with the collector outlet temperature T_0. If $T_0 \leq T_i$, then no heat is being collected by the collector, and air should not be circulated through the collectors. If $T_0 > T_i$, then there is usually energy available, and the controller will turn on the blower and the hot-water pump (HWP) and control the dampers to direct the flow through the storage.

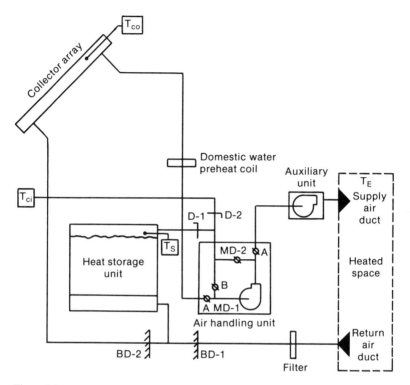

Figure 4.4
Schematic of solar air heating system.

If the zone requires heat, the controller determines whether heat can be supplied from either the collector or from storage. These determinations are again made by comparing temperatures. If $T_0 \leq T_i$, then heat cannot be supplied from the collector; therefore, the next question is whether the heat can be supplied from storage. This is determined by comparing the storage temperature T_s with the reference temperature T_{sr}, which is nominally set at 100°F (38°C). If $T_s \leq T_{sr}$, then storage cannot provide heat. If, however, $T_s > T_{sr}$, then the controller will turn on the blower and control the dampers to direct the flow through the storage container and to the zone, thereby providing heat to the enclosure. It is possible, of course, that T_e may continue to decrease because of high heat losses from the zone. If T_e drops below the reference temperature T_{r2} then it is clear that still more heat is required. In this case the controller will turn on the auxiliary heater.

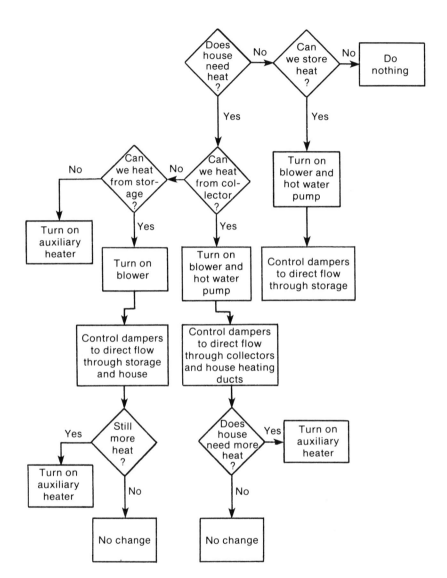

Figure 4.5.
Flow chart of control strategy.

If the controller determines that heat could be supplied from the collector, i.e., if $T_0 > T_i$, then the controller will turn on the blower and the hot-water pump and control the dampers to direct the flow through the collectors and then to the house heating ducts. It is still possible that this may not supply adequate heating to the zone, and consequently the controller must compare the enclosure temperature with the second reference temperature to determine whether or not the auxiliary heater must be turned on.

Table 4.1 is a truth table for this system. The left side of the table shows temperature comparisons. An entry of one in the table indicates that the statement is true, whereas a zero indicates a false statement. For example, if $T_e < T_{r1}$, then a one is entered in the first column. The X's represent the "don't care" situation. The middle portion of the truth table shows the various operations. For example, a one in the B_{main} column indicates that the main blower is turned on, whereas a zero indicates that the blower is turned off. Also, a one in the MD-1 column indicates that port A on motorized damper number one is open and port B on motorized damper number one is closed. The right side of the table indicates the mode of operation. For example, consider the first line in the truth table. This line corresponds to the mode where heat is supplied from storage. If T_e is less than T_{r1} but greater than T_{r2}, then we would enter a one in the first column and a zero in the second column. Furthermore, if $T_0 \leq T_i$, then we enter a zero in the third column. Suppose, however, that the storage temperature T_s is greater than the reference temperature T_{sr}; we would want the controller to turn on the blowers and direct the flow through the storage to the house. This flow is directed by controlling motorized dampers one and two. When the blower comes on, it will draw air into the return air ducts, shown in figure 4.4, through the filter, through backdraft damper number one, and then into the lower plenum on the storage unit. The air will then flow up through the storage unit, being heated in the process, and exit the storage unit at the top of the plenum. At MD-1, B must be open and A closed so the flow can reach the main blower. The air will exit the main blower and go through MD-2; at MD-2, A must be open and B closed so that the flow of air can be directed to the supply air ducts. Since we assumed that $T_e > T_{r2}$, the auxiliary unit does not need to be turned on, and therefore a zero is entered in the gas column in the truth table. Also, since the collectors are not being operated, the hot-water pump will not be turned on and, therefore, a zero is entered in the HWP

Table 4.1
Truth table for control strategy

1	2	3	4	1 + 3	1	3	1	$1\cdot2 + 1\cdot\bar{3}\cdot\bar{4}$	3	
$T_e < T_{r_1}$	$T_e < T_{r_2}$	$T_o > T_i$	$T_s > T_{sr}$	B_{main}	B_{aux}	MD-1	MD-2	Gas	HWP[a]	Mode
1	0	0	1	1	1	0	1	0	0	Heating from storage
1	1	0	1	1	1	0	1	1	0	Heating from storage plus auxiliary
1	×	0	0	1	1	0	1	1	0	Heating from auxiliary
1	0	1	×	1	1	1	1	0	1	Heating from collector
1	1	1	×	1	1	1	1	1	1	Heating from collector plus auxiliary
0	×	1	×	1	0	1	0	0	1	Store heat
0	×	0	×	0	0	×	×	0	0	Do nothing

[a] HWP = hot-water pump.

column of the truth table. The remaining modes of operation shown on the flow chart of the control strategy are illustrated in the truth table. The symbols at the top of the middle portion of the truth table represent the logic to be implemented; for example, above the B_{main} column we observe the notation $1 + 3$. This indicates that the main blower is to be turned on whenever there is a one in column 1 or a one in column 3; similarly, the auxiliary blower is to be turned on whenever there is a one in column 1, port A on $MD-1$ is to be opened whenever there is a one in column 3, port A on $MD-2$ is to be opened whenever there is a one in column 1, and the gas is to be turned on whenever there is a one in columns 1 and 2 or a one in column 1 and a zero in columns 3 and 4.

Figure 4.6 shows a circuit diagram with discrete components that may be used to implement the control logic just developed. The comparators on the left compare the various temperatures throughout the system. These signals are then sent through *AND* gates, *OR* gates, and inverters to

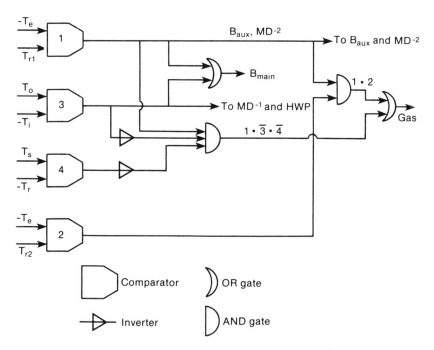

Figure 4.6
Circuit diagram.

generate the signals that are sent to the motorized dampers, blowers, and auxiliary unit.

The solar system designer would not ordinarily be concerned with this level of detail but would purchase a control unit that accomplishes the task; however, the designer and the installer should understand the functions of the control unit to properly check the system and ensure that it operates as desired.

4.2.2 Proportional Control

A proportional controller can reduce overshoot and cycling problems in bang-bang controllers. A proportional controller varies the output of the controlled device so that the output is proportional to the difference between the controlled variable and the set point; e.g., to control the temperature in a building, the proportional controller varies the output of the heating device so that the output is proportional to the difference between the desired temperature and the actual temperature at any instant. Therefore, the output of the heating plant is small if the error signal is small, and large if the error signal is large. Overshoot may still occur with a proportional controller, but not to the extent that it occurs with a bang-bang controller.

4.2.2.1 Zone Temperature Control

Figure 4.7 illustrates the proportional controller (with saturation) for zone control.

The time variation of T_e for proportional control is again found by taking the inverse Laplace transform of equation (4.3), using

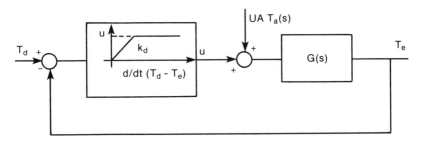

Figure 4.7
Block diagram of a proportional controller with saturation.

$$U(s) = k_p[T_d(s) - T_e(s)].$$ (4.5)

The solution is, again for constant T_a,

$$T_e(t) = T_d \frac{k_p}{UA + k_p} \{1 - e^{-((UA+k_p)/C_E)t}\}$$

$$+ \frac{UAT_a}{UA + k_p} \{1 - e^{-((UA+k_p)/C_E)t}\} + T_e(0)e^{-((UA+k_p)/C_E)t},$$ (4.6)

where k_p is the gain for the proportional controller, which may be adjusted to vary the system response.

4.2.2.2 Collector Loop Control

A proportional controller may also be used to control the fluid mover in the collector loop of a solar energy system. To do this, the controller varies the mass flow rate of fluid through the collector array, decreasing the flow rate when the temperature difference between the collector outlet and the storage outlet is small and increasing the flow rate when this tempera-ture difference increases. This results in eliminating the cycling problem that exists in bang-bang controllers. DOE has sponsored several studies (Kent, Winn, and Huston 1980; Kovarik and Lesse 1976; Lunde 1982; McDonald, Farris and Melsa 1978; Orbach, Rorres, and Fischel 1981; Taylor 1978; Wang and Dorato 1983; Winn 1983a) concerning the advan-tages of proportional controllers versus bang-bang controllers for col-lecting energy in solar energy systems. These studies are discussed in the next section.

The proportional controller normally has both an offset and a satura-tion value. The offset value is required to prevent the controller from operating its control device continuously. Without an offset value, the controller could turn the fluid mover on when the difference between the collector outlet temperature and storage temperature is too small to make energy collection worthwhile. The saturation value represents the maxi-mum output of the controlled device. In the case of the fluid mover in the collector loop, the maximum flow rate is attained at this value.

Lunde (1982) proposed the stagnation-point controller, a pseudo-proportional controller. Its rate of energy collection is directly proportional to the difference between a collector temperature at a no-flow condition and the storage temperature at any instant in time, with the proportionality constant being the modified collector heat-loss coefficient $F_R U_L$. However, the stagnation-point controller is actually a bang-bang controller; it turns the fluid mover on whenever the no-flow collector temperature increases to the storage temperature and turns the fluid mover off whenever the no-flow collector temperature decreases to the storage temperature. This controller is discussed later.

4.2.3 Integral Zone Temperature Control

Greater control efforts may be needed when building systems do not respond quickly enough to demands for increased temperature. An integral controller provides this extra control. The integral controller is designed to increase the controller output in proportion to the time integral of the error; i.e., the longer the building temperature remains below the desired value, the greater will be the output from the heating plant. An integral controller results in a slow response but does have a stabilizing effect on system response.

Figure 4.8 illustrates an integral control system. The integral controller is represented by

$$u(t) = k_i \int_{t_0}^{t} [T_d(\tau) - T_e(\tau)]dt, \tag{4.7}$$

where k_i is the gain for an integral controller. The Laplace transform of $u(t)$ is

$$u(s) = \frac{k_i}{s}[T_d(s) - T_e(s)]. \tag{4.8}$$

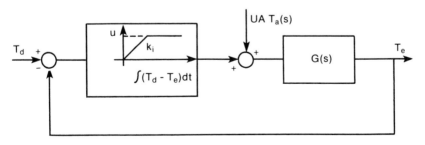

Figure 4.8
Block diagram of an integral control system.

Inserting equation (4.8) into equation (4.2) and taking the inverse transform of $T_E(s)$ results in

$$T_e(t) = \left\{ \frac{k_i}{C_E} \frac{1}{s_1 s_2} + \frac{1}{s_1 - s_2} \left[\frac{1}{s_1} e^{s_1 t} - \frac{1}{s_2} e^{s_2 t} \right] \right\} T_d$$

$$+ \left(\frac{U A T_a}{C_E} \right) \frac{1}{s_1 - s_2} [e^{s_1 t} - e^{s_2 t}]$$

$$+ T_e(0) \frac{1}{s_1 - s_2} [s_1 e^{s_1 t} - s_2 e^{s_2 t}], \tag{4.9}$$

where

$$s_{1,2} = -\frac{1}{2} \left(\frac{UA}{C_E} \right) \pm \frac{1}{2} \sqrt{ \left(\frac{UA}{C_E} \right)^2 - 4 \frac{k_i}{C_E} }. \tag{4.10}$$

The controller gain k_i may be adjusted to give a desired response. Note that if k_i is sufficiently large, the system response will be oscillatory.

4.2.4 Derivative Zone Temperature Control

Derivative controllers may be used to achieve a faster response. A derivative controller leads to a response that is proportional to the time derivative of the error and is represented analytically as

$$u(t) = k_d \frac{d}{dt} [T_d(t) - T_e(t)], \tag{4.11}$$

where k_d is the gain for a derivative controller.

The system response may be varied by adjusting the gain. Derivative controllers result in quick responses but sometimes cause instabilities to occur. They are also strongly affected by small fluctuations in temperatures.

The derivative controller is represented in figure 4.9. The system response to derivative control is obtained by transforming $u(t)$, substituting this value into equation (4.2) and taking the inverse transform. The transform of the derivative controller is

$$U(s) = k_d \{ [s T_d(s) - T_d(0)] - [s T_e(s) - T_e(0)] \}. \tag{4.12}$$

Substituting into equation (4.2) solving for $T_e(s)$, and taking the inverse transform results in

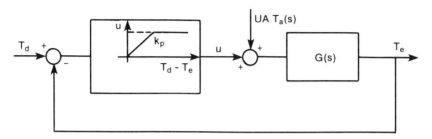

Figure 4.9
Block diagram of a derivative control system.

$$T_e(t) = T_e(0)e^{-(UA/(C_E+K_d))t}$$
$$+ T_a[1 - e^{-(UA/(C_E+k_d)t)}]. \qquad (4.13)$$

Note that this would not be a desirable controller for zone control, since if the time rate of change of the error signal approaches zero, the control will approach zero and, because of the loss terms, the enclosure temperature will approach the ambient temperature. Therefore, the derivative controller should be used in conjunction with a proportional controller, resulting in proportional plus derivative (PPD) control.

4.2.5   PID Zone Temperature Controllers

The previously discussed control types may be combined to give a "proportional-integral-differential controller," known as a PID controller. The control law is expressed as

$$u = k_p(T_d - T_e) + k_i \int_{t_0}^{t} (T_d - T_e)\, dt + k_d \frac{d}{dt}(T_d - T_e), \qquad (4.14)$$

where k_p is the gain for a proportional controller.

A PID controller can reduce energy consumption in three ways. First, it can reduce overshoot in heating the space (an improved method of thermostat "anticipation"). Second, if needed to implement a proportional actuator, the PID controller could increase the average steady-state efficiency of a given size combustion heat exchanger by operating the burner at less than nominal rating most of the time. Third, the PID controller can increase transient efficiency by reducing on/off cycling and reduce flue losses by causing the heat exchanger to be at a temperature that is well below the nominal design temperature at the end of each "on" cycle.

The system response may be determined in the same manner as before. The gains may be adjusted in order to attempt to obtain some desired response.

4.2.6 Design of Controllers

There are two basic approaches to the design of controllers. The first is trial-and-error method, and the second is the analytical design method.

4.2.6.1 Trial-and-Error Method

Figure 4.10 illustrates the trial-and-error method, which is the most often used method in designing control systems. In the classical trial-and-error design procedure, one selects a set of performance specifications and then attempts to adjust the controller gains to satisfy them. Performance specifications for a zone heating system typically include the desired temperature, overshoot, and time delay. The trial-and-error design process tends to be inefficient and expensive. Moreover, when one selects a set of performance specifications, one has no way of knowing whether or not those performance specifications can be satisfied.

Some general guidelines, drawn from experience, can ameliorate these problems somewhat by suggesting a reasonable initial control strategy. Kutscher (1983) developed some design guidelines at the Solar Energy Research Institute (SERI) as part of a DOE project on solar industrial process heat systems. His guidelines can be used to select controllers. In general, he recommends using proportional control when cycling action (due to on/off control) is undesirable, setpoint changes are small or infrequent, and the steady-state deviation between the set point and the controlled variable (i.e., the offset) can be tolerated. He recommends using integral control when the offset must be reduced or eliminated and the set point changes are frequent, but not when start-up overshoot must be eliminated and the process can be controlled with high-gain proportional control. Finally, he suggests adding derivative action to proportional control when a faster response is required, but not if the distance-velocity lag is significant or the process is noisy.

4.2.6.2 Analytic Design

Some designers may incur considerable costs for computer simulations that specify a set of gains for selected controllers to meet a set of performance specifications. However, they may find it more cost-efficient to use an analytical design process, which eliminates guesswork and which can

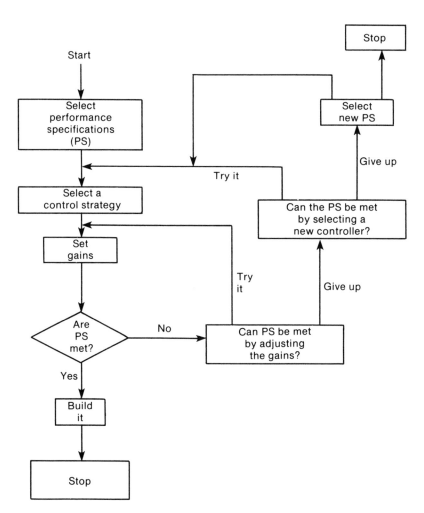

Figure 4.10
Trial-and-error design procedure.

determine directly the control strategy that will result in optimizing system performance. In the analytical design process one selects a cost function to be optimized. Hence, the process is often referred to as optimal control.

A typical formulation of the optimal-control problem is the linear regulator problem. This type of control problem applies to situations where the objective is to bring a controlled quantity, such as $(T_d - T_e)$, to zero. A linear regulator for this application could be designed to minimize the weighted sum of the root-mean-square deviation from the set point and the mean energy input

$$J = \int_0^{t_f} \{A[u(t)]^2 + B[E(t)]^2\}\, dt, \tag{4.15}$$

where J represents the performance index, $E(t)$ is the deviation from the set point, and $u(t)$ is the controlled energy input. $E(t)$ is related to $u(t)$ (and to other factors, such as weather and occupant behavior, which are both treated as noise) by a set of differential equations that govern the dynamic behavior of the heated space and backup heating plant. The chief parameters in these governing equations are the room parameters, the resistance to heat loss (R) and the thermal storage capacitance (C_E).

One issue raised by the integration of backup heaters with solar-heated buildings is the interaction of controls. In particular, passive-solar-heated buildings present a number of problems. First, the time constants of rooms in such buildings are usually much longer than in conventional buildings, because both R and C are much greater. The thermal capacitance effect alone has a major significance: any overshoot of the set point by the backup heater controller represents a relatively large quantity of energy absorbed by thermal capacitance elements. This overshoot will then persist (decay very slowly) because of the large time constant of the heated space. It is therefore apparent that significant energy savings can be realized by reducing overshoot and thermostat dead bands.

The quadratic criterion represented by equation (4.15) helps reduce overshoot, because increasing deviations $E(t)$ are penalized with progressively increasing severity. Since the penalties are integrated over time, the resulting control law will involve some integral feedback, which will tend to reduce final error. Another useful feature of the optimal regulator is reduced peak demand. The quadratic penalty on $u(t)$ tends to limit parasitic (auxiliary) power.

4.3   Controllers for Collecting Solar Energy

This section describes controllers that are used for collecting solar energy. First, fluid flow controllers that are used to control the fluid flow rate through collector arrays are briefly described. Bang-bang controllers, proportional controllers, stagnation point controllers, and optimal controllers are discussed in detail.

DOE has sponsored a number of projects involving controllers for regulating the fluid flow rate through collector arrays in solar energy systems. DOE has sponsored projects involving bang-bang, proportional, optimal, and adaptive control. The major results from these studies are discussed in the following paragraphs.

4.3.1   Bang-Bang Control

On/off operation of a fixed flow rate collector pump is the most widely used pump and control configuration. The make/break power switching in the controller is familiar to the installer, electrician, etc. On/off control can be used to switch any motor of appropriate voltage and current rating. Manufacturers use both electromechanical (relays) and solid-state outputs in these controllers.

The preferred on/off temperature set points to be used in bang-bang control depend on many characteristics that affect the system. These include collector performance; heat exchanger features; thermal capacitance of the collector and piping; the system flow rate; system loss characteristics; and probe placement, accuracy, and precision. Collector probe placement is particularly critical in its effect upon the system's transient thermal performance. Analytical results, generally confirmed by field experience, show that for efficient operation the collector probe should be near the collector outlet at approximately 90% of the length along the flow path (Herczfeld, Fischl, and Konyk 1980). If the sensor is located too near the exit, it turns the fluid mover on too late, and if it is located too near the entrance, excessive cycling occurs. An exception to this general rule is that a more downstream location may be preferred for systems in climates that experience large amounts of radiation to the night sky.

The dead band (hysteresis) between the off and on temperature settings minimizes pump "hunting"; i.e., frequent switching between on and off, during morning start-up and afternoon shutdown or intermittent cloud conditions. The "off" setting must be sufficiently high to accommodate normal sensor inaccuracies but low enough to allow maximum collection of solar energy while assuring that collection stops when parasitic power for the pumps exceeds the value of the solar energy gain (i.e., if the coefficient of performance (COP) ≤ 1, collection should be stopped). The experience of the controller industry is that if an on-to-off temperature ratio of from 4/1 up to 6/1 is maintained with the ΔT_{off} temperature ranging from $1°$ to $2°C$ ($1.8°$ to $3.6°F$), satisfactory operation will result in most types of flat-plate liquid transport systems. Usually, the higher ratio of ΔT_{on} to ΔT_{off} is applied when ΔT_{off} is at its lower range of values. That is, if ΔT_{off} is approximately $1°C$ ($1.8°F$), then ΔT_{on} would be approximately $6°C$ ($10.8°F$). Similarly, the lower value of the ratio of on-to-off temperature differentials is applied when ΔT_{off} is near its upper range of values. Thus the hysteresis in liquid systems is normally in the range from $6°$ to $8°C$ ($10.8°$ to $14.4°F$), with some systems as high as $11°C$ ($19.8°F$).

Air transport systems normally require ΔT_{off} temperatures of $5°C$ ($9°F$) and as high as $11°C$ ($19.8°F$) with an $8°$ to $14°C$ ($14.4°$ to $25.2°F$) hysteresis. This slightly wider range for hysteresis is required to compensate for the effects of the relatively low specific heat of air, the relatively low efficiency of blowers (as compared to pumps), the relatively high difference in temperature between absorber plate and fluid, and the relatively high temperature increase of the fluid passing through the collector.

Solar designers and installers should note that while pump cycling often arouses concern in homeowners, it causes very little (if any) equipment degradation if limited to a small number (approximately 6 cycles or fewer) during a 10–15 minute start-up or shutdown period. Cycle rates of this magnitude have little impact on motor winding temperatures, on the pump's bearing life, or on the controller output switches, which are generally rated for more than 150,000 closures. Conversely, pump cycling indicates a longer collection duty cycle and hence more stored energy in many cases. Nonetheless, should the designer or installer wish to limit the cycling in an on/off controlled installation, the system operation is generally best served if it is done by lowering the flow rate. Controllers with manually adjusted or timing delays are recommended only for experimental installations because of the high field failure rates of components required for this type of application. Controllers with time delays, additional circuit logic, or variable flow rate outputs should be considered in installations where cycling is chronic.

4.3.1.1   Effects of Temperature Settings on Cycling Rates

Many investigators (Herczfeld et al. 1978; Lewis and Carr 1978; Kent, McGavin, and Lantz 1978; Winn and Winn 1981; Schiller, Warren, and Auslander 1980) have studied the problem of selecting appropriate values for ΔT_{on} and ΔT_{off} for bang-bang controllers. If thermal capacitance of the collectors is negligible and if ΔT_{on} is set too high, the amount of useful energy collected is reduced, and if ΔT_{on} is set too low or ΔT_{off} is set too high, excessive cycling of the fluid mover occurs. Clearly, it is not worthwhile to operate the fluid mover to collect energy if the rate at which energy is collected is less than or equal to the power required to collect the solar energy. This is a useful criterion for determining the appropriate value for ΔT_{off}. The value for ΔT_{on} may then be determined by setting the ratio of ΔT_{on} to ΔT_{off} to a value based on the radiation and number of cycles acceptable. The following paragraphs describe the development of these relationships for the temperature settings.

The relationship for ΔT_{off} is well known (Kent, Winn, and Huston 1980; Schiller, Warren, and Auslander 1980; Alcone and Herman 1981). It is

$$\Delta T_{off} \geq k P_m / \dot{m} c_p, \tag{4.16}$$

where P_m is the power delivered to the motor of the fluid mover, k is a constant coefficient determined by the value of the energy being displaced, $\dot{m}$ is the mass flow rate, and c_p is the specific heat of the fluid. For electric resistance heating, one would normally choose k to be unity. The values for ΔT_{off} as determined from equation (4.16) are typically lower than those used in practice. In practice, the values are higher to account for sensor inaccuracies and drift; they are typically on the order of $1°$–$2°C$ ($2°$–$4°F$) for liquid-based systems.

It has also been shown (Kent, Winn, and Huston 1980; Schiller, Warren, and Auslander 1980; Duffie and Beckman 1980) that if

$$\Delta T_{off} \leq (F_R U_L A_c / \dot{m} c_p) \Delta T_{on}, \tag{4.17}$$

where F_R is the collector heat removal factor, U_L is the collector loss coefficient, and A_c is the area of the collector, then cycling will not occur. However, equation (4.17) is not necessary and sufficient to prevent cycling; it is very conservative and leads to extremely high values for ΔT_{on}. Previous studies (Herczfeld et al. 1978; Schiller, Warren, and Auslander 1980) relating to this problem have been based on numerical simulations and have provided useful results. An analytical development is presented next.

4.3.1.2   Analytical Formulation

More realistic values for ΔT_{on} than those given by equation (4.17) may be obtained from Winn (1983b). When the fluid mover is "on," the rate of energy collection may be represented by

$$\dot{Q}_u = \dot{m}c_p(T_0 - T_i) = F_R A_c[G_T(\tau\alpha) - U_L(T_i - T_a)], \tag{4.18}$$

where G_T is the instantaneous radiation on a tilted collector. Let

$$T_0 = T_i + \Delta T_{on} \tag{4.19}$$

and

$$T_i = T_s. \tag{4.20}$$

Then equation (4.18) may be written as

$$\dot{m}c_p\Delta T_{on} = F_R A_c[G_T(\tau\alpha) - U_L(T_s - T_a)]. \tag{4.21}$$

Solving for $T_s - T_a$ results in

$$T_s - T_a = F_R(\tau\alpha)G_T/F_R U_L - (\dot{m}c_p/A_c)\Delta T_{on}/F_R U_L. \tag{4.22}$$

4.3.1.3   Phase Plane Representation

Equation (4.22) represents a family of straight lines in a "phase space" of $T_s - T_a$ and G_T, as illustrated in figure 4.11. The slope of each line is given

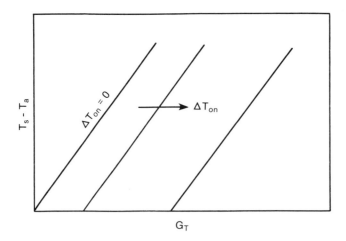

Figure 4.11
Phase plane representation.

by the ratio between $F_R(\tau\alpha)$ and $F_R U_L$. Now consider just two members of this family of lines: the first for $\Delta T_{on} = \Delta T_{off}$ and the second for $\Delta T_{on} = (\dot{m}c_p/A_c)\Delta T_{off}/F_R U_L$. These two lines define a cycling region, as illustrated in figure 4.12. To the left of the $\Delta T_{on} = \Delta T_{off}$ line, the fluid mover remains off, whereas to the right of the line for ΔT_{on} as determined from equation (4.17), the fluid mover will remain on. The cycling rate is constant along the straight lines within the cycling region.

4.3.1.4 Cycling Rate

The variation in fluid temperature along the length of the collector may be approximated by a linear relationship. The maximum value for radiation $G_{T_{NC}}$ for which cycling will not occur is that value for which the outlet temperature T_0 is just equal to $T_s + \Delta T_{off}$. This determines the slope of the T_0 versus x line. An increase in G_T will increase the slope. To determine $G_{T_{NC}}$, consider a steady-flow problem for which an energy balance equation results in

$$G_{T_{NC}} = [\dot{m}c_p \Delta T_{off} + F_R U_L A_c (T_s - T_a)]/[F_R(\tau\alpha) A_c]. \tag{4.23}$$

The time to "empty" the collector of the fluid initially in the collector at the time the fluid mover is turned on is given by

$$t_E = m/\dot{m}; \tag{4.24}$$

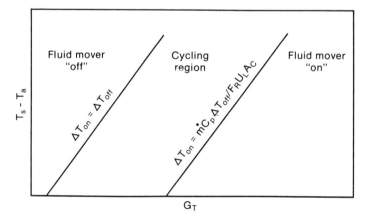

Figure 4.12
Cycling region.

the collector outlet temperature at the time t_E is given by

$$T_0 = T_s + [F_R(\tau\alpha)G_T - F_R U_L(T_s - T_a)](A_c/\dot{m}c_p); \tag{4.25}$$

and the time required to raise the collector fluid outlet temperature from T_0 to $T_s + \Delta T_{\text{on}}$ may be determined from an energy balance equation for the "no flow" condition. The result is

$$(\dot{m}c_p/A_c)(T_s + \Delta T_{\text{on}} - T_0)/t_H = A_c[G_T(\tau\alpha) - U_L(T_s - T_a)], \tag{4.26}$$

where t_H represents the time to heat the fluid. Substituting for T_0 results in

$$t_H = \left(\frac{m}{\dot{m}}\right)\left[\frac{(\dot{m}c_p/A_c)\Delta T_{\text{on}}}{G_T(\tau\alpha) - U_L(T_s - T_a)} - F_R\right]. \tag{4.27}$$

The total cycle time T is given by

$$T = \frac{m}{\dot{m}}\left[\frac{(\dot{m}c_p/A_c)\Delta T_{\text{on}}}{G_T(\tau\alpha) - U_L(T_s - T_a)} - F_R + 1\right]. \tag{4.28}$$

This expression is valid for $G_{T_{\text{min}}} < G_T < G_{T_{NC}}$, where

$$G_{T_{\text{min}}} = U_L(T_s - T_a)/(\tau\alpha) \tag{4.29}$$

and

$$G_{T_{NC}} = [\dot{m}c_p\Delta T_{\text{off}} + F_R U_L A_c(T_s - T_a)]/[F_R(\tau\alpha)A_c]. \tag{4.30}$$

The cycling rate ω is given by

$$\omega = 1/T. \tag{4.31}$$

4.3.1.5 Number of Cycles

The number of cycles that occur while the radiation value remains within the cycling region may be obtained by integrating the equation for the cycling rate with respect to time. That is,

$$N = \int_{t_0}^{t_f} \omega(t)\,dt, \tag{4.32}$$

where N represents the number of cycles and t_0 and t_f represent the times corresponding to $G_{T_{\text{min}}}$ and $C_{T_{NC}}$, respectively. In order to perform this integration analytically, it is necessary to have an analytical representation for the radiation G_T as a function of time. Schiller, Warren and Auslander (1980) performed numerical studies using the following model:

$$G_T(t) = G_{T_M} \sin\left(\frac{\pi}{12} t\right).$$ (4.33)

This results in

$$N = \left(\frac{m}{\dot{m}}\right)$$

$$\cdot \int_{t_0}^{t_f} \left[\frac{-U_L(T_s - T_a) + G_{T_M}(\tau\alpha)\sin(\pi t/12)}{(\dot{m}c_p/A_c)\Delta T_{\text{on}} - (1 - F_R)U_L(T_s - T_a) + (1 - F_R)(\tau\alpha)G_{T_M}\sin(\pi t/12)} \right] dt,$$ (4.34)

which may be integrated in closed form to give

$$N = \text{INT}\, \frac{(\dot{m}c_p/A_c)}{(mc_p/A_c)} \frac{12}{\pi} \frac{1}{(1 - F_R)}(x_f - x_0) - \frac{2(\dot{m}c_p/A_c)\Delta T_{\text{on}}}{(1 - F_R)\sqrt{B}}$$

$$\times \left\{ \tan^{-1}\left[\frac{C\tan(x_f/2) + E}{\sqrt{B}}\right] - \tan^{-1}\left[\frac{C\tan(x_0/2) + E}{\sqrt{B}}\right] \right\},$$ (4.35)

where

$$x_f = \sin^{-1}\left\{ \frac{1}{G_{T_M}} \left[\frac{(\dot{m}c_p/A_c)\Delta T_{\text{off}}}{F_R(\tau\alpha)} + \frac{F_R U_L}{F_R(\tau\alpha)}(T_s - T_a) \right] \right\},$$ (4.36)

$$x_0 = \sin^{-1} \frac{1}{G_{T_M}} \frac{F_R U_L}{R_R(\tau\alpha)}(T_s - T_a),$$ (4.37)

$$B = \left[\frac{\dot{m}c_p}{A_c}\Delta T_{\text{on}} - (1 - F_R)U_L(T_s - T_a) \right]^2 - \left[(1 - F_R)G_{T_M}(\tau\alpha) \right]^2,$$ (4.38)

$$C = \frac{\dot{m}c_p}{A_c}\Delta T_{\text{on}} - (1 - F_R)U_L(T_s - T_a), \text{ and}$$ (4.39)

$$E = (1 - F_R)G_{T_M}(\tau\alpha).$$ (4.40)

This analysis may be used to determine the number of cycles and the energy collected as a function of ΔT_{on} and ΔT_{off}. Note that rather than use $\dot{m}/m$ in equation (4.35), the ratio has been expressed in terms of fluid capacitance and collector capacitance per unit collector area. If the term (mc_p/A_c) represents fluid plus collector mass and specific heat, then the equation for t_E is not exact. However, it will be shown that the error involved in this approximation is small for representative collectors.

4.3.1.6 Comparisons with Earlier Results

Schiller, Warren, and Auslander (1980) used a numerical simulation program to determine the number of cycles that would occur as a function of
ΔT_{on} for various values of G_{T_M} and for given values of ΔT_{off} and collector
parameters. The parameter values used were as follows:

High-Gain, High-Flow, Clear Day

$(\tau\alpha) = 0.84$
$F_R = 0.937$
$(mc_p/A_c) = 14.3$ kJ/m^2 °C (0.7 Btu/ft^2.F)
$(\dot{m}c_p/A_c) = 511$ kJ/h m^2 °C (25 Btu/h.ft^2.F)
$\Delta T_{\text{off}} = 1.7$°C (35.1°F)
$G_{T_M} = 946$ W/m^2 (300 Btu/h.ft^2.F)
$T_s - T_a = 25$°C (77°F)
$U_L = 3.97$ W/m^2 °C (0.7 Btu/h.ft^2.F)
$\Delta T_{\text{on}} = 5$°C (41°F)

High-Gain, Low-Flow, Clear Day

$(\tau\alpha) = 0.84$
$F_R = 0.93$
$(mc_p/A_c) = 14.3$ kJ/m^2 °C (0.7 Btu/ft^2.F)
$(\dot{m}c_p/A_c) = 306$ kJ/h m^2 °C (15 Btu/h.ft^2.F)
$\Delta T_{\text{off}} = 1.7$°C (35.1°F)
$G_{T_M} = 946$ W/m^2 (300 Btu/h.ft^2.F)
$T_s - T_a = 25$°C (77°)
$U_L = 3.97$ W/m^2 °C (0.7 Btu/h.ft^2.F)
$\Delta T_{\text{on}} = 5$°C (41°F)

Low-Gain, High-Flow, Clear Day

$(\tau\alpha) = 0.84$
$F_R = 0.937$
$(mc_p/A_c) = 14.3$ kJ/m^2 °C (0.7 Btu/ft^2.F)
$(\dot{m}c_p/A_c) = 511$ kJ/h m^2 °C (25 Btu/h.ft^2.F)
$\Delta T_{\text{off}} = 1.7$°C (35.1°F)
$G_{T_M} = 473$ W/m^2 (150 Btu/h.ft^2.F)
$T_s - T_a = 36.1$°C (97°F)
$U_L = 3.97$ W/m^2 °C (0.7 Btu/h.ft^2.F)
$\Delta T_{\text{on}} = 5$°C (41°F)

High-Gain, High-Flow, Clear Day

$(\tau\alpha) = 0.84$
$F_R = 0.93$
$(mc_p/A_c) = 14.3$ kJ/m^2 °C (0.7 Btu/ft^2.F)
$(\dot{m}c_p/A_c) = 306$ kJ/h m^2 °C (15 Btu/h.ft^2.F)
$\Delta T_{\text{off}} = 1.7$°C (35.1°F)
$G_{T_M} = 473$ W/m^2 (150 Btu/h.ft^2.F)
$T_s - T_a = 36.1$°C (97°F)
$U_L = 3.97$ W/m^2 °C (0.7 Btu/h.ft^2.F)
$\Delta T_{\text{on}} = 5$°C (41°F)

The cycles reported for each case were 10, 2, 61, and 10, respectively. The
analytical solution presented in equation (4.35) leads to 8, 2, 37, and 7
cycles for each time the value for the incident radiation passes through the
cycling region for each case. This agreement between the analytical and
numerical results is sufficiently satisfactory that the analytical representation may be used for purposes of analysis. For example, for a given collector, one may develop graphs of the number of cycles as a function of ΔT_{on},
as illustrated in figure 4.13. The analytical solution enables one to conduct
sensitivity studies without having to resort to (possibly) lengthy numerical
solutions using a computer. This is illustrated in the next section. Note
that for the high-gain, high-flow, clear-day case, cycling ceases at a value
for ΔT_{on} of approximately 14°C (25°F), whereas using equation (4.17) re-

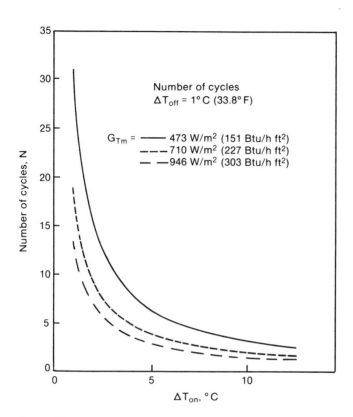

Figure 4.13
Number of cycles versus ΔT_{on}.

sults in a value of $\Delta T_{on} = (511 \text{ kJ/h m}^2 \text{ °C}) (1°\text{C})/(0.937) (3.97 \text{ W/m}^3 \text{ °C})$
$(3.6) = 38°\text{C}$.

4.3.1.7 Sensitivity Analysis

The equation for the number of cycles is of the form

$$N = N\left(\frac{\dot{m}c_p}{A_c}, \frac{mc_p}{A_c}, F_R, \tau\alpha, U_L, \Delta T_{on}, \Delta T_{off}, T_s - T_a, G_{T_M}\right). \tag{4.41}$$

The sensitivity of N to any of the parameters may be determined analytically by partial differentiation with respect to that parameter. For multiple parameter variations, one has

$$\Delta N = \sum_{j=1}^{n} \left(\frac{\partial N}{\partial x_j}\right) \partial x_j, \tag{4.42}$$

where ∂x_j represents the variation in parameter x_j. Since uncertainties exist in some of the parameter values, particularly in ΔT_{off} and the collector capacitance term, there will be a corresponding uncertainty in N. The most likely value for the uncertainty in N is given by

$$\Delta N = \left[\sum_{j=1}^{n} \left(\frac{\partial N}{\partial x_j} \partial x_j\right)^2\right]1\sqrt{2}. \tag{4.43}$$

The sensitivity of N to ΔT_{off} is given by

$$\partial N/\partial \Delta T_{\text{off}} = \frac{(\dot{m} c_p/A_c)(12/\pi)}{(1 - F_R)\sqrt{1 - A^2 G_{T_M} F_R(\tau\alpha)}}$$

$$\times \left\{1 - \frac{(\dot{m} c_p/A_c)\Delta T_{\text{on}} C}{[B + C\tan D + E)^2]\cos^2 D}\right\}, \tag{4.44}$$

where

$$A = \frac{1}{G_{T_M}}\left\{\frac{(\dot{m} c_p/A_c)\Delta T_{\text{off}}}{F_R(\tau\alpha)} + \frac{F_R U_L}{F_R(\tau\alpha)}(T_s - T_a)\right\}, \tag{4.45}$$

$$D = 0.5\sin^{-1}(A), \tag{4.46}$$

and B, C, and E are given by equations (4.38), (4.39), and (4.40), respectively.

The sensitivity to collector capacitance is simply

$$\frac{\partial N}{\partial(m c_p/A_c)} = -N/(m c_p/A_c). \tag{4.47}$$

A method for estimating the collector capacitance is presented in Huang and Lu (1982).

The sensitivity to ΔT_{off} for the case considered previously and for various values of ΔT_{on} is illustrated in figure 4.14. It is apparent that the number of cycles is strongly sensitive to ΔT_{off} when ΔT_{on} is relatively near ΔT_{off}; hence, the location of the collector outlet sensor is very important in determining the number of cycles.

4.3.1.8 Energy Collected
The amount of energy collected as a function of the number of cycles may be determined from the above analysis (Kahwaji and Winn 1985). The

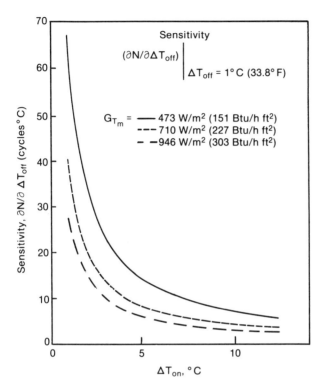

Figure 4.14
Sensitivity to ΔT_{off}.

times corresponding to the no-flow and no-cycling conditions may be determined and then the relative increase in collected energy (RICE) may be calculated from

$$\text{RICE} = \frac{Q_C - Q_{NC}}{Q_{NC}}, \tag{4.48}$$

where Q_C and Q_{NC} represent the energy collected under cycling and no-cycling conditions, respectively, and where

$$Q_C - Q_{NC} = \int_{t_0}^{t_f} Q_u(t)\, dt. \tag{4.49}$$

Figure 4.15 shows a graph of RICE as a function of ΔT_{on} for a low-gain, high-flow system.

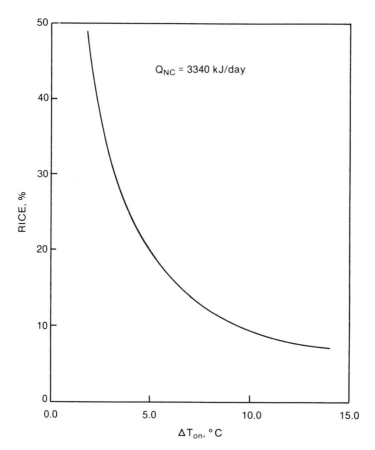

Figure 4.15
RICE versus ΔT_{on}.

4.3.1.9 Collector Capacitance Effects

The above results were based on the Hottel-Whillier model, for which the capacitance of the collector is neglected. If the collector capacitance is large, then a different model should be used. Such a model has been proposed (Klein 1973; Orbach, Rorres, and Fischl 1981) and is given by

$$\frac{\partial T_f(x,t)}{\partial t} + \frac{\dot{m}c_p}{WC_c}\frac{\partial T_f(x,t)}{\partial x} + \frac{F'U_L}{C_c}T_f(x,t) = \frac{F'U_L}{C_c}\left[\frac{(\tau\alpha)}{U_L}G_T + T_a\right], \qquad (4.50)$$

where C_c is the collector capacitance per unit area and T_f is the steady-state temperature.

This model has been validated by comparison with experimental results obtained from Middlebury College under contract to DOE (Wolfson 1979). During periods for which there is constant flow through the collector array, the steady-state temperature rise across the collector is given by

$$\Delta T_{fc} = \frac{\tau_{tc}}{\tau_c} G_T(\tau\alpha)/U_L - (T_s - T_a). \tag{4.51}$$

During periods for which there is no flow, the difference between the collector temperature and the storage temperature may be obtained from equation (4.50) and is given by

$$T_{fc}(1,t) - T_s = F(t) + T_{fc}(1,0)e^{-t/\tau_c} - T_s, \tag{4.52}$$

where

$$F(t) = \frac{1}{\tau_c} \int_0^t \left[\frac{(\tau\alpha)}{U_L} G_T(\tau) + T_a(\tau) \right] e^{-(t-\tau)/\tau_c} d\tau. \tag{4.53}$$

If one assumes that the flow rate has been turned on, then equation (4.51) applies. The temperature difference across the collector will be a function of G_T, as indicated by equation (4.51), and if G_T is not sufficiently large, the temperature difference may decrease and reach ΔT_{off}, at which point the controller would shut off the flow rate through the collector. If ΔT remains greater than ΔT_{off}, then no cycling will occur. One could compute from equation (4.51) the time t_{off} at which the temperature difference reaches ΔT_{off}. This value for t_{off} could then be substituted into equation (4.52) to determine the limiting value for ΔT_{on} to prevent cycles. An analytical solution may be obtained if one assumes that C_c is approximately equal to the fluid capacitance per unit collector area, $\dot{m}c_p/A_c$, and that $T_s - T_a$ is constant. The resulting equation for ΔT_{on} (Herczfeld, Fischl, and Konyk 1980) is

$$\Delta T_{on} = \{F - \omega\tau_c[(G_T(\tau\alpha)/U_L)^2 - F^2]\}/(1 + \omega^2\tau_c^2), \tag{4.54}$$

where

$$F = \Delta T_{off}\tau_c/\tau_{tc} + (T_s - T_a), \tag{4.55}$$

and

$$\omega = 2\pi/24 \text{ (rad/h)}. \tag{4.56}$$

The values for ΔT_{on} that will prevent cycles, as obtained from this equation, are in very close agreement with those obtained from equation (4.17) when applied to actual systems.

Herczfeld, Fischl, and Konyk (1980) have used the plug flow model described previously to conduct parametric sensitivity analyses. They determined the sensitivity of the energy collected and the number of cycles to various parameters classified as control, system, and climatic parameters, defined as follows:

$$\underline{\phi} = \begin{cases} \underline{\phi}_c & \text{(vector of control parameters)} \\ \underline{\phi}_s & \text{(vector of system parameters)} \\ \underline{\phi}_w & \text{(vector of climatic parameters)} \end{cases}$$

where

$$\underline{\phi}_c = \begin{cases} \Delta T_{\text{off}} & \text{(turn-off set point, °C)} \\ \Delta T_{\text{on}} & \text{(turn-on set point, °C)} \\ \dot{m} & \text{(collector fluid flow rate, kg/h)} \end{cases}$$

$$\underline{\phi}_s = \begin{cases} \tau_{tc} & \text{(system time constant for } \dot{m} \neq 0 \text{ (h))} \\ \tau_c & \text{(system time constant for } \dot{m} = 0 \text{ (h))} \end{cases}$$

and

$$\underline{\phi}_w = \begin{cases} G_{T_M} & \text{(maximum solar irradiance (kJ/h m}^2) \\ T_s - T_a & \text{(storage to ambient temperature difference, °C).} \end{cases}$$

The sensitivity in system performance was investigated by varying the parameters of $\underline{\phi}$ about the nominal values shown in table 4.2. The results are shown in figure 4.16a. The difference between these results and those presented earlier simply reflects the difference between the two models used. Note from figure 4.16b that as the collector capacitance decreases, the number of cycles and the total accumulated energy increase, which is in agreement with the results presented earlier.

4.3.2 Proportional Control

The proportional controller operates by controlling pump speed or by using a throttling valve with a constant or multiple-speed pump (Kutscher et al. 1982). For the variable-speed pump, it increases power to the pump

Table 4.2
Baseline system parameter values

Nominal values of parameters		Climate	Set points
$C_c = 10.22 \dfrac{kj}{C-m^2}$	$U_L = 18.65 \dfrac{kj}{m^2 - C - r}$	$T_a(t) = 0°C$	$\Delta T_{on} = 5°C$
	$F' = 0.95$		
$W = 1.0$ m	$(\tau\alpha) = 0.8281$	$I(t) = I_o \sin \omega(t - 6)$	$\Delta T_{off} = 2°C$
	$L = 2.0$ m		
$\dot{m}_c = 227.19 \dfrac{kg}{h}$	$c_p = 4.19 \dfrac{kj}{kg - C}$	$I_o = 3000 \dfrac{kj}{m^2 - h}$	$\Delta T_h = 3°C$
	$A_c = 2.0$ m^2		

C. Byron Winn

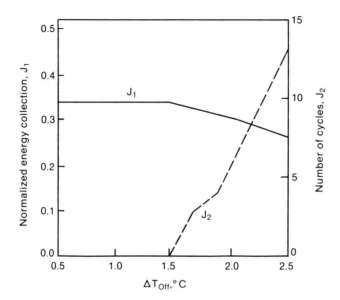

Figure 4.16a
Energy collected and number of cycles as a function of ΔT_{off}. Source: Herczfeld et al. 1980.

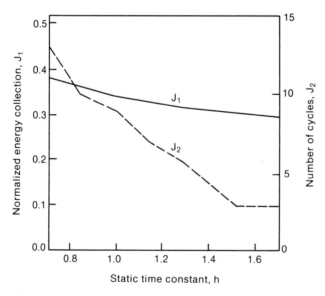

Figure 4.16b
Energy collected and number of cycles as a function of system static time constant. Source: Herczfeld et al. 1980.

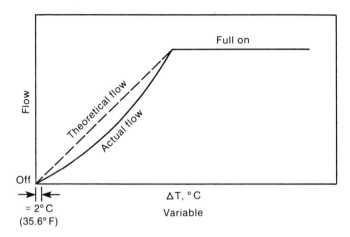

Figure 4.17
Variable-flow controller characteristics.

based on the collector-to-storage temperature difference, thereby increasing collector fluid flow rate. The main advantage of the variable output controller is its elimination of the typically large ΔT_{on}, which allows a longer collection duty cycle in intermittent cloudy weather without cycling. This advantage is achieved with low-speed pump operation and without pump cycling that would occur in an on/off system with a low turn-on temperature and small dead band (hysteresis).

In variable flow, the electrical power to the pump is interrupted, on a cycle-to-cycle basis, to generate the flow versus ΔT characteristics of figure 4.17 (Kent, Winn, and Huston 1980). The theoretical flow on figure 4.17 is not generally achieved because of the third power relationship of current to flow, nonlinear I^2R losses in the pump motor, and highly variable system head characteristics. The power is interrupted with a Triac™ output that may be used either to chop the input power's individual half-wave or to modulate the number of wave forms being delivered to the pump. The duty cycle of the interruption is a ramp function of collector-to-storage temperature difference, typically from 1° to 2°C (2° to 4°F) for low setting to 6°C to 9°C (11° to 16°F) for high setting (varies with manufacturer). This means that there will be no flow until the temperature difference between collector and storage exceeds the lower value, say 1°C (16°F). At that point, the fluid will begin to flow at a low rate. As the temperature difference increases, the flow rate will increase until, at the high limit, say 9°C (20°F), the fluid will be circulating at its maximum rate. There will be no further increase in the flow rate, even though the temperature difference may increase. The low fluid flow rate that is initiated through the collector array as soon as the low turn-on temperature is reached allows for a longer period of collection than would be realized with a bang-bang controller with a high ΔT_{on}. Thermal shock to the collectors is also reduced because of the reduced flow rate; also, the requirement for a large on/off hysteresis is eliminated. Some manufacturers claim (Schlesinger 1977) an increase in energy collected when compared with high offset temperature on/off systems with a low thermal capacitance collector (a lightweight collector) by as much as 6%–8% in cloudy and overcast conditions.

The stagnation temperature controller is an example of a pseudoproportional controller (Lunde 1982; Williams, White, and Lunde 1980). Again, consider the Hottel-Whillier model for useful energy collected, as given by

$$\dot{Q}_u = F_R A_c [G_T(\tau\alpha) - U_L(T_s - T_a)]. \tag{4.57}$$

The threshold value for G_T for which useful energy can be collected is given by setting $\dot{Q}_u = 0$; thus

$$G_{T_{th}} = \frac{F_R U_L}{F_R(\tau\alpha)}(T_s - T_a). \tag{4.58}$$

Figure 4.18 plots G_T and $G_{T_{th}}$ versus time. The controller turns the fluid mover on at time t_{on} and off at time t_{off}. Unfortunately, a pyranometer would be needed to implement this control. This is obviously too expensive but may be circumvented by considering the collector stagnation temperature. From equation (4.57), the stagnation temperature is given by

$$T_{stag} = T_a + \frac{F_R(\tau\alpha)}{F_R U_L} G_T. \tag{4.59}$$

Figure 4.19 plots the stagnation temperature and the storage temperature versus time.

These results are similar to those in figure 4.18. The fluid mover should be turned on when $T_{stag} \geq T_s$ and off when $T_{stag} \leq T_s$. The rate of energy collection is given by

$$\dot{Q}_u = F_R A_c [G_T(\tau\alpha) - U_L T_A]. \tag{4.60}$$

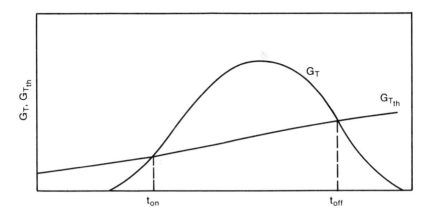

Figure 4.18
Qualitative variation of radiation and radiation threshold versus time.

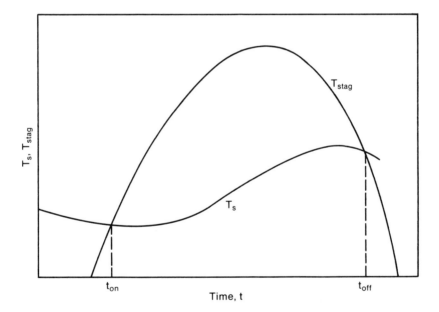

Figure 4.19
Qualitative variation of stagnation and storage temperature versus time.

However, from the equation for the stagnation temperature, one obtains

$$F_R(\tau\alpha)G_T = F_R U_L(T_{\text{stag}} - T_a). \tag{4.61}$$

Hence, on substituting into the useful energy equation, one obtains

$$\dot{Q}_u/A_c = F_R U_L(T_{\text{stag}} - T_s). \tag{4.62}$$

At any time then, the rate of energy collected per unit of collector area is directly proportional to the difference between the collector stagnation temperature and the storage temperature, with the proportionality constant being $F_R U_L$. Lunde (1982) discusses implementation of this controller.

4.3.3 PID Control

Control of the outlet temperature of a molten-salt solar thermal central receiver is very important for satisfactory operation. A central receiver, which is usually mounted on a tower, is simply a heat exchanger that is used to absorb solar energy. The central receiver absorbs solar radiation reflected by heliostats, which are automatically controlled tracking reflectors. The outlet temperature from the central receiver must be controlled to within about $5°C$ ($10°F$), even though the solar energy input can vary significantly because of cloud cover, dust on the collector, etc. DeRocher, Melchier, and McMordie (1982) examined this problem and presented results for various control strategies, in particular for a PID type of controller. They varied the flow rate as the solar energy input changed in order to maintain the outlet temperature within $5°C$ ($10°F$) of its set point [$\sim 567°C(-1053°F)$]. They selected control gains for the PID controller through a trial-and-error design process; these were $k_p = 1$ kg/s $°C$, $k_i = 0.01$ kg/s^2 $°C$, and $k_d = 10$ $kg/°C$. Figure 4.20 shows results from simulation studies of the system using the PID controller.

It is clear from figure 4.20 that the PID type of controller causes the system to respond satisfactorily to step disturbances. The outlet temperature remains within $2.5°C$ ($4.5°F$) of the desired value on the high side and within $1.4°C$ ($2.5°F$) on the low side. The overshoot and settling time for flow are relatively small; however, the temperature response is somewhat oscillatory. Improved performance could probably be obtained by either optimizing the gains for the PID controller or by using the analytical design process to determine the optimal control law, which may be different from PID control.

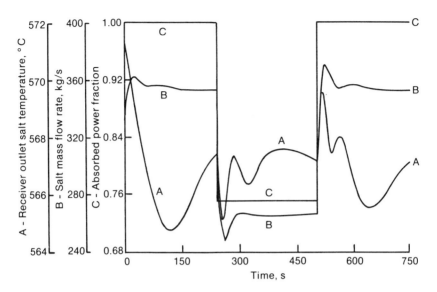

Figure 4.20
System response to step function disturbances in absorbed flux. Source: De Rocher,
Melchior, and McMordie 1982.

4.3.4 Optimal Control

The rate at which energy is collected by a solar heating system can be
varied by varying the flow rate through the collectors. Increasing this flow
rate, however, increases the power required to drive the fluid mover, as
illustrated in figure 4.21. Ideally, one would choose the flow rate $\dot{m}_{opt}$ that
maximizes the difference between the rate at which energy is collected and
the parasitic power. This is quite different from maximizing the energy
collection rate, which would be at $\dot{m}_{max}$ but which would not provide any
net benefits.

Kovarik and Lesse (1976) first approached the problem of maximizing
the difference between solar power and fluid moving power. Their approach resulted in a two-point boundary value problem that was solved
numerically. Their solution, however, could not be implemented in a practical controller, because it was not a function of measurable states of the
system. Later, Winn and Hull (1979) presented an approximate analytical
solution to the problem that could be implemented. Their simulation results were in close agreement with those of Kovarik and Lesse. Winn
and Winn (1981) developed a controller that implemented the optimal

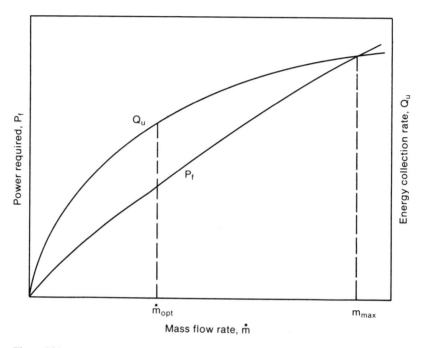

Figure 4.21
Power required and rate of energy calculated as functions of the flow rate.

strategy and tested it in Solar House II at Colorado State University. These studies were all based on using the Hottel-Whillier model for the collector. Orbach, Rorres and Fischl (1981) considered the case where the collector capacitance is included in the model and obtained slightly different results. Also, if a stratified storage tank is included in the system model, still different results for the optimal flow rate are obtained.

Winn and Winn (1981) show that to maximize energy collection, bang-bang control is best. However, others (Pontryagin et al. 1962), using the Pontryagin Maximum Principle, show that the optimal control is not bang-bang if one wishes to maximize the difference between the rate at which solar energy is collected and the parasitic power required. Winn and Winn (1981) present details of the derivations.

4.3.4.1 Distributed Lumped Model
Orbach, Rorres, and Fischl (1981) incorporated the dynamic behavior of the collector into the system dynamics by using a distributed parameter

collector model represented by a bilinear partial differential equation. The equation models the temperature profile $T_f(x, t)$ of the collector fluid as a function of distance along the collector (x) and time (t). The Hottel-Whillier model is a limiting case for this model when the collector is operating under equilibrium conditions or when the thermal capacitance of the collector is ignored (see equation [4.50]). The initial and boundary conditions are

$$T_f(x, 0) = T_0, \quad 0 < x \le L, \tag{4.63}$$

$$T_f(0, t) = T_s(t), \quad t_0 \le t \le t_f. \tag{4.64}$$

The storage tank was modeled as a single-node tank governed by the following equation:

$$\frac{dT_s}{dt} = \frac{\dot{m}c_p}{C_S}[T_f(L, t) - T_s(t)] - \frac{(UA)_s}{C_S}[T_s(t) - T_{as}(t)], \tag{4.65}$$

where C_S is the thermal capacitance of the storage.

The problem was formulated to maximize the following objective function:

$$J = \int_{t_0}^{t_f} \{[\dot{m}c_p[T_f(L, t) - T_s] - P(\dot{m})]dt\}. \tag{4.66}$$

That is, given the initial conditions, the solar irradiance as a function of time and the system parameters, determine the flow rate during the time interval $t\varepsilon(t_0, t_f)$ that maximizes the net energy $J(\dot{m})$ subject to the above state equations for the fluid temperature in the collector and the temperature in the storage and subject to the control constraint $0 \le \dot{m} \le M$. The pump power function $P(\dot{m})$ was assumed to satisfy the following three conditions:

i) $P(\dot{m})$ is piecewise continuous for $0 \le \dot{m} \le M$,

ii) $P(0) = 0$,

iii) $P(\dot{m}) > \dot{m}$; for $0 < \dot{m} < M$.

These conditions were imposed to ensure that the optimal control strategy would be a bang-bang controller. If these conditions are not satisfied, for example, if $P(\dot{m}) = C \cdot \dot{m}^3$, then the possibility of some intermediate flow rate being optimal exists.

Since one of the state equations is a partial differential equation, the maximum principle must be modified to determine the optimal control for this problem. Using the methods of optimal control for distributed parameter systems (Sage 1968), it may be shown that the optimal control is a bang-bang controller, satisfying

$$\dot{m}(t) = \begin{cases} 0, & \text{if } S(t) \leq k \\ M, & \text{if } S(t) > k' \end{cases} \tag{4.67}$$

where

$$S(t) = WC_c[t_f(L, t) - T_s(t)] - D(t), \tag{4.68}$$

and

$$D(t) = \int_0^L \left[p(x, t) - \frac{WC_c}{C_s} \lambda(t) \right] \frac{\partial T_f(x, t)}{\partial x} dx, \tag{4.69}$$

and where $p(x, t)$ and $\lambda(t)$ are adjoint variables satisfying

$$\frac{\partial p(x, t)}{\partial t} = -\dot{m}c_p \frac{\partial p(x, t)}{\partial x} + \frac{F'U_L}{C_c} p(x, t), \tag{4.70}$$

$$\frac{\partial \lambda(t)}{dt} = \left[\frac{\dot{m}c_p}{C_S} + \frac{(UA)_s}{C_S} \right] \lambda(t) + [WC_c - p(o, t)]\dot{m}c_p/WC_c, \tag{4.71}$$

subject to the following conditions:

$$p(x, t_f) = 0, \quad 0 \leq x < L,$$

$$p(L, t) = WC_c[1 + \lambda(t)/C_S], \quad t_0 \leq t \leq t_f,$$

$$\lambda(t_f) = 0.$$

The state equations, the adjoint equations, and the control equation represent a system of equations that must be solved to determine the switching times for the optimal controller. To do this, the solar irradiance $G_T(t)$ must be known for the time interval $t\varepsilon(t_0, t_F)$. Since this cannot be known for future values of time, it is not possible to implement this open-loop optimal control policy. If, however, it is assumed that the optimal control has exactly two switches, then the control reduces to a feedback control strategy that may be implemented. The two-switch policy is represented by

$$\dot{m}(t) = \begin{cases} 0, & t_0 \leq t \leq t_1 \\ M, & t_1 < t < t_2, \\ 0, & t_2 \leq t \leq t_f \end{cases} \tag{4.72}$$

where t_1 is the turn-on time and t_2 is the turn-off time. The times, t_1 and t_2, may be determined from the necessary conditions for optimality. By applying the necessary conditions, it may be shown that

$$\Delta T_{on} = \frac{k}{WC_c(1 - e^{-F'U_L/MC_c})} + \frac{kM}{(UA)_s}, \tag{4.73}$$

and

$$\Delta T_{off} = \frac{k}{WC_c}. \tag{4.74}$$

The fluid mover is then turned on the first time at which the temperature rise across the collector ΔT_c is equal to ΔT_{on} and is turned off the first time after t_1 at which $\Delta T_c = \Delta T_{off}$. It turns out that ΔT_{on} calculated by equation (4.73) gives results that are almost identical to those obtained by calculating ΔT_{on} from equation (4.17). For example, for the case considered earlier (in section 4.3.1), it was found that ΔT_{on} as calculated from equation (4.17) was equal to 38°C (68°F). Using the same values for system parameters and $k = 40$ J/m, $F'U_L/MC_c = 2.1 \times 10^{-4}$, $M = 2.5$ m/s, and $(UA)_S = 4.46$ J/s°C, one finds from equation (4.73) hat ΔT_{on} is equal to 36°C (65°F).

This optimal strategy will apply in cases where the solar irradiance increases rapidly in the morning and decreases rapidly in the evening, and the ambient temperatures do not vary significantly. In this case, the two-switch policy will apply. Orbach, Rorres, and Fischl (1981) suggest that this is also a reasonable suboptimal control strategy that approximates the true optimal control strategy so long as the durations between successive switches are large compared with the transit time of fluid in the collector.

Two difficulties remain with respect to this analysis. The first is that the load has not been considered in the system models. The second is that the storage tank has been modeled as a one-node tank. Results of recent work at the National Bureau of Standards (NBS) and the University of Wisconsin suggest that the optimal control strategy may be quite different if these effects are included in the analysis. This work is described next.

4.3.4.2 Stratified Storage Model

Work conducted jointly at the University of Wisconsin-Madison and NBS (Fanney and Klein 1983) indicates that significant improvements in the performance of solar domestic hot-water systems may be realized by substantially decreasing the collector fluid flow rate to achieve a greater degree of thermal stratification in the liquid storage tank. The increase in collector efficiency, resulting from lower collector inlet temperatures, more than offsets the decrease in the collector heat removal factor that results from the lower flow rates. Flow rates of less than one-fifth the normally recommended flow rates have been found to lead to significant improvements in system performance. The lower flow rates also result in lower parasitic losses.

In the NBS tests, tank temperature profiles within each storage tank for six different solar domestic water heating systems were measured by cop-

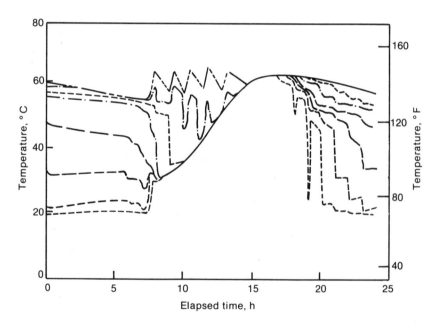

Figure 4.22
Single tank direct storage tank temperature profiles versus time. Source: Fanney and Klein 1983.

per constantan thermocouples positioned every 152 mm (6 in.) along the vertical axis of each tank. Figure 4.22 shows the results for a single-tank direct system. The curves on the figure represent the temperature histories at each vertical position within the tank for a 24-hour period. The results show that during periods of no energy collection, the tank is well stratified; however, when circulation through the storage tank commences, the lower half of the tank becomes well mixed. The top curve on the figure indicates cycling caused by an auxiliary heating element in the upper portion of the tank, which cycles on and off as a result of mixing between the lower, solar-heated portion of the tank and the upper auxiliary-heated portion of the tank. The load that was used in the experimental program was the Rand profile.

The results for this single-tank direct system are in rather sharp contrast with those for a thermosyphon system, as shown on figure 4.23. Here, it is apparent that stratification within the thermosyphon storage tank is excellent at all times, regardless of whether or not energy is being collected. The

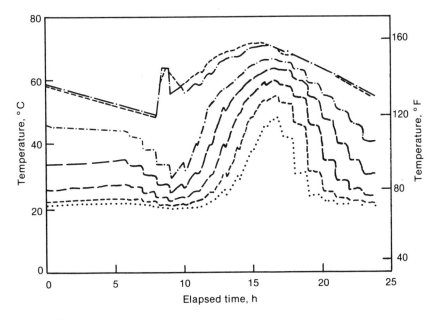

Figure 4.23
Thermosyphon storage tank temperature profiles versus time. Source: Fanney and Klein 1983.

lack of mixing within the thermosyphon storage tank during periods of solar energy collection is a result of the very low flow rates that existed through the solar collector array. Also, during long-term testing, the thermosyphon system required the least amount of auxiliary energy. It was suggested that the most important reason for this was that the low collector fluid flow rates occurring in the thermosyphon system result in a high degree of thermal stratification in the storage tank, which, in turn, results in a higher collector efficiency.

Unresolved problems remain in this area. Although these results indicate that system performance may be improved by decreasing the collector flow rate, results from the Solar Federal Building Program (SFBP) on larger systems indicate otherwise. Several systems were operated at 15%–20% reductions in the flow rates and all had reduced performance (Pekrul 1985). Possible reasons include unbalanced flow in large collector arrays at low flow rates, different load profiles for the SFBP systems, mixing in the storage tanks caused by high flow rates on the load side, and decreased heat-transfer coefficients at the lower flow rates. Until these issues are resolved, it will remain unclear as to the optimal control of the collector flow rate in solar systems.

4.3.5 Adaptive Control

DOE has sponsored research on adaptive controllers for controlling collector fluid flow rates. In one project, Kent, McGavin, and Lantz (1978) developed a high-performance solar controller that offered novel, adaptive control for collecting solar energy. The adaptive control feature reduced the flow rate as load demand was satisfied to increase the system coefficient of performance. The strategy was to use a two-speed flow control. A simulation study indicated that an adaptive algorithm with a two-speed collection mode could approach the operating efficiency of an optimal variable-speed controller. The two-stage bang-bang controller reduced operating costs by more than 50% over a single-stage controller. Moreover, the study indicated that an adaptive anticipator, based on outdoor temperature, could significantly increase the comfort level within a conditioned space by virtually eliminating the lag and overshoot problems that were discussed earlier. The controller was developed and tested under DOE contract and was marketed between 1979 and 1983.

Another DOE-sponsored study involving adaptive control was conducted at Los Alamos National Laboratory and involved the solar heated and cooled National Security and Resources Study Center. The adaptive control aspects involved the heating and cooling of the building and bang-bang control was used for the collector flow rate. This project is described in section 4.4.1.

4.3.6 Laboratory Tests of Controllers

SERI tested controllers for solar domestic water heating systems as part of the Systems and Materials Effectiveness Research Program. The objective of the program was to encourage manufacturers, designers, and researchers of solar domestic hot-water control systems to improve the reliability and performance of these systems. Only bang-bang controllers were tested. Seven controllers were selected for testing under controlled laboratory conditions (Farrington and Myers 1983). The tests revealed problems caused by sensor inaccuracy, sensor failures under stagnation conditions, heat and humidity, and, in particular, reliability. Additional studies (Argonne National Laboratory 1981) have indicated that many of the problems encountered with respect to solar heating systems have been related to the controllers. Detailed discussions of controllers are presented in Farrington and Myers (1983) and Kent, Winn, and Huston (1980).

Results of Farrington and Myers's tests of controllers are summarized as follows:

1. Controllers sometimes signaled for pump operation even though the storage temperature was greater than the collector temperature. This was a result of sensor drift, sensor inaccuracies, controller inaccuracies, or inaccuracies caused by measuring surface temperatures instead of fluid temperatures.

2. Two of six controllers tested under elevated temperature and humidity conditions failed a later test.

3. All sensors using 3000-ohm thermistors and one 10,000-ohm thermistor failed a stagnation temperature test. Two of the remaining 10,000-ohm thermistors were affected by the high temperatures.

4. The performance of three of four freeze protection switches was affected by the stagnation temperature tests, resulting in sensor drifts up to 5.7°C (10.3°F).

5. Thermistor resistance-temperature responses from the manufacturers differed by as much as 2°C (4°F) at low temperatures.

6. Thermistor self-heating was as high as 1°C (33.8°F).

7. Controllers from the same manufacturers and of the same model often differed significantly in performance.

4.4   Controllers for Distributing Energy

This section discusses projects sponsored by DOE and others on energy distribution in building heating or cooling systems. First, conventional solar heating systems are discussed, followed by a discussion of controllers for off-peak storage systems.

4.4.1   Conventional Systems

Winn and Hull (1978) analyzed control strategies for building heating systems and determined optimal control strategies to maximize solar energy use and minimize auxiliary energy use while maintaining building temperature within desired limits. They also conducted sensitivity analyses to determine the effect of building parameters on the performance of the optimal controllers. Figure 4.24 illustrates the system they modeled. They examined the control problem of determining the controls u_1, u_2, and u_3 illustrated in figure 4.24 to minimize the following objective function:

$$J = \int_{t_f}^{t_0} \{C_1(T_e - T_d^2) + C_2 u_2^2 - C_4 \dot{Q}_u + C_5 u_1^2\} dt. \tag{4.75}$$

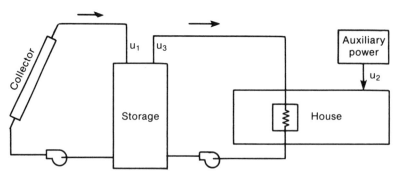

Figure 4.24
Schematic of solar space heating system.

They modeled the building and the storage tank as single-node systems and determined the resulting optimal control strategies by applying the methods discussed previously. To successfully implement these control strategies, system parameters had to be determined. Hays (1978) addressed this problem and reported significant improvements in performance for the optimal controllers compared with conventional bang-bang controllers.

In a similar study, Taylor (1978) examined a two-stage approach to cost-effective controllers through adaptive optimization. The objective of this study was to evaluate the cost-effectiveness of sophisticated controllers for solar heating systems and offset their increased cost with the energy savings they would produce. Figure 4.25 illustrates the system that was examined. Taylor developed a suboptimal control strategy and performed simulation studies. In the simulations, storage was assumed to be unstratified.

Wang and Dorato (1983) examined the same system in the presence of periodic disturbances in ambient temperature and solar radiation, and

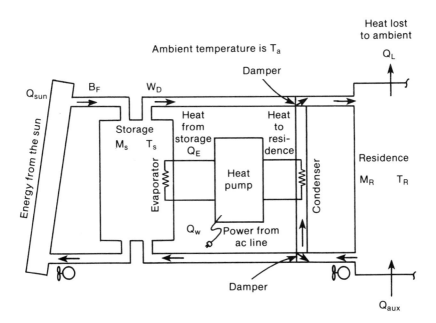

Figure 4.25
System schematic. Source: Taylor 1978.

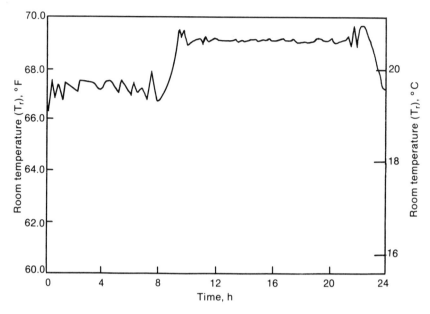

Figure 4.26a
Room temperature variation with a conventional controller. Source: Wang and Dorato 1983.

applied a modified gradient technique to determine the solution for the periodic optimal control problem. The residence was modeled as a single node, as was the storage. The solar radiation and ambient temperature were modeled as periodic functions. The performance index was

$$J = \int_{t_0}^{t_f} \{(T_e - 68)^2 + C_1(Q_w - Q_{aux})\} dt, \tag{4.76}$$

where Q_w was the power input to the heat pump and Q_{aux} was the auxiliary heating as illustrated in figure 4.25. The coefficient C_1 is a weighting coefficient for the trade-off between energy use and discomfort, and was selected as 0.002. Optimal control theory was applied to determine the optimal control strategy that minimizes the objective function J.

Figures 4.26 and 4.27 give results obtained from the conventional controller and the optimal controller, respectively. Figure 4.26 shows results for the conventional controller with a dead band of 4°F (2°). The room temperature variation is shown in figure 4.26a, the energy to the heat

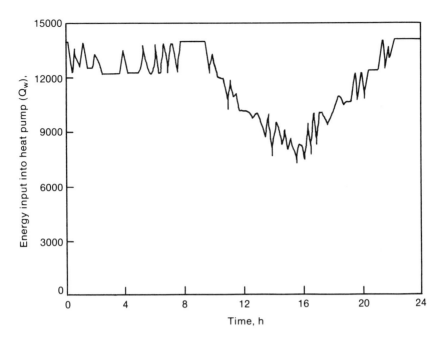

Figure 4.26b
Variation of energy to heat pump with a conventional controller. Source: Wang and Dorato 1983.

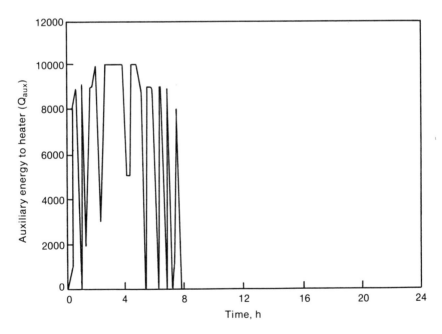

Figure 4.26c
Variation of energy to auxiliary heater with a conventional controller. Source: Wang and Dorato 1983.

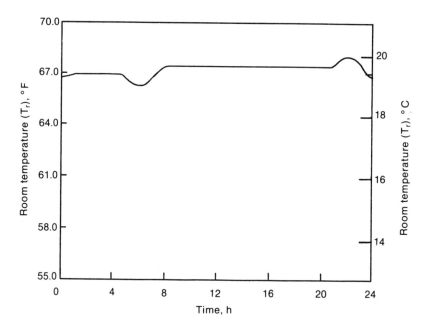

Figure 4.27a
Room temperature variation with optimal periodic controller. Source: Wang and Dorato 1983.

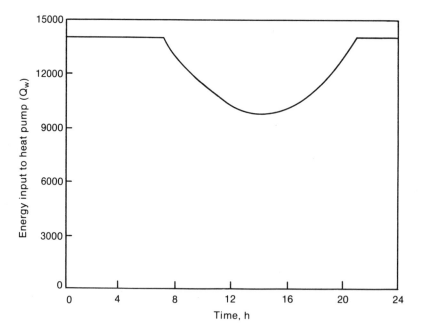

Figure 4.27b
Variation of energy to heat pump, with an optimal periodic controller. Source: Wang and Dorato 1983.

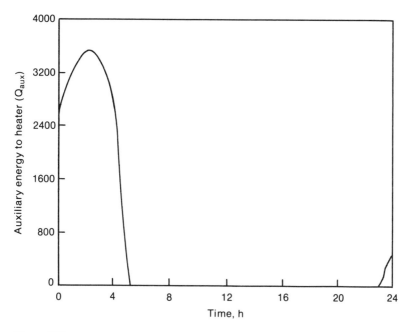

Figure 4.27c
Variation of energy to auxiliary heater, with an optimal periodic controller. Source: Wang and Dorato 1983.

pump is shown in figure 4.26b, and the energy to the auxiliary heater is shown in figure 4.26c. For the optimal controller, the room temperature results are shown in figure 4.27a, the energy to the heat pump in figure 4.27b, and the energy to the auxiliary heater in figure 4.27c. A comparison of these two sets of figures shows that the auxiliary energy is reduced with the optimal controller. In addition, the variables are much more uniform under the action of the optimal controller.

McDonald, Farris, and Melsa (1978) also examined adaptive controllers for solar heated and cooled buildings and applied adaptive and optimal control techniques to HVAC systems in large buildings for the National Security and Resources Study Center at the Los Alamos National Laboratory. They performed computer simulations for the adaptive optimal controller and compared them with simulations for a conventional controller. They found that the adaptive optimal controller resulted in approximately 30% savings in the auxiliary energy for heating and 20% for

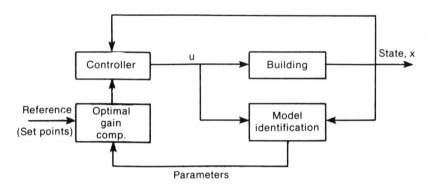

Figure 4.28
Block diagram of an adaptive optimal controller.

cooling when compared with the results obtained from a conventional controller.

Figure 4.28 shows a block diagram for the adaptive optimal controller. The concept of adaptive optimal control is to first identify a linearized model of the system and then apply linear optimal control theory to determine an optimal control strategy. In reality, however, the building and HVAC system represent a nonlinear system with operating points that can vary over a wide range. Since the linearized model is valid only in a region about an operating point, identifying the linearized model must be an ongoing process. The optimal controller is then adapted for each new linearized model or system operating point.

4.4.2 Off-Peak Storage Systems

Solar systems using conventional bang-bang controllers tend to cause reduced load factors for utility companies, because the systems provide energy during the midday and demand peak energy in phase with the peaks in the utility's systemwide load curve. Figure 4.29 illustrates this; the top shows a utility systemwide load curve, and the bottom shows the load curve for a particular building. The reduced load factors result in higher unit costs of production for the utility. This problem has been addressed in several studies. Debs (1978) examined the management of electrical backup demand for solar heating and cooling buildings and outlined some basic approaches to the problem with emphasis on the utility customer relations. Lorsch, Oswald, and Crane (1978) also examined the problem

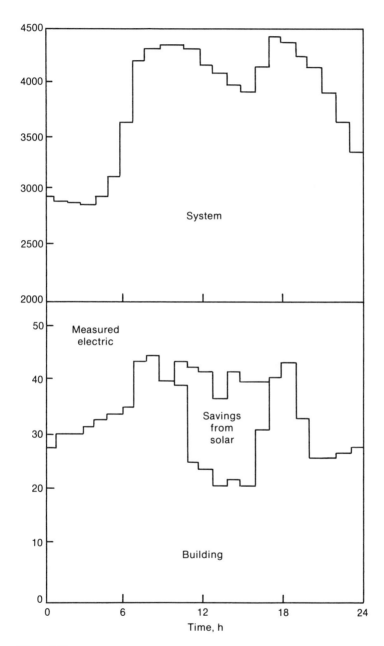

Figure 4.29
Load profile for a delicatessen. Source: Winn 1983.

with an emphasis on control strategies for electrical peak shaving. They
presented an excellent qualitative discussion but did not develop any optimal control algorithms. Winn and Duong (1978) tried to develop an optimal control strategy for peak-load reduction in solar heated buildings. In
this study, they developed a short-term weather forecasting model based
on Kalman filtering. The forecasting model estimated the amount of auxiliary energy required for the next day. They also used a linear programming scheme to determine the optimal electric energy consumption
sequence and an optimal linear regulator to minimize the temperature
deviations in the conditioned space.

In another study, sponsored by the Electric Power Research Institute,
Winn (1984) developed control strategies for off-peak storage systems and
then analyzed these for nine separate utility service areas. He calculated

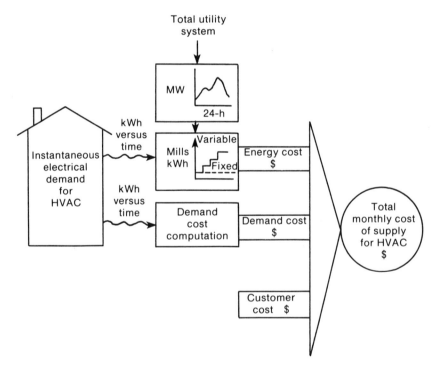

Figure 4.30
Methodology for computing the cost of supply. Source: Winn 1983.

the cost of supply for each utility according to the methodology illustrated in figure 4.30.

To determine the best control strategy for the discharge of solar and off-peak storage, dynamic optimization may be employed. The optimization problem is formulated to determine the on-peak power draw to minimize

$$J = \int_{t_0}^{t_f} [\dot{Q}_{\text{on-peak}}(t)]^2 dt, \tag{4.77}$$

subject to the dynamic equations of the enclosure and the storage,

$$C_E dT_e/dt = \dot{Q}_{\text{on-peak}} + \dot{Q}_{ST} - \dot{Q}_{\text{load}}, \tag{4.78}$$

$$C_S dT_s/dt = \dot{Q}_{\text{sol}} - \dot{Q}_{ST} - \dot{Q}_{\text{loss}}, \tag{4.79}$$

where $\dot{Q}_{\text{on-peak}}$ is on-peak resistance heat rate, W (Btu/h), $\dot{Q}_{ST}$ is the rate of energy removal from storage, W (Btu/h), $\dot{Q}_{\text{sol}}$ is the rate of supply of solar energy to storage, W (Btu/h), $\dot{Q}_{\text{load}}$ is heating load, W (Btu/h), and $\dot{Q}_{\text{loss}}$ is the rate of energy loss from storage, W (Btu/h).

Minimizing J will minimize the product of coincident demand ($\dot{Q}_{\text{on-peak}}$) and on-peak energy consumption ($\dot{Q}_{\text{on-peak}} dt$) while satisfying the dynamic equations that describe the thermal performance of the residence and the storage. The dynamic equations used in this problem formulation assume a well-mixed storage and a uniform (but not constant) enclosure temperature.

Several reasonable assumptions make this a very straightforward problem. First, in a well-designed system, the storage losses are small and can safely be ignored in formulating a control strategy. Second, the enclosure temperature is held nearly constant during the on-peak period by the thermostat in the enclosure. Certainly, when diversity is considered, the utility sees the enclosure temperature as constant, and, therefore, the time derivative of enclosure temperature is effectively zero. Third, the boundary conditions on the storage temperature can be specified. For heating, the storage temperature at t_0 is the maximum storage temperature $T_{s_{\max}}$, which is determined by reasonable design practice. The temperature at t_f is the minimum storage temperature $T_{s_{\min}}$ below which the pump will be deactivated. With these assumptions, the problem may be solved using Pontryagin's Maximum Principle as described earlier.

Let the control u be defined as

$$u = \dot{Q}_{\text{on-peak}}. \tag{4.80}$$

Minimize

$$J = \int_{t_0}^{t_f} u^2(t)dt, \tag{4.81}$$

subject to

$$C_S \frac{dT_s}{dt} = \dot{Q}_{\text{sol}} - \dot{Q}_{\text{load}} + u. \tag{4.82}$$

Let

$$x_1 = T_s, \tag{4.83}$$

$$\dot{x}_2 = u^2, \quad x_2(0) = 0. \tag{4.84}$$

Then we want to maximize the scalar function

$$\phi = -x_2(t_f), \tag{4.85}$$

as this is equivalent to minimizing J. The Hamiltonian function is constructed as

$$H = \frac{\lambda 1}{C_S}(\dot{Q}_{\text{sol}} - \dot{Q}_{\text{load}} + u) + \lambda_2 u^2, \tag{4.86}$$

where the adjoint variables are defined by

$$\dot{\lambda}_1 = -\frac{\partial H}{\partial x_1} = 0, \tag{4.87}$$

and

$$\dot{\lambda}_2 = -\frac{\partial H}{\partial x_2} = 0. \tag{4.88}$$

The optimal control is found by

$$\frac{\partial H}{\partial u} = 0 = \frac{\lambda_1}{C_S} + 2\lambda_2 u, \tag{4.89}$$

or

$$u_{\text{opt}} = \frac{-\lambda_1}{2\lambda_2 C_S}. \tag{4.90}$$

To find λ_1 and λ_2, use

$$\lambda^T(t_f) = \frac{\partial \phi}{\partial x_f} - u^T \frac{\partial \psi}{\partial x_f}, \tag{4.91}$$

where ψ represents terminal conditions on the states; then

$$\frac{\partial \phi}{\partial x_f} = (0 \quad -1), \tag{4.92}$$

and

$$\frac{\partial \psi}{\partial x_f} = (1 \quad 0). \tag{4.93}$$

Therefore,

$$\lambda_1 = -\mu, \tag{4.94}$$

$$\lambda_2 = -1, \tag{4.95}$$

and

$$u_{\text{opt}} = \frac{-\mu}{2C_S}. \tag{4.96}$$

To evaluate μ, use the constraint equation

$$C_S \dot{x}_1 = \dot{Q}_{\text{sol}} - \dot{Q}_{\text{load}} - \frac{1}{2C_S} \mu. \tag{4.97}$$

Integrating,

$$C_S(T_{s_{\min}} - T_s) = \int_t^{t_f} \left[\dot{Q}_{\text{sol}} - \dot{Q}_{\text{load}} - \frac{1}{2C_S} \mu \right] dt. \tag{4.98}$$

Solving for μ and substituting gives

$$u_{\text{opt}} = \frac{1}{t_f - t} \int_t^{t_f} [\dot{Q}_{\text{load}} - \dot{Q}_{\text{sol}}] dt - C_S(T_s - T_{s_{\min}}), \tag{4.99}$$

or

$$\dot{Q}^*_{\text{on-peak}} = \overline{\dot{Q}}_{\text{load}} - \overline{\dot{Q}}_{\text{sol}} - \frac{C_S(T_s - T_{s_{\min}})}{(t_f - t)},$$
(4.100)

where the overbar indicates an average over the entire on-peak period and the asterisk represents the optimal value. The optimal rate of energy delivery from storage is

$$\dot{Q}^*_{ST} = \dot{Q}_{\text{load}} - \overline{\dot{Q}}_{\text{load}} + \overline{\dot{Q}}_{\text{sol}} + \frac{C_S(T_s - T_{s_{\min}})}{(t_f - t)}.$$
(4.101)

Implementing this strategy requires a knowledge of the current heating load, the total heating load during the on-peak period, and the total amount of solar energy delivered to storage during the on-peak period. For a simple off-peak storage heating system, $\dot{Q}_{\text{sol}} = 0$.

A proportional discharge strategy results if one assumes that the heating load is constant throughout the day and no solar energy will be collected. In this case,

$$\dot{Q}^*_{ST} = \frac{C_S(T_s - T_{s_{\min}})}{(t_f - t)}.$$
(4.102)

Implementing the optimal discharge strategy for solar storage is not easy. In addition to needing an estimate of the heating load for the entire on-peak period, the optimal discharge of solar storage requires advanced knowledge of the amount of solar energy to be collected. This points up the need for accurate short-term weather predictions. The Prototype Regional Observing and Forecasting Service (PROFS) program may prove useful in this regard.

Table 4.3 compares supply costs for the bang-bang, proportional, and optimal controllers. These costs were developed for the Public Service Company of New Mexico for the year 1990. In the comparison, the baseline heating system was considered to be an electric central heating system.

4.5 Recommendations for Further Research

The foregoing literature review identified a number of areas for further research. Some of the suggested research involves following through with previous research topics for which the tasks are fairly well defined. Other research topics involve inquiries where the magnitude of improvements in cost or performance are not well defined. Investigators venturing into these areas of research will be embarking on an uncharted course, defining new tasks and abandoning unpromising control concepts or analytical methods as their work progresses. Table 4.4 summarizes recommended research activities.

Table 4.3
Comparison of solar storage discharge strategies

	Baseline heating system	Conventional strategy	Proportional strategy	Optimal strategy
Annual peak HVAC coincident demand	13.32 kW	13.07 kW	10.29 kW	9.59 kW
Coincident HVAC Annual demand				
cost of supply[a]	2551	2227	2020	1905
% of baseline	100	87	79	75
Annual HVAC fuel				
cost of supply[a]	1396	705	730	691
% of baseline	100	51	52	49
Annual HVAC total				
cost of supply[a]	3946	2932	2749	2596
% of baseline	100	75	70	96

[a] in 1990 dollars

Table 4.4
Summary of recommended research activities

	Controller function/type			
	Collecting		Distributing	
Recommended research	Bang-Bang	Prop. and PID	Bang-Bang	Prop. and PID
Quantify parasitic losses	×	×	×	×
Investigate controller based on stagnation temperature	×			
Determine the effects of cycling on system performance and cost	×		×	
Develop charts for specifying ΔT_{on}	×			
Determine partial excitation characteristics of pumps and blowers		×		×
Examine multirate circulators				
Test relationship for differential control thresholds	×	×		
Compare costs and performance	×	×		
Test the use of the PID input state in mixed storage systems	×			
Test the performance of systems at very low flow rates	×	×		
Test controllers that reduce overshoot				×
Test off-peak storage controllers			×	×
Examine the use of weather prediction in controls			×	×
Develop control systems simulated program	×	×	×	×
Test for reliability	×	×	×	×

Nomenclature

A_c	area of collector
C_1	weighting coefficient for the trade-off between energy use and discomfort
C_E	heat capacitance of a conditioned space (kJ/°C)
C_S	thermal capacitance of the storage [kJ/°C (Btu/°F)]
C_c	collector capacitance per unit area
c_p	specific heat (kJ/kg °C)
$E(t)$	deviation from set point
F_R	collector heat removal factor
$F_R U_L$	modified collector heat-loss coefficient
G_T	instantaneous radiation on a tilted collector (W/m²)
$G_{T_{th}}$	threshold value for G_T
$G_{T_{NC}}$	maximum value for radiation
G_{T_M}	minimum value for radiation
H	Hamiltonian function
J	performance index
$J(\dot{m})$	net energy
k	constant coefficient
k_d	gain for a derivative controller
k_i	gain for an integral controller
k_p	gain for a proportional controller
L	loss function
m	mass
$\dot{m}$	mass flow rate
$\dot{m}_{\max}$	maximum flow rate
m_{opt}	optimal flow rate
mc_p/A_c	fluid capacitance per unit collector area
N	number of cycles
P_f	power required

P_m power to a pump motor (W)

Q_{aux} auxiliary heating

Q_C energy collected under cycling condition

Q_{load} heating load [W (Btu/h)]

Q_{loss} rate of energy loss from storage [W (Btu/h)]

Q_{NC} energy collected under no-cycling condition

$Q_{\text{on-peak}}$ on-peak resistance heat rate [W (Btu/h)]

$\dot{Q}_{\text{sol}}$ rate of supply of solar energy to storage [W (Btu/h)]

$\dot{Q}_{ST}$ rate of energy removal from storage [W (Btu/h)]

$\dot{Q}_{ST}^*$ optimal rate of energy delivery from storage

$\dot{Q}_u$ energy collection area

Q_W power input to the heat pump

R resistance to heat loss

RICE   relative increase in collected energy

s Laplace transform variable

t time (s)

t_0 time at the beginning of the on-peak period

t_f time at the end of the on-peak period

T total cycle time (s)

t_E time to empty collector

T_a ambient temperature (°C)

T_d desired temperature (°C)

T_e temperature of an enclosed space (°C)

T_f steady-state temperature

T_{fc} steady-state temperature rise across the collector

T_i collector inlet temperature (°C)

T_0 collector outlet temperature (°C)

ΔT_{off} turn-off set point temperature (°C)

ΔT_{on} turn-off set point temperature (°C)

T_r room temperature

T_s storage temperature (°C)

T_{sr}	reference temperature (°C)
T_{stag}	stagnation temperature (°C)
u	control variable
$u(t)$	controlled energy input
UA	area-conductance product for a conditioned space (W/°C)
U_L	collector loss coefficient (W/m²°C)
u_{max}	maximum value of the control function, $u(t)$
$U(s)$	Laplace transform of the control function, $u(t)$
W	p. 28
x	static vector
ω	cycling rate
δx_j	variation in parameter x_j
ϕ_c	vector of control parameters
ϕ_s	vector of system parameters
ϕ_w	vector of climatic parameters
τ	p. 12
$\tau\alpha$	p. 20
τ_c	system time constant for $m = 0$ (h)
τ_{tc}	system time constant for $m \neq 0$ (h)

References

Alcone, J., and R. Herman. 1981. Simplified methodology for choosing controller set-points. *Proc. ASME Solar Energy Division 3d Annual Conf. on SSEA and Operational Results.* Reno, NV.

Argonne National Laboratory. 1981. *Final reliability and materials design guidelines for solar-domestic hot water systems.* ANL/SDP–11, Solar/0909–81/70. Argonne, IL: Argonne National Laboratory.

Baz, A., A. Sabry, A. Mobarak, and S. Marcos. 1983. On the tracking error of a self-contained tracking system. Submitted to *J. of Solar Energy Engineering.*

Debs, A. S. 1978. *Management of electric back-up demand for solar heating and cooling applications.* DOE/CS/31595-11. Washington, DC: Department of Energy.

De Rocher, W. L., Jr., D. K. Melchior, and R. K. McMordie, 1982. Control simulations of a molten salt solar thermal central receiver. Proc. ASME, pp. 403–408.

Diamarogonas, A., and A. Mouriki. 1980. Synthesis of four bar linkage for solar tracking. *Solar Energy* 25(3):195–99.

Duffie, J. A., and W. A. Beckman. 1980. *Solar engineering of thermal processes*. New York: Wiley.

Fanney, A., and S. Klein. 1983. Performance of solar domestic hot water systems at the National Bureau of Standards: measurements and predictions. *Proc. ASME Solar Energy Division 5th Annual Conf*. Orlando, FL, April 1983, pp. 200–211.

Farber, E., C. Morrison, and H. Ingley. 1976. A self-contained solar tracking device. ASME Paper No. 76–WA/HT-26. ASME Winter Annual Meeting, New York, NY, December 1976.

Farrington, R., and D. Myers. 1983. *Evaluation and laboratory testing of solar domestic hot water control systems*. SERI/TR–254–1805. Golden, CO: Solar Energy Research Institute.

Hays, D. K. 1979. Optimal control and identification in space heating systems. M.S. thesis, Department of Mechanical Engineering, Colorado State University, Fort Collins, CO.

Herczfeld, P., R. Fischl, H. Klafter, and A. Orbach. 1978. Study of pump cycling in the control of solar heating and cooling systems. *Proc. 1st Workshop on the Control of Solar Energy Systems for Heating and Cooling*. Newark, DE: American Section of the International Solar Energy Society.

Herczfeld, P., R. Fischl, and S. Konyk, Jr. 1980. Solar flat plate collector control system sensitivity analysis. Proc. SSEA. San Diego, CA, pp. 69–74.

Huang, B. J., and H. J. Lu. 1982. Performance test of solar collector with intermittent output. *J. Solar Energy* 28(5).

Kahwaji, G., and C. B. Winn. 1985. Effect of the cycling rate on energy collection for bang-bang controllers. Accepted by *J. Solar Energy Eng*.

Kent, T. B., M. J. McGavin, and L. Lantz. 1978. Development of a novel controller. *Proc. 1st Workshop on the Control of Solar Energy Systems for Heating and Cooling*. Newark, DE: American Section of the international Solar Energy Society.

Kent, T. B., C. B. Winn, and W. G. Husto. 1980. *Controllers for solar domestic hot water systems*. Contract No. AH–9–8189–1, SERI/TR–9819–1A. Golden, CO: Solar Energy Research Institute.

Klein, S. 1973. The effects of thermal capacitance upon the performance of flat-plate solar collectors. M.S. Thesis, University of Wisconsin-Madison.

Kovarik, M., and P. F. Lesse. 1976. Optimal control of flow in low temperature solar heat collectors. *Solar Energy* 18.

Kutscher, C. F., R. L. Davenport, D. A. Dougherty, R. G. Gee, P. M. Masterson, and E. K. May. 1982. *Design aproaches for solar industrial process heat systems*. SERI/TR–253–1356. Golden, CO: Solar Energy Research Institute.

Lewis, R., Jr., and J. B. Carr. 1978. Comparative study of on/off and proportionally controlled systems. *Proc. 1st Workshop on the Control of Solar Energy Systems for Heating and Cooling*. Newark, DE: American Section of the International Solar Energy Society.

Lorsch, H., R. Oswald, and R. Crane. 1978. Control strategies for electrical peak shaving. *Proc. 1st Workshop on the Control of Solar Energy Systems for Heating and Cooling*. Newark, DE: American Section of the International Solar Energy Society.

Lunde, P. 1982. Control of active systems using the stagnation temperature. *Solar Age* 69–70.

McDonald, T. M., D. R. Farris, and J. L. Melsa. 1978. Energy conservation through adaptive optimal control for a solar heated and cooled building. *Proc. 1st Workshop on the Control of Solar Energy Systems for Heating and Cooling*.

Orbach, A., C. Rorres, and R. Fischl. 1981. Optimal control of a solar collector loop using a distributed-lumped model. *Automatica* 17(3):535–39.

Pontryagin, L. S., V. G. Boltyanskii, R. V. Gambrelidge, and E. F. Mischenko. 1962. *The mathematical theory of optimal processes.* New York: Interscience.

Sage, A. P. 1968. *Optimum systems control.* Englewood Cliffs, NJ: Prentice-Hall.

Schiller, S. R., M. L. Warren, and D. M. Auslander. 1980. Comparison of proportional and on/off solar collector loop control stratgies using a dynamic collector model. *J. Solar Energy Eng.* 102(4).

Schlesinger, R. J. 1977. Preliminary comparison of proportional and full on/off control systems for solar energy applications. *Proc. 1977 Annual Meeting, AS/ISES.* Orlando, FL.

Taylor, T. 1978. *Two-stage approach to cost effective controllers via adaptive optimization.* Singer Corporation.

Wang, Y., and P. Dorato. 1983. *Optimal periodic control of a bilinear solar system.* Albuquerque, NM: Department of Electrical and Computer Engineering, University of New Mexico.

Williams, G., J. White, and P. Lunde. 1981. The solar energy threshold method. *Proc. 1981 Annual Meeting, AS/ISES.* Philadelphia, PA.

Winn, C. B. 1983a. Controls in solar energy systems. *Advances in solar energy.* American Solar Energy Society, pp. 209–240.

———. 1983b. The effects of temperature settings on cycling rates for bang-bang controllers. *J. Solar Energy Eng.* 105:277–80.

Winn, C. B., and N. Duong. 1978. An optimal control strategy for peak load reduction applied to solar structures. *Proc. 1st Workshop on the Control of Solar Energy Systems for Heating and Cooling.* Newark, DE: University of Delaware.

Winn, C. B., and D. E. Hull. 1978. *Optimal control studies of solar heating systems.* COO–451901. Washington, DC: Department of Energy.

———. 1979. Singular optimal solar controller. *Proc. 1978 IEEE Conference on Decision and Control.* San Diego, CA.

Winn, R. C., and C. B. Winn. 1981. Optimal control of mass flow rates in flat plate solar collectors. *J. Solar Energy Eng.* 103(2).

Wolfson, R. L. T., and H. S. Varvey. 1980. Control strategies for dual temperature solar hot water systems: an experimental comparison. *Proc. 2d Annual Systems Simulation and Economic Analysis Conference.* San Diego, CA.

5 Optimum System Design Techniques

Jeff Morehouse

5.1 Introduction

Optimization of a given system design is generally based on system performance parameters and the economics of the prevailing financial situation. As discussed below, this tie between system performance and finances/economics relative to design optimization is compounded by the many ways of analyzing the financial situation, which, in turn, is affected by the system owner's situation (e.g., residential versus commercial, tax bracket, local energy costs). The object of this chapter is to present these various system performance and economic factors and considerations as they lead into the techniques used to optimize solar system designs.

5.1.1 The Optimization Process

The optimization process can be grouped into three steps (Michelson 1982):

1. Choosing the optimizing criterion for either a thermal/technical basis (e.g., minimum auxiliary energy use) or an economic basis (e.g., life-cycle costs);

2. Calculating the thermal performance of the system using the parametric values and variations of concern;

3. Using the optimization method (which could be a single variable type, contour type, or multivariant type method) to determine the optimum variable(s) value.

Currently, most optimization processes involve an economic basis criterion, thermal performance calculations from simulations or correlations, and a single variable type optimization.

5.1.2 Noneconomic-Based Designs

It should be mentioned that there are designed solar systems where system economics is *not* considered a major factor. The primary reason for ignoring economic factors is the desire to insure a supply of energy against the possibility that other energy sources become unavailable (DOE Solar Working Group 1978). The major factor driving the design is to have the system produce the needed energy. However, even this approach to system design can be said to be economic-based in terms of what one is willing to pay for this "insurance."

All systems designed by competent practitioners involve many subsystem and component sizing and optimum decisions that are not primarily economic in nature. The choice of the piping size, pump size, auxiliary energy equipment size, storage size, and many other components is determined first by the thermal and mechanical requirements of designing a workable system. Only then does the cost of acquiring these components, and trade-offs among various styles and vendors, become an optimization consideration at the component and subsystem level. Thus, it should be remembered that there are design optimizations primarily involved with the thermal/mechanical workings of any system; these thermal/mechanical factors are expanded on in section 5.2.

Another reason why solar systems are designed and constructed without major emphasis on economics is related to a solar system being considered a "symbol." The solar system can be used to advertise a business (and thus may be economically linked) and/or to act on expressed environmental concerns (pollution, resource depletion). Systems designed under this rationale are constructed only when "excess" funds are available.

5.1.3 Financial/Economic-Based Designs

The techniques used for economic-based optimized system designs either maximize or minimize an economic "figure of merit" relative to an investment analysis. This economic figure of merit may be stated in purely financial terms (cash flow, payback, life-cycle cost, marginal cost) or it may involve an energy-dollars mix ($/Btu). There are many methods of investment analysis and these are discussed in detail in chapter 2 of volume II of this series, *Solar Resources*, which includes extensive references.

While the method and figure of merit chosen for optimization may differ, almost all optimizations involve two generic factors: (1) system costs (capital and operating) and (2) thermal performance (energy displaced). In fact, most solar system optimizations can be generalized as attempts to displace conventional energy at the lowest cost. Thus, this chapter discusses the thermal design and economic factors associated with solar systems that are used in optimization techniques employing various economic approaches and figures of merit.

5.1.4 Chapter Outline

As mentioned earlier, section 5.2 treats the basic solar system thermal sensitivities involving sizing and performance of components and subsystems. Section 5.3 discusses the various energy flows and deals with the concept of net energetics and the energy displaced or saved. Also, the definition of the primary economic and financial factors involved in system costs is presented.

In section 5.4, the optimization techniques are described in terms of the thermal and economic data required to use them, and the output and display from each technique is presented. The techniques are compared, and their limitations, weaknesses, and strengths are summarized in section 5.5.

5.2 System Design Sensitivities

In order to optimize a given system design, it is generally necessary to have information concerning the system thermal performance and cost over a range of system sizes and/or configurations. In turn, this system performance/cost sensitivity depends on the size and choice of appropriate and available subsystems and components.

5.2.1 System Configuration

The basic solar system configuration can be viewed in terms of the "sensitivity" or "optimization" opportunities associated with the major subsystem operations: collection, storage, conversion, and distribution.

As seen in figure 5.1, these major subsystems are not always found in every design, or they may be combined in function. Also, it should be noted that almost all solar system designs include a conventional system, able to meet the design load, as backup to (or incorporated in) the solar system.

As mentioned earlier, the amount of conventional energy displaced is a major factor in optimization. The amount of solar energy collected is of first concern; that this collected energy is then delivered to the load is equally important. Collection and delivery of energy are not unrelated activities, mainly due to the need to store energy against solar energy versus load mismatches (nights, cloudy days). However, the size of the collector subsystem generally governs the fraction of the conventional

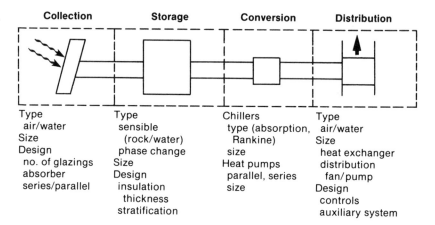

Figure 5.1
Basic solar system configuration sensitivity/optimization opportunities.

energy able to be displaced. Thus, the collector area is considered the primary system sensitivity parameter.

Storage size sensitivity is often considered the second most important factor. However, the more "mundane" design issues associated with pump and fan sizing leading to parasitic energy consumption, along with control system design, are often of greater importance in displacing energy. Also, the thickness of insulation on pipes, storage, etc. can be a major factor in optimum design, as can be appropriately sized heat exchangers.

For a given design configuration, the sizes and types of components are varied to determine the system thermal performance of various combinations. As shown in figure 5.2 (Hughes and Morehouse 1979), the system thermal performance is usually presented in terms of fraction of the load met by solar versus collector size for various other component variations. Thus, the basic system energy displacement information needed for the optimization is generated. Further discussion of how to determine the energy displaced is given in section 5.3.1.

5.2.2 Subsystem/Component Design Considerations

In the following paragraphs, the specific selection and sizing considerations associated with the collectors, the storage, and the conversion equipment are discussed. System thermal performance effects and optimization are discussed as influenced by these subsystems and their components.

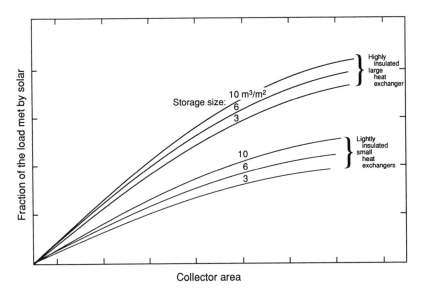

Figure 5.2
Presentation of component variations and resultant conventional energy displacement.

5.2.2.1 Collectors

The initial selection of a collector is based on whether a liquid or air system is being designed. The choice between liquid- and air-based systems is usually dependent on the experiences of local contractors and dealers, the type of *conventional* heating employed, and the installation layout available for the particular system. Overall efficiencies and costs of air versus liquid systems appear generally comparable.

The major design question to be answered with the collectors is the size of the collector array. However, the collector performance and collector array area together determine the amount of solar energy collection, which is the primary system-level variable to be optimized. Thus, collector type and area are major trade-off variables.

EFFICIENCY VERSUS COST.   The major sensitivity/trade-off that must be made with collectors is in efficiency versus cost. The adage that "you get what you pay for" is valid in that higher cost usually buys higher collector efficiency. This higher efficiency is bought by reducing the collector thermal losses, either by suppressing the conductive/convective losses or by reducing the radiative losses.

The glazing is the predominant path for the loss of thermal energy. The conductive/convection loss to the ambient can be reduced by using more glazing layers, but at increased cost and with a reduction in the insolation reaching the collector's absorber plate. The use of a vacuum between the glazing and absorber is also used to suppress convection losses, but evacuated tube collectors are more expensive than the flat-plate types. The radiative loss from the hot absorber plate can be reduced by using a selectively emitting surface. These selective surfaces are generally more expensive, slightly less absorbing, and some types may be more prone to degradation than conventional black surfaces. Other trade-offs are available with the glazing material itself, including various glass and plastic candidates having a variety of solar transmittances, thermal windows ("greenhouse" transmittance), antireflective properties, and lifetime degradation modes.

The efficiency-versus-loss sensitivity between different collectors can be graphically seen in collector performance curve plots, as shown in figure 5.3. A given collector's performance is plotted in terms of efficiency (η = energy collected/energy incident) for various insolation (I) and ambient/ collector temperature differences (ΔT = absorber plate temperature − ambient temperature) conditions. Figure 5.3 illustrates the collector trade-off that most commonly must be made: line A represents a collector with high efficiency at low temperatures but a steep slope (high losses/unit

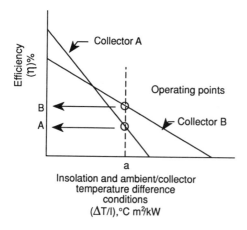

Figure 5.3
Collector performance curves for efficiency-versus-loss sensitivities.

area); and line *B* represents a lower slope (lower losses/unit area) but with lower efficiency at lower temperatures. Collector B (double glazing and/or selective surface) would generally cost more per unit area than collector A (single glazing, black surface); thus, the efficiency of B must be greater for it to be the more cost-effective. The "operating point" plotted, dashed line *a* in the figure, illustrates that the temperature/insolation parameter must be high enough (high absorber fluid temperatures and/or low ambient temperatures and/or low insolation values) before collector B is the more efficient.

To determine the collector's annual efficiency performance, ambient conditions and other subsystem effects (mainly storage temperature profiles) must be known and taken into account in order to determine an integrated "operating point" $(\Delta T/I)$ for a given system and collector combination; this is usually done by use of a computer simulation. The final determination as to which collector to use would be made by evaluating the incremental energy saved by the more efficient collector system and relating it to the additional collector cost (Warren and Wahlig 1985; Scholten 1984; O'Dell et al. 1984; Barnes 1981).

TILT AND AZIMUTH ANGLE.   The collector array can be oriented through tilt and azimuth angle to maximize the insolation incident on the collectors. The tilt angle is defined as the angle between the plane of the collectors and the horizontal, while the azimuth angle is the angle between due south and the direction the collector is facing (see figure 5.4).

The appropriate tilt-angle determination is based on the local latitude the seasonal use of the collectors (using the varying solar declination: $\pm 23.5°$). The geometric relationship between these variables is shown schematically in figure 5.5. As can be seen, the collectors should be tilted at an angle greater than the latitude angle for winter time (heating season). Figure 5.6 (Colorado State University 1977) illustrates the quantitative effect on collector heating season performance around the latitude-plus-10-degrees optimum value. For annual performance, as for a solar water heater, the rule of thumb is for the tilt to equal the latitude, while the cooling season optimum angle is close to latitude minus 10 degrees. However, each system design with its particular characteristics should be checked for tilt-angle sensitivity, especially if unusual seasonal load conditions are involved.

The effect of collector azimuth angle (collectors oriented east or west of due south) is not as severe as that of tilt angle. As seen in figure 5.7

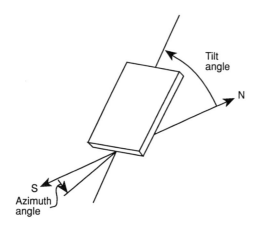

Figure 5.4
Collector tilt and azimuth angle geometry.

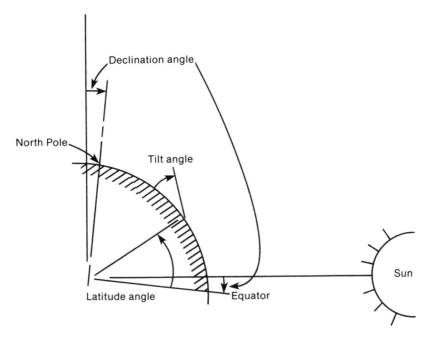

Figure 5.5
Geometry of tilt, latitude, and declination angles.

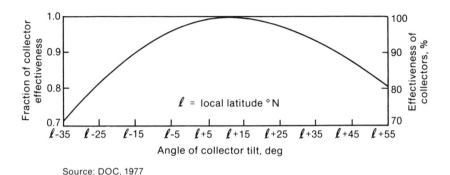

Source: DOC, 1977

Figure 5.6
Effect of solar collector tilt on annual heating performance.

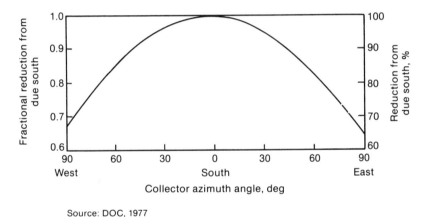

Source: DOC, 1977

Figure 5.7
Effect of solar collector orientation on annual heating performance.

(Colorado State University 1977), a 30° variation from south results in a performance degradation of only 4% to 5%. However, local conditions of weather (e.g., consistent morning fog) or time-of-day loads to be met (e.g., large late-morning hot water usage) could mean that the collector azimuth angle should be designed off-south for optimum system performance.

5.2.2.2 Storage

The sensitivity of system performance to storage size is due to two factors: first, the use of storage to meet loads when direct solar energy is unavail-

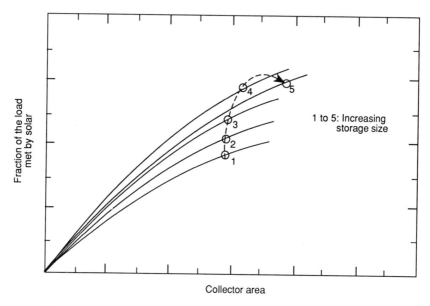

Figure 5.8
Overall system performance variation for increasing storage size.

able; second, the effect of storage temperature on system efficiency. These two factors are related and should be examined when sizing storage (Winn and Baer 1984; Braun et al. 1981; Collares-Pereira et al. 1984).

The use of storage to meet the load when solar energy is unavailable generally shows the system performance indicated in figure 5.8 as storage size increases, and at some large enough storage size the system performance can actually begin to decrease. This "diminishing returns" effect occurs because the storage can be made so large that the storage temperature cannot be raised to usable levels to meet the load. System performance is also degraded when storage is undersized and the storage temperatures would increase rapidly during the day, forcing the collectors to operate at low efficiency (high $\Delta T/I$; see figure 5.3 [Hughes and Morehouse 1979]) for a major portion of the day.

For a given collector size, the thermal performance optimization of storage size can be determined, as illustrated in figure 5.9. Rules of thumb for sizing storage are provided below (Colorado State University 1977), but again, caution should be exercised if any "unusual" loads, weather, or system designs are being considered:

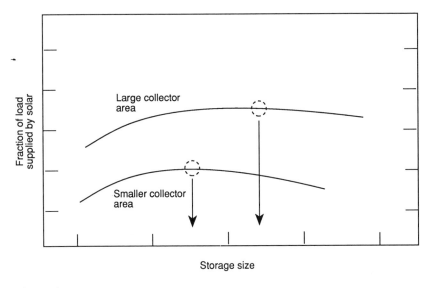

Figure 5.9
System performance at fixed collector size and varying storage size.

1. *Heating System*
Air: 1/2 to 1 ft^3 of 3/4 in. to 1 in. pebbles per ft^2 of collector;
Hydronic: 1.5 to 2.5 gallons per ft^2 of collector;

2. *Domestic hot water*
Preheat tank: 1.5 to 2.0 times the size of the DHW auxiliary tank (conventional water heater).

5.2.2.3 Conversion Equipment
The conversion equipment used in solar systems is generally associated with either cooling systems or heat pump systems. The sizing of this equipment is usually based on load, not on the solar system characteristics. This sizing by load occurs because the conversion equipment is usually also the backup system.

However, sizing of the solar subsystem for input of thermal energy into the conversion equipment is a major design problem. For a series-connected heat pump system, the sizing of the collectors and storage is critical to ensure that the heat pump is always supplied with energy within the proper temperature range. Similarly, for a solar-driven chiller system, the solar thermal input capacity and temperature are critical parameters in designing for efficient operation. Often the optimized solar cooling system will be designed to have a solar-driven capacity of only 50% to 75% of the conversion equipment capacity, with conventional energy used for full-load capacity situations (Warren and Wahlig 1985; Anand et al. 1984).

5.3   Thermal and Economic Factors

As described earlier, the optimization techniques are based on economic maximizing or minimizing as related to energy saved. The subsections below discuss the energy used and saved in solar systems, and the definition of the primary financial elements used in the economic evaluations of solar systems.

5.3.1   Net Energetics Balance

The determination of the energy saved by a given solar system is always in relation to a conventional system. For a valid comparison and determination of energy saved, the solar and conventional systems must do the same job, which usually means meeting the same thermal load. If the solar and conventional systems are comparable, the energy saved is defined as the difference in total energy purchased for the two systems:

energy saved = conventional system purchased energy

− solar system purchased energy.

As can be readily imagined, the only "energy saved" with economic value is that which would have been purchased; thus, the emphasis here is to account for all purchased energies.

A net energetics balance is based on accounting for both purchased and "free" energy flows into and out of a system. The "net" in net energetics usually refers to determining the total purchased energy consumption. The energy flows into and out of a system must "balance" in the long term. Figure 5.10 illustrates the general types of input and output energy flows that go into an energy balance, which can be expressed as:

energy in = energy out;

"free" inputs + purchased inputs = output to load + thermal losses.

The "free" inputs refer to the solar energy collected and/or the ambient energy (heat from the atmosphere, ground, or water) used. The purchased

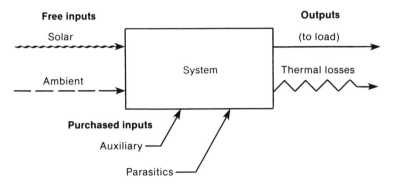

Figure 5.10
Energy flows for energy balance.

inputs can be divided into the auxiliary energy used to power the conversion and/or backup equipment and the parasitic energy, which is the energy used to power the pumps, fans, motorized valves, controllers, etc. that are used in the overall system. These parasitic energy flows are often mistakenly ignored in net energy calculations for energy savings, but they can often be a very significant factor, as, for example, fans in air systems (Hughes and Morehouse 1979; Warren and Wahlig 1985).

5.3.1.1 Load versus Energy Consumption
One of the major system figure of merits is the system coefficient of performance (COP). The system COP is defined similarly to an efficiency:

$$\text{COP system} = \frac{\text{energy delivered to load}}{\text{energy purchased}}.$$

A conventional electrical resistance heating system would have a COP = 1, while a gas or oil furnace system COP might be in the 0.5 to 0.9 range. A conventional heat pump system COP will vary from 1.5 to 2.5, depending on the geographic location. Solar heating and hot water systems generally have system COPs in the 2 to 5 range, which roughly corresponds with meeting 40% to 90% of the load with solar and accounting for parasitics.

5.3.1.2 Auxiliary Systems
The auxiliary system may be an integral part of the solar system, or it may be purely a backup (which could be the conventional competition). In any case, an important item in solar system optimization is that the amount of

energy purchased to supply both the auxiliary and the conventional competition system must be accurately determined. Thus, it is very important to know the efficiency and COP of the auxiliary equipment in order to relate between load met and energy purchased. The magnitude and type of energy purchased must be known so that costs can be calculated for the optimization process.

5.3.1.3 Parasitic Energy

As mentioned, the parasitic energy is often not accounted for; unless included, the system COP will be overstated. A "solar delivery" COP is sometimes used to indicate the magnitude and effective use of parasitic energy needed to deliver solar energy to the load:

$$\text{solar delivery COP} = \frac{\text{solar energy delivered to load}}{\text{parasitic energy purchased}}.$$

Solar delivery COPs are often greater than 10 in hot water systems and in well-designed solar space heating systems.

5.3.2 Economic Assumptions

The economics of a solar system are based both on the capital cost of the system and the planned savings due to reducing purchased energy. While these two factors seem straightforward, the financial situation is complicated by taxes and credits, fuel cost escalations, and interest and inflation, all representing unknown future variables about which some assumptions must be made (Percino 1979; Ruegg 1975).

5.3.2.1 Money Costs by Present Worth

The costs and savings associated with the solar system lead to the determination of the cost or worth of money. Generally, the cost of money is discussed in terms of its "present worth." If, for example, one has to pay $5,000 in 5 years, then only $4,000 cash has to be obtained today and put in the bank to grow with interest to $5,000. Thus, the present worth of $5,000 in 5 years is only $4,000. The present worth of a future sum is determined by both the length of time until the future payment is due and the interest rate during that period. Usually, a "discount rate" is used to represent the interest rate that could be obtained from an alternate investment. The calculation of the present worth of a single future sum is given by (Mitchell 1983):

present worth of a single payment = (future sum) × (present worth factor), (5.1)

where the present worth factor $PWF(n, d)$ is given for n years at d discount (or interest) rate:

$$PWF(n, d) = \frac{1}{(1 + d)^n} \qquad (5.2)$$

For the situation of determining the present worth of a series of annual payments over n years with d interest rate prevailing, and including an annual escalation rate f (such as could happen with rising fuel prices), then the calculation would be (Mitchell 1983):

Present worth of a series of annual payments = (annual sum in first year)

× (present worth factor), (5.3)

where for a series of annual payments:

$$PWF(n, f, d) = \frac{1}{(d - f)}\left[1 - \left(\frac{1 + f}{1 + d}\right)^n\right], \qquad (5.4)$$

unless $d = f$, then:

$$PWF(n, f = d) = \frac{n}{1 + d}. \qquad (5.5)$$

The accounting for a general inflation rate in the overall economy may be handled (a bit simplistically) by using "real" discount/interest and escalation rates. Real rates mean the rates are those expected *above* the general inflation rate. The use of real rates provides a consistent, understandable picture of money costs without predicting inflation rates.

Several items should be noted. First, in order to determine the present worth of future expenditures and revenues, an assumption must be made concerning the expected interest/discount rate. Secondly, a simplifying assumption can be made that the capital cost of a system *is* the present worth of the system, even if a loan is taken out and annual payments are made. This is possible since any loan assures itself of the "proper" present worth (system cost plus banker's fee) through its own interest rate. However, tax advantages may accrue via the loan as discussed below.

5.3.2.2 Energy Costs

The annual energy costs and/or energy savings are the primary concern when determining present worth figures for a solar energy system. The cost of future energy depends on the present fuel cost and the predicted fuel cost escalation rate. The use of equation (5.3) for a series of annual fuel payments is appropriate for most situations, where f represents the expected annual fuel escalation rate. If more than one fuel is purchased, present worth calculations of costs or savings for each fuel are needed.

5.3.2.3 Tax Effects

The situation for a given solar system's economics is further complicated, though usually financially benefitted, by considering the effects of federal and state taxes. Tax credits (direct reductions in payable taxes) may be available for businesses and individuals who install solar systems, from both the federal and state governments. State governments may also offer tax deductions (a reduction in the amount subject to tax) as an incentive to install solar systems. Both the tax credit and the tax deduction act to reduce the cost of a solar system to residential and commercial buyers.

However, there is one tax effect for businesses that is not favorable for solar systems. Since fuel for businesses is an expense, fuels for businesses already rate as a tax deduction. Thus, fuels for businesses may be viewed as really only costing one-half the market price, or solar systems displacing a certain value of fuels can be viewed as displacing an equal value of tax deductions. Either way, this business-tax situation leads to a lowering of solar system savings.

5.4 Optimization Techniques

The optimization techniques may be generally grouped into two types: energy cost minimization methods and financial-based methods. The techniques involving energy cost minimizations are intended to find the lowest ratio of system cost to solar energy used (e.g., $/Btu). The financial-based techniques involve optimization of life-cycle costs, payback time, return on capital, and/or a cash flow analysis (Percino 1979; Ruegg 1975).

5.4.1 Financial-Based Methods

Most of the federally sponsored optimization methods use life-cycle costing in some manner. However, most commercial and individual buyers and sellers of solar equipment use some form of payback, return on capital, or cash-flow analysis for selection of system and size. It should be noted, as will be demonstrated later, that the various methods do not always lead to the same answer for a given situation.

5.4.1.1 Life-Cycle Costing

The life-cycle cost of a system is the sum of all the costs needed to buy and operate the system throughout the system's expected lifetime. These costs incurred by the system through acquisition and operation over the period of analysis are referenced to present worth values. The system cost elements included in most life-cycle cost analyses are:

1. System acquisition costs: initial investment costs including design, delivery, installation, building modification, value of system, occupied space, and tax credits (negative);

2. System repair and replacement costs: cost of repairing or replacing system parts, exclusive of routine maintenance, and net of insurance reimbursement and parts salvage;

3. Maintenance costs: cost of routine upkeep, maintenance, labor, and parts;

4. Operating costs: cost of all fuels and parasitic energy used in operating the system including primary and auxiliary equipment (backup conventional fuel cost included);

5. Insurance costs: cost of insuring the system;

6. Tax costs: incremental property tax costs, federal/state income tax reductions due to interest paid and depreciation, and tax reduction due to fuel expenses;

7. Salvage: the salvage value of the system at the end of the period of analysis, net of removal and disposal costs.

There are two approaches to life-cycle cost comparisons between alternative systems: the total cost of each alternative or the cost differences between alternatives. The total cost approach is the simplest to formulate, but the determination of all the system costs is often a difficult task. The difference-in-costs, or net-benefits, approach involves determining only those costs that are not common between the compared systems. The difference approach has usually fewer items that must be cost-analyzed, thus allowing a simpler statement and formulation. Assuming life-cycle costs and salvage value costs, a comparison between systems A and B would be calculated as follows (Mitchell 1983):

$$\text{net benefits} = \text{LCC}(A) - \text{LCC}(B), \tag{5.6}$$

where

$$\text{LCC}(A) = \text{first costs}(A) + \text{annual costs}(A) \times PWF(n, f, d)$$

$$- \text{salvage costs}(A) \times PWF(n, d),$$

with a similar formulation for system B.

It can be seen that the net-benefits method is appropriate for a number of reasons. First, solar systems are almost always compared directly with a given conventional energy system, and the net-benefits method illustrates either a positive or negative benefit in present dollars between the compared systems. Second, the net-benefits method is suitable for using either total or difference costing, noting that the benefits format is especially suited for difference costing. Third, the present-value total costs can be tabulated for each system independently and the net benefits determined for a series of comparisons. Last, the presentation of net benefit (sometimes referred to as "solar savings") itself is easily understood and accepted by a wide variety of solar system investigators (Collares-Pereira et al. 1984; Percino 1979; FCHART 1978; SOLCOST 1978; Schroder and Reddemann 1982; Manson and Mitchell 1982; Drew 1985; Suzuki et al. 1985; Brooks and Duchon 1983).

The optimization of solar system sizing using life-cycle costs is based on calculating the life-cycle costs for a variety of system sizes. As system size increases, the first costs will increase, but the annual conventional fuel costs should decrease (see figure 5.8). Thus, different size systems will have differing life-cycle costs.

For solar system optimum sizing, the accepted collector sizing techniques are to minimize the solar system life-cycle costs (LCC) or maximize the system's life-cycle net benefits against the collector area used in the system. Figure 5.11 illustrates the two possible situations that can arise when LCC and net benefits analyses are employed. Figure 5.11a represents the case when the LCC reaches a minimum and the net benefit reaches a maximum (even if the net benefit is negative as shown). In this situation, the optimum system collector size is identified and the system can be designed.

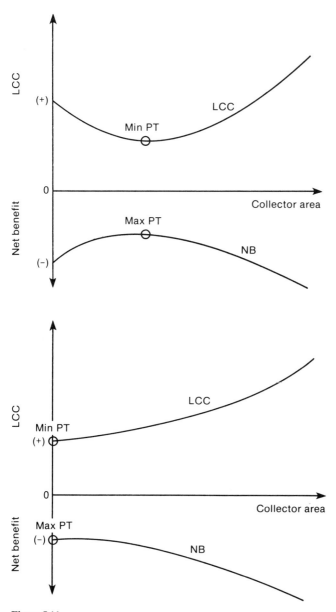

Figure 5.11
Two cases for sizing analysis by life-cycle costs or net benefits. (a) LCC and NB analyses have finite collector area minimum/maximum points. (b) LCC and NB analyses have *no* finite collector area minimum/maximum analysis.

The second case, figure 5.11b, arises when no maximum or minimum is present at finite collector area. The LCC continually increases with more collector area, and the net benefit continually decreases. Zero collector area (no solar system) is the optimum economic size. In this case, the energy cost minimization methods must be used to size the system, as will be discussed later.

5.4.1.2 Payback Period

The payback period is the length of time needed to recoup the additional investment required by a solar system, with the solar system's lower annual costs providing the "payback" cash. The "simple" payback period is the additional investment required divided by the annual savings generated by the solar system (Michelson 1982). However, more rigorously, the "discounted" payback period is determined using life-cycle costs by solving for the number of years in the present worth formula (equation [5.4]) when the net benefit is set equal to zero.

In an approach similar to life-cycle costing optimization, the payback periods for a range of system sizes can be calculated, and the system size with the minimum payback period is chosen as the optimum design size (Schroder and Reddemann 1982). However, the correct financial approach is to utilize marginal payback criteria as the system size is changed (Lunde 1982). A situation can arise that any calculated payback period is longer than the systems expected lifetime; again, other sizing methods are needed in this case.

5.4.1.3 Return on Capital

The return on capital is determined by the annual net savings relative to the required extra investment for the solar system. Similar to the payback analysis, a "simple" return on capital would take the annual cash savings and divide them by the first cost. However, the "discounted" return on capital accounts for the time value of money, where the life-cycle cost formulation in equation (5.6) would be solved for the "discount rate" in the present worth factor (Schroder and Reddemann 1982; Gordon and Rabl 1982).

Optimization would be approached by finding the highest return on capital for the range of system sizes and costs being investigated or, more correctly, by finding the minimum allowable marginal return point. It should be noted that the various system sizes and costs needed for a life-cycle cost/net benefits optimization analysis will also be the values needed for the payback and return on capital analyses (Schroder and Reddemann 1982).

5.4.2   Energy Cost Minimization Methods

Another approach to determining optimum system size is to determine what system size and configuration gathers, delivers, or displaces energy at the least cost. An advantage of this approach is that an optimum is always found, no matter how unfavorable the solar economics may be. A disadvantage is that the method, by itself, may lead the user to an uneconomical decision.

Again using collector area as the primary sizing variable, it can be seen that the ratio of system costs to system energy ($/Btu) will be infinite at both zero collector and infinite collector area. At zero collector area, while system costs are finite, no system energy is gathered, delivered, or displaced, and the $/Btu ratio approaches infinity. As the system gets larger, system costs keep increasing, but the system energy term approaches a finite value (see figure 5.8). Thus, as the collector area goes to infinity, so

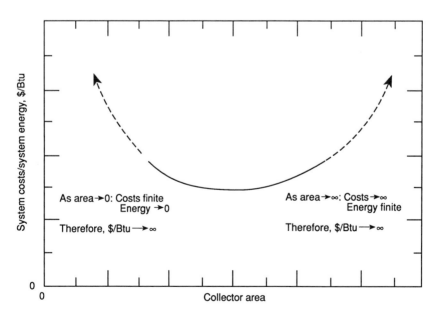

Figure 5.12
Necessity for a minimum $/Btu ratio.

does the $/Btu ratio. Figure 5.12 illustrates the necessity of the $/Btu minimum ratio.

5.4.2.1 System Cost/Energy Collected

One of the most widely used energy cost ratios is that of the initial system cost to solar energy collected. This ratio is the simplest to determine, because only the total system cost and the estimated annual collector array performance are needed. The ratio can be based on either the annual collection of energy or the system lifetime collection of energy. This initial cost/energy collected ratio is also used by passive solar designers as an optimization method. The obvious problem with this approach is that life-cycle economics are completely ignored.

Another related method (Löf 1983) is the solar-or-no-solar decision, which is made by comparing the present price of the energy being displaced with the first year interest cost of the loan for various size solar systems. Unless some size solar system provides heat at lower costs than the present price of the displaced conventional fuel, the optimum is a conventional system without solar. This method is often used for screening purposes.

5.4.2.2 Life-Cycle Cost/Energy Delivered

The ratio of the system life-cycle costs to the solar energy delivered to the load is also widely known and used (Sillman 1981; Lund 1984). This ratio requires a more detailed analysis, as the life-cycle costs must be determined along with the thermal analysis of how much of the collected solar energy is ultimately delivered to the load. This ratio is more sensitive to proper design of a solar system than the $/Btu-collected ratio described previously. The disadvantage of this ratio (and the next one described below) is that *estimates* are needed for the LCC parameters, which can greatly affect the ratio, depending on the estimates used.

5.4.2.3 Life-Cycle Cost/Energy Displaced

The ratio with the greatest sensitivity to system design is that of the life-cycle cost to purchased energy displaced (Drew 1985; Sodha et al. 1984). The purchased energy displaced is a "net energetics" calculation and requires the most detailed knowledge of the overall system energy flows. This thermal information, combined with the life-cycle cost details, provides an indicator that will lead to a thermally optimum design based on the rationality of the LCC approach.

5.5 Comparisons among Techniques

To illustrate the advantages, disadvantages, and differences in the various optimization approaches, examples of the application of the techniques are provided below using the thermal performance data and financial data listed below for an assumed active solar system design.

ASSUMED THERMAL PERFORMANCE DATA. A conventional system meets an annual heating and hot water load of 27.6×10^6 Btu (8087 kWh). A solar system is designed that has the annual thermal performance as a function of collector area as seen in table 5.1.

ASSUMED FINANCIAL DATA. The conventional system will have an initial cost of $1,500. The solar system will have an initial cost of $4,000 in fixed costs plus $300/m^2 of collector in variable costs. The salvage value of the solar system is $1,000 at the end of a 20-year life, while the conventional system has zero salvage value. The fuel escalation rate is assumed equal to the general inflation rate; thus $f = 0$. The real discount/interest rate assumed is 8%. With these assumptions the present worth factors are:

$PWF(n = 20, d = .08) = 0.215$, for single payment,

and

$PWF(n = 20, f = 0, d = .08) = 9.818$, for annual payments.

Two fuel costs are examined: $50/10^6$ Btu (17¢/kWh) and $18/10^6$ Btu (6¢/kWh). With these two fuel rates, the annual cost of fuel for the conventional system is $1,380 and $496, respectively. The LCC of the conventional system is $1,500 plus 9.818 times the annual fuel cost, resulting in

Table 5.1
System annual thermal performance for collector subsystem sizes area

Collector area (m^2)	Solar energy collected (10^6 Btu)	Solar energy delivered to load (10^6 Btu)	Parasitic energy (10^6 Btu)	Purchased energy displaced (10^6 Btu)
5	8.8	7.9	1.0	6.9
10	16.0	14.4	1.5	12.9
15	21.0	18.9	2.0	16.9
20	23.0	20.7	2.5	18.2
25	24.6	22.1	3.0	19.1

Table 5.2
System costs at two different fuel rates for various collector areas

Collector area (m^2)	Initial costs ($)	$50/10^6 Btu fuel		$18/10^6 Btu fuel	
		Net fuel costs ($)	LCC ($)	Net fuel costs ($)	LCC ($)
5	5,500	1,035	15,447	373	8,943
10	7,000	755	14,198	272	9,454
15	8,500	535	13,538	193	10,176
20	10,000	470	14,399	169	11,446
25	11,500	424	15,458	153	12,787

$15,048 for $50/10^6$ Btu fuel and $6,370 for $18/10^6$ Btu fuel. (For clarity, tax credits and deductions were ignored, as were maintenance and other miscellaneous costs.)

5.5.1   Financial Based Methods

5.5.1.1   Life-Cycle Cost and Net Benefit
The life-cycle costs of the solar system are determined by summing the initial cost and the present worth of the annual fuel costs, minus the present worth of the salvage value ($215). The costs for the two fuel rates are contained in table 5.2.

Note that for the $50/10^6$ Btu fuel cost case, a minimum LCC exists, while at $18/10^6$ Btu, the LCC values steadily increase. These two fuel costs illustrate the two situations possible with LCC and net-benefit analyses, where no optimization is possible in one case.

The net benefit (solar savings) is calculated by subtracting the solar system LCC from the conventional system LCC ($15,048 for $50/10^6$ Btu and $6,370 for $18/10^6$ Btu).

Figure 5.13 shows the LCC and net-benefit curves for this example with the optimum being around 15 m^2 for $50/10^6$ Btu, and no optimum obtainable with LCC or net benefit if the fuel cost is $18/10^6$ Btu.

5.5.1.2   Payback and Return on Capital
The payback and return-on-capital indicators are calculated from the solar systems' initial cost and the fuel cost savings. Table 5.3 shows the results.

While the discounted paybacks and returns on capital are "worse" than the "simple" values, note that they minimize and maximize around the

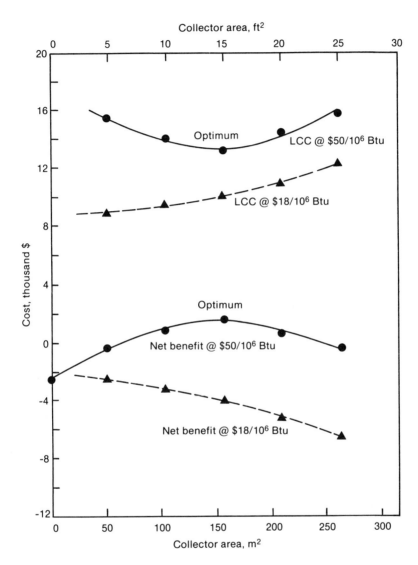

Figure 5.13
LCC and net benefit versus collector area for two fuel prices.

Table 5.3
Payback and return at two different fuel rates for various collector areas

	$50/10^6$ Btu fuel				$18/10^6$ Btu fuel	
Collector area (m^2)	Simple payback (years)	Simple return on capital (%)	Discounted payback (years)	Discounted return on capital (%)	Simple payback (years)	Simple return on capital (%)
5	15.9	6.3	100+	2.3	44	2.2
10	11.2	8.9	30	6.3	31	3.2
15	10.1	9.9	22	7.7	28	3.6
20	11.0	9.1	28	6.5	30	3.3
25	12.1	8.3	43	5.4	33	3.0

same collector area. However, this situation (discounted being worse and/ or obtaining same optimum collector area) will not always be the case (Schroder and Reddemann 1982; Gordon and Rabl 1982).

5.5.2 Energy Cost Methods

As discussed earlier, the energy cost minimization methods always give a minimum value. For the data in this example, the three energy cost ratios are plotted in figure 5.14, (a) for $50/10^6$ Btu and (b) for $18/10^6$ Btu. The minimums for $50/10^6$ Btu fuel are in the 15-to-20 m^2 range, while the minimums for $18/10^6$ Btu fuel are slightly less than 15 m^2.

5.5.3 Summary Comparison

The example above illustrates the use of life-cycle costs and net-benefit optimization to size a solar system when the solar system was an economi- cally viable alternative to a conventional system (the $50/10^6$ Btu fuel cost case). The lower fuel cost case ($18/10^6$ Btu) was found to be uneconomic against the conventional system alternative, and energy cost methods have to be used to get a designed size via optimization.

In examining the LCC and net-benefit $50/10^6$ Btu curves in figure 5.13, it can be seen that the maximum and minimum values are in a range where little variation in $ amounts is noted with substantial collector area variation (see 10-to-20 m^2 range). This situation of obtaining a broad optimum is the normal situation and allows the designer good flexibility (Hughes and Morehouse 1979; Warren and Wahlig 1985; Collares-Pereira et al. 1984; Lund 1984).

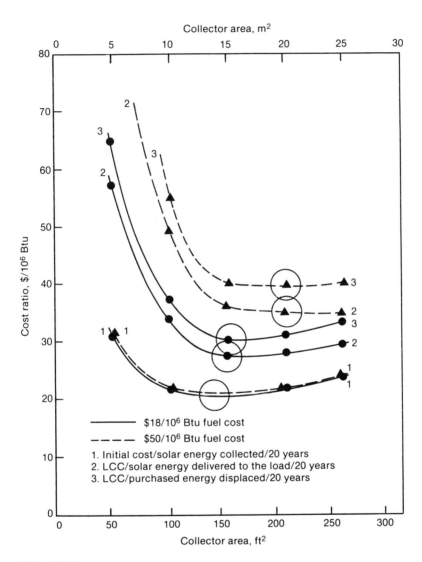

Figure 5.14
Energy cost ratio minimizations.

The example also illustrates *the* major problem encountered in trying to find an optimum solar system size using the life-cycle cost or savings method—zero size is found for the uneconomic case (the $18/10^6$ Btu situation). While this is the correct and rational answer from a purely economic view, the number of assumptions that must be made (future fuel prices, maintenance costs, inflation and interest rates, etc.) lends a degree of uncertainty that may invalidate the calculation. The use of the minimum/maximum payback/return-on-investment approaches (instead of marginal rate calculations) and the $/Btu methods is often needed when no finite optimum size is obtainable with LCC or net benefit. However, the overall economic competitiveness of solar versus conventional systems is being ignored in order to justify designing for a given size.

It should also be mentioned that "simple" calculations, those not involving present worth determinations, are only approximations, but their ease of use often permits an economic evaluation that otherwise might not be attempted by the designer. However, for major investments, it should be obvious that LCC and marginal analysis are the methods to use for optimization.

This chapter deals specifically with the optimization of active solar systems themselves; the system thermal performance and system costs lead to curves that yield a minimum or maximum point indicating a component or subsystem optimum size. In summary, it should be realized that the solar system designer is faced with having to perform a "global" optimization, which includes energy conservation, backup system size and type, and passive/hybrid system design (Lunde 1982; Kirkpatrick and Winn 1985; Evans et al. 1985), plus the trade-offs due to other "noneconomic" factors (aesthetics, construction schedules, etc.). Thus, the global optimization process is usually much more complicated and involved than just "finding the dips in a couple of curves."

References

Anand, D. K., et al. 1984. Second law analysis of solar powered absorption cooling cycles and systems. *J. Solar Energy Eng.* 106:291–98.

Barnes, P. R. 1981. The optimization of solar heating systems. *Solar Energy* 26(4):375–76.

Braun, J. E., et al. 1981. Seasonal storage of energy in solar heating. *Solar Energy* 26(5):403–11.

Brooks, D. M., and C. E. Duchon. 1983. A comparison of two procedures for optimum sizing of a solar heating system. *Solar Energy* 29(2):129–35.

Collares-Pereira, M., et al. 1984. Design and optimization of solar industrial hot water system with storage. *Solar Energy* 32(1): 121–33.

Colorado State University. 1977. *Solar heating and cooling of residential buildings: sizing, installation and operation of systems.* Chapter 14. Washington, DC: U.S. Department of Commerce.

DOE Solar Working Group. 1978. *Solar energy research and development: program balance.* Washington, DC: U.S. Department of Energy.

Drew, M. S. 1985. A ranking procedure for active solar heating systems. *Solar Energy* 35(3): 291–93.

Evans, B. L., et al. 1985. A design method for active-passive hybrid space heating systems. *Solar Energy* 35(2): 189–97.

FCHART users' manual. 1978. EES Report 490–3, University of Wisconsin.

Gordon, J. M., and A. Rabl. 1982. Design, analysis and optimization of solar industrial process heat plants without storage. *Solar Energy* 28(6): 519–30.

Hughes, P. J., and J. H. Morehouse. 1979. *Comparison of solar heat pump systems to conventional methods for residential heating, cooling, and water heating.* Chapters 4 and 7 of final report. DOE Contract No. DE–8CO4–78CS034261. McLean, VA: Science Applications.

Kirkpatrick, A., and C. B. Winn. 1985. Optimization and design of zone heating systems, energy conservation, and passive solar. *J. Solar Energy Eng.* 107: 64–69.

Löf, G. O. G. 1983. Solar heating economic—a simple and rational basis for replacing fuel with solar energy. *Proc. 8th Biennial Congress of the ISEE.* Solar World Congress, Perth, Australia.

Lund, P. D. 1984. Optimization of a community solar heating system with a heat pump and seasonal storage. *Solar Energy* 33(3/4): 353–61.

Lunde, P. J. 1982. A simplified approach to economic analysis of solar heating and hot water systems and conservation measures. *Solar Energy* 28(3): 197–203.

Manson, B. E., and J. W. Mitchell. 1982. A regional comparison of solar, heat pump, and solar-heat pump systems. *J. Solar Energy Eng.* 104: 158–64.

Michelson, E. 1982. Multivariate optimization of a solar water heating system using the SIMPLEX method. *Solar Energy* 29(2): 89–99.

Mitchell, J. W. 1983. *Energy engineering.* New York: Wiley.

O'Dell, M. P., et al. 1984. Design methods and performance of heat pumps with refrigeration-filled solar collectors. *J. Solar Energy Eng.* 106: 159–64.

Percino, A. M. 1979. *A methodology of determining the economic feasibility of residential or commercial solar energy systems.* SAND 78–0931. Sandia Laboratories.

Ruegg, R. T. 1975. *Solar heating and cooling in buildings: methods of economic evaluation.* NBS IR 75–712. Gaithersburg, MD: National Bureau of Standards.

Scholten, W. B. 1984. A comparison of energy delivery capabilities of solar collectors. *J. Solar Energy Eng.* 106: 490–92.

Schroder, M., and B. Reddemann. 1982. Three different criteria to evaluate the economics of solar water heating systems. *Solar Energy* 29(6): 549–55.

Sillman, S. 1981. Performance and economics of annual storage solar heating systems. *Solar Energy* 27(6): 513–28.

Sodha, M. S., et al. 1984. A simple cost analysis for hybrid systems. *Solar Energy* 32(4): 561–62.

SOLCOST space heating handbook. 1978. DOE/CS–0043/3. Washington, DC: U.S. Department of Energy.

Suzuki, M., et al. 1985. An analysis of solar heating systems that use vapor-compression cycles. *Solar Energy* 34(1):43–57.

Warren, M. L., and M. Wahlig. 1985. Cost and performance goals for commercial active solar absorption cooling systems. *J. Solar Energy Eng.* 107:136–40.

Winn, C. B., and C. A. Baer. 1984. Off-peak storage size analysis. *J. Solar Energy Eng.* 106:483–89.

II SOLAR WATER HEATING

6 Solar Water Heating—Introduction and Summary

George Löf

6.1 Perspective

Solar water heating is the only solar application that already had a commercial existence prior to the establishment of federal programs in solar energy in the early 1970s. This situation resulted in federal involvement substantially different from that in solar heating, solar cooling, solar electricity generation, and other applications that had no commercial use at that time.

Federal government emphasis in solar water heating during the NSF, ERDA, and DOE programs was on *demonstration* rather than the research and development aspects. Under these programs, funding was provided for the design, construction, and, in some cases, the testing of residential and commercial solar water heating systems. Manufacturers of solar equipment, mainly the collectors, teamed with architects, engineers, and plumbing contractors to design, install, and provide to the user solar water heating systems of many varieties, sizes, and capabilities for demonstration in various U.S. climatic regions.

Although the federal government supported research and development in solar collectors, including improvements in materials and designs, nearly all of the developments in solar water heating *systems* were undertaken by manufacturers and designers without direct federal funding. (In chapter 7, the roles of these several participants are outlined and the results delineated.) The government did, however, provide support for the development of test methods and standards for solar water heating systems and their components.

6.2 Historical Background

6.2.1 Solar Water Heaters and Their Residential Use prior to 1970

6.2.1.1 Hot Water Requirements and Solar Availability

Residential solar water heating is intrinsically a logical application for solar energy. A typical American family of 4 persons uses 50 to 100 gallons of hot water per day throughout the year, requiring about 50,000 Btu per day, or 15 to 20 million Btu per year. In a favorable sunny climate, a solar water heater of the best quality should provide 500 Btu per average day for each square foot of solar collector, so a water heater based on 60 square feet of collector should supply, on average, about 30,000 Btu per day. Daily and seasonal variation will, of course, occur. This contribution to the total hot water heating requirements, from a comparatively modest-sized unit, can thus readily exceed 50% and in some circumstances reduce the need for auxiliary energy (usually natural gas or electricity) to a minor fraction of the total energy required.

6.2.1.2 Commercial Developments—Types and Areas

Shortly before the beginning of this century, solar water heaters made their commercial appearance in southern California. First in the form of roof-mounted tanks, and a few years later as glazed tubular solar collectors and elevated storage tanks in a thermosyphon arrangement, several thousand systems were sold for residential use by a commercial firm and its successor (Butti and Perlin 1980).

Usage declined in southern California in the 1920–1930 period but nearly simultaneously began to develop in the Miami area of Florida. Several companies supplied thermosyphon solar water heaters for domestic use in southern Florida in the 1920–1950 era. It has been estimated that over 50,000 were made and sold in this period. Increased availability of natural gas and low-priced electricity in combination with the cost and inconvenience of repair and maintenance of the solar water heaters contributed to their virtual demise in the 1950s (Butti and Perlin 1980).

6.2.1.3 Decline in U.S. Application and Growth in Other Countries

Although the manufacture and use of solar water heaters declined substantially in the 1940–1970 period, commercial development in other countries, particularly in Israel, Japan, and Australia, became important (U. N. Conference on New Sources of Energy 1961; Löf and Close 1967; Morse 1955). Stimulated by high residential electricity prices, thermosyphon solar water heaters became commonplace in multifamily Israeli dwellings in the 1950s and 60s. Government policy contributed to this growth through encouragement and requirement of this application in new housing projects.

Developments in Japan and Australia were also active in this period, but the degree of commercialization did not approach that in Israel until the 1970s (Löf and Close 1967; Morse 1955).

6.2.2   Development and Commercialization of Residential Solar Water Heaters in the United States since 1970

6.2.2.1   Growth of the Industry

Following the oil embargo and sharp rise in price of petroleum in 1973, manufacture and sale of solar water heaters began anew. In 1974, more than 20 companies started the production of flat-plate solar collectors in the United States, most of which were coupled to pumps and storage tanks in active residential water heating systems. Plumbing contractors and solar water heating retailers sold and installed direct heating recirculation systems in Florida and other southern states, and indirect heaters employing antifreeze solutions were used in colder climates. By 1979, annual installation of collectors for solar hot water had grown to 3 million square feet, corresponding roughly to about 50 thousand residential solar water heaters per year. By 1984, these figures had increased two-to-three fold (DOE 1984).

Although the main thrust of commercial activity in solar water heating has been in the residential sector, large installations for commercial establishments such as laundries, restaurants, and factories were also made and sold during this period.

6.2.2.2 Influence of Improvements in Solar Collectors

As indicated in other sections of this volume, there was active federal support of research and development on solar collectors during this period. Progress by manufacturers and developments in research institutions supported by federal funds led to the use of radiation selective surfaces, corrosion resistant (copper) absorbers, and high transmission glass. These and other developments contributed to improved performance of solar hot water systems.

6.2.2.3   Research and Development on Solar Water Heaters

Because industry moved rapidly into commercialization of solar water heating in 1974 and subsequent years, government-funded research was directed mainly toward development of test methods and comparative evaluation of various system types and configurations rather than toward system development. As indicated in subsequent chapters, theoretical and experimental evaluations of efficiencies of thermosyphon and pumped direct and indirect types, and of single-tank and dual-tank storage were made (Fanney and Liu 1980, Fanney and Klein. 1983). Comparative performance of representative manufactured systems was measured under controlled laboratory conditions by state energy offices and government-operated electric utilities (Löf and Karaki 1983). Reliability and maintainability of solar water heaters were also evaluated in the federal DHW program. Approximately twenty residential and commercial solar hot water systems were designed, fabricated, installed and monitored under a joint NASA-DOE program to develop new concepts.

6.2.2.4   The Solar Heating and Cooling Demonstration Program

Although there was only a limited federal program of research and development on solar water heating systems, several thousand commercially manufactured solar hot water systems were funded by DOE and HUD for use in the solar heating and cooling demonstration program (Vitro Corporation 1979–1983). Systems were designed by manufacturers, architects, and heating and plumbing specialists for specific buildings, the owners of which were funded by the federal agencies for the installation of solar water heating systems. Most of these systems were built between 1977 and 1980. Mixed results were derived from this program. The installations ranged from poorly designed systems assembled from inferior hardware to sound designs employing high-quality components. Monitoring of over 100 of these systems indicated performance that ranged from unacceptably low efficiencies to levels much as expected. Reports on many of the monitored systems were useful to manufacturers in appraising the effectiveness and reliability of specific system designs and in discarding those that were found troublesome or ineffective (Vitro Corporation 1979–1983). A State Initiative Program under the direction of HUD and DOE, administered by the Northeastern U.S. state energy offices, resulted in the installation of over four thousand systems. Requirements of minimum performance to meet or exceed 50% of the DHW load and participation in collector thermal performance and durability/reliability tests were mandatory. Experience in this and other test programs helped provide a basis for developing a product certification program.

6.2.2.5   Indirect Benefits of DOE Solar Space Heating R&D Program

The federal R&D program in solar heating and cooling had indirect benefits in the development of solar water heating. Many systems included water heating with space heating (and sometimes with cooling), and designs for combined functions were developed and evaluated under this program (Karaki and Löf 1981). Particularly in the residential sector, a solar space heating system was seldom built without the water heating capability, so the development of the combination has been important. Models for predicting performance of these combined systems were developed, experimental installations were made and monitored, and various alternative designs were compared. The solar heating and cooling demonstration program also included many systems in which water heating was an essential part. Housing developers and the manufacturers of solar water heaters were given an opportunity to work together for more integrated solar system design and to gain experience in financing and marketing solar-equipped homes. Indirectly, therefore, the federal solar space heating program contributed substantially to the improvement in solar water heaters, particularly those used in combination with space heating.

6.2.3 Application of Technology to Industrial and Commercial Hot Water Requirements

The use of solar-heated water for commercial and industrial purposes was initiated in the federal demonstration programs during the 1976–1980 period. Federal funding of several hundred commercial hot water installations and a dozen industrial applications of solar hot water for process heat were implemented. With a few exceptions, monitoring of these systems showed disappointing performance, for reasons ranging from poor design to improper installation and inadequate maintenance (Löf and Karaki 1983; Vitro Corporation 1979–1983). There was no "standard" system, each one being designed for the installation at hand, often by people who had no experience with solar. There were, however, some well-designed and capably installed systems that clearly demonstrated the commercial and industrial usefulness of solar water heating.

In addition to the federally funded demonstration projects involving solar water heating, many privately financed installations have been made, particularly under conditions that involved third-party ownership of the solar water heating system. Favorable tax treatment of this type of investment has made it possible for manufacturers to provide the components and systems, for investors to purchase and install the systems, and for users to pay for the hot water by use of a formula that guarantees savings relative to the cost of fuel otherwise incurred. Installations for large hot

water supplies for process purposes and for domestic sanitary use in apartments and multifamily installations have benefited from the experience
derived by manufacturers from the federal solar demonstration programs.

6.3   Summary of Current Status

6.3.1   Extent of Application in Home and Industry

Solar water heating is the most widely practiced solar technology today. It
has been variously estimated that there are 0.5 to 0.8 million residential
solar water heaters in the United States and that commercial solar hot
water installations total over 5 million square feet of solar collectors. Penetration of these markets is, however, extremely small as a percentage.
Less than one percent of dwellings and of commercial establishments are
provided with solar heated water.

6.3.2   Types of Systems in General Use

As indicated in chapter 7, active circulating systems of several types are in
use in the United States. Two indirect types, involving heat exchangers,
predominate. The drain-back system involves circulation of pure water in
a closed loop through the collector and heat exchanger; the antifreeze type
employs a nonfreezing glycol solution or other liquid for transfer of heat
from the collector to the exchanger. In the drain-back design, freeze protection is provided by automatic drainage of the water from the collector
into a small reservoir when the circulating pump stops.

There are two direct heating types: the drain-down type, in which freeze
protection is provided by the operation of valves to drain the collector
when freezing threatens, and the recirculation type, in which freezing is
prevented by continual operation of the circulating pump during cold
sunless periods. Freeze protection reliability afforded by both these systems is now recognized to be inadequate in most sections of the United
States.

In the milder climates in the United States, the use of two other types of
solar hot water heaters is growing: the thermosyphon type, in which water
circulates from a collector to an elevated storage tank, without a pump,
and the tank or integral storage collector (ISC) type, in which one or more
water tanks or containers are heated directly by the sun. The risk of freez-

ing, particularly in piping to and from these heaters, places limitation on the areas of their reliable use.

In commercial and industrial applications, large solar water heating systems usually involve the circulation of nonfreezing collector fluids, freeze protection by drainage not usually being practical because of collector size.

6.3.3 Economic Viability

As indicated in chapter 10, the cost of solar heated water is a function of system efficiency, solar climate, the capital investment in the system, and the cost of money. To a lesser extent, the cost of solar hot water also depends on the cost of operating and maintaining the system.

Although the typical cost of manufacturing a high-quality residential solar water heater with a 64-square-foot collector, storage tank, and all other necessary components is well below $1,000, the costs of management, marketing, distribution, and installation usually bring the total installed price to about $4,000. With the cost of money at 10% per annum and the delivery of about 10 million Btu per year in a good solar climate, the cost of solar heat for hot water becomes approximately $40 per million Btu. Without subsidies or other incentives, the cost of solar heat is clearly above the typical $5 to $10 cost of one million Btu of heat from natural gas and $15 to $30 from electricity. Other factors, however, motivate home owners to purchase solar water heaters, with a perceived advantage beyond the question of cost.

It may be expected that as demand for solar water heating increases, marketing and distribution costs will decrease and total installed system prices will go down. Some solar specialists have claimed that substantial manufacturing cost reductions are possible, but experienced manufacturers do not consider such changes to be either likely or of significant benefit. Conversely, any sacrifice in product quality that might be required by a manufacturing cost reduction (in materials or labor) is believed to be a deterrent to the purchase of solar water heaters and, therefore, a contribution to increased prices.

Commercial use of solar water heating does not involve the high marketing costs experienced in the residential sector. At a substantially higher production level than now exists, large systems can be expected to require an investment of $25 to about $40 per square foot of collector, which, at

10% annual interest, should provide 200,000 Btu per year at a cost $2.50 to $4. The equivalent cost of solar heat is thus $12.50 to $20 per million Btu. In the absence of subsidies, tax benefits, and other incentives, heat at these prices is not competitive with other sources available to the average commercial and industrial user. It is believed, therefore, that where decisions are dictated by costs only, as is usual in the commercial and industrial sectors, solar water heating will not be a significant factor until conventional energy prices have doubled or trebled, or unless the cost of money (interest rates) may have decreased to a level near 5%.

6.3.4 Trends and Prospects

It can be expected that commercial manufacture and sale of solar water heaters, mainly in the residential sector, will continue at current levels. It may also be expected that improvements in product quality developed by manufacturers and installers will lead to greater system reliability, appeal, and user confidence. As confidence improves, demand will grow, and marketing costs should decrease. The price of solar systems relative to that of conventional water heaters may well become less decisive in the choice.

6.4   Contents of Solar Water Heating Section of This Book

Chapter 7 is a description of systems for supplying solar heated water. It covers numerous types that have been used or proposed and emphasizes those that are in most extensive use.

In chapter 8, performance of systems is discussed, particularly with reference to several monitoring programs involving large numbers of operating systems in the field.

Chapter 9 contains an analysis of reliability and operational characteristics of solar water heating systems.

Finally, chapter 10 contains extensive data on costs of solar water heating systems, particularly in large installations of the federal solar heating and cooling demonstration program.

References

Butti, K., and J. Perlin. 1980. *A golden thread*. New York: Van Nostrand, Reinhold.

Karaki, S., and G. O. G. Löf. 1981. Active space heating and hot water supply with solar energy, *Solar Technology Assessment Project*, vol. 3. Cape Canaveral: Florida Solar Energy Center.

Löf, G. O. G., and D. Close. 1967. Solar water heaters. In *Low temperature engineering applications of solar energy*, edited by R. C. Jordan, p. 61. New York: ASHRAE.

Löf, G. O. G., and S. Karaki, 1983. System performance for the supply of solar heat. *Mechanical Engineering* 105(12):32.

Morse, R. N. 1955. Solar water heaters. *Proc. World Symposium on Applied Solar Energy*, 1st ed., p. 191. Menlo Park, CA: Stanford Research Institute.

Fanney, A. H., and S. A. Klein. 1983. Performance of solar domestic hot water systems at the National Bureau of Standards—measurements and predictions. *ASME J. Solar Energy Engineering, 105* p. 311.

Fanney, A. H., and S. T. Liu. 1980. Comparison of experimental and computer-predicted performance of six solar domestic hot water systems. *ASHRAE Trans. 86*, p. 823.

Proceedings of the U.N. Conference on New Sources of Energy. Vol. 5, 1964. United Nations. New York.

U.N. Conference on New Sources of Energy. *See Proceedings of the U.N. Conference on New Sources of Energy*.

U.S. Department of Energy. 1985. Solar Collector Manufacturing Activity 1984. Washington, DC: DOE.

Vitro Corporation. 1979–1983. *Comparative Report: Performance of Solar Hot Water Systems*. Reports Solar/0024-80/41, Solar/0024-82/41, Solar/0024-83/41. Washington, DC: DOE.

7 Hot Water Systems: System Concepts and Design

Elmer Streed

7.1 Introduction and Background

7.1.1 General Considerations

Solar domestic hot water (SDHW) systems are proving to be a cost effective application when properly designed for specific climatic conditions and use requirements. Because the SDHW system performance depends upon its installed configuration and because of the variety of operating conditions and the deviation in reported data for similar systems, it has been difficult for a potential user to select the most appropriate system from a performance and economic standpoint.

Experience has shown that the relatively simple process of transforming the solar energy striking the collector into heated water delivered to an end use involves the following factors:

- Climatic variations including available solar energy and ambient temperature variation
- Load demand profile including rate, time, and temperature
- Source water temperature variations
- Heat transfer media type—liquid, air, etc
- Method of media circulation—forced or thermosyphon
- Method of heat transfer—direct or indirect
- Effectiveness of heat exchanger (if used)
- Collector type, area, array design, and location
- Flow rate through collectors
- Pump or blower efficiency and pressure characteristics
- Storage tanks—number, size, and location
- Thermal insulation for piping and storage
- Controller characteristics
- System degradation—corrosion, dust, temperature, environmental exposure
System reliability—sensors, valves, controller, pumps
- Monitor instrumentation—type and location

Systems research and development studies of each of these factors for DHW systems were not feasible, nor were they the intent of government sponsored programs (ERDA 1976). System design and development was conducted primarily by the solar industry with government support to conduct demonstration and performance evaluation. Limited prototype system design and development was performed on advanced systems with emphasis on testing and evaluation to provide valid test data for comparison with operational performance data obtained during later demonstration.

A directed R&D program which emphasized areas of particular interest for SDHW systems included the following:

• Materials characterization, development and testing

• Component development—collectors, storage, heat exchangers, piping

• Systems and controls—integrated active systems, system design, controllers, passive and hybrid systems

• Nonengineering aspects—economic analysis, architecture, and construction

This chapter contains design data, system descriptions, and summaries of experience reported or indirectly developed as the result of the applied research and demonstration of SDHW systems. Since considerations important in designing water heating systems are also basic to other solar heating and cooling applications, there may be some duplication or omission of information found in other chapters describing components of systems.

An excellent summary of solar hot water heating designs and experience in Australia has been reported by Morse and Close (1977).

7.1.2 Operating Considerations

Initial government efforts to establish design requirements for health, safety, and acceptable levels of performance were developed under Public Law 93–409, Solar Heating and Cooling Demonstration Act of 1974. Publication of Interim Performance Criteria in January 1975 (NASA 1975; GPO 1975) provided guidelines for the industry to design, build, install, service, and evaluate solar energy applications related to building and domestic hot water heating. Because of the national scope of the program, the documents covered broad design aspects related to climatic, load, and reliability requirements for all areas of the country.

Interim documents as well as later publications (NBSIR 1978) were prepared for use in the demonstration portion of the 1974 act, which was performed under the direction of the Department of Housing and Urban Development (HUD) and the Department of Energy. Therefore, the building system operating characteristics had to conform to similar standards in the HUD Minimum Property Standards (NBSIR 1977) for nonsolar residential buildings. The most significant of these operating requirements for SDHW systems were as follows:

1. Hot water shall be provided at a tap temperature of 60°C (140°F) at the draw and recovery rates required for conventional domestic hot water heating systems.

2. Hot water storage shall have a volume capacity not less than conventional domestic hot water heating systems.

3. An auxiliary heating system shall be provided to furnish 100 percent backup for the solar contribution with the same degree of reliability and performance as a conventional system.

4. Liquid heat transfer systems employing nonpotable water shall use a heat exchanger design with a minimum of two walls or interfaces between the nonpotable liquid and the potable water supply.

5. Systems components and heat transfer fluids shall be protected from or resistant to damage by freezing when used in climates prone to freezing temperatures.

Similar requirements were specified for commercial building service hot water applications where appropriate (NBSIR 1976). Additional discussion of guidelines to aid in the design, development, and evaluation of commercial systems is presented by Hunn (1977). Changes, based upon field experience, were made in some of these considerations as described in section 7.4.

7.1.3   Climatic Design Variables

A comparison of the design variables for residential solar water heating applications illustrates some major differences. Ambient temperature is not an important factor, however, because the system operating temperature tends to adjust itself to maintain relatively consistent collection efficiency. Residential hot water demand, which directly influences the load, is not greatly affected by weather or season changes. System and heat loss data for specific applications are presented in section 7.3.

One seasonal load factor that must be considered is the source water temperature. Variations in the average monthly makeup water temperature for fourteen U.S. cities (Mutch 1974) can be as high as 39°F between winter and summer. The calculated average daily domestic hot water heating load for single family residences ranged from 50,000 to 108,000 Btu for January and from 50,000 to 87,000 Btu for July (NBSIR 1977). The winter loads in the colder climates are considerably higher when the available solar energy is less. Therefore, system collector sizing is often based upon providing a minimum of 40% of the load requirement for the month with a minimum of solar radiation (usually January).

A comparison of the measured seasonal performance of several collector and system factors for SDHW systems in single family residences, residential multifamily, and commercial systems is shown in table 7.1 (Cramer, Spears, Pollack 1980). An upward-pointing arrow indicates better (not necessarily higher) values for the factor occurred in the summer, while a down-pointing arrow indicates the winter value was better. The data are representative of three to nine systems of various designs installed in widely different climatic regions.

For single family residences, with SDHW only, the overall and operational collector efficiencies for wintertime operation were higher (30% and 41% compared to summer values of 15% and 29%, respectively). Increased collector subsystem COP and lower threshold losses are also shown. These improvements in the winter-time are primarily the result of lowered collector inlet temperatures permitting higher collector efficiency and longer operation of the fluid transfer device with subsequent greater solar energy collection.

Factors that had better summer values included the solar fractions and the AUXFR (ratio of purchased energy to the ideal energy required). Only the air system solar fraction was higher in the winter, attributed to the low air-to-water heat exchanger efficiency. The summer average values for liquid hot water solar fraction was 85.3% compared to the air hot water solar fraction of 52.9%.

Generally, the seasonal variations for commercial and residential multifamily hot water systems were the same as for single family residential systems. The higher overall collection efficiency in the summer for com-

Table 7.1
Comparison of SDHW system performance factors for summer and winter (Cramer, Spears, Pollack 1980)

System application	Overall collector efficiency	Operational collector efficiency	System solar fraction	Hot water solar fraction	Ratio of auxiliary to hot water demand	Ratio of hot water load to hot water demand	Ratio of solar energy used to collected	Ratio of operational to total incident	System COP	Hot water COP	Collector COP	
Residential single-family DHW only Active Liquid collector (6 systems)	→	→	←	←	←	→	→	→	→	no change	←	Summer / Winter
Residential single-family DHW with SH Active Liquid collector (9 systems)	→	→	←	←	←	→	→	→	→	→	→	Summer / Winter
Residential single-family DHW with SH Passive Liquid collector (3 systems)	→	NA	←	←	←	→	←	NA	NA	NA	NA	Summer / Winter
Residential single-family DHW with SH Active Air collector (5 systems)	→	→	→	←	←	→	→	→	→	→	→	Summer / Winter

Table 7.1 (continued)

System application	Overall collector efficiency	Operational collector efficiency	System solar fraction	Hot water solar fraction	Ratio of auxiliary to hot water demand	Ratio of hot water load to hot water demand	Ratio of solar energy used to collected	Ratio of operational to total incident	System COP	Hot water COP	Collector COP
Residential multifamily DHW only Active Liquid collector (5 systems)	→	→	←	←	←	←	→	←	→	→ (Summer) ← (Winter)	→ (Summer)
Commercial DHW only Active Liquid collector (4 systems)	←	→	←	←	→	→	←	←	→	→ (Summer) ← (Winter)	→ (Summer)

mercial systems is attributed to better control logic. The lower value for the ratio of auxiliary to hot water demand for commercial systems in the winter indicated the use of less auxiliary energy per gallon of water per degree increase. Additional comparisons between residential single family, multifamily, and commercial systems are presented in section 7.2, System Descriptions.

7.1.4 Hot Water Demand Profiles

The influence of demand profile on SDHW systems performance is related primarily to the coincidence of the demand and collected solar energy, and the storage thermal insulation effectiveness. The commonly accepted RAND profile for residential domestic water consumption (see figure 7.1) exhibits both a morning and evening peak in hot water usage. However, measured usage reported by the National Solar Data Network (NSDN) (Raymond 1981) for eight residential systems exhibited larger morning use contrary to the greater evening use characteristic of RAND (see figure

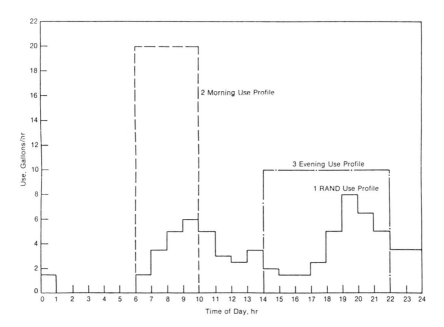

Figure 7.1
Assumed SDHW demand profiles, described in terms of use vs. time of day (Farrington, Murphy, Noreen 1980).

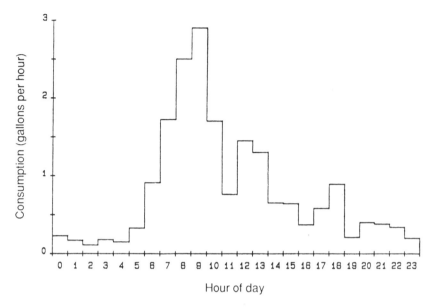

Figure 7.2
Typical 24-hour hot water consumption profile for residential SDHW systems in NSDN (Raymond 1981).

7.2). As intuitively expected, simulation studies (Buckles and Klein 1980; Fisher and Fanney) show that afternoon or evening use provides the greatest SDHW system performance. Data relating solar fraction to three usage profiles for well-insulated storage tanks and poorly insulated storage tanks are presented in figure 7.3a and 7.3b, respectively. A study has shown that day-to-day load variations can have a significant effect on system performance (Buckles and Klein 1980).

The load profile for commercial systems may be even more significant because of unique facility operation. Because of these systems' greater size and investment costs, system simulation studies with real load and climatic data are necessary to size the major components—collector and storage—and to evaluate variations with insulation and heat transfer media type and flow rate.

7.1.5 Economic Design Trade-Offs

Design trade-offs that will result in the most economical system require life-cycle cost analysis as described in chapter 5. Rules of thumb based

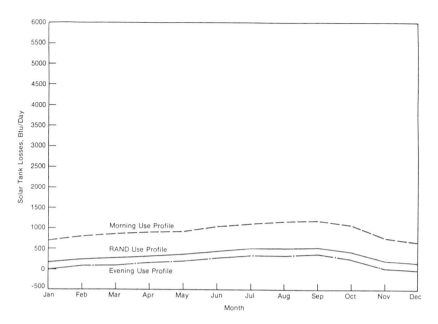

Figure 7.3a
Solar tank losses as a function of month for three demand use profiles, assuming
well-insulated storage tanks (Farrington, Murphy, Noreen 1980).

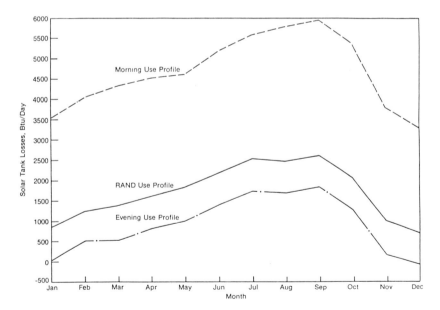

Figure 7.3b
Solar tank losses as a function of month for three demand use profiles, assuming poorly
insulated storage tanks (Farrington, Murphy, Noreen 1980).

upon experience with similar applications and climatic regions are often used to provide the initial values for sizing and economic comparison calculations. Individual residential SDHW system optimizing cannot generally be justified by running a complex life-cycle cost program. Therefore, simplified methods provided by design handbooks (Balcomb 1975) and computerized relationships based upon long-term average characteristics (Beckman, Klein, and Duffie 1977) are available to perform economic design trade-off calculations.

As an example of these simplified methods for a SDHW system, the rule of thumb for storage capacity is about 10 to 15 Btu per degree Fahrenheit per square foot of collector (Btu/°F ft²) or 200 to 300 kilojoules per square meter per degree Celsius (KJ/m² °C). Referring to

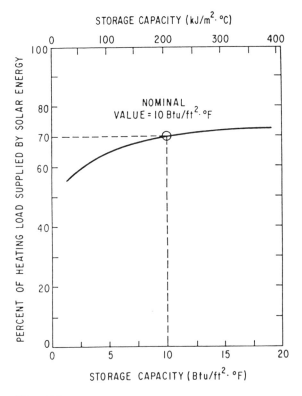

Figure 7.4
Effect of varying storage capacity upon percentage of heating load supplied by solar energy (Balcomb 1975).

figure 7.4, illustrating the relationship between the storage capacity and the percent heating load supplied by solar energy, the 10 Btu/°F ft² will provide about 70% of the load for the average system operated under average conditions. It should be noted that the rules of thumb do not apply to commercial or industrial systems where use patterns and system characteristics may differ significantly.

The collector is usually the most costly system component. Although its cost varies significantly with type, a rule of thumb that is commonly used in sizing residential SDHW systems is that array be sized to meet about 40% of the load under winter conditions. Again applying the rule of thumb that the total storage capacity should meet one to two days of hot water supply, a nominal collector area can be determined. By the use of pre-selected collector cost, the collector thermal efficiency equation,

$$\eta = F_R(\tau\alpha) - F_R U_L [(t_{f,r} - t_a)]/I, \tag{7.1}$$

the rule of thumb for storage size, a range of collector sizes, and the other system operational characteristics in an analytical program, a relationship between life cycle savings and collector area can be determined as illustrated in figure 7.5. Various collector thermal efficiencies can also be com-

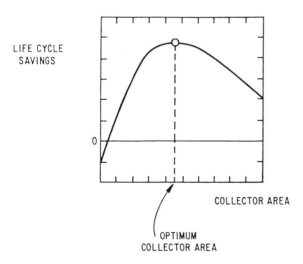

Figure 7.5
Determination of optimum collector area (Cole, Nield, Rohde, Wolosewicz 1980).

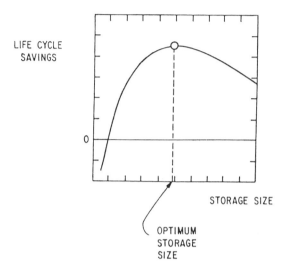

Figure 7.6
Determination of optimum storage size (Cole, Nield, Rohde, Wolosewicz 1980).

pared by this method. A similar study to optimize storage size life-cycle savings can be performed as illustrated in figure 7.6.

Analysis of the experimental studies of five different SDHW systems employing the same collector type under identical climatic and operating conditions resulted in the widely varying collector efficiencies shown in figure 7.7 (Farrington, Murphy, and Noreen 1980). Field experience with operating systems monitored by the NSDN has shown average collector operating efficiencies that differ from those of single-panel tests performed under steady state conditions. A comparison of the efficiency of a residential-sized array (45 ft^2) with test measurements is illustrated in figure 7.8. A similar comparison for a 1,798 ft^2 collector array in a residential multifamily system showed closer agreements between test and operational efficiency (figure 7.9). At a commercial site with a 9,124 ft^2 array collection efficiency was about 10 percentage points less than obtained in single panel tests (figure 7.10).

A similar analysis of NSDN collector efficiency data on operating systems (McCumber 1979) contains an assessment of some of the causes of differences in array efficiency and single-panel efficiency. Fluid flow rate and panel flow distribution in large arrays, piping thermal losses, operat-

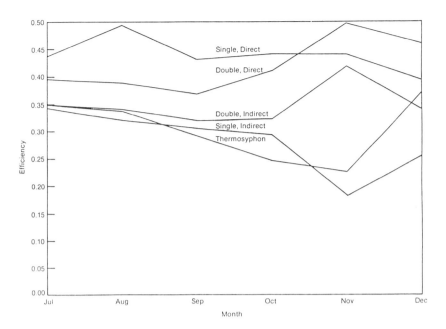

Figure 7.7
Collector efficiencies as a function of system configuration and season (Farrington, Murphy, Noreen 1980).

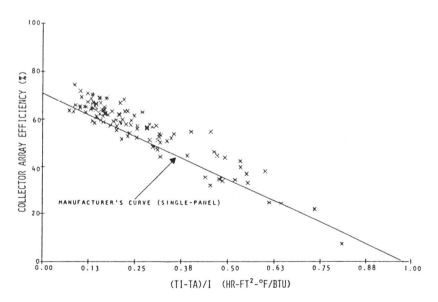

Figure 7.8
Average collector efficiency for residential SDHW system (Raymond 1982).

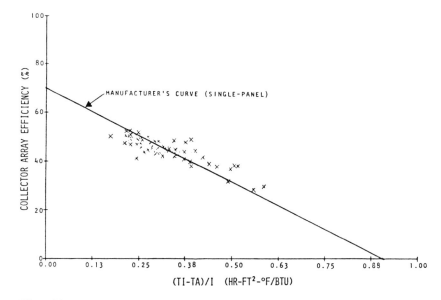

Figure 7.9
Average collector efficiency for multi-family SDHW system (Raymond 1982).

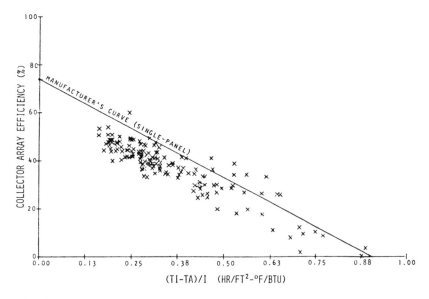

Figure 7.10
Average collector efficiency for commercial SDHW system (Raymond 1982).

ing wind and temperature conditions, and deterioration of the cover transmittance or absorptivity of the absorber were cited as potential factors.

7.1.6   Early U.S. Designs

Early U.S. SDHW system designs during the time period of 1890 to 1930 were based upon the batch heater or thermosyphon principle for use in warm climates of southern California, Florida and Arizona (Butti and Perlin 1980). Initial systems consisted of one or more blackened tanks mounted in a box that was located in an unrestricted sunny site. The system design was purely functional to provide enough warm or hot water for afternoon requirements; no thermal insulation was used. Improvements to the basic "bread-box" design included the use of a glass cover over the box and operation in conjunction with a wood stove to maintain a supply of hot water regardless of weather conditions. In 1909, design changes patented by William J. Bailey (Butti and Perlin 1980) helped revolutionize the southern California industry and led to a basic domestic hot water system design that is conceptually similar to some present-day systems. Bailey separated the collection and storage of heat by employing a "flat-plate" glass-covered collector with a parallel grid of copper pipes welded to a blackened copper absorber plate. In addition, he insulated the storage tank, which was mounted in a separate elevated location. Combining this system with an auxiliary heater allowed users to have hot water day and night with an efficient solar contribution. The system was later provided with freeze protection by use of an antifreeze fluid in the collector loop with a heat exchanger coil in the hot water tank. Early design requirements evolved by meeting consumer needs and desires at a realistic price. Systems were characterized as reliable, capable of providing hot water twenty-four hours, and meeting up to 75% of annual hot water heating bills with an investment payoff in one to three years.

Between 1920 and 1930, large discoveries of natural gas in southern California resulted in a cheap fuel that replaced solar energy use as rapidly as the gas distribution systems were installed. The sale of solar water heaters in large urban areas decreased drastically after 1926.

About this same time, rapid increases in population in Florida combined with the high cost of gas fuel and electricity led to the introduction of the "Sun Coil" for hot water heating. Modifications included the use of steel pipe soldered to a copper absorber plate and an insulated metal collector housing instead of the wood frame use in California. Enlarged systems comprising 14 ft-by-4 ft collectors connected to storage tank of 300 to 2,500 gallons were installed on apartment houses and hotels. The use of solar hot water systems flourished in Florida until 1941 when a government freeze on the nonmilitary use of copper severely affected the fabrication of a suitable absorber plate.

After World War II, the expanded use of dishwasher and washing machines in a more affluent society increased hot water requirements beyond the capabilities of prewar systems. Enhanced corrosion problems arose from the use of copper pipe and steel tanks and absorbers with higher water temperature requirements for the appliances. Additional economic convenience advantages with mass-produced electric water heaters and reduced electric power costs provided an appealing package to residential building developers. Solar water heating decreased as maintenance costs on existing solar water heating systems increased.

7.1.7 System Development and Demonstration

Initial use of site-specific selection and installation of available components and systems was employed early in 1976 in the National Solar Demonstration Program to gain experience in performance and to identify problems in existing residential and commercial systems. The success of these first cycle demonstrations was limited, however, by attempts to improperly adapt materials, components, or the early systems to different user requirements and climates. Subsequent development and residential demonstrations emphasized integrated system projects in which builders, developers, and solar energy system manufacturers and distributors combined to provide complete residential building systems.

The design, installation, and operation of large multifamily and commercial systems were largely site-specific, and except for components such as collectors and controllers, system designs were unique and assembled or built on site.

The integrated system philosophy enabled manufacturers to design packaged systems with particular climate, load, control, and installation features. By working with the building developer, installation and operational problems could be identified at an early date. This approach built confidence and understanding at levels important for long-term successful marketing. It is important to note that domestic solar hot water heating systems can be efficiently retrofit much more readily than space heating systems, and therefore designs evolved that are packaged but not constrained to installation during the building construction.

To provide consumer confidence, the production and marketing of these packaged systems was accompanied by design and performance criteria and standards (NBSIR 1980). Establishing the criteria involved the identification and development of standard test procedures to evaluate materials and components and to test system performance. The criteria were also used as a basis for the preparation of model or local building codes (DOE 1980) and federal specifications.

7.2   Generic System Types

Although domestic hot water heating is a relatively simple energy requirement, numerous system designs have evolved to meet thermal performance, durability/reliability, freezing, and building and site needs in all areas of the United States. Five basic types can be categorized as follows: passive (including integral collector storage, thermosyphon, and phase change), direct heating (service water heated in collector), indirect heating (service water heated in heat exchanger outside of collector), and air collector. (Phase change systems may also be considered indirect types because water is heated in an exchanger.) These categories can be further broken down to one- and two-tank storage configurations. The use of toxic antifreeze solutions requires double-walled heat exchangers for all DHW systems and some process heating applications. The method of freeze protection may use special controller functions to drain and vent frost-prone system plumbing.

Most SDHW systems are assemblies of the same types of components. Solar collectors and water tanks (sometimes in combination) and auxiliary heat sources are used in systems of all types. In all active systems, controllers and pumps are also used. The characteristics of these components seldom differ from one system to another, some of them are not even exclusively used in solar systems. Water storage tanks, auxiliary heaters, and controllers may therefore be described in general without associating them with any specific system type. Modifications and variations required for their use in particular systems may then be identified with those generic types.

7.2.1 Storage Tank Options

Individual performance characteristics of system components are described in chapter 8 and are discussed here only in general terms related to system design options. The influence of some storage options is summarized (Fanney et al. 1982) as follows:

- single tank vs. double tank configuration
- stratified vs. mixed tank performance
- nominal vs. improved thermal insulation

A reference direct heating SDHW system comprising a 4.1 m^2 (41.4 ft^2) flat-plate, single glazed, nonselective collector was used in a detailed computer simulation study to evaluate options in storage configuration. All storage tanks were 0.352 m^3 (80 gal) capacity with tank heat loss coefficients of 0.74 W/m^2°C (0.13 Btu/hr°F ft^2) and 0.403 W/m^2°C (0.071 Btu°F ft^2) corresponding to R–7 and R–13 thermal insulation blankets.

A hot water load of 0.308 m^3 (70 gal) per day with a heating range from 13°C (55°F) to 60°C (140°F), an ambient tank temperature of 15°C (60°F), and an hourly draw profile were assumed. Noon and evening peak draw profiles were used in the simulated studies.

Table 7.2 shows the results of simulations on eight combinations of tanks, insulation, and stratification. The data represent one complete year for a system located in Madison, Wisconsin. The influence of improved thermal insulation is illustrated by run 1B for a nonstratified double tank system, showing a 6.8% increase in the solar fraction. Examination of the results for the R–7 insulation case (typically used in electric storage tanks prior to 1980) reveals that twice as much heat was lost from the auxiliary tank as from the preheat tank and that the heat loss was equal to about 38% of the solar energy collected. Improvement in system performance can be achieved by minimizing thermal losses and, since auxiliary tank capacity has little effect on collector efficiency, the auxiliary tank size should be commensurate with hot water requirements. Stratification in the two-tank configuration resulted in 11.4% and 15.8% increases in solar fraction for the R–7 and R–13 cases, respectively. The primary effect of stratification is to reduce the collector inlet temperature and thus increase the collector efficiency.

The major improvement in single-tank system performance is obtained by achieving thermal stratification. As exhibited by run 3C with stratifica-

Table 7.2
Variations in thermal performance of reference SDHW system due to storage tank stratification and heat loss (Fanney, Thomas, Scarborough, Terlizzi 1982)

Run no.	No. tank	Preheat tank stratification	Heat loss coefficient Btu/hr°F ft²	Total energy reqd. 10^6 Btu	Solar fraction of total	Storage heat loss fraction			Storage tank Q_{Loss}/Q_{Solar}	Collector efficiency
						Preheat tank	Aux. tank	Total		
1A	2	nonstrat.	.130	21.28	.500	.063	.127	.190	.377	.474
1B	2	nonstrat.	.071	19.96	.534	.039	.076	.115	.213	.465
1C	2	nonstrat.	0.0	17.69	.586	0.0	0.0	0.0	0.0	.453
2A	2	stratified	.130	21.74	.552	.055	.127	.182	.321	.534
2B	2	stratified	.071	20.90	.618	.032	.072	.104	.178	.529
3A	1	nonstrat.	.130	20.87	.323	0.0	.136	.136	.405	.303
3B	1	stratified	.130	19.45	.600	0.0	.082	.082	.135	.513
3C	1	stratified	.071	18.80	.614	0.0	.047	.047	.075	.508

tion and R–13 insulation, the best overall performance is achieved in terms of total energy required and minimum heat loss. An investigation of methods to enhance stratification in a single tank with a single heating element located in the upper portion was demonstrated with the use of a special inlet tube (Fanney et al. 1982). The tube, shown in figure 7.11, has alternate, equally spaced holes down the inner and outer tubes. It was experimentally shown that a stratified tank with this tube produced system fractional energy savings 8.7% higher than a tank with a plain tube to the same tank depth. The slower temperature decay rate in the upper portion of the tank results in lower auxiliary energy usage for the system when the tank is stratified.

Figures 7.12a and 7.12b illustrate the excellent thermal stratification obtained in a closed-loop indirect heating residential SDHW system using silicone as the heat transfer fluid (Jackson 1979). The 120-gallon preheat storage tank was equipped with a finned tube heat exchanger and operated at a flow rate of 5 gallons/minute (19 l/min).

7.2.2   Gas Backup Heater Development

Achieving stratification in a conventional single tank heated by gas or oil auxiliary is not normally possible because these tanks are constructed with the heat input at the tank bottom; free convection results in circulation of the heated water and a fully mixed tank. An analytical and experimental study of methods to obtain an efficient backup gas-fired heater was conducted by Morrison (1980). A schematic of an engineering prototype design comprising two storage tanks, a gravity return water heat pipe, and a unique side-flue type combustion chamber assembled into a thermally insulated container is shown in figure 7.13.

Stratification in the tanks was enhanced by using an offset "split-tube" connecting pipe and by the natural heat input into the demand tank by the heat pipe. Temperature differences as high as 32°C (90°F) were obtained between the bottom and the top of the tank. Burner efficiencies of about 80% were measured under typical DHW loads and operating conditions.

7.2.3   Controller Strategies

The primary function of the controller in a SDHW system is the efficient collection of solar energy by the controlling pump or blower settings. Other functions involve operation of components for freeze protection, high temperature protection, and special functions such as auxiliary en-

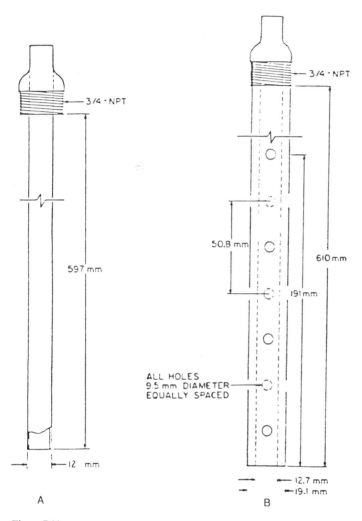

Figure 7.11
Comparison of standard hot water tank return tube (A) with modified tube (B) used to
enhance temperature stratification (Fanney, Thomas, Scarborough, Terlizzi 1982).

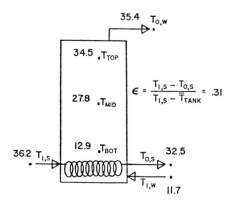

Figure 7.12a
Stratification and heat exchanger perfomance (°C) for February (Jackson 1979).

JUNE

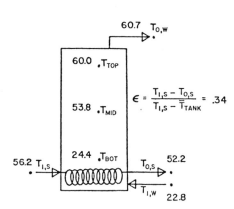

Figure 7.12b
Stratification and heat exchanger performance (°C) for June (Jackson 1979).

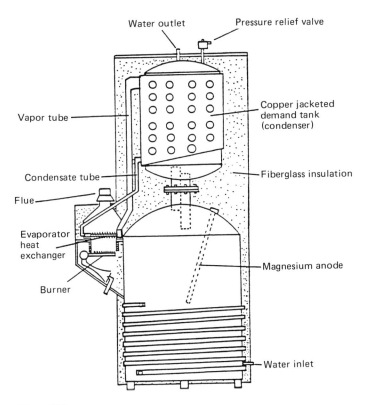

Figure 7.13
Schematic of gas back-up heater for solar water heating (Morrison 1980).

ergy supply, indication display, and optional manual control for energy conservation. A brief discussion of flow control technique is presented because of its impact on system efficiency and parasitic energy cost. Controller functions for 18 generic SDHN systems and complete description of strategy and equipment adequate for system design, procurement, installation, and operation are presented in SERI (1981).

SDHW flow control is achieved by two types of differential controller outputs. The most commonly used "bang-bang" or on/off controller operates the collector pump or blower either full on or off by switching the power on and off. Less common is the variable-flow controller, which modulates the power supplied to the pump or the position of a valve during the "on" time to make the flow rate a function of the difference in

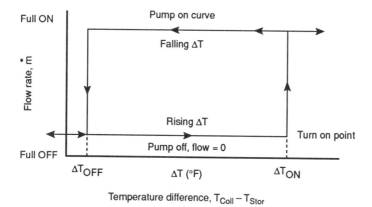

Figure 7.14a
ON-OFF controller characteristics (SERI 1981).

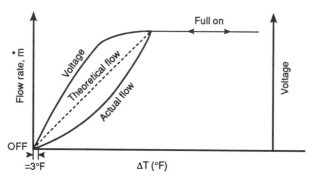

Figure 7.14b
Variable flow controller characteristics (SERI 1981).

temperatures of collector and storage. The flow characteristics of each type of controller are shown in figures 7.14a and 7.14b.

The trade-off in selecting the on/off temperature set points used in a "bang-bang" control depends on several system characteristics. These include collector performance, heat exchanger effectiveness, piping thermal capacitance system flow rate, system thermal losses, temperature transducer placement, accuracy, and precision. The placement of the temperature transducer (probe) in the collector is particularly important. Analytical results, generally confirmed by field experience, show the

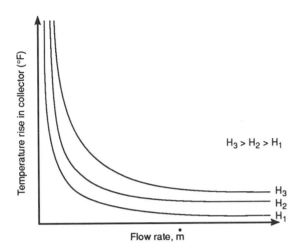

Figure 7.15
Qualitative representation of temperature rise as a function of flow rate and solar radiation, H. (SERI 1981).

collector probe should be near the collector outlet at approximately 90% of the length along the flow path in the collector.

The dead band (hysteresis) between the temperature off and on setting, shown in figure 7.14a, minimizes pump cycling or "hunting" during morning startup and afternoon shutdown or intermittent cloud conditions. The "off" setting must be high enough to accommodate normal sensor inaccuracies but low enough to allow collection of as much solar energy as possible. Collection should cease when the parasitic power for the pump exceeds the solar energy gain. Controller industry experience has shown that if an on-to-off temperature ratio of 4/1 to 6/1 is maintained with the ΔT_{off} temperature ranging from 2°F to 4°F, efficient operation will result in most types of flat-plate liquid transport systems. It is found that if ΔT_{off} is at approximately 2°F, then ΔT_{on} would be approximately 12°F. Similarly, the lower value of the ratio of on-to-off temperature differentials is applied when ΔT_{off} is near its upper range of values. The hysteresis range for most liquid systems is in the range of 12°F to 16°F, with some as high as 20°F.

The qualitative relationship between mass flow rate ($\dot{m}$) and solar radiation (H) on the ΔT choices is illustrated in figure 7.15. At low mass flow rates, the rise in fluid temperatures is greater than at high flow rates, resulting in larger values of ΔT_{on} and ΔT_{off}. Similarly, in regions of high solar radiation, the ΔT values could be increased. The values should remain within the limits stated above.

Air transport systems normally require ΔT_{off} temperatures at $10°F$ and as high as $20°F$ with a $15°F$ to $25°F$ hysteresis. The wide hysteresis is required to compensate for the effects of relatively low air specific heat and the lower efficiency of blowers compared to pumps.

The variable-flow (proportional) controller's major advantage is derived from the essential elimination of pump cycling and the use of a longer collection duty cycle during cloudy weather than can be obtained with low speed pump or blower operation. The theoretical flow, shown in figure 7.15b, is generally not achieved because of the third power relationship between electric current and fluid flow rate, the nonlinear I^2R losses in the pump motor, and highly variable system head characteristics. Therefore, the actual flow rate usually varies with motor voltage, as shown in figure 7.14b. The motor duty cycle interruption used to vary the voltage is a ramp function of the collector-to-storage ΔT with a ΔT of $2°F$ or $3°F$ as a low setting, and $10°F$ to $16°F$ as a high setting.

The low fluid flow rate initiated at low turn-on temperature allows for a longer period of solar energy collection than would be achieved with a "bang-bang" controller set at a high ΔT_{on}. Thermal shock to susceptible components such as vacuum tube collectors is reduced. Better temperature stratification in storage and more uniform hot water delivery temperature can usually be obtained. Claims have been made for increased energy collection of about 7% for a low thermal capacitance collector under partly cloudy conditions.

The main advantage attributed to the variable flow rate controller is achieved by increasing the difference between useful energy collected and pumping energy used. The increase in electric COP results from the nonlinear relationship between parasitic losses and flow rate. As the flow rate increases, the parasitic losses typically increase as a function of the flow rate to the x power, where x is between 2 and 3, as illustrated in figure 7.16. The increase in power with flow rate is depicted by the curve with increasing slope, and the rate of useful energy collection is shown by the concave downward curve. The most efficient flow rate, i.e., the rate corresponding to the highest electric COP, would be at the point of greatest separation between the curves.

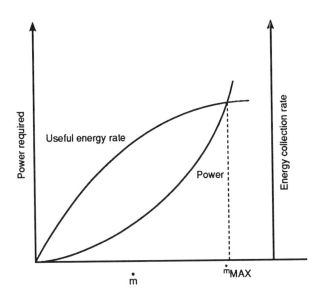

Figure 7.16
Power required and rate of energy collected as functions of flow rate (SERI 1981).

The main disadvantage of variable flow rate systems is the greater expense of the pump and power control. Only pumps with a series wound motor are safe for operation. Other drawbacks include the reduced reliability of field wired output devices and the generation of electromagnetic interference by the chopped-wave modulated controller, which can interfere with radio and television reception. The full-wave modulated output is not recommended for air systems because of a high occurrence of pulsed duct noise.

Parasitic energy consumption can be substantial and must be considered for energizing automatic valves, controller activation, as well as pump or blower operation. An analysis of the parasitic energy consumed by six different systems operated under identical load and climatic conditions resulted in significant power usage by all systems. Table 7.3 shows that a liquid double-tank direct system and an air system required seven times as much electric energy as a thermosyphon system. Table 7.4 shows that parasitic energy used over a 20-year period can add only 2.8% to the cost of hot water in a thermosyphon system but as much as 44% in an air system.

Table 7.3
Parasitic energy for various SDHW systems

System	Pump operation (h)	Calculated[a] energy consumed by pump (kWh)	Days of testing	Calculated[b] energy consumed by solenoid valves (kWh)	Calculated total (kWh)	Measured total (kWh)
Single tank direct	681.7	68.2	127	91.4	159.6	125.3
Double tank direct	882.9	88.3	127	91.4	179.7	145.3
Single tank indirect	690.2	69.0	127	NA	69.0	68.5
Double tank indirect	870.5	87.1	127	NA	87.1	71.0
Air system	644.9	64.5 (+48.2 Fan)[c]	127	NA	112.7	145.0
Thermosyphon	NA	NA	121 (45 days w/freeze protection)	32.4	32.4	19.4

Source: Farrington, Murphy, Noreen 1980.
a. Energy consumed by pumps = hours of pump operation × 100 W.
b. Energy consumed by each solenoid valve = days of testing × 15 W × 24 h.
c. Energy consumed by fan in air system = hours of pump operation × 75 W.

Table 7.4
Influence of parasitic energy on energy cost for 6 SDHW systems over 20 years for collector cost of $162/m² ($15.00/ft²)

	Without parasitics			With parasitics		
System	$/GJ	($/MBtu)	Relative ranking[a]	$/GJ	($/MBtu)	Relative ranking[a]
Thermosyphon	11.93	(12.58)	1.00	12.25	(12.92)	1.00
Single tank direct	15.11	(15.94)	1.27	18.70	(19.73)	1.53
Single tank indirect	20.37	(21.49)	1.71	22.63	(23.87)	1.85
Double tank direct	21.07	(22.23)	1.77	27.18	(28.67)	2.22
Double tank indirect	25.37	(26.76)	2.13	28.62	(30.19)	2.34
Air system	41.56	(43.84)	3.48	59.95	(63.23)	4.89

Source: Farrington, Murphy, Noreen 1980.
a. Normalized system costs with lowest system cost equal to 1.00.

7.2.4   System Types—Residential

7.2.4.1   Passive Designs

7.2.4.1.1   INTEGRAL STORAGE COLLECTOR   Integral storage collector (ISC) designs range from a single black tank laid horizontally in an insulated box (sometimes referred to as a bread box or batch heater) with a transparent cover, to a series of interconnected tanks with a total storage capacity of 0.075 m³ (20 gal.) to 0.227 m³ (60 gal.) of directly heated water, which can be used as a preheat tank for a conventional DHW tank as depicted in figure 7.17. Simple designs can be used year around in warm climates directly or as preheaters with cold weather drainage. Limited freeze protection is achieved by integral roof mounting, double or triple glazing, automatic drip valve, freeze tolerant tank construction, or integral electrical heating. Practical considerations restrict their use in cold climates, since they generally have no positive freeze protection other than by manual draining.

7.2.4.1.2   THERMOSYPHON   The simplicity of installation, operation, and maintenance has resulted in a resurgence of thermosyphon systems in mild U.S. climates. Although similar in concept to the early designs employed in the United States between 1910 and 1930, the availability of new materials for selective absorbers, improved insulation, and better seals have contributed to improved efficiency and durability. Thermosyphon systems used predominantly in Australia, Japan, and Israel are described in Morse

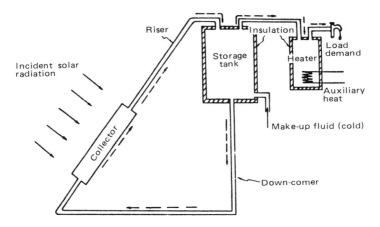

Schematic of Two-Tank Thermosyphon SDHW System

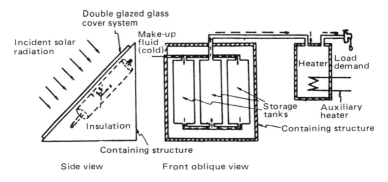

Schematic of Integral SDHW System

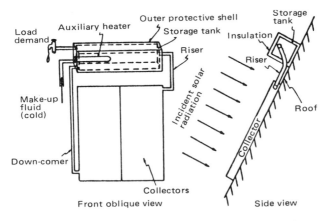

Schematic of One-Tank Thermosyphon SDHW System

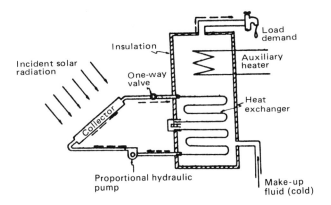

Schematic of One-Tank Pumped SDHW System

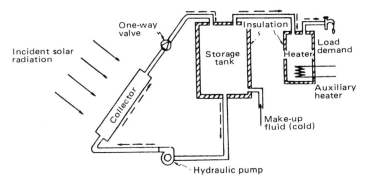

Schematic of Two-Tank Pumped SDHW System

Figure 7.17
Schematics of integral, thermosyphon and pumped SDHW systems (Farrington, Murphy, Noreen 1980).

and Close (1977). The thermosyphon heating system utilizes the change in water density with temperature to induce circulation of heated water from the collector to storage. As illustrated in figure 7.17, the storage tank must be elevated at least 45 cm (18 in.) above the collector to obtain adequate circulation when solar energy is absorbed. Properly sized headers and a minimum of tubing bends can produce circulation flow rates of about 0.011 kg/s m^2 (15 lb/hr ft^2) with a temperature rise of about 8°C (15°F) per pass through the collector. The system can be operated either at main pressure or in a gravity flow mode.

The storage tank can be separate from the collector and located outside or inside the building, or can be combined and mounted adjacent and above the collector. Typical combined units employ about 1.40 m^2 (15 ft^2) to 2.79 m^2 (30 ft^2) of collector area and 0.075 m^3 (20 gal.) to 0.113 m^3 (30 gal.) storage capacity. Multiple units can be installed in series or parallel to provide faster recovery, higher temperature water, or greater capacity to meet the hot water requirements. Temperature stratification in the storage tank is desirable for improved collector efficiency. Judicious design of the collector inlet and outlet and the hot and cold water line locations and configuration are important.

Some type of freeze protection is desirable even in the lower latitude areas of the United States. Complete draining of the collector can be achieved by utilizing solenoid valves, controller, syphon breaker and air vent. One solenoid valve closes the collector inlet and drains the collector when actuated by the controller at about 3°C (38°F) exterior temperature. A check valve in the piping leading from the collector to the tank prevents reverse thermosyphoning as well as tank drainage. The added system complexity and dependence upon electrical power for solenoid operation has resulted in the use of antifreeze loops with heat exchanger or thermostatically controlled drain valves as preferred methods. Freeze protection for colder climates has been achieved in thermosyphon systems by using a freeze-resistant fluid in the collector, which also circulates through a thermally sealed jacket around the outside of the storage tank as shown in figure 7.18. The jacket serves as a heat exchanger and added thermal insulation to reduce the heat loss from storage. An electrical heating element is provided in the storage tank for use as a single-tank system.

In all systems used in cold climates, water piping exposed to outdoor conditions (supply to the SDHW system connections between collector

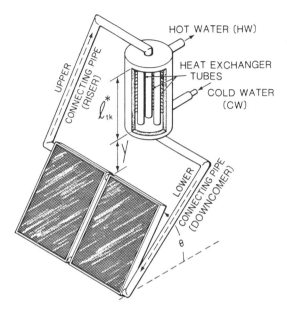

Figure 7.18
Thermosyphon SDHW system with a heat exchanger in the storage tank (Mertol, Place, Webster, Greit 1981).

and tank, and hot water delivery piping) must be protected from freezing by use of electric heat tapes or by automatic drip valves to maintain at least a low flow of supply water through the system during cold, sunless periods.

7.2.4.1.3 PHASE CHANGE   A nonfreezing and noncorroding refrigerant with a boiling point in the range of 4°C (39°F) to 23.7°C (75°F) and a freezing point below -94°C (-137°F) can be used as the heat transfer fluid in a solar water heating system. Figures 7.19a and 7.19b show systems used in the United States. The system shown in figure 7.19a consists of a flat-plate collector and storage tank with an inside heat exchanger (Best 1981). The absorbed solar radiation causes the refrigerant (R–114 in this case) to vaporize in the collector and rise into the heat exchanger where the vapor delivers heat by condensing, the liquid flowing by gravity back into the collector. An adjustable pressure limiter valve is set to a value of about 607 kPa (88 psi) equivalent to stored water temperature of 68°C (155°F).

REFRIGERANT SOLAR COLLECTOR SYSTEM

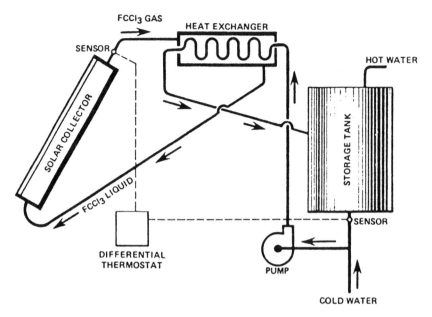

Figure 7.19a
Phase-change type thermosyphon solar heater with heat exchanger in storage tank
(Best 1981).

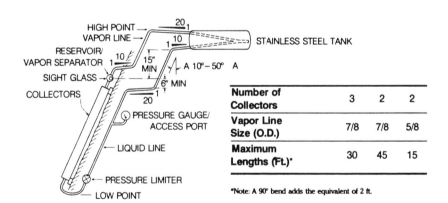

Number of Collectors	3	2	2
Vapor Line Size (O.D.)	7/8	7/8	5/8
Maximum Lengths (Ft.)*	30	45	15

*Note: A 90° bend adds the equivalent of 2 ft.

Figure 7.19b
Refrigerat (phase-change) type collector with external heat exchanger (Schreyer 1981).

The system shown in figure 7.19b utilizes R–21 (trichlorofluoromethane) in the collector loop (Schreyer 1981). A differential thermostat control measures the refrigerant vapor temperature at the top of the collector and compares it with the water temperature in a constant temperature reservoir. When the vapor temperature is greater than the reservoir temperature by 6°C (11°F), the circulating pump turns on; when the temperature differential falls below 1.66°C (3°F), the pump is turned off. Water is circulated through the heat exchanger at a rate of 3.78 l/min (1 gpm).

As in other thermosyphon type systems, the collector must be located at least 38 cm (15 in.) below the storage tank or the heat exchanger. Piping must be designed and installed without liquid or vapor traps. Tight silver soldered joints and cleanliness prior to and during system fabrication and charging with refrigerant are necessary. Concern about the compatibility of the refrigerant and ozone or other environmental effects must be recognized.

7.2.4.2 Pumped Circulation Systems

7.2.4.2.1 DIRECT HEATING SYSTEMS

Single and Double Tank Designs. One and two-tank pumped SDHW systems without provisions for freeze protection are also shown in figure 7.17 for comparison with the integral tank and thermosyphon systems. Modifications to these relatively simple systems are described in the following Sections for applications made in colder US climates and commercial systems.

Single Tank, Draindown Design. In a single-tank, direct heating system, depicted in figure 7.20, potable water is circulated through the collector to a single storage tank. Efficient operation is dependent upon developing thermal stratification in the storage tank; consequently, the auxiliary heating source (normally an electric element) is located in the upper portion of the tank. Cold water is supplied to the collector from the bottom outlet of the tank as solar heated water enters the tank near its center. Valves for relief, balancing, venting, and direction control are used as required. The control starts the circulator when temperature in the upper portion of the collector is about 9°C (16°F) above the lower portion of the storage tank. Circulation is stopped when the temperature difference decreases to about 2°C (4°F).

228

Elmer Streed

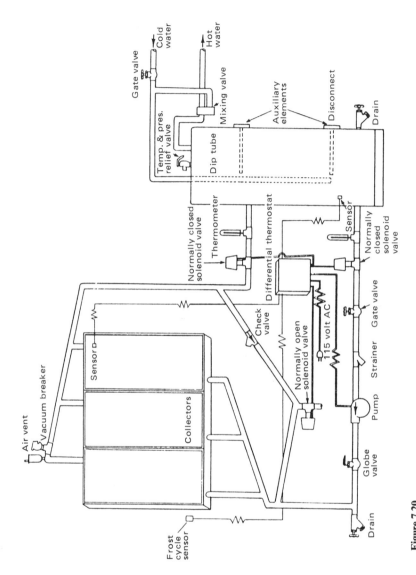

Figure 7.20
Schematic for direct heating draindown system (Franklin Res. Ctr. 1979).

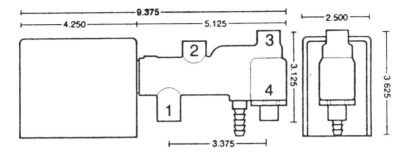

Figure 7.21
Illustration of motorized unit to control three valves simultaneously (Sunspool Corp 1976).

Freeze protection can be provided by controlling motorized (usually solenoid operated) valves, as illustrated in what is commonly termed a "draindown" design. One valve closes the supply to the collectors, another opens the drain valve, and the third closes the line from the collectors to the storage tank, thereby maintaining pressure in the tank as the collectors drain. Air enters the collectors through an automatic vacuum breaker valve. A commercially available spool valve serving all three functions simultaneously is illustrated in figure 7.21. Air vent and vacuum relief valves are located at the highest points in the system to facilitate filling and draining. An unpressurized open loop drainback system has an air space in the top of the vented solar tank. When the pump stops, the collector drains into this space as the air displaces the water in the collector. A booster pump is used in the hot water service line to provide service pressure.

Solar Tank, Auxiliary Tank, and Automatic Draindown. This system consists of collectors connected in parallel or series, a solar storage tank, an auxiliary hot water heater, flow control valves, an on/off differential temperature controller with freeze protection circuitry, and a circulating pump. This system is shown schematically in figure 7.22. Its operation is not as dependent on thermal stratification as the single-tank system, but collector efficiency improves with stratification in the preheat (solar) tank. The system incorporates provision for freeze protection by automatically draining the water from the collector and exposed lines as in the single-tank system. The freeze control system may use an exterior-mounted thermostatic switch to sense the ambient temperature, or several sensors may

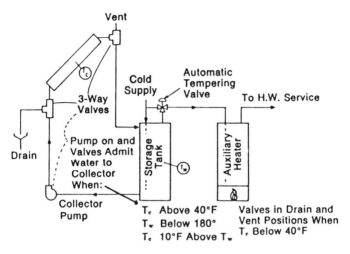

Figure 7.22
SDHW system in series with convential water heater (DOE 1978).

be attached to the first collector supply manifold and the last collector outlet manifold. The sensors activate normally closed valves, which then open with the proper signal or when there is a power failure. The collector piping must be corrosion resistant, or, in some instances, large systems may be automatically supplied with an inert gas (nitrogen) during drain-down to minimize oxidation effects.

To eliminate the possibility of overheated water at points of use, a mixing (tempering) valve is usually installed in the hot water line. A typical mixing valve, shown in figure 7.23, blends cold water with hot water to a maximum temperature of about 140°F.

Recirculation Systems (Single Tank and Auxiliary Heater Types). In mild climates where freezing temperatures seldom occur, solar collectors and water piping may be protected by circulating water from the storage tank through the collector during unusually cold weather. Even though the water loses heat during those periods, the simplicity of these systems and the low frequency of freezing conditions justify the use of these recirculation designs in appropriate climates, such as in Florida and southern California.

The equipment in recirculation systems is essentially the same as in the draindown types except for controls and valves. Instead of freeze sensors

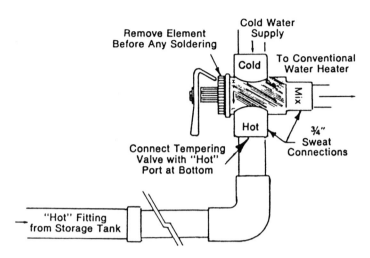

Figure 7.23
Typical tempering value assembly (DOE 1978).

actuating drain valves, check valves, and vacuum relief valves, the control maintains circulation through the collector when ambient temperatures or temperatures in the collector fall below preset levels.

As with draindown systems, the reliability of this freeze-protection technique is subject to the accuracy and durability of the controller and freeze sensors, and, in addition, to the continuous availability of electric power for pump operation. Power outages, sometimes coincident with severely cold weather, are a risk factor that must be considered.

A variation on the recirculation design, which is sometimes used in mild climates, provides a low flow of water from the mains through collectors and piping during freezing weather. The low temperature sensor and controller activate a drip valve that allows sufficient water flow to waste to prevent ice formation in the collector piping. A valve directly activated by low temperature can avoid dependence on electric supply to provide protection.

7.2.4.2.2 INDIRECT HEATING SYSTEMS   Indirect heating SDHW systems were developed to overcome the lack of reliability associated with draining liquid collectors and piping to avoid damage from freezing. Instead of circulating the service water through the collector, another fluid, such as nonfreezing liquid or air, is heated in the collector and pumped through a heat exchanger where the energy is transferred to potable water. Another indirect type is the drainback system, in which water in a closed loop is the collection medium from which heat is transferred to potable water in an exchanger; freezing is avoided by draining the collector into a small reservoir.

Indirect Anti-Freeze Systems.   Examples of single-tank and double-tank indirect solar water heaters in which a nonfreezing liquid is circulated through the collector and heat exchange coil (in tank) are shown schematically in figures 7.24 and 7.25. External counterflow heat exchangers (requiring a second pump for circulating water from the solar tank through the exchanger) are also used. Pumping requirements, heat transfer characteristics, maintenance, and initial costs are factors to be considered in selection of the heat transfer fluid and freeze protection. Glycol solutions with corrosion inhibitors are commonly used, and systems employing mineral oils and silicone oil have been commercialized. An external heat exchanger with pressure relief valve, vent valve, pressure gauge, pump, and controls can be obtained as a prepackaged unit as shown in figure 7.26.

Expansion of the heat transfer liquid when heated in a closed loop must be accommodated by use of an air cushion tank. In a diaphragm-type air cushion tank, the fluid is separated from the air, thereby providing a relatively maintenance-free system. The diaphragms are usually large enough to cover the bottom and one-half of the sides of the tanks as shown in figure 7.27. Expansion tank volumes in typical residential-size systems are 3 to 5 gallons. The tanks are usually precharged on the air side of the diaphragm to a pressure of about 6 psig. Typical tank sizes as a function of the number of gallons of fluid in the loop are available from SMACNA (1978). If an antifreeze solution such as ethylene glycol is used with the water, expansion requirements are larger. Correction factors for various concentrations are also available from SMACNA (1978).

Drainback Systems.   In drainback systems, the heat transfer medium is pure water in a closed loop, automatically draining by gravity into a partially filled tank. A heat exchanger is necessary because the city water pressure on the potable water side would otherwise prevent draining. An indirect system was developed by Embree (Best 1982) to provide freeze protection without the use of check valves and electrically operated sole-

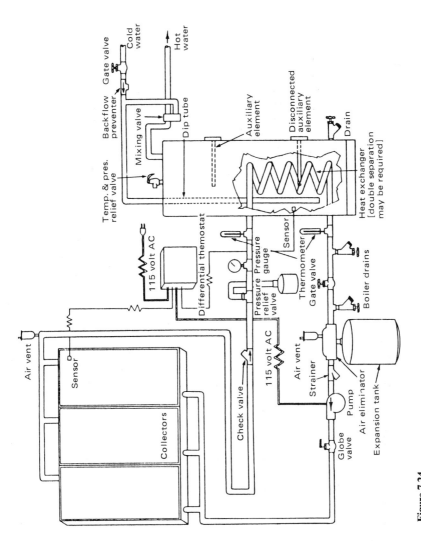

Figure 7.24
Single tank indirect system (Franklin Res. Ctr. 1979).

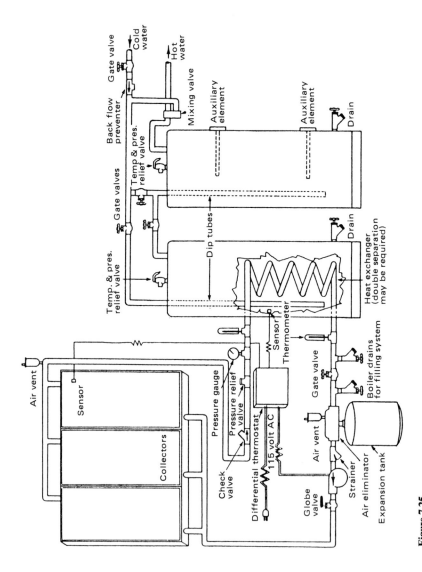

Figure 7.25
Double tank indirect system (Franklin Res. Ctr. 1979).

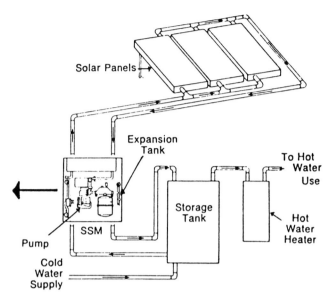

Figure 7.26
Prepackaged solar heat transfer module for indirect system (DOE 1978).

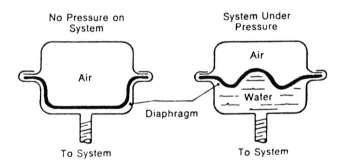

Figure 7.27
Diaphram air cushion (DOE 1978).

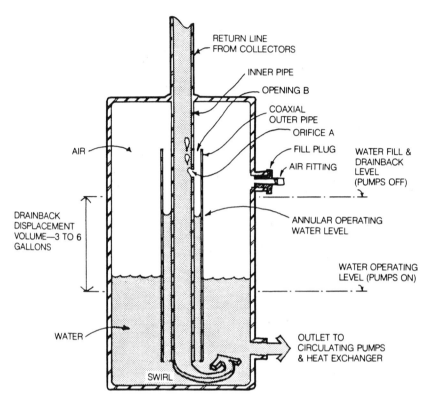

Figure 7.28
Air/water reservoir (sump tank) for use in drainback system (Best 1982).

noid valves. The pressurized closed loop technique illustrated in figure
7.28 employs a small, partially filled reservoir, or sump tank, with a capacity of about $0.0227m^3$ (6 gal.), located in the return line from the collectors.
The collector return line extends below the operating water level in the
reservoir. Spill flow from orifice A is transferred to the annular operating
water level within the coaxial outer pipe. Heat is delivered to the service
water in an exchanger coil either in the reservoir or in the solar hot water
storage tank. When the controller shuts off the circulating pump, air flows
from the tank through opening B and orifice A to break the vacuum in
the inner pipe and in the collector return line. Air rises to the collector
through the return line, and water drains from the collector into the reservoir. By maintaining a closed system, problems with corrosion, venting,

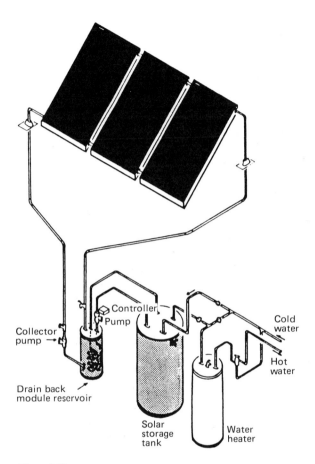

Figure 7.29
Indirect drainback system with separate reservoir and heat exchanger container (Solaron Corp. 1980).

and vaporization are reduced and quiet efficient operation results. Positive drainage not requiring electric power or temperature sensing provides total protection from freezing.

In the most widely used preassembled systems, the heat exchanger coil is located in the drainback reservoir, as shown in figure 7.29. The advantages of this system are:

• completely closed system

• efficient high thermal capacitance water for heat transfer

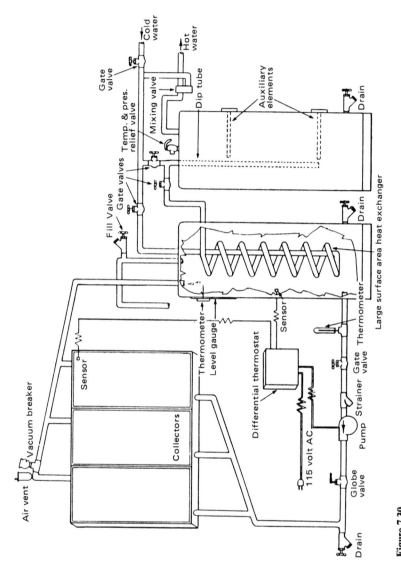

Figure 7.30
Indirect drainback system (Franklin Res. Ctr. 1979).

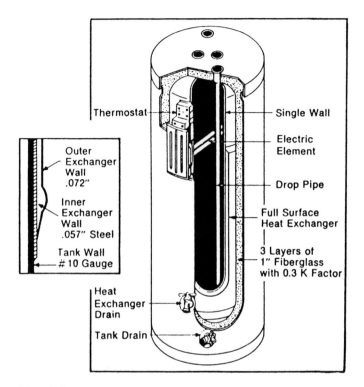

Figure 7.31
Coil inside storage tank and wrap around heat exchangers (DOE 1978).

- failsafe protection from freezing and overheating
- versatile installation and adaptability to space heating

Another drainback design, illustrated in figure 7.30, involves a solar water tank of typical size (80 to 120 gal.) containing water in a closed loop and an air space at least as large as the volume of water in the collector and associated piping. The tank also contains a large heat exchanger coil through which city water enters the system. The pump circulates the heat-transfer water through the collector and the partially filled tank; when hot water is drawn from the system for use, cold water flows into the exchanger coil, where it is heated. When the pump stops, air from the top of the tank displaces the water, which drains from the collector into the tank.

If nonpotable materials (corrosion inhibitors, antifreeze solutions) are added to the water in the collector loop, most plumbing codes require a double-walled heat exchanger, such as a shell-and-tube type in the collector loop or a "wrap-around" heat exchanger on the storage tank, as illustrated in figure 7.31.

Air Systems Replacement of the liquid heat transfer fluid and collector with an air collecting and circulating system offers advantages in freezing climates. However, liquid problems, such as corrosion, freezing, leaks, and valving, must be weighed against blower power costs, heat exchanger efficiencies, and air tightness. A typical air system schematic is shown in

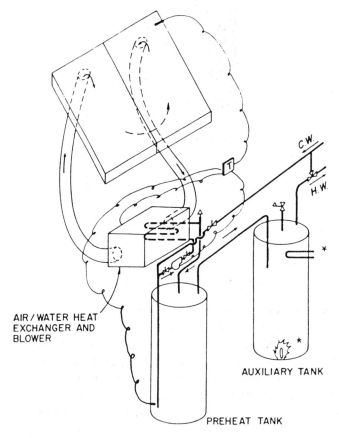

Figure 7.32
Indirect solar heating with air collector and air/water heat exchanger (Fanney, Klein 1983).

figure 7.32. The hot air passes through an air-to-water heat exchanger. Water is circulated through the heat exchanger and back into the preheat tank when sufficient energy is available as sensed by a solar radiation sensor or a collector temperature sensor. A preheat tank system can be used as shown with gas, oil, or electric auxiliary, or a single tank system may be used with electric auxiliary.

7.3 Operating Systems

7.3.1 General

This section is devoted to describing solar hot water systems that were designed, installed, and operated in various parts of the United States by the solar industry. Systems are categorized into single-family residential, multifamily residential, and commercial. Their performance has been monitored with varying amounts of data reported. The descriptions include site location, anticipated load, and SDHW components. Unique features, such as special controls, heat loss, and loads, are discussed where pertinent and available.

7.3.2 Residential, Single-Family

7.3.2.1 Integral Storage Collector

This passive-type SDHW system (Cramer, Spears, Pollack 1980; Nash, Cunningham 1979) consists of two 30-gallon uninsulated steel tanks. The tanks were painted black, positioned in a south-facing wall and enclosed with double glazing having a gross area of 44 ft^2. Domestic hot water is preheated in the tanks and, upon demand, flows to a conventional gas-fired domestic hot water tank. Reflective insulation around the sides of the enclosure enhances the collection of energy. Active freeze protection is provided by small beads of foamed plastic blown into the space between the two glass covers. The system was site-built in Longmont, Colorado.

On-site solar and temperature measurement made during one year of testing, indicates about one-third of the incident solar energy was collected. Approximately 15% of the collected energy was used for preheating service hot water, whereas the principal use was for space heating. The losses from this system are relatively low because of low operating tank temperatures.

7.3.2.2 Direct Heating

7.3.2.2.1 SINGLE-TANK SYSTEM   One of the simplest types of active direct heating single-tank systems, is located in Kaneohe, Hawaii. The four-panel selective black, single-cover collector array mounted in two rows has a gross collecting area of 77 ft^2. The panels face 10 degrees east of south at a tilt angle of 22 degrees. The water-circulating pump operates with a flow rate of 2.3 gallons per minute. Potable solar heated water is stored in a conventional 120-gallon hot water tank with a 4-kWh electrical heater. No freeze protection is required at this site.

The energy-flow measurements indicate that about 29% of the solar radiation was collected and that subsequent heat losses were about 25% of the total collected. The system was designed to provide 95% of the annual DHW load and is actually providing about 91%.

7.3.2.2.2 TWO-TANK SYSTEMS   Solar DHW systems with two tanks have experienced the greatest use because of the large number of retrofit installations in which the conventional DHW storage tank was not satisfactory for solar preheat but could provide the auxiliary function.

A two-tank direct heating SDHW system is installed at Rancho Santa Fe, California. The open-loop system contains a 68 ft^2/flat, black collector array, a 120-gallon preheat tank, and a conventional 40-gallon gas-heated hot water tank. The collector array faces 25 degrees west of south at a tilt angle of 34 degrees. Energy-flow measurements for operation of this system from October through December 1981 showed collection efficiency of 46%, the system provided 46% of the energy required for water heating.

Two prototype SDHW systems developed under the NASA Advanced System Development Program were installed in San Diego, California, and Tempe, Arizona (Elcam 1980). These systems featured a cascade control to minimize the use of auxiliary energy. The operational characteristics and performance of six systems were initially analyzed. Two systems with heat exchangers, external and internal, with and without dump tanks, a direct dump option, and one direct heating system without freeze protection were compared. A draindown system with a direct dump unit, three-port valve, pressure switch, vacuum relief valve, check valve, and air vent valve, shown in figure 7.33, was selected for installation. Other system components included 64 ft^2 of flat black collector facing south at a 30 degree tilt angle. The total storage capacity of the two tanks was 104 gallons. Performance results are presented in Elcam (n.d.).

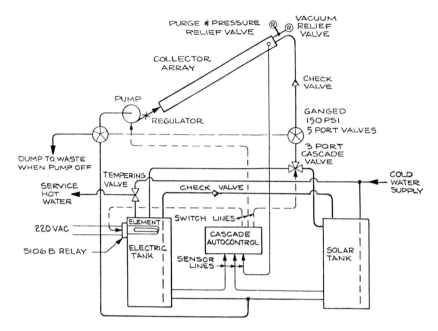

Figure 7.33
Direct heating system with cascade auto-control and direct dump (Elcam Inc. 1980).

7.3.2.3 Indirect Heating-Drainback System

A drainback system is installed on a single-family residence located in East Meadow, Long Island, New York. It consists of 48 ft² of flat, black double-cover collectors and a conventional electric hot water tank which is equipped with a solar hext exchanger. A drainback reservoir is located between the collector and pump. The collectors face south at a tilt angle of 36 degrees.

Off-peak power is used from midnight to 7:00 A.M. for auxiliary heating in the upper two-thirds of the tank. The energy measurements show that during four fall months, collection efficiency averaged 44%; 64% of the hot water demand was met by the solar supply.

7.3.3 Residential, Multifamily

7.3.3.1 Thermosyphon

The Hei Wai Won site is a four-story apartment building in Honolulu, Hawaii. Three separate thermosyphon systems operating at city water pressure are used to provide hot water for the building. This description is for one of these systems, which supplies hot water to six apartments and one laundry room.

The collector array consists of 807 ft^2 of flat, black single-glazed panels facing due south at a tilt of 24 degrees. The collected solar energy is stored in a 1,230-gallon tank mounted on a tower above the collectors. No freeze protection is needed in this climate. Conventional 30-gallon electric hot water tanks are provided in each apartment, and an 85-gallon tank with electric auxiliary is used in the laundry. Experience has shown that the system is overdesigned for the current DHW demand, so the auxiliary tanks are turned off during most of the year.

Energy measurements for system operation from September 1980 through April 1981 indicates a solar collection efficiency of 28%. The SDHW system has supplied about 98% of the hot water energy requirements. Heat losses from the solar energy systems transport and storage are about 20% of the collected energy; losses in the distribution system are an additional 27% of the collected energy.

7.3.3.2 Direct Heating

7.3.3.2.1 SINGLE-TANK SYSTEM   A relatively simple single-storage-tank system with an in-line gas-fired boiler is integrated into a 101-unit, four-story apartment building in Albuquerque, New Mexico, (Cramer, Spears, and Pokack 1980). The SDHW system is operated independently from a solar hot water space heating system. A 2,478-ft^2 tracking concentrator collector array provides heated water to a 1,740-gallon storage tank. Domestic hot water is continuously circulating through the building.

Energy measurements for the 5-month period from October 1980 through February 1981 indicated that about 27% of the radiation on the aperture of the collectors was used: 17% was delivered to storage and 10% to the hot water distribution system. Approximately 35% of the energy supplied to the hot water is lost in the distribution system.

7.3.3.2.2 TWO-TANK SYSTEM   SDHW systems for larger applications generally employ two or more storage tanks because of distribution or freeze considerations. The direct heating type has advantages of simplicity, thermal effectiveness, and lower cost. Disadvantages include freeze-protection complexity, corrosion susceptibility, and potential water contamination. A large, well-insulated storage with good thermal stratification serving as a preheat tank can be operated with the collector array to maintain low collector inlet temperatures and good collection efficiencies.

Multifamily buildings such as apartment houses or condominiums in mild climates are good applications for these systems. A 31-unit, three-story condominium located in San Diego, California, has a direct heating system with a 520-ft^2 flat, black single-glazed collector array facing south at a tilt angle of 42 degrees. A main preheat 1,000-gallon glass-lined storage tank is located underground. The preheated water is supplied, on demand, to 31 conventional electric storage tanks in each dwelling unit.

The energy measurements for 12 months of operation during 1980 shows a solar collection efficiency of 43%. Total energy loss from storage and the hot water subsystem was about 29% of the total solar and auxiliary energy supplied. A system solar fraction of 36% was achieved for the year.

7.3.3.3   Indirect Heating

7.3.3.3.1   IN-LINE COLLECTOR HEAT EXCHANGER   An indirect SDHW system is used in a ten-story, 101-unit apartment building in Burlington, Vermont. The 2,342-ft^2 collector array is oriented 42 degrees west of due south at a tilt angle of 45 degrees. Freeze protection is achieved by pumping a 67% propylene glycol/water solution through the collectors and an in-line water heat exchanger. The collected solar energy is stored in a 2,699-gallon water tank in a closed loop (figure 7.34). The system also has an auxiliary 865-gallon tank with internal heating coil supplied from two gas-fired boilers.

A two-mode control system is used for effective solar energy use:

Mode 1, collector-to-storage.   Pumps in the collector loop and the storage loop are activated when the collector outlet temperature exceeds the solar thermal storage temperature by 10°F. Water in storage is heated in heat exchanger HX-1. When the temperature difference decreases to 3°F or less, the pumps are turned off.

Mode 2, storage-to-preheat.   Pumps in the solar thermal storage to DHW preheat exchanger loop are activated when the temperature of the solar thermal storage exceeds the temperature on the discharge side of the DHW preheat heat exchanger (at HX–3) by 10°F. The city water is preheated, on demand, as it passes through the DHW preheat heat exchangers HX–2 and HX–3. When this temperature difference decreases to 3°F,

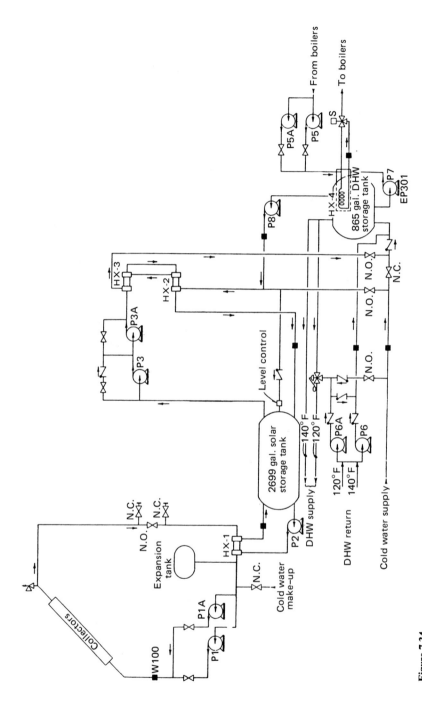

Figure 7.34
Double tank SDHW system for multi-family residential use in Burlington, Vermont (Raymond 1982).

the pumps are inactivated. Final heating of the DHW supply is provided by auxiliary energy in the 855-gallon tank, thermostated at 140°F.

An energy-flow diagram for the system operation from July 1981 to December 1981 indicated the solar collection efficiency was measured at 31 percent; transport losses reduced delivery to the storage subsystem to 27% of incident radiant energy. Additional heat losses from the DHW subsystem substantially reduced useful delivery and increased auxiliary requirements.

7.3.4 Commercial Systems

7.3.4.1 Direct Heating

7.3.4.1.1 SINGLE-TANK SYSTEM A direct heating, single-tank SDHW system supplies hot water to a 271-room luxury hotel located in Honolulu, Hawaii. The 160 flat-plate collector panels are arranged in 80 parallel rows of two series-connected panels for a total gross area of 3,307 ft². They lie flat on the roof. A 4,591-gallon storage tank insulated with 3 inches of fiberglass is also mounted on the roof.

Solar-heated water is transferred from storage, on demand, to the DHW recirculation loop through two gas-fired boilers for auxiliary heating as necessary. A system schematic is depicted in Figure 7.35.

Energy measurements showed a solar collection efficiency of 37% and solar heat delivery to the hot water circulating system of 28% of incident solar radiation. Substantial energy losses from the hot water recirculation loop are characteristic of such plumbing systems.

7.3.4.1.2 TWO-TANK SYSTEMS A solar hot water heating system was designed to provide 60% of the 1,100-gallon daily requirement of an industrial laundry in Dallas, Texas. The 43-panel flat-plate collector has a flat, black painted absorber and one plastic glazing. The collector array has a gross area of 1,011 ft² and is mounted one degree west of due south at a tilt angle of 27 degrees.

The 2,000-gallon storage tank is insulated with one inch of polyurethane and is located outdoors. There is no provision for automatic freeze protection. As shown in figure 7.36, preheated water is supplied to storage from a wastewater heat recovery system and from the solar collector. Heated water is delivered from storage to two 100-gallon conventional gas-heated hot water tanks operating at 150°F. Additional heating by

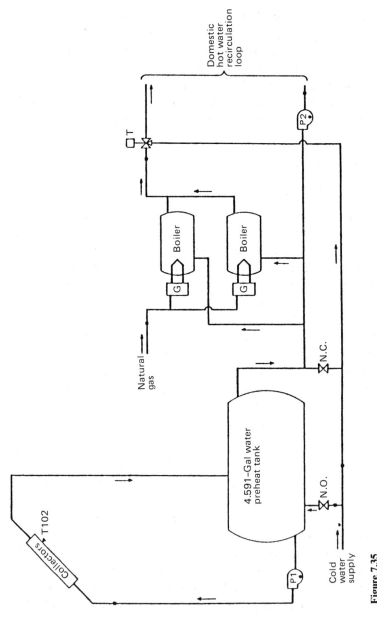

Figure 7.35
Boiler assisted SDHW system for motel in Honolulu; Hawaii (Cramer, Spears, Pollack 1981).

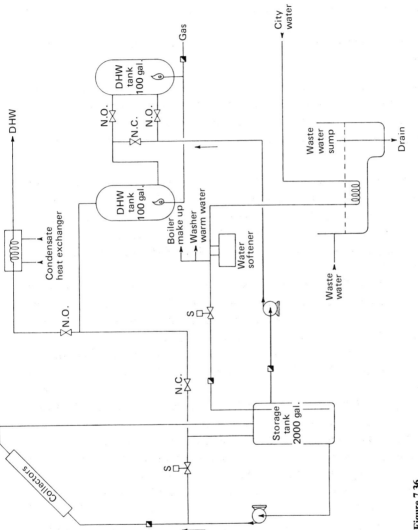

Figure 7.36
Commercial type SDHW application in Dallas, Texas (Raymond 1982).

steam boiler condensate and heat exchanger is also available for final heating of laundry water.

The system has two modes of operation for solar collection and process water:

Mode 1, collector-to-storage. Water is pumped from solar thermal storage through the collector array and back to storage. The pump is activated when the collector array temperature is 18°F above the storage temperature and continues until the temperature difference drops below 2°F.

Mode 2, hot water demand. When there is a demand for laundry water, city water enters a 100-foot pipe coiled in the 275-gallon sump pit, where it is preheated by exchange with waste water. The city water temperature is raised to a level between 80°F and 95°F before reaching the solar thermal storage. Water drawn from the solar thermal storage is then heated by the auxiliary systems to 150°F.

The energy-flow measurements diagram for 10 months of operation indicated the collection efficiency was only 19%, thus only 17% of the laundry heat demand for hot water was supplied. The system design objective was 60% of the laundry hot water heat demand.

A similar two-tank system is installed in a six-story university dormitory building in New Orleans, Louisiana. Potable water is used in an open-loop system for collector-to-storage heat transfer and for thermal storage. A 4,590-ft² (gross area) collector system is arranged in 15 arrays mounted 16 degrees west of south at a tilt angle of 83 degrees.

The 5,000-gallon storage tank is located on the ground outside the building. Freeze protection is provided by pumping warm water from storage through the collectors when the outside temperature is less than about 38°F. Upon demand, stored preheated water flows to two existing 1,500-gallon hot water tanks. Auxiliary energy is supplied by the central heating plant to provide about 9,000 gallons of water per day at 140°F.

Energy measurements for six months of system operation indicated a solar collection efficiency of 29%. High thermal losses in the hot water subsystem necessitated substantial auxiliary use.

7.3.4.2   Indirect Heating

7.3.4.2.1   TWO-TANK SYSTEMS   An indirect, two-tank SDHW system is located in a school building in Leesburg, Virginia. The 1,225-ft² flat-plate collector array faces 15 degrees east of south at a tilt angle of 37 degrees. Silicone oil is circulated through the collector and a heat exchanger coil in a 2,056-gallon water tank located inside the building.

Preheated water is supplied from the storage tank to a 1,000-gallon DHW tank for additional heating to 140°F as required. A timed recirculation loop from the DHW tank to service hot water lines operates during normal use hours.

Energy measurements for operation of the system from August 1980 to February 1981, indicates a collection efficiency of only 17% and a solar contribution to the hot water subsystem of approximately 60%.

An indirect system with antifreeze in the collector loop and an external heat exchanger was installed in a restaurant located in Washington, D.C. The system is designed to heat about 10,000 gallons of water per day to a temperature of 150°F. The 6,254-ft^2 collector array faces southwest at a tilt angle of 55 degrees.

Potable water is heated in a liquid-to-liquid heat exchanger by a propylene glycol solution circulated through the solar collector. Two insulated 5,000-gallon tanks are used for DHW storage. Auxiliary energy is supplied from a gas-fired boiler in the recirculation loop.

Energy-measurements reveal a relatively high collector efficiency of 35% and a hot water subsystem solar fraction of 33%.

An indirect system employing a ground-mounted collector array for process water heating is located in Sioux Falls, South Dakota (Raymond 1981). Freeze protection and energy transfer are provided by a 50% water/ silicone antifreeze solution in the collector loop. The 510 single-glazed, selective flat-plate collectors have a gross area of 9,124 ft^2. The south facing panels are tilted 35 degrees from horizontal.

Solar heated water from the heat exchanger is stored in a 18,790-gallon tank from which heat is delivered to the DHW in a second exchanger supplying a 500-gallon tank. Auxiliary energy is obtained from two steam heaters. The system schematic is shown in figure 7.37. Four operational modes are used as follows:

Mode 1, collector-to-heat exchanger #1. Pump PIA and pump PIB are activated when the collector plate temperature exceeds the temperature at the bottom of the solar heated storage tank by 18°F.

Mode 2, heat exchanger #1-to-storage. When the collector fluid temperature at the heat exchanger exceeds the temperature at the bottom of the

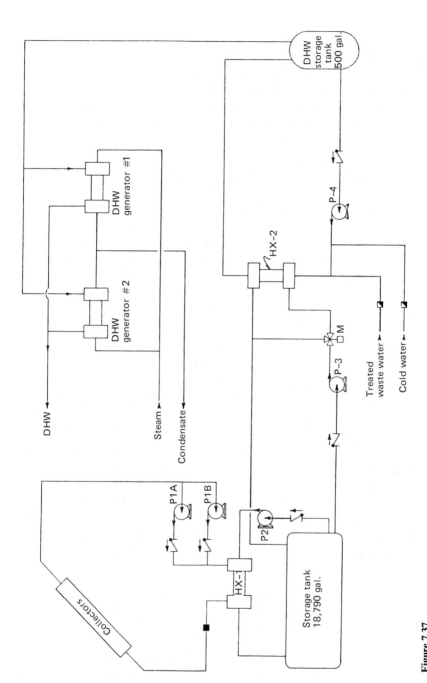

Figure 7.37
Commercial ground-mounted collector system in Sioux Falls, S. Dakota (Raymond 1982).

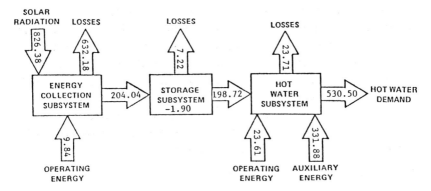

Figure 7.38
Energy flow diagram for commercial system in Sioux Falls, S. Dakota (Cramer, Spears, Pollack 1981).

solar storage tank by 18°F, pump P2 is activated to circulate water from heat exchanger #1 to storage.

Mode 3, storage-to-heat exchanger #2-to-HW load. When the temperature at the top of the solar storage tank exceeds the temperature of the water in the DHW tank by 6°F, pump P3 is activated to circulate water from solar storage to heat exchanger #2, and pump P4 is activated to circulate water from the hot water tank to heat exchanger #2.

Mode 4, hot water consumption. When DHW is used, city water or treated wastewater flows into the hot water storage tank and is preheated. The water then passes through the auxiliary steam heaters where its temperature is increased to 140°F.

An energy-flow diagram, figure 7.38, shows a solar collection efficiency of 25% and a solar fraction of 37%.

7.4 System Field Experience

7.4.1 General

The design criteria and rationale for selection, fabrication, installation, and operation of solar domestic hot water systems are intended to conform to the provisions of existing building codes and licensed building contractor practices. Reasonable protection of the public health and safety is considered when publishing recommended requirements in the area of electrical, building, mechanical, and plumbing (DOE 1980). In general, materials are fabricated and installed so that the system shall provide the following: (a) adequate structural strength, (b) adequate resistance to weather, moisture, corrosion, and fire, and (c) acceptable durability and economy of maintenance. The elevated temperature and moisture exposure encountered in overall normal and no-flow operation of hot water systems has resulted in the development of over twenty ASTM standards, related to durability and reliability.

7.4.2 Residential System Problems

Early experience in SDHW systems was gained in the New England area with 100 installations of systems in single-family residences (Little 1977). The installations were made in a freezing climate to demonstrate the suitability of commercially available designs (1976), to meet the energy needs of the area, and to evaluate future integration of SDHW systems into utility operations. A summary of initial design problems by generic system type is shown in table 7.5. The predominance of improper installation of thermal insulation, control malfunction, and freezing problems plagued the demonstration program.

Additional monitoring and evaluation of installed residential (Boone, Leopoid, Shipp 1980) and commercial systems (Cash 1978) has illustrated problem areas. A breakdown of problems categorized into major subsystems for residential use is shown in table 7.6. It should be noted that these problems include both SDHW and space heating applications in a

Table 7.5
Problems experienced in New England SDHW demonstration

System type Design problems	Water draindown	Closed loop		Air
		Water	Antifreeze	
Inadequate insulation	F	F	F	F
Malfunctioning controls	F	F	F	—
Freeze-up	F	F	IF	—
System air lock	IF	F	IF	—
Noisy equipment	F	—	—	IF
High circulating power	IF	—	—	IF

Source: Little, Inc. 1977.
Code: F—Frequent, IF—Relatively Infrequent, - Did not occur.

Table 7.6
Summary of reported problems by major subsystems, cycles I, II, III, IV, and IVa

	Cycle I		Cycle II		Cycle III		Cycle IV		Cycle IVa		Total	
	No.[a]	%[b]	No.	%	No.	%	No.	%	No.	%	No.	%
Collector subsystem	75	31.8	120	30.4	129	38.2	9	50.0	4	66.7	337	33.8
Storage subsystem	28	11.9	56	14.2	46	13.6	5	27.8	1	16.7	136	13.6
Energy transport subsystem	61	25.9	103	26.8	64	18.9	2	11.1	1	16.7	231	23.2
Control subsystem	37	15.7	75	19.0	44	13.0					156	15.6
Auxiliary subsystem	7	3.0	11	2.8	9	2.7	2	11.1			29	2.9
Building sybsystem	7	3.0	9	2.3	18	5.3					34	3.4
Unclassified problems	21	9.0	21	5.3	28	8.3			4		74	7.4
Total	236		395		338		18		10		997	

Source: Boone, Leopold, W. Shipp 1980.
a. Number of problems within each cycle by solar subsystems.
b. Percent of problems to total problems within each cycle.

Table 7.7
Comparison of the number of system problems per system as demonstration program progressed

Problems Cycle	No. of systems	No. of problems	Ratio prob/system
I	107	236	2.2
II	217	395	1.82
III	308	338	1.09
IV	34	18	0.53
IVa	51	10	0.20

Source: Boone, Leopold, Shipp 1980.

total of 717 solar energy systems. 71% were retrofit installations, of which about 75% were SDHW; of the remaining 29% systems installed in new building units, about half were for water heating.

Of the 997 recorded technical problems, one-third were attributed to the collector subsystem. The energy transport system had about one-fourth of the total problems, and the control and storage subsystems accounted for a majority of the other problems.

Table 7.7 shows that the number of problems per system decreased dramatically as the Demonstration Program progressed from cycle I to cycle IVa. Designers, installers, and users gained experience that resulted in substantial improvements in SDHW technology and application.

The analysis of these residential system problems was used to generate new requirements and revisions in existing requirements for SDHW performance criteria and the HUD Solar Intermediate Minimum Property Standards. Some of these changes are summarized as follows:

1. *Collector rain test*
The leakage of rain into collectors has been a recurring problem. The excessive moisture inside collectors resulted in absorber deterioration, reduced thermal insulation effectiveness, and cover glass condensation. The recommended rain test provides data on collector resistance to thermal shock and water spray penetration (Waksman et al. 1979).

2. *Requirements for heat exchangers and circulation loops*
The leakage of heat exchangers due to freezing, improper installation, and lack of proper fluid control resulted in the contamination of potable water.

Heat exchanger requirements were strengthened to specify a minimum of two physically separated walls or interfaces between the potable water and the nonpotable liquid. The level of protection must be directly related to the toxicity potential of the heat transfer fluid used. Guidelines relating the degree of toxicity and the extent of protection have been studied (Metz and Orleski 1978).

3. *Waterproofing of insulation for exterior tanks, pipes, and ducts*
The relatively large thermal loss from storage tanks and heat transfer distribution systems was due, in part, to inadequate or damaged thermal insulation. Maximum rate of heat loss from storage tanks was therefore specified at 10% per 24 hours, requiring a thermal conductance not to exceed 0.075 Btu/hr ft^2 °F (0.40W/cm^2 °C). R–4 thermal insulation is recommended for pipes less than 1 inch in diameter, and R–6 insulation is recommended for pipes 1 to 4 inches in diameter. All exterior or underground pipes and ducts shall be water proofed to stop water leakage into the components and insulation.

4. *Draindown system freeze protection*
Freeze-related problems for these systems were attributed to the failure of temperature sensors, solenoid valves, vent valves, or pressure relief devices. The hardware elements reported damaged due to freezing include: absorber assemblies, collectors, header connectors, flow regulators, heat exchangers, pipe assemblies, and storage tanks. Emphasis was added to check the concentration of antifreeze solutions for adequacy in the operating geographic location.

Automatic air vents, vacuum relief valves, and draindown valves require protection from freezing by the proper combination of sensing and control mechanisms. Draindown systems must be designed and installed so that all lines are sloped a minimum of 1/4 inch per foot (21mm/meter) and that lines are adequately supported and leak-tight. Bottle gas (nitrogen) or air dryers should be used for air-assisted draindown systems so that moisture cannot be drawn into control or vent valves and freeze.

Fail-safe freeze protection is desirable, because in the event of a power failure, methods involving pumps, syphons, or electrical heating tapes will be ineffective. Extensive experience with such systems has led most manufacturers and system designers to replace the draindown type of solar water heater with indirect types, such as the drainback and antifreeze systems.

7.4.3 Commercial System Problems

A compilation of eighty-five commercial system problems (Cash 1978) did not distinguish between types of systems but indicated that control, sensor, freezing, and collector or collector array distribution problems dominated. Because of special load requirements and the integration of the solar system with auxiliary heating equipment, large systems generally require more complex controllers than usually employed.

As experience accrued, significant improvement in installer skills, product familiarity, and use of numeric temperature readout meters resulted in virtual elimination of installer checkout problems (Kent and Winn 1979). However, probe accuracy, tracking accuracy, optimum differential offset, and hysteresis are factors that continue to require attention in system control design and installation.

7.4.3.1 Freeze Protection
Particular concern with freeze protection for large-scale systems resulted in a comparative evaluation of antifreeze and draindown protection methods (Gunardson 1979). The study concluded that water draindown freeze protection has a strong economic advantage. The economic advantage of draindown system results from lower initial system cost, no heat exchanger, less plumbing, no toxic fluid contamination concern, and lower operating cost for pumping. Net solar energy delivered by the water draindown system is 10% to 20% higher than the output for antifreeze systems. However, careful attention must be given to detailed design of the system to provide fail-safe protection with no traps and proper sloping of the drain lines. It should be noted, however, that SDHW systems for single-family residence are seldom installed with the precautions that can be afforded in large commercial systems, nor is preventive maintenance usually performed. The performance advantages of draindown types are obtainable in small residential systems only at a much higher risk of freeze damage.

7.4.3.2 Sensor Installation
Freezing and control problems are often related to improper sensor type, location, or mounting. Normally, the temperature difference between the sensor and the point of desired measurement should not exceed the precision range of the sensor. Thermosyphon or backflow in an idle loop can

cause sensors to give erroneous readings of system operation. The sensor should be mounted in direct contact with the measured object or fluid. If sensors for sensing fluid temperatures cannot be immersed in the fluid, they should be held in direct contact with the containment device by use of a heat transfer paste and mechanical clamp. Thermal insulation of the sensor and surrounding area may be necessary to shield the area from extraneous heat flows. Details of sensor selection, location, and mounting are discussed in SERI (1981), DOE (1978), and Waite et al. (1979).

7.4.3.3 Collector Array Performance
System performance is influenced by collector array orientation, fluid flow uniformity, fluid flow rate, and heat loss resulting from inadequate insulation or leakage.

7.4.3.3.1 COLLECTOR ORIENTATION For maximum annual output, collectors in SDHW systems should face due south and have a tilt angle approximately equal to the latitude. Experience has shown that tilt deviation of 20 degrees from latitude and deviation of 30 degrees from true south will decrease performance by less than 10% in most locations. Highrise buildings or small or unfavorable roof orientation may require special architectural consideration. A typical relationship between collector tilt and solar space heating performance is shown for Columbus, Ohio, in figure 7.39. A similar typical plot relating collector orientation and solar heating performance is presented in figure 7.40. It should be noted that these two graphs apply to typical solar space heating systems in which

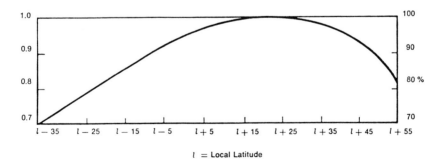

Fig. 7.39
Effect of solar collector tilt on annual solar heating performance (DOE 1978).

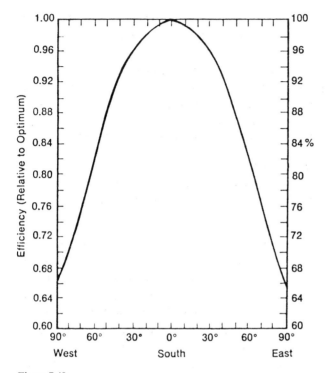

Figure 7.40
Effect of solar collector orientation on solar heating performance (DOE 1978).

optimum collector tilt angle is higher than for SDHW systems because of seasonal load variation. Unique site conditions, such as shading, prevailing cloud patterns, or load profile should be considered in emphasizing morning or afternoon solar gains.

7.4.3.3.2 PANEL ARRAY PIPING Data were presented in figure 7.10 illustrating collector array efficiency lower than single panel collector efficiency. Array piping arrangements can result in flow imbalance and overheating of sections of the array. Three collector array piping designs are illustrated in figure 7.41.

Use of parallel flow-direct return design may result in flow and temperature difference between parallel paths because of pressure drops in the supply and return headers. Sizing of the headers for minimum pressure

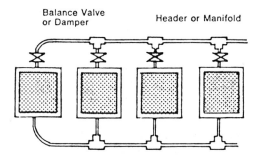

Parallel flow-direct return

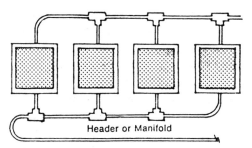

Parallel flow-reverse return

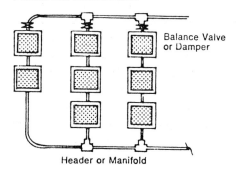

Series flow

Figure 7.41
Collector array piping circuits (DOE 1978).

drop and the use of balancing valves can equalize the flow. Provision should be made to measure pressures so that individual collector flow rates can be adjusted.

Parallel flow-reverse return designs are usually preferable to the direct return design. Provisions for flow balancing may still be required in some reverse piping systems, depending upon the overall size and type of collector.

Series flow may be used in large planar arrays to reduce the amount of piping required by allowing collector assemblies to be served by the same supply and return manifolds. This arrangement can be used to increase collector output temperature or to allow placement of collectors on non-rectangular surfaces. More than three collectors in series are seldom used because of increased pressure requirements and higher temperatures of fluid entering collectors following the first. In large arrays, flow differences of up to 3 to 1 in individual collectors may be acceptable, provided that none of the collectors operates at a flow below a practical minimum. The influence of flow rate and fluid property on the efficiency of collectors containing a variety of absorber plates is given in Thomas (1980) and Youngblood, Schultz, and Barber (1979), and the effect of flow rate on a typical air-type collector is presented in Jones, Shaw, and Löf (1979).

The detection of fluid circulation problems in large arrays can become tedious and time-consuming even with flow balance provisions. Infrared thermographic techniques have been used to show temperature distributions that indicated flow imbalance, air blockage, and broken collectors in large arrays (Eden 1980). Temperature variations over a range of 5° to 10°C for the entire thermograph can be observed with a sensitivity of about 1°C. Figure 7.42 illustrates thermographs showing a blocked cluster in one array and normally operating collectors in another array.

7.5 Installation, Servicing, and Monitoring

7.5.1 General

The installation, checkout, and servicing of solar hot water systems are as important as the proper system design in assuring good performance. There has been considerable evidence that improper or inadequate atten-

 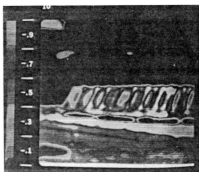

Blocked Cluster   Normally Operating Collectors

Figure 4.2
Thermograph of blocked solar cluster and of collectors operating normally (Eden 1980).

tion to these steps is a contributing factor to poor performance, accelerated deterioration, and system damage (Boone, Leopoid, Shipp 1980; Cash
1978). The difficulties of mandating and enforcing criteria and practices for
proper installation and checkout are acknowledged in the industry by the
formation of contractor associations and state requirements for permits
and licensed operations. It is essential that education of designers, installers, and consumers continue and that authoritative handbooks and
manuals be used to alleviate these problems.

Handbooks containing design and installation guidelines for control
systems (Sharp et al. 1980a) and procedures for checking the operation of
different types of installed systems are available (Sharp et al. 1980b). A
summary of the primary requirements for evaluation of the installation,
operation, and maintenance is presented in the following sections.

7.5.2 System Installation

Instructions that describe the installation and removal (where required) of
each solar energy system component in a step-by-step fashion shall be
provided. The instructions shall describe the overall system operation and
the relationship and interconnections between components and their interface with the building and site. Areas to consider include (Franklin
Research Center 1979; Florida Solar Energy Center 1980):

1. COLLECTOR   2. PIPING
   Orientation and tilt   Leaks
   Shading   Supports
   Mounting   Insulation
   Roof support   Valves and vents
   Stagnation protection   Heat transfer fluid
                                   Pressure/temperature relief
                                   Penetrations

3. CONTROL   4. STORAGE TANK
   Sensors   Location
   Freeze protection   Insulation
   Weather resistance
   Wiring

7.5.3 Checkout Procedure and Operation

Instructions that describe the operation of the system in all modes including startup and shutdown under normal and emergency conditions shall be provided. Description of critical temperature, flow, and pressure information for checkpoints shall be given. Upon completion of the installation, the total system operation shall be inspected and tested for acceptable performance as specified in the design.

Operation of the control system shall be verified with respect to temperature set points, pump or blower operation, freeze protection, and fail-safe operation. Fluid level, flow direction and rate, and fluid composition (if antifreeze is liquid) shall be checked. All piping, valves, and connections shall be inspected for leaks. Air collectors and ducting shall be inspected for leaks. Air collectors and ducting shall be inspected for excessive leakage or sagging when operating at expected temperature and pressure. Damper operation shall be checked to insure positive air seals when closed to prevent thermosyphon losses or freezing in the air-to-water heat exchanger. Piping shall be installed with the proper slope to facilitate drainage or thermosyphon action. Controls, sensors, dampers, and valves should be marked to identify their function and pipes marked to show flow direction.

7.5.4 Maintenance

A maintenance manual with a schematic diagram showing the location of critical components, wiring, and fluid flow shall be provided. The manual shall include a plan for periodic servicing of components or testing fluids and flushing of liquid passages, if required. A description of hazards that could arise during maintenance and precautions required shall be provided. Provision shall be made for system monitoring, and instructions for the use of simple sensors or indicators to reveal proper operation shall be given.

7.5.5 System Monitoring

The large-scale use of SDHW systems is predicated upon the proper operation of the system without the use of sophisticated monitoring equipment. Experience in the residential part of the National Demonstration Program indicated that the performance of noninstrumented installations can differ greatly from one installation to another because of undetected individual system operational problems. The presence and effects of these problems can be minimized by providing the system operator with equipment capable of detecting improper operation and degraded performance.

The availability of on-site maintenance personnel and their familiarity with measurement instrumentation at large multifamily or commercial solar installations has resulted in microprocessor monitoring techniques, instrumentation, and data analysis for reporting the performance of solar water heating (Conference 1978).

Two considerably different monitoring methods have been employed in the National Demonstration Program for SDHW systems. The National Solar Data Network (NSDN) carried out detailed monitoring of 70 systems that have the means of solar water heating (Nash and Cunningham 1979; DOE 1977), and about 200 SDHW systems in the Northeast were evaluated by use of simple instrumentation and manual data logging.

In the NSDN program, the analysis of solar hot water heating subsystems required calculation of the following performance factors (Cramer, Spears, and Pollack 1980; Raymond 1981):

Acronym	Performance factor	Units
HWDM	Hot Water Demand	Btu
HWL	Hot Water Load	Btu
HWSE	Hot Water Solar Energy Used	Btu
HWAT	Auxiliary Thermal Energy Used	Btu
HWOPE	Hot Water Subsystem Operating Energy	Btu

HWOPE	Solar Specific Hot Water Operating Energy	Btu
HWCSM	Hot Water Consumption	gallons
THW	Hot Water Temperature	°F
TSW	Supply Water Temperature	°F
HWSFR	Hot Water Supply Fraction	%

These parameters, in addition to parameters relating to solar collection and storage, allow intersystem comparison of the performance of SDHW systems. The data system was initially used to determine proper system operation by comparing temperature setpoints, operating energy, and actual load to design values.

Each site contained standard industrial instrumentation adapted to the particular site. Sensors measured temperatures, flows, solar radiation, electric power, fossil fuel use, and other parameters as shown in the typical system schematics presented in section 7.2. The sensors were wired into a junction box, which was connected to a microprocessor data logger, called a Site Data Acquisition System (SDAS). The SDAS could read up to 96 different channels—one for each sensor. The SDAS converts analog volt-age input to each channel to a 10-bit word. The SDAS samples each channel and records the values on a cassette tape at 30-second intervals.

Each SDAS was connected through a modem to a telephone line trans-mitting the data to a central computer facility. Typically, the central com-puter collected data from each SDAS six times per week, although the tape could hold three to five days of data, depending upon the number of channels used. The digital data were then processed by software and con-verted into Engineering Units. These detailed measurements were tabu-lated on a weekly basis for the site analyst and transformed into plots, graphs, and processed reports summarizing the system performance.

In contrast to the relative complexity of the NSDN, a simple moni-toring technique was devised to help evaluate the performance of about 200 of the approximately 10,000 SDHW systems installed in the Northeast States Solar Water Initiative (Erhartic 1979). The objectives of the moni-toring program were:

1. evaluate the effects of equipment certification HUD warranties, instal-ler training, and solar industry maturation on the quality of the installed product

2. establish maintenance and service costs

3. establish the reliability, safety, and useful life of solar water heaters

The low-cost monitoring system (LCS) consisted of the following instrumentation:

- two bimetal thermometers
- watt-hour meter
- elapsed time indicator
- water flow meter
- pressure gauge

The monitoring equipment was installed by the solar system installer, and the data were logged daily by the homeowner. A typical installation is shown in figure 7.43. The LCS provided data that included:

- the amount of hot water consumed
- temperatures of inlet cold water and delivered hot water
- the amount of purchased electrical energy used for water heating
- the daily operational time of the solar water heating system

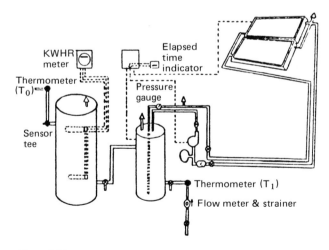

Figure 7.43
Instrumentation schematic for SDHW system (Erhartic 1979).

Two difficulties with the LCS arose from the simple instrumentation and manual data logging. Since only overall energy parameters were measured, component heat loss characteristics could not be determined. Manual once-per-day recording of temperatures assumes constant temperatures for storage inlet and outlet. While the inlet temperature should be relatively constant, the outlet temperature depends on many variables and may be difficult to correlate with actual day-long performance. Provisions were made to install a Btu meter, if required, later to verify the accuracy of the LCS.

References

Balcomb, D. 1975. *Solar heating handbook for Los Alamos.* Report UC–5967–CONF–75027–1. Los Alamos, NM: Los Alamos Scientific Laboratory.

Beckman, W. A., S. A. Klein, and J. A. Duffie. 1977. *Solar heating design by the f-chart method.* New York: Wiley.

Best, D. 1981. What you should know about phase-change water heaters. Pub. *Solar Age,* Vol. 6, 12 *Solar Vision Inc.,* Harrisville, N.H.

Best, D. 1982. Innovations with real expectations. *Solar Age* vol. 7 No 3 (March 1982): Solar Vision Inc., Harrisville, N.H.

Buckles, W. E., and S. A. Klein. 1980. Analysis of solar domestic hot water heater. *Solar Energy* 25 No. 5:417–24.

Boone, T., S. Leopold, and W. Shipp. 1980. *Problem identification of solar systems used in the HUD residential solar heatng and cooling demonstration program.* Quarterly report #4. Washington, DC: National Bureau of Standards.

Butti, K., and J. Perlin. 1980. *A golden thread.* New York: Van Nostrand Rheinhold, p. 129.

Cash, M. 1978. *Hardware problems encountered in solar heating and cooling systems.* DOE/NASA TM–78172. Huntsville, AL: NASA, MSFC.

Cole, R. L., K. J. Nield, R. Rohde and R. M. Wolosewicz. 1980. *Design and installation manual for thermal energy storage.* ANL–89–75. Argonne National Laboratory. NTIS, Springfield, VA.

Conference on Performance Monitoring Techniques for Evaluation of Solar Heating and Cooling Systems. Washington, DC, April 1978.

Cramer, M. A., J. W. Spears, and F. O. Pollack. 1980. *Comparative report—performance of solar hot water systems 1980–1981.* Solar/0024–82/4, DE83–000069. Springfield, VA: NTIS.

DOE. 1977. *National Solar Data Network.* SOLAR/003–77–17. Washington, DC: Department of Energy.

DOE. 1978. *Fundamentals of solar heating.* HCP/M 4038–01, SMACNA 0224, DOE cont. no. EG–77–C–01–4038–2. Washington, DC: Department of Energy.

DOE. 1980. *Recommended requirements to code officials for solar heating, cooling and hot water systems.* DOE/CS/34281–01. Washington, DC: Department of Energy.

Eden, A. 1980. *The analysis of solar collector array systems using thermography.* SERI/TR–351-494. Golden, CO: SERI.

Elcam Inc. N.d. *Solar energy system evaluation—final report for Elcam—Tempe.* DOE/NASA CR-161492. Huntsville, AL: MSFC.

Elcam Inc. 1980. *Design package for solar domestic hot water system.* DOE/NASA CR-161558. Huntsville, AL: MSFC.

ERDA 36-6. 1976. *National program for solar heating and cooling of buildings.* Washington, DC: Energy Research and Development Administration, Division of Solar Energy.

Erhartic, P. 1979. Interim performance results from the Northeast States Solar Water heater Performance Monitoring Program. SERI/TP-245-430, SOLAR/0500-80/00. *Proc. Solar Heating and Cooling Systems Operational Results. Conf.* Colorado Springs, CO, November 1979, p. 33.

Fanney, A. H., and S. A. Klein. 1983. Performance of solar domestic hot water systems at the National Bureau of Standards-Measurements and Predictions", *J. Solar Energy Eng.* 105.

Fanney, A. H., W. C. Thomas, C. A. Scarborough, and C. P. Terlizzi. 1982. *Analytical and experimental analysis of procedures for testing solar domestic hot water systems.* BSS-140. Washington, DC: National Bureau of Standards.

Farrington, R. B., L. M. Murphy, and D. L. Noreen. 1980. *A comparison of six generic solar domestic hot water systems.* SERI/RR-351-413. Golden, CO: Solar Energy Research Institute.

Florida Solar Energy Center. 1980. *Florida standard practice for design and installation of solar domestic water and pool heating systems.* Draft standard. Cape Canaveral, FL: Florida Solar Energy Center.

Franklin Research Center. 1979. *Installation guidelines for solar DHW systems in one- and two-family dwellings.* HUD-PDR-407. Washington, DC: U.S. Government Printing Office.

Gunardson, H. H. 1979. Evaluation of freeze protection systems for large scale solar energy system. *Proc. 3d Solar Heating and Cooling Demonstration Program Contractors Review,* Norfolk, VA., December 1979.

Hunn, B. D. 1977. *ERDA facilities design handbook.* LA-UR-77-186. Los Alamos, NM: Los Alamos Scientific Laboratory.

Interim performance criteria for solar heating and combined heating/cooling systems and dwellings. 1975., Stock No. 0324-01043. Washington, DC: Government Printing Office.

Interim performance criteria for commercial solar heating and combined heating/cooling systems and facilities. 1975. Doc. no. 98M10001. Huntsville, AL, NASA, George Marshall Space Flight Center.

Jackson, D. R. 1979. Performance of a solar hot water system installed in a interstate highway visitor center—results of one year of monitoring. *Proc. Solar Heating and Cooling Systems Operational Results Conf.,* Colorado Springs, CO: SERI, p. 69.

Jones, D. E, L. E. Shaw, and G. Löf. Air leakage effects on active air-heating solar collector system. SERI/TP-245-430, SOLAR/0500-80/00. *Proc. Solar Heating and Cooling Systems Operational Results Conf.,* Colorado Springs, CO, November 1979, p. 131.

Kent, T. B, and C. B. Winn. 1979. Performance of controllers in solar heating systems. *Proc. Solar Heating and Cooling Systems Operational Results Conf.* SERI/TP-245-430, SOLAR/0500-80/00. Colorado Springs, CO., p. 311.

Little, A. D., Inc. 1977. *Interim report on New England electric residential solar water heating experiment.* Westborough, MA: A. D. Little, Inc.

Mutch, J. J. 1974. *Residential water heating, fuel consumption, economics and public policy.* Rand Corporation, R-1498-NSF.

McCumber, W. H. 1979. Collector array performance of instrumented sites in the National Solar Data Network. *Proc. Solar Heating and Cooling Systems Operational Results Conf.*, Colorado Springs, CO, November 1979.

Metz, F. A., and M. J. Orleski. 1978. *State-of-the art study of heat exchangers used with solar assisted domestic hot water systems* (*potential contamination of potable water supply*). NBSIR 78–1542. Washington, DC: National Bureau of Standards.

Mertol, A., Place, W., Webster, T., Greif, R. 1981 Detailed loop model (OLM) analysis of liquid solar thermosyphons with heat exchangers. LBL-10699 Rev., Lawrence Berkeley laboratory, Berkeley, Colif. 94720, DOE Contract No. W-7405-Eng-48.

Morrison, D. J. 1980. *Development of a gas-backup heater for solar domestic hot water systems.* DOE/CS/34696–1, DOE cont. no. DE–AC02–78CS34696. Santa Cruz, CA: Altas Corporation.

Morse, R. N, and D. J. Close. 1977. Applications of solar energy for heating and cooling of buildings. Chap. 9 in *Solar water heating.* Atlanta, GA: ASHRAE.

Nash, J. M., and G. W. Cunningham. 1979. Thermal performance of DHW systems in the National Solar Data Network during the 1978–1979 heating season. *Proc. Solar Heating and Cooling Systems Operational Results Conf.*, Colorado Springs, CO, November 1979, p. 8.

NBSIR 76–1187. 1976. *Interim performance criteria for solar heating and cooling systems in commercial buildings.* Springfield, VA: National Bureau of Standards.

NBSIR 77–1272. 1977. *Intermediate standards for solar domestic hot water system/HUD initiative.* Washington, DC: National Bureau of Standards.

NBSIR 78–1562. 1978. *Interim performance criteria for solar heating and cooling systems in residential buildings.* Springfield, VA: National Bureau of Standards.

NBSIR 80–2095. 1980. *Performance criteria for solar heating and cooling systems in residential buildings.* Gaithersburg, MD: National Bureau of Standards.

Raymond, M. G. 1981. *Comparative report—performance of solar hot water system—1980–1981.* Solar/0024–83–41. Springfield, VA: NTIS.

Schreyer, J. M. 1981. Residential application of refrigerant-changed solar collectors. *Solar Energy* 26 No. 4:307–12.

SERI. 1981. *Controllers for solar domestic hot water systems.* SERI/TR–98189–1A. Fort Collins, CO: Solar Environmental Engineering.

Sharp, K., J. Lefler, B. Winn, and T. Kent. 1980a. *Systems checkout handbook for solar domestic hot water systems.* Golden, CO: SERI.

———. 1980b. *Handbook for controllers used in solar domestic hot water systems.* Golden, CO: SERI.

Thomas, W. C. 1980. *Effects of test fluid composition and flow rates on the thermal efficiency of solar collectors.* NBS-GCR 80–250. Washington, DC: National Bureau of Standards.

Waite, E., et al. 1979. *Reliability and maintainability evaluation of solar control systems.* ANL/SDP–TM79–5. Argonne National Laboratory, Argonne, IL.

Waksman, D., E. R. Streed, L. W. Reichard, and L. E. Cattance. 1979. *Provisional flat plate solar collector testing procedures: first revision.* NBSIR 78–1305A. Washington, DC: National Bureau of Standards.

Youngblood, W. W., W. Schultz, and R. Barber. 1979. *Solar collector fluid parameter study.* NBS-GCR 79–184. Washington, DC: National Bureau of Standards.

8 The Performance of Residential Solar Hot Water Systems in the Field

William Stoney

8.1 Introduction

This chapter summarizes the results of investigations the principal purpose of which was to define the performance of commercially developed and sold solar systems. The discussion centers on the results of field tests on solar domestic hot water (SDHW) systems conducted by government and public utilities. Field testing has often been conducted in the form of programs of sufficient size that statistically valid data were obtained. As will be seen, these tests provide a variety of usable and unique data on the performance and operations of the commercially produced domestic hot water systems bought and used by the public. However, the data may sometimes prove frustrating, because most of the field test programs do not have enough measurements or are not of sufficient size to permit the performance results to be used to compare the effectiveness of different system types and/or sizes, or to attempt to validate theoretical calculations.

8.1.1 History

The history of SDHW field test programs started in the mid-1970s when a utility, the New England Electric System, offered a program to its customers that involved obtaining data on the performance of SDHW systems. That experience was a troubled one, beset with problems that were mostly caused by the inexperience of the technicians with the details of installing these "simple" but new products. That effort, however, aided the next attempt at SDHW field testing, which came about in the same area as a result of the federal rebate program of 1978–79. The program provided $400 rebates to SDHW purchasers in the New England states (included in that group were Pennsylvania, New Jersey, New York and Delaware) and in Florida. These states were chosen because their conventional hot water systems depended, for the most part, directly on oil— or on electricity generated by oil. This program was administered by the U.S. Department of Housing and Urban Development (HUD) and the participating states. HUD set up performance standards that SDHW systems had to meet in order to qualify for the rebate program. The standards resulted in the creation of collector performance and reliability testing programs in several states. The programs and the requirements of the federal solar tax credits legislation then led to the collector rating and certification program in which nearly all manufacturers of solar collectors have participated.

It is less well known that two field test programs were also initiated as a direct result of the HUD program. The largest, funded by the U.S. Department of Energy (DOE) and managed by the Northeast Solar Energy Center (NESEC), included over 100 systems located mostly in the Northeast. A smaller effort involving 20 systems was initiated by the Florida Solar Energy Center, which was in charge of the HUD-related program in Florida. These tests were designed to answer the basic question: do commercially available SDHW systems meet the HUD performance requirement of 50% solar contribution?

The programs were very useful but immediately raised more questions. The Florida effort, which included inspection of the installed system, showed that the homeowners usually had no way of knowing if their solar system was working at all, much less how well it performed. The New England tests indicated a similar situation. Because of the much larger number of subjects and the fact that daily measurements of both water use and backup energy were made, the NESEC program provided the first really comprehensive look at the very wide variability that characterized the use of hot water by the public. The lack of user knowledge of system performance raised concerns as to the information on actual field performance available to the installer, dealer, and manufacturer as well. The realization that industry had limited means of assessing the performance of its products led to the creation in 1980 of an extensive field test program (approximately 150 systems) by the Southern Solar Energy Center (SSEC).

DOE had long recognized the importance of gathering field test data and to this end also created the National Solar Data Network (NSDN). This program is discussed in detail in several other sections of this book. It is mentioned here, because its basic purpose, the understanding of solar system performance in the field, is the same as that of the programs noted above. However, its test population contained very few stand-alone SDHW systems and thus can contribute little to the statistical information discussed in this section. NSDN methodology involved extensive and detailed measurements of all system components and functions, and the interested reader should consult the available reports for examples of the way in which systems operate under the dynamic conditions of varying loads, weather, etc.

During the early 1980s, other organizations, principally public utilities and a few state energy offices, began field test programs with similar objectives. ESG, Inc., while completing the test program started by the Southern Solar Energy Center, became aware of these activities and with DOE sponsorship conducted a review meeting with these organizations. The proceedings of this meeting, held in Atlanta in July 1982, contain program descriptions and results from the 14 organizations in attendance (ESG 1982). Although there appeared to be a great diversity in the way the programs were conducted and their data analyzed and presented, there was a great deal of similarity in the overall results. The common conclusion of all the studies was that the very wide range in measured performance simply indicates the wide variability of SDHW operating conditions, system size, and system types that exist in the field. Installation variables, such as collector tilt and orientation shading factors, climate and weather, and water use practices, all have effects on system performance.

A list of the programs and their major characteristics is provided in table 8.1. As the table shows, there are nearly a thousand SDHW systems represented in the programs listed.

Table 8.1 lists "other" DHW systems as part of certain field test programs (fourth column). Category "other" is made up principally of conventional electric backup hot water systems, but it also includes heat recovery units, hot water heat pumps, and large storage tanks with timers. It is important to realize that while many utilities have had large-scale statistical test programs measuring the energy used by hot water heaters, most of the data do not include measurement of the water quantity used in conjunction with the energy used. As discussed below, the lack of such information causes a significant problem in determining the energy savings that widespread use of SDHW systems would provide.

Two publications resulted from DOE-sponsored workshops at the Florida Solar Energy Center (FSEC). The first conference, held in December 1980, considered only the performance monitoring of SDHW systems. Two years later, DOE, through the Solar Energy Research Institute (SERI), sponsored a second conference to include the monitoring of all active solar systems and to discuss the information available from the field on solar system reliability and maintainability. This conference was held at FSEC in January 1983. Both proceedings contain considerable discussion of field test instrumentation options and the inevitable trade-offs be-

Table 8.1
Summary of field test programs and key measurements

Organization	Location	Sites SDHW	Sites Other HW	Energy meas. Water used	Back-up	Parasitic	Delivered Btu
SSEC & ESG, Inc.	FL, AL, GA, NC, SC, TX, VA	152		×	×	×	
NESEC	New England States & NY, NJ, DE, PA	124		×	×	×	
Baltimore Gas & Light	Baltimore, MD	30		×	×	×	×
Long Island Lighting	Long Island, NY	160		×	×		
Florida Solar Energy Center	Florida	20	60	×	×		×
Kansas Power & Light	Kansas	142		×	×		
Oregon State Dept. of Energy	Oregon	37	8	×	×	×	
Carolina Power and Light	North Carolina	20	66	×	×		×
Tennessee Valley Authority	Tennessee	140		×	×(T)		×
San Diego Gas & Electric	San Diego, CA	87	18	×	×	×(20)	×(12)
Portland General Electric	Portland, OR	31	18	×	×	×	
Public Service Company of New Mexico	New Mexico	30	30		×(T)		
EPRI with Local Utilities	CO, MS, TN, OR, AR, TX, CA, NM, NY, WI	0	110	×	×		
Illinois Power	Illinois	18		×	×		
TOTALS		991	310				

Notes: W = Weekly   T   = Total (includes backup and parasitic energy)
        D = Daily   TIN = Temperature at inlet (groundwater)
        M = Monthly   TDEL = Temperature of delivered water

Table 8.1 (continued)

Temperatures				Meas.	
TIN	TDEL	TSET	TAMB	freq.	Comments
×	×	×		W	Water temperature from hand-held thermo-meters, parasitic energy from timer on pump
×	×	×	×	D	Same as above
				M	Area ground-water temperature available
				M	Parasitic energy estimated
		×		1/4 hr.	Degree day weather data available
				M	Analysis support by University of Kansas Center for Research
×	×			M	Tank standby losses measured, certain test sites chosen near weather stations
				M	Program also included heat pumps, heat recovery units, and large capacity superinsulated tanks
×	×		×	1/4 hr.	
		×		W	12 systems with Btu meters and 15-minute record intervals
×	×			M	Systems periodically operated without solar input to determines losses
		×		1/4 hr.	Water usage not measured
×		×		M	
		×		1/4 hr.	Analysis support by University of Illinois

TSET = Temperature setting for controller or thermostat
TAMB = Temperature of ambient air at tank (relates to tank loss)

tween more data on few sites vs. less data on many sites. The proceedings of the second conference (Yarosh 1983) contain valuable material on the field experience with different types of instrumentation, including discussions on the pros and cons of the use of Btu meters in large-scale field tests.

This section has briefly described the history of SDHW field test programs that were initiated primarily to determine the performance of systems being sold to the public. As noted above, the earliest tests quickly revealed that there were numerous problems with the first installations and that the problems were often not detected by either the homeowner or the installer. These findings led to adding the objective of determining installation quality and system reliability to subsequent field tests, and to the initiation by DOE of specific programs concerned with reliability and maintenance of systems in the field. As these programs were implemented and as designers, manufacturers, and installers acquired field experience, the quality of solar water heating systems improved, and performance, reliability, and durability showed significant gains.

8.2 Field Test Program Descriptions

8.2.1 Objectives

Organizations perform field test programs for a variety of reasons. For example, the purpose of the work by the public utilities, as reported in ESG (1982), included load prediction, capacity prediction, consumer assistance, regulatory compliance, validation of a loan or rebate program, and determination of typical demand profiles. As a result, it comes as no surprise that the testing techniques of the programs differ in the instruments used, the quantities measured, types and numbers of systems tested, and in the way the data are reduced, analyzed, and presented. However, since the purpose of this chapter is to provide an overview of the data available, it is necessary to rationalize a procedure that allows the varied data to be compared. This is best done by defining a common question that all or most of the program data can answer.

During discussions among people involved in these test programs, it was usually agreed that whatever the programmatic reason for the test, the technical objectives of the field test programs could be divided into attempting to answer two basic questions: (1) how much energy will solar and other unique water heating systems save compared with conventional systems; and (2) what are the effects of nonconventional systems on the demand load profile. Another closely related question, answered in a number of the test programs, how much hot water is consummed by residential users. Since only four of the smallest of the thirteen data sets discussed in this chapter measured the daily load profiles, the available data are only statistically useful on a national level in answering the first question on energy savings. That is a primary focus in this chapter.

While the stated purposes of field test programs rarely included system reliability data as an original goal, many of the programs added questions to their reporting forms to obtain information on system problems that could affect the measured performance.

8.2.2   Test Methods

Table 8.1 provides an overview of the measurement *systems* used. The measurement *methods* can be divided into several basic types. In all of the programs except the one by the Public Service Company of New Mexico, the hot water used by each system along with the backup energy required were measured. These two measurements together with measured or assumed temperature values taken at daily to monthly intervals and analyzed over one-month periods can provide a common basis for comparing all of the different test results. The tests varied widely with respect to the measurement and use of temperature data; some specific comparisons between results require the use of data such as generic ground water and ambient temperatures in the area.

In three programs, and partially in two others, delivered water energy was measured directly with Btu meters. This measurement provided valuable additional information, but its higher cost limited its use to smaller test samples.

In four of the programs and part of a fifth, recording devices were used to obtain data every 15 minutes, thereby providing daily profiles of the measured quantities. These data are especially useful for utilities concerned with the effect of the time of the use of the backup energy relative to their system's peak energy periods.

The above differences in measurement techniques result in different analytical potentials for the collected data as discussed in the following section.

8.2.3   Interpretative Methods and Assumptions

Large-scale tests are created to produce statistically valid figures of the average or mean response of the total population. Statistical sampling is a complex process that involves strict rules for the selection and size of the test populations. With the possible exception of the New Mexico program, the solar system data reported herein are not statistical in this sense. In fact, most of the questions on the data reduction and interpretation techniques treated below result from having to determine the savings generated by the test system relative to a conventional system, the performance characteristics of which are either assumed from laboratory tests or derived from other field measurements not related to the solar system performance test.

8.2.3.1   Comparison Assumptions

In any comparison of solar system performance with conventional or other advanced hot water systems, the fundamental decision is to choose a basis for the comparison. There are four equivalency relationships that appear useful. These are described and compared below. All are based on evaluating the expression for the annual energy saved, $AES = EP_C - EP_S$, in accordance with different equivalency assumptions, where EP_C is the energy purchased for a conventional hot water system and EP_S is the purchased energy for a solar hot water system. The four relationships are:

AES_{UD} annual energy saved based on equivalent user demographics

AES_G annual energy saved based on equivalent gallons used

AES_{BW} annual energy saved based on equivalent Btu of hot water delivered

AES_{BT} annual energy saved based on equivalent Btu of total load delivered (hot water plus tank losses).

Other terms in subsequent equations are:

EP energy purchased

EHW delivered hot water energy

EL system heat losses

k conversion factor constant

g gallons of water used per period

TIN   inlet water temperature

TSET   controller temperature setting

Q_U energy supplied by the solar collector

Subscript *C* applies to conventional hot water systems, and subscript *S* applies to solar hot water systems.

EQUIVALENT USER DEMOGRAPHICS, AES_{UD} The annual energy saved can be expressed as the difference in the mean energy used by the two populations, i.e., $AES_{UD} = EP_C - EP_S$, where EP_C and EP_S are population averages.

The theoretical justification for this comparison is the assumption of equivalent demographic characteristics for each test population. This assumption requires that equivalent sets of typical households can be found for solar systems and conventional systems. Accuracy can be achieved if the populations are large (relative to the other methods) and demographically identical. The data would seem to be suitable for load production studies if the above two criteria are met.

However, of the test programs noted in table 8.1, only one involved this method. This was the program by the Public Service Company of New Mexico. One reason why this was the only test program relying on this comparison technique is that the method has another basic limitation. Analyses of data concerning the relative savings of individual systems cannot be performed because data on water use in individual systems are not required and thus not obtained by this method. It would also be necessary to include variations in system type, system size, and installation practice. The effects of load and load profile are also important.

This shortcoming led to the inclusion in the other programs of data on the amount of water used or on the thermal energy content of the delivered water (in Btu). Either additional measurement makes possible the calculation of savings defined in the three different ways discussed below.

WATER QUANTITY AND PURCHASED ENERGY METHODS—AES_G, AES_{BW}, AND AES_{BT} In order to compare conventional and solar hot water system performance on the basis of equivalent gallons used, Btu of hot water delivered, or Btu of total load, it is necessary to measure the number of gallons used or the delivered Btu in each solar system being tested.

The relationships for the annual energy saved for the three equivalencies are easily derived from the basic equation, $AES = EP_C - EP_S$, where it should be noted that we can now deal with a comparison of an individual solar system and an equivalent conventional system.

For electric backup systems, EP_C is simply the sum of the delivered hot water energy, EHW_C, and the system losses, EL_C. The savings equation can thus be easily expressed in terms of the three possible equivalencies by using the proper subscripts to denote the relationship of the conventional energy calculation to the solar system quantities. Note that the $EHW = kg(TSET - TIN)$, where k is a constant, g is gallons used per period, and TIN and $TSET$ are the inlet and controller temperatures, respectively. With the need for all comparisons to be based on the same ground water temperature ($TIN_C = TIN_S$), the three variations become:

1. $AES_G = kg_S(TEST_C - TIN_S) + EL_C - EP_S$ (equivalent gallons)
2. $AES_{BW} = kg_S(TEST_C - TIN_S) + EL_C - EP_S$ (equivalent Btu of hot water)
3. $AES_{BT} = kg_S(TEST_C - TIN_S) + EL_C - EP_S$ (equivalent Btu of total load)

8.2.3.2 Rationalizing the Choice between Equivalencies

The choice should depend, in part, on the use to which the answers will be put. For a number of purposes, it does not matter which equivalency is used, as long as the choice is clearly indicated and consistently and explicitly used in the derivative values and conclusions drawn from the data. One important principle to be considered in making a choice is to select the equivalency that requires the least number of assumptions to be made in reducing the data.

AES_{BT} is by definition the useful energy supplied to the system by the solar collector. (Note that the energy supplied by the solar collector, Q_U, is $Q_U = EHW_S + EL_S - EP_S$). Its use in some form is required to make direct comparisons of the field test results with theoretical values calculated by programs such as FCHART. However, except where the delivered energy is measured by Btu meters, its use involves either making a number of assumptions about each system being tested or increasing the test complexity by attempting to measure the solar system's losses as was done in the Oregon program. Probably more significant numerically is the assumption that the system set point temperature accurately represents the temperature of the delivered water. In the programs that did not involve Btu measurements, this value was obtained either by intermittent (daily, weekly, or monthly) spot measurements or by assuming that the set point switch of the backup system was accurate. Either method has its obvious problems and is inevitably the cause of some of the variability exhibited by the results.

It is sometimes important to examine the differences between the results by use of the various approaches. Evaluations of these differences can be obtained by using the Btu measurements where available and by theoretical estimates of maximum solar energy. Comparisons are shown in the system performance section of this chapter, and they indicate that the difference can be quite significant. An important issue is how much of the hot water consumed by a residence results from constant volume draws or energy draws. For example, a dishwasher or washing machine uses a volume draw, whereas a person taking a shower uses an energy draw in the sense that the outlet temperature is modulated to the comfort of the individual user. If the storage temperature is reduced significantly, greater amounts of hot water are consumed. On the other hand, for constant volume draws, the set temperature has no bearing on the amount of water consumption. Also, the seasonal changes in the supply water temperature are important in determining the monthly energy consumption or the monthly energy used for heating the water.

8.2.3.3 Comparison Procedure

The critical data reduction relationships are indicated by the three equations in section 8.2.3.1. As noted above, the annual energy saved, expressed either as a total yearly value or the daily average of that value, is the difference between the calculated conventional purchased energy, EP_C, and the measured solar system purchased energy, EP_S.

Note that the inlet water temperature, TIN, is a measured quantity of the test site, and thus it is not an arbitrary variable in the calculation of EP_C. The constant, k, in each equation, is chosen to give the values in the desired units and time period and is equal to .00244 for energy calculated in kWh/day and for g given in gallons/day.

The absolute value of the savings derived in this way depends directly on the values chosen to define $TSET$ and EL_C. The magnitude of the differences that result between systems is greatly influenced by the set point temperature chosen for the conventional system and is also proportional to the gallons used by the solar system being tested. Below 40 gallons per day, the value chosen for EL_C can be relatively significant also.

Table 8.2
Summary of EPRI data on conventional systems

Location	g (gals/day)	EP_C (kWh/day)	TSET °F	TIN °F	TSET-TIN) °F	EL_C (kWh/day)
Warm						
Corpus Christi, TX	63.7	11.2	131.5	73.5	58	2.2
Moderate						
Los Angeles, CA	85.8	18.6	146.0	70.8	75.2	2.9
Little Rock, AR	53.4	13.4	138.1	64.3	73.8	3.9
Albuquerque, NM	83.0	16.9	142.6	64.1	78.5	1.0
Chattanooga, TN	59.7	14.1	134.8	60.2	74.6	3.2
San Francisco, CA	87.2	20.3	144.7	66.0	78.7	3.6
Cold						
Portland, OR	70.7	16.9	138.1	56.0	82.1	2.7
Denver, CO	53.9	14.9	139.5	55.2	84.3	3.8
Rochester, NY	57.1	16.6	139.3	54.0	85.3	4.7
Madison, WI	64.9	16.4	143.0	53.5	89.5	2.2
Minneapolis, MN	50.0	14.6	141.0	52.6	88.4	3.8
Average-all sites	66.3	15.8	139.9	60.9	79.0	3.1
Average-warm	63.7	11.2	131.5	73.5	58.0	2.2
Average-moderate	73.8	16.7	141.2	65.1	76.2	2.9
Average-cold	59.3	15.9	140.1	54.3	85.9	3.4

Table 8.2 is a summary of the best data known to the author that provide statistical insight into establishing *TSET* and EL_C for conventional systems. EPRI requested 11 utilities to monitor 10 sites each for a year. This program resulted in the data shown in the first four columns. The equation $EL_C = EP_C - kg\,(TSET - TIN)$ provides the values for the last two columns. As shown in the table, the data set can be divided into three categories according to groupings of the individual inlet water temperatures. This arrangement corresponds somewhat to the climate, which is referred to as warm, moderate, and cold. It is interesting to note that the southernmost site, Corpus Christi, Texas, which has the highest average ground water temperature of 73.5°F, also has the lowest *TSET*, 131.5°F.

8.3 System Performance

8.3.1 Operational Variability

Table 8.3 shows the result of comparing the data collected from a number of programs using equations 1 and 3. Original data from the various tests

Table 8.3
Comparison of calculated annual energy savings

Location	AES average kWh/month	Number of sites analyzed	Data reduction methods			Comments
			Equation used	TSET °F	EL_C kWh/day	
Florida (ESG)	219	87	1	130	2.1	
Florida (FSEC)	215	16	1	130	2.1	
	191	16	3	N/A	N/A	Delivered load measured with Btu meters
NC, VA, AL	243	37	1	130	3.1	
Pennsylvania	228	40	1	140	2.9	
Massachusetts	289	23	1	140	3.7	
New York (L.I.)	172	167	1	140	3.1	
Maryland (Baltimore)	306	23	1	140	3.1	Btu also from solar input to auxiliary tank
	167	23	3	N/A	N/A	
Oregon (State)	192	36	1	140	3.1	T_{out} and T_{in} measured monthly. Average $EL_C = 3.2$
	150	36	3	Meas.	Meas.	
Oregon (Portland)	156	38	1	140	Meas.	Average $EL_C = 2.6$
	138	31	3	Meas.	Meas.	Average $EL_C = 2.3$
Kansas	342	73	1	140	3.1	Data reduced using a constant 4.9 gal/kWh for the conventional system. This equates to 67 gallons using TSET = 140° and $EL_C = 3.1$ kWh/day
Tennessee (Nashville)	277	17–28	1	140	3.1	
	169	17–28	3	N/A	N/A	Delivered load measured with Btu meters

were reworked using either the gallon-equivalency method or the total-load-equivalency method (if the required Btu data were available). It was possible to derive both AES_G and AES_{GT} for several of the programs: Florida (FSEC), Tennessee, Oregon, and Baltimore. The $TSET$ and EL_C values noted allow quantitative comparison between the two methods. The results show that the savings produced by the gallon-equivalency method are consistently higher than those for the total-load method, and by significant amounts. The differences between conventional and single-tank solar system losses are, on the average, not likely to be much over 1 kWh/day or 30 kWh/month. (If separate solar and conventional tanks are used, total losses are typically double the conventional losses, resulting in differences of 2 to 4 kWh/day between solar and conventional systems.) Also, note that the two Oregon programs produced heat loss measurement averages of 3.2 and 2.3 kWh/day, which span the 2.7 average measured for the conventional systems in Portland (shown in table 8.2). Thus, most of the differences in energy savings calculated by the two methods must be due to average delivery temperatures lower than the $TSET$ used as the delivery temperature in the AES_G calculations.

In the Baltimore AES_{BT} calculations (assuming the Btu meters to be accurate), the average delivered temperature was 118°F. This is significantly lower than the $TSET$ of 140°F used in the AES_G calculations. For a few systems, the Btu measurements indicated yearly average temperatures as low as 104°F. However, an average temperature that low would seem to indicate the probability of errors in the solar input measurement, since showers and baths feel chilly just below this temperature. The calculated average temperatures of the Tennessee Valley Authority (TVA) tests are over 121°F, and thus their Btu measurements seem more reasonable.

Whatever we should believe about the accuracy of the average delivered water temperatures, there is no doubt that the field tests show that the solar users saved considerable amounts of energy by the acceptance of temperatures lower than the 130°F to 140°F level at which the conventional systems are set (Stoney and Jones 1983). This is an extremely important message for the solar industry, since it points out the significance of user operational patterns in the achievement of energy savings. This observation could have both design and educational implications. Preliminary results of tests in North Carolina (ESG 1982) indicate that the use of timers on the backup power increased the savings by 20%. Similar findings were shown in the same reference for the Florida tests. An increase in solar savings fraction from .64 to .82 was achieved by Florida users who kept their backups entirely off year round, compared with those who kept them on. This practice translates into a 28% increase in savings.

8.3.2   Usage Variability

Another useful finding of the field test programs relates to the wide variability exhibited in water use patterns, daily or yearly. While there has been a remarkable confirmation of the much used ASHRAE mean value of the rate of 20 gallons per day per person, the tests have simultaneously shown that the probability of encountering values between 10 and 30 gallons per day per person can be almost as high as that of encountering the value of 20 gallons per day per person. Figure 8.1 illustrates the variation in water usage shown by typical field measurements.

From the solar system designer's point of view, the extremely wide variations of hot water demand between days, weeks, and months is very important and should be considered when the system capacity is being determined. Also, it is clear that test programs for performance certification should be configured to test the full range of use rates in order to be realistic in their results.

8.3.3   Weather Variability

While the operational conditions discussed above are a major reason for performance variability, several other factors are at work also. Most obvious, of course, are the variations caused by the climate differences as exhibited in the data by the month-to-month differences. Several examples of the monthly variations are presented in figure 8.2. There is a tendency for southern locations to produce a fairly flat or somewhat bimodal monthly pattern with the summer peak on savings more pronounced in the colder climates.

These effects may also relate to annual weather patterns. This is illustrated in figures 8.3a and 8.3b. The Pennsylvania data (figure 8.3a) over a 2-year period illustrate an expected variability. Data from the Florida test sites for the winters of 1981–82 and 1982–83 (figure 8.3b) show that the severe cold in 1982–83 caused the savings to decrease 27% in February, while the whole winter's reduction was a significant 14%.

An indirect effect of climate and seasonal weather differences is the variation in supply water temperature and the resulting differences in energy requirements for the hot water supply. Average surface water tem-

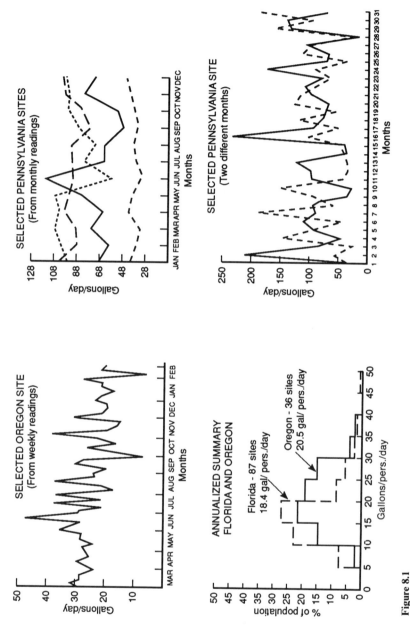

Figure 8.1
Examples of the variability in water usage shown by typical field test data.

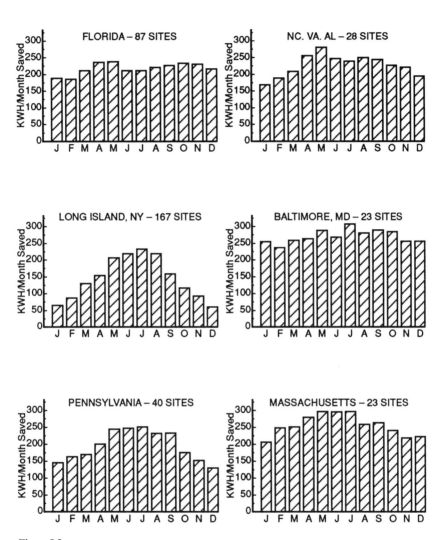

Figure 8.2
Variation in monthly energy savings at different locations in the eastern united states.

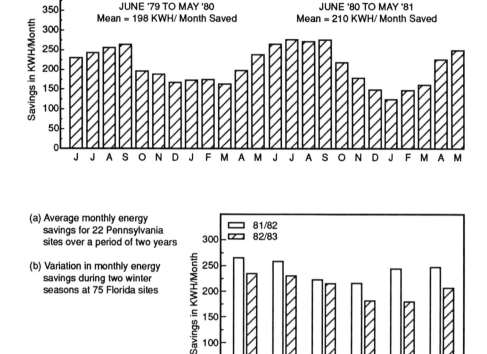

Figure 8.3
Examples showing typical annual and seasonal variations in energy savings.

peratures are higher in regions having warm climates, and differences in winter and summer supply water temperatures can exceed 20°F, particularly in the northern regions of the United States. A 25% difference in heat requirements can thus be experienced.

Since there are many other variables that could be involved in the above differences, it is interesting to find a data set that can illustrate the climate effect with relative lack of ambiguity. Table 8.4 presents the details of the systems and the data from 23 Florida sites illustrated in figure 8.3b. These sites were all under test over two slightly overlapping annual periods. The principal measure of user variability, hot water use, was precisely the same for each period, and the second user indicator, the percentage of time the backup was kept on, produced very similar averages for the two periods. Since the majority of the systems were new, it is not likely that age caused the decrease. Thus, we are left with only the weather to account for the 9% decrease in performance experienced by these 23 systems.

8.3.4 System Variability

Field tests seldom involve a sufficient number of installations to provide statistically valid data for comparing performance of as few as two different designs, much less for assessing the effects of difference among a large number of designs. Performance differences resulting from widely varying operating conditions, water usage patterns, and weather conditions conceal the effects of design differences unless a very large number of systems can be examined. Although hundreds of systems were involved in the studies analyzed here, sample sizes were not large enough to permit determination of the effects of design on system performance. These relationships can best be measured in laboratory setting, where only the design is varied and all other factors are constant (Gutierrez et al. 1974; Fisher and Fanney 1983). Investigations of this type have shown that system performance is strongly dependent on size, type, design, and quality of the system and its components, which is particularly true of the solar collector (Fanney and Klein 1983). It should be noted again, however, that laboratory testing should be designed to provide comparative measurements over the entire range of operating conditions shown by actual field monitoring programs.

There is, however, one field program that has some claim to statistical validity in terms of defining system performance differences between solar, conventional, and two other alternative system types. The Florida Solar Energy Center (FSEC) has conducted a program to monitor twenty systems of each of the following four types: solar systems, heat recovery units, heat pump water heaters, and conventional electric systems. While the primary emphasis of this program was on time-of-day energy demand, the data are also useful for total-load analysis. The data on the solar systems are included in table 8.3, line 2, and further information is in Merrigan 1983.

When the data for 16 of the solar systems are reduced by the AES_G method, the average savings are 215 kWh per month. This is very close to the value of 219 kWh per month derived from the much larger sample of

Table 8.4
Annual performance comparisons for twenty-three Florida sites

Site code	Annual energy saved (kWh) (1)	(2)	Hot water usage (gal/day) (1)	(2)
01013	2654	2569	69	74
01021	1536	1243	27	35
01022	3898	3499	86	87
01025	2561	2340	47	45
01131	3364	3092	64	67
01132	4971	4016	112	100
01212	1692	1965	25	35
01233	3012	2974	55	57
01235	3484	3325	70	76
01242	1666	1437	78	77
01311	1773	1627	26	24
01314	4331	4506	133	145
01324	1060	1310	75	73
01325	2393	2549	52	60
01411	1182	1349	26	26
01413	4590	3935	141	137
01511	3065	2915	67	69
01513	2008	1693	48	50
01514	1768	1611	43	36
01531	2707	2562	55	55
01533	5489	4253	124	112
01534	1949	1703	30	30
01535	4849	3938	108	97
Averages	2870	2627	68	68

Period (1)—August 1981 to July 1982.
Period (2)—April 1982 to March 1983.

Table 8.4 (continued)

% Back-up available		Gallons per purchased kWh		Solar savings fraction	
(1)	(2)	(1)	(2)	(1)	(2)
14	11	16	14	63	58
4	22	17	10	73	50
16	26	26	19	76	68
2	12	32	24	83	78
65	100	46	25	87	76
100	100	33	23	80	72
2	25	48	32	90	83
2	1	76	53	92	88
14	14	55	31	88	79
88	100	11	10	39	34
5	19	46	33	90	86
54	27	19	18	62	60
96	60	9	9	25	31
100	100	24	22	75	72
86	67	12	15	59	68
100	100	18	15	62	54
21	25	26	21	76	71
74	100	16	12	65	53
35	37	15	14	63	64
4	5	28	23	79	74
7	8	34	20	81	68
27	36	38	21	87	77
37	30	34	23	81	72
41	45	30	21	73	67

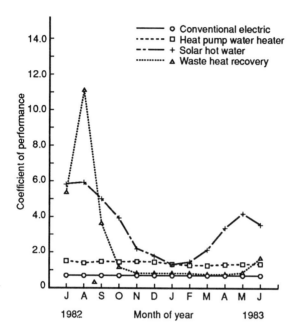

Figure 8.4
Monthly coefficient of performance data from tests of four types of water heating systems in Florida.

the ESG (1982) test program in Florida and thus provides some confirmation of the statistical applicability of the small sample.

The FSEC performance data were reduced to coefficients of performance (COP), defined as the ratio of energy delivered in the hot water divided by the energy used by the system for backup and operation of pumps, valves, and controllers. Using this coefficient, the measurements provided the data for figure 8.4, which shows the monthly values for the four systems. The average COP loads are: 0.8 for the conventional electric; 1.5 for the heat pump heater; 2.5 for the solar; and 1.1 for the heat recovery units.

The above values can be translated into savings by noting that $AES_G = EHW$ $(1/COP_C - 1/COP_S)$ and that $EHW = COP_C \times EP_C = COP_C \times ETOT_C$ (where ETOT is the total energy acquired by the system). Thus, the savings can be expressed in terms of the total energy required by the conventional systems, or $AES = ETOT_C (1 - COP_C/COP_S)$. This relationship is often defined as the solar fraction but can be used more generally as simply a savings fraction. Thus, the three systems compare with the measured conventional values as follows, $AES/ETOT$ is 0.68 for the solar systems, 0.46 for the heat pump, and 0.27 for the waste heat recovery units.

It appears that the above technique would be a reasonable way to evaluate system effectiveness. However, such comparisons may still be subject to the nonlinear effects that use patterns have on such savings fractions. These effects can easily be seen from the variability in the COP values of the tests. The conventional sites had a standard deviation of 0.09 in COP relative to a mean value of 0.8, which compared to a standard deviation of 1.83 relative to a mean of 3.29 for the solar systems. The usage in gallons per day by both sets averaged 52.5 and 55.0 with standard deviations of 21 and 23 for the conventional and solar sets, respectively. Thus, the solar values are exhibiting the greater variance found in other tests. For comparable accuracy, the solar sample should therefore be much larger than its conventional counterpart.

8.3.5 Individual System Results

In the previous sections, field test data have been analyzed statistically. This section illustrates some of the data provided by the individual test sites.

Figure 8.5 shows individual site data from the ESG test program. Each of the charts applies to the sites of a single installer. This form of the data has been of the most interest to the test participants since it provides a basis for comparing the individual systems with each other. This information makes it possible for the installer to evaluate the performance characteristics of his systems and thus to develop a sense of which features are useful and which are not. Unfortunately, the operational variable can mask the effect of the system variables, which are the only ones under control of the installers.

The wide variation in the performance of nearly every installer's set of systems is somewhat surprising. Some of this variation may be caused by system problems that are marked by sudden drops in an otherwise consistent set of monthly data. The reasons for the sharp increases in several of the systems are not immediately obvious and may in some cases (e.g., Charleston) be errors in reading the weekly values. These monthly values are more useful when examined in conjunction with the annual values and physical characteristics given in table 8.5.

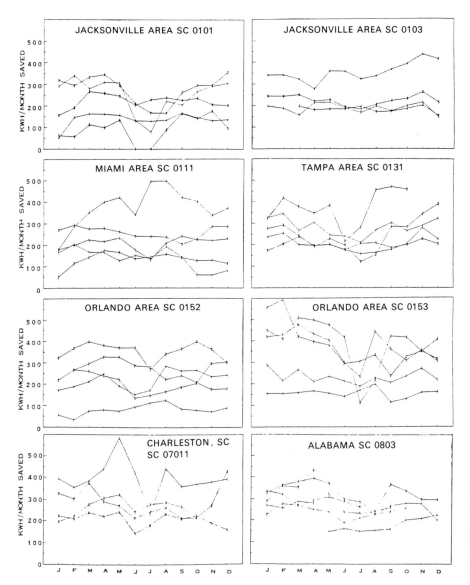

Figure 8.5
Typical data from field test program with results shown for the sites of eight installers
located in southeastern United States. Site codes (SC) are those noted on Table 8.5.

8.4 Conclusions

Results of the field tests show individually and collectively that system performance varied over a wide range. Identification of the principal causes of the variation has been an important objective of the program and has provided important information on the practicality and usefulness of solar water heating systems.

Performance variability appears to result mainly from two operational variables. The first, and most important, is the large difference in hot water temperatures consciously or unconsciously accepted by users. The second is the wide range of hot water demand and consumption. Both of these factors strongly affect solar collection efficiency and overall system performance.

Laboratory studies have shown that differences in performance of solar water heaters also result from differences in system and component design, but determination of these influences in the field test program was not possible.

Average energy saved by solar water heaters with typical collector areas of 40 to 60 square feet were about 200 to nearly 400 kWh per month in nearly all geographic areas studied. The better performers in the more favorable solar climates provided 60% to 90% of the domestic hot water requirements with annual savings of 4000 to 5000 kWh.

The field test data reduction issue draws attention to the need to define the purpose of the test before assuming what measurements need to be made and what coefficients should be derived from them.

Finally, the field test data provide ample proof that solar systems work well and save considerable energy on average. If system designs and operational methods are improved to yield the savings shown in the higher ranges achieved in many of the tests, solar systems should provide much better performance in the future.

Table 8.5
Annualized data summary for thirty-eight sites in the southeastern United States

Rank no.	Site code	Solar savings fraction projected	Annual energy saved proj.	Return on invest proj.	Data period	System type
1	01011	81	3337	12	8109–8303	drainback
2	01012	80	3470	13	8111–8303	drainback
3	01013	59	2562	12	8107–8303	drainback
4	01014	60	1485	8	8110–8303	drainback
5	01015	32	1009	4	8112–8303	drainback
6	01031	73	2237	10	8110–8303	drainback
7	01032	75	2109	17	8203–8303	recirculation
8	01033	75	2661	13	8111–8303	drainback
9	01034	93	4133	32	8109–8303	drainback
10	01111	89	3275	16	8109–8301	recirculation
11	01112	100	4326	19	8108–8210	thermosiphon
12	01113	88	3001	13	8108–8301	recirculation
13	01114	81	2592	17	8110–8303	recirculation
14	01115	74	2307	14	8109–8303	recirculation
15	01311	87	1695	11	8108–8303	recirculation
16	01312	53	1514	10	8111–8303	recirculation
17	01313	82	3161	21	8110–8303	recirculation
18	01314	59	4309	23	8108–8303	recirculation
19	01315	88	2406	16	8110–8303	recirculation
20	01521	30	898	12	8109–8303	recirculation
21	01522	61	2817	18	8110–8303	draindown
22	01523	57	4008	18	8110–8303	draindown
23	01524	76	2549	15	8112–8303	draindown
24	01525	88	2743	19	8202–8303	draindown
25	01531	75	2656	29	8108–8303	recirculation
26	01532	88	4867	23	8109–8212	recirculation
27	01533	73	4947	30	8108–8303	recirculation
28	01534	80	1812	11	8108–8303	recirculation
29	01535	75	4389	18	8108–8303	closed loop
30	07011	63	4664	14	8201–8303	closed loop
31	07012	75	3715	14	8111–8302	recirculation
32	07013	63	3160	12	8111–8303	recirculation
33	07015	68	2362	10	8201–8303	closed loop
34	08031	79	2616	14	8111–8303	drainback
35	08032	71	3199	18	8111–8210	drainback
36	08033	73	3939	26	8112–8303	drainback
37	08034	70	3150	11	8201–8303	drainback
38	08035	76	3651	14	8201–8209	drainback

Note: Monthly data from these sites is presented in figure 8.5.

Table 8.5 (continued)

Coll. area sqft	Tanks gal/gal	Gal per day	Gal per purch. kWh	Pump hours per day	Actual occup.	Percent back-up avail.	System cost after tax credit
78	120/0	65.0	30	4.2	4.1	7	$1833
65	120/0	69.0	28	7.0	4.0	17	$1804
52	80/0	70.9	15	6.4	2.9	13	$1454
52	82/0	33.7	12	4.6	2.0	25	$1200
64	120/0	46.4	8	5.4	2.1	75	$1667
32	100/0	45.0	20	6.4	1.8	5	$1500
40	82/0	40.7	21	4.3	2.0	100	$ 960
52	120/0	54.1	22	6.9	3.5	8	$1380
64	120/0	72.9	90	6.0	5.1	100	$1020
64	82/0	57.4	51	3.0	4.0	6	$1380
64	120/40	71.6	$$$$	0.0	3.8	0	$1557
60	82/0	50.9	45	4.6	3.7	59	$1560
60	82/0	48.7	29	3.7	3.0	13	$1137
40	82/0	47.1	22	5.0	2.4	7	$1200
36	82/0	25.5	37	3.8	1.9	14	$1152
40	82/0	45.2	12	3.0	2.7	84	$1176
36	82/0	66.4	35	5.7	3.5	4	$1140
36	82/0	141.8	17	6.4	4.7	47	$1500
36	82/0	42.5	47	1.8	4.1	100	$1152
30	52/0	46.1	8	3.1	2.2	10	$ 480
40	82/0	78.6	16	5.0	3.7	45	$1020
64	120/0	124.8	15	8.3	5.0	98	$1410
40	82/0	52.3	24	5.3	5.9	69	$1260
40	82/0	48.0	45	5.5	2.1	2	$1080
40	82/0	56.3	23	6.3	4.0	5	$ 690
64	120/0	97.9	52	4.0	5.8	10	$1497
80	82/0	122.3	24	5.9	4.5	8	$1260
40	82/0	30.2	25	2.5	2.4	36	$1114
80	120/0	103.8	26	6.4	4.7	29	$1838
64	82/52	105.1	14	5.4	4.2	18	$2014
42	82/0	60.9	18	6.7	2.4	7	$1530
60	82/0	63.4	12	5.5	3.8	45	$1594
42	66/40	38.6	13	4.9	3.6	12	$1440
69	65/0	34.7	18	5.0	1.9	2	$ 945
69	80/0	54.5	15	7.5	2.1	8	$ 840
69	80/0	81.5	21	7.5	3.6	10	$ 980
104	120/0	59.5	16	5.7	3.7	9	$1470
87	120/0	68.0	22	7.2	6.3	9	$1316

References

ESG, Inc. 1982 *Proceedings of the Solar Hot Water Field Test Technical Review Meeting, July 14–15, 1982.* DOE/CH/10122–7. Prepared for the U.S. Department of Energy by ESG, Inc., Atlanta, GA.

Fanney, A. H, and S. A. Klein. 1983. Performance of solar domestic hot water systems at the National Bureau of Standards—Measurements and Predictions. *J. Solar Energy Eng.* 105.

Fisher, R. A., and A. H. Fanney. 1983. Thermal performance comparisons for a solar hot water system subjected to various hot water load profiles. *ASHRAE Journal.*

Gutierres, G., F. Hincaple, J. A. Duffie, and W. A. Beckman. 1974. Simulation of forced circulation water heaters; effects of auxiliary energy supply, load type and storage capacity. *Solar Energy* 15: 287.

Merrigan, T. 1983. *Residential conservation demonstration, domestic hot water, final report.* FSEC–CR–90–83. Prepared for the Florida Public Service Commission. Cape Canaveral, FL: Florida Solar Energy Center.

Stoney, W. E., and W. M. Jones. 1983. *A method for comparing solar system test data with theory utilizing the computer program FCHART 4.0.* ESG, Inc., Energy Systems Group WP–106–83. Atlanta, GA: ESG.

Yarosh, M., ed. 1983. *Conference proceedings, Performance Monitoring of Active Solar Energy Systems Second National Workshop, January 20–21, 1983.* FSEC–CR–78–83 (EI). Cape Canaveral, FL: Florida Solar Energy Center.

9 Mechanical Performance and Reliability

R. M. Wolosewicz

9.1 Introduction

This chapter consolidates the information available on the reliability of solar energy systems used for space heating or domestic hot water. The accepted definition of reliability is the probability that a component or a system will perform its required function under the specified conditions for a specified period of time. Maintenance and instrumentation requirements as well as suggested troubleshooting guidelines are presented. In addition, corrosion and scaling in these alternative energy systems are discussed. This chapter does not include system-sizing criteria specified by simulation codes such as SOLCOST, TRNSYS, or f-chart.

Argonne National Laboratory made some attempts to gather reliability and maintainability information on federally sponsored solar energy systems. These data were gathered as a secondary effort to the collection of system performance data through the National Solar Data Network (NSDN); they are summarized in Chopra, Cheng, and Wolosewicz (1978); Chopra and Wolosewicz (1978); Mavec et al. (1978); and Waite et al. (1979). These data, which provided information on the number of problems that occurred with a particular component or subsystem, were useful to system designers; however, they did not yield hard reliability data on failure rates of the components used in solar energy systems. Recent reports by Jorgensen (1983) and Kendall et al. (1983) summarize available reliability data on components used in solar energy systems.

Solar energy system reliability studies based on component reliability data from nonsolar sources were performed at Argonne National Laboratory. These studies evaluated domestic hot water and space heating systems (Chopra and Wolosewicz 1980; Vresk et al. 1981; Wolosewicz and Vresk 1982, 1983a, and 1983b).

Although applying reliability engineering techniques to estimate the service performance of solar energy systems is new, the techniques are used daily in some industries. They are used in aerospace and nuclear industry to ensure system safety, in electric power to obtain consistent system operations, and in consumer products to reduce warranty costs or to anticipate potential manufacturing problems.

Figure 9.1 illustrates a typical failure-rate curve for a system assembled from components. When the system is first put on line (the break-in

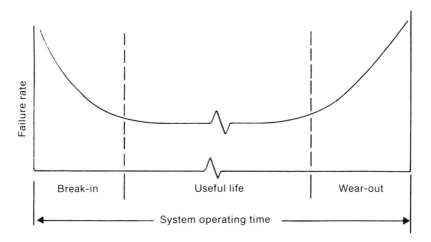

Figure 9.1
Typical system failure-rate curve.

period), the failure rate is high because of design errors, omissions, or operator errors. After the break-in period, the useful life portion of the system life cycle begins. The failure rate drops, then remains virtually constant. Any malfunctions or failures that occur are random and result primarily from the degradation of material structures by fatigue, creep, or poor maintenance.

After the system has been in service for a certain length of time, wear begins to affect its performance, and the failure rate increases (the wear-out period). At this point, a decision must be made to either overhaul the system or abandon it.

Solar energy system decisions are based on the service to be provided. Although solar energy systems are similar, they are seldom identical. Each system has specific features, and a system designed and installed in New England may not resemble a system intended for use in Florida. Despite these differences, reliability engineering can be applied to any system to reduce downtime and increase productivity.

9.2 Component Reliability

One of the major parameters of system reliability is the estimate of the mean time between failure (MTBF) or the mean life (ML) during the useful portion of the component or system life cycle. As illustrated in figure 9.1, the useful life of a component generally is characterized by a virtually constant failure rate. Component reliability is characterized by an exponential distribution function (Von Alven et al. 1964; Green and Bourne 1972; DOD 1979) and can be expressed as

$$R = \exp(-\lambda t), \tag{9.1}$$

where

R is reliability or probability that a component will operate without malfunction or failure for a specified period of time under the stated operating conditions;

λ is the component failure rate, usually expressed as the number of malfunctions or failures per unit of time;

t is the time during which the component is subjected to the operating conditions.

To evaluate equation 9.1, the component failure rate must be known. The best sources for component failure rates are field data or the results of accelerated testing. If these data are not available, component failure rates may be estimated from the failure rates of the units that comprise the component or subsystem.

For components with constant failure rates, equation 9.1 applies. Here several reliability measures, such as mean life (ML), mean time between failure (MTBF), and mean time to failure (MTTF), are used interchangeably (Von Alven et al. 1964). Although each of these reliability measures has a slightly different meaning, an accepted definition of ML, MTBF, and MTTF is the total operation time of a number of identical components, divided by the number of failures during the measured time period. In reliability testing, ML and MTBF are used interchangeably and indicate the average life of the components being tested. In this section, ML, MTBF, or MTTF will also indicate the average life of the components or systems. Because of the assumption of constant failure rates, Von Alven et al. (1964), Green and Bourne (1972), and DOD (1979) indicate that

$$ML = MTBF = MTTF = 1/\lambda. \tag{9.2}$$

The time, t, used to compute component reliability in equation 9.1 is the continuous operating time. When components such as pumps or powered valves in solar energy systems do not operate continuously, the value of t must be modified and replaced by $t \times d$, where d is the period of operation (duty cycle) factor expressed as a ratio of the operating time to the total mission time. The component ML, MTBF, or MTTF estimates should be adjusted accordingly, by increasing their value by a factor of $1/d$ (DOD 1979).

9.3   Component Failure Rates

Component failure rates are the number of failures within a specified number of operating hours. These failure rates imply that the component functioned continuously under normal operating conditions until failure occurred. The failure rate values generated by testing under normal operaing conditions are defined as base failure rates.

9.3.1   Base Failure Rates

Field or laboratory data provide the most accurate failure-rate information, but they are usually not in the public domain. Manufacturers regard these data as confidential, because they are expensive to generate and their publication would reveal design criteria. In the absence of failure data, however, failure-rate estimates can be generated from warranty data or from reliability analysis.

Assume that a component has a manufacturer's warranty that extends for one year from the date of installation. It is the judgment of the engineer specifying the component that this piece of equipment will operate on the average of H hours per day. With this background information, Vresk et al. (1981) indicate that an upper-bound estimate for the base failure rate, λ_{bu}, can be computed from

$$\lambda_{bu} = 0.0274/WH, \tag{9.3}$$

where W is the warranty period in years.

Manufacturer warranties are conservative. A lower bound on the base failure rate can be estimated by assuming that the component will operate continuously for at least as long as the warranty period. For this case, Vresk et al. (1981) indicated that:

$$\lambda_{bL} = 0.000057/W. \tag{9.4}$$

Table 9.1
Comparison of calendar time, operating time, and failure rates

Calendar time		Operating time (h)	Estimated failure rates (no. of failures/h)
Years	Hours		
1.25[a]	10,950	2,281[b]	4.4×10^{-4}
2.50	21,900	21,900	4.6×10^{-5}

Source: Wolosewicz and Vresk 1982.
a. Assumed warranty period.
b. Component operates five hours daily.

Table 9.1 summarizes these results and indicates the differences between operating time and calendar time.

If failure-rate data and warranty data are not available, component failure rates can be estimated from the failure rates of the elements that make up the component. For a component that has a number of series-connected elements that must function for the component to function, the overall failure rate is (Von Alven et al. 1964; Green and Bourne 1972; DOD 1979):

$$\lambda_b = \lambda_1 + \lambda_2 + \cdots + \lambda_n, \tag{9.5}$$

where λ_1 through λ_n are the failure rates of the individual elements that comprise the system.

More complicated components can be represented as a combination of series- or parallel-connected elements (Von Alven et al. 1964; Green and Bourne 1972; DOD 1979).

9.3.2 Operational Failure Rates

Base failure rates obtained from field data, warranty information, or reliability studies are based on the assumption that the component operates under normal conditions. However, operational parameters—temperature, voltage, time of day, etc.—modify the base failure rate values. Depending on the component and available testing data, DOD (1979) suggests the following functional expression for the failure rate of a component:

$$\lambda_p = \lambda_b(f_1 \times f_2 \times f_3 \times \cdots \times f_n), \tag{9.6}$$

where

λ_p is the component operational failure rate;

λ_b is the component base failure rate under normal operating
         conditions;

$f_1 \ldots f_n$ are environmental parameters affecting the failure rate.

For most of the components used in solar energy space heating or
domestic hot water systems, the various environmental parameters used in
equation 9.6 are unavailable at the present time. However, data from the
National Solar Data Network indicate that certain components have duty
cycles of only part-day operation (Kendall et al. 1983; DOE 1979c).

In this situation, the failure rate for a component in a solar energy
system can be estimated from DOD (1979):

$$\lambda_p = \lambda_b[d + (1 - d)a], \tag{9.7}$$

where

λ_b is the component's base failure rate;

d is the duty cycle per day (operating h/24);

a is a parameter to account for degradation during nonoperating
     periods.

The values assigned to the degradation parameter by the military range
from zero to one-half (DOD 1979). When the value of a is zero, the component is not degraded during nonoperational periods. If the degradation
parameter is assigned a value, say one-half, then the component degrades,
under nonoperating conditions, at one-half of the fully operational rate.

9.3.3 Solar Component Operational Failure-Rate Summary

The component failure rates used in previous reliability studies of solar
energy systems were based upon nonsolar data sources (Chopra and
Wolosewicz 1980; Vresk et al. 1981; Wolosewicz and Vresk 1982, 1983a,
and 1983b). A recent report by Jorgensen (1983) identified other studies
(Kendall et al. 1983; Jones 1983; Goldberg 1978) that gave failure-rate
data on components used in solar energy systems.

These data are presented in table 9.2. The table also summarizes data on
similar components used for nonsolar applications.

A review of the information in table 9.2 indicates that base failure rates
(λ_b) of components in solar and nonsolar applications are similar. With

Table 9.2
Failure rate ranges for solar system components

Component	Base failure rate (λ_b)[a] failures, 10^{-6} h	Base failure rate (λ_b)[b] failure, 10^{-6} h	Duty-cycle parameter[c] d	Assumed degradation parameter[d] a
Single flat-plate collector panel	11.4–114[e]	—	0.25	0.0–0.5
Single tubular collector module	23–73[f]	—	0.25	0.0–0.5
Control system	5.7–28.5	27.9	1.0	0.0
Storage tank or expansion tank	7.6–23	3.8–10.3	1.0	0.0
Polymeric hose	23–38	13.8	1.0	0.0
Piping system	0.02–5	—	1.0	0.0
Pump	8–150	3.8	0.25	0.2–0.4
Check valves	5.7–11.4	—	1.0	0.0
Pressure relief valves	5.7–11.4	8.4	1.0	0.0
Air vent or air separator	14–200	—	1.0	0.0
Heat exchanger	2.3–14	7.1	0.25	0.2–0.4
Heat exchanger in storage tank	11.4–23	—	1.0	0.0
Fan	22–44	—	0.25	0.0

a. Base failure rate from Vresk et al. (1981) and Wolosewicz and Vresk (1982).
b. Base failure rate from Jorgensen (1983), Jones (1983), and Goldberg (1978).
c. Duty-cycle parameter from DOE (1979).
d. Degradation parameter is based on engineering judgement and parameter ranges from 0 to 0.5.
e. Calculated failure rate from Vresk et al. (1981).
f. Calculated failure rate from Wolosewicz and Vresk (1982).

the exception of data on the control systems and pumps, these more recent data represent an average of the previous nonsolar information (Vresk et al. 1981; Wolosewicz and Vresk 1982).

The control system base failure-rate range of 5.7 to 28.5×10^{-6} was based on calculations presented in Waite et al. (1979). The upper failure-rate value of 28.5×10^{-6} agrees well with the more recent field data value of 29.7×10^{-6} failures per hour. The lower bound of 5.7×10^{-6} failures per hour is the failure rate that would be indicative of control systems using integrated circuits instead of discrete components.

The operational failure rates λ_p for the components used in solar energy systems can be computed from equation (9.7) using the information in table 9.2.

9.4 System Reliability

In developing a reliability model for a specific system, the main objective is to include a degree of complexity appropriate for the accuracy required and the data available. An overly complex model results in analytical difficulties and insufficient data; an overly simplified model can lead to inaccurate conclusions and difficulties in substantiating the assumptions.

One of the tasks in deriving a reliability formula is to prepare a functional diagram of the system describing how the input and output elements are related. System reliability therefore reflects the successful operation of one or more of the component parts. Conversely, system malfunction is represented by one or more component malfunctions. The components can be combined in series, so that the system fails if any one component fails, or they can be combined in parallel, so that when one component fails, another is available to perform the same function.

A more complex configuration consists of system components operating in series and parallel combinations. System reliability is computed by entering the block reliabilities and failure rates in the system-reliability formula and evaluating the resulting equation for the time periods of interest. Computation of system reliability provides an estimate of the MTBF.

To evaluate the reliability of a system, consider the block diagram in figure 9.2. In this system, all seven components must operate for the system to function. The reliability of this system is expressed by the following:

$$R_s = \exp(-\lambda_s t), \tag{9.8}$$

where

λ_s is $\lambda_1 + \lambda_2 + \lambda_3 + \lambda_4 + \lambda_5 + \lambda_6 + \lambda_7$;
R_s is the system failure rate;
t is time.

The failure rates for the components are presented in table 9.3. The system failure rate λ_s, which is the sum of the individual failure rates, is

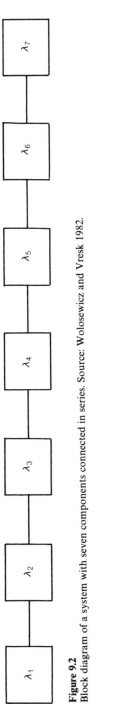

Figure 9.2
Block diagram of a system with seven components connected in series. Source: Wolosewicz and Vresk 1982.

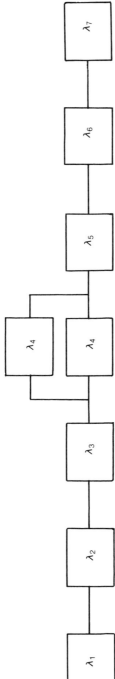

Figure 9.3
Block diagram of a system with one redundant component. Source: Wolosewicz and Vresk 1982.

Table 9.3
Component failure rate data

Component	Assumed failure rate (10^{-5} failures/h)
1	1.2
2	1.06
3	1.4
4	3.5
5	0.3
6	0.3
7	0.3

Source: Wolosewicz and Vresk 1982.

8.06×10^{-5}. Component 4 has the highest failure rate and accounts for 43% of the system failure rate, which could be reduced by using a component of higher quality. If that is not possible, a parallel combination as shown in figure 9.3 could be considered.

For the parallel combination, the system reliability can be expressed as

$$R_{sp} = [2\exp(-\lambda_s t) - \exp(-\lambda_4 + \lambda_s)t], \tag{9.9}$$

where R_{sp} is the system reliability with the parallel component.

Once the system reliability equation has been defined, the MTBF can be computed as shown in Von Alven et al. (1964), Green and Bourne (1972), and DOD (1979):

$$\text{MTBF} = \int_0^\infty R_s(t)\, dt. \tag{9.10}$$

Substituting equations (9.8) and (9.9) into equation (9.10), the MTBFs are

$$\text{MTBF}_s = 1/\lambda_s,$$

$$\text{MTBF}_{sp} = (2/\lambda_s) - [1/[\lambda_4 + \lambda_s]].$$

Using data in table 9.3, the MTBFs for these two systems are

$$\text{MTBF}_s = 12{,}407 \text{ h},$$

$$\text{MTBF}_{sp} = 16{,}163 \text{ h}.$$

By using the parallel combination, the effect of the component with the highest failure rate is reduced. The system with the redundant component has MTBF 1.3 times greater than that of the original system.

Solar energy systems used for space heating or domestic hot water use multiple solar panels that are generally connected in parallel. Chopra and Wolosewicz (1980) assessed the reliability of domestic hot water systems, and Wolosewicz and Vresk (1982) assessed the reliability of a combined domestic hot water and space heating system.

The reliability equations for solar energy systems that contain more than two collectors are lengthy and similar in form to equation (9.9). These equations can be generated by first using the concept of partial redundancy to estimate the collector array reliability (Wolosewicz and Vresk 1982; Von Alven et al. 1964). Because all of the remaining elements are in series, their overall failure rate is the sum of the individual failure rates. The reliability of these remaining system elements can then be expressed as

$$R_r = \exp(-\lambda_r t).$$

The overall system reliability can then be expressed as

$$R_s(t) = \left[\sum_{x=R}^{m} \frac{m! r_c^x (1 - r_c)^{m-x}}{x!(m-x)!} \right] \exp(-\lambda_r t), \tag{9.11}$$

where

k is the minimum number of collector modules or panels that must be operational;

m is the total number of collectors;

r_c is the reliability of a collector module or panel; can be expressed as $\exp(-\lambda_c t)$, where λ_c is the failure rate of a single collector module or panel;

λ_r is the overall failure rate of all system elements excluding the collectors;

t is time.

The MTBF of a particular system can then be computed from equation (9.10) when the reliability expression $R(t)$ is obtained from equation (9.11).

The preceding equations for estimating system reliability and MTBF are based on the assumption that the same duty cycle applies for each day of the year. If the components undergo n different duty cycles (n different values of d) through a period of time t (the number of sunny days in one year), system MTBF may be estimated by computing an averaged daily duty-cycle factor from

$$d = 1/365 \left[\sum_{i=1}^{n} d_i t_i \right], \tag{9.12}$$

where d_i is the i-th duty-cycle factor and corresponds to the number of hours per day that sufficient insolation is available, and t_i is the number of days in a year having a duty cycle of d_i.

The value of d is then used to compute λ_r and λ_c. With the new failure-rate values for the collector and the remaining system elements, the new values of MTBF can be computed from the preceding equations.

9.5 Comparison of System Reliabilities

The reliability and MTBF values for six generic domestic hot water systems were presented in Vresk et al. (1981). Wolosewicz and Vresk (1982) presented the MTBF of a drainback space heating system. The conditions analyzed were no hot water, no space heat, overheating, and freezing.

9.5.1 No Hot Water

Figure 9.4 illustrates the reliability block for a draindown domestic hot water system. Reliability block diagrams for other systems are similar and contain either more or fewer component blocks. The results of these analyses, based on the failure rates in table 9.2 and a 6-h/day duty cycle, are summarized in table 9.4.

One design represents a system with two collector panels. If the two panels are connected in a series, both collector panels must function for the system to operate. This is the two-out-of-two collector case. The same mathematical expression also applies to two collectors connected in parallel when both collectors are required to meet the hot water load.

The other designs contain four, six, or eight collectors connected in parallel. When two or more of the four collectors are working, the collector array supplies at least 50% of the load. When three out of four collec-

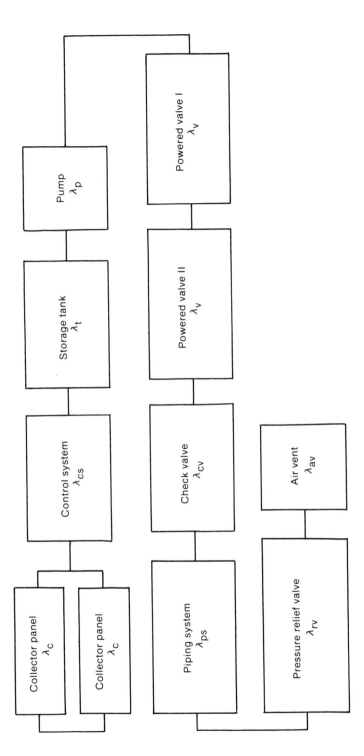

Figure 9.4
Reliability block diagram of a solar draindown DHW system with two collector panels. Source: Vresk et al. 1981.

Table 9.4
Summary of mean time between failures of generic solar DHW systems[a]

Number of collectors in system	Percentage of rated capacity	Range of MTBFs for different systems (days)[b]				Indirect	
		draindown	Circulating water	Thermoshypon	Drainback	Liquid	Air
2	50	80–947	97–1144	320–2418	174–1742	232–1418	—
2	100	64–846	72–975	174–1880	116–1434	141–1206	110–764
4	50	81–950	92–1149	304–2424	171–1749	223–1424	—
4	75	82–922	83–1078	213–2172	138–1623	170–1349	—
4	100	53–760	60–883	109–1500	87–1195	96–1039	81–94
6	50	83–950	95–1150	294–2431	177–1752	225–1427	—
6	67	71–940	90–1130	229–2278	152–1685	186–1267	—
	100	47–690	50–790	81–1251	67–1027	75–913	64–634
8	50	83–952	96–1152	288–2437	176–1756	223–1430	—
8	75	70–926	83–1106	183–2120	132–1607	155–1346	—
	100	41–632	42–714	63–1072	55–901	59–813	52–585

a. Based on Vresk et al. (1981).
b. Assumes 6-h/day duty cycle.

tors are working, at least 75% of the load is met. Similar conditions apply to the remaining configurations.

System manufacturers can use the data in table 9.4 to estimate how long a solar domestic hot water system should last before major maintenance is required. For example, a draindown system with two panels in parallel would be expected to operate from 64 to 846 days at full load before major maintenance would be required. If the system performance is allowed to deteriorate to no lower than the 50% level, the expected system MTBF can be extended approximately 12%. Similar conclusions can be drawn for the other systems and collector designs.

The large estimated MTBF values for the thermosyphon system are based on its design. This system has simple controls, does not require a pump, and has only one powered valve for overheat protection. Although this system is self-regulating, the state-of-the-art designs do not include built-in freeze protection. If users of thermosyphon systems could be relied upon to drain the system when freezing conditions were expected, these systems could be used in all climates.

The small MTBF values for the air system are due to the number of components in the system. The two motorized dampers have the same effect as the powered valves in other systems. In addition, the air vent accounts for approximately 25% of the system failure rate. If higher-quality air vents and dampers were installed, then the MTBF of the two-collector system would approach 1000 days.

Indirect liquid (antifreeze) systems have estimated MTBFs approximately 15% greater than those of draindown systems. The larger MTBF values for glycol/water systems over draindown systems occur because the former do not have vent valves and vacuum breakers. However, the glycol/water-system estimates are based on the assumptions that the glycol/water-solution pH and inhibitors are checked regularly, that the solutions buffers are added when required, and that the solution is changed when necessary.

Estimated MTBFs in table 9.4 are based on an assumed 6-hour operating day. Although a 6-hour day is typical of systems in the National Solar Data Network, data from DOE (1979c) can be used with equation (9.12) to estimate an average duty cycle factor for solar energy systems. If these data are inappropriate, more localized data can be obtained from the U.S. Weather Service.

Data concerning estimated MTBFs for the generic systems indicate:

1. Dual storage tanks reduce the mean life of a system by approximately 15% compared with a single-tank system;

2. Flexible hose interconnections reduce system mean life by approximately 50% compared with tubing that has soldered or brazed connections;

3. An order-of-magnitude difference exists between minimum and maximum estimated system mean lives.

This order-of-magnitude difference reflects the spread in the component failure-rate data. These data, summarized in table 9.2, represent the best available data in the open literature.

The MTBF estimates in table 9.4 could be used by system manufacturers to estimate warranty periods and to set up maintenance schedules. However, this information is appropriate only for the generic systems presented in this section. System manufacturers must develop comparable data based on the performance of their systems.

9.5.2   Heating System Reliability Assessment

In assessing the reliability of heating systems with a domestic hot water option, figure 9.5, the active time for the various components was estimated from data in DOE (1979). The active time of components (time under load) is summarized in table 9.2, which also indicates the values for the duty-cycle parameter to be used for computing the operational failure rates for the various components.

The duty-cycle parameter for the majority of the components listed in table 9.5 is the ratio of the active time to the number of hours in a day. The exceptions to this straightforward calculation are for pump P-2 and the water-to-air heat exchanger.

The latter two components provide the heating for the home and are not required during the warmer months. As a result, the minimum value of the duty-cycle parameter was set at 0.25. The maximum value assigned to the pump and heat exchanger duty cycle was 0.40. This latter value will apply to a solar-assisted heating system located in the northern sections of the United States.

The MTBF of five collector configurations was evaluated by using data in table 9.2 and the reliability equations. These calculations are summa-

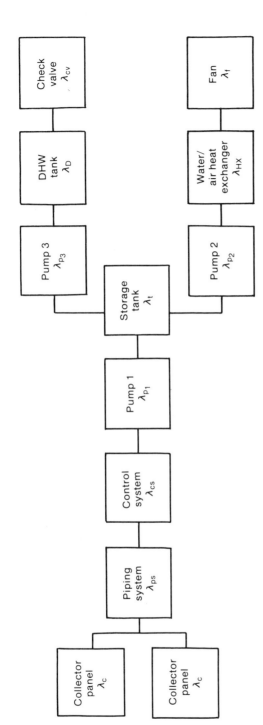

Figure 9.5
Reliability block diagram of a solar drainback system for spaceheating and domestic hot water. Source: Wolosewicz and Vresk 1982.

Table 9.5
Summary of the average active time of components used in solar space heating systems

Component	Average active time, h/day	Duty-cycle parameter
Collector	6	0.25
Storage	24	1.0
DWH tank	24	1.0
Control system	24	1.0
Pump 1[a]	6	0.25
Pump 2[a]	12–18	0.25–0.40
Pump 3[a]	12–18	0.25
Fan	12–18	0.50
Water-to-air heat exchanger	8–10	0.25–0.40
Piping system	24	1.0
Air vent	24	1.0

Source: Wolosewicz and Vresk (1982).
a. See Figure 9.5.

rized in table 9.6. This table also summarizes system MTBF values for those configurations in which the domestic hot water option or the space heating option is not included.

The lack of variation in the MTBF values as a function of the size of the collector array and the number of operational collectors indicates that the collectors have little effect on the system MTBF. For these systems, the MTBF of the system without any collectors is approximately 802 days and the MTBF at 50% load is 792 days. However, the economic implications of replacing collectors (price and installation charges for four collectors versus their life expectancy) during the expected life of the entire system must be considered in the selection and the specification of the collectors.

If the collector system must supply only space heat, not both space heat and domestic hot water, the system MTBF increases by approximately 1.26. If only domestic hot water is to be supplied, the system MTBF is approximately 1.4 times greater than if both space heat and domestic hot water are to be supplied. This increase in the system MTBF is a direct consequence of reducing the number of components in the system.

Table 9.6
Estimated mean time between failures (days) for a drainback space heating and DHW system

Collector modules	Percentage of rated capacity	Complete system		Heating system only		DHW system	
		Min	Max	Min	Max	Min	Max
6	50	113	800	146	1011	147	1128
	67	110	792	139	996	130	1109
	100	80	602	102	715	60	773
8	50	113	800	146	1011	147	1128
	75	110	785	139	993	115	1084
	100	69	558	88	650	50	700
10	50	113	800	146	1011	147	1128
	70	110	785	139	989	118	1080
	100	66	508	80	595	45	638
12	50	113	800	146	1011	147	1128
	75	110	785	139	989	112	1080
	100	58	475	73	555	38	587
14	50	113	800	146	1011	147	1128
	70	110	785	139	989	118	1060
	100	55	445	69	518	33	543

Source: Wolosewicz and Vresk (1983).

9.5.3 Overhead Protection Reliability

If hot water demand is low and insolation is high, domestic hot water systems can overheat. Vresk et al. (1981) presented three ways to prevent overheating. In one method, as overheated water is discharged from storage, cold makeup water enters the storage tank. As the storage tank cools down, the solar collection cycle is started and cool water enters the collectors. A second technique locks out the collector loop pump, which protects the storage tank while subjecting the collectors to stagnation conditions. The third method relies on a heat exchanger, valves, and a pump to dissipate collector loop heat, which protects the collectors from stagnation and the storage tank from high temperatures.

The reliability of these overheat protection techniques may be assessed by reliability block diagrams. Table 9.7 summarizes these results. Data to compute these reliabilities were taken from table 9.2.

Table 9.7
First-year overheating probabilities for solar DHW systems

| | Probability range (%) | | |
System	Water-discharge method	Pump-lockout method	Exchanger method
Draindown	9–99	5–53	14–100
Circulating water	5–40	5–53	13–100
Thermosyphon	5–40	NA[a]	NA[a]
Drainback	10–40	2–15	9–99
Indirect (antifreeze)	5–40	5–40	13–100
Air	5–40	5–40	14–100

Source: Vresk et al. (1981).
a. NA—not applicable.

Discharging heated water for overheat protection relies on a pump in a draindown system, or on makeup water in the other systems. Data in table 9.7 indicate that the makeup water method is approximately twice as reliable as the pump method.

Overheating protection by locking out the collector loop pump is as reliable as introducing cold makeup water. However, the solar collectors must be able to withstand wet stagnation or dry stagnation in drainback or air systems.

The third overheat protection method, which uses a heat exchanger, is the least reliable. This method is the most sophisticated of the overheat protection techniques because it uses additional valves and pumps as well as a heat exchanger. This third method is not recommended for residential domestic hot water systems.

9.5.4 Freeze Protection Reliability

Solar energy systems that use water as the collector coolant are attractive because water has the highest heat capacity and the lowest viscosity of all of the solar heat transfer fluids. These properties permit low flow rates to be used to extract the heat from the collectors so that the collector loop pump does not have to overcome the larger frictional forces that would be present if other heat transfer fluids were used. In addition, where allowed by local sanitary codes, a single-wall heat exchanger can be used for the preheating of domestic hot water.

Some of the disadvantages of using the antifreeze solutions are their cost and their lower heat capacity as compared to water. To compensate for the reduced heat capacity, the collector flow rates in these systems should be approximately 20% greater than in comparable water systems. Because the antifreeze fluids are more viscous than water, pipe diameters should be from 30% to 50% greater than for water systems (ERDA 1977).

The cost of antifreeze solutions dictates their use only in the solar system collector loop. A heat exchanger is the thermal link between the collector loop and the storage container. The presence of the heat exchanger requires higher collector operating temperatures, and these lower the collector's efficiency.

The remaining type of solar system uses air solar collectors. If only space heating is required, then the solar system is immune to freezing. However, when a domestic hot water option is included, the air-to-water heat exchanger must be freeze-protected by using appropriately installed and maintained dampers. Another possibility would be to locate the air-to-water heat exchanger in a duct through which cold air cannot flow (NSDN 1982; Solaron 1976; SMACCNA 1977).

The flow chart format in figures 9.6, 9.6a, and 9.6b was chosen to present a composite picture of the reasons why some of the solar energy systems experienced freezing problems (Vresk et al. 1981). Many of these failures were documented, and other failure modes were deduced by examining the system schematics (Chopra, Cheng, and Wolosewicz 1978; Chopra and Wolosewicz 1978; Mavec et al. 1978; Waite et al. 1979). It must be emphasized that all of these failures did not occur at a specific installation. At some of the solar demonstration sites, one failure led to other operational difficulties.

A review of the information presented in figures 9.6, 9.6a, and 9.6b indicates that some of the problems arise from a lack of attention to the engineering details. Other problems result from a lack of knowledge of the specific requirements of solar systems. For example, in conventional heating and ventilating systems, a valve does not usually require special sealing or operational characteristics. Therefore, a valve manufacturer's statement that his valves have a "tight closure" does not necessarily imply airtightness or bubbletightness.

Considering the high cost of solar systems, it is inadvisable to attempt to "save" money with less expensive valves or pumps. The result of a failure can be a freeze-up, and repairing the resulting problem will be more

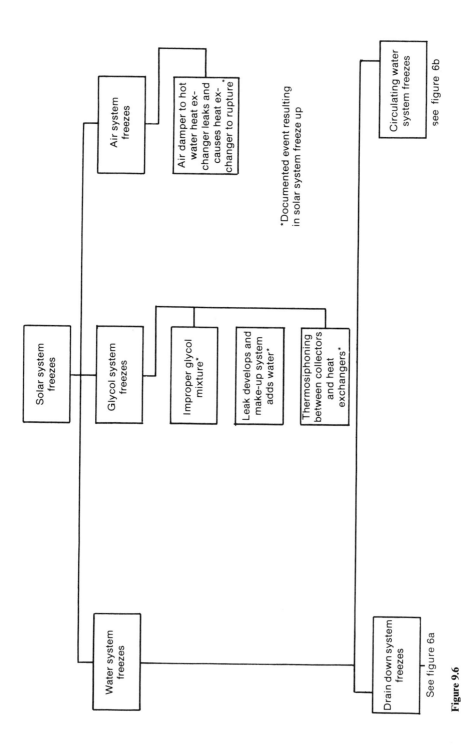

Figure 9.6
Component problems leading to freeze-ups in solar energy systems. Source: Vresk et al. 1981.

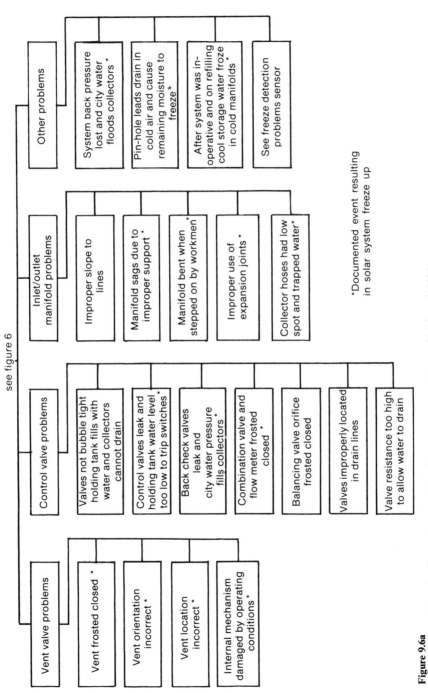

Figure 9.6a
Component problems leading to freeze-ups in solar energy systems. Source: Vresk et al. 1981.

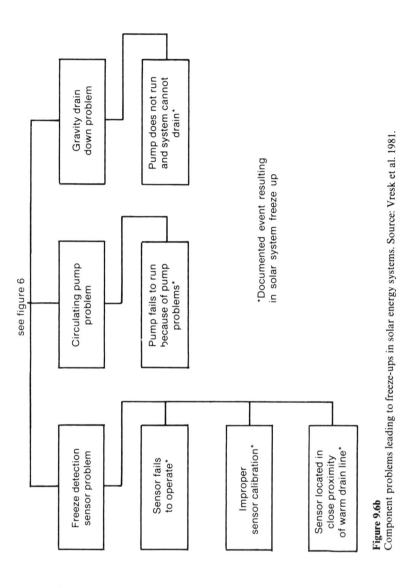

Figure 9.6b
Component problems leading to freeze-ups in solar energy systems. Source: Vresk et al. 1981.

Table 9.8
Freezing probabilities of solar DHW systems

System	Probability range (%)		
	After first year	After second year	After third year
Drainback	8–28	15–48	21–63
Circulating water	15–81	28–96	40–100
Draindown	34–48	57–900	71–100

Based on Vresk et al. (1981).

costly than the simple replacement of a single component. Wolosewicz and Vresk (1982) discussed various freeze-protection techniques and presented a conservative estimate for the temperature at which a solar collector array could freeze. Section 9.7.1 discusses the selection of freeze-protection setpoints.

If water-cooled solar energy systems are to be widely accepted in the continental United States, the systems must not freeze up in cold weather. Reliability of freeze-protection methods for draindown, drainback, and circulating-water systems was derived from reliability block diagrams and generic system schematics. The results are presented in table 9.8.

Table 9.8 does not include results for the indirect or thermosyphon systems. Indirect systems that use the correct glycol-water solution concentration should not freeze. If freezing occurs, either the glycol concentration was inadequate or the system thermosyphoned. These problems are documented in figure 9.6.

Glycol systems can also freeze if the control system malfunctions and the glycol circulates through the collector loop. This situation is similar to the system thermosyphoning and causing the water side of the collector-loop heat exchanger to freeze.

Based on data presented in table 9.8, the drainback system appears to have the lowest probability to experience a freezing problem. Drainback systems should be approximately twice as reliable as draindown systems because control valves are not required in the collector loop. If an air vent is present in a drainback system, it is not exposed to outdoor ambient conditions, nor does the air vent have to function for freeze-protection.

Data presented in figure 9.6a indicate that valves in draindown systems can malfunction and cause the system to freeze. This condition is also supported by data in table 9.8. In addition, if the piping system is not properly sized or if the manifold slopes are incorrect, the time required to drain the system could become excessive and the system could freeze.

Although the Sunbelt does not have prolonged subfreezing temperatures, during December 1983, temperatures as low as 14°F (-10°C) were reported in Tallahassee, Florida. A survey conducted by the Florida Solar Energy Center indicated that approximately 10% of the circulating water systems throughout the state froze during December 1983. The three major causes of failure were loss of electrical service, controller/pump/sensor failure, and frozen vent valves (Stutzman 1984).

Data in table 9.8 indicate that a circulating water system should be more reliable than a draindown system. For this reliability figure to be achieved, the freeze-protection sensors must be in the proper location (Kimball 1981), the sensors must be calibrated (Vresk et al. 1981; Wolosewicz and Vresk 1983a, 1983b), and the electrical utility service must not fail.

Although Kimball (1981) does not give detailed data on the reasons for pump/controller/sensor failures, from engineering judgment it is reasonable to assume that the massive freezing failure in Florida in December 1983 was probably caused by improper sensor location or sensor calibration. If the sensors are improperly located and are not properly calibrated, the controller does not receive the proper input to activate the pump to prevent freezing damage. If circulating water systems are to be used for freeze protection, the sensor calibration must be verified before each winter season.

While the controller and the pump could have failed during the freezing conditions, Kimball (1981) does not indicate if the pumps and controllers were operational while the freeze damage was being repaired. It is possible that some of the controller and pump failures could have been caused by brownout conditions or by voltage spikes on the power lines when full power was restored. Failures of air-conditioning or refrigerator compressors due to brownout or voltage spike conditions were common in the home appliance industry until protective devices were developed and installed on compressors.

Thermosyphon systems are installed in regions where freezing temperatures either do not occur or occur seldom. When freezing temperatures do occur, some method of freeze protection must be provided. The usual technique is to install electric heaters in the collector array. This technique is not recommended in colder climates because the cost of the electricity to power the electric heaters will exceed the benefits derived from solar energy.

If heaters are installed in thermosyphon systems, the reliability of the freeze-protection method depends on the location of the temperature sensors and on the reliability of the local electric utility. Vresk et al. (1981) discuss the location of the freeze-protection sensors.

The hot water loops of well-designed air systems should not freeze, although, as shown in figure 9.6, air systems can freeze if the air dampers leak. Freeze protection for an air-to-water heat exchanger can be similar to the protection described for circulating-water systems. The reliability of this freeze-protection method should be the same as shown on center line in table 9.8.

9.6   System Maintenance

For solar energy domestic hot water or space heating systems to meet the goals of reducing the need for fossil fuels, these systems must have instrumentation to detect potential problems or when a problem has occurred. In addition, system maintenance must be performed on a regular basis. The following sections present recommendations for system instrumentation as well as condensed troubleshooting guidelines.

9.6.1   Instrumentation for System Monitoring

Temperature sensors and differential thermostats are required to operate solar energy systems. Additional instruments are needed for system monitoring, checkout, and troubleshooting.

For residential systems, minimum instrumentation includes thermometers, pressure gauges, and sight glasses. If feasible, thermometers are installed close to the temperature sensors. If thermometers are not installed permanently, test plugs or thermowells should be provided so that thermometers can be installed when required.

Pressure gauges or gauge cocks should be installed across pumps, fans, and heat exchangers. An unusually large pressure drop across a pump or heat exchanger indicates that flow passages are blocked.

Flowmeters, anubars, pilot tubes with pressure indicators, or circuit setters may be installed to monitor fluid flows. However, if fluid measurements are not needed, sight glasses in major loops should be adequate.

A sophisticated monitoring system provides readouts for temperature sensors, valve or damper position, pump or fan operating modes, and possible fluid flow rates. At present, the expense of such sophisticated instrumentation may not be justified. However, significant cost reductions are possible with mass-produced integrated circuits, so sophisticated monitoring systems may be cost-effective within a few years. Farrington and Myers (1983) suggest several system configurations for detecting system faults.

9.6.2 System Startup Testing

After the system has been installed, all components cleaned, the system purged, and all sensors calibrated, but before the insulation has been installed, the system must be started and tested. Testing is necessary to detect leaks, to verify the control sequence, to locate any defective pumps or valves, and to identify and eliminate installation or design errors.

Testing procedures are specific to each system. Vresk et al. (1981) and Wolosewicz and Vresk (1982) outlined the procedures to be followed for domestic hot water and space heating systems.

1. Inspect pressure relief valves on tank(s) and on the collector array to be sure that valves operate and exit ports are not plugged.

2. Check systems periodically for: insulation deterioration; system leaks; correct thermostat setting to make certain the setting has not drifted from the position established during startup and testing; proper operation of draindown valves (these valves are closed during summer months and may stick and not open when needed to prevent a freeze-up); and correct operating mode.

3. Verify that the pump does not operate at night, using a flowmeter, sight glass, or pressure gauges if such gauges are installed.

4. Verify draining down by measuring the temperature in the vicinity of the freeze-protection sensor. Thermowells similar to those shown in figure 9.7 are recommended. If water temperature is below 40°F (4°C) and the system has not drained, drain the system manually and call a service person.

5. On a sunny day, verify pump operation by noting if the flowmeter or sight glass indicates fluid motion. In addition, check the temperature dif-

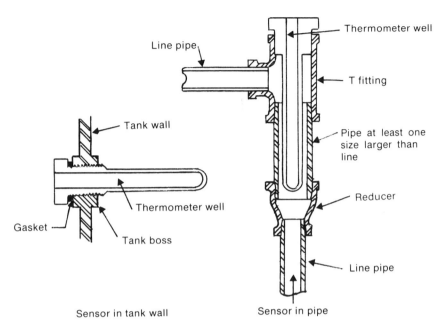

Figure 9.7
Thermowells. Source: Vresk et al. 1981.

ferential between collectors and storage by using thermometers or thermowells. If the temperature differential is above the high-temperature differential setpoint and the system has not started, be sure the pump and
controller are energized. If power is available and the system does not
start, call a service person. Similarly, if the temperature differential setpoint is below the low-temperature differential setpoint and the system
is running, call a service person.

6. Check the pump lubrication as per the manufacturer's specifications.

9.6.3   Solar System Preventative Maintenance

To ensure that the solar system will perform reliably and meet the MTBFs
that are presented in tables 9.4 and 9.6, basic preventative maintenance is
required. With the exception of sensor calibration, the remaining preventative maintenance actions can be performed by the user.

The following preventative maintenance should be performed at least
once a year:

1. Inspect the flashing and collector mounting hardware. Tighten, replace, and recaulk as required.

2. If hose connections are used on the collector array, they should be inspected and tightened as necessary. Replace degraded hoses.

3. Verify that sensors are still in the correct locations. Check the sensor calibration.

4. Inspect hand-operated valves for leaks. Adjust the packing. Open and close valves to verify proper operation.

5. Drain and flush all tanks. Clean strainers and filters.

6. Inspect and operate pumps, control systems, and automatic valves before the heating season begins.

7. Inspect pressure relief valves, air vents, and vacuum breakers on tanks and collector loop, verifying valve operation. Check that the exit ports are not plugged.

8. Check the water level in the storage tanks. If necessary, add makeup water or drain the water to the proper level and check the isolation valves.

9.6.4 System Troubleshooting

Although the exact configuration of solar energy systems depends on the manufacturer or the system designer, the troubleshooting information in tables 9.9 and 9.10 should help to identify and correct problems. Additional troubleshooting guidelines can be found in Vresk et al. (1981) and Wolosewicz and Vresk (1982).

9.7 Control Systems

Control systems and their associated sensors initiate solar collection cycles, activate auxiliary heaters, and activate freeze-protection and/or heating sequences. All of these operating conditions must be considered during the design phase of a control system so that hot water and/or space heat is available on demand.

Control system problems affected the performance of many of the systems installed as part of the HUD cycles of the Commercial Solar Demonstration Program (Bartlett 1978a, 1978b; Abrams 1978; Sparks and Raman 1978a, 1978b; Winn and Parkinson 1978; Moore 1981; DOE 1979b, 1979c). Some of the reasons why these control systems failed to

Table 9.9
Troubleshooting indirect drainback systems

Problem	Components	Possible causes	Corrective action
System does not start	Power supply	1. Tripped on overload.	1. Determine cause and replace fuse or breaker.
		2. Open circuit breaker.	2. Check and close.
		3. Defective transformer.	3. Replace.
		4. Line voltage fluctuating.	4. Inform power company.
		5. Brownout.	5. Provide brownout protective device and inform power company.
		6. Control switch on "off" position.	6. Turn to "auto" position.
	Thermostat T-1	1. High and low temperature differential setpoints too high.	1. Reset according to specifications and/or results obtained during system startup and testing.
		2. Defective component.	2. Replace thermostat.
		3. Loose contacts.	3. Tighten wires.
		4. Thermostat is out of calibration.	4. Recalibrate.
	Sensor	1. Defective.	1. Replace.
		2. Improper installation.	2. Reinstall.
		3. Defective control cable.	3. Replace.
		4. Sensor out of calibration.	4. Recalibrate.
	Pump	1. Motor failure.	1. Check brush holders, throwout mechanisms, centrifugal switches, or other mechanical components that may be loose, worn, dirty, or gummy. Replace worn components and reassemble.
		2. Overload protection switch shuts down pump motor.	2. Determine cause of overloading; check if balancing valve is in proper position.
		3. Defective shaft; impeller or coupling.	3. Replace.
		4. Defective bearings.	4. Replace.
	Control circuitry	1. Circuit continuity lost.	1. Check and repair.
		2. Bad contacts.	2. Check and correct.
System starts but cycles	Thermostat	High and low temperature differential setpoints are too close together.	Reset according to specification and/or results obtained during system startup and testing.
	Control circuitry	1. Circuit continuity lost.	1. Check and repair.
		2. Bad contacts.	2. Check and correct.
		3. Pump cycles on internal overload.	3a. Check voltage. b. Check pump flow. c. On shaded pole motor, check if shading pole ring is open and replace.

Table 9.9 (continued)

Problem	Components	Possible causes	Corrective action
Pump runs but water does not flow to collectors	Isolation pump valve(s)	Valve(s) closed.	Open valve(s).
	Pump impeller	Impeller broken or separated from shaft.	Replace impeller and/or shaft assembly.
	Blocked liquid flow passage	Pipe damaged.	Replace damaged section.
System runs continuously	Thermostat T-1	1. Low temperature differential setpoint set too low. 2. Defective component. 3. Thermostat is out of calibration.	1. Reset according to specifications or settings determined during startup. 2. Replace thermostat. 3. Recalibrate.
	Sensor(s)	1. Defective sensor(s). 2. Sensor(s) is out of calibration. 3. Incorrect sensor in circuit.	1. Replace. 2. Recalibrate. 3. Replace.
	Control circuitry	Bad contacts.	Check and correct.
System leaks	Pipe joints	1. Thermal expansion and contraction. 2. Joint improperly made.	1. Provide flexibility and reassemble. 2. Reassemble leaky joint.
	Hose connection	Clamp does not hold tightly.	Tighten up the hose clamp, replacing clamp or hose if necessary.
	Relief valve	1. Improper pressure setting. 2. Defective component.	1. Check pressure setting and correct. 2. Replace.
Poor solar energy collection	Collector array	1. Undersized collector area. 2. Collectors shaded. 3. Flow rate too high or too low. 4. Heat transfer surface covered with scale deposits. 5. Leaks.	1. Install more collector area. 2. Remove obstacle or install collectors in sunlit location. 3. Rebalance flow. 4. Flush collector loop. 5. Repair.
	Piping	1. Insufficient insulation. 2. Improper weather protection. 3. Insulation damaged.	1. Add additional insulation. 2. Provide proper weather protection. 3. Repair.
	Heat exchanger	1. Undersized. 2. Clogged.	1. Install proper size unit. 2. Clean heat exchanger.

Table 9.9 (continued)

Problem	Components	Possible causes	Corrective action
System noisy when operating	Pump cavitation	1. Restricted pump suction line. 2. Air in the system. 3. Low fluid level in reservoir.	1. Remove restrictions. 2. Check pipe installation at pump inlet. 3. Refill to proper level.
	Pump bearings	1. Bearing worn. 2. Bearing damaged owing to misalignment.	1. Replace bearing. 2. Align pump and motor shaft.
	Piping	1. Air locked in the piping. 2. Piping vibrates.	1. Check pipe installation. 2. Provide adequate pipe support.
No hot water	Makeup water shutoff valve	Valve closed.	Open valve.

	Heater failed to actuate	1. Electric heater, single-tank systems. No power to electric heater (single-tank system only).
		2. Gas heater, two-tank systems only.
		  a. Failure to ignite (gas off).
		  b. Safety switch malfunctioning.
		  c. Defective thermo-couple and/or automatic pilot valve.
		  d. Pilot won't stay lit.
		    (1) Too much primary air.
		    (2) Dirt in pilot orifice.
		    (3) Pilot valve defective.
		    (4) Loose thermocouple connection.
		    (5) Defective thermo-couple.
		    (6) Improper pilot gas adjustment.
		3. Both systems.
		  a. Thermostat defective.
		  b. Bad contacts.
			Check overload protection and correct.
			
			a. Open manual valve.
			
			b. Check and replace.
			
			c. Check and replace.
			
			(1) Adjust pilot shutter.
			(2) Open orifice.
			(3) Replace.
			(4) Tighten.
			
			(5) Replace.
			
			(6) Adjust.
			
			a. Replace.
			b. Correct.

	Temperature modulating valve	1. Valve defective. 2. Sensor defective.	1. Replace. 2. Replace.
Hot water temperature not high enough	Hot water thermostat	1. Thermostat setting too low. 2. Thermostat out of calibration.	1. Set thermostat higher. 2. Recalibrate or replace thermostat.

Table 9.9 (continued)

Problem	Components	Possible causes	Corrective action
Hot water temperature not high enough	Auxiliary heater	Heater undersized for hot water demand.	Replace when heater fails.
	Safety switch	Set too low.	Check and reset.
	Burner, two-tank system only	1. Burner clogged. 2. Undersized burner orifice.	1. Clean. 2. Provide correct size orifice.
	Temperature modulating valve	1. Sensor out of calibration. 2. Temperature set too low. 3. Valve spring too weak.	1. Recalibrate. 2. Reset. 3. Replace.
Water temperature too high	Hot water temperature thermostat	1. Thermostat setting too high. 2. Thermostat out of calibration. 3. Bad contacts.	1. Reset. 2. Recalibrate. 3. Correct.
	Sensor	Out of calibration.	Recalibrate or replace.
	Temperature modulating valve	1. Sensor out of calibration. 2. Temperature set too high.	1. Recalibrate. 2. Reset.
System overheats	Overheat protection valve V-1	1. Valve sticks. 2. Return spring failed.	1. Cycle valve by jumping control system or replace actuator. 2. Replace.
	Makeup water supply valve	Valve closed.	Open the valve.
	Overheat protection thermostat	1. Thermostat setting too high. 2. Defective component. 3. Thermostat is out of calibration. 4. Loose contacts.	1. Reset in accordance with specifications. 2. Replace thermostat. 3. Recalibrate. 4. Tighten wire.
	Control circuitry	1. Circuit continuity lost. 2. Loose contacts. 3. Coil contacts burned out.	1. Check and repair. 2. Tighten contacts. 3. Replace.
	Sensor(s)	Out of calibration.	Recalibrate.
System discharges continually	Overheat protection valve V-1	1. Valve leaks. 2. Valve sticks.	1. Replace. 2. Cycle valve by jumping control system or replace actuator.
	Overheat protection thermostat T-1	1. Defective. 2. Out of calibration. 3. Setpoint too low.	1. Replace. 2. Recalibrate. 3. Reset.
	Control circuitry	Bad contacts.	Check and replace.

Source: Vresk et al. (1981).

Table 9.10
Troubleshooting glycol or oil system

Problem	Components	Possible causes	Corrective action
System does not start	Power supply	1. Tripped on overload.	1. Determine cause and replace fuse or breaker.
		2. Open circuit breaker.	2. Check and close.
		3. Defective transformer.	3. Replace.
		4. Line voltage fluctuating.	4. Inform power company.
		5. Brownout.	5. Provide brownout protective device and inform power company.
		6. Control switch on "off" position.	6. Turn to "auto" position.
	Thermostat T-1 and T-2	1. High and low temperature differential setpoints too high.	1. Reset according to specifications and/or results obtained during system startup and testing.
		2. Defective component.	2. Replace thermostat.
		3. Loose contacts.	3. Tighten wires.
		4. Thermostat is out of calibration.	4. Recalibrate.
	Sensor	1. Defective.	1. Replace.
		2. Improper installation.	2. Reinstall.
		3. Defective control cable.	3. Replace.
		4. Sensor out of calibration.	4. Recalibrate.
	Pump	1. Motor failure.	1. Check brush holders, throw-out mechanisms, centrifugal switches, or other mechanical components that may be loose, worn, dirty, or gummy. Replace worn components and reassemble.
		2. Overload protection switch shuts down pump motor.	2. Determine cause of overloading; check if balancing valve is in proper position.
		3. Defective shaft; impeller or coupling.	3. Replace.
		4. Defective bearings.	4. Replace.
	Control circuitry	1. Circuit continuity lost.	1. Check and repair.
		2. Bad contacts.	2. Check and correct.

Table 9.10 (continued)

Problem	Components	Possible causes	Corrective action
System starts but cycles	Thermostat	High and low temperature differential setpoints are too close together, or insufficient overlapping of thermostat T-1 low and thermostat T-2 high setpoints.	Reset according to specification and/or results obtained during system startup and testing.
	Time-delay relay	Time delay "times out" too soon (freeze protection sensor does not heat up in time).	Increase time-delay relay setting.
	Control circuitry	1. Circuit continuity lost. 2. Bad contacts. 3. Pump cycles on internal overload.	1. Check and repair. 2. Check and correct. 3a. Check voltage. b. Check pump flow. c. On shaded-pole motor, verify that shading-pole ring is open and replace.
Pump runs, but coolant does not flow to collectors	Valve V-1 closed	1. Actuator defective. 2. No power to actuator. 3. Valve sticks.	1. Replace actuator. 2. Check wiring. 3. Cycle valve by jumping control system or replace actuator.
	System air locked	Air vent(s) jammed closed.	Replace vent(s).
	Pump impeller	Impeller broken or separated from shaft.	Replace impeller and/or shaft assembly.
	Blocked liquid flow passage	Pipe damaged.	Replace damaged section.
System runs continuously	Thermostat T-2	1. Low temperature differential setpoint set too low. 2. Defective component. 3. Thermostat is out of calibration.	1. Reset according to specifications or results obtained during startup. 2. Replace thermostat. 3. Recalibrate.
	Sensor(s)	1. Defective sensor(s). 2. Sensor(s) is out of calibration. 3. Incorrect sensor in circuit.	1. Replace. 2. Recalibrate. 3. Replace.
	Control circuitry	Bad contacts.	Check and correct.

Table 9.10 (continued)

Problem	Components	Possible causes	Corrective action
System leaks	Pipe joints	1. Thermal expansion and contraction. 2. Joint improperly made.	1. Provide flexibility and reassemble. 2. Reassemble leaky joint.
	Hose connection	Clamp does not hold tightly.	Tighten up the hose clamp, replacing clamp or hose.
	Relief valve	1. Improper pressure setting. 2. Defective component.	1. Check pressure setting and correct. 2. Replace.
Poor solar energy collection	Collector array	1. Undersized collector area. 2. Collectors shaded. 3. Flow rate too high or too low. 4. Heat transfer surface covered with scale deposits. 5. Leaks.	1. Install more collector area. 2. Remove obstacle or install collectors in sunlit location. 3. Rebalance flow. 4. Flush collector loop. 5. Repair.
	Piping	1. Insufficient insulation. 2. Improper weather protection. 3. Insulation damaged.	1. Add insulation 2. Provide proper weather protection. 3. Repair.
	Collector fluid	1. Antifreeze/water concentration in collector loop fluid is improper. 2. Collector loop degraded.	1. Provide proper concentration. 2. Follow in manufacturer's recommendations.
System noisy when operating	Pump cavitation	1. Restricted pump suction line. 2. Air in the system.	1. Remove restrictions. 2. Manually vent the system if automatic air vents are not adequate.
	Pump bearings	1. Bearing worn. 2. Bearing damaged due to improper alignment.	1. Replace. 2. Properly align pump and motor shaft.
	Piping	1. Air locked in the piping. 2. Piping vibrates.	1. Air vent the system. 2. Provide adequate pipe support.
	Air vents	1. Improperly sized. 2. Plugged.	1. Install proper air vents. 2. Clean and/or replace.

Table 9.10 (continued)

Problem	Components	Possible causes	Corrective action
No hot water	Makeup water shutoff valve	Valve closed.	Open valve.
	Heater failed to actuate	1. Electric heater, single-tank systems. No power to electric heater (single-tank system only).	Check overload protection and correct.
		2. Gas heater, two-tank systems only.	
		a. Failure to ignite (gas off).	a. Open manual valve.
		b. Safety switch malfunctioning.	b. Check and replace.
		c. Defective thermocouple and/or automatic pilot valve.	c. Check and replace.
		d. Pilot won't stay lit.	
		(1) Too much primary air.	(1) Adjust pilot shutter.
		(2) Dirt in pilot orifice.	(2) Open orifice.
		(3) Pilot valve defective.	(3) Replace.
		(4) Loose thermocouple connection.	(4) Tighten.
		(5) Defective thermocouple.	(5) Replace.
		(6) Improper pilot gas adjustment.	(6) Adjust.
		3. Both systems.	a. Replace.
		a. Thermostat defective.	b. Correct.
		b. Bad contacts.	
	Temperature modulating valve	1. Valve defective.	1. Replace.
		2. Sensor defective.	2. Replace.
Hot water temperature not high enough	Hot water thermostat	1. Thermostat setting too low.	1. Set thermostat higher.
		2. Thermostat out of calibration.	2. Recalibrate or replace thermostat.
	Auxiliary heater	Heater undersized for hot water demand.	Replace when heater fails.
	Safety switch	Set too low.	Check and reset.
	Burner, two-tank system only	1. Burner clogged.	1. Clean.
		2. Undersized burner orifice.	2. Provide correct size orifice.
	Temperature modulating valve	1. Sensor out of calibration.	1. Recalibrate.
		2. Temperature set too low.	2. Reset.
		3. Valve spring too weak.	3. Replace.

Table 9.10 (continued)

Problem	Components	Possible causes	Corrective action
Water temperature too high	Hot water thermostat	1. Thermostat setting too high. 2. Thermostat out of calibration. 3. Bad contacts.	1. Reset. 2. Recalibrate. 3. Correct.
	Sensor	Out of calibration.	Recalibrate or replace.
	Temperature modulating valve	1. Sensor out of calibration. 2. Temperature set too high.	1. Recalibrate. 2. Reset.
Overheating of the system	Overheat protection valve V-2	1. Valve sticks. 2. Return spring failed.	1. Cycle valve by jumping control system or replace actuator. 2. Replace.
	Makeup water supply valve	Valve closed.	Open the valve.
	Collector loop valve V-1 closed	1. Actuator defective. 2. No power to actuator. 3. Valve sticks.	1. Replace actuator. 2. Check wiring. 3. Cycle valve by jumping control system or replace actuator.
	Overheat protection thermostat	1. Thermostat setting too high. 2. Defective component. 3. Thermostat is out of calibration. 4. Loose contacts.	1. Reset in accordance with specifications. 2. Replace thermostat. 3. Recalibrate. 4. Tighten wire.
	Control circuitry	1. Circuit continuity lost. 2. Loose contacts. 3. Coil of the contacts burned out.	1. Check and repair. 2. Tighten contacts. 3. Replace.
	Sensor(s)	Out of calibration.	Recalibrate.
System drains continually	Overheat protection valve V-2	1. Valve leaks. 2. Valve sticks.	1. Replace. 2. Cycle valve by jumping control system or replace actuator.
	Overheat protection thermostat T-3	1. Defective. 2. Out of calibration. 3. Setpoint too low.	1. Replace. 2. Recalibrate. 3. Reset.
	Control circuitry	Bad contacts.	Check and replace.

Source: Vresk et al. (1981).

perform as expected were due to the use of complex operating modes and improper interfacing of the solar energy subsystems with the auxiliary heating equipment. Additional factors that degraded solar energy system performance included failure of the control unit; not accounting for time-dependent insolation rates and the response of the storage subsystem; and control decisions made on the basis of small temperature differences that were affected by improperly located temperature sensors.

Control system anomalies were and still are more difficult to identify than outright failure of a controlled subsystem. The only apparent symptom is higher-than-expected utility bills. In these cases, the system problems were traced to incorrectly placed temperature sensors, improperly chosen setpoints, miswired control units, or subtle owner modifications that were difficult to detect (Waterman 1981).

9.7.1 Control System Setpoint Selection

Control setpoints determine system efficiency and reliability. If the setpoints are too close together, the pump cycles excessively and may fail prematurely. If the setpoints are too far apart, or otherwise improperly selected, the system may not start, or it may start too late or run too long (Waite et al. 1979; Chopra and Wolosewicz 1980; Farrington and Myers 1983; Waterman 1981; Barer et al. 1980; Winn and Winn 1980; Kent and Winn 1978; Alcone and Herman 1981).

Selection of the high-temperature differential setpoint is not critical from the standpoint of energy collected. In fact, for each operating day, there is a different optimum high-temperature setpoint. The daily optimum depends on insolation, ambient temperature, storage temperatures, and wind velocity.

If the high-temperature setpoint is set too high, the system starts later than it should. However, a compensating effect occurs, because the collector fluid is at a higher temperature than it would be for an earlier startup. Conversely, if the high-temperature setpoint is set too low, the system starts too early and cycles excessively (because the difference between high and low setpoints is too small). In either case, total energy collected is the same if the collectors are well insulated and heat losses are small compared to the total energy available. For reliability and low maintenance costs, the high-temperature setpoint must be far enough away from the low-temperature setpoint to prevent any unnecessary cycling that wears out pump motors.

There is no fixed procedure for selecting the high-temperature differential setpoint. Although the usual value for liquid systems is 20°F (11°C), setpoints as low as 15°F (7°C) have been used. The usual high-temperature differential setpoint for air systems is 25° to 45°F (14° to 25°C). If in doubt, it is better to be conservative and set the system to start only when adequate insolation is available.

Selecting the low-temperature differential setpoint is more critical than selecting the high temperature setpoint. If the low-temperature setpoint is too low, the collector loop pump runs either continuously or too late into the day. In either case, the electrical energy required to power the collector loop pump or fan may exceed the available solar energy.

Proper selection of the low-temperature setpoint requires a comparison between the value of the energy collected and the cost of collecting the energy. Vresk et al. (1981) and Alcone and Herman (1981) present techniques for estimating the low-temperature setpoint. These techniques can be used for custom solar energy systems to minimize the possibility of encountering a negative value for the low-temperature differential setpoint, with a resulting energy loss.

In prepackaged solar domestic hot water systems, manufacturers can compensate for nonlinearities in the control-sensor package, sensor tolerances, and differential-thermostat hysteresis. Through proper testing, the manufacturer can establish the correct low-temperature differential setpoint for his particular system.

Water at 120°F (49°C) is hot enough for most household purposes, although higher temperatures may be required for laundry and dishwashing. However, the higher the setting of the auxiliary-heater thermostat, the lower the seasonal efficiency of the solar domestic hot water system. In fact, it is more economical to use an electrical booster heater with a dishwasher than to maintain 140°F (60°C) or higher water temperatures on an auxiliary-heater/storage tank. Most commercially available domestic hot water heaters have temperature settings marked medium and high. The recommended temperature for the medium setting is 120°F (49°C) and 140°F (60°C) for the high setting.

A conservative estimate of the ambient temperature at which a collector array could freeze due to combined effects of nocturnal radiation and convention is presented in figure 9.8. In developing these results, it was assumed that the back and sides of the collector array are well insulated, and conductive losses are neglected (Vresk et al. 1981).

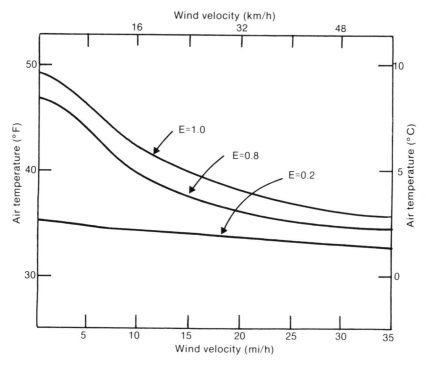

Figure 9.8
Conservative estimates for the air temperatures that cause solar collectors to freeze. Source:
Vresk et al. 1981.

In figure 9.8, three values of absorber-plate emissivities are considered.
The highest curve characterizes a perfect black body and is shown for
reference. Commercially available collectors with nonselective surfaces
have emissivities that are approximately 0.8. Theory predicts that collectors of this type can begin to freeze at an air temperature of 47°F (8°C) in
a 5-mi./h (8-km/h) wind. As the wind velocity increases, the ambient temperature at which freezing occurs decreases.

Selectively coated absorber plates have emissivities that are approximately 0.2. As shown in figure 9.8, collectors of this type can begin to
freeze at 35°F (4°C) in a 5-mi./h (8-km/h) wind. As a guide, the freeze-
detection sensors should be set to detect water temperatures of at least
40°F (4°C). The system designer must evaluate the accuracy of the instrumentation and the suitability of the sensor location before selecting a
setpoint.

The lowest system temperature does not necessarily occur only in the collectors. On a cold, partly cloudy day when water is not circulating through the collectors, marginal insolation may keep water in the collectors from freezing. However, any exposed collector loop piping may be losing heat to the environment. Thus a combination of weather, piping insulation, and time can reasonably be assumed to lead to freezing in the piping even though water in the collectors is above the freezing point. Consequently, either the freeze-protection thermostat setting must be high enough to protect the coldest portions of the piping system, or two sensors (one on the absorber plate and the second in the inlet manifold) could be used. Control system modifications are required to accommodate the use of two freeze-protector sensors.

9.7.2 Sensors

Sensor selection and installation are perhaps the most important, and least appreciated, aspects of solar energy control systems. The control system cannot produce correct outputs from inaccurate, unreliable sensor inputs. In solar energy systems, temperature measurement is the important sensor function. A collector loop sensor is subjected to a wide range of temperatures ranging from subzero to stagnation. Before selecting a sensor, the designer must confirm that it can survive the anticipated operating conditions without physical damage or loss of accuracy. Also low-voltage sensor circuits must be located away from 120/200-V AC lines to avoid electromagnetic interference (Waite et al. 1979; Farrington and Myers 1983).

Sensor calibration is critical to system performance. A routine calibration maintenance schedule is essential. The heating/ventilating/air-conditioning maintenance contractor or trained solar heating specialist is best qualified to establish and maintain sensor calibration. The average building superintendent or homeowner cannot be expected to have the needed equipment or skills.

9.7.3 Control-System Failure Modes

State-of-the-art controls for solar energy systems are designed around integrated circuit chips. Other control system components, each of which is made in several versions by more than 50 manufacturers, are triacs, relays, and transformers. The large number of variations precludes any description of each control system design.

Table 9.11
Portion of failure modes and effects analysis for DHW control system

Component	Failure mode	Direct effect	Effect of system	Failure rate $(/10^6 \text{ hr})$
Resistor R1	Open	Sensor S_1 signal becomes erratic.	System operates in an erratic manner.	0.003
	Shorted	Sensor S_1 signal rises to saturation level.	Pump does not start when solar energy is available.	0.003
Resistor R9	Open	Integrated-circuit reference voltage is out of range, and amplifier cannot turn on.	Pump cannot operate in automatic mode.	0.003
	Shorted	Integrated circuit reference input is too low, and amplifier is on.	Pump runs continuously.	0.003

Source: Waite et al. (1979).

DOD (1979) and NRC (1975) give failure rates for the electrical components in solar energy control systems. Using schematics from several sources and the control schematics in Russell (1978), DOE (1976), and DOE (1978), a failure modes and effects analysis (FMEA) can be used to estimate the failure rate of control systems. FMEA is a reliability analysis technique that is well suited to solar energy systems, because the control systems tend not to have any redundancies to avoid single-failure modes. In the FMEA technique, each component is examined to determine the ways in which it can fail, and the consequences of each failure mode are assessed with respect to the total system. The probability of failure is then computed by summing the failure probabilities that contribute to a specific failure. Table 9.11, a portion of an FMEA for a control system, illustrates the technique (Waite et al. 1979). DOD (1979) and Wang et al. (1979) present other applications.

9.8 Corrosion and Scaling

In solar energy systems, corrosion must be controlled because, otherwise, the system's reliability is reduced and premature failure can occur. In some of the early solar energy systems that were erected during the HUD Cycles or as part of the Commercial Solar Demonstration Program, electrochemically active materials were joined together without dielectric couplings, or the inhibitor levels in the glycol solutions were incorrect (Chopra, Cheng, and Wolosewicz 1978; Sparks and Raman 1978a, 1978b; Moore 1981).

9.8.1 Parameters Affecting Corrosion in Liquid Systems

Because at least six parameters affect the rate of metal corrosion in aqueous systems, it is difficult, if not impossible, to quantify the reduction in a system's reliability due to corrosion. These six parameters are:

1. Heat transfer solution pH
2. Aeration
3. Impurities and corrosion products
4. Electrical conductivity of the solutions
5. Temperature
6. Flow rate

Given sufficient testing time and data, the effects of these parameters could be quantified in a form similar to equation (9.6). The equations for a component's failure rate could then be used in the reliability equations in section 9.4 to assess the system reliability.

9.8.1.1 Heat Transfer Solution pH
The corrosion rate of a metal is highly dependent on the acid or alkali concentration in the fluid. Metals generally corrode more rapidly in acidic solutions. That is, the lower the pH, the greater the corrosion rate. Aluminum is the exception. The optimum pH for aluminum is around 6.5. The optimum pH for copper is near neutrality, or around 7.0 (Wranglen 1976).

For a three-metal system composed of aluminum-copper-iron, the corrosion potential diagram (Pourbax) indicates that there is only a narrow range of pH, between 7.5 to 8.5, where corrosion will not occur in uninhibited aqueous solutions at room temperature (Wranglen 1976). This narrow pH range is difficult to maintain, and suitable inhibitors are required to expand the region where corrosion will not occur. The inhibitors that are used must combine two functions: to seal and reinforce the protective films that form on aluminum, and to reduce the deleterious effects of heavy metal ions from the ionic dissolution of copper and iron on aluminum (Bombara and Bernaboi 1976).

9.8.1.2 Aeration

Oxygen dissolved in water is a cathodic stimulant and is one of the major contributors to corrosion in aqueous-based solar energy systems. Chlorine from the chlorination of potable water also acts as a cathodic stimulant. Dissolved oxygen in ethylene-glycol-water and propylene-glycol accelerates the decomposition of glycol into glycolic and/or formic acid (Cowan and Wintritt 1976).

If oxygen is excluded from the system, the rate of corrosion is reduced. In closed-loop solar domestic hot water or space heating systems, there is usually no problem in excluding oxygen. However, the system should be designed to prevent the entrainment of air, which can occur in turbulent areas beneath a dead-air space or by suction at a leaking pump seal.

9.8.1.3 Impurities and Corrosion Products

Raw tap water contains a number of dissolved impurities that can accelerate the corrosion process. These impurities include chlorides and sulphates, calcium and magnesium carbonates, copper and iron ions, and other dissolved solids. Each impurity causes a different corrosion problem, but the solution to each problem is the same: use deionized water with inhibitors or inhibited glycols.

Aggressive anions such as chlorides and sulphates increase corrosion rates in water or glycol-water solutions. These ions lower the electrical resistivity of corrosion cells and penetrate and break down protective films that may form on the metal surfaces.

Traces of copper or iron in heat transfer fluids can initiate pitting corrosion of aluminum collectors. Complexing inhibitors can be used to deactivate such noble ions in the solution. The addition of a getter tube or column in the collector loop can also reduce the effects of these ions.

A getter tube contains a large area of less noble metal. As the heat transfer fluid passes through the getter tube, the noble metal ions are displaced by sacrificing the less noble metal, The material in the tube must be replaced periodically as a result of this ion displacement. Although a getter tube does not eliminate corrosion in the liquid-based systems, corrosion is reduced and the getter tube provides insurance in the event of inhibitor breakdown.

Another possibility is to use a sacrificial anode. While this technique provides corrosion protection, the sacrificial anode must be sized and located properly. Unless these precautions are taken, one component in

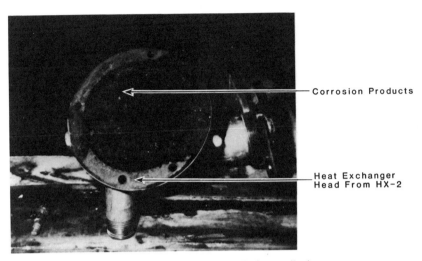

(a) Corrosion Products Accumulated at the Heat-Exchanger Head

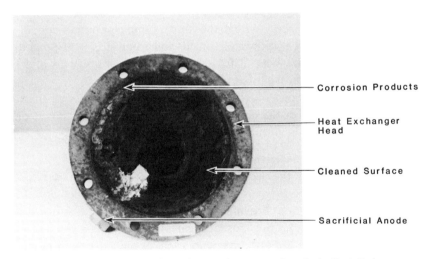

(b) Sacrificial-Anode-Prompted Local Corrosion Protection of the Heat-Exchanger Head

Figure 9.9
Corrosion products accumulated in heat exchanger head. Source: Cheng et al. 1981.

the system will be protected at the expense of the remaining portions of the solar energy system.

Precipitates from soluble metal ions in water or from system metal dissolution accelerate inhibitor depletion rates by providing large surfaces for inhibitor absorption. Green et al. (1948) found a hundredfold increase in iron corrosion in the presence of hydrated iron oxide and a lesser increase in corrosion of aluminum and solder. Mercer and Wormell (1958) reported that suspensions of ferric hydroxide rapidly depleted phosphate inhibitor but had less effect on benzoate and none on borate.

Cheng et al. (1981) discussed the failure analysis of a metal system that had stagnated. The inhibitor in the propylene-glycol was depleted, and the system had corroded to the extent that it had to be replaced.

The corrosion products consisting of iron oxide contaminated with zinc had accumulated in the head of the collector-loop heat exchanger. Although the heat exchanger was protected because of a sacrificial anode, as shown in figure 9.9, the remainder of the collector loop had severe corrosion problems.

On the same solar energy system, the copper tubing from the secondary loop heat exchanger was connected directly to carbon steel (1008) piping. As illustrated in figure 9.10, a dielectric coupling was not used and a galvanic cell was established that pitted the steel pipe.

The deepest corrosion pit extended to 38% of the original wall thickness. A cross section of the pit area is shown in figure 9.10c. Based on linear extrapolation, the pitting rate was calculated to be 35 mil/yr.

9.8.1.4 Electrical Conductivity of the Fluid

In multimetal systems, the electrical conductivity of the heat transfer fluid affects corrosion rates. In general, the greater the fluid conductivity, the higher the corrosion rate. Providing a dielectric separation between dissimilar metal connections prevents local galvanic corrosion.

For example, there are instances with some of the early solar energy systems where aluminum collectors corroded because they were connected to copper piping without using a dielectric coupling. Similarly, electrochemical incompatibility between copper pipe and a connecting steel flange accelerated local corrosion of the more active steel flange face. The leaking joint was eliminated by replacing the steel flange with a brass flange.

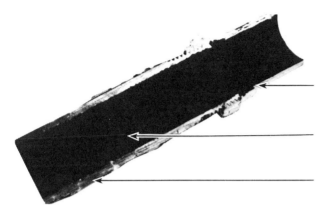

(a) Section of Iron/Copper Joint Near Heat Exchanger HX-1, Showing Accumulated Corrosion Products

(b) Cleaned Iron/Copper Joint Seen in Fig. 11-a, Showing Pitting Corrosion

(c) Cross Section of Iron Pipe Seen in Fig. 11-a and -b, Magnified 12X and Showing Pitting Corrosion

Figure 9.10
Corroded copper-iron joint showing pitting corrosion. Source: Cheng et al. 1981.

9.8.1.5   Temperature

The effect of temperature on corrosion is best summarized by an accepted rule of thumb. This rule indicates that for each 20°F (10°C), the corrosion rate doubles. Temperatures in flat-plate collectors should not exceed 185°F (85°C) when the system is in operation. During shutdown, fluid temperatures in a closed-loop system can exceed 212°F (100°C). As a safety measure, a pressure relief valve should be provided. Ethylene glycol at these temperatures can break down into corrosive acids especially if oxygen is present. High temperatures can also break down some inhibitor systems, forming precipitates that further aggravate corrosion.

To avoid these problems, fluid overheating must be prevented. One technique is to drain the fluid when the system is shut down for prolonged periods. If this is not possible, the quality of the glycol solution should be checked at regular intervals (Grover 1979).

9.8.1.6   Flow Rates

Flow rates through a solar collector must be carefully controlled and the manufacturer's recommendations must be followed in order to avoid turbulence, which removes protective oxide films. Partial restrictions of channels by debris can produce high local flow velocities that cause erosion/corrosion.

A failure analysis presented in Cheng et al. (1981) indicated that the perforation of the outlet tube of the collector originated at the inside surface (see figure 9.11) and not from the outside surfaces of the soldered boss and outlet tubes. This perforation can be attributed to erosion/corrosion resulting from the heat transfer fluid flowing through the narrow passage inside the tube clogged with corrosion products (see figure 9.11).

The corrosion products inside the outlet tube are the result of intergranular corrosion of the steel by the ethylene-glycol solution. This solution had an insufficient nitrate inhibitor concentration of 680 ppm. The recommended nitrate concentration should have been between 2000 and 4000 ppm (Cheng et al. 1981).

9.8.2   Glycol-Water Solutions

Commercial antifreeze solutions contain several important ingredients, including corrosion inhibitors, foam suppressants, and dyes. Antifreeze solutions also contain buffers to keep the heat transfer fluids alkaline by neutralizing acids formed by the oxidation of the fluid.

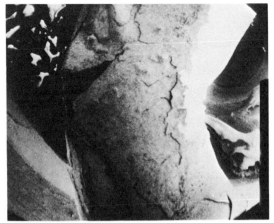

(a) Side View of Corrosion Failure in
    Outlet Tube

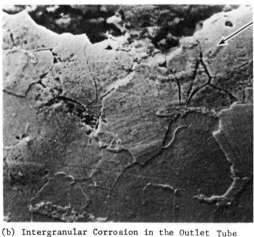

Tube
Matrix

(b) Intergranular Corrosion in the Outlet Tube
    Matrix

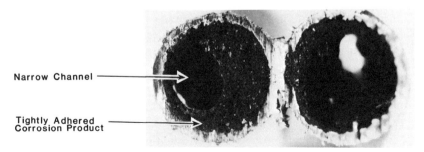

Narrow Channel

Tightly Adhered
Corrosion Product

(c) Cross Section of Clogged Outlet Tube

Figure 9.11
Tube perforation caused by erosion/corrosion. Source: Cheng et al. 1981.

There are three inhibitor categories: pH buffers that neutralize acid decomposition products, providing an optimum pH to minimize corrosion rates; film formers that react with the metal to form a protective layer against pitting; and complexing agents that tie up heavy metal ions in solution. Vresk et al. (1981) discussed the characteristics of several inhibitors that can be added to antifreeze solutions to reduce their corrosivity.

Foam suppressants are included in antifreeze solutions to prevent formation of foam caused by entrainment of air. Foam in a closed system can erode the pump and reduce the heat transfer at the absorber plate, causing hot spots. Hot spots can in turn degrade the glycol solutions and reduce the life expectancy of the fluid.

9.8.3 Calcium Carbonate Scaling

The American Society of Testing Materials lists 32 common deposits found in water handling systems (ASTM 1980). The nature of the deposits depends on the chemical composition of the water, the thermal conditions under which they are formed, and other factors. These deposits affect system performance by reducing the carrying capacity of the pipes, reducing the efficiency of the heat exchanger, localizing corrosion attack, and raising operating costs due to inefficiency, increased downtime, and maintenance.

Chemicals likely to form scale in aqueous solutions are calcium carbonate, magnesium hydroxide, calcium silicate, and calcium sulphate. Calcium carbonate has the greatest potential for deposition of scale, which forms mostly on the collector's internal absorber surfaces. The appearance of scale at this location is due to the fact that calcium carbonate becomes less soluble as the temperature increases (Singh et al. 1978).

Calcium bicarbonate in hard water gives rise to an insoluble calcium carbonate film. This film or scale may prevent corrosion, but it also reduces pipe diameters, impairs surface heat exchange properties in absorber tubes and heat exchangers, and blocks collector inlet and outlet passages.

In waters supplied by some municipalities, temporary hardness caused by calcium and magnesium bicarbonates may degrade solar system performance because of corrosion and scale formation. Permanent hardness may also be important if chlorides and sulphates or calcium and magnesium are present in significant concentrations. Water containing fewer than 50 ppm of total hardness is considered soft. If the hardness exceeds 250 ppm, the water is considered hard.

Water treatment techniques can minimize the operational problems associated with mineral scaling. For those techniques to be effective, the scaling potential of the water supply must be estimated. Vresk et al. (1981) present a simplified technique for estimating the scaling potential of water.

9.9   Conclusions

Reliability block diagrams have been presented and used in conjunction with available failure-rate data from the open literature to estimate the reliability and mean-life range for six generic solar domestic hot water systems as well as a drainback space heating system with a domestic hot water option. Designers can use these techniques and data and apply them to their specific designs to estimate the time between major maintenance efforts.

Using the troubleshooting guidelines, engineers and system designers can prepare detailed troubleshooting checklists for their specific design. These checklists could be tailored for experienced service personnel or the homeowner.

The reliability techniques presented in this chapter, when used in conjunction with the corrosion and scaling information, should be useful to help system manufacturers in establishing warranty periods or to suggest major service intervals for their systems. The techniques presented indicate that solar energy systems can be made reliable and cost-effective as long as engineering details are developed and then implemented in accordance with established engineering principles.

References

Abrams, D. W. 1978. Evaluation of solar energy control systems. *Proc. Solar Heating and Cooling Operational Results Conf.* SERI/TP-49-063. Golden, CO: Solar Energy Research Institute.

Alcone, J. M., and R. W. Herman. 1981. Simplified methodology for choosing controller set-points. *Solar Engineering.* Ed. R. L. Reid et al. New York: American Society of Mechanical Engineers.

ASTM (American Society for Testing Materials). *1980. 1979 Annual handbook*, part 30 Philadelphia, PA: ASTM.

Barer, C. A., et al. 1980. *Enhancement of performance of active systems by optimal control and system identification techniques.* SERI/TP–351–431. Golden, CO: Solar Energy Research Institute.

Bartlett, J. C. 1978a. Comparative evaluation of the thermal performance of solar energy systems. *Proc. Solar Heating and Cooling Operational Results Conf.* SERI/TP–49–063. Golden, CO: Solar Energy Research Institute.

————. 1978b. Evaluation of solar energy control systems. *Proc. Solar Heating and Cooling Operational Results Conf.* SERI/TP–49–063. Golden, CO: Solar Energy Research Institute.

Bombara, G., and U. Bernaboi. 1976. Corrosion of aluminum in multimetal water systems. *British Corrosion Journal* 11(I):25.

Cheng, C. F., et al. 1981. Corrosion problems associated with solar fluid and components. *Solar Engineering.* Ed. R. L. Reid et al. New York: American Society of Mechanical Engineers.

Chopra, P. S., C. F. Cheng, and R. M. Wolosewicz. 1978. *Preliminary evaluation of selected reliability, maintainability, and materials problems in solar heating and cooling systems.* Argonne National Laboratory Report SDP-TM-78–2. Dept. of Energy Report SOLAR/O900–78/70.

Chopra, P. S., and R. M. Wolosewicz. 1978. *Reliability and maintainability evaluation of freezing in solar systems.* Argonne National Laboratory Report SDP-TM-78–3, Dept. of Energy Report SOLAR/0901–78/70.

————. 1980. Application of reliability, maintainability, and availability engineering to solar heating and cooling systems. *Proc. IEEE Annual Reliability and Maintainability Symposium,* San Francisco, CA.

Cowan, J. C., and D. H. Weintritt. 1976. *Water former scale deposits.* Houston, TX: Gulf Publishing Company.

DOD. *see* U.S. Department of Defense.

DOE. *see* U.S. Department of Energy.

ERDA (U.S. Energy Research and Development Administration). 1977. *ERDA facilities solar design handbook.* ERDA 77–65.

Farrington, R. E., and D. Myers. 1983. *Evaluation and laboratory testing of solar domestic hot water control systems.* SERI/TR–254–1805. Golden, CO: Solar Energy Research Institute.

Green, A. H., and A. J. Bourne. 1972. *Reliability technology.* NY: Wiley-Interscience.

Green, D. H., et al. 1948. Requirements of an engine antifreeze and methods of evaluation. *American Society for Testing Materials Bulletin* 154:57.

Goldberg, M. D. 1978. Solar performance and component reliability. *Proc. Solar Heating and Cooling Operational Results Conf.* SERI/TP–49–063. Golden, CO: Solar Energy Research Institute.

Grover, M., ed. 1979. *Corrosion prevention in solar heating systems.* Ottawa, Canada: Natural Research Council of Canada, Solar Information Series.

Jones, W. M. 1983. *Solar water heater performance monitoring program of SDHW systems.* Cape Canaveral, FL: Florida Solar Energy Center.

Jorgensen, G. J. 1983. *An assessment of historical reliability and maintainability data for solar hot water and space conditioning systems.* Draft Report SERI/TR–253–2120. Golden, CO: Solar Energy Research Institute.

Kendall, P. W., et al. 1983. *A reliability and maintainability study of select solar energy system components in the National Solar Data Network.* SERI/TP–254–1948. Golden, CO: Solar Energy Research Institute.

Kent, T. B., and C. B. Winn. 1978. Performance of controllers in solar heating systems. *Proc. Solar Heating and Cooling Operational Results Conf.* SERI/TP–49–063. Golden, CO: Solar Energy Research Institute.

Kimball, B. A. 1981. How to get reliable freeze protection for solar collectors. *Solar Engineering,* pp. 24–27.

Mavec, J., et al. 1978. *Reliability and maintainability evaluation of solar collector and manifold interconnections.* Argonne National Laboratory Report SDP–TM–79–4. Dept. of Energy Report SOLAR/0902–79/70.

Mercer, A. D., et al. 1958. Research and experience with sodium benzoate and sodium nitrite mixtures as corrosion inhibitors in engine coolants. *SCI The Protection of Motor Vehicles from Corrosion.* SCI Monograph no. 4. London: Society of Chemical Industry.

Moore, D. C. 1981. Lessons learned from the HUD solar demonstration program. *Proc. 1981 Annual Meeting of the American Section of the International Solar Energy Society,* vol. 4.1.

NRC (U.S. Nuclear Regulatory Commission). 1975. *Reactor safety study: An assessment of accident risks in the U.S. commercial nuclear power plants.* NRC Report WASH–1400 (NUREG 75/104). Washington, DC: NRC.

NSDN (National Solar Data Network). 1982. *The effects of air-damper leaks on solar energy system performance.* NSDN Report SOLAR/0012–78/29.

Russell, T. 1978. For your solar water heater, build a temperature control. *Popular Science* 214 (3):134–38.

Singh, I., et al. August 1978. *Prediction of calcium carbonate scaling from water in solar energy systems.* Argonne National Laboratory Report ANL/SDP-TM-79-9. U.S. DOE Report SOLAR/0905–79–70.

SMACCNA (Sheet Metal and Air Conditioning Contractors National Association). 1977. *Heating and air conditioning system installation standards for one and two family dwellings and multi-family housing, including solar.* Tysons Corner, VA: SMACCNA.

Solaron. 1976. *Application engineering manual.* Commerce City, Colorado: Solaron Corporation.

Sparks, H. R., and K. Raman. 1978a. Lessons learned on solar system design problems from the HUD solar residential demonstration program. *Proc. Solar Heating and Cooling Operational Results Conf.* SERI/TP–49–063. Golden, CO: Solar Energy Research Institute.

————. 1978b. Lessons learned on solar system installation, operation and maintenance problems from the HUD solar residential demonstration program. *Proc. Solar Heating and Cooling Operational Results Conf.* SERI/TP–49–063. Golden, CO: Solar Energy Research Institute.

Stutzman, G. W. 1984. *Survey of the solar industry after the December 1983 freeze.* Florida Solar Energy Center Report FSEC–RM–15–85. Cape Canaveral, FL: Florida Solar Energy Center.

U.S. Department of Defense. 1979. *Reliability prediction of electronic equipment, military standardization handbook.* MIL–HDBK—217c. Washington, DC: DOD.

U.S. Department of Energy. 1976. *Preliminary design package for sunspot domestic hot water heating system.* DOE/NASA Contractor Report CR–150605.

————. 1978. *Solar control design package.* DOE/NASA Contractor Report CR–150771.

————. 1979a. *Solar energy system performance evaluation for the J.D. Evans home.* DOE Report, SOLAR/1013–79–14.

————. 1979b. *Solar energy system performance evaluation for the J.D. Evans home.* DOE Report SOLAR/1012–79–14.

————. 1979c. *Environmental data from sites in the National Solar Data Network.* DOE Report SOLAR/0010–79–11.

Vresk, J., et al. 1981. *Final reliability and materials design guidelines for solar domestic hot water systems.* Argonne National Laboratory Report ANL/SDP–11. DOE Report SOLAR/0909–81/70.

Von Alven, W. H., et al. 1964. *Reliability engineering.* NY: Prentice-Hall.

Waite, E., et al. 1979. *Reliability and maintainability evaluation of solar control systems.* Argonne National Laboratory Report SDP–TM–79–4. DOE SOLAR/0903–79/70.

Wang, P. Y., et al. 1979. *Reliability, maintainability and availability engineering for integrated community energy systems.* Argonne National Laboratory Report ANL/CNSV–6.

Waterman, R. E. 1981. Analysis, simulation and diagnosis of solar energy control system anomalies. *Solar Engineering.* Ed. R. L. Reid et al. New York: American Society of Mechanical Engineers.

Winn, C. B., and B. W. Parkinson. 1978. Experiences realized with air and liquid space heating systems during the past three years. *Proc. Solar Heating and Cooling Operational Results Conf.* SERI/TP–49–063. Golden, CO: Solar Energy Research Institute.

Winn, R. C., and C. B. Winn. 1980. *Implementation of an optimal controller of the second kind.* SERI/TP–351–433. Golden, CO: Solar Energy Research Institute.

Wolosewicz, R. M., and J. Vresk. 1982. *Reliability and maintainability design guidelines for combined space heating and domestic hot water systems.* Argonne National Laboratory Report ANL/SDP–12.

————. 1983a. Reliability and maintenance of solar domestic hot water systems. ASHRAE Paper DC–83–16. *ASHRAE Trans,* vol. 89, part 2A, 2B.

————. 1983b. Protection against freezing in solar energy systems. ASME Paper 83–WA/SOL–14. New York: American Society of Mechanical Engineer.

Wranglen, G. 1976. *An introduction to corrosion and protection of metals.* Frome & London: Butler & Tunner, p. 70.

10 Costs of Commercial Solar Hot Water Systems

Thomas King and Jeff Shingleton

10.1   Introduction

Cost is certainly an important factor affecting the rate of adoption of solar energy systems. However, the cost of these systems has been difficult to quantify accurately to permit meaningful evaluation of cost effectiveness. In addition, comparisons between systems are made difficult because the term "cost" means different things to different people. Possible definitions are:

1. Construction cost of the system
2. Life cycle cost of the system
3. Cost of energy delivered by the system

Option 3, the cost of energy delivered by the system (dollars per million Btu), is a complete way of comparing one energy system to another. This cost would include such factors as:

- construction cost
- reliability/downtime
- maintenance costs
- system lifetime
- operating energy costs
- energy delivered
- tax credits and other tax effects
- interest rate

Various life-cycle cost parameters can be developed using the same list of factors that will permit comparison of a solar energy system to any other investment or potential purchase. The major difficulty in dealing with options 2. and 3. is that all the factors that make up these costs tend to be very site- and system-specific. For instance, the amount of energy produced is affected by climate and system efficiency. In turn, system efficiency is a strong function of the nature of the load at the site, shading, collector mounting angle, control settings, and numerous other factors. The same system installed in two different sites could yield significantly different costs of energy or life-cycle costs. In addition to being very site-specific, the nature of some variables makes them very difficult to measure. For instance, maintenance costs have been identified accurately for very few systems because these costs are difficult to track over time even though they can be significant.

In very few cases have all cost variables been quantified in operating systems with sufficient accuracy to permit an evaluation of true cost effectiveness using life-cycle-cost or cost-of-energy approaches. Thus, one must be satisfied with a less than complete way of looking at system costs, if data based on real systems are to be used. Installation or construction cost is the best choice, if not the only choice, remaining.

Construction cost is not without merit as a measure of system cost. Certainly, it is usually the major factor in total system life-cycle cost. Also, construction costs of a number of *nonresidential* systems have been identified in a consistent, accurate, detailed manner.

This chapter presents construction costs of *commercial* (nonresidential)-sized systems in terms of dollars per square foot of collector area, which does not account for variations in efficiency of different systems. However, when the data limitations are recognized, very useful conclusions can be drawn.

The purpose of this chapter is to provide a perspective on the construction cost of solar water heating system and on the factors that influence cost, based on data generated by various programs. Emphasis is placed on costs at the subsystem level. In addition to reporting these costs in terms of dollars per square foot of collector area, subsystem costs are presented in terms of percent of total system cost.

The focus of the chapter is the construction cost of hot water systems in nonresidential applications. Comparable data for residential systems are not available. However, data for other system types (heating and cooling) are also presented because many subsystems are comparable across system types. By examining a larger data base that includes other system types, better conclusions can be drawn for water heating systems at the subsystem level. Also, comparisons of the total cost of different system types are useful in placing water heating systems in perspective.

Chapter 15 is similar to this chapter, but it focuses on the cost of solar space heating systems. The raw data presented in chapter 15 are also contained in this chapter, but chapter 15 specifically limits its discussion to conclusions that can be made about space heating system costs only.

Data have been obtained for presentation here from a number of sources. Most of the discussion centers on actual construction cost data collection from installed systems through detailed review of construction records. The programs under which the systems were funded were the following:

Program	Number of systems analyzed	Reference
The National Solar Heating and Cooling Demonstration Program (SHAC)	23	King et al. (1979)
The Industrial Process Heat Field Test Program (IPHFT)	10	Shingleton and King (1981)
The Tennessee Valley Authority (TVA) Commercial and Industrial Solar Water Heating Project	11	Burrows and Ira (1982)
The Solar Federal Buildings Program	8	Mueller Associates (1982)

In addition to projects from the above programs, one privately funded project designed by Mueller Associates, Inc., is included in the analysis (Flesher 1982).

The projects from the SHAC Demonstration Program are the oldest of the projects investigated, dating between 1976 and 1979. The eight projects from the Solar Federal Buildings Program were all designed by Mueller Associates, Inc., and construction costs were obtained from bid data (Mueller Associates 1982). The TVA, utilizing similar collection and analysis procedures as those used in the other demonstration programs, has provided detailed data for eleven of their projects (Burrows and Ira 1982). The data collection procedures included site visits, discussions with construction contractors and subcontractors who actually installed the system, and detailed review of their records. The primary aim is determination of accurate total system costs. However, in all cases, subsystem level data have also been obtained for the following subsystems:

• collector array
• support structure

• energy transport (piping and pipe insulation)
• storage
• electrical and controls
• general construction

For the early demonstration program projects, the "demonstration," or research effect on costs, was minimized in the analysis through the following methods:

• No costs for instrumentation other than normal control instrumentation costs were included. Many of the early systems were instrumented heavily through the National Solar Data Network (DOE 1979).
• Many different types of organizations were involved in the construction of the systems. To permit better comparisons, the effects of the widely varying overhead of these organizations was eliminated by identifying "bare" construction costs and multiplying the bare costs by an overhead and profit margin typical for the construction industry.
• The costs presented do not include the design costs.
• Auxiliary energy system costs are not included.

The systems included in this analysis were constructed between 1976 and 1982. All the costs are presented in terms of 1981 or 1982 dollar values by utilizing the construction cost index to account for inflation during the period. The result of these cost data collection efforts is that a significant body of data is now available on the cost of solar thermal systems in nonresidential applications. This is particularly true for hot water systems with 27 systems reported.

In addition to the detailed construction cost data obtained for individual installations, this chapter also discusses several other projects that involved cost analysis. For instance, the DOE Active Program Research Requirements Program (APRR) included work to develop a procedure for estimating the cost of projected solar energy systems (Science Applications 1983). This procedure resembles conventional engineering cost estimating procedures. The background data for the procedure were obtained from a large number of sources.

Another source of data discussed in this chapter is the Active Solar Installation Survey performed by the DOE Energy Information Administration (EIA) (Applied Management Sciences 1983). The survey data were obtained from contacts during 1980 and 1981 with solar-related firms that provided information, including cost data, on the installations with which they were involved.

Section 10.2 consists of the presentation and analysis of detailed construction cost data collected by on-site surveys, including discussions of subsystem costs and cost factors. This is followed by presentation of additional cost data from several other sources (section 10.3). Section 10.4 presents cost-estimating information derived from the compiled cost data that can be used by system designers.

10.2   Presentation and Analysis of Cost Data

This section provides a description of the characteristics of the systems for which cost data have been collected. Also presented are the detailed cost data for each system along with various analyses of system and subsystem cost trends for hot water and steam solar thermal systems. Similar analyses of the detailed cost data for space heating solar thermal systems are presented in chapter 15, "System Costs of Solar Space Heating Systems."

The data are presented here in two forms. Tables are provided that contain all the system descriptive and cost data for each system, organized according to system application and collector type. Data trends are identified and discussed. The availability of these detailed data makes it possible for the reader to perform independent analyses. In addition to the tables, data are provided here in the form of bar graphs that illustrate the range of costs associated with various systems, subsystems, and applications. The bar graphs are intended to facilitate the visualization of trends and are not particularly precise. The reader is cautioned to rely only on the detailed data tables for specific cost data.

10.2.1   Presentation of Data Tables

Table 10.1, Summary of system characteristics, provides descriptive information, while table 10.2, Summary of system costs, provides the basic cost data that have been used in the analysis. The systems are organized in these tables according to their application: hot water, steam, space heating, and combined space heating and cooling. The systems for each application are further arranged in these tables according to the program that, provided funding to each project, generally in order of increasing total

Table 10.1
Summary of system characteristics

Project number	System name	Funding program	Solar application	New/retro	Area (ft^2)	Collector type	Reference
	Hot water systems:						
1	Aratex	SHAC	HW	R	6350	FPL	(1)
2	Iris Images	SHAC	HW	N	600	FPL	(1)
3	Ingham	SHAC	HW	N	9374	FPL	(1)
4	Campbell	IPHFT	HW	R	7335	FPL & PTC	(2)
5	York	IPHFT	HW	N	8960	SLAT CONC.	(2)
6	LaFrance	IPHFT	HW	R	5861	ET	(2)
7	Hogate's	SHAC	HW	R	5840	FPL	(1)
8	Loudon County	SHAC	HW	R	1169	FPL	(1)
9	Union General	TVA	HW	R	2400	S. POND	(3)
10	The Connection	TVA	HW	R	390	FIN TUBE	(3)
11	Univ. of South	TVA	HW	R	560	FPL	(3)
12	Appalachian CV	TVA	HW	R	1025	FPL (SB)	(3)
13	Driftwood Inn	TVA	HW	R	74	FPL	(3)
14	Mtn. Heritage	TVA	HW	R	237	PTC	(3)
15	Cracker Barrel	TVA	HW	R	545	FPL	(3)
16	Univ. of Tenn.	TVA	HW	R	384	FPL	(3)
17	Siskin Memorial	TVA	HW	R	174	FPL	(3)
18	Villa Maria	TVA	HW	R	2424	FPL	(3)
19	Mayland Tech.	TVA	HW	R	1200	ET	(3)
20	Chesapeake Col.	PRIVATE	HW	R	3924	FPL	(4)
21	Fairfield	FED	HW	R	3000	FPL	(4)
22	Delaware Valley	SFBP	HW	R	4032	FPL	(4)
23	Westover AFB	SFBP	HW	R	2304	FPL	(4)
24	San Diego	SFBP	HW	R	1728	FPL	(4)
25	Shreveport	SFBP	HW	R	2016	FPL	(4)
26	Turner	SFBP	HW	R	1728	FPL	(4)

		SFBP					
27	E. C. Clements	SFBP	HW	R	1255	FPL	(4)
	HW SYSTEMS	All	HW		2774		
	Process heat systems:						
28	Tropicana	IPHFT	IPH (STEAM)	R	10000	PTC	(2)
29	W. Pepperell	IPHFT	IPH (STEAM)	R	7500	PTC	(2)
30	J & J	IPHFT	IPH (STEAM)	R	11520	PTC	(2)
31	Home Laundry	IPHFT	IPH (STEAM, HW)	R	6496	PTC	(2)
	AVERAGES	IPHFT	IPH		8879		
	HW/STM AVGS.				3561		
	Space heating and hot water systems						
32	Moseley	SHAC	SH, HW (LIQ)	R	376	FPL	(1)
33	Telex	SHAC	SH, HW (LIQ)	R	10700	FPL	(1)
34	Billings	SHAC	SH, HW (LIQ)	R	1660	FPL	(1)
35	Charlotte	SHAC	SH, HW (LIQ)	R	3950	FPL	(1)
36	Blakedale	SHAC	SH, HW (LIQ)	N	928	FPL	(1)
	AVERAGES	SHAC	SH, HW (LIQ)		3525		
	Air space heating systems						
37	Howard's Grove	SHAC	SH (AIR)	R	2357	FPA	(1)
38	DuCat	SHAC	SH (AIR)	R	7000	FPA	(1)
39	Aberdeen	SHAC	SH, HW (AIR)	R	1260	FPA	(1)
40	Scattergood	SHAC	SH, HW (AIR)	R	2240	FPA	(1)
41	Concord	SHAC	SH (AIR)	R	1737	FPA	(1)
	AVERAGES	SHAC	SH (AIR)		2919		
	SH AVERAGES				3221		
	Heating and cooling systems						
42	Page Jackson	SHAC	SH & C	N	10943	FPL	(1)
43	Irvine	SHAC	SH & C & HW	R	5000	ET	(1)
44	Trinity	SHAC	SH & C & HW	R	15633	FR. LENS	(1)

Table 10.1 (continued)

Project number	System name	Funding program	Solar application	New/retro	Area (ft²)	Collector type	Reference
45	Mt. Rushmore	SHAC	SH & C	R	1725	FPL	(1)
46	North Hampton	SHAC	SH & C	R	3660	FPL	(1)
47	Columbia Gas	SHAC	SH & C & HW	R	2978	ET	(1)
48	Radian	SHAC	SH & C	R	350	FR. LENS	(1)
49	Reedy Creek	SHAC	SH & C	N	3840	SLAT CONC.	(1)
	AVERAGES				5516		
	SH & C AVERAGES	SHAC	SH & C		4241		

Notes:

Funding program: SHAC—Solar Heating and Cooling Demonstration Program
IPHFT—Industrial Process Heat Field Test Program
TVA—Tennessee Valley Authority
FED—Misc. Federal Programs
SFBP—Solar in Federal Buildings Program

Solar Application: HW—service hot water
IPH—industrial process heat
SH—space heating
C—space cooling

Collector type: FPL—flat-plate liquid
FPA—flat-plate air
ET—evacuated tubular
FR. LENS—Fresnel lens
SLAT CONC.—slat concentrator
S. POND—solar pond
PTC—parabolic trough concentrator
FIN TUBE—finned tube
SB—site-built

References: (1)—King et al. (1979).
(2)—Shingleton and King (1981).
(3)—Brrows and Ira (1982).
(4)—Mueller Associates (1982).

Table 10.2
Summary of system costs (1981 dollars)

Project number	System name	Cost $/ft²	Collector array		Support structure		Energy transport		Thermal storage		Electrical & controls		General construction	
			$/ft²	%	$/ft²	%	$/ft²	%	$/ft²	%	$/ft²	%	$/ft²	%
	Hot water systems													
1	Aratex	35.80	11.70	32.68	3.80	10.61	14.40	40.22	3.30	9.22	1.60	4.47	1.40	3.91
2	Iris Images	37.90	10.50	27.70	10.00	26.39	5.60	14.78	2.90	7.65	2.50	6.60	6.10	16.09
3	Ingham	46.10	10.30	22.34	15.60	33.84	15.60	33.84	1.70	3.69	2.40	5.21	1.10	2.39
4	Campbell	47.50	22.40	47.16	6.20	13.05	7.40	15.58	6.40	13.47	8.00	16.84	2.60	5.47
5	York	56.50	37.50	66.37	7.20	12.74	4.70	8.32	0.00	0.00	7.80	13.81	0.00	0.00
6	LaFrance	61.80	26.90	43.53	8.30	13.43	18.40	29.77	2.60	4.21	5.10	8.25	0.00	0.00
7	Hogate's	62.70	18.50	29.51	12.50	19.94	18.50	29.51	6.70	10.69	1.50	2.39	4.80	7.66
8	Loudon County	75.90	21.80	28.72	23.90	31.49	17.00	22.40	7.50	9.88	4.60	6.06	1.10	1.45
9	Union General	25.80	6.70	25.97	4.70	18.22	12.70	30.62	2.80	10.85	4.70	18.22	0.00	0.00
10	The Connection	40.20	20.30	50.50	0.00	0.00	7.90	19.65	12.10	30.10	0.50	1.24	0.00	0.00
11	Univ. of South	44.90	12.80	28.51	11.60	25.84	14.10	31.40	3.50	7.80	2.50	5.57	0.00	0.00
12	Appalachian CV	46.50	17.60	37.85	8.80	18.92	14.60	31.40	3.10	6.67	2.40	5.16	0.00	0.00
13	Driftwood Inn	57.40	24.10	41.99	2.20	3.83	14.40	25.09	12.60	21.95	4.10	7.14	0.00	0.00
14	Mtn. Heritage	60.50	14.50	23.97	19.50	32.23	22.40	37.02	5.70	9.42	7.40	12.23	0.00	0.00
15	Cracker Barrel	60.90	23.00	37.77	5.80	9.52	18.70	30.71	9.80	16.09	3.60	5.91	0.00	0.00
16	Univ. of Tenn.	62.20	18.30	29.42	8.70	13.99	24.60	39.55	7.00	11.25	3.50	5.63	0.00	0.00
17	Siskin Memorial	67.00	16.60	24.78	15.60	23.28	21.30	31.79	3.70	5.52	9.80	14.63	0.00	0.00
18	Villa Maria	73.80	17.30	23.44	20.40	27.64	24.10	32.66	7.40	10.03	4.60	6.23	0.00	0.00
19	Mayland Tech.	90.00	22.10	24.56	10.40	11.56	32.30	35.89	18.20	20.22	7.10	7.89	0.00	0.00
20	Chesapeake Col.	45.00	14.80	32.89	6.50	14.44	15.20	33.78	2.40	5.33	5.80	12.89	0.50	1.11
21	Fairfield	45.20	16.10	35.62	4.60	10.18	12.40	27.43	2.80	6.19	4.20	9.29	4.70	10.40
22	Delaware Valley	48.40	13.30	27.48	9.10	18.80	15.90	32.85	3.40	7.02	4.50	9.30	2.20	4.55
23	Westover AFB	55.90	16.90	30.23	9.30	16.64	16.90	30.23	3.50	6.26	7.00	12.52	2.30	4.11
24	San Diego	58.30	16.00	27.44	9.80	16.81	19.30	33.10	1.40	2.40	8.90	15.27	3.00	5.15
25	Shreveport	59.40	15.10	25.42	12.00	20.20	21.40	36.03	3.10	5.22	5.30	8.92	2.50	4.21
26	Turner	68.80	18.20	26.45	9.50	13.81	23.80	34.59	5.30	7.70	8.90	12.94	3.00	4.36

Table 10.2 (continued)

Project number	System name	Cost $/ft²	Collector array		Support structure		Energy transport		Thermal storage		Electrical & controls		General construction	
			$/ft²	%	$/ft²	%	$/ft²	%	$/ft²	%	$/ft²	%	$/ft²	%
27	E. C. Clements	75.40	17.40	23.08	3.00	3.98	33.80	44.83	4.30	5.70	12.40	16.45	3.70	4.91
	HW SYSTEMS AVERAGE	55.92	17.80	32.42	9.59	17.09	17.13	30.11	5.30	9.43	5.21	9.30	1.44	2.81
	Industrial process heat system													
28	Tropicana	55.20	31.40	56.88	11.10	20.11	10.30	18.66	0.00	0.00	2.40	4.35	0.30	0.54
29	W. Pepperell	62.80	23.70	37.74	7.00	11.15	13.20	21.02	0.00	0.00	18.20	28.98	0.00	0.00
30	J & J	74.50	28.50	38.26	10.70	14.36	13.70	18.39	5.30	7.11	13.00	17.45	3.30	4.43
31	Home Laundry	123.60	38.20	30.91	19.20	15.53	38.60	31.23	1.30	1.05	16.40	13.27	10.20	8.25
	AVERAGES	79.03	30.45	40.95	12.00	15.29	18.95	22.32	1.65	2.04	12.50	16.01	3.45	3.31
	Space Heating and hot water systems													
32	Moseley	47.40	13.80	29.11	4.60	9.70	16.00	33.76	5.10	10.76	6.40	13.50	1.10	2.32
33	Telex	53.30	17.30	32.46	9.20	17.26	20.70	38.84	2.10	3.94	2.80	5.25	1.30	2.44
34	Billings	56.80	20.40	35.92	10.80	19.01	19.80	34.86	2.50	4.40	3.50	6.16	0.00	0.00
35	Charlotte	77.50	18.00	23.23	8.20	10.58	35.10	45.29	4.60	5.94	4.60	5.94	7.10	9.16
36	Blakedale	86.00	12.00	13.95	21.40	24.88	24.20	28.14	11.80	13.72	11.80	13.72	0.10	0.12
	AVERAGES	64.20	16.30	26.93	10.84	16.29	23.16	36.18	5.22	7.75	5.82	8.91	1.92	2.81
	Air space heating systems													
37	Howard's Grove	26.20	13.40	51.15	3.30	12.60	4.40	16.79	3.80	14.50	1.30	4.96	0.00	0.00
38	DuCat	41.60	21.90	52.64	12.90	31.01	4.90	11.78	0.00	0.00	1.90	4.57	0.00	0.00
39	Aberdeen	59.30	24.80	41.82	1.90	3.20	18.40	31.03	7.80	13.15	4.70	7.93	1.30	2.19
40	Scattergood	61.00	24.50	40.16	16.00	26.23	11.20	18.36	5.30	8.69	3.80	6.23	0.30	0.49
41	Concord	87.80	18.60	21.18	35.70	40.66	17.10	19.48	7.20	8.20	8.60	9.79	0.60	0.68
	AVERAGES	55.18	20.64	41.39	13.96	22.74	11.20	19.49	4.82	8.91	4.06	6.70	0.44	0.67
	SH AVERAGES	59.69	18.47	34.16	12.40	19.51	17.18	27.83	5.02	8.33	4.94	7.81	1.18	1.74

Heating and cooling ssystems

42	Page Jackson	60.70	15.90	26.19	22.00	36.24	16.30	26.85	2.10	3.46	0.40	0.66	4.20	6.92
43	Irvine	68.20	28.30	41.50	10.70	15.69	20.90	30.65	0.00	0.00	6.70	9.82	1.70	2.49
44	Trinity	73.80	29.10	39.43	7.60	10.30	22.40	30.35	5.80	7.86	4.00	5.42	5.00	6.78
45	Mt. Rushmore	97.10	21.20	21.83	19.30	19.88	45.50	46.86	5.00	5.15	9.20	9.47	7.00	7.21
46	North Hampton	137.70	21.40	15.54	8.80	6.39	50.10	36.38	12.40	9.01	22.50	16.34	18.60	13.51
47	Columbia Gas	143.80	53.00	36.86	14.20	9.87	23.80	16.55	6.30	4.38	20.30	14.12	26.30	18.29
48	Radian	152.80	37.40	24.48	21.80	14.27	60.40	39.53	10.70	7.00	13.60	8.90	6.70	4.38
49	Reedy Creek	226.60	150.50	66.42	6.50	2.87	32.60	14.39	15.20	6.71	7.70	3.40	14.50	6.40
	AVERAGES	120.09	44.60	34.03	13.86	14.44	34.00	30.19	7.19	5.45	10.55	8.52	10.50	8.25
	SH & C AVERAGES	86.53	30.08	34.10	13.05	17.26	24.66	28.88	5.98	7.05	7.43	8.12	5.32	4.82

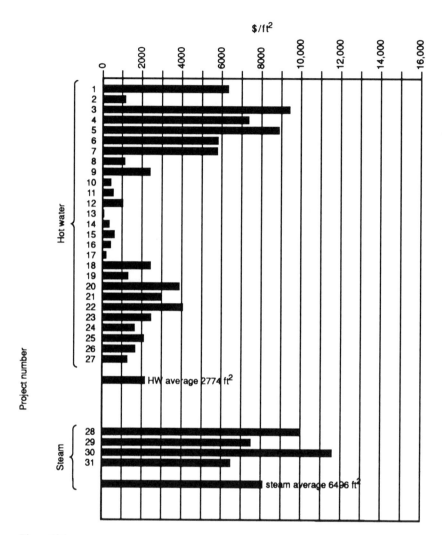

Figure 10.1
Array size vs. application.

system cost within each group of projects. As described in section 10.1, the systems for which cost data are available have been funded for construction by various federal programs, including: the Solar Heating and Cooling Demonstration Program (SHAC) (King et al. 1979); the DOE Solar Industrial Process Heat Field Test Program (IPHFT) (Shingleton and King 1981); programs of the Tennessee Valley Authority (TVA) (Burrows and Ira 1982); and miscellaneous federal programs (FED) (Flesher 1982). In addition, included are a number of systems designed by Mueller Associates, Inc., in Baltimore for the Solar Federal Buildings Program (SFBP) (Mueller Associates 1982) and for a private client (PRIVATE) (Mueller Associates 1982).

Table 10.1 and figure 10.1 show that the systems in the data set for hot water solar thermal systems exhibit a broad range of array sizes, from less than 100 to more than 9,300 square feet. Hot water (HW) systems in the data set include those that use-flat plate liquid (FLP), evacuated tubular (ET), and parabolic trough concentrating (PTC) collectors. Also included are systems that use slat concentrating and shallow solar pond collectors, although the vast majority of the HW systems use FPL collectors. The steam systems all use PTC collectors and, with an average size of more than 8,800 square feet, they are generally larger than the hot water systems.

Figure 10.2, a bar graph of array size versus collector type, illustrates a large range of system sizes for nearly every collector type. (This figure includes many systems whose balance-of-system costs are analyzed primarily in chapter 15 on the cost of solar thermal space heating systems.) The flat-plate liquid (FPL), flat-plate air (FPA), and evacuated tubular collector (ETC) arrays cluster in a size range between 2,900 to 5,900 square feet, facilitating cost comparisons between these competing collector types. The systems with parabolic trough concentrating (PTC), slat concentration (SLAT CONC), and Fresnel lens (FR. LENS) concentrating collectors generally have larger arrays, averaging 7,170 square feet.

10.2.2 Total System Costs

It is important to reiterate here that most of the data provided in this chapter and in chapter 15 are for *commercial* and *industrial*, and not *residential* solar energy systems. The cost methodology that was applied to the collection and analysis of SHAC and IPHFT systems consisted of

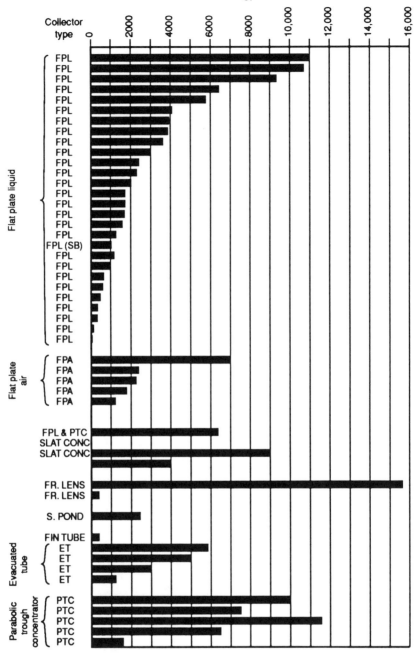

Figure 10.2
Array size vs. collector type.

determining the bare cost of construction within each of several defined subsystem categories. These cost data were adjusted to a common year of construction (1981) using historic construction cost inflation factors, while industry-standard overhead and profit factors were applied to the bare costs. The premise is that this methodology results in a close approximation of the cost of these systems, had they been entirely funded by private sector.

Figure 10.3 shows a bar graph of the total system cost data from table 10.2 for HW and steam systems. HW systems averaged $44.92/ft^2 and steam systems were $79.03/ft^2 on average. The figure shows that there is very good agreement on total system costs between the three data sets for HW systems. The fact that the TVA projects' total system costs, which were direct quotes from private contractors in a competitive environment, agree so closely with those costs for the SHAC, MAI, and SFBP systems, confirms the validity of both the data on HW system costs and of the cost analysis techniques employed.

Total system cost data were examined to determine whether economy-of-scale effects exist. Economy-of-scale effects of interest in solar systems cost analyses are those trends that suggest that large solar thermal systems have a lower $/ft^2 cost than similar small solar thermal systems. The utility and accuracy of the economy-of-scale factors are dependent on the quality and range of the data set. Using the full data set for total system costs in table 10.1 shows a relatively limited economy-of-scale effect. That data set includes 27 relatively similar systems that represent system sizes from 74 to 9,374 square feet.

One of the hopes of the solar community is that solar energy construction costs will be reduced as the benefits of experience in design and installation and in mass production of solar components are realized. Therefore, the total unit costs were tested to determine whether the year in which a project was constructed was a significant cost factor. Since all the costs were expressed in 1981 dollars, inflation was not a factor in the statistical test. Unfortunately, the results showed no significant evidence that costs are coming down.

In a similar manner, total unit costs of a subset of data were tested to determine whether regional variations in labor and materials costs were an important cost factor. The regional cost factor,s used were taken from the city cost index for mechanical construction in *Means Mechanical and*

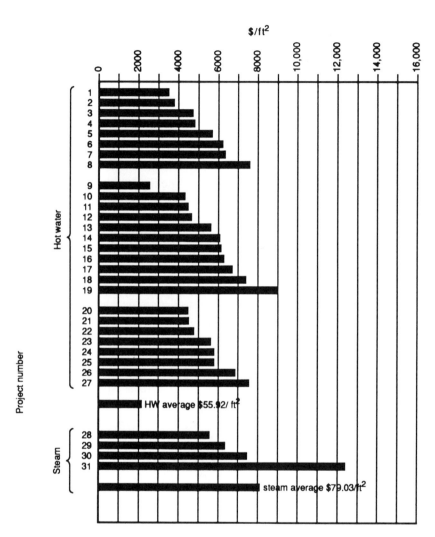

Figure 10.3
Total system cost vs. application.

Electrical Cost Data 1981 (Means 1981). Regional variations in materials and labor costs were not shown to be a significant factor in the cost of these systems.

10.2.3 Collector Array Costs

One would anticipate that collector array costs are relatively independent of system application. These costs can be shown to be more specifically dependent on the collector type. Table 10.2 lists every system in the data set, including space heating and cooling (SH&C) systems, organized according to application and, generally, according to funding program.

Figure 10.4, a bar graph of the collector cost of these systems, shows general trends in collector cost for each collector type. The least expensive collector array was for the single shallow solar pond collector on the TVA Union General system at $6.70/ft^2.

The collector array costs for the flat-plate liquid (FPL) collector arrays in all the federal and private systems were remarkably consistent at an average installed cost of $16.73/ft^2. The flat-plate air (FPA) collector array costs averaged $20.64/ft^2. Evacuated tubular (ET) collector arrays with an average of $32.58/ft^2 showed a consistently higher cost than FPL or FPA arrays installed. If the one ET array that is so much more expensive than the others is removed from the analysis, the ET array average cost is $25.77/ft^2. The two slat concentrator arrays cost $37.50 and $150.50/ft^2, while the PTC arrays, at $27.26/ft^2 interestingly, showed an average installed cost nearly equivalent to that for the ET arrays. The one finned tube collector array, at $20.30/ft^2, was more expensive than the average flat-plate array.

10.2.4 Support Structure

Support structure costs tend to be very dependent on the specific characteristics of the site, the collector characteristics, and whether the support structure is designed to perform more than simple structural functions in the installed solar system. For systems where the collectors rest directly on an existing roof or where the collector by design does not require significant additional structural elements to support it, the support structure costs for the solar thermal system can be very low or zero. Such is the case for the slat concentration collectors, which are integrated into the building roof structural system.

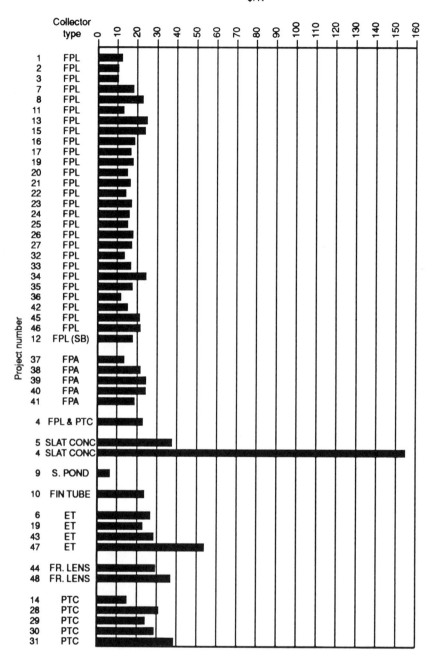

Figure 10.4
Collector cost vs. collector type.

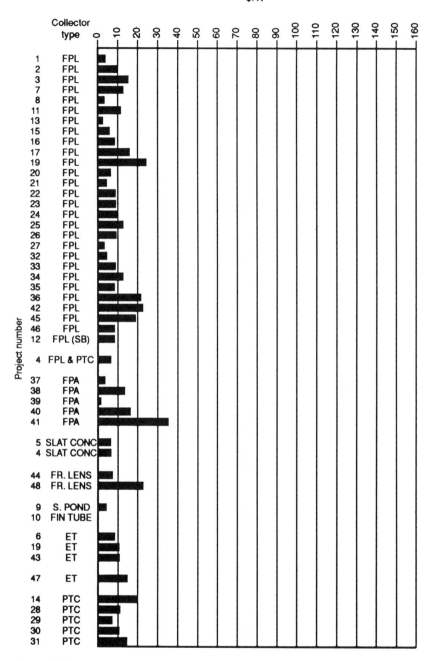

Figure 10.5
Support structure cost vs. collector type.

Figure 10.5 is a bar graph of structural costs for systems organized according to collector type. Structural costs for FPL systems averaged $11.03/ft^2 and were within a fairly narrow range. Structural costs for FPA systems were similar to costs for FPL systems, as one might expect. Structural costs for ET arrays, at $10.90/ft^2, also compare favorably with those for FPL and FPA arrays. Structural costs for PTC arrays ranged broadly, but had an average near other collector types at $13.50/ft^2.

Figure 10.6 illustrates the results of an analysis of the correlation of structural system costs to the number of functions that a specific structure performs. This analysis, performed on a majority of the systems in tables 10.1 and 10.2, indicates that, on average, an extra $6.00/ft^2 is paid for support structures that perform more than simple structural functions in an installed solar thermal system. Some of the functions that may be performed are:

- house system equipment
- improve system appearance
- provide reflectors
- act as roof/building structure
- act as roof waterproof membrane

In addition, it was determined that, based on the data available, there is little cost difference between roof-mounted and ground-mounted structures.

10.2.5 Energy Transport Costs

An analysis of energy transport costs reveals that these costs are related more to system application than to collector type. Energy transport costs for HW system are discussed here, while the reader is referred to chapter 15 for a discussion of energy transport costs for those systems.

Figure 10.7 shows a bar graph of energy transport costs organized according to system application and generally according to federal program, and in order of increasing total system cost. An inspection of the data for HW systems from various programs reconfirms the validity of the cost analysis methodology and the quality of the data obtained. Generally, it can be seen that energy transport costs account for an average of between $12.70 and $19.84/ft^2 in HW systems. Within each federal program data set, the data also interestingly suggest that systems with the highest energy transport costs also exhibited higher total system costs.

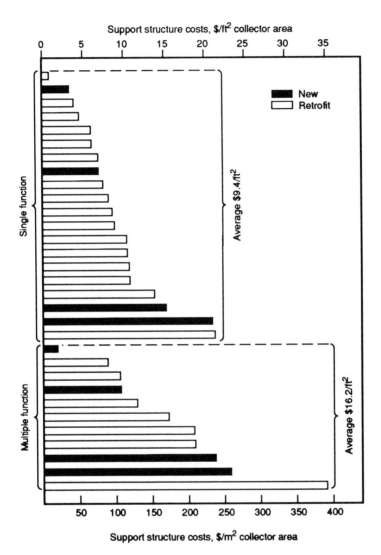

Figure 10.6
Support structure costs, $/ft^2 collector area.

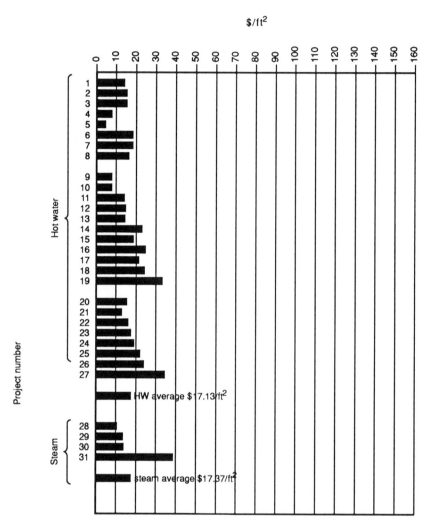

Figure 10.7
Energy transport cost vs. application.

10.2.6 Storage Costs

Because of the wide variation in storage capacity per collector area encountered among the projects, storage costs are expressed in terms of storage capacity rather than collector area. Statistical tests indicated that storage costs are dependent on both type and location of the vessel.

Figure 10.8 represents the storage subsystem costs by vessel type and by location for the majority of systems listed in table 10.1. Unpressurized steel tanks were the least expensive, followed by fiberglass tanks, rock bins, and pressurized steel tanks. In terms of location, buried storage was least expensive, followed by exterior and then interior storage. These costs do not include the cost of extra piping or ductwork to and from buried and exterior storage. Nevertheless, the results are surprising, since most estimation techniques would lead one to believe that interior tanks, which do not require waterproofing, would be less expensive.

Storage costs, on the average, represent only 8% of total system cost, and storage performance (particularly heat loss) has a large effect on system thermal performance. Therefore, in general, design measures that aim to reduce storage system cost but which may reduce storage system performance are probably unwise.

10.2.7 Electric and Control Costs

Electric and control costs tend to be primarily dependent on system application, not collector type or size. Figure 10.9 presents a bar graph of electric & controls cost organized by application and by federal program. These costs for the SHAC and TVA HW systems are very similar with an average of \$4.19 and \$4.56/ft^2. The electric & control costs for the MAI and SFBP projects are slightly higher, at \$7.12/ft^2, because those systems incorporate instrumentation for performance measurement also in this cost category.

The electric & controls costs of the steam systems, at an average \$12.50/ft^2, are generally higher than the HW systems.

Figure 10.10 is a bar graph of the percentage of total system cost represented by the electric & controls for HW and steam systems, which demonstrates less variability than these costs on a \$/ft^2 basis. HW and steam systems electric & controls cost accounted for averages of 9.3 and 16.0 percent of total system costs.

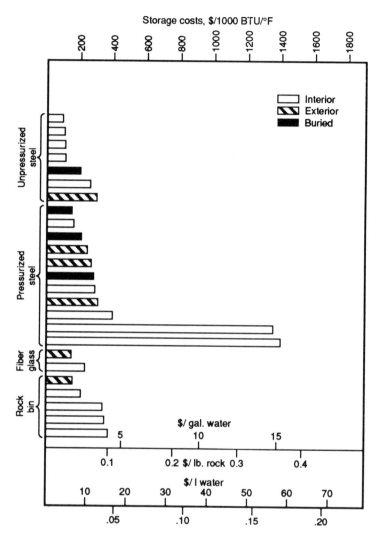

Figure 10.8
Storage costs, $/1,000 Btu.

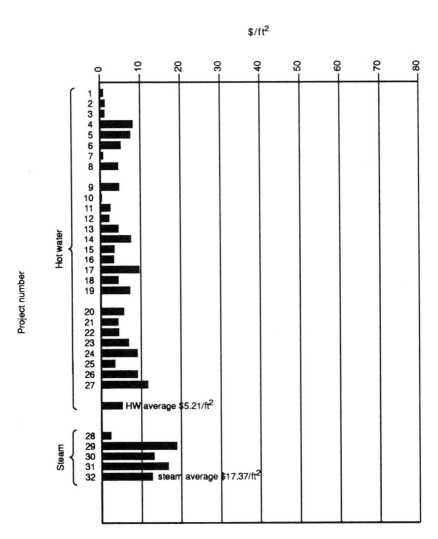

Figure 10.9
Electric and controls cost vs. application.

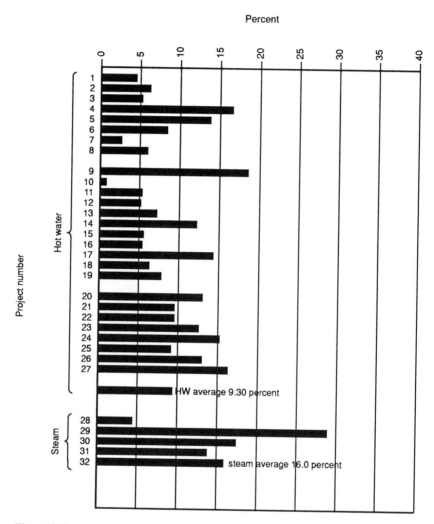

Figure 10.10
Electric and controls cost percentage of total vs. application.

10.2.8   General Construction Costs

The category of general construction costs includes any construction that is not directly related to a specific subsystem in a solar thermal system but which is necessary for the installation of that solar thermal system at a particular site. These costs are therefore very site-specific and show no identifiable trends. For those systems with general construction costs, these costs are generally less than $4.00/ft^2 and account for less than 5 percent of total system costs.

10.3   Other Detailed Cost Data

10.3.1   Active Program Research Requirement (APRR) Project

10.3.1.1   Background
During 1982, the DOE Office of Solar Heat Technologies sponsored a project to identify the research priorities in active solar thermal energy systems. That project, known as the APRR project (Science Applications 1983), took a "top-down" approach and centered around system concepts in order to identify areas of potential cost or performance improvement in solar thermal energy systems. Research activities were defined and evaluated with respect to their potential for contributing to the development of one or more cost-competitive active solar thermal systems.

10.3.1.2   Approach
In order to facilitate the systems approach that was adopted, it was necessary to develop, as part of the APRR project, consistent and detailed systems- and subsystems-level construction cost estimates for a range of solar thermal system types. One of the major efforts of the APRR project was the development and cataloging of these cost data for both current (year 1982) and advanced (year 2000) solar thermal systems (Science Applications 1983).

Construction cost estimates were developed for commercial and residential solar thermal space heating, combined space heating and hot water, and space cooling systems. Systems were evaluated that utilize flat-plate air (FPA), flat-plate liquid (FPL), and evacuated tubular (ET) collectors. Advanced systems included those that utilize thin-film collectors, heat pumps, Rankine, absorption, and desiccant cooling devices. The economic viability of these systems was evaluated in comparison with conventional conditioning systems with a variety of conventional fuel sources.

Cost goals were set in accordance with system economic requirements, and preferred research activities were identified that showed the greatest potential to achieve these system cost goals.

10.3.1.3 Results

The APRR current system cost estimates were developed mainly from manufacturer and installer surveys and conventional construction cost-estimating manuals rather than from actual construction cost data from completed systems. The APRR results are interesting for a number of reasons.

First, the APRR cost estimates compare favorably with total system cost data from actual systems that are presented in section 10.2. This is illustrated in table 10.3, which contrasts summary cost estimates from the APRR project (organized according to the cost categories previously used) with average cost data for similar system types developed from actual construction cost data from section 10.2. The APRR total system cost estimates for combined space heating and hot water systems, ranging from $49.54/ft^2 to $66.94/ft^2, compare well to the average of total system cost for system, $59.69/ft^2 presented earlier. Table 10.3 shows similar comparisons between estimated and actual subsystem costs, particularly in the collector array and energy transport subsystems categories.

Second, the APRR project results provide a thorough and consistent set of cost estimates for use in various parallel program research activities. The effect is to establish a yardstick by which to measure specific research activity progress and potential.

Last, the APRR project results present researchers' best estimates of the cost reductions that are possible through continued advanced materials and systems research. Table 10.3 presents a summary of the cost reductions that are projected for advanced residential and commercial combined space heating and hot water solar thermal systems. It is interesting to note that the largest cost reductions projected are in the collector and energy transport subsystems. The APRR reports that these cost reductions can be achieved primarily through the use of thin-film collectors and plastic piping. There are still significant materials-related durability problems that must be solved in order for these approaches to realize widespread use in solar thermal systems.

Table 10.3
Summary of selected APRR project results vs. actual construction cost data

Application (1981 dollars)	APRR case #	Area ft²	Total $	Cost $/ft²	Collector type	Collector and support structure $/ft²	%	Thermal storage $/ft²	%	Energy transport $/ft²	%	Electric & controls $/ft²	%
Residential SH & HW	1.00	250.00	13295.00	53.18	FPA	26.25	49.36	7.74	14.55	17.08	32.12	2.10	3.95
Advanced		250.00	9395.00	37.58	FPA	11.25	29.94	7.74	20.60	17.08	45.45	1.50	3.99
Residential SH & HW	2.00	250.00	16736.00	66.94	ET	38.25	57.14	8.29	12.38	16.60	24.80	3.80	5.68
Advanced		225.00	10557.00	46.92	ET	21.25	45.29	9.21	19.63	13.35	28.45	2.80	5.97
Residential SH & HW	3.00	250.00	14236.00	56.94	FPL	28.25	49.61	8.28	14.54	16.60	29.15	3.80	6.67
Advanced		250.00	6120.00	24.48	FPL	7.92	32.35	4.41	18.01	9.34	38.15	2.80	11.44
Commercial SH & HW		800.00	39634.00	49.54	FPL	33.23	67.07	4.90	9.89	10.22	20.63	1.19	2.40
Advanced		800.00	19370.00	24.21	FPL	18.25	75.37	4.90	20.24	6.20	25.61	1.19	4.91
Averages for 10 Commercial SH (& HW) systems in table 10.1		3221.00		59.69	FP	30.87		5.02		17.18		4.94	

Source: Science Applications (1983).

10.3.2   Active Solar Installation Survey

10.3.2.1   Background

The Office of Solar Heat Technology and the Energy Information Administration sponsored the EIA-701 program, known as the Active Solar Installation Survey (Applied Management Sciences 1983). The EIA and contractors performed surveys of the activities of private solar thermal sys tem installation contractors during 1980 and 1981. In addition to publishing the results of the individual year surveys, the EIA published results of a detailed comparison made of both years' survey data (Applied Management Sciences 1983). The discussion here is concerned primarily with the results of the 1981 survey.

10.3.2.2   Approach

Detailed questionnaires were mailed to every known solar thermal system installation contractor in the United States. The questionnaires solicited information concerning level of business activity with several categories of solar thermal systems. The questions were structured so that the EIA could perform statistical analyses of the results to determine state, regional, and national patterns in system installation size, cost, and in numbers of installations for residential and commercial pool heating, hot water, space heating, and cooling systems. These categories were further subdivided to show trends in both single-family and multifamily hot water residential applications.

10.3.2.3   Results

Of particular interest are the results reported concerning average system size and cost. Tables 10.4 and 10.5 (Applied Management Sciences 1983) provide summaries of these results.

   The survey is one of the few efforts that had been made to specifically ascertain the cost of privately constructed systems and to collect data for residential systems. (As discussed previously, the data in section 10.2 represent commercial systems only.) There are limitations in the use of the data from the survey. In order to make effective comparisons between the costs shown in the survey and the detailed cost data of section 10.2, one must have comparable detailed information on the characteristics of the systems reported in the survey. However, these system characteristics, such as collector type, storage quantity, etc., are not available for the surveyed systems. Also, since detailed subsystem cost data were not a part

Table 10.4
Shift in the average size of active solar installations between 1980 and 1981 in square feet of collector area

| | Market sector | | | | | | | | |
| | Single-family | | | Multifamily | | | Commercial | | |
Applications	1980	1981	%	1980	1981	%	1980	1981	%
Pool heaters	367	376	+2	888	820	−8	1,208	1,212	0
Water heaters	59	61	+3	143	280	+96	388	517	+33
Space heaters	146	287	+97	167	312	+87	702	340	−52
Combined (water & space)	358	224	−37	360	390	+8	1,185	1,154	−3
Cooling	N/A	22	N/A	N/A	2,641	N/A	N/A	3,851	N/A
Other[a]	167	164	+1	1,800	1,137	−37	3,108	3,159	+2

Source: Applied Management Science (1983).
N/A—not available.
a. Cooling system data were included in calculations of average system size for 1981 to make figures more compatible with the 1980 data, which includes cooling.

Table 10.5
Shift in the average cost of active solar installations between 1980 and 1981 in dollars per square foot

| | Market sector | | | | | | | | |
| | Single-family | | | Multifamily | | | Commercial | | |
Applications	1980	1981	%	1980	1981	%	1980	1981	%
Pool heaters	9.20	9.21	0	7.66	9.17	+20	11.07	12.39	+12
Water heaters	49.02	53.03	+8	39.95	29.18	−27	30.86	25.57	−17
Space heaters	24.60	32.71	+33	16.68	20.95	+26	15.57	33.01	+112
Combined (water & space)	22.57	43.14	+91	30.83	36.65	+19	40.44	29.85	−26
Cooling	*	46.13	*	N/A	63.99	N/A	N/A	63.30	N/A
Other[a]	23.60	45.16	+91	27.23	60.87	+124	40.31	49.46	+23

Source: Applied Management Science (1983).
N/A—not available.
a. Cooling system data were included in calculations of average prices for 1981 to make figures more compatible with the 1980 data, which includes cooling.

of the survey, it is not possible to compare subsystem costs in section 10.2 and those in the survey or to determine what caveats apply to these data, such as hidden costs to the owner or installation subcontractors, which might have gone unreported.

The data of table 10.5 show very large changes in some system types between 1980 and 1981. For instance, the cost of commercial space heating systems increased from \$15.57 to \$33.01 per square foot. Combined water and space heating systems showed a change from \$40.44 to \$29.85 per square foot. The magnitude of these variations from year to year causes concern about the accuracy with which the data reflect actual total installed costs. However, if these data are accurate, they indicate significantly lower average installed costs than the data of table 10.2 and the costs generated in the APRR project. The cause of these differences is open to conjecture. Under-reporting of total costs by survey participants is a strong possibility.

10.4   Cost Estimating Guidelines

The ability to make reasonable, accurate estimates of the costs associated with constructing solar energy systems is important at every phase of the design process. Simple cost estimates are required at the conceptual phase for the initial economic feasibility and system-sizing procedure. For this purpose, cost estimates can be derived from the construction cost data presented in previous sections. During the design phase, designers must be able to assess the cost impact of various schematic design choices. Finally, detailed design cost estimates are usually performed by contractors bidding for the job and are valuable to the designer or owner when assessing bids for the construction contract.

In developing the guidelines presented in this section, the authors have relied heavily on the cost data presented in the section 10.1. However, since some of the projects probably no longer represent the state of the art, allowance has been made for those cost data that are not considered to be representative. The authors have also relied on their considerable experience with actual costs, cost estimating, and the design process.

Engineering and design fees vary widely and are difficult to estimate. Engineering fee data were not collected for the demonstration projects, but some general guidelines for estimating these fees for solar energy systems are provided here.

The engineering fees referred to include the cost of conceptual design, detailed design and specification, bid supervision and evaluation, and construction supervision. Designing a commercial-scale solar energy system requires at least three man-months if complete detailed specifications and drawings are to be provided. Design costs will increase with the solar construction contract size and system complexity. Design costs will also be higher if the services of an architect or structural engineer are required in addition to the mechanical design.

Cost-estimating manuals commonly indicate a percentage of the mechanical construction cost to estimate the mechanical engineer's fee, which ranges from 4 to 12 percent (Means 1987). For solar design, the percentage is somewhat higher, based on our experience as solar designers, typically by 6 to 15 percent, due to increased complexity. Only small systems (less than 1,000 square feet or 92.9 m^2) or very complex systems would normally experience engineering fees greater than 15 percent. Only very large and straightforward system designs would normally experience fees less than 8 percent (true until solar energy system design becomes more commonplace).

10.4.1   Useful Data Sources

A number of manuals are commonly used to estimate mechanical and electrical construction costs (Means 1987; Dodge 1987; National Construction Estimator 1987). These manuals list typical material and labor costs required to perform construction tasks that are common to mechanical and electrical construction projects. Data from these manuals can be very useful for estimating the cost of most of the construction tasks for an active solar project, particularly in the following areas:

- piping
- ductwork
- insulation
- liquid storage
- electric
- controls

Manufacturers' representatives and dealers are also useful data sources for material cost of specific components and equipment. Obtaining price lists or simply calling dealers can provide accurate material cost estimates for the following types of equipment:

- collector panels
- heat exchangers
- pumps and fans
- control values and mechanical dampers
- storage tanks

   Other references are available that provide insights into solar thermal system costs (Burrows and Ira 1982; King et al. 1979; Kutscher et al. 1982; Shingleton and King 1981).

10.4.2   Cost-Estimating in the Schematic Design Phase

During the schematic design phase, the designer needs an estimate of construction costs in order to perform feasibility and system-sizing studies. At this point, the designer has, at best, a rough concept of the system design; therefore, his cost estimate can only be approximated. Tables 10.6 and 10.7 are guidelines for making initial estimates for use in the feasibility and sizing analysis. These previously unpublished tables were developed by Mueller Associates, Inc., based on our experience in designing systems and on the detailed data described earlier in this chapter.

   These tables list the subsystems that make up a solar energy system, the design categories for that subsystem, the range of costs that might be expected, and what the authors consider to be typical costs. Note that unlike the cost data presented earlier in 1981 dollars, all costs are expressed here in either 1983 dollars/unit net aperture collector area or in 1983 dollars/gallon or ton of rocks (alternately 1983 dollars/liter or kilogram of rocks) for the storage subsystem. No value is given for a fixed cost component of the total system because statistical tests of the demonstration projects' cost suggest that for large systems (>300 ft^2 or 27.8 m^2 of collector), this cost component is negligible. Therefore, system costs can be described accurately solely as a variable cost per collector area within the size-range typical of commercial buildings.

   To use the table to arrive at a total system cost estimate, add up the typical subsystem costs located under the appropriate design category. The designer can use the range figures by applying judgment. For example, if the designer anticipates that an expensive support structure will be

Table 10.6
Conceptual phase cost-estimating guide[a] (English units)

Subsystem	Category	Range	Typical	Units
Collector array	Site-built	6–13	11	$/ft²
	Liquid flat plate	13–21	16	collector area
	Air flat plate	15–25	19	
	Evacuated tubes	26–32	30	
	Tracking concentrator	21–43	32	
Support	Single function	3–15	9	$/ft²
structure	Multiple function	7–25	15	collector area
Energy	Hot water	9–19	16	$/ft²
transport	Liquid-to-air heat	18–34	25	collector area
	Air heating	6–20	11	
	Steam	11–34	20	
	Heating and cooling	22–56	39	
Storage (liquid)	Unpressurized steel tank	1.2–3.2	1.9	$/gal.
	Pressurized steel tank	2.2–7.5	3.2	
	Fiberglass tank	1.9–2.8	2.2	
Storage (air)	Rock bin	111–222	154	$/ton rocks
Electric and	Hot water	2–6	3	$/ft²
controls	Liquid-to-air heat	3–10	6	collector area
	Air heating	2–9	4	
	Steam	2–20	13	
	Heating and cooling	4–22	11	
General				$/ft²
construction	—	0–17	4	collector area

Source: Mueller Associates (1982).
a. Cost includes materials and labor and contractor's overhead and profit. Costs are in 1983 dollars.

required because of building roof structure design constraints or if long piping runs will be required, then he should use values from the upper end of the ranges for the support structure and energy transport subsystems. However, the designer should avoid being overly optimistic or pessimistic when choosing values from the range column. Table 10.8 describes the design factors that determine which end of the range is appropriate. The table is previously unpublished work of Mueller Associates, Inc.

10.4.3 Cost-Estimating during the Design Phase

As the designer performs detailed design work, decisions are constantly made that will impact the project construction cost. Many of these decisions involve cost/performance trade-offs. For example, using higher effi-

Table 10.7
Conceptual phase cost-estimating guide[a] (metric units)

Subsystem	Category	Range	Typical	Units
Collector array	Site-built	64–140	118	$/m^2
	Liquid flat plate	140–226	172	collector area
	Air flat plate	161–269	204	
	Evacuated tubes	280–344	323	
	Tracking concentrator	226–463	344	
Support	Single function	32–161	97	$/m^2
structure	Multiple function	75–269	161	collector area
Energy	Hot water	97–204	172	$/m^2
transport	Liquid-to-air heat	194–366	269	collector area
	Air heating	64–215	118	
	Steam	118–366	215	
	Heating and cooling	237–602	420	
Storage (liquid)	Unpressurized steel tank	.32–.84	.50	$/liter
	Pressurized steel tank	.58–1.98	.84	
	Fiberglass tank	.50–.74	.58	
Storage (air)	Rock bin	.12–.24	.17	$/kg
Electric and	Hot water	22–64	32	$/m^2
controls	Liquid-to-air heat	32–108	64	collector area
	Air heating	22–97	43	
	Steam	22–215	140	
	Heating and cooling	43–237	118	
General				$/m^2
construction	—	0–183	43	collector area

Source: Mueller and Associates (1982).
a. Cost includes materials and labor and contractor's overhead and profit. Costs are in 1983 dollars.

ciency panels will improve thermal performance but may also increase costs. Analysis of cost/performance trade-offs frequently requires consideration of only differential costs of the more expensive options and incremental energy output increases that they provide. Other decisions may not affect thermal performance that much (for example, support structure design). By maintaining an attitude of cost consciousness and applying value engineering concepts, the designer will help keep construction costs down.

Table 10.8 contains some of the design-related factors that impact construction costs. These factors may be useful to the designers in two ways. First, they provide guidance for deciding which values to use in the range column in table 10.6. The designer can thereby refine his initial cost esti-

Table 10.8
Design cost factors

Subsystem	Cost factor	Comments
Collector array	Manifold type	Internally manifolded collectors cost about the same as externally manifolded collectors but require less than half the labor to connect the collectors.
	Panel size	Larger panels may require less labor per collector area for mounting and connecting.
Support structure	Roof imposed structural constraints	In some retrofit situations, the support structure must span large distances, requiring larger, more expensive structural members.
	Flat/sloped roof	In general, support structures for sloped roofs require less material and labor.
	# penetrations through roof membrane	Roof penetrations are expensive and should be kept to a minimum.
	Reflectors	Reflectors add materials costs.
	Aesthetic constraints	"Cosmetic" details for support structures add significantly to costs.
Energy transport	Long pipe/duct runs	Long runs to and from an isolated collector array or storage vessel add to materials and labor costs.
	Freeze protection	Direct systems are probably less expensive. The type and size of heat exchanger used in an indirect system can significantly affect energy transport costs.
	Complexity/# modes	The simpler the system, the more opportunity for cost savings.
Storage	Size	The most important cost factor is size (storage capacity per collector area).
Electric and control	# modes/# actuators	The more complex the control system, the more cost will be incurred.
	Monitoring equipment	Instrumentation and performance monitoring equipment can be very expensive.

Source: Mueller Associates, Inc.

mate as he refines his conceptual design. Second, the list may help the designer maintain an attitude of cost-consciousness. While the list is by no means complete, the factors listed indicate the more important cost decisions facing the designer.

10.4.4   Cost-Estimating during the Postdesign Phase

After the system has been designed, complete with drawings and specifications, a detailed cost estimate is normally prepared by the bidder or by the contractors who will bid on the project.

Often, the designer will also prepare a cost estimate, although, usually, this estimate is less detailed than that of the contractor. This design estimate can be useful in assessing the bids. If all the bids are much higher than the estimate, perhaps additional contractors should be invited to bid. If the lowest bid is much lower than the estimate, perhaps the contractor's ability to perform should be questioned.

Contractors differ in the process they use to prepare the cost estimate, but the following description is representative of what normally happens. The contractor uses the design drawings and the specifications to formulate a construction plan as a series of tasks. He then prices the major components by contacting suppliers and subcontractors, estimating the

Table 10.9
Detailed cost-estimating guide[a] (English units)

Item	Units	Material	Labor	Labor hrs.	Total	Total, including subcontractor
Mount and connect collector panels (not including cost of panels)	$/ft^2 Collector area	.90	1.12	.10	2.02	2.52
For air collectors, add		0.00	1.12	0.10	1.12	1.40
For tracking collectors, add		1.00	.56	.05	1.56	1.95
For external Manifold, add		0.00	1.57	.10	1.57	1.96
Collector support structure		4.30	2.90	.25	7.20	9.00
Maximum		7.20	4.80	.40	12.00	15.00
Minimum		1.20	1.20	.10	2.40	3.00
For reflectors, add		1.80	0.00	0.00	1.80	2.25
For multiple function, add		3.00	1.80	0.15	4.80	6.00
For sloped roof, add		3.00	2.00	.20	5.00	6.25
Rock bin	$/ton rocks	67	56	5.0	123	154
Maximum		100	78	7.0	178	222
Minimum		50	39	3.5	89	111

Source: Mueller Associates, Inc.
a. Costs are in 1983 dollars.

material and labor cost required to complete each task. He will obtain estimates or firm bids from any subcontractors he plans to use on the job. He will then add his overhead, profit rate, and a contingency to arrive at a bid value. The labor and material cost estimates are often obtained from cost-estimating manuals, although many contractors rely instead on cost data compiled from previous experience.

The usefulness of the data contained in the cost-estimating manuals and of information from manufacturers' representatives and dealers has been discussed above. However, these sources are generally insufficient in themselves to prepare a detailed estimate for a large, commercialized solar construction job. Tables 10.9 and 10.10, based on previously unpublished

Table 10.10
Detailed cost-estimating guide[a] (metric units)

Item	Units	Material	Labor	Labor hrs.	Total	Total, including subcontractor
Mount and connect collector panels (not including cost of panels)	$/m^2 Collector area	9.68	12.06	.10	21.74	27.18
For air collectors, add		0.00	12.06	.10	12.06	15.08
For tracking collectors, add		10.76	6.04	.05	16.80	21.00
For external Manifold, add		0.00	16.90	.10	16.90	21.12
Collector support structure		46.29	31.22	.25	77.51	96.89
Maximum		77.50	51.67	.40	129.17	161.46
Minimum		12.92	12.92	.10	25.84	32.30
For reflectors, add		19.38	0.00	0.00	19.38	24.22
For multiple function, add		32.30	19.38	.15	51.68	64.60
For sloped roof, add		32.30	21.53	.20	53.83	67.29
Rock bin	$/1,000 kg of rocks	74	62	5.0	136	170
Maximum		110	86	7.0	196	245
Minimum		55	43	3.5	98	122

Source: Mueller Associates, Inc.
a. Costs are in 1983 dollars.

work of Mueller Associates, Inc., provide additional cost information for the following solar construction tasks:

- collector panel mounting (labor)
- collector array manifold (labor)
- support structure (labor and materials)
- rock bin storage (labor and materials)

10.4.5 Nondesign-Related Cost Factors

Cost factors that are not directly related to system design but may impact system cost include the following:

Number of bidders for construction contract. By ensuring that a large number of contractors are invited to bid, one can have greater confidence that the low bid will be reasonable.

Number of contractors in area/general economic conditions. Contract bid values are sensitive to the laws of supply and demand. Contract bids will be lower if there are many contractors and little current construction in the area. The converse is also true.

Transportation requirements. Transportation of materials to an isolated site may boost material costs.

Time constraints. A tight completion deadline may require the use of overtime labor that will boost labor costs.

Material delivery delays. Construction schedules may be interrupted if delivery of important materials and equipment must be delayed, leading to increased costs.

Weather/season. Inclement weather may interrupt construction schedules. Labor productivity may be reduced during cold winters and hot summers.

Regional variations in labor and material costs. While this was not a significant factor in the demonstration projects, costs for a particular project may be affected by local cost variations.

Legal and union restrictions. Local building-code restriction and local union regulations may affect system cost.

Contingency. The contingency allowance added to the contractor's bid will depend heavily on the extent and nature of his experience with solar and government projects. A contractor's contingency allowance can often be reduced by carefully planned prebid conferences and detailed drawings and specifications.

In-house construction supervision vs. construction management vs. general contractor. Exploring all possible contractual arrangements for the solar system construction can sometimes help reduce total construction cost.

10.4.6   Example Cost-Estimating—Fairfield

The Fairfield solar energy system was designed in 1979 and 1980, and constructed during 1980 (Flesher 1982). The system provides hot water for a nursing-care facility located near Baltimore, Maryland. The solar energy system was retrofit onto the existing building, but the original building was constructed with steel tubing stub-ups extending from the roof so that a solar collector support structure could be added later.

Early in the design phase, the following design parameters were determined:

• Collectors. Space available for about 144 liquid flat-plate collectors.

• Support structure. The stub-ups would enable a lightweight steel support structure to be constructed relatively inexpensively.

• Energy Transport. Freeze protection would use a glycol solution in the collector loop separated from the rest of the system by a heat exchanger. A relatively long piping run would be required to connect the collector array to the mechanical room.

• Storage. A pressurized steel tank with a capacity of about 5,000 gallons would be used.

At this point, a preliminary cost estimate could be prepared using the values in table 10.6 as follows:

Collector area—144 panels $\times$ 20 ft^2/panel = 2,880 ft^2

Collector array—$16/ft^2 $\times$ 2,880 ft^2 = $ 46,080

Support structure—$6/ft^2 $\times$ 2,880 ft^2 = $ 17,280

(A figure from the small end of the range column would be used because of the stub-ups.)

Energy transport—$18/ft^2 × 2,880 ft^2 = $ 51,840

Energy transport—\$18/ft^2 × 2,880 ft^2 = \$ 51,840

(A figure from the higher end of the range column is used because of the long pipe runs.)

Storage—\$3.2/gallon × 5,000 gallons	= \$ 16,000
Electric and controls—\$3/ft^2 × 2,880 ft^2	= \$ 8,640
General construction—\$4/ft^2 × 2,880 ft^2	= \$ 11,520
Total	= \$151,360

During the detailed design phase, the design was revised with the following changes:

The collectors selected had a net aperture area of 20.80 ft^2 per panel for a total net aperture area of 20.80 × 144 = 3,000 ft^2.

The storage tank selected had a capacity of 4,000 gallons.

Using these final figures, a total construction cost estimate can be calculated as follows:

Collector array—\$16/ft^2 × 3,000 ft^2	= \$ 48,000
Support structure—\$6/ft^2 × 3,000 ft^2	= \$ 18,000
Energy transport—\$18/ft^2 × 3,000 ft^2	= \$ 54,000
Storage—\$3.2 gallon × 4,000 gallons	= \$ 12,800
Electric and controls—\$3/ft^2 × 3,000 ft^2	= \$ 9,000
General construction—\$4/ft^2 × 3,000 ft^2	= \$ 12,000
Total	= \$153,800

10.4.7   Residential and Packaged Systems

Nearly all of the information in this chapter is based on the costs of large commercial solar water heating systems. Residential-size solar water heaters have not been included. Currently, most of the solar energy systems being installed are residential hot water systems, and the majority of these are prepackaged systems. Generally, one contractor will be responsible for all aspects of construction and his crews are experienced enough with his particular type of system that he can quickly provide an owner a price for a specific home based solely on this experience. A large component of the price of a residential solar water heater is the cost of marketing, which can approach fifty percent of the installed price.

10.5 Conclusions

The availability of accurate construction cost data is essential to the research and development as well as the design of any technology but especially of solar thermal energy systems that require new combinations of skills and materials.

Gaps do exist, however, in the availability of cost data for solar thermal systems. These gaps include:

Information on more recent systems
The systems described in sections 10.2 and 10.3 are at least several years old at the time of this writing. Similar cost analyses of more recent solar thermal systems might identify some new trends. For instance, "learning-curve" information may be more easily discerned. Many of the systems studied were designed by individuals who were relatively inexperienced in the design of solar thermal systems. Also, many of the systems were built by contractors who were inexperienced in their construction. A comparison of the costs of these systems with costs of systems designed and constructed by experienced designers and contractors would be worthwhile.

Information on residential systems
Although the installation survey indicated the level of activity in the private sector, it did not tell enough about the characteristics of these systems or important caveats on their reported costs.

Information on larger solar thermal systems
There are a number of large-scale Industrial Process Heat Field Test projects for which detailed cost data could be obtained. In addition, a number of very large privately-funded, third-party-financed projects have been constructed and would be valuable to evaluate. Cost data on these large systems would be useful to investigate economy-of-scale effects further.

References

Applied Management Sciences. 1983. *Key trends in active solar installation activities between 1980 and 1981, final report.* DOE/CE/30692–1.

Burrows, D., and S. Ira. 1982, *Cost savings through first cost reduction.* Tennessee Valley Authority.

Dodge manual for building construction pricing and scheduling. 1987. New York: McGraw-Hill.

DOE. *see* U.S. Department of Energy.

Flesher, A. R. 1982. Retrofit solar hot water system for a nursing home. *Proc. Mid-Atlantic Energy Conference and Exposition*, December 1982.

King, T. A., et al. 1979. Cost effectiveness—An assessment based on commercial demonstration projects. *Proc. 2d Annual Solar Heating and Cooling System Operational Results Conference*, Colorado Springs, CO, November 1979.

Kutscher, C. F., et al. 1982. *Design approaches for solar industrial process heat systems.* SERI/TR–253–1356; Golden, CO: Solar Energy Research Institute.

Means. *see* R. S. Means Company, Inc.

Mueller Associates, Inc. 1982.

National Construction Estimator. 1987. Carlsbad, CA: Craftsman Book Company.

R. S. Means Company, Inc. 1981. *1981 Means mechanical and electrical cost data.* Kingston, MA: R. S. Means.

R. S. Means Company, Inc. 1987. *1987 Means mechanical cost data.* Kingston, MA: R. S. Means.

Science Applications, Inc. 1983. *Active program research requirements, final report.* Washington, DC: U.S. Department of Energy.

Shingleton, J. G., and T. A. King. 1981, *The analysis of construction costs of ten solar industrial process heat systems.* SERI/TR–09144–1. Golden, CO: Solar Energy Research Institute.

U.S. Department of Energy. 1979. *instrumented solar demonstration systems project description summaries.* DOE Solar/0017–79/34.

III SOLAR SPACE HEATING

11 Space Heating Systems—Introduction and Summary

George Löf

11.1 Introduction

Approximately one-fifth of the energy consumed in the United States is used for heating buildings. Although space heating demands are high in seasons of lowest sunshine, nearly all buildings in the United States intercept more winter solar energy than their heat requirements.

These circumstances offer incentive and potential for providing a large fraction of the nation's space heating requirements with solar energy. Limitations on solar collection efficiency and variations in solar availability and heating demand require that dependable auxiliary heat be available for correcting the mismatch in solar supply and heat demand. In most applications, the disadvantages of day-to-night solar fluctuation are mitigated by short-term (6 to 18 hours) thermal storage.

11.2 Historical Background

11.2.1 Solar Space Heating prior to 1970

The development of active solar heating of buildings began in the 1938–1945 period with the design and construction of active experimental systems in four residential type buildings in the United States. Two installations included large active solar water heaters employing roof-mounted flat-plate collectors and interior storage tanks for solar heated water. Two systems were based on air-heating collectors and heat storage in a bin of small rocks or in containers of a fusible salt. The performance of these systems was monitored and reported in the technical literature (Hottel and Woertz 1942; Telkes 1955; Löf 1955).

Development activity in the subsequent two decades (1945–65) was limited. Low energy prices provided little incentive for reducing fuel use. Fourteen solar heating systems were built and tested during this period, including three in Japan, one in Italy, and one each in Australia, New Zealand, and Canada. Three systems in the United States were of the air type, and another four were based on liquid-heating collectors. Funding came from personal financing by the investigators, the manufacturers of materials and equipment, university sources, and governments. Performance data on some of the systems were procured and published (Löf 1964).

Although most systems provided solar heat, and useful design information was obtained, commercial development was precluded by the disparity in the relatively low cost of fuels and the considerably higher capital cost of the solar equipment. Most of the experimental systems provided proof that they could be used to provide a substantial portion of the annual heat requirements of single-family dwellings, but no commercial interest was created.

11.2.2 Development and Commercialization of Residential Solar Space Heating since 1970

Reacting to shortages and major price increases of petroleum early in the next decade, 1973 to 1974, solar energy research and development programs were established by departments of the U.S. government. Several program plans organized by the National Science Foundation, the National Aeronautic and Space Administration, and the Department of Housing and Urban Development included the solar heating of buildings. Through a new agency, the Energy Research and Development Administration (ERDA), funding of research and development in solar heating was initiated in universities, research centers, and some of the national laboratories. Projects on solar collectors, materials, heating systems, controls, and other topics resulted in substantial increase in technical information in this field.

In 1975, ERDA also initiated the Solar Heating and Cooling Demonstration Program. Its purposes were the establishment of public understanding and confidence in solar heating, identification of practical system designs, and, indirectly, the stimulation of equipment manufacture and installation by industry. Funds were provided to builders and to the owners of existing buildings for purchase and installation of solar heating systems, which were specified in the proposals accepted for support. The grants for residential application were administered by the Department of Housing and Urban Development, and for commercial and institutional use, by ERDA. The solar research and demonstration activities of ERDA were transferred to the newly established Department of Energy in 1978.

In the 1975–1980 period, extensive R&D activity and vigorous demonstration programs resulted in the formation of many commercial firms to manufacture solar collectors and to assemble and install solar heating systems. Although the principal market was for solar water heaters, the demonstration programs provided a market for hundreds of solar space heating systems. The primary new components in these systems, solar collectors, were made by over two hundred companies of greatly varying size and capability.

Further government incentives for solar heating use resulted from the establishment of tax credits for purchasers of residential and commercial solar heating systems starting in 1980. The program reached its full potential in 1984, when 2.4 million square feet of solar collectors were installed in active solar space heating systems. A few leading manufacturers, some of which were solar departments in companies making other products, sold many residential and most of the commercial systems throughout the United States.

As long as the federal solar tax credit program resulted in attractive net consumer costs, solar heating systems were salable, but when the program ended in December 1985, demand dropped sharply, and nearly all U.S. manufacturers discontinued their solar activities.

11.3 Current Status of Solar Heating

11.3.1 Residential and Commercial Application

Although the current market for active solar heating is small, it is estimated that about 40 thousand houses are being partially heated with active solar systems in the United States. No accurate statistics are available, but there is indirect evidence that collectors exceeding 800 square feet in area are used in the heating of about one thousand commercial and institutional buildings. It is estimated that about 80 percent of the heating systems employ liquid collectors and water tank heat storage, 10 percent air type with pebble-bed storage, and 10 percent air type with no special solar heat storage.

Solar heating installations are distributed widely over the United States, with a higher per-capita use in areas with cold, sunny winters. The Rocky Mountain states, parts of New England, and the northwestern states are favored. Higher population densities in the mid-Atlantic and Midwest regions have led to substantial sales in those areas. Although winter solar energy is favorable in the southern states, mild winters limit the practical use of solar heating in that part of the country.

11.3.2   Types of Systems

Flat-plate liquid heating collectors with heat storage in tanks of water are used in the majority of residential solar heating systems and in nearly all institutional and commercial buildings. Freeze-protection is usually secured through use of a nonfreezing collector fluid (commonly aqueous antifreeze solutions) and heat exchange with water for heat storage. Also widely used are drainback systems in which water is the heat collection fluid that drains from the collector into the storage tank when the pump is not operating. The building is usually heated by circulating hot water to radiator-convector units, fan-coils, or baseboard finned tubes in the rooms, or by heat exchange to air in forced warm-air systems.

Collectors comprising banks of glass or metal tubes surrounded by glass-enclosed evacuated spaces are used to a limited extent in liquid space heating systems.

The other principal type of space heating system employs a collector for direct heating of air, which is circulated to the rooms or supplied to a pebble bed for short-term (half-day) heat storage. When no solar energy is being received, room air is heated by circulation through the pebble bed.

Air and liquid space heating systems are also usually provided with heat exchange coils for domestic hot water supply, a particularly useful summer function when space heating is not needed. Since the solar contribution is usually less than half the total annual space heating requirements, and even less in the coldest months, auxiliary heat in the form of a conventional supply (natural gas or oil furnace, electric coil, or heat pump) is necessary.

11.3.3   Economics of Solar Space Heating

With only a few exceptions, active solar space heating is more expensive than heating with conventional fuels and methods. Nearly all the cost is associated with the capital investment in the system and the interest and depreciation on that investment. Well designed and constructed systems require little maintenance and operating expenditures. Many complete systems, particularly of the air type, have been installed at total prices of $30 to $50 per square foot of collector (approximately $300 to $500 per square meter), requiring annual debt service charges, at 10 percent per annum, of $3 to $5 per square foot ($30 to $50 per square meter). In a favorable climate with a long heating season and above-average solar radiation, annual deliveries of 150,000 Btu of usable heat per square foot of collector (approximately 1,500 MJ/m^2) can be obtained if a combination space heating and water heating system is conservatively sized, i.e., designed to supply up to one-half the annual space heating requirements. The effective cost of solar heat is therefore $20 to $35 per million Btu (approximately $20 to $35 per GJ).

Natural gas is available in most sections of the United States at an effective cost of about $7 to $10 per million Btu. Fuel oil at $1 per gallon will supply heat in a typical furnace for $10 per million Btu. Electric resistance heat at 7 cents per kWh is $20 per million Btu delivered.

The above simplified comparison shows that even with favorable assumptions for costs and performance of the solar heating system, heating with conventional fuels is less expensive. In some localities, however, electricity price is much higher than in the foregoing example, as high as 15 cents/kWh, or $40 per million Btu. Solar is competitive with electric heating in such circumstances if the capital cost, interest rates, and annual energy production are comparable to the above example.

Of the previously cited installed system price of $50 per square foot of collector, factory costs of the collector, storage unit, controls, pumps, fans, heat exchangers, and piping rarely exceed $20 per square foot of collector. Low sales volume, lack of efficient marketing channels, installer inexperience, and other problems not related to manufacturing costs have resulted in distribution costs as high as 50 percent of the total installed system price.

The manufacture of solar collectors has usually been on a sufficient scale, however, to permit minimum prices for materials and fabrication, even of the very best quality. Significant reduction in manufacturing cost is unlikely unless unacceptably low quality materials and workmanship are chosen.

It is therefore evident that substantial reduction in solar heating cost depends on reducing marketing, distribution, and installation costs, which requires, in turn, a more conventional pattern of selling a major home appliance in new and retrofit markets. This pattern should result from a sizable increase in demand as consumer confidence and preference are established through maintenance and improvement in quality of components, systems, installation, and service.

The idealized circumstances in the foregoing example are only occasionally encountered, so solar space heating (and, generally, water heating also) is not usually cost-competitive. When tax credits were available, that subsidy often reduced the net cost of the system sufficiently for an economic choice of solar to be made. Since costs of operating existing systems of good quality are generally low (for maintenance and electric power) and since the investments have already been made, their operation can be expected to continue for the life of the system. New systems are not likely to be widely installed, however, until economic factors or public opinion change.

11.3.4   Problem Areas

As seen in the foregoing subsection, solar heat can seldom be provided at the costs of conventional energy sources. In addition, user experience with many systems has been unsatisfactory due to costly repairs, poor performance, and disappointing fuel savings. These factors have created widely held views that solar heating is a poor choice, even if costs were comparable. Many satisfied users provide only partial, and often less visible, support for this relatively new technology.

Much of the capital cost of residential solar heating systems has been in marketing, distributing, and servicing. A problem related to the quality-confidence factor is the multiplicity of solar heating system designs used in dwellings and commercial buildings. New and untested ideas were frequently built into systems, with widely varying success. There was no generally accepted "code of practice" for solar space heating, and even the more common solar water heating systems were of at least five substantially different types. Efforts to standardize practice through recognition of a limited number of "generic" systems known by experience to provide efficient, dependable performance are needed.

11.3.5   Industry Infrastructure

In the mid-1980s, the "solar space heating industry" comprised: (1) solar collector manufacturers and the suppliers of their materials; (2) suppliers (manufacturers) of accessories such as tanks, controls, pumps, and other components often identical with items used in conventional (nonsolar) equipment; (3) solar *system* suppliers (often collector manufacturers) who provided complete "packages" of components for a solar space heating system; (4) solar system designers (usually HVAC engineers with experience in conventional heating systems) who generally were involved in the design of large commercial and institutional solar space heating systems; (5) solar system installers (HVAC contractors and plumbing and heating contractors, for large systems and some residential systems), and special-ized *solar* system suppliers not in the conventional HVAC and plumbing-heating installer business; (6) solar system maintenance firms (usually the installers, but also HVAC customer service firms and solar maintenance and repair specialists).

The foregoing organizations all participated to varying extent in the solar heating industry. Universities and R&D organizations undertook much of the major portion of research, development, and testing of com-ponents and systems.

All of the types of organizations in the industry were involved to various degrees in sales and marketing. If a collector manufacturer or solar sys-tems supplier did not sell directly to users, there would be at least a distrib-utor (who usually warehoused the product and handled other products also), and sometimes also a wholesaler, in the marketing chain. It is evi-dent that the cost of distributing solar heating systems to the residential market must cover the handling of the products by several types of firms, or if more direct marketing is used, comparable (or sometimes higher) costs must be absorbed by the original supplier or manufacturer.

In comparison with the typical marketing path of residential solar sys-tems, commercial and institutional systems have usually been sold by the collector manufacturer or system supplier directly to the HVAC installing contractor or the building owner on specifications by the system design engineer. Competitive bidding is often used in the selection process.

11.3.6   Trends and Prospects

The 1980s have seen a major expansion, followed by nearly a total col-lapse, of the solar heating industry. Built largely on R&D projects and demonstration programs of the U.S. Department of Energy, the total val-ue of installed solar space heating systems had surpassed a billion dollars by 1985. The market disappeared when tax credits expired, and, almost simultaneously, petroleum price dropped to half its former level. Natural gas prices also softened as availability exceeded demand. Excess electric generating capacity interrupted the steady rise in electricity prices. Public concern over energy availability and cost ended nearly as suddenly as it began.

At the present writing (1989), there appears to be a slight increase in interest in domestic use of solar energy. Environmental concerns (pollu-tion from fossil fuels, the "greenhouse" effect, nuclear waste problems, oil spills, and other accidents) are stimulating the interest. Although a clear trend is not yet evident, there are prospects for a slow rebuilding of the industry. Initial costs of solar heating are not likely to decrease, but creative marketing strategies, high quality products, government incentives, and public concern for the environment could make solar heating the system of choice for users in favorable circumstances

11.4   Contents of Space Heating Section of This Book

Chapter 12 contains descriptions of the most widely used types of space heating systems and their operating principles and characteristics. Their applicability and limitations in various types of buildings are also indicated.

The performance of solar space heating systems in a number of well-instrumented experimental houses is detailed in chapter 13. Comparisons of reported efficiencies are drawn and opportunities for improvements are shown.

Chapter 14 contains the results of monitoring 20 solar heating systems of widely different types in the DOE program for demonstrating solar heating in commercial and government buildings. Performances are compared and conclusions are drawn.

In chapter 15, the costs of solar heating are analyzed. The results are based on actual experience in the DOE solar heating demonstration program.

European experience with long-term storage of solar heat for use in district heating systems and studies of the applicability and potential of this technology in the United States are presented in chapter 16.

The final chapter in this section, chapter 17, is a review of comparatively new solar heating technology that resulted primarily from research and development activities supported by DOE. Also presented are concepts that, although not currently used, may become important in the future.

References

Hottel, H. C., and B. B. Woertz. 1942. Performance of flat plate solar collectors. *Trans. ASME* 64:91.

Löf, G. O. G. 1955. House heating and cooling with solar energy. *Solar energy research*. Ed. F. Daniels and J. A. Duffie. Madison, WI: University of Wisconsin Press, p. 33.

————. 1964. The use of solar energy for space heating. *Proc. U.N. Conference on New Sources of Energy*, E–35, GrS14, Rome.

Proc. World Symposium on Applied Solar Energy. 1956. Menlo Park, CA: Stanford Research Institute.

Space heating with solar energy. 1954. M.I.T. 1950 Symposium Proceedings. Cambridge, MA: MIT Press.

Telkes, M. 1955. Solar heat storage. *Solar energy research*. Ed. F. Daniels and J. A. Duffie. Madison, WI: University of Wisconsin Press, p. 57.

Trans conference on the use of solar energy—The scientific basis. 1958. Tucson, AZ: University of Arizona Press.

Zarem, A. M., and D. D. Erway, eds. 1963. *Introduction to the utilization of solar energy*. New York: McGraw-Hill.

12 Space Heating: System Concepts and Design

S. Karaki

12.1 General Requirements

A solar system for space and water heating can be designed to suit any particular application, residential or commercial, new or retrofit. It is technically feasible to design a solar system that can provide 100% of the heating needs of a building, but generally it is uneconomical to do so. Practical solar heating systems are designed to displace up to about 50% of conventional fuel needs and require auxiliary heating systems that are fully capable of supplying the total heating load when no solar energy is being collected and when stored solar energy has been depleted.

When space heating systems serve occupied buildings, it is economical to include solar heating of domestic hot water (DHW) in the system. For residential systems during the summer, when there is no space heating load, the entire solar system can be devoted to water heating so that solar can supply a substantial portion, if not all, of the DHW heating needs. However, for large commercial systems, summer operation to provide a relatively small DHW demand may not be worthwhile.

The size of a solar system (primarily the collector area and storage volume) for a particular building depends on the portion of the total load the system is expected to provide. Size also depends on climate and location. The type of system, whether air- or liquid-based, depends in part on the application and to a large extent on the designer's choice. Very often, large systems may be subdivided into several smaller systems to better suit the application. Smaller systems may be easier to control, provide better overall efficiency, and cost collectively less than a system with one large collector array and storage.

12.1.1 Application and Environment

12.1.1.1 Residential or Commercial

There are many more residential applications of solar heating systems (to 1985) than commercial applications, primarily because of economic reasons. Solar heating systems are as adaptable technically for commercial applications as for residential applications, except that for very large buildings, available area for collector placement may constrain system size. For most residential applications, selecting an air- or liquid-based system is largely a matter of designer or owner preference. Air-based solar systems can be as effective as liquid-based systems, and costs for equally sized systems are essentially the same. For large commercial applications where collectors are not part of the buildings served, liquid systems have an advantage over air systems of smaller-size fluid conduits in the transport subsystem. Heat transport in large air ducts is more expensive than in liquids through pipes.

12.1.1.2 New Building or Retrofit

There is considerable freedom in selecting a solar system type for a new building and in designing that system to provide an arbitrary (but economical) fraction of the total load. Orientation of the building, slope of the roof for mounting collectors, location and space for thermal energy storage, and type of heat distribution system can be selected according to system choice. Options are fewer for retrofit applications. Building orientation is fixed, roof slope may not be suitable for collector mounting, choice of storage location and size may be limited, and some interior changes may be necessary to accommodate the heat transport subsystem. The existing heating and distribution system may also restrict the selection of solar system type and design.

12.1.1.3 System Size

The size of the solar system is logically based on economics. Among the economic figures of merit are total capital investment (first cost), least cost for energy, life-cycle cost, life-cycle cost savings, payback, and return on investment (ROI). Discussion of economic considerations related to system selection and sizing is included in chapter 15. While positive life-cycle cost savings and short-term payback are criteria most readily understood by homeowners, a high ROI is usually the criterion for commercial systems.

While other, noneconomic, factors can affect choice of system size for residential applications, using solar systems that can provide from 30% to perhaps 50% of the total annual heating requirements is a practical choice. Sizes of commercial systems, while also dictated by economic constraints, also depend on availability of suitable locations for collectors.

12.1.1.4 Climate

Solar radiation availability and heating demand are inversely related; in sunny winter climates, the need for space heating is low, while in areas with prevailing clouds in winter, temperatures are generally low and heating demand is high. It is commonly assumed that solar heating systems are, therefore, more suitable for climates between these two extremes, in a region covering a considerable portion of the area of the United States between northern latitudes of 30 to 45 degrees, and including most of the U.S. population. It is not clear, however, that solar systems are less effective in displacing fossil fuels in the northern climates of Europe and Canada, where the much longer heating season means a considerably greater heating load. While solar radiation is least during the coldest periods of winter, the fall and spring periods may provide substantial opportunity for displacement of conventional heating fuels.

Although beyond the scope of this chapter, there are systems using annual storage schemes that collect thermal energy during warm periods for use in the cold periods. For such systems, summer as well as winter climate is very important.

12.1.1.5   Freezing Considerations

System type selection for space and water heating and its specific design are greatly affected by concerns for freezing. Air systems are clearly superior in this regard, but a number of commercially available nonfreezing liquids are suitable for use in liquid-based systems. Early water-based collection systems, which depended on detection of freezing temperatures to drain the system, suffered considerable freeze damage when the controls failed to detect onset of freezing (Chapra and Wolosewicz 1978). Improved equipment, simpler system designs, and better control strategies have greatly reduced occurrences of freezing in collectors, but, nevertheless, damage occasionally is reported, particularly with first-generation system designs.

Because the consequences of freezing can be catastrophic to the system, or at least very expensive to repair, considerable attention should be given to selecting the type of system and important design aspects that prevent freezing.

12.1.2   Supplementary Requirements

12.1.2.1   Thermal Energy Storage

Thermal energy storage is needed if a solar heating system is to provide heat overnight or during cloudy periods of the day. A variety of materials can be used for heat storage, and the desirable characteristics are the following: (1) easy to transfer heat from the heated fluid to the storage material; (2) readily recoverable and supplied to the heating system; (3) subject to little internal losses or losses to the environment; (4) inexpensive; and (5) requiring minimal floor area or volume.

The two principal materials used for storage are water for liquid-based systems and rock pebbles for air-based systems. Water has a high heat capacity and rocks have one-fifth as much (one-third per unit volume), and both are inexpensive. The other liquids and solids that may be used for sensible-heat storage are almost always more expensive than rock or water per unit of heat stored.

Another method of storing heat involves using phase-change materials such as eutectic and hydrated salts. These materials store latent as well as sensible heat. The principal advantage of these materials is that a smaller volume (or mass) of material is needed than with water or a pebble bed. Considerable research has been devoted to phase-change materials, and a comprehensive discussion of them can be found in volume 5 of this series.

12.1.2.2 Auxiliary Heating System

Practically sized solar systems require auxiliary heating to supply the portion of the heating load that the solar systems cannot supply. System controls are commonly designed to use solar energy for heating and to rely on the auxiliary heating system whenever solar energy is not available and storage has been depleted. Even if the control strategy included provisions to share the load between the solar and auxiliary heating systems during delivery, there is no assurance that an appropriate share of solar energy would always be available. The auxiliary heater must, therefore, be fully capable of delivering heat at a rate sufficient to maintain comfort conditions during the coldest nights of winter.

Without an adequately sized auxiliary heating system, it would be difficult to receive approvals for solar heating systems from building officials. It would also be difficult to secure financing for the building or solar system, or to sell the building.

12.1.2.3 Temperature

The temperature of energy collection and delivery from a solar system depends on system design and operating control strategy. For domestic hot water, any solar heat supplied at a temperature above water temperature in the city mains is useful, as is solar heat used for space heating to raise and maintain room air temperature within the comfort zone. Temperature influences the size of the heat transfer equipment selected for a specific design, as well as flow rates of the heat transfer fluid used in the system.

For efficient collection of solar energy, fluid temperature delivered to the collector should be as low as possible, but for effective heating, high temperatures are preferred. It is necessary, therefore, to consider storage volume in relation to collector area when designing a solar system. Hydronic heating systems, for example, circulate hot water at high temperatures (80°C [176°F] or higher), but liquid systems with heat delivery through fan coils can operate very effectively at more moderate temperatures (below 60°C [140°F]), and radiant panels can operate at still lower temperatures. Hot-air furnaces usually deliver air at 50°–60°C (122°–140°F). The temperature of energy delivery from solar systems, however, depends on collector area, fluid flow rate, and volume of storage. Generally, a control strategy that insists on constant high temperature delivery from a solar system will collect less useful energy during a given period than a control strategy that allows for variable temperatures in the system (Kovarik and Lesse 1976).

12.1.2.4   Load Profile

Time distribution of space and water heating demand (load profile) depends to a large extent on building function and occupant habits. Load profiles, in turn, affect system performance and selection of system size (Karaki et al. 1985). Although the DHW load profile tends to have minor influence on performance of an integrated space- and water-heating system, the space heating load profile can influence decisions concerning collector area, particularly for systems with no storage. The selection of the auxiliary heating unit size may depend on the load profile. Selection of the load heat exchanger for a liquid-based system and air flow rate for an air-based system will generally determine the maximum rate of heat delivery to the rooms. Building heat loss rates higher than delivery rates from the solar system will necessitate using auxiliary heat to maintain comfort conditions.

12.1.3   Combined Space and Water Heating Systems

12.1.3.1   Rationale and Load Features

With the exception of a few special applications, buildings that require space heating also require domestic water heating. Because it is seasonal, compared to domestic water heating that is needed throughout the year, it is economically advantageous to integrate a DHW subsystem with a solar space heating system. Some attention is required in an integrated design to assure adequate domestic water heating while space heating needs are being met.

12.1.3.2   General Methods and System Types

A number of design alternatives can be arranged to integrate domestic water heating with space heating. A common approach for a liquid-based residential-size system is to provide a DHW subsystem using solar storage as the heat source (Ward and Löf 1975; Duffie and Beckman 1980). For a large collection system with a small DHW load, the electric energy required to operate the solar system during nonheating periods may be more than the equivalent thermal energy needed to supply the DHW load. In such a case, using a separate DHW system may be a wiser choice. With air systems, an air-to-water heat exchanger is commonly used to heat water while storing thermal energy in a pebble bed (Ward et al. 1977; Solar Energy Applications Laboratory 1980a).

To prevent the possibility of potable water becoming contaminated by a nonpotable liquid during the heat transfer process, a double-wall heat exchanger should be used. Two single-wall heat exchangers, one between the collector and storage, and another between the storage and the potable water, are not acceptable as a substitute for a double-wall heat exchanger. Systems that use potable water in the collector will not require a heat exchanger.

For ease of maintenance and replacement, heat exchangers should be installed where they will be readily accessible. Generally, this means mounting the heat exchanger externally to the storage tank, but there are designs that include long lengths of copper coils (single wall) suspended or supported in storage tanks (NBS 1977a). Such designs may use a pump to circulate water through the coil, but often they are connected to the cold water supply with delivery directly to an auxiliary DHW heater, thereby eliminating the need for a pump. The possible reduction in capital and operating cost for such a design is offset by the poor heat transfer rate from hot water in storage to the domestic hot water.

The air-to-water heat transfer coil in an air system should be located in the collector outlet duct (leading to storage). The size and type will depend on the desired heat transfer rate (effectiveness). A high heat transfer rate results in large heat extraction and temperature drop in the air stream. Since lower temperature heat will then be stored, the ability of the solar system to meet the space heating load will be reduced. A balanced design is generally needed in an integrated system to permit maximum utilization of solar energy and minimum requirement for auxiliary energy.

There are some air-heating system designs that place the water-heating coil inside the storage unit (SMACNA 1981). Disadvantages in this approach are that (1) heating will be ineffective without air movement through storage and (2) heat would have to be stored during summer in order to heat water. If the storage unit is inside the building, storing heat during summer and consequent heat losses will significantly overheat the rooms or increase the cooling load. As with liquid systems, heat exchangers located externally to storage are preferred for ease of inspection and maintenance.

12.2   Liquid Systems

The many concepts and varieties of designs for liquid-based solar heating systems can be classified into two basic types, direct and indirect heating systems (Solar Energy Applications Laboratory 1980a). Systems in which the storage fluid is circulated through the collector are direct heating systems, and systems that use a heat exchanger to transfer heat from collector fluid to storage fluid are indirect heating systems. Although many fluids are suitable for storing heat, water is commonly used because of its availability and low cost. Water is also used as the collecting fluid, but antifreeze compounds must be added or the collector and exposed piping must be drained to prevent freezing. Nonfreezing organic liquids may also be used as the collector heat transfer fluid.

12.2.1   Direct Heating Systems

The first of two basic designs for direct heating solar systems is a draindown type in which only water is used to both collect and store thermal energy. When freezing or boiling conditions are approached, the collector and pipes exposed to the atmosphere are automatically drained to the sewer through a sensor-controlled drain valve.

In the second design, which is a drainback type, water in the collector drains back into the storage tank or a smaller vessel. This water can be pretreated and additives can be used to prevent corrosion. Additives can also be used in draindown systems, but they would become diluted with makeup water, and small amounts would need to be added periodically to maintain an appropriate concentration level. In space heating systems, when the collector is above the level of the storage tank, the liquid can drain back from the collectors and piping into the storage tank, but in systems where the storage tank is completely filled with water or the water level in storage is above the collector array, more complex designs have to be used to drain the collector and associated piping.

12.2.1.1 Draindown

A draindown system is depicted in figure 12.1. Although freeze protection is often cited as the principal reason for this concept and design, protection from boiling and overpressure is also provided. To drain the collector, the headers, and the connecting piping, an inlet valve on the supply line, and another on the return line from the collector array are closed. Two drain valves are then opened to drain the collectors and exposed piping. An alternative is to use a single-action multiport valve (shown as inset in figure 12.1) to close the connecting lines to the pressurized solar storage tank and to open the collector lines to the drain. As water drains from the collector, suction causes the siphon breaker valve to open so that air is drawn in to replace the water leaving the collector.

All pipelines are sloped to assure complete drainage, and a syphon breaker and air vent are located at the high point of the collection subsystem. After drainage, water must be added to the system when collection is resumed. In most instances, makeup water is provided directly from the mains, often through a pressure-reducing valve. Where the solids content of the supply water is high, treatment of the replacement water is advisable to prevent solids deposition in collector passages and foreshortening the useful life of the solar collectors.

Draindown systems appear to have little use in space heating systems, and then only in those designed mainly for the DHW supply. Daily filling and draining of the collectors during the heating season seems impractical. The main use has been in DHW systems in climates where freezing temperatures occur infrequently and where pressurized storage and collectors are needed. Even in these situations, excessive freeze failures are resulting in a trend toward the drainback design.

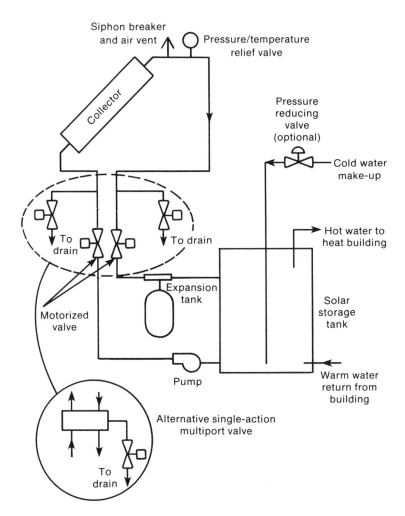

Figure 12.1
Draindown liquid solar space-heating system.

12.2.1.2 Drainback

A basic drain-back system is depicted in figure 12.2, and a variation is shown in figure 12.3. Water in storage is heated directly in the collector, and both freezing and boiling are prevented because water drains back into the storage tank when the pump stops.

As in the draindown design, all segments of the fluid loop must be sloped adequately to drain. If water accumulated in a pipeline in an un-heated zone, it could freeze. Probably the most reliable design is one in which there is no specific control for collector drainage. When pump oper-ation stops, whether by electric power interruption or on signal from the controller, water drains from the collector back through the pump into the tank, while air rises into the collector from the top of the storage tank through the return pipe. If there are no water traps in the collector piping and if the return (down-flow) pipe is large enough, this system is virtually fail-safe for freeze protection. However, the nightly exposure of cold col-lector tubing and system piping to warm humid air rising from the storage tank may result in significant heat loss from storage. Exposure of metal collector tubing and system piping to humid air also increases corrosion rates.

A disadvantage of the system depicted in figure 12.2 is that water must be circulated against full static head losses as well as friction head losses in the supply piping and through the collector. Energy consumption for pumping can be minimized by using the syphon return design in figure 12.3. Static head between the water level in storage and the top of the collector is recovered in the water-filled down-flow return line. An auto-matic air inlet valve at a high point in the piping opens when the collector temperature approaches a preset low temperature (or if there is power outage). The open-air inlet permits drainage of the collector and piping into the storage tank through the two connecting pipes while the collector fills with air from the atmosphere. When collector temperature again rises, pumping resumes and air is forced out of the collector into the storage tank, and out of the storage vent. Although this system requires less pumping power than the non-syphoning type, dependence on a control sensor and automatic valve, even though opening without power, in-creases the risk of freeze damage.

The design variation of the drainback design shown in figure 12.3 uses a motorized water drain valve on a branch of the return pipe (shown in

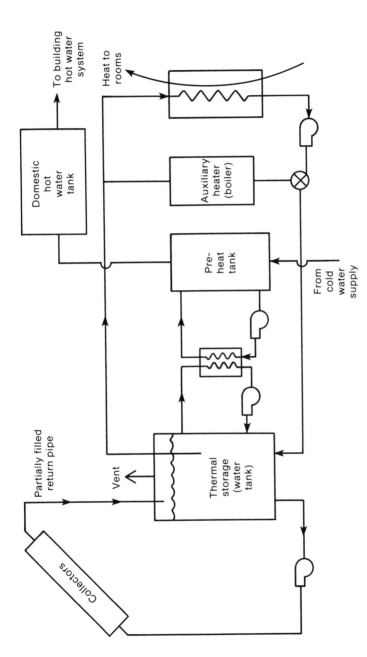

Figure 12.2
Schematic diagram of a drainback space heating and water heating system (gravity return).

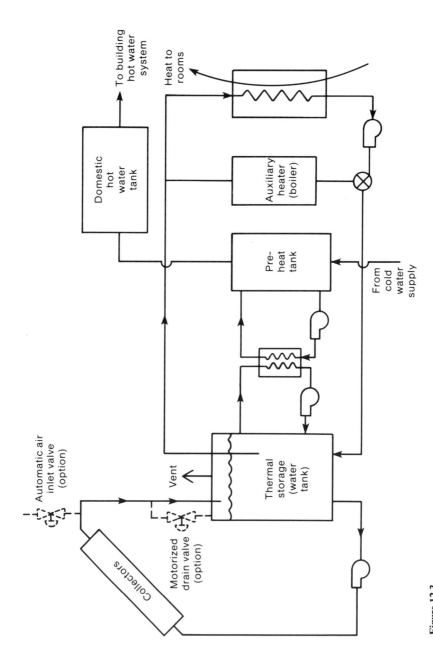

Figure 12.3
Schematic diagram of a drainback space heating and water heating system (syphon return).

dotted outline) instead of an electrically actuated air inlet valve. When a sensor in the collector shows a temperature near the freezing point (or if there is loss of electric power), this valve opens. Because the branch pipe is open to the air space in the tank, the weight of water in the return line causes a negative pressure at the top of the collector. This negative pressure causes the automatic air inlet valve to open and the collector to drain. When pumping is restarted, both valves close, air is pushed into the storage tank, and water circulation is reestablished.

The most common drainback design includes the best features of the systems shown in figures 12.2 and 12.3 and eliminates the risk of freeze controller failure. The motorized drain valve in figure 12.3 is replaced by a small-diameter tube, and the automatic air inlet valve is eliminated. There are no valves in the collector loop, and a syphon return develops as soon as the loop is filled with water. To ensure syphoning, the return pipe from the collector must be small enough to provide an adequate liquid velocity. When the pump stops, air can move up through the small tube into the collector return pipe, bubbling through the water in that pipe until drainage occurs.

12.2.1.3 Control Strategies

The basic control strategy for space and water heating solar systems is to maximize solar energy collection and utilization and to minimize electrical energy use for collection and distribution. While a theoretically optimum strategy for each system with a specific load in a particular climate can be prescribed (Winn, Johnson, and Moore 1974), practical control devices are generally not available to implement the optimum strategy for each system. Fortunately, even a simple controller performs reasonably well without inflicting severe penalties on the system. The simplest controller is a differential thermostat that starts or stops an electric motor that operates a pump or valve. More complex controllers regulate motor speeds. A variation of this type is a photovoltaic device that provides electric energy to a DC motor operating a centrifugal pump (Cromer 1985). Pump speed changes according to available solar radiation.

A set of simple controls for a direct-heating solar system is shown in figure 12.4 (Solar Energy Applications Laboratory 1980a). The space heating control system consists of temperature sensors S1 and S2 and differential thermostat no. 1 (which controls pump 1), and a two-stage room thermostat TR1–TR2 that controls pump 2, the circulation fan, and auxiliary heater. A second differential thermostat no. 2 controls pumps 3 and 4 for heating domestic water. Because the collector contains no water when the pump is not running, sensor S1 must be in contact with the absorber (back side). When the temperature at S1 is greater than the temperature at S2 by a preset amount (5°–10°C [9°–18°F]), pump 1 starts, and fluid circulates through the collector. Pump 1 will operate until the temperature at S1 is only slightly higher (about 2°C [4°F]) than at S2. Domestic water heating is controlled in similar manner, using sensor S3 at the top of the solar storage tank and S4 at the bottom of the preheat tank.

A two-stage thermostat is the simplest way to control space heating in a solar heating system (Solar Energy Applications Laboratory 1980a). The first stage, TR1, provides solar heating by controlling pump 2 and the circulation fan, while the second stage, TR2, controls an auxiliary heater. Pump 2 can circulate solar-heated water while auxiliary heat is also being supplied, or an interlock can be provided to prevent pump 2 from operating whenever the auxiliary heater is activated. If solar heat is to be used at the same time as auxiliary heat, the load heat exchanger should be arranged to permit preheating the air to the auxiliary heater. A parallel arrangement for using either solar or auxiliary heat requires that the circulation pump be disabled whenever auxiliary heating is in demand. Using sensor S3 at the top of storage, or an aquastat, pump 2 can be controlled to extract solar heat from storage whenever the temperature at top of storage is above a minimum level, whether or not the second stage of the thermostat is engaged. Such a strategy enables maximum use of solar energy for space heating.

Control strategies to maximize solar energy collection include using variable-speed or multispeed pumps (motors) (Winn and Hull 1979; Karaki, Brisbane, and Löf 1985). When the solar radiation level is low, a low flow rate through the collector permits continued operation. When the solar radiation level is high, a high flow rate maximizes energy collection. Many types and designs of pumps are available, and reasonable care should be exercised in selecting high-performance pumps for either variable-speed or fixed-speed applications. When pump selection is limited for a specific system design and multiple speed is desired, another alternative is to use two pumps in series. Control of pump speed (or of multiple pumps) is easiest with a set temperature difference between S1 and S2. When the difference is less than a preset amount, pump speed is reduced,

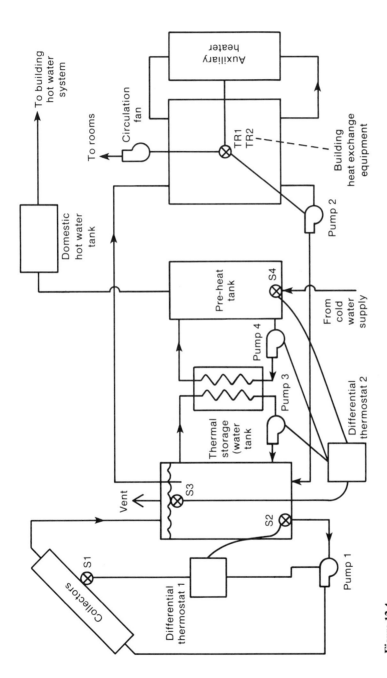

Figure 12.4
Schematic diagram of a single-liquid (drainback) space heating and water heating system (gravity return). Source: Solar Energy Applications Laboratory (1980a).

and when it is greater than the preset value, pump speed is increased. A dead band about the preset temperature difference prevents instabilities.

Control strategy for preheating domestic water is similar to that for solar space heating. However, because flow rates are much less, variable-speed strategies are generally not utilized. If DHW use constitutes a substantial portion or all of the load, then the main storage tank becomes the preheat tank and the preheat tank shown in figure 12.4 is not used in the design.

12.2.2 Indirect Heating

A widely used indirect liquid-heating solar system design is shown in figure 12.5 (Duffie and Beckman 1980: Löf and Tybout 1974: Ehrenkrantz Group 1979). The system collects solar heat in a nonfreezing liquid, delivers the heat to water storage through a liquid-to-liquid heat exchanger, and supplies heat from storage to the space heating and DHW equipment. When solar heat cannot meet DHW or space heating demands, conventional units supply auxiliary heat. A solar DHW heater can be completely separate from a solar space heating system, or it may be more convenient and economical to combine them. For residential applications, an integrated system can be used throughout the year, providing essentially all of the DHW needs during summer. However, for large systems with small DHW needs, operating a large collection system during the summer may not be economical, and a smaller, separate DHW system may be more practical.

12.2.2.1 Antifreeze Solutions
A commonly used liquid in the collector loop is a mixture of water and either ethylene glycol or propylene glycol. The advisable glycol concentration depends on the minimum temperature expected in the region. A 50–50 mixture provides maximum protection to $-30°C$ ($-22°F$). A centrifugal pump, usually at the lowest position in the loop, circulates the liquid solution through the collectors and heat exchanger, typically at about 0.02 g min^{-1} ft^2 (0.8 l min^{-1} m^{-2}) of collector.

In many dual-liquid systems, the collector loop is pressurized (to a moderate 12–15 psi [83–103 kPa] pressure). The likelihood of boiling is thereby reduced, positive pressure at pump inlet is assured, and fluid loss is minimized. A pressurized expansion tank (like those in conventional hydronic heating systems) is usually mounted in the line near the circulat-

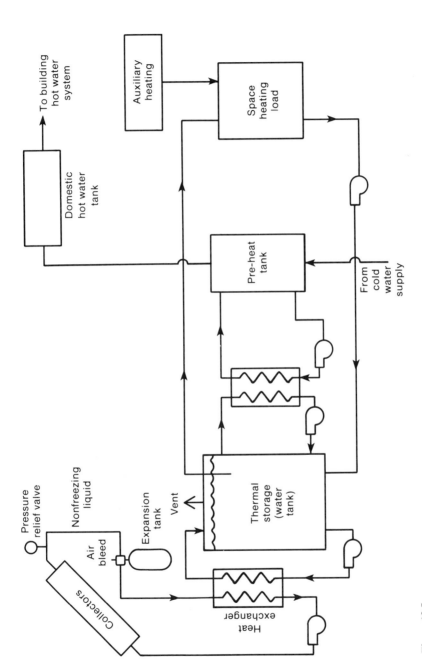

Figure 12.5
Indirect heating of storage fluid (water) space and DHW heating system.

ing pump. The expansion tank can be a bladder type or air-filled type with vent valves. An automatic air vent and a pressure relief (safety) valve are also provided in each loop of the collection system.

12.2.2.2.   Other Heat Transfer Fluids

Some systems use nonfreezing oil as the heat collection fluid. Therminol®, Bray-oil®, and silicones are examples. In addition to providing complete protection from freezing, these liquids will not boil at temperatures attainable in flat-plate collectors and are noncorrosive. They are more expensive than antifreeze solutions, however, and their heat transfer properties are poorer.

A heat exchanger and water heat storage are always used in systems with a nonfreezing liquid because of the excessive cost of filling a thermal storage tank with such expensive liquids. The heat exchanger is usually a commercial tube-and-shell type through which water is circulated from storage by a centrifugal pump at a volumetric rate about 50% greater than in the collection loop. Water is supplied to the pump from the bottom of the tank and returned to the tank near its top after being heated in the exchanger.

12.2.2.3   Control Strategies

Strategies to collect solar energy and to control its distribution for space and water heaters are basically the same for indirect- and direct-heating systems (Solar Energy Applications Laboratory 1980a). The most significant difference is in protecting the system from boiling. In drainback and draindown designs, a high-temperature sensor or aquastat at the top level in storage can be used to interrupt electrical service to the pump to prevent fluid circulation and further heating. A similar strategy may be used to stop collection if a nonboiling liquid is used in the collector. However, if a water-glycol mixture is the collector fluid, a possible alternative is to allow the vented storage tank to boil and provide makeup water, either manually or automatically, to maintain adequate water level in the storage tank. Such a design is not encouraged where frequent boiling can be expected. Increased water hardness and scaling and corrosion of heat exchanger and tank surfaces will reduce the life of these components.

With some additional cost, a heat rejector can be included in the collector loop to protect against boiling. This unit is usually an air-cooled, water-to-air heat exchanger. The heat rejector capacity must be as large as the peak energy delivery rate from the collector. Thermostatic control, using either collector outlet temperature or top-of-storage temperature and a wide dead band, is appropriate for controlling the heat rejector.

The room thermostat controls fluid circulation through the load heat exchanger. Sophisticated load-management controllers (Warren et al. 1981) are available, but a simple two-stage thermostat can adequately control room temperature to maintain comfort levels with a temperature swing of only a few degrees. On first-stage contact, solar-heated water is first used to heat the rooms. If the water in storage is not sufficiently warm to meet the heating demand, the room air continues to cool down until the second-stage contact of the thermostat activates the auxiliary heater. Since the auxiliary heater size is adequate to meet the heating load for the coldest night during the winter, the room air temperature will be restored to the set temperature. The principal difficulty with a two-stage thermostat is the temperature swing between set temperature and second-stage contact. With a dead band of $1.5°–2°C$ ($2.7°–3.6°F$) at each contact point, there is potentially a $6°C$ ($10.8°F$) swing when the rooms are being heated by the auxiliary system. The dead bands can be reduced to minimize the temperature swing, or additional controls can be used to initiate auxiliary heating whenever the top level in storage drops to a preset temperature ($30°–34°C$ [$86°–93°F$]). Since average storage water temperature may be higher, this strategy can adversely affect collector efficiency in a minor way and perhaps reduce annual solar energy collection and use.

12.2.3   Heating of Domestic Hot Water

A typical arrangement for preheating domestic hot water in an integrated liquid space-heating system is shown in figure 12.6 (Solar Energy Applications Laboratory 1980b). Hot water in solar storage is used to heat potable water in a smaller tank through a heat exchanger. If water in solar storage is nonpotable because of anticorrosion additives, or if initially potable water could become contaminated, a double-wall heat exchanger must be used. In the arrangement shown in figure 12.6, the capacity of the preheat tank is selected to be adequate for a typical daily demand. For a residential application, this means a 60–80 gal. (250–300 l) tank. A conventional (auxiliary) hot water heater supplements the solar-heated water as required. As hot water is used, water preheated by solar energy replaces the quantity drawn from the auxiliary hot water heater. If the temperature of the preheated water is less than the preset value, auxiliary fuel or electricity is used to raise the temperature to the desired level. If the preheated

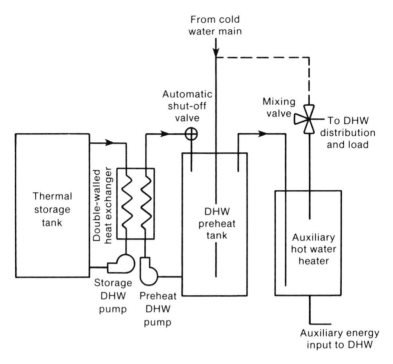

Figure 12.6
Schematic arrangement of a solar DHW subsystem. Source: Solar Energy Applications
Laboratory (1980b).

water can be hotter than the desired level, for safety considerations a
tempering (mixing) valve should be used to prevent delivery of excessively
hot water. Alternatively, controls can be used to prevent overheating in
the preheat tank, thereby avoiding the need for the tempering valve.

Instead of using both a preheat tank and a smaller auxiliary tank in the
DHW subsystem, they may be combined in a single tank when auxiliary
heating is by electricity. Only the upper electric element of the combined
tank is used for auxiliary heating, and the return line from the heat exchanger is extended below the level of the heating element. If the circulation rate through the heat exchanger is sufficiently low, electrically heated
water in the upper portion of the tank will not mix with the solar-heated
water in the lower part of the tank. A principal advantage of the one-tank
system is reduced heat losses from storage (and reduced capital invest-

ment). Depending on the size and effectiveness of the heat exchanger, the solar contribution from the one-tank system can be nearly the same as from the two-tank system (Löf, Duff, and Hancock 1981).

12.2.3.1 Heat Exchanger Types and Locations

To prevent contamination of household water and water mains, double-walled heat exchangers should be used for transfer of heat from solar storage to domestic water. Depending on the heat exchanger design, they can be internal or external to the DHW tank. An important feature of double-walled heat exchangers must be that leaks through either wall are readily detectable so that the unit may be repaired or replaced.

One type uses a tube wrapped around the outside of the preheat tank (NBS 1977b). A Roll-Bond® sheet is particularly suitable for this purpose. Other designs involve tube in tube, with one liquid inside the inner tube and the other liquid outside the outer tube (Karaki and Westhoff 1982). The annulus between the tubes is vented so that leaks can be detected. This type can be located either inside or outside the tank, with the external location preferred because of inspection and maintenance. Other external heat exchanger designs include a pair of tubes wrapped around a steel shell in helical fashion, with potable water flowing through one tube and nonpotable water in the other. Another design involves parallel tubes cross-connected with many short heat transfer fins to conduct heat from the hot tubes to the cold tubes (Solar Energy Applications Laboratory 1980a).

Double-wall heat exchangers are not as commonly available as single-wall shell-and-tube heat exchangers, and size selections may be limited. Typical flow rates for both hot and cold water are 2–3 gal min^{-1} (7–11 l min^{-1}) with temperature drops of 3°–5°C (5°–9°F) across the heat exchanger and temperature differences up to 10°C (18°F) between hot and cold water. For single-tank systems, a higher heat recovery rate may be desirable than for dual-tank systems, in which case a larger heat exchanger should be selected.

12.2.3.2 Operating Strategies

During the summer months, the DHW subsystem in an integrated solar heating (and/or cooling) system can supply virtually all of the hot water needs of most buildings. Danger of overheating should be prevented in the design, either with appropriate controls or by including a tempering valve on the delivery line. In either event, there is no specific operating strategy different from winter operation that will impact on the solar contribution to the total load.

During the winter months, when storage water temperatures are lower and solar will not meet the combined demands, priorities must be considered. If space heating is relatively more important than DHW heating, then the differential thermostat can be set to provide water heating only when storage temperature is greater than a large preset difference above mains temperature. Normal strategy is to set the temperature difference between the top of solar storage and bottom of the DHW tank to about 6°C (11°F) to start and 1°C (2°F) to stop fluid circulation through the heat exchanger.

12.2.4 Comments and Recommendations

There are advantages and disadvantages with each type of liquid system discussed in this section. Direct-heating systems have some cost advantages over indirect-heating systems but are more subject to corrosion damage than closed-loop indirect-heating systems. Drainback designs offer simplicity in operation, although the annual operating energy requirement is inherently greater than with draindown designs. Reliance on temperature sensors for freeze protection of draindown systems has proved costly for many installations when systems failed to drain and collector absorbers and pipes burst after freezing. Nonaqueous, nonfreezing liquids in indirect-heating systems have proved to be relatively maintenance-free compared to aqueous antifreeze solutions; however, heat exchangers and pumps add to initial system costs. In terms of performance, any system can be designed to provide comparable annual quantities of thermal energy to a load, but the cost per unit of energy delivered can vary considerably from system to system.

For simplicity, a direct-heating drainback system with nonsyphon return is recommended among liquid-heating systems. There is considerable merit in keeping systems as simple as possible, because they are easy to control, reliable in operation, and require little need for maintenance. Where collector designs are not suitable for use in a drainback system, an indirect-heating system with a nonfreezing, nonboiling, noncorrosive heat transfer fluid is recommended. Heat exchangers external to the storage tank are recommended for both collection and delivery of solar energy to the load, mainly because of ease of inspection and maintenance.

12.3 Air Systems

There are probably as many varieties in designs of air-based solar heating systems as there are liquid systems; however, all air systems are of the direct-heating type. With air systems, freezing and boiling are not concerns, and air leakage is the main issue for design and installation. Unlike leaks in liquid systems, an air leak does not physically damage the building or its contents.

12.3.1 Basic System

The main components of a basic air-heating system are shown in figure 12.7. Their primary functions are the same as in a liquid collection and

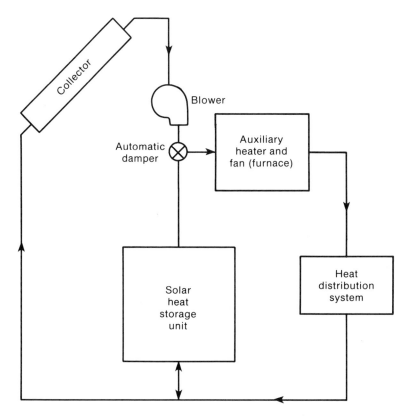

Figure 12.7
Schematic diagram of a basic air-heating solar system.

storage system. An important advantage of the air system is its capability
for the direct supply of heat from the collector to the living space. Liquid
collection and storage systems involve transfer of heat from water to air,
whereas room air is always used for both collection and distribution in air
systems. Auxiliary heat is nearly always supplied as a supplement or boost
in air systems (figure 12.7), but in liquid systems having hot water distribution (figure 12.1), the auxiliary is in parallel rather than in series with the
solar supply.

The most economical and effective type of heat storage for an air system
is a pebble bed or crushed rock through which air is circulated and in
which heat is stored and transferred (Solar Energy Applications Laboratory 1980a). This key component serves not only as a heat storage unit but
also as a heat exchanger during the storing cycle, and also when transferring stored heat to room air. The use of a pebble bed results in beneficial
temperature stratification in the storage unit and a substantially higher
solar collector efficiency than if transferred to a nonstratified storage unit.
Phase-change materials may also be used, but their costs are higher than
pebbles and the isothermal nature of such a storage unit may result in a
lower collector efficiency than that obtained with a stratified pebble bed
unit.

12.3.1.1 Blower Locations

The circulation blower for the collector can be located in the cold-air duct
leading to the collector or in the hot-air duct from the collector. An important advantage in locating the blower in the hot-air duct is that the collectors will operate under subatmospheric pressure, and if there is any air
leakage in the collectors, cold air will enter the array and be heated by the
collector. Air at room temperature and of a quantity equal to the cold air
admitted through the collector array will be forced out of the building.
With the blower in the cold-air inlet duct leading to the collector, the
collector array will be pressurized, and solar-heated air at temperatures
warmer than room air will leak out of the array to the atmosphere. Cold
air at a mass flow rate equal to the leakage rate will infiltrate the building,
thus adding to the heating load. Both analytical and experimental studies
(Close and Yusoff 1978; Karaki and Hawken 1982) have shown that it is
technically better to place the blower in the hot-air side of the collector.
Depending on location of the leak, thermal efficiency of the collector array
is improved with entrance of cold air from the atmosphere.

With direct-drive blowers, locating the motor in the hot-air stream may cause operational difficulties if the motor winding becomes hot enough to trip the overheat breaker. Using a pulley-driven blower with the motor external to the duct or a motor with a higher operating temperature limit is advisable.

Whether the collector circulation blower is located in the supply or return duct, a separate distribution blower is desirable. For zone-controlled distribution systems, multiple blowers or a multispeed blower can be used. With separate collector and distribution blowers, selection of each blower size and type may be based on their respective air flow rate and pressure head requirements. Air flow rate through the collector can then be independent of air flow rate through the distribution system whether the blowers operate independently or simultaneously.

An optional design is to use one blower for air circulation on both the collection and distribution loops. When the air flow rate through the collector is nearly the same as that required for the distribution system, using one blower instead of two (or more) may reduce the capital cost of the system, but additional automatic dampers are required. With a one-blower system, there is little flexibility to control heat delivery to the rooms, and heat will be delivered to all the rooms when required at the central thermostat location.

12.3.1.2 System Operation with Storage

A common type of air-heating solar system with storage is shown in figure 12.8 (Solar Energy Applications Laboratory 1980a). The components include: (1) a flat-plate air-heating solar collector; (2) a pebble-bed heat storage unit to and from which heat is transferred by circulating air through the bed; (3) a control unit that includes the sensors and control logic necessary to collect and store solar heat when available and to automatically maintain comfort conditions at all times; (4) an air-handling module with filters and blowers; (5) a solar hot-water heater consisting of an air-to-water heat exchanger and a preheat storage tank connected to an auxiliary hot water heater; and (6) an auxiliary heating unit (usually a warm-air furnace) to provide 100% backup space heating when storage temperatures are insufficient to meet demands or when the solar system is not operating.

In an air-heating system, the collector absorbs solar radiation and converts it to heated air. Air is circulated from the solar system to the building

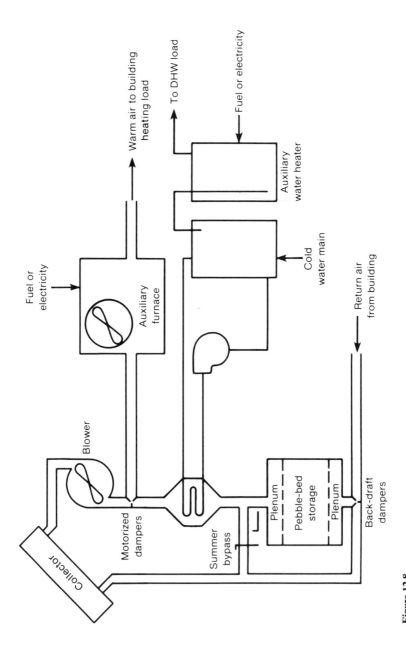

Figure 12.8
Air-heating solar system. Source: Solar Energy Applications Laboratory (1980).

in the same manner as in most modern air-heating systems. As air flows from one end of the collector to the other, its temperature normally rises from 20°C (68°F) to 55°C (131°F) or 65°C (149°F) during the middle of the day. As shown in figure 12.9, the building is heated directly by air from the collector whenever heating is needed during sunny periods. Cool air from the building is returned to the collector for reheating.

Heat is stored by using the heat-exchange and heat-storage characteristics of dry pebbles, the most practical storage medium for use with air-heating collectors. When heat is not needed in the building, solar-heated air is routed through the storage unit as in figure 12.10, thereby heating the pebbles; cool air, usually near 20°C (68°F), returns to the collector for reheating. Temperature stratification in the storage unit (higher near the top, lower near the bottom) assures maximum heat recovery from the solar air collector and beneficial air supply temperatures to the rooms.

In the evening and nighttime hours, heat is delivered to the rooms by circulating air from the building through the pebble bed, as in figure 12.11. Because of temperature stratification in the storage unit, this mode supplies heat to the rooms at the highest available temperature. The system automatically provides auxiliary heating from fuel or electricity when the solar supply is insufficient to meet requirements.

Domestic hot water can be heated by an air-to-water heat exchanger in the hot air duct from the collector. As shown in figure 12.10, when solar heat is being transferred to storage, heat is also being supplied to the hot water system if the water pump is operating. The temperature of the air passing through the water-heating coil is thus usually reduced by a few degrees. When warm air is being supplied from storage to the rooms (see figure 12.11), no water heating occurs because there is no water circulation through the coil. Other coil locations have been used, but this position has proved to be best in protecting the coil from freezing.

So that solar-heated water can be available in the summer when no space heating is needed, a bypass duct is opened, as shown in figure 12.12, and air flow to storage and to the building is blocked. Nearly 100% of typical hot-water requirements can thus be met by solar energy in the 4–5 warm months of the year, and unwanted heating of the building by losses from storage is avoided.

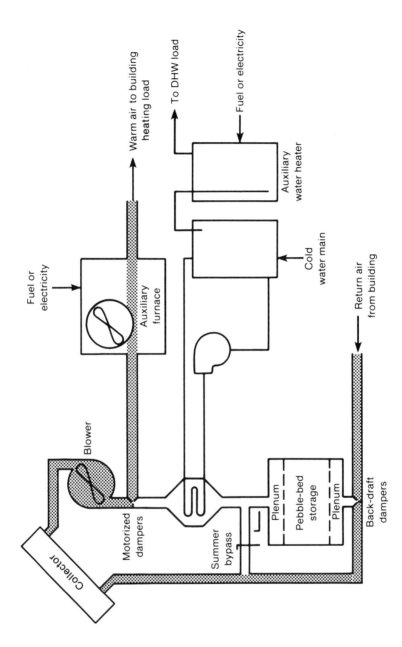

Figure 12.9
Heating from collector.

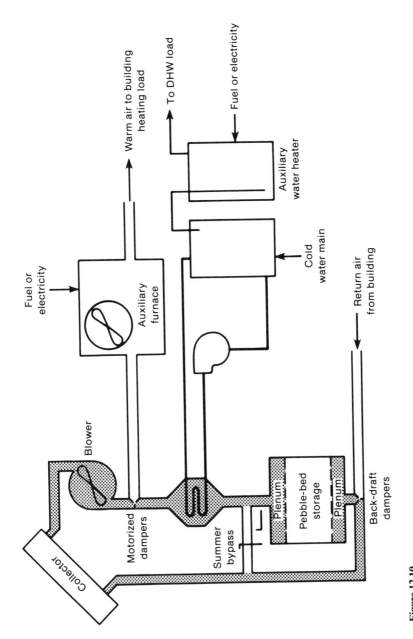

Figure 12.10
Storing heat from collector.

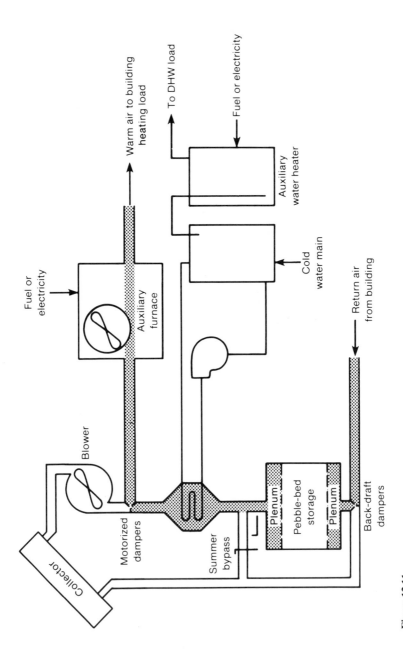

Figure 12.11
Heating from storage.

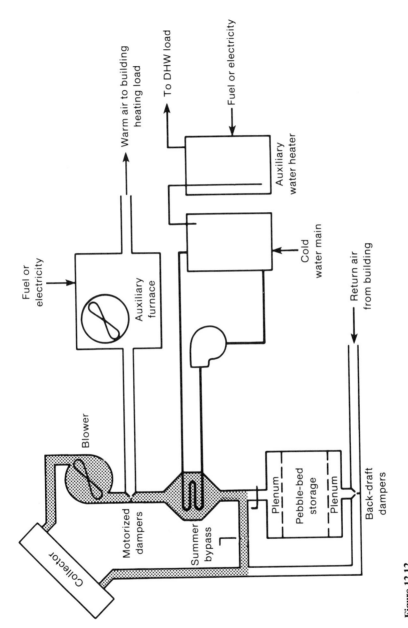

Figure 12.12
Summer heating of domestic hot water.

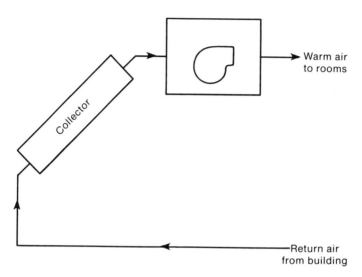

Figure 12.13
Solar system without storage.

12.3.1.3   Systems without Storage

Collector area for the type of system discussed in the previous section is generally more than that of one or two modules. It is expected that more heat will be collected than required by the load during the collection period, hence the excess thermal energy is stored. If the collector area is relatively small with respect to the heating load, collected solar energy can usually be used immediately during the winter months, and a separate storage unit becomes unnecessary. This daytime solar heating can also be effective on cool days in spring and fall. Such a simple system is depicted in figure 12.13. For various commercial applications with daytime occupancy and requirements for fresh air ventilation, high efficiency no-storage systems can be designed to use virtually all of the thermal energy provided by the collectors. During the warmer months, the collector can either be covered or allowed to stagnate.

12.3.1.4   Control Strategies

A basic set of sensors for control of an air-heating solar system is shown in figure 12.14. Sensor S1 in the collector is compared to S2 in a duct near the bottom of storage, and when the temperature difference is greater than a preset amount, about 10°C (18°F), blower B1 starts. Depending on the

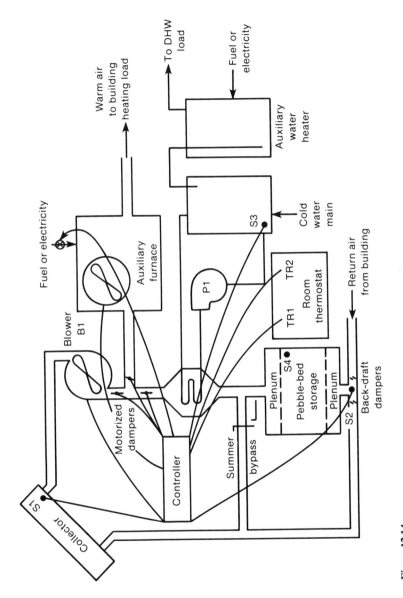

Figure 12.14
Control for air-heating solar system.

status of the two-stage room thermostat, TR1, TR2, the motorized dampers will direct the warm air flow to either the rooms or storage. If room heating is required, the blower in the furnace will also be activated by the controller; if not, air flow will be directed through storage. A comparison of temperatures at S1 and S3 will also determine whether water pump P1 should be activated. In this basic design, domestic water cannot be heated while warm air is being delivered to the rooms from the collector, and because the comparison is between S1 and S3, there will be no water heating (usually) when warm air is being delivered from storage to the rooms, even though the heat exchanger is located in a position where heat transfer is possible. The strategy is simply that during the coldest part of the heating season, it does not matter whether stored solar heat is used for heating space or heating water. That strategy could be easily changed by adding a fourth sensor, S4, to the top layers of the pebble bed and then controlling P1 by the temperature difference between S4 and S3, whenever S1 is less than S3 and the furnace blower is running. In that event, annual solar energy utilization could be increased slightly by using stored solar heat to heat water during the warmer months of spring and fall when heat in storage is greater than the space heating demand.

Room heating from storage is controlled by elements TR1 and TR2 in the room thermostat. When the first-stage contact point, TR1, is reached, air circulation is reversed through storage, with return air from the rooms entering at the bottom of storage and warm air exiting from the top for supply to the rooms through the auxiliary furnace. The second-stage contact, TR2, controls the fuel valve (or electrical switch) of the auxiliary furnace.

Control of heating systems without storage is basically simple. If the temperature in the collector, as determined by S1, is greater than a preset value, about 30°C (86°F), the blower circulates room air through the collector. On warm days, this simple control may result in overheating, so an interlock with the room thermostat is usually employed to prevent blower operation unless the TR1 contact is also closed.

Advanced control strategies can be adopted for air systems as well as liquid systems. For example, a variable or multispeed blower, B1, can modestly improve overall collection efficiency (Karaki, Lantz, and Waterbury 1982). Sensor S4 can be used as an anticipator so that if storage is below a set temperature (such as 30°C [86°F]), the auxiliary furnace will be used immediately to prevent a large temperature swing in the rooms. As with liquid systems, a single-tank DHW subsystem can also be used effectively, provided electricity is the auxiliary source.

12.3.2   Heating of Domestic Hot Water

Domestic water heating with an air system is accomplished by an air-to-water, cross-flow heat exchanger in the hot-air duct from the collector. When solar-heated air is being delivered to storage, heat is also supplied to the hot water system by circulating water through the heat exchanger. The temperature of the air passing through the water-heating coil is therefore reduced by a few degrees depending on the size of the heat exchanger. In a typical solar air system, water is heated only during collection hours, thus the size of the heat exchanger and the air-temperature drop at the heat exchanger depend on the daily quantity of water to be heated. During the heating season, available solar heat will be shared between space and water heating in accordance with size of heat exchanger and control strategy. During the summer when no space heating is needed, the entire collector array can be devoted to DHW heating.

12.3.2.1   Heat Exchanger Types and Locations

The heat exchanger for preheating DHW should be located where it cannot freeze. By placing the heat exchanger in the heated space inside the building, there is reasonable protection from freezing, but if a collector damper were open and cold air from the collector were drawn through the coil at night by the furnace fan, water in the coil could freeze. A safe location for the heat exchanger is between the inlet to storage and the room distribution duct as shown in figure 12.8. Even if the damper in the collector return duct were not tightly closed, cold air could not come in contact with the heat exchanger in this location.

Some designs of air heating systems include finned copper tubes or plain copper tubes placed in the top layers of the pebble bed to preheat the potable water flowing from the mains to the preheat tank (SMACCNA 1981). Although a large area of coil surface would assure reasonable heat transfer, it would be difficult to inspect for leaks, and maintenance could be expensive. Since the heat transfer rate from storage to the coil is limited unless there is air movement through storage, there is no inherent advantage in placing a heat exchanger inside the pebble bed container compared to a more conventional placement in the duct.

12.3.2.2 Operating Strategies

A basic strategy for preheating DHW in an integrated air system is to operate the water circulating pump only while collecting. This strategy assumes that the heat exchanger is large enough to provide the desired preheating while solar energy is being collected, or that it is more important to use solar energy for heating space than for heating water. An alternative strategy to permit DHW preheating while heating the rooms is discussed in section 12.3.1.4.

12.3.3 Comments and Recommendations

Air-heating systems offer a simple and direct approach to space heating with solar energy, and by including DHW heating capability, the solar system can be used throughout the year. There are no freezing, boiling, or significant corrosion problems, and the working fluid in the system is the medium to be conditioned. Collectors made of high-quality materials, including tempered glass covers and metal absorbers with quality coatings, can be expected to provide long and effective service. With proper maintenance, air-heating systems should have a long life, perhaps as long as the building itself. Heat exchangers for domestic water heating and domestic water tanks may require infrequent replacement. Preventive maintenance of parts of the system requiring service can greatly extend useful life.

Air leakage in ducts and through closed dampers can degrade system performance, so they should be carefully installed. Ducts must be insulated, because heat leakage from ducts can reduce effectiveness of controlled delivery of heat. There are no significant technical barriers to successful wide-scale implementation of solar air systems, particularly for residential use. For large commercial applications, large ducts may be required to carry large amounts of air to and from the collectors. Compared to liquid systems with large collector arrays, substantially greater heat losses through duct walls are likely than through the walls of much smaller pipes. However, large systems, particularly those without storage, can be subdivided into smaller, independently controlled systems with much smaller ducts.

When properly designed and installed, air-heating solar systems can be durable and operate effectively and reliably for a very long time. They are very well suited to residential applications.

For certain commercial uses, solar systems can be used effectively during the winter even without storage. For example, large spaces like terminal buildings for transportation vehicles, which require continuous ventilation, can use all of the solar heat provided by the system for pre-heating ventilating air. Other commercial buildings, such as large ware-houses, can use the contents for heat storage, so that simple air systems can be used.

12.4 Auxiliary Heating

12.4.1 Liquid-Based Systems

12.4.1.1 Boilers and Furnaces

There are several methods for supplying auxiliary heat in a liquid-based solar system for space heating. In virtually all practical solar heating designs, auxiliary heat is supplied to the fluid stream in which heat is dis-tributed to the various zones in the building. Hydronic distribution (base-

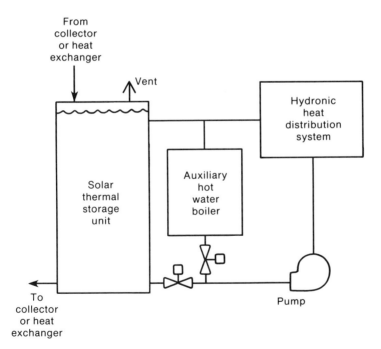

Figure 12.15
Auxiliary heat supply in hydronic heat distribution system. Source: Solar Energy Applications Laboratory (1980a).

board strips, individual fan coils, imbedded tubing, cast radiators) uses a
fuel-fired or electrically heated hot water boiler, as indicated in figure
12.15 (Solar Energy Applications Laboratory 1980a). If a central heat
exchanger and ducted warm air are employed, auxiliary heat is most economically and practically supplied in a furnace or electric heater through
which the solar-heated air passes to the rooms, as shown in figure 12.6
(Solar Energy Applications Laboratory 1980a).

In hydronic distribution systems, the hot water boiler is best used in
parallel with the solar supply from storage rather than in series with it, so
that only one source is used at a time (Solar Energy Applications Laboratory 1980a). The series design is seldom used because some of the heat
supplied to the water passing through a thermostated auxiliary boiler
would flow on through the load exchangers and accumulate in the solar

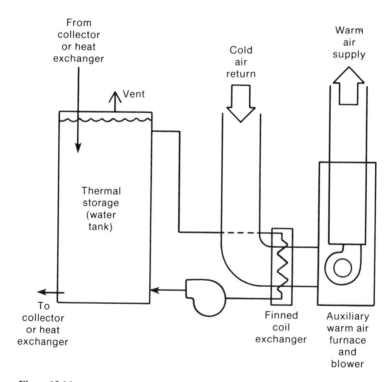

Figure 12.16
Solar heating with auxiliary furnace. Source: Solar Energy Applications Laboratory
(1980a).

heat storage tank. The resulting temperature rise in solar storage would reduce collector efficiency and capacity for solar heat storage. As shown in figure 12.15, a single pump with automatic valves is satisfactory for circulating hot water from either solar storage or the auxiliary boiler. The controller regulates the system so that when the demand cannot be met by solar, auxiliary is used.

If heat is distributed as warm air heated by exchange with solar-heated water, auxiliary heat is usually supplied to the air in a warm-air furnace located downstream from the solar coil, as shown in figure 12.16. Auxiliary heat cannot affect solar storage in this design, so this type of arrangement is advantageous. Using stored solar heat even at a low temperature of 25°C (77°F) is thus made possible by further heating of the tepid air to useful temperatures greater than 50°C (122°F).

12.4.1.2   Heat Pumps

An electrically driven heat pump is a special type of auxiliary heater for a solar space heating system (Mitchell, Freeman, and Beckman 1978). The heat pump uses electric energy to extract heat from a low-temperature source and delivers the heat at a higher temperature. By this process, heat delivery at useful temperature may be several times the electric energy input. Depending on the source, annual performance factors (heat delivery per unit electric input) between 2 and 3 are commonly achieved (Andrews, Catan, and LeDoux 1982). With air-source heat pumps in colder climates, a somewhat lower factor generally prevails (Karaki 1983), and with water-source heat pumps and source temperatures above 10°C (50°F), a slightly higher factor can be achieved (Karaki and Westhoff 1982). The principal advantage in using a heat pump as an auxiliary heater during the winter is its use as an air conditioner during summer.

One method of heat pump use in a warm-air distribution system, illustrated in figure 12.17, involves the condenser coil (heating coil) of an air-to-air heat pump in the air circuit downstream from the solar-to-air exchanger. Because of pressure limitations in the heat pump, the flow of water through the solar coil is usually interrupted when auxiliary heat is required so that air is not supplied to the heat pump evaporator coil at excessive temperatures (e.g., not above 35°C [95°F]). As in conventional heat pump installations, electric resistance coils are used during periods when outdoor temperatures are so low that the capacity of the heat pump is insufficient to meet the heat demand. The resistance coil is also used

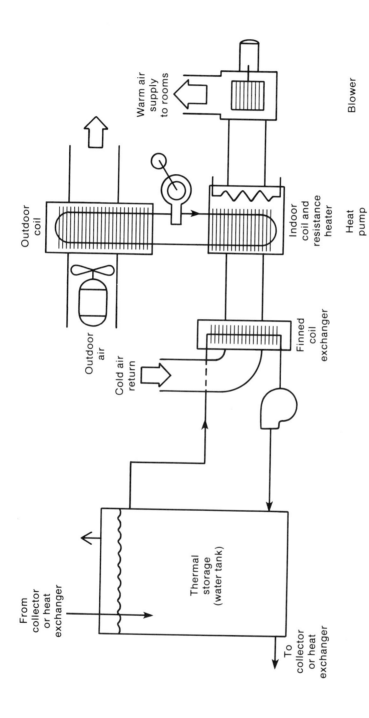

Figure 12.17
Solar heating with auxiliary air-to-air heat pump.

when the outdoor coil is being defrosted, and for some types of heat pumps when outdoor air temperature drops below a set temperature (e.g., $-20°C [-4°F]$).

Another alternative is an air-to-water heat pump arranged as shown in figure 12.18. Replacing the boiler in figure 12.15 with the heat pump condenser coil and backup electric resistance heater assures adequate heat supply. Heat distribution may be hydronic or air.

12.4.2 Air-Based Systems

12.4.2.1 Warm-Air Furnace

Auxiliary heat is usually supplied in solar air systems by using a warm-air furnace in series with the collector and storage units, as shown in figure 12.19 (Solar Energy Applications Laboratory 1980a). This design permits maximum supply of solar energy by using the solar system to preheat the air when heat requirements are greater than solar heat availability. Essentially all of the collected and stored solar heat can thus be used, even if at low temperature.

Warm-air furnaces can be fueled with gas, oil, or propane, or they can be supplied with electricity. A dual-stage house thermostat usually controls the auxiliary fuel or electricity, with the lower temperature contact activating the fuel valve or electric switch.

Electric motors for blowers in warm-air furnaces are frequently mounted in the cabinet through which air passes, and are thereby air cooled. When used as an auxiliary heater in a solar air system, this motor usually operates in a warm air stream, occasionally at high temperatures. Motors able to operate at such temperatures should therefore be used, rather than the type usually supplied with the furnace.

12.4.2.2 Heat Pumps

An air- or water-source heat pump may also be used to supply auxiliary heat to an air system. An air-to-air heat pump is shown in figure 12.20. The condenser coil of the heat pump heats the house air as the outdoor evaporator coil extracts energy from ambient air. Electric resistance backup is also necessary for meeting high heating demands. Summer cooling can be supplied by reversing the evaporator and condenser functions so that electrically driven vapor-compression air-conditioning is provided in a conventional manner.

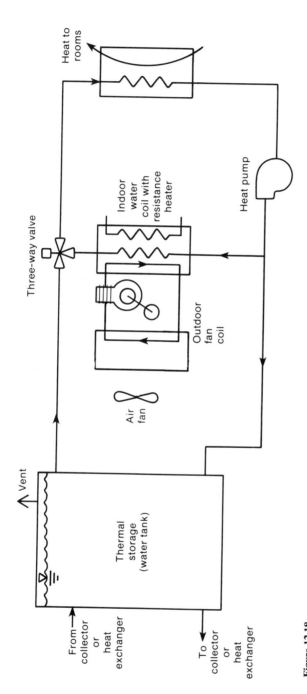

Figure 12.18
Auxiliary heat from air-to-water heat pump.

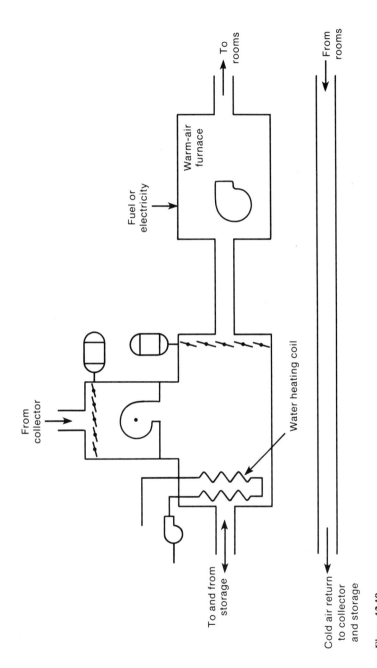

Figure 12.19
Solar heating system with warm air furnace. Source: Solar Energy Applications Laboratory (1980a).

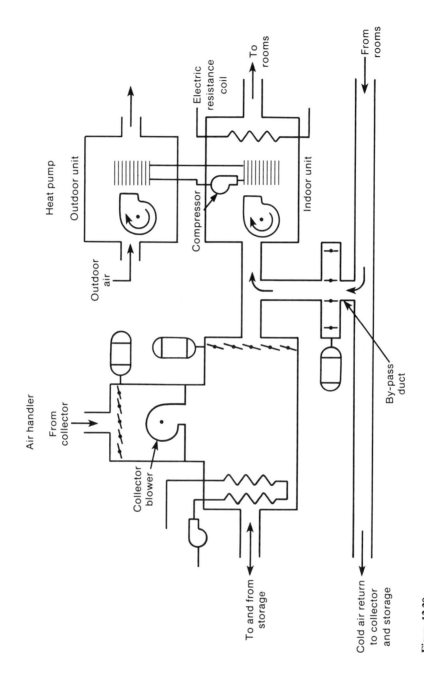

Figure 12.20
Solar heating system with air-to-air heat pump auxiliary.

Because of pressure limitations in commercially available heat pumps, they are not usually operated as temperature boosters for solar-heated air. An air temperature of 38°C (100°F) may not be sufficient for heating the building, but supply of air at that temperature to the condenser coil would cause excessive pressure in the heat pump. For this reason, air is supplied to the heat pump from the house air return via the bypass shown in figure 12.20 rather than from the solar system. A motor-operated damper in the bypass duct opens when this operating mode is required. This damper and the heat pump compressor and fans are controlled by the low-temperature contact in the two-stage house thermostat, which simultaneously closes the damper at the air handler outlet. The electric resistance backup coil is actually controlled by a temperature sensor in the heat pump system.

12.4.3   Comments and Recommendations

Using an auxiliary heating system to maximize use of collected (and stored) solar heat is technically and economically advantageous to the entire heating system. There are constraints, however, especially in retrofit installations, that may prevent placing the auxiliary heating unit in the preferred locations suggested in this section. When such occasions arise, the impact on overall performance should be carefully assessed.

12.5   Combination Systems

In principle, most heat pump systems are solar assisted, because the air or water that supplies heat to them derives its thermal energy content from the sun. However, in this chapter, solar-assisted heat pumps are those for which specific design efforts are made to supply solar thermal energy actively or passively in air or water to the evaporator of the heat pump. Dual-source heat pumps use both natural sources and energy from man-made solar heating systems. Ground-coupled heat pumps extract thermal energy from the ground through extended lengths of buried pipe. All three types are discussed in this section as combination systems.

12.5.1   Solar-Assisted Heat Pump

12.5.1.1   Liquid-Based Systems
The concept of a solar-assisted heat pump is the collection of solar heat at a comparatively low temperature and the supply of stored solar heat to

456   S. Karaki

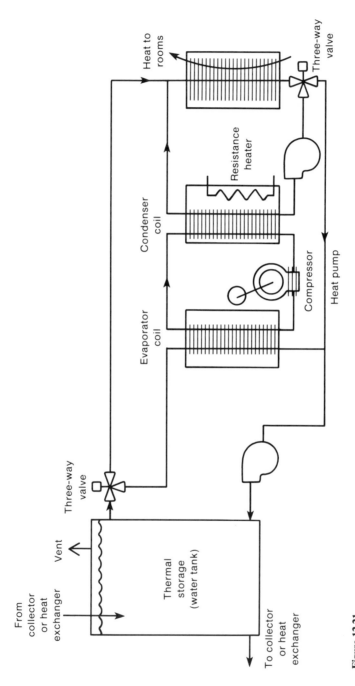

Figure 12.21
Solar-assisted heat pump (liquid-to-liquid). Source: Solar Energy Applications Laboratory (1980a).

the evaporator of the machine at a temperature higher than that of the outdoor ambient air. The coefficient of performance (COP) of the heat pump can then be higher than if outdoor air were the source, and collector efficiency can be higher than if operated at higher conventional temperatures.

When water in solar storage in a liquid-based system is not hot enough to heat room air directly, it may be used as a source of heat to the heat pump evaporator coil, as illustrated in figure 12.21 (Solar Energy Applications Laboratory 1980a). Heat is then supplied to the building by heating water in the condenser coil, with circulation to the living space or by exchange with air. In multizone distribution systems where individual water-to-air heat pumps in each zone are supplied with low-temperature water, using solar preheated water as the heat pump supply may reduce electricity use, provide heated air where and when needed, and minimize heat distribution cost. Because the fluid temperature delivered from the collectors may be as low as $25°-30°C$ ($77°-86°F$), collection efficiency can be higher than that obtained in a solar system supplying heat directly to load at water temperatures of $50°-85°C$ ($122°-185°F$). When solar storage approaches the freezing point in midwinter as a result of decreased solar collection and large withdrawals of heat for heat pump supply, electric resistance elements are called on to meet the demand.

12.5.1.2 Air-Based Systems

A solar-assisted heat pump in an air-based solar system is illustrated in figure 12.22 (Karaki and Hawken 1982). When no solar heat is being collected, heat is supplied to the building by circulating room air through the condenser coil; when heating the rooms directly from the collector, the condenser coil is bypassed. If the storage temperature is sufficiently high to heat the rooms, hot air will flow from storage through the condenser coils, but the heat pump will not be allowed to operate simultaneously. When the storage temperature drops below a level useful for direct heating, additional heat is withdrawn from storage by circulating air from the top of storage through the evaporator coil to the bottom of storage. As a practical matter, $-20°C$ ($-4°F$) is a reasonable lower limit for extracting thermal energy from storage because the COP of the heat pump will approach unity at such a low temperature. With very cold storage temperatures, collection efficiency increases, but air delivery temperature from

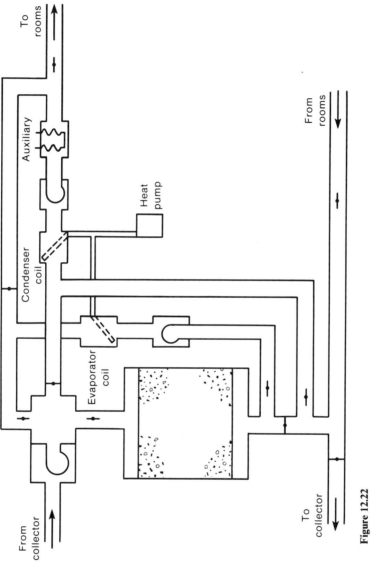

Figure 12.22
Solar-assisted heat pump with an air-based solar system. Source: Karaki and Hawken (1982).

the collector will be low unless the flow rate is controlled by a variable-speed or multispeed blower.

12.5.2   Dual-Source Heat Pumps

Dual-source heat pump systems use energy from a natural source as well as solar heat that has been collected and stored. The evaporator in water-to-water or water-to-air heat pumps is supplied with surface or ground water, and atmospheric air is used as the energy source in air-to-air heat pumps. Dual-source heat pump systems (Alcone 1981) are uncommon; in general, if the natural source is adequate as an energy source when collected solar energy has been depleted, it will also be an adequate source when solar energy is available. A more common design employs solar for direct heating and the heat pump, with energy extracted from a natural source, as auxiliary (Karaki 1983).

12.5.3   Ground-Coupled Heat Pumps

These systems are modifications of water-to-water or water-to-air heat pumps (Bose et al. 1980; Backlund 1981). Lengths of (usually) plastic pipe are buried in the ground, and by circulating an antifreeze liquid through this ground loop, heat is extracted from the soil and supplied to the evaporator of the heat pump. Since thermal energy in the soil is provided either by the sun or by slowly moving ground water, this system is a type of solar-assisted heat pump.

12.5.4   Comments and Recommendations

Technically, combination systems can be made to perform reasonably well. Compared to conventional solar systems, additional components are required, and there is greater complexity in control equipment and control strategy. With greater system complexity, there naturally follows some reduction in reliability. Adding many more automatic valves and dampers is an invitation to future maintenance problems. A well-designed, ground-coupled heat pump could be reasonably simple in design and perform well in selected regions where the ground can remain warm through the heating period. In very cold regions, long lengths of buried pipe would have to compensate for low rates of heat transfer.

A general disadvantage of many solar-assisted heat pump systems is that the unit cannot readily be used for air-conditioning (cooling) during the summer. A heat pump in parallel with a conventional solar system could be used quite simply for space cooling during the summer. An automatic feature of standard heating/cooling thermostats can easily accommodate the functions as seasons change.

12.6   System Integration and Factory Assembly

Solar heating systems have been generally designed or sized to fit a specific application, and selection of components has often been left to the installer. Success or failure of systems often depends on familiarity of installers with various available components. It is not surprising that mismatching of components has occurred, resulting in poor system performance.

Solar DHW systems suffered many failures during early stages of development, but common mistakes were gradually eliminated by developing packaged systems. Matching components were preselected by manufacturers of residential DHW systems and premounted (generally on the hot water tank) so that only the collectors and piping were field-installed. Pumps, valves, heat exchangers, controls, and sensors were factory-installed. The performance of DHW systems steadily improved over the years, and by 1985, nearly all residential solar DHW systems being marketed were packaged systems.

Packaged space-heating and integrated systems could lead to fewer mistakes in design and installation and result in better-performing systems. However, variations in applications are so numerous that it is difficult for manufacturers to commit to standard designs. The considerably large size and spatial separation of components in residential heating systems compared to those in DHW systems also preclude the manufacture, transport, and installation of fully assembled systems. After many years of serious marketing by virtually hundreds of manufacturers, only a handful are marketing fully integrated space-heating systems, and probably none of these is a "package" in the usual sense. The U.S. Department of Energy has in recent years encouraged manufacturers to produce packaged solar heating systems, but with little success.

Standardization of residential-size solar heating systems, to the limited extent that has occurred, has been based on tax credits. Systems were sized, hence priced, to fit the constraints of federal and state tax credits. After the collector and storage, the two largest components in the system, have been sized to fit a particular budget, the sizes of pumps, pipes, ducts, and heat exchangers are fixed, so that the total cost, including installation, is in a "marketable" cost range. Not surprisingly, costs of systems in 1985 were commonly about $10,000, which was the maximum outlay eligible for federal and several state tax credits for residential systems. These systems were typically 4 to 6 liquid collector panels with a total area of 120–200 ft^2 (37–61 m^2), 120–240 gal (454–908 l) of hot water storage, heat exchangers for domestic hot water and warm air, and the necessary pumps, valves, piping, and controls. Systems sizes for commercial buildings were not constrained by tax incentives, so there were no standard packages for commercial buildings.

The market for solar systems is not sufficiently secure (in 1985) for individual companies in the industry to commit to standard packages. Much more emphasis is placed on marketing approaches than on packaging. While, in concept, it should be easy to decide on one or more standard-size systems made up of specific numbers of collectors, a fixed volume of storage, and other selected standard components that could deliver a "rated" amount of thermal energy annually, the difficulties in marketing such a system has caused manufacturers to retreat to relatively safer grounds of making only the components and depending on others to design, market, and install systems. As the industry matures, packaged systems of standard sizes may become more commonplace.

References

Alcone, J. M. 1981. Ground coupling and heat pump compatible solar collectors. *Premeeting Proc. of the Annual DOE Active Solar Heating and Cooling Contractors' Review Meeting*, Washington, DC. CONF.810912–PRELIM.

Andrews, J., M. A. Catan, and P. LeDoux. 1982. *Solar energy and heat pumps: evaluation of combined systems for heating and cooling of buildings.* BNL51603 under contract DE–AC02–76CH00016. Upton, NY: Brookhaven National Laboratory.

Backlund, J. C. 1981. Development and analysis of a ground coupled solar assisted heat pump system for residential application. *Premeeting Proc. of the Annual DOE Active Solar Heating and Cooling Contractors Review Meeting*, Washington, DC. CONF.810912–PRELIM.

Bose, J. E., et al. 1980. Earth coupled and SAHP system. *Proc. 5th Annual Heat Pump Technology Conference*, Oklahoma State University, 14–15 April, 1980. OSU Mechanical and Aerospace Department Engineering Extension, pp. 5.1–5.11.

Chapra, P. S., and R. M. Wolosewicz. 1978. Freezing problems in operational solar demonstration sites. *Proc. Solar Heating and Cooling Systems Operational Results Conference*, Colorado Springs, CO. Golden, CO: Sol'ar Energy Research Institute.

Close, D. J., and M. B. Yusoff. 1978. The effect of air leaks in solar air collector behavior. *Solar Energy* 2:459.

Cromer, C. J. 1985. The effect of circulation control strategies on the performance of open loop solar DHW systems. *Proc. Joint ASME/ASES Solar Engineering Conference. Solar Engineering 85*, ASME.

Duffie, J. A., and W. A. Beckman. 1980. *Solar engineering of thermal processes.* New York: Wiley.

Ehrenkrantz Group, and Mueller Associates, Inc. 1979. *Active solar energy system design practice.* Solar/0802–79/01. U.S. DOE National Solar Data Program.

Karaki, S., T. E. Brisbane, and G. O. G. Löf 1985. *Performance of the solar heating and cooling system for CSU Solar House III.* SAN–11927–15. Fort Collins, CO: Colorado State University, Solar Energy Applications Laboratory.

Karaki, S., G. O. G. Lof, T. E. Brisbane, and D. Wiersma. 1985. *Performance of a solar heating and cooling system with phase-change storage, CSU Solar House I.* SAN–11927–16. Fort Collins, CO: Colorado State University, Solar Energy Applications Laboratory.

Karaki, S., and P. Hawken. 1982. *Performance evaluation of a solar/solar assisted heat pump system in CSU Solar House II.* COO–30122–27. Fort Collins, CO: Colorado State University, Solar Energy Applications Laboratory.

Karaki, S., T. G. Lantz, and S. Waterbury. 1982. *Performance evaluation of a solar air-heating system with multi-speed blower control for solar energy collection and off-peak use of heat pump for cooling.* Annual report under contract DE–AC02–79–CS–30122. Fort Collins, CO: Colorado State University, Solar Energy Applications Laboratory.

Karaki, S., and M. Westhoff. 1982. *Performance evaluation of a state-of-the-art liquid-based solar system with auxiliary heat pump.* COO–30122–29 under contract DE–AC02–79–CS–30122. Fort Collins, CO: Colorado State University, Solar Energy Applications Laboratory.

Karaki, S. 1983. *Performance of a drain-back solar heating and hot water system with auxiliary heat pump.* SAN–30569–33 under contract DE–AC03–81–CS–30569. Fort Collins, CO: Colorado State University, Solar Energy Applications Laboratory.

Kovarik, M., and P. F. Lesse. 1976. Optimal control of flow in low temperature solar heat collectors. *Solar Energy* 18:431.

Löf, G. O. G., and R. A Tybout. 1974. Design and cost of optimized systems for residential heating and cooling by solar energy. *Solar Energy* 16:9.

Löf, G. O. G., W. S. Duff, and C. E. Hancock. 1981. Development and improvement of liquid systems for solar space heating and cooling—CSU Solar House I. *Proc. ASME Solar Energy Division Conference*, Reno, NV, p. 68. ASME.

Mitchell, J. W., T. L. Freeman, and W. A. Beckman. 1978. Heat Pumps Solar Age 3:7.

NBS. 1977a. *Intermediate minimum property standards for solar heating and domestic hot water systems.* NBSIR 77–1226. Washington, DC; NBS.

————. 1977b. *Intermediate standards for solar domestic hot water systems/HUD initiative.* NBSIR 77–1272. Washington, DC: NBS.

SMACNA (Sheet Metal and Air Conditioning Contractors National Association). 1981. *Installation standards for heating, air conditioning and solar system.* 5th ed. Merrifield, VA: SMACNA.

Solar Energy Applications Laboratory. 1980a. *Solar heating and cooling of residential buildings—design of systems.* S/N 003–001–00089–5. Washington, DC: Government Printing Office.

————. 1980b. *Solar heating and cooling of residential buildings—sizing, installation and operation of systems.* S/N 003–001–00088–7. Washington, DC: Government Printing Office.

Ward, D. S., and G. O. G. Löf. 1975. Design and construction of a residential solar heating and cooling system. *Solar Energy* 17:13.

Ward, D. S., C. C. Smith, G. O. G. Lof, and L. Shaw. 1977. Design of solar heating and cooling system for CSU Solar House II. *Solar Energy* 19:79.

Warren, M., S. Schiller, M. Wahlig, G. Sadler, C. Vilmer, and C. Weaver. 1981. *Experimental and theoretical evaluation of control strategies for active solar energy systems.* Solar Energy Program Annual Report 1980. LBL–11984. Berkeley, CA: Lawrence Berkeley Laboratories.

Winn, C. B., and D. E. Hull. 1979. Singular optional solar controller. *Proc. 1978 IEEE Conference on Decision and Control*, San Diego, CA.

Winn, C. B., G. R. Johnson, and J. B. Moore. 1974. Optimal utilization of solar energy in the heating and cooling of buildings. Paper presented at the ISES Annual Meeting, Colorado State University, Fort Collins, CO.

13 System Performance

Charles Smith

13.1 Introduction

Much early work in solar heating was concerned with components, such as collectors and storage. It has become recognized, however, that solar heating is the product of a collection of components comprising a system and needs to be studied as such. Because of the interaction of components, optimal system performance occurs under conditions different from those for optimal behavior of each component. For example, optimal collection efficiency would not necessarily be coupled with least auxiliary usage.

The performance of a large number of systems in public use has been monitored as part of the National Solar Data Network (the results of that study are presented in a separate chapter of this volume). Direct experimental research into system operation and the factors affecting system performance has also been undertaken. This chapter reports the outcome of such studies, sponsored primarily by the Department of Energy. The results are divided into sections on liquid systems, air systems, comparisons between the two, and solar-assisted heat pump systems.

Component performance, which is addressed in other chapters, has advanced beyond that of solar system performance. That difference is due to two basic reasons. Except for the collectors and heat storage units, and perhaps the control hardware, components are the same as found in nonsolar HVAC or hot water installations. Secondly, the components are built, tested, and inspected under controlled factory conditions. Solar systems, particularly those for space heating, are assembled under conditions involving a wide range of worker skills, quality control, and inspection capabilities. The results of these differences have become evident in comparisons between many commercially installed systems and R&D installations. The experimental projects have had the advantage of design and installation know-how obtained from experience with previous installations and of continuity of design improvements under consistent conditions. Capable selection of components can go unrewarded if the installation is poorly done. In this chapter, the most common installation factors that affect performance are discussed.

13.1.1   Experimental Research in Solar Systems

Principal activity in experimental solar systems research, development, and design evaluation has taken place at universities and scientific laboratories. The Colorado State University Solar Houses are representative of this group. The Solar Houses are three identical residential-size buildings, each with a living area of 133 m^2 (1350 ft^2) in addition to a full-heated basement. Solar systems provide heating in the winter, cooling in the summer, and domestic hot water all year. Two of the houses (I and III) have liquid-based systems, and the other (II) has an air-based system. Systems are modified annually to incorporate new components and configurations. The first system began operation in 1974. Since that time, over 35 different system configurations have been operated, including 14 for liquid-based heating, 6 for air-based heating, and 15 for cooling.

Other institutions have also made significant contributions in solar space heating system experimental research. They include the University of Delaware, University of Pennsylvania, University of Wisconsin, George Washington University, New Mexico State University, the University of Tennessee (in conjunction with the Tennessee Valley Authority), Trinity University, and the Los Alamos Scientific Laboratory. The Department of Energy has also sponsored research undertaken by private companies. The work done by these institutions and companies and the results obtained are discussed in subsequent sections of this chapter.

13.1.2   Measures of System Performance

There are four generally accepted measures of solar system performance:

1. *Collection efficiency* applies to the performance of the solar energy collection subsystem. It is the energy collected, divided by the radiation incident upon the collectors. The radiation may be the total radiation, or it may be only that incident while the collection subsystem is operating.

2. *System efficiency*, or solar heating performance factor, is the solar heat delivered to the load, divided by the total radiation incident upon the collector. It is similar to the collection efficiency but also takes into account heat loss from pipes and storage. The net system efficiency is the solar heat delivered less the electrical inputs to the system, divided by the incident solar radiation. This figure must be used, however, with the caution that the delivered heat may not have the same economic value as the equivalent energy in the form of electricity.

3. *Solar fraction* is the fraction of the total heat requirement that is met by solar energy. The figure relates the output of the solar system to the size of the load, and is dependent on the size of the system as well as on its efficiency.

4. *Electrical coefficient of performance* is the solar heat delivered to the load, divided by the electrical energy used to operate the system.

These measures may be evaluated on a daily, monthly, or seasonal basis.

13.2   Liquid Systems

Work sponsored by the Department of Energy in solar heating system research spans a range from the development of commercially viable technologies to studies undertaken in order to increase understanding of system performance. Colt, Inc., developed a commercial single-family sized system (Colt 1980). The system used paraffin oil for the collector fluid and water tanks with internal heat exchangers for heat storage. The collectors were single-glazed and employed a nonselective absorber coating.

Wormser Scientific Corporation developed a solar heating system that utilized reflective pyramidal optical focusing collection (WSC 1980). At George Washington University, the Solaris system consisting of a trickle-water collector and a combined water tank/rock bed storage was evaluated (GWU 1977). Although it is chiefly associated with passive solar and phase-change material studies, the University of Delaware solar house (Solar One) was also used for experiments with liquid-based collection systems (UD 1978).

The Los Alamos National Security and Resource Study Center at the Los Alamos Scientific Laboratory has been used as an experimental facility to study the control of solar heating and cooling systems (Murray, Traump, and Hedstrom 1977; Hedstrom, Hurray, and Balcomb 1978). Trinity University in San Antonio, Texas, had a large solar heating and cooling system used as a research facility (Crum 1978). At New Mexico State University, a solar heated and cooled building was extensively monitored (Fenton et al. 1978).

The systems mentioned above have been used to investigate several different aspects of system performance. Significant results from these investigations are included in the summary of results at the end of this section. Two experimental systems have been selected for more comprehensive discussion: the experiments with liquid systems in Colorado State University Solar Houses I and III, and with air systems in Solar House II. This is not a selection based upon superior performance; the projects have been chosen because they truly conform to the concept of experimentation; they are examples of both liquid and air systems in buildings of nearly identical design, and they have yielded the most extensive and definitive performance data available.

Solar House I was designed in 1973 as a typical custom residence of that period with a few special features to accommodate uses of the building for experimental purposes. Originally, site-built flat-plate collectors were installed, because commercially manufactured collectors were not available. All of the mechanical equipment, with the exception of a cooling tower for the chiller, was placed within the building enclosure. In 1976, a test bed for collectors was erected to the south of the building and a few years later, the solar equipment, except for the load coil and the DHW subsystem, was moved to a separate enclosed space adjacent to the house.

Solar Houses II and III were constructed in 1975–76. The architecture of these buildings conformed substantially with Solar House I, but different solar systems were installed. Solar House II was provided with an air-based system, while a liquid-based system was used in Solar House III. Initially, Solar House III was to be fitted with more advanced or more experimental solar equipment than that used in Solar Houses I and II. In the progression of experiments, this distinction became less pronounced as both Solar House I and III received several state-of-the-art components. Solar House II served to advance the design of air systems as well as to provide data for comparing the performance and operation of air versus liquid-based systems. Table 13.1 provides a brief chronology of liquid solar systems installed in Solar Houses I and III that pertain to space heating. In addition to references cited in the table, other reports (Löf and Ward 1974; Löf et al. 1974; Ward, Löf, and Smith 1975; Duff 1977; Leflar and Duff 1977) contain information on these systems and their performance.

From 1977 to 1980, two sets of collectors were used alternately to supply energy to the solar heating and cooling system in House I. One set consisted of evacuated-tube collectors installed on an adjacent test bed, and the other consisted of flat-plate collectors on the building roof. In conjunction with seven different collectors and four heat storage units, there have been four types of absorption chillers, four kinds of auxiliary

Table 13.1
Liquid solar heating systems in Solar Houses I and III at Colorado State University

Year	Solar House I Collector	Storage	Collector-to-storage transfer	Solar House III Collector	Storage	Collector-to-storage transfer	Reference
1974–1975	Site-built, F.P.[a] double glazed non-selective absorber. Area = 71 m² (67)[b]	4000 liters, water, non-stratified, light weight site-built tank	Double loop (glycol anti-freeze)				Ward and Löf (1976)
1975–1976	Site-built, F.P. double-glazed	"	Glycol Solution	Owens-Illinois evacuated tube, 41 m², reflective back surface	4540 liters, horizontal cylindrical steel tank	Single-loop water, non-draining	Ward, Löf, and Smith (1976) Ward, Ward, and Oberoi (1979)
1976–1977	Site-built, F.P. double-glazed	"	Glycol	"	"	"	Conway et al. (1978) Ward, Ward, and Oberoi (1979)
1977–1978	a. Corning evacuated tube Area = 50 m² (40)[b] b. Site-built F.P. double-glazed	a. 4000 liters, water, commercial tank b. 4000 liters, water, site-built tank	a. Glycol b. Glycol	Chamberlain flat-plate, 59 m² single-glazed, slective absorber	"	Double-loop (glycol anti-freeze)	Conway et al. (1978) Löf and Duff (1979) Ward, Ward, and Oberoi (1979)
1978–	a. Philips	a. 4000 liters,	a. Glycol	Chamberlain	"	"	Duff and Löf (1981)

Table 13.1 (continued)

	Solar House I			Solar House III			
Year	Collector	Storage	Collector-to-storage transfer	Collector	Storage	Collector-to-storage transfer	Reference
1979	evacuated tube, Mk IV Area = 44.7 m² (40.6)[b] b. Miromit, Area = 56 m² (49)[a] single-glazed slective, F.P.	water, site-built stratified b. 4000 liters, water, direct contact, liquid exch.	b. Heat transfer oil	flat-plate, 59 m²			Ward, Ward, and Oberoi (1980)
1979–1980	a. Philips E.T. MkIV 44.7 m² b. Miromit E.P.	a. 4000 liters, water, commercial tank b. 4000 liters, rectangular, sectional stratified, plastic liner	a. Glycol b. Heat transfer oil	Chamberlain flat-plate 59 m²	"	"	Colorado State University (1980) Karaki and Westhof (1982)
1980–1981	Philips E.T.[c] heat pipe, MkI, VTR141 59.6 m²	4000 liters, rectangular sectional, stratified, plastic liner	Glycol	Chamberlain flat-plate	"	"	Löf et al. (1982)
1981–1982	Philips evacuated tube, heat pipe, VTR141	4000 liters, rectangular, sectional, stratified, plastic liner	Glycol	Chamberlain flat-plate	"	"	

1982–1983	Philips E.T. heat pipe VTR361 54.5 m²	4000 liters, pressurized water in steel tank	Glycol	a. Brookhaven plastic film F.P. b. Chamberlain F.P. reinstalled	4540 liters horizontal stratified steel tank	water (drain-back)	Karaki, Vattano, and Brothers (1984) Karaki (1984)
1983–1984	Philips E.T. heat pipe VTR361, ripple reflector, 54.5 m²	4000 liters, pressurized water in steel tank	Glycol	Revere flat-plate 48.7 m² double-glazed selective black chrome absorber	"	"	Karaki et al. (1985a) Karaki, Brisbane, and Löf (1985)
1984–1985	Sunmaster TRS-81 evacuated tube, 25.7 m²	4000-liter tank with stratification manifolds	Water (drainback)	Revere F.P. 48.7 m²	"	"	Karaki et al. (1985b) Karaki et al. (1985c)

Source: Löf, Duff, and Hancock (1981).

a. F.P.—flat-plate collector.

b. Absorber area.

c. E.T.—evacuated tube collector.

heaters, three DHW subsystems, and a number of different ancillary components. Table 13.1 shows the major components tested in House I. Also listed are the solar heating components used simultaneuosly in House III. Collectors included an evacuated-tube type, an unsuccessful plastic film prototype, and two commercial flat-plate systems. Conventional storage and heat transport arrangements were used. Extensive investigations with many system combinations provided valuable information about effective system designs and a wealth of performance data that could be obtained only from physical experiments.

Beginning in 1977, there was considerable emphasis on the development of systems that utilized evacuated-tube collectors, particularly for enhancing solar cooling performance. The original configuration of a liquid-based solar heating system at Colorado State University involved the use of site-built, flat-plate collectors, heat collection in an antifreeze solution, heat exchange to water, and heat storage in an unstratified tank. Improvements that have been developed include the use of high-performance evacuated-tube collectors, heat collection in water with freeze-proof, drainback design, and thermal stratification in storage. Most of these modifications have resulted in performance improvement.

To illustrate and compare the performance of systems, average monthly data on a number of systems operated at Colorado State University Solar Houses I and III are presented in the following sections. Detailed analyses have been published in a series of papers and reports cited in the list of references. The principal results were also presented in a review publication (Löf and Karaki 1983). The systems examined provided solar heating, solar cooling, and solar heated domestic hot water.

13.2.1 Thermal Performance

Due to the wide range of systems tested in the three structures, it is useful to organize the results first according to collector type and subsequently according to other variables. The most notable distinction between collectors in the Colorado State studies is the general type: liquid flat-plate (FP), liquid evacuated-tube (ET), and air flat-plate. Figure 13.1 presents results of performance of solar systems employing several types of liquid collectors. (Section 13.3 addresses the comparison between liquid and air flat-plate collector systems.) This classification permits direct comparison under the same weather conditions but omits the effects of other system

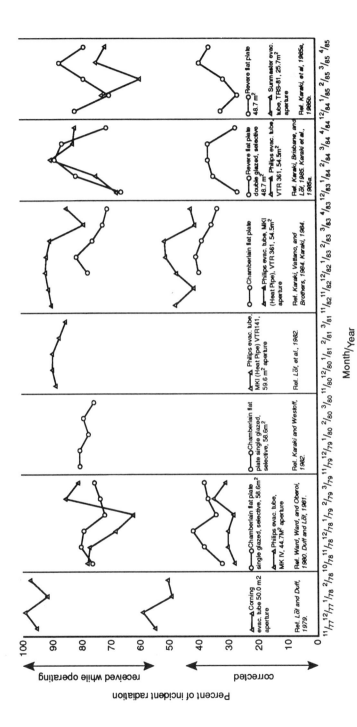

Figure 13.1
Performance of liquid solar heating systems at Colorado State University.

variables, such as double-loop versus drainback, or temperature stratification versus nonstratification in storage.

Among the important conclusions that may be drawn from the performance comparisons shown in figure 13.1 is the substantial month-to-month variation in average efficiency of each system. The average monthly efficiency of a system employing one type of flat-plate collector ranged form 32 to 44 percent. The efficiency of another flat-plate collector covered a range form 26 to 39 percent. Variation in efficiency of one evacuated type was 49 to 59 percent; another, 28 to 35 percent, a third, 28 to 32 percent, another ranged from 41 to 51 percent, and a fifth type varied from 29 to 42 percent. Although not always the case, efficiencies were usually higher in mid-winter months when systems were fully utilized; collection and storage temperatures were therefore lower and thermal losses were less. An international comparison of buildings in various climatic regions showed that evacuated-tube collectors usually outperformed flat-plate types in months of low solar radiation (Duff 1982).

Another conclusion is that there are large differences in performance of systems with different collectors, particularly of the evacuated-tube types. Flat-plate collectors provided higher system efficiency in three seasons from 1978 to 1981 than did two types of evacuated-tube collectors, whereas an evacuated-tube system outperformed two flat-plate systems in 1982–1984 operations. In the winter of 1977, an evacuated type also showed superior performance, but in 1984–1985, the average efficiency of another evacuated type was about the same as a commercial flat-plate collector system.

Evacuated-tube collectors would be expected to deliver heat to storage when solar intensities are lower than those necessary for useful delivery form flat-plate types. The upper lines in figure 13.1 parially support this expectation, but they also show considerable departure from this concept. Although usually utilizing more than 80 percent of the total incident radiation (one evacuated-tube type consistently utilized more than 90 percent), there were periods when some of the evacuated-tube types did not operate when flat-plate collectors were delivering heat. Other system variables, such as average collection temperatures and the set points of controllers, also affect these differences.

For conventional space heating, the use of most types of ET collectors showed a relatively small margin of benefit over the FP type. The highest performance was achieved with an experimental ET collector that has not been commercially manufactured in the United States. The major attraction of ET collectors remains for solar cooling applications where higher operating temperatures are a substantial advantage.

Recognizing from fundamental considerations as well as from practice that collection efficiency depends on the difference in temperature between storage and the outdoor air, figure 13.2 is presented to show the influence of this variable on ET and FP liquid systems (Karaki, Brisbane, and Löf 1985; Karaki et al. 1985a). It is seen that there is less incentive to maintain a low storage temperature level (more specifically, the return temperature of fluid supplied to the ET collector) than in systems with FP collectors. This difference naturally leads the designer of FP systems to strive for temperature stratification in thermal storage tanks. Storage stratification in systems with ET collectors, although beneficial, has less influence on collection efficiency; in fact, it is difficult to maintain stratification in systems delivering heat to air conditioners or hydronic heating loops, which are candidates for such collectors.

The principal advantage of the better ET collectors over the FP type for conventional forced-air heating or domestic water heating is their higher collection efficiency at low solar radiation levels and low atmospheric temperature. In overcast climatic areas, efficient ET collectors can usually supply solar heat under conditions at which FP types are ineffective.

A development trend that is noteworthy, especially with liquid systems, is operating the collectors at reduced flow rates and increased stratification of temperatures in storage tanks. High flow rates were initially used in order to obtain satisfactory heat transfer rates at small temperature differences in the collectors. Larger piping, pumps, and fill volumes of liquid in the lines, as well as minimal stratification in storage, resulted in a number of performance penalties that can be minimized by use of optimum flow strategies.

13.2.2 Fraction of Load Supplied by Solar Energy

The extreme variability in solar radiation (day to night, one day to another because of cloudiness; month to month because of day length) coupled with high variability in hourly, daily, and seasonal heat demands makes it a practical impossibility to provide 100 percent solar heating, even with substantial heat storage capacity. With such variation, it is obvious that a solar system designed to meet a high fraction, say, 90 percent, of the annual heating requirements will have to be large enough

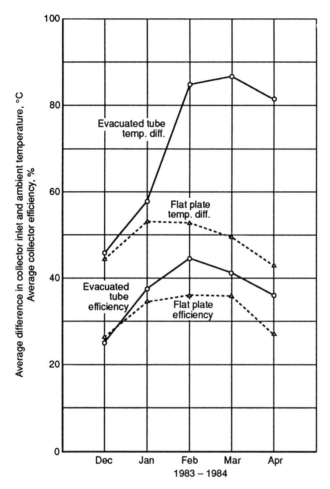

Figure 13.2
Average efficiencies and temperature differences for space heating with flat plate and evaculated tube collectors.

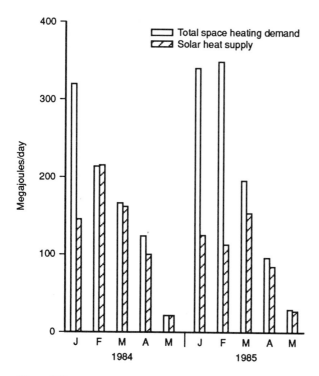

Figure 13.3
Average heat demand and solar heat supply in Solar House I, Colorado State University.

to meet most of the heavy midwinter demands; such a large system will have a capacity much greater than needed through the rest of the year. The economic result is an excessive first cost (because of large size) and a substantial waste of available solar heat except during the coldest weather. Because of the the need for annual utilization of the solar system to be high enough to justify its purchase, high solar fractions typical of many early designs gave way to smaller systems that typically provided solar fractions about half the annual loads.

The percentage of total annual heat requirements that a solar system is designed to supply is an economic consideration. Most of the residential systems installed in the 1970s and 1980s met 30 to 70 percent of the total demand with solar. If the annual fraction is 50 percent, for example, daily fractions vary from zero to 100 percent, and monthly averages range from comparatively low fractions in midwinter to nearly 100 percent in late spring and early fall when loads are low and solar energy is usually high. Figure 13.3 illustrates how the monthly solar fraction varies during two heating seasons (Karaki et al. 1985a, 1985b). This experimental system was designed to provide cooling as well as heating, forcing use of a system larger than would have been used only for heating.

The significance of this range in solar fraction relates to the need for auxiliary capacity and any differential in auxiliary energy costs for variable demand. Figure 13.3 shows that auxiliary energy demand for space heating ranged from zero to 236 MJ/day (224,200 BTU/day), with an average demand of 103 MJ/day (97,800 BTU/day). In the case of electric auxiliary, this variation can indicate the desirability of off-peak energy storage if there are substantial demand charges or time-of-day pricing; in the case of fuel auxiliary, the need for good efficiency at part load capacity is indicated. The data in figure 13.3 show that solar provided nearly all the heat required in spring months, indicating that during this period, more solar energy was available than could be used.

Figure 13.4 is the diagram of a system that was used in 1982–83 to measure the effects of solar system capacity on fraction of load provided by solar (Karaki, Vattano, and Brothers 1984). The winter of 1982–83 was mild, particularly in January and February. As a consequence, the average daily heating load was less than 200 MJ/day (190,000 BTU/day), as seen in figure 13.5. With the large, 54 m² array of high efficiency collectors (Philips evacuated tube, heat pipe), substantially all of the heating requirement was provided by solar energy. To obtain data when solar energy provides a smaller part of the heating load, the lower array of collector tubes was removed during the last week of February. reducing collector area by 50 percent to 27 m² (293 ft²) aperture area. As a result, solar contribution to space heating was reduced to 58 percent in March (when it would have been essentially 100 percent if at the original size). When solar contribution for space heating is high, storage temperature is usually high and collection efficiency decreases. With high-efficiency ET collectors, however, storage temperature has only moderate influence on collector efficiency. The decrease in average storage temperature resulting from reducing collector size in March did not produce an increase in collector efficiency; in fact, with cloudier conditions, average efficiency dropped from 51 percent in February to 41 percent in March.

A portion of the solar energy collected in the large system during the December-February period was not needed for heating. Losses from stor-

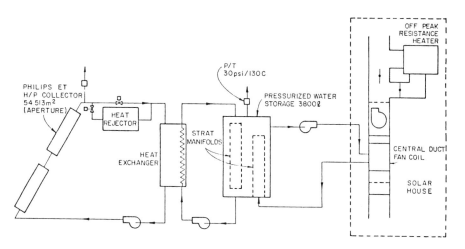

Figure 13.4
Diagram of heating system in Solar House I, Colorado State University.

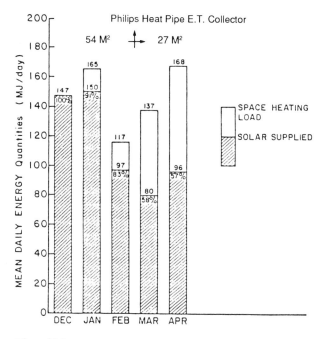

Figure 13.5
Influence of collector size on fraction of heating load carried by solar energy.

age and intentional heat rejection in a "dump" heat exchanger accounted
for the excess. If the full array had been used during March, an even
greater percentage of the collected thermal energy would have been un-
used. It may be concluded that the solar heating system as originally used
in December through February was oversized for the heating load encountered during the 1982–83 season.

Although it is evident that the collector array of 54.5 m^2 (586 ft^2) Philips
VTR 361 heat-pipe collectors was unnecessarily large for the heating demand in Solar House I, the extra capacity provided flexibility to under-
take a wide range of experiments in an experimental laboratory facility
where solar cooling was being developed.

13.2.3 Space Heating System with Low-Cost Collectors

In the summer of 1982, low-cost plastic collectors were installed in a system provided with a solar-assisted (series) heat pump (Löf, Westhoff, and
Karaki 1984). Inability of plastic materials in the collectors to withstand
stagnation temperatures resulted in their failure soon after installation,
thereby precluding testing and performance evaluation. The issues concerning low-cost/low-temperature collectors are treated elsewhere in this
series in chapters devoted to collectors. However, it should be noted that
solar heating systems must be installed to the same standard of quality
and reliability as conventional heating equipment. If the collectors themselves are not up to the same standard, the entire installation becomes of
inferior quality.

13.2.4 Comparison between Drainback and Double-Loop System

In most climatic regions where solar heating is of interest, the collector
array must be protected from freezing during nonoperating hours. (Freeze-
tolerant collectors are of a special design, which is addressed in volume 5
of this series, Solar Collectors, Energy Storage, and Materials.) Double-
loop antifreeze systems and drainback systems have been studied and
compared at Colorado State University. Drainback systems have been
operated in Solar House I and Solar House III (Löf et al. 1982; Karaki
1984). The best comparison between the two types is based on measurements with the same collector array (on Solar House I) used successively
for both systems (Goumaz 1981). This is a unique study because the operating systems were essentially identical except for the method of delivering
heat to storage. Data from operation of the two systems are not directly

Table 13.2
Seasonal performance of double-loop antifreeze systems and drainback systems employing the same collector and storage unit

System	Efficiency, including storage losses, as useful energy (%)	Efficiency, not including storage losses, as useful energy (%)
Dual liquid ($\varepsilon = 0.66$)	24.3	22.1
Dual liquid ($\varepsilon = 0.80$)	24.9	22.6
Drainback, open return	28.8	26.2
Drainback, open return without stratification	27.4	24.9
Drainback syphon return[a]	26.8	24.4

Source: Goumaz (1981).
Note: ε is the effectiveness of the liquid-to-liquid heat exchanger.
a. Limiting case.

comparable, however, because heating loads and solar radiation differed. The data were therefore used to validate a computer simulation model, which was then used in a comparison of the systems with identical weather and load patterns. System efficiency results are shown in table 13.2.

Conclusions drawn from the comparison study and from general experimental operating experience can be summarized as follows:

• Drainback systems can operate at significantly higher efficiency than equivalent dual-liquid systems. Based on the parameters of CSU Solar House I, an open return, drainback system provided 17.1 percent more heat through a full season than that delivered by the double-loop system.

• The major factors in the improved performance of the drainback system are the absence of a heat exchanger and of the morning startup capacitance effect.

• A syphon-return drainback system has the lowest pumping power requirement. In the comparison study, the double-loop system had a pumping power requirement twice that of the syphon-return drainback system, and the open-return drainback system had a pumping power requirement 4.6 times that of the syphon-return drainback system.

• A significant source of heat loss in drainback systems may be vapor migration to the collector when the collector loop pump is not running. Heat loss due to condensation in the empty collectors may be avoided by employing a liquid trap in the collector line.

13.2.5 Thermal Storage

To achieve seasonal solar heating fractions of at least 30 percent, part of the heat collected in the daytime must be stored for later use. On mild sunny days, very little heat may be needed until nighttime; practical storage systems should thus have the capacity to accumulate essentially all the energy that can be collected on a clear day for use during the following night. Larger storage, for more than one day's collection, has not proved cost-effective. Typical storage volumes are about 100 liters per square meter of collector (about 2 gallons per square foot of collector).

Research and testing on thermal storage have been directed toward container design and materials, maintenance requirements, and water temperature stratification (Michaels 1979). Storage quantity is also an important design consideration, but its effect on system performance can be reliably evaluated by calculations employing simulation models and hourly weather and solar data.

Storage tank materials and construction are selected on the basis of corrosion resistance and allowable temperature and pressure (Ward and Oberoi 1980). Even low-temperature solar systems have the potential of generating high temperatures in storage during the equipment lifetime. Steel tanks with various inert, nonmetallic interior coatings represent a proven approach to durable liquid (water) storage containers. Tank connections involving dissimilar materials should be avoided because of thermal expansion and corrosion hazards.

Thermal losses from storage were found in several studies to be more significant than originally projected. Probable causes were faulty insulation applications, moisture penetration and other damage to insulation, vapor flow into connecting piping, and conduction through supports and connections.

If heat lost from the tank is into the conditioned space, it is a useful contribution to the space heating requirements except when no heating is needed. (If solar cooling is being provided, this heat loss can be a significant penalty, adding up to 30 percent more cooling demand). Recognizing that storage heat losses do occur, one solution is to have the storage tank in a space where the loss can be retained in the house in winter, but vented to the outdoors in summer.

Maintenance demands of conventional tanks were found to be minimal, although a failure occurred with a built-up rectangular bin lined with a plastic sheet. Leakage resulted from failure of seams at corners and other stress points.

Thermal stratification in heat storage tanks is beneficial to system operation. With a low water temperature near the bottom of the tank, the temperature at the collector inlet is lower than if the tank were fully mixed; heat losses from the collector are therefore reduced, and collection efficiency is improved. The temperature of the heated water supplied to the load is higher than it would otherwise be, thereby increasing the value and usefulness of the energy. Several types of stratification enhancement devices have been tested in systems (Karaki et al. 1985c; Löf et al. 1982), and a general design technique for such devices is available (Gari 1983). In operation, the stratification enhancement devices can maintain temperature differences of 7–10°C (12–18°) between the top and bottom of the tank if flow rates through the collector and load loops are low enough to provide such differences in inlet and outlet temperatures. Temperature differences as high as 15°C (27°F) have been observed. Temperature stratification permits a 5 to 7 percent increase in energy collection over that which would be obtained if the tank were fully mixed (Löf et al. 1982).

13.2.6 System Efficiency

Not all of the energy delivered by a solar collector is actually utilized in a building. While solar collection efficiency is often 30 percent or higher, seasonal system efficiency (the ratio of solar heat utilized in the building to the incident solar energy on the collectors) has frequently been less than 20 percent. The difference in these quantities is the heat lost from collector loop piping, the storage tank, and from load loop piping. In some circumstances, collected heat may also be intentionally discharged from collectors or heat exchanger coils to avoid overheating. Table 13.3 illustrates system efficiency values found in various U.S. solar demonstration projects (Löf and Karaki 1983). Although some systems operated efficiently, others suffered from the effects of design faults, operating problems, accidents, and poor control.

Experimental solar heating systems have shown substantially better performance than most of the systems in demonstration projects. Better

Table 13.3
System performance of solar heating and DHW demonstration projects

Site	Collector area	System type[a]	Loc.	System efficiency η_s (%)	Energy savings of collector per ft² Year MJ/m² (Btu/ft²)	Day MJ/m² (Btu/ft²)
Saddle Hill Lot 36	315	HX-G	MA	25	1351 (119,000)	3.69 (325)
Oakmead Industries	2022	HX-G	CA	21	1373 (121,000)	3.78 (331)
Montecito Pines	950	DB	CA	18	1044 (92,000)	2.86 (252)
RORU Clemson	388	A	SC	13	749 (66,000)	2.04 (180)
John Byram	312	A	KS	13	726 (64,000)	1.99 (175)
Scattergood	2496	A	IA	13	635 (56,000)	1.76 (155)
M. F. Smith	512	DD	RI	12	692 (61,000)	1.91 (168)
Summerwood M	378	DD	CT	12	647 (57,000)	1.76 (155)
Design Construction	792	DD	MT	12	499 (44,000)	1.38 (122)
Lawrence Berkeley			CA	11	670 (59,000)	1.84 (162)
J. D. Evans A	350	DD	MD	11	579 (51,000)	1.60 (141)
First Manufactured #9	288	A	TX	10	601 (53,000)	1.65 (145)
J. D. Evans E	350	W	MD	10	522 (46,000)	1.42 (125)
Helio Thermics #8	416	DD	SC	8	397 (35,000)	1.08 (95)
First Manufactured #10	288	A	TX	7	431 (38,000)	1.18 (104)
Perl Mack			CO	7	408 (36,000)	1.11 (98)

Summerwood G (Attic)	448	A-W	CT	6	329 (29,000)	0.90 (79)
Sir Galahad	640	HX-G	VA	5	306 (27,000)	0.84 (74)
Matt Cannon	597	W	FL	5	238 (21,000)	0.66 (58)
Helio Thermics #6 (Attic)	416	A + W	SC	5	227 (20,000)	0.61 (54)
Maulden Corp.	704	A	IN	3	136 (12,000)	0.36 (32)
Washington Natural Gas	591	A	WA	3	79 (7,000)	0.23 (20)
Brookhaven	2500	HX-G	NY	2	79 (7,000)	0.22 (19)
Average				10	556 (49,000)	1.52 (134)

Source: Löf and Karaki (1983).
a. A—Air; DD—Draindown; HX-G—Heat exchange, glycol solar; W—Water, no freeze protection; DB—Drainback.

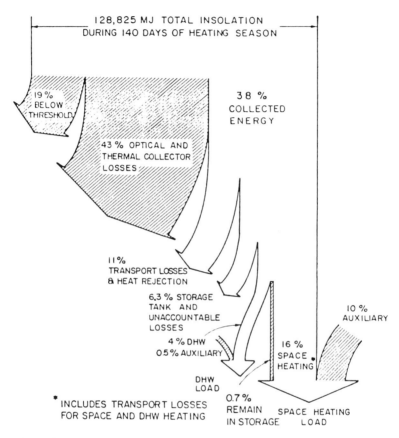

Figure 13.6
Energy flow for heating season, CSU Solar House I.

design, capable engineering and fabrication, good control of operation, and diligent maintenance have yielded superior performance.

Figure 13.6 is a typical energy flow diagram for a liquid solar heating system (Karaki et al. 1985a). It shows that 38 percent of the incident solar energy was collected and that about half of that quantity (16 percent plus 4 percent) was delivered to the space heating system and the water heating system. About half of the collected energy (17 percent of the incident solar energy) escapes from piping and storage inside the building, and therefore contributes to the space heating supply in an uncontrolled manner. Heat that escapes from piping and storage when needed (as during the winter months) reduces auxiliary requirements and constitutes part of the solar fraction. Another portion must be considered a loss because it escapes when not needed (as during mild weather) and tends to overheat the building. In the example shown in figure 13.6, it is estimated that about half of the "losses" inside the building were useful, thereby making solar space heating about 25 percent of incident solar energy; including hot water, the solar fraction was about 29 divided by 29 plus 10 auxiliary, equal to 0.74. A system design with reasonable attention to energy losses should provide performance similar to that depicted in figure 13.6. Significant departure from this distribution could indicate the need for more attention to reducing heat losses. Such attention is not only a matter of better thermal insulation but also of the proper sizing of components, reduction of cycling loss due to excessive thermal mass, and improved control strategy.

13.2.7 Parasitic Power Requirement

The electric power required to collect solar heat is expressed as the electrical coefficient of performance (COP). It is the solar energy collected, divided by the electric energy needed to operate pumps, fans, and controls.

In early solar heating testing programs, little attention was given to electric power consumption. High liquid flow rates were used, and complex piping networks often required excessive pumping power. Flow meters and special valves required for procurement of test data added flow resistance. Typical electrical COP levels in the Colorado State University testing program were about 15, with the highest examples near 20. More recent system designs, including syphon return from the collector and reduced flow rates, have permitted significant reductions in pump power consumption. Electric requirements have decreased to levels corresponding to COP values as high as 50, making electricity an almost negligible operating-cost item.

13.2.8 Domestic Hot Water

At small additional cost, a solar system designed to supply part of the space heating requirements of a building can also be used to meet a substantial portion of the domestic hot water (DHW) demand. This feature is an economic advantage, because it permits use of the solar equipment and reduction in purchased energy during spring, summer, and fall, when space heating demand is small or completely absent. Solar DHW has been incorporated in all of the space heating systems designed and tested at Colorado State University.

The portion of the total collected solar energy that is supplied to DHW depends on the solar system capacity and the demand for space heating. Since heating demand is highly variable, the energy available for DHW also varies. In the system to which figure 13.6 applies, about 10 percent of the collected energy is used for water heating.

In most of the systems tested at Colorado State University, DHW was heated by exchange with solar-heated water circulated from the main heat storage tank. Another arrangement included an exchanger in the return line from the collector, and in a third design, the DHW exchanger was in the load-loop circuit. Among the three configurations, use of main storage as the source for DHW heating provides the largest delivery of solar-heated water. A heat exchange coil or jacket in the DHW tank or an external exchanger with two small circulating pumps can be used. A single DHW tank, with electric auxiliary in the upper portion, has been found more efficient than a separate solar tank and auxiliary water heater because of lower external heat loss (Löf and Karaki 1983).

13.2.9 Summary

Since 1974, numerous systems have been tested and evaluated. They have shown a wide range of performance. Systems with factory-built collectors have proven superior to those employing site-built collectors. Well designed and operated heating systems employing single-glazed flat-plate collectors with selective absorber surfaces show average annual heating system efficiencies of 30 to 35 percent. The development of evacuated-tube collectors has led to solar heating systems having efficiencies a few percent higher, with the greatest advantage at low solar radiation levels.

Drainback systems show evidence of having better efficiencies and higher electrical COP than the equivalent double-loop antifreeze system; 17 percent higher heat delivery and 50 percent less electricity consumption are possible.

The use of stratification-enhancing devices in storage has been shown to provide great benefits. Consistent temperature difference of 7–10°C (12–18°F) between the top and bottom of storage can be obtained. This stratification leads to improvement in collector efficiencies of 5 to 7 percent.

The cost of solar heat is inversely proportional to system efficiency, provided that maintenance, repair, and durability factors do not cause significant difference in operating costs. A capably designed solar heating system employing durable, efficient components, controlled and operated to maximize thermal delivery, will provide solar heat at minimum cost.

Over the course of the DOE-sponsored research into liquid-based solar heating systems, the efficiency of such systems—the efficiency with which solar radiation can be converted to heat delivered to DHW and the conditioned space—has more than doubled.

13.3   Air Systems

A residential air-based solar heating system was constructed for experimental purposes at Colorado State University (Solar House II) in 1975. The performance of this system has been presented in reports and papers, including Karaki (1978) and Karaki and Hawken (1982). Other experimental air-system work supported by the U.S. Department of Energy includes a Mobile/Modular Home Unit at Los Alamos Scientific Laboratory (Hedstrom, Moore, and Balcomb 1978), the Arlington Solar House of the University of Wisconsin-Madison (Erdmann and Persons 1978), the Colorado State University solar residence/greenhouse, and the Löf (Denver) solar house (Ward and Löf 1976).

The Los Alamos Mobile/Modular Home Unit was designed to be a relatively low-cost system. It had a unique thermal storage system consisting of small jars of water around which air flow was directed. The solar system supplied about 70 percent of the house heating requirements, with a solar system efficiency of 17 percent.

The Arlington Solar House was provided with rock-bed heat storage and a unique evacuated-tube, air-heating collector. The rock bed was sized to store auxiliary energy as well as solar heat; off-peak electricity was used as the auxiliary energy source.

The Department of Energy also sponsored system development work by solar companies. Contemporary Systems, Inc., developed a system-which included site-built collectors integrated into roof construction (Contemporary Systems 1979). Fern Engineering developed a system in which two-pass air heating collectors and water storage were used (Fern Engineering 1978). The Solar Engineering and Equipment Co. developed systems using prefabricated steel building material and collectors employing transparent Tedlar® film (Solar Engineering and Equipment Co., 1978).

The General Electric Company developed a self-contained solar air heating system with an evacuated tube collector (SEEC 1978). The unit, known as a Solarsource Furnace, consisted of two groups of air collector tubes with a common center manifold, a variable size rock-bin storage compartment, and built-in circulation fans. The system was factory-built and transported by truck or forklift for placement on the ground near the house. The unit was evaluated on a mobile home, a house, and in an installation for providing DHW.

A unique study dealt with long-term performance of a solar air-heating system (Ward and Löf 1976; Ward, Löf, and Hadley 1977). The Löf House, built in 1957, has one of the earliest air-heating systems still in use. Performance data spanning a 15-year period showed a decline in output of 1.1 to 1.8 percent per year, mainly due to a gradual decrease in solar collector efficiency resulting from some deterioration of absorber surfaces. Annual maintenance costs were calculated to be between 0.6 and 1.6 percent of the original system cost.

In 1975, a combination residence/greenhouse structure was built and tested at Colorado State University. Air was preheated in the greenhouse before it entered flat-plate solar collectors, from which it passed through a rock-bed heat storage unit (Smith 1977). This scheme effectively utilized the excess solar heat from a well-built attached greenhouse, but the preheating of air before it entered the collectors decreased collection efficiency by 5 to 10 percent. While the greenhouse aspect of this study set it apart from the normal space heating comparisons, some useful information was obtained. The electrical COP, sometimes assumed to be lower with air systems, was surprisingly high—approaching 20. This performance was obtained by good rock-bed storage design and control strategy. It was also found that fan motors (high temperature "B" class) could operate in the heated air stream without damage. This capability permits negative pressure in the collectors, thereby avoiding hot air leakage, improving thermal efficiency, and permitting lower-pressure duct construction.

Solar House II was first provided with a site-built air collector (Karaki 1978) and then with factory-built, internally manifolded Solaron® air heating collectors (Karaki et al. 1980). Improvements have been made periodically to the air system, and major modifications, such as the installation of an air-to-air heat pump in parallel with the solar system, were tested (Karaki et al. 1980; Karaki, Lantz, and Waterbury 1982). The performance of liquid and air systems in Solar Houses I, II, and III has been compared and reported (Westhoff and Karaki 1981; Duff, Karaki, and Löf 1981).

Areas that have received attention in experimental system studies of air-based solar heating include the effect of variable air-flow rates, the effect of air leakage, and the characteristics of pebble-bed heat storage units.

13.3.1 Thermal Performance

Performance of an air-based solar heating system is illustrated with data from the operation of Colorado State University Solar House II. The system includes 55.8 m^2 (600 ft^2) of single-glazed, selective surface air collectors and 10.2 m^3 (360 ft^3) of pebble bed storage. Auxiliary heating is provided by a 3-ton, air-to-air heat pump, backed up with in-duct electric resistance heating for use when the heat pump capacity is insufficient. Domestic hot water is solar-heated in an air-to-water heat exchanger.

Table 13.4 (Karaki et al. 1980), contains a summary of the heating season performance in the form of monthly averages of the daily totals of the principal energy measurements. The more important results are presented in bar-graph form in figures 13.7, 13.8, and 13.9.

Figure 13.8 shows the monthly average daily solar radiation, solar radiation while collectors were operating, and the solar heat collected. The collection efficiency for the season was 33 percent, and the collectors were operating during the period in which 87 percent of the total radiation was received.

The monthly average daily loads (the total of space heating and hot water) and the solar contribution are shown in figure 13.8. The solar fraction ranges from 48 to 85 percent, with a seasonal value of 65 percent. The space heating load, examined in more detail in figure 13.9, accounts for 89 percent of the total; solar energy supplied 67 percent of this load, while the heat pump and the electric resistance heating supplied 16 and 17 percent, respectively.

The system efficiency, i.e., the solar energy utilized divided by the radiation incident upon the collectors, ranged from 19 to 36 percent, with a seasonal average of 29 percent.

13.3.2 Parasitic Power Requirements

In the system under discussion, the electric energy used for solar heat collection was 3.9 percent of the energy collected, and the electric energy used to distribute heat from storage was 3.2 percent. The electrical coeffi-

Table 13.4
Summary of key monthly mean daily quantities and performance measures for a solar air system

Number	Item	Units	1978 DEC	1979 JAN	FEB	MAR	APR	MAY	Season
1	Total incident solar insolation	(MJ/m²·day)	16.74	15.96	18.71	17.77	20.56	12.51	17.55
2	Insolation solar while collecting	(MJ/m²·day)	14.16	14.30	16.52	15.28	18.32	9.84	15.28
3	Thermal energy collected	(MJ/m²·day)	6.31	5.57	6.25	5.61	5.87	3.86	5.77
3a	Thermal energy collected	MJ/day	365.29	322.52	362.18	324.86	340.11	223.33	334.43
4	Collector operating efficiency	3/2	0.45	0.39	0.38	0.37	0.32	0.39	0.38
5	Collector daily efficiency	3a/1	0.38	0.35	0.33	0.32	0.29	0.31	0.33
6	Total heating load	MJ/day	594.00	614.78	447.03	377.81	265.12	337.33	451.90
6a	Space heating load	MJ/day	551.19	557.70	382.20	306.57	193.43	280.72	390.82
6b	DHW load	MJ/day	42.81	57.08	64.83	71.24	71.69	56.61	61.08
7	Solar energy utilized	MJ/day	353.53	297.91	340.06	285.79	223.60	223.65	294.42
7a	Space heating	MJ/day	331.68	271.91	300.96	247.82	176.59	203.60	261.16
7b	DHW preheating	MJ/day	21.85	26.00	39.10	37.97	47.01	20.05	33.26
8	Auxiliary thermal energy	MJ/day	240.47	316.87	106.97	92.02	41.52	113.67	157.49
8a	Heat pump space	MJ/day	74.28	129.42	51.40	48.40	15.09	51.91	63.11
8b	Elec. resistance space	MJ/day	145.23	156.37	29.84	10.35	1.75	25.21	66.56
8c	Elec. resistance, DHW	MJ/day	20.96	31.08	25.73	33.27	24.68	36.55	27.82
9	Elec. energy to collect	MJ/day	17.70	12.32	12.04	11.19	13.11	8.11	12.95
10	Elec. energy to dist. solar	MJ/day	12.77	10.75	10.10	9.12	3.57	11.75	9.42

11	Elec. energy for heat pump	MJ/day	50.70	84.88	27.33	22.35	7.88	24.38	37.83
12	Elec. energy for H.P. blower	MJ/day	8.59	13.13	3.91	2.86	1.03	3.15	5.75
13	Solar system efficiency	$7/(1 \cdot A_c)$	0.36	0.32	0.31	0.28	0.19	0.31	0.29
14	COP to collect solar	3a/9	20.64	26.18	30.09	29.03	25.94	27.50	26.36
15	COP to distribute solar	7/10	27.69	27.71	33.67	31.34	62.63	19.03	31.25
16	COP overall solar	$7/(9+10)$	11.60	12.91	15.36	14.07	13.41	11.26	13.78
17	COP heat pump	8a/11	1.47	1.52	1.88	2.17	1.91	2.13	1.67
18	COP auxiliary heating	$8a + 8b/(8b + 11 + 12)$	1.07	1.12	1.33	1.65	1.58	1.46	1.18
19	COP overall heating	$6/(8b + 8c + 9 + 10 + 11 + 12)$	2.32	1.99	4.10	4.24	5.09	3.09	2.82
20	Number of days of measurement		31	30	27	31	30	11	160
21	Solar fraction total	7/6	0.60	0.48	0.76	0.76	0.85	0.66	0.65
22	Solar fraction space	7a/6a	0.60	0.49	0.79	0.81	0.91	0.73	0.67
23	Solar fraction, DHW	7b/6b	0.51	0.46	0.60	0.53	0.66	0.35	0.54
24	F-chart fraction		0.55	0.45	0.72	0.78	1.00	0.61	0.64
25	Difference (percent)	$(24 - 21)/21$	−8.3	−6.2	−5.2	2.4	17.6	−7.9	−1.5

Source: Karaki et al. (1980).

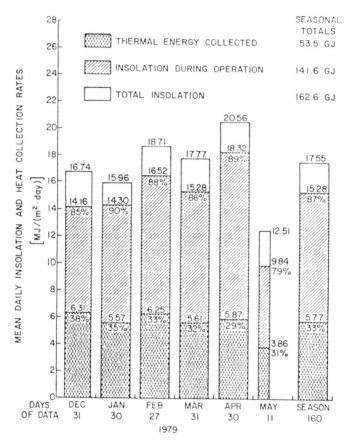

Figure 13.7
Insolation and thermal energy collected in a solar air system.

cient of performance for collection was 26, while the overall electrical coefficient of performance was 14.

13.3.3   Air-Flow Rates

Initial studies in CSU Solar House II (Karaki 1978) led to a recommended air-flow rate of 10 l/sec m^2 (2 cfm/ft^2) of collector area. To achieve higher air and storage temperatures, collector flow rates were sequentially reduced in the winter of 1978–79 to 7 and 3 l/sec · m^2 (1.4 and 0.6 cfm/ft^2)—Karaki et al. (1980). When the flow rate was lowered from 10 to 7 l/sec · m^2 (2 to 1.4 cfm/ft^2), the higher operating temperatures resulted in a heat collection decrease of about 15 percent.

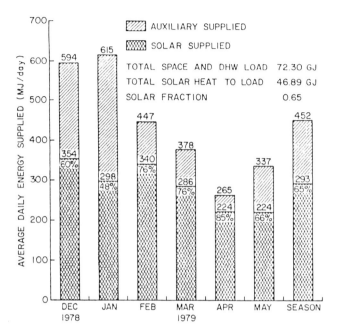

Figure 13.8
Average daily energy supply for hot water and space heating in a solar air system.

In the period 1979–80, multiple-speed blower control was used (Karaki, Lantz, and Waterbury 1982). The multiple-speed blower is used to match air rate to collection, with a low flow rate early and late in the day, and a maximum rate at noon. It was found that use of variable flow can increase the quantity of energy collected compared to collection at constant flow rate with on/off control. The magnitude of the increase depends on the flow rate used. If the air flow rates can differ by a factor of 3 between low and high speed, the increased collection may be as much as 20 percent.

13.3.4 Air Leakage

The effect of air leakage on the performance of solar air-heating systems has been investigated (Ward and Löf 1976; Karaki et al. 1980; Close and Yusoff 1978). Air leakage into the collector is found to have a positive benefit. Systems should always be configured so that the collector array is at lower pressure than the ambient; thus, there is no leakage of heated air from the collector. In measurements on side-by-side arrays (Karaki et al.

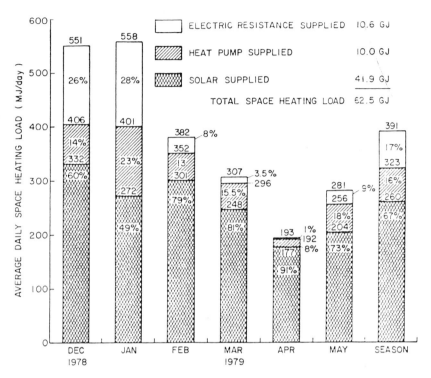

Figure 13.9
Monthly average daily space heating load for a solar air system.

1980), one had a leakage rate of 7 percent, and the other a leakage rate of 15 percent. With equal inflow rates, the array with a leakage rate of 15 percent delivered 10 percent more thermal energy than the array with smaller leakage rate.

The positive benefit from leakage comes from the fact that the ambient air is usually at a lower temperature than the collector inlet air, and the total collector flow rate is higher. Infiltration reduces the operating temperature of the collector and increases its efficiency. If the leakage is excessive, however, lower delivery temperature may result in decreased capacity of the heat storage unit.

13.3.5 Pebble-Bed Characteristics

Pebble-bed heat storage performance has been monitored in the Arlington House in Wisconsin (Erdmann and Persons 1978) and CSU Solar House

II (Karaki et al. 1980). In both installations, considerable variation in air velocity across the pebble bed was observed. Even when gross short-circuiting and channeling were avoided, there was a tendency for flow to be greater near the walls of the storage bin than through the middle of the bed.

In 1979, the storage unit in CSU Solar House II was opened and examined in the course of remodeling the system. Even though the rock had been screened and washed before first being installed, excessive fines were found, particularly in the central "core" of the bed. Fracturing of the rock during loading, and size segregation by rolling of larger pebbles on a conical heap probably caused the poor distribution. When the CSU pebble-bed storage unit was reconstructed, vertical dividers were used. The remodeled storage showed a more uniform flow distribution pattern (Karaki et al. 1980; Karaki, Lantz, and Waterbury 1982).

The pebble bed at CSU was also used for testing its cold storage potential (Karaki, Lantz, and Waterbury 1982; Karaki, Armstrong, and Bechtel 1977). Day-night exchange was employed, with added off-peak operation of the heat pump. Even though condensation occurred in parts of the bed, no unusual biological or bacteriological growth was found on the pebbles.

Pebbles about 4 cm (1.5 in.) in diameter have been found a practical size for effective heat transfer and low pressure drop. The size is not critical provided it is reasonably uniform. Uniformity is important because of its influence on void fraction and air flow characteristics. The crushing and screening at gravel yards is normally satisfactory, but careful transport and distribution in the bin are also necessary.

Pebble-bed geometry is important for avoiding excess air pressure drop while at the same time promoting temperature stratification. Vertical flow through a bed 1 to 2 m deep has been found most practical. The lateral dimensions follow from the storage mass requirement as dictated by the overall system size.

13.3.6 Summary

System performance characteristics are dependant on the equipment used and the relative size of the load. A well-designed solar air heating system has operated with an all-season solar fraction of 65 percent, a collection efficiency of 33 percent, an overall system efficiency of 29 percent, and an electrical COP of 14. The solar fraction is, of course, a characteristic of the relative size of the solar installation and the heating requirements of the building.

The performance of the systems is sensitive to the air-flow rate through the collectors and to the control temperature settings used. Multiple-speed blowers are capable of providing performance enhancement. The development of "smart" controllers and efficient variable speed motors will facilitate the practical application of variable-flow, constant-temperature solar air heating systems.

Moderate air leakage into air collectors has been found to have a positive benefit. This does not mean that leakage should be encouraged, but it need not be scrupulously avoided. Collectors should not be operated at positive pressure relative to the atmosphere and the inside of a building, however, as leakage of hot air from the Collector is a performance penalty.

Pebble-bed design is important. Air-flow channeling should be avoided. Careful loading of the bed with clean, uniform gravel, and vertical baffles or partitions in the storage improve air flow distribution. Pebble beds should be located inside the building so that moisture does not leak into or condense on the material, thereby avoiding the development of odor-causing deposits and growths.

13.4   Comparisons between Air and Liquid Systems

The performance of systems employing flat-plate and evacuated-tube, liquid-heating collectors in Colorado State University Solar Houses III and I is compared, in figure 13.10 with the performance of the system in which flat-plate, air-heating collectors are used in CSU Solar House II (Duff, Karaki, and Löf 1981). Space heating and domestic hot water were provided in both systems. While nearly identical conditions prevailed in these side-by-side air/liquid solar system comparisons, some conditions differed. Commercial collectors were used in each system; as previously indicated, the buildings were the same size and adjacent to each other. The heating demand was 30 percent higher in Solar House III (the liquid system), mainly because of some heating of a garage area.

The performance of the two systems employing flat-plate collectors during four months of the heating season is shown in figure 13.11 (Westhoff and Karaki 1981); table 13.5 contains daily average values of the principal conditions and operating results. Although air is inferior to liquid as a heat transfer medium, the system interaction and effective thermal stratifi-

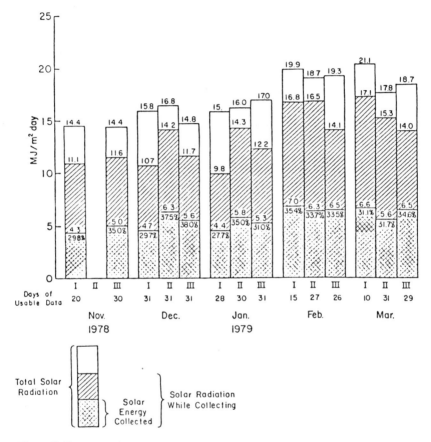

Figure 13.10
Seasonal performance of three types of solar heating systems at Colorado State University.

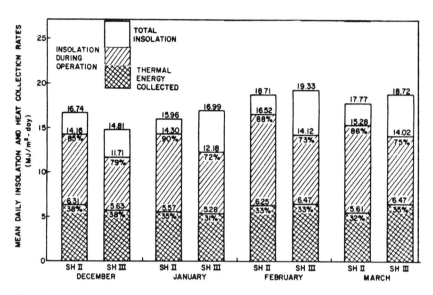

Figure 13.11
Average solar collection efficiency of air system (SH II) and liquid system (SH III).

cation in the heat storage unit serve to overcome this handicap. Lower fluid supply temperature to the collector lessens heat loss to the atmosphere and reduces the threshold level of useful solar radiation. Although the efficiency of the collector *during operation* was 39.5 percent for air and 46 percent for liquid, the mean *daily* collection efficiency for both was found to be identical at 34 percent. It was concluded that the overall efficiency of well designed and operated air systems is essentially the same as that of liquid systems of comparable quality.

It was found that the air system required less electric energy than the liquid system. The air system electric COP was 29 percent higher than the COP of the liquid system in spite of a 13-percent longer collector running time. Neither power requirement is considered typical, however, because there had been no effort to minimize or optimize power use.

Since air and liquid systems have approximately equal solar heating capability, selection should be based on other characteristics. Air collectors usually cost less than the liquid type, but air ducting may be more expensive than liquid piping. If a forced warm air heating system is used, solar air heating has an advantage because a liquid solar system requires a heat exchanger. Freezing, boiling, corrosion, and leakage are avoided in

Table 13.5
Performance comparison of air system in Solar House II and liquid system in Solar House III, Colorado State University, December 1978–March 1979

Line	Item	Units/Definitions	December		January		February		March		Period	
			Air	Liq.	Air	Liq.	Air	Liq.	Air	Liq.	Air	Liq.
1	Total insolation	MJ/m²-day	16.74	14.81	15.96	16.99	18.71	19.33	17.77	18.72	17.26	17.38
2	Insolation while collecting	MJ/m²-day	14.16	11.71	14.30	12.18	16.52	14.12	15.28	14.02	15.02	12.95
3	Thermal energy collected	MJ/m²-day	6.31	5.63	5.57	5.28	6.25	6.47	5.61	6.47	5.93	5.94
3a	Thermal energy collected	MJ/day	365.3	329.9	322.5	309.4	362.2	379.1	324.9	379.1	343.3	347.9
4	Collector operating efficiency	Line 3 ÷ Line 2	0.45	0.48	0.39	0.43	0.38	0.46	0.37	0.46	0.39	0.46
5	Collector daily efficiency	Line 3 ÷ Line 1	0.38	0.38	0.35	0.31	0.33	0.33	0.32	0.35	0.34	0.34
6	Total heating load	MJ/day	594.0	734.5	614.8	847.5	447.0	570.8	377.8	620.9	509.6	698.7
7	Solar energy utilized	MJ/day	353.5	322.9	297.9	299.5	340.0	363.0	285.8	359.8	318.8	334.9
8	Auxiliary thermal energy	MJ/day	240.5	411.6	316.9	548.0	107.1	207.8	92.0	261.1	190.8	363.8
9	Elec. energy to coll. and dist.	MJ/day	30.47	34.64	23.07	38.91	22.14	31.45	20.31	31.45	24.07	34.25
10	Solar system efficiency	Line 7 ÷ (Line 1 · A_c)	0.36	0.37	0.32	0.30	0.31	0.32	0.28	0.33	0.32	0.33
11	Overall solar COP	Line 7 ÷ Line 9	11.60	9.32	12.91	7.70	15.36	11.54	14.07	11.44	13.24	9.78
12	Solar fraction	Line 7 ÷ Line 6	0.60	0.44	0.48	0.35	0.76	0.64	0.76	0.58	0.62	0.48
13	Number of days of data	Days	31	31	30	31	27	27	31	29	119	118

Source: Westoff and Karaki (1981).

air systems, but bulkier and heavier heat storage is a disadvantage. These factors point toward the greater applicability of liquid systems in retrofit and commercial use, and air systems in new residential construction.

13.5 Solar-Assisted Heat Pumps

The use of a heat pump in conjunction with a solar energy system has been widely studied. A number of configurations are possible for the use of a solar-assisted heat pump (SAHP) in both heating and cooling applications. Heating applications can be divided into series and parallel types. In the series SAHP design, the solar part of the system supplies comparatively low temperature energy to a heat pump specially designed for this application. In the parallel configuration, a conventional heat pump serves as the auxiliary heat supply for a solar heating system. The series system has the advantages that: (1) the collector array operates at lower temperatures, leading to increased collection efficiencies and the possibility of reduced collector cost, and (2) the heat pump operates at higher supply temperatures than in a conventional application, leading to higher heat pump efficiencies. The principal disadvantage is that electric power is needed for heat pump operation whenever heat is required, even though the solar supply could be adequate by itself in sunny weather. In the parallel SAHP design, the heat pump is used only when solar cannot meet the demand. But the heat pump usage tends to occur when ambient temperatures and heat pump efficiencies are low and backup electric resistance heaters may have to be used.

In the cooling season, the heat pump can be used to cool the storage medium at night in order to provide a supplementary source of cooling during the day. Such a practice has the advantages of heat pump operation at higher efficiency during the cooler hours of the night and the use of cool storage to avoid peak daytime utility demand charges.

The series configuration SAHP improves the collection efficiency of a solar energy system. However, a number of economic studies (Hwang and Bessler 1979; Hughes, Morehouse, and Swanson 1979; Chandrashekar, Sullivan, and Hollands 1979) showed that the series SAHP system was not economically competitive. The parallel SAHP was compared with alternative gas and electric backup heating and found to be economically advantageous in relatively mild climates, especially when the heat pump can also be used for cooling.

The SAHP program administered by the Department of Energy included the development of low-cost collectors suitable for SAHP systems, investigation of ground-coupled storage, and the development of heat pumps specifically designed for SAHP applications.

System performance investigations can be separated into two groups, performance monitoring of systems in the field and experimental systems operation at research institutions. E-Tech monitored six ground-coupled SAHP systems in four geographical areas (E-Tech 1980). Two concepts were used. In the first, solar heat is stored in the ground, and the heat pump uses the ground as its heat source. In the second concept, solar heat is added only to a storage tank, and the heat pump uses either the storage tank or the ground as its source. System performance was found to depend strongly on climate. In warmer climates, ground temperature was high enough and stable enough that the addition of solar heat provided minimal benefit. In colder climates, heating of the ground by solar heat was beneficial.

In the Phoenix House in the city of Colorado Springs, a SAHP system operated with ground-coupled seasonal storage (Jardine 1980). The system provided an annual solar fraction ranging from 60 percent to 80 percent, depending on the ratio of heating load to cooling load. However, depletion of stored heat resulted in high electrical usage.

Two side-by-side systems were monitored at the Twin Crown Realty building in Denver (Hays 1980). These were solar-augmented heat pump systems; an air-collection solar system was used in one, and a liquid-based system in the other.

The University of Toledo operated a liquid-based SAHP system (Eltimsahy et al. 1980) in which solar heat was collected in a tank that was used as a source for the heat pump to deliver higher temperature heat to a second tank from which heat was supplied to the house heating system. The house was heated during the 1978–79 heating season without the heat pump and during the 1979–80 heating season with the heat pump. Monthly average results for the 1979–80 heating season are summarized in table 13.6. The solar collection efficiency of this system with the heat pump had a seasonal average of 33.9 percent. This result may be compared with the 17 percent solar collector efficiency during the previous season when the heat pump was not used. Use of the heat pump doubled the collection efficiency but at a cost, of course, for the electric energy supplied to the heat pump.

Table 13.6
Solar-assisted heat pump monthly average data for the 1979/80 heating season in the University of Toledo Solar House

		November 1979	December 1979	January 1980	February 1980	4 Month Average
Average ambient temperature	°C	7.85	3.2	−1.2	−2.7	1.9
	F	46	38	29.8	27	35.4
Total insolation	kWh/m²	69	64.9	65.9	88.6	71.8
	Btu/ft²	21,922	20,618	20,901	28,074	22,788
Electric energy supplied to heat pump	kWh/day	55.6	55	34	41.7	46.6
Electric energy supplied to auxiliary devices	kWh/day	23	29	28	34	28.4
Usable energy of solar source	kWh/day	96.7	127	102.9	146.3	117.7
Solar collection COP		4.22	4.38	3.9	4.3	4.20
Heat pump COP		2.37	3.03	3.75	4.23	3.37
Solar collectors efficiency	%	34.5	33.5	37.5	30	33.9
Percent of load supplied by solar	%	70.5	64	44	51	57.5
Percent cost of load supplied by solar	%	91	88	62	73	78.6

Source: Eltimsahy et al. (1980).

The University of Tennessee and the Tennessee Valley Authority tested an air-based SAHP system (Bedinger et al. 1981). The heat pump was operated in different configurations; as a parallel SAHP, as a series SAHP, and as a conventional air-to-air heat pump. The overall thermal performance factor of these systems is defined as the total heat delivered to the conditioned space, divided by the total purchased electric energy required to operate the system. The system performance factors are shown in table 13.7. These data indicate that in a mild climate such as in Tennessee, the series SAHP required the least electricity purchase, followed by the stand-alone heat pump and the parallel SAHP system.

At Colorado State University, an air-based SAHP was operated in parallel and series arrangements (Karaki and Hawken 1982; Karaki et al.

Table 13.7
System performance factors for solar-assisted heat pump systems at the University of
Tennessee/VTA Solar House

System type	System performance factor
Parallel SAHP with off-peak heating	1.66
Series SAHP without off-peak heating	2.40
Air-to-air heat pump without off-peak heating	1.99

Source: Bedinger et al. (1981).

1980; Westhoff and Karaki 1981). When the heat pump was operating in
a series configuration in March, average collection efficiency was 52 percent, while in April, when the heat pump did not operate, average collection efficiency was 40 percent. When operated in series, the average heat
pump COP was 2.16, and in the parallel configuration, the average was
1.5. However, the use of auxiliary energy in the two systems was nearly the
same, mainly because in the series configuration, the heat pump could
seldom meet the auxiliary heat demand, so electric resistance heating was
frequently needed. The total quantity of purchased energy is thus dependent not only on SAHP system design, but also on the climate and the size
of the solar system relative to the load.

A parallel SAHP system was also operated in conjunction with a liquid-
type collector at Colorado State University (Karaki and Westhoff 1982).
Little difference was found in the performance of air- and liquid-based
systems with parallel heat pump auxiliary heating units.

The air-based SAHP system at Colorado State was also used to cool the
pebble-bed storage unit at night during the summer (Karaki et al. 1980).
Use of the heat pump to provide cooling capacity in storage for subsequent daytime room cooling was found to be a viable mode of operation
if off-peak electric rates are available and where peak rates are at least
double the base rates.

From a relatively limited amount of operating experience with SAHP
systems, the following conclusions can be drawn:

1. Series SAHP systems usually require less purchased energy than fuel-
assisted and electric-resistance-assisted solar heating systems and stand-
alone heat pump systems. However, economic studies indicate that the
high initial cost of series SAHP systems places them at an economic
disadvantage.

2. An air-to-air heat pump in a parallel SAHP system has a lower average COP than the heat pump in a stand-alone heat pump system, because it operates mainly during severe weather when its capacity must be augmented by electrical resistance heaters. These systems may be practicable, however, if cooling is also needed and if gas-fired heating is not available.

3. Cold-charging of a rock bed by a heat pump may be practical if there is a sufficient difference in on-peak and off-peak electric prices. Provision must be made for moisture removal from storage.

References

Bedinger, A. F. G., J. J. Tomlinson, R. L. Reid, and D. J. Chaffin, 1981. Performance of a solar augmented heat pump, *Proc. ASME Solar Energy Division, 3d Annual Conference on Systems Simulation, Economic Analysis, Solar Heating and Cooling Operational Results*, Reno, NV.

Chandrashekar, M., N. T. Le, H. F. Sullivan, and K. G. T. Hollands, 1979. A comparative study of solar assisted heat pump systems for Canadian locations, *Proc. ISES International Congress*, Atlanta, GA, p. 782.

Close, D. J., and M. B. Yusoff, 1978. The effects of air leaks on solar air collector behavior. *Solar Energy* 20(6): p. 459.

Colorado State University Solar Energy Applications Laboratory, 1980. *Six solar heating and cooling program tasks, Solar House I, Solar House II, and Solar House III.* Progress Report C00-30122-12, Colorado State University to Department of Energy.

Colt, Inc., 1980. *Development of an integrated commercial or single family size solar heating and hot water system using paraffin oil in the solar loop.* Final report, contract EY-76-C-01-2283 to Department of Energy.

Contemporary Systems, Inc., 1979. *Development of an integrated solar space heating system.* Final report contract NAS-8-32243 to Department of Energy.

Conway, T. M., W. S. Duff, R. B. Pratt, G. O. G. Löf, and D. B. Meridith, 1978. *Evaluation of high performance evacuated tubular collectors in a residential heating and cooling system: Colorado State University Solar House I.* Progress report C00-2577-14 to Committee on the Challenges of Modern Society (CCMS), Department of Energy.

Crum, J. W., 1978. Performance evaluation of the Trinity University Solar Energy System, *Proc. Solar Heating and Cooling Operational Results Conference*, Colorado Springs, CO, p. 195.

Duff, W. S., 1977. *Evaluation of the Corning and Philips evacuated tubular collectors in a residential solar heating and cooling system.* Final report C00-4012-1 from Colorado State University to Energy Research and Development Administration, Washington, D.C.

Duff, W. S., 1982. *Eight evacuated collector installations.* Interim report Colorado State University to the International Energy Agency, Fort Collins, CO.

Duff, W. S., S. Karaki, and G. O. G. Löf, 1981. Performance of three well instrumented residential solar energy systems. *ASHRAE Transactions* 87(2): p. 611–629.

Duff, W. S., and G. O. G. Löf, 1981. *The performance of evacuated tubular solar collectors in a residential heating and cooling system.* Report C00-2577-20 Colorado State University to Department of Energy.

Eltimsahy, A. H., R. L. Barnes, R. G. Molyet, and M. L. Jones, 1980. Performance evaluation of a solar house utilizing a high temperature heat pump. *Proc. Annual Meeting of the American Section of the International Solar Energy Society*, Phoenix, AZ, p. 319.

Erdmann, D. R., and R. W. Persons, 1978. The Arlington Solar House of the University of Wisconsin—Madison. *Proc. Solar Heating and Cooling Systems Operational Results Conference*, Colorado Springs, CO, p. 75.

E-Tech, Inc., 1980. *Solar assisted heat pump field performance evaluation.* Final report DOE/SF/10549/TI to Department of Energy.

Fenton, D., D. La Plante, R. L. San Martin, S. Diamond, C. Packard, H. Shaw, and W. Stevens, 1978. Performance and operating experience for the New Mexico Dept. of Agriculture Solar Heated and Cooled building. *Proc. Solar Heating and Cooling Operational Results Conference*, Colorado Springs, CO, p. 233.

Fern Engineering, Inc., 1978. *Development of residential space and DHW solar system using air heating collector.* Final report contract NAS-8-32246 to Department of Energy.

Gari, H. W. K, 1983. Stratification enhancement in solar liquid thermal storage tanks: Analysis and test of inlet manifolds. Ph.D. diss. Mechanical Engineering Department, Colorado State University.

General Electric Co., 1977. *Solar heater test program.* Final report contract EY-76-02-2705 (GE Document No. 77SD34262) to Department of Energy.

George Washington University, 1977. *Evaluation of a solar heating and cooling system using a trickle collector and combined water rockbed storage.* Final report contract EY-76-S-02-2589 to Department of Energy.

Goumaz, J. Y., 1981. Comparison drain-back and dual liquid solar heating and domestic hot water systems. M. S. thesis, Colorado State University.

Hays, D. K., 1980. Instrumentation and monitoring of the Twin Crown Realty buildings in Denver, Colorado. *Proc. Annual DOE Active Solar Heating and Cooling Contractors Review Meeting*, Incline Village, NV., p. 4–34.

Hedstrom, J. C., S. W. Moore, and J. D. Balcomb, 1978. Performance of Los Alamos Solar Mobile/Modular Home Unit II. *Proc. Solar Heating and Cooling Systems Operational Results Conference*, Colorado Springs, CO, p. 67.

Hedstrom, J. C., M. S. Murray, and J. D. Balcomb, 1978. Solar heating and cooling results for the Los Alamos Study Center. *Proc. Solar Heating and Cooling Systems Operational Results Conference*, Colorado Springs, CO, p. 155.

Hughes, P. J., J. H. Morehouse, and T. Swanson, 1979. Comparison of combined solar heat pump systems to conventional alternatives. *Proc. ISES International Congress*, Atlanta, GA, p. 772.

Hwang, B. C., and W. F. Bessler, 1979. Economics of solar assisted heat pump systems for residential use. *Proc. ISES International Congress*, Atlanta, GA, p. 767.

Jardine, D. M., 1980. Phoenix/City of Colorado Springs solar assisted heat pump project—Phase III. *Proc. Annual DOE Active Solar Heating and Cooling Contractors Review Meeting*, Incline Village, NV, p. 4–15.

Karaki, S., 1978. *Performance evaluation of the first solar air-heating and nocturnal cooling system in CSU Solar House II.* Final report C00-2868-5 Colorado State University to Department of Energy.

Karaki, S., 1984. *Performance of a drain-back solar heating and hot water system with auxiliary heat pump.* Final report SAN-30569-33 Colorado State University to Department of Energy.

Karaki, S., P. R. Armstrong, and T. N. Bechtel, 1977. *Evaluation of a residential solar air heating and nocturnal cooling system.* U.S. Department of Energy special format report C00-2868-3, Colorado State University to Department of Energy.

Karaki, S., T. E. Brisbane, and G. O. G. Löf, 1985. *Performance of the solar heating and cooling system for CSU Solar House III, summer season 1983 and winter season 1983–84.* Final report SAN-11927-15, Colorado State University to Department of Energy.

Karaki, S., T. E. Brisbane, S. S. Waterbury, and T. G. Lantz, 1980. *Performance evaluation of a state-of-the-art solar air heating system with auxiliary heat pump.* Annual report C00-30122-4, Colorado State University to Department of Energy.

Karaki, S., and P. Hawken, 1982. *Performance evaluation of a solar/solar assisted heat pump system in CSU Solar House II.* Final report SAN-30569-12, Colorado State University to Department of Energy.

Karaki, S., T. G. Lantz, and S. S. Waterbury, 1982. *Performance evaluation of a solar air-heating system with multiple-speed blower control for solar energy collection and off-peak use of heat pump for cooling.* Annual report C00-30122-28, Colorado State University to Department of Energy.

Karaki, S., G. O. G. Löf, T. E. Brisbane, D. Wiersma, G. Cler, 1985a. *Performance of solar heating and cooling systems employing evacuated tube collectors.* Final report SAN-11927-28, Colorado State University to Department of Energy.

Karaki, S., G. O. G. Löf, T. E. Brisbane, D. Wiersma, G. Cler, 1985b. *A liquid-based flat-plate drain-back solar heating and cooling system for CSU Solar House III.* Report SAN-11927-29, Colorado State University to Department of Energy.

Karaki, S., P. Vattano, and P. Brothers, 1984. *Performance evaluation of a solar heating and cooling system consisting of an evacuated tube heat pipe collector and air cooled lithium bromide chiller.* Final report SAN-30569-32, Colorado State University to Department of Energy.

Karaki, S., and M. Westhoff, 1982. *Performance evaluation of a state-of-the-art liquid-based solar system with auxiliary heat pump.* Final report C00-30122-29, Colorado State University to Department of Energy.

Leflar, J. A., and W. S. Duff, 1977. *Solar evacuated tube collector-absorption chiller systems simulation.* Report C00-2577-13, Colorado State University to Department of Energy.

Löf, G. O. G., and W. S. Duff, 1979. *Comparative performance of two types of evacuated tubular solar collectors in a residential heating and cooling system.* Progress report C00-2577-19, Colorado State University to Department of Energy.

Löf, G. O. G., W. S. Duff, and C. E. Hancock, 1981. Development and improvement of liquid systems for solar space heating and cooling—CSU Solar House I. *Solar Engineering—1981. Proc. ASME Solar Energy Division 3d Annual Conference in System Simulation, Economic Analysis/Solar Heating and Cooling Operational Results,* Reno, NV, p. 68.

Löf, G. O. G., W. S. Duff, C. E. Hancock, and D. Swartz, 1982. *Performance of eight solar heating and cooling systems in CSU Solar House I, 1980–1981.* Final report SAN-30569-14, Colorado State University to Department of Energy.

Löf, G. O. G., and S. Karaki, 1983. System performance for the supply of solar heat. *Mechanical Engineering* 105(12): p. 32.

Löf, G. O. G., and D. S. Ward, 1974. *Solar heated and cooled buildings.* Progress report C00-2577-1, Colorado State University to National Science Foundation/RANN, Washington, D.C.

Löf, G. O. G., D. S. Ward, J. C. Ward, and C. C. Smith, 1974. *Design and construction of a residential solar heating and cooling system.* Progress report C00-2577-4 Colorado State University to National Science Foundation/RANN, Washington, DC.

Löf, G. O. G., M. A. Westhoff, and S. Karaki, 1984. *Performance evaluation of an active solar heating and cooling system utilizing low-cost plastic collectors and an evaporatively cooled absorption chiller.* Report SAN-30569-30, Colorado State University to Department of Energy.

Michaels, A. I., 1979. Central storage for building heating-panel chairman summary report. DOE Report CONF-790328-P3. *Proc. Solar Energy Storage Options,* San Antonio, TX, p. 5.

Murray, M. S., M. S. Traump, and J. C. Hedstrom, 1977. Solar heating and cooling results for the Los Alamos Study Center. *Proc. 1977 Annual Meeting of the American Section of the International Solar Energy Society,* Orlando, FL, p. 9-1.

Smith, C. C., 1977. Solar space heating of a house and attached greenhouse. *Proc. 1977 Annual Meeting of the American Section of the International Solar Energy Society,* Orlando, FL, p. 33-9.

Solar Engineering and Equipment Co., 1978. *Development of integrated single family or light commercial solar assisted heating system using air collectors made from prefabricated steel building materials.* Final report, Contract NAS-8-32247, Department of Energy.

University of Delaware, 1978. *Performance of advanced systems and subsystems experiments on University of Delaware Solar One.* Final report, Contract EY-76-S-02-2589, to Department of Energy.

Ward, D. S., G. O. G. Löf, 1976. *Design, construction, and testing of a residential solar heating and cooling system.* Progress report C00-2577-10, Colorado State University to Committee on the Challenges of Modern Society (CCMS), Energy Research and Development Administration, Washington, DC.

Ward, D. S., G. O. G. Löf, C. C. Smith, 1975. *Performance of a residential solar heating and cooling system.* Progress report C00-2577-9, Colorado State University to Department of Energy.

Ward, D. S., G. O. G. Löf, C. C. Smith, 1976. *System modifications and refinements for a residential solar heating and cooling system.* Interim report C00-2577-11, Colorado State University to Department of Energy.

Ward, D. S., and H. S. Oberoi, 1980. *ASHRAE handbook of experiences in the design and installation of solar heating and cooling systems.* Colorado State University, Fort Collins, CO, p. 36.

Ward, D. S., J. C. Ward, and H. S. Oberoi, 1979. *Residential solar heating and cooling using evacuated tube solar collectors—CSU Solar House III.* Final Report C00-2858-24, Colorado State University to Department of Energy.

Ward, D. S., J. C. Ward, and H. S. Oberoi, 1980. *CSU Solar House III—solar heating and cooling system performance.* Report C00-30122-11, Colorado State University to Department of Energy.

Ward, J. C., and G. O. G. Löf, 1976. Long term performance of a residential solar heating system. *Solar Energy* 18(4): p. 301.

Ward, J. C., G. O. G. Löf, and L. N. Hadley, 1977. *Maintenance costs of solar air heating systems.* Final report C00-2830-1, Colorado State University to Energy Research and Development Administration, Washington, DC.

Westhoff, M., and S. Karaki, 1981. Performance comparison of air and liquid solar heating systems for CSU Solar Houses II and III. *Proc. Meeting of the American Solar Energy Society*, Philadelphia, PA, p. 461.

Wormser Scientific Corporation, 1980. *Multi-unit solar heating and hot water system using a pyramidal optical concentrating collector*. Final report contract NAS8-32250 to Department of Energy.

14 Performance of Residential and Commercial Systems

S. M. Embrey

14.1 Introduction

Operational performance monitoring of state-of-the-art solar energy systems is ongoing throughout the United States, through efforts by government agencies and private contractors. One of the primary sources of data produced by this monitoring is the National Solar Data Network (NSDN), organized through the U.S. Department of Energy (DOE) (Vitro 1980).

This chapter summarizes background information on monitoring efforts, NSDN analyses, major results, and key limitations to solar performance.

In 1974, Congress passed the Solar Heating and Cooling Demonstration Act. This act established a program of research, development, and demonstration directed toward reducing the nation's dependence on nonrenewable resources by stimulating the development and use of solar energy systems. Initially, the Energy Research and Development Administration (ERDA) and then DOE managed the federal Solar Energy Research, Development and Demonstration Program. ERDA and DOE were assisted in the demonstration portion of the program by the Department of Housing and Urban Development (HUD), the National Bureau of Standards (NBS), and other federal agencies and private contractors.

The demonstration program was divided into two parts: a residential program (for which HUD had prime responsibility) and a commercial program (directed by DOE). In both programs, funds were allocated for new and retrofit building projects in a variety of climatic and geographic regions throughout the United States. One objective of these projects was to demonstrate the economic viability of solar energy systems for hot water heating, space heating, and space cooling. Another objective was to provide data on the technical performance of solar energy systems. Data were collected in two ways: manually (noninstrumented data) and electronically (instrumented data).

On-site owner questionnaires were the source of the noninstrumented data, both technical and nontechnical, which were entered into the data base. Included were data describing the demonstration projects and their solar energy systems as well as data on system cost, fuel savings, reliability, etc.

In order to assess the impact of the Solar Demonstration Program, instrumentation was installed at over 100 of the approximately 5,000 federally funded projects. This effort began in FY 1977, when NSDN was officially created under the Solar Demonstration Program.

After a data collection system was developed, the monitoring of about 30 sites began in late 1977. Sites were added to the network at a rate yielding over 165 sites by early 1980. Not all systems were considered reportable, for various reasons, including system problems, unoccupied buildings, poor system performance, data collection failure, and analysis problems.

A series of publications was developed to disseminate the information gathered through NSDN, and was made available through the DOE Technical Information Center (TIC), in Oak Ridge, Tennessee. These publications were:

1. *Monthly Performance Reports (MPR)*. These reports give detailed technical data on the monthly performance of the various solar systems. The reports describe the solar system and give daily values for system performance factors. The reports also include computer-generated tables and critical operating information.

2. *Solar Energy System Performance Evaluations (Seasonal Reports)*. These reports compile monthly-level data into a comprehensive seasonal system evaluation. They present data in tables and figures and include in-depth analyses of system interactions.

3. *Environmental Data Reports*. These reports compile environmental data from every reporting site in the network; they present insolation, temperatures, and wind velocity, along with other environmental parameters.

4. *Comparative Reports*. These reports compare thermal performance and are published after sufficient data are available for comparison.

5. *Solar Project Description Reports*. These reports document "as-built" site configurations at commercial demonstration sites.

6. *Solar Project Cost Reports*. These reports document actual costs at commercial demonstration sites.

7. *Program Information Reports*. These reports describe the NSDN program on an "as-needed" basis.

8. *Reliability and Materials Assessment Reports*. These reports assess reliability of solar systems and performance of various materials used in solar systems.

Late in 1979, the Solar Demonstration Program was transferred to Vitro Corporation, Silver Spring, Maryland, and new goals were established. One of Vitro's major new goals in FY 1980 was to improve the quality and dissemination of information. They made several changes during this period.

The MPRs reported essentially the same information as in previous years; however, Vitro improved their format, added an energy flow diagram, and revamped the report to improve readability. The reports provided more interpretive analyses, and Vitro improved dissemination channels to increase the number of potential users. An effort to improve support documents was also undertaken in order to improve the information and the data base of performance results. In addition, seasonal performance reports were expanded and improved in format and readability to be used by a large audience. The reports were revamped using contacts in the solar community to better address the needs of the potential users. An advisory board of independent experts was created, with representatives from the solar industry, professional societies, the media, and solar and architectural groups. These respected experts in the solar field were invited to examine and comment on the new NSDN directions. As a result, comparative reports were expanded to include more sites, better interpretation, and significant findings.

Several new reports and procedures were created to fill the gaps in the dissemination process, including a special one-page *Performance Bulletin*, which was tailored to the general public. Major technical presentations were made at professional society conferences, with an emphasis on interpreting the performance results. These efforts continued through FY 1981.

In FY 1981, however, significant changes took place in the NSDN program. The number of reporting sites was reduced. Privately funded solar systems were monitored to assess unsubsidized solar system performance as opposed to the federally subsidized projects. Monthly reports were written but were no longer published. At the end of FY 1981, approximately 48 sites remained in the network. The level of funding was reduced below the levels originally planned, necessitating a reduced level of contractor effort.

Significant changes also occurred in FY 1982, including a reduction in NSDN site activity and a further reduction in level of effort. The principal changes included the development of a Portable Instrumentation Package

(PIP) to analyze noninstrumented sites in a nonintrusive manner (six sites were analyzed using PIP), the continuation of reports, although the number of sites was reduced, a reduction in dissemination activities, equipment repair on an "as-convenient" basis, suspension of computer analysis during the "off" season, and a further reduction in the NSDN program for FY 1983, with only 12 sites being monitored.

The NSDN-archived data base represents a significant resource for the solar community. No effort has been made to organize and update the data base, which is a tremendous reservoir of data for developing models, verifying simulation codes, assessing environmental resources, evaluating designs, and other useful activities. The tremendous quantity of NSDN data was scheduled for review in FY 1984.

14.2   Overview of the Sites

14.2.1   Introduction

NSDN monitored various residential and commercial solar systems. This section presents results from those sites that performed best and that produced the most useful and valid data; they include heating sites and combined hot water and heating sites.

Heating sites have space heating capability, with no additional system loads supplied by solar energy, with at least one heating season of actual performance data. Two systems, Telex Communications and Bell Telephone of Pennsylvania, were selected as the best performers in the space heating category.

Combined hot water and heating sites have both hot water heating and space heating capabilities. NSDN monitored more sites of this type than any other. For this study, 18 systems with at least a full year of data were included. This criterion was selected because the efficiency of these systems depends on the seasonal load, which is large in the winter and small in the summer.

The rest of this section discusses the collector array efficiencies for the selected systems and the individual site categories. The following section examines the composite performance of the sites. Individual sites are discussed, particularly when performance was limited because of unique problems. Table 14.1 presents the site characteristics of selected systems in the NSDN.

Table 14.1
Characteristics of selected systems and sites in the NSDN

Site / Location	Collector type / Collector size / Collector fluid, Freeze protection / Storage type / Storage capacity	Building type / Heated floorspace / Storage location	Auxiliary 1 / Auxiliary 2 / Solar delivery	Comments
Bell Telephone of Pennsylvania, Westchester Co, Pennsylvania	Flat-plate, 196 m² (2112 ft²), Water, draindown, Water, steel tank, 22,740 l (6000 gal.)	Commercial, 910 m² (9800 ft²), Mechanical room	Heat pump solar (20), Electric boiler, Duct heat exchanger	2-story brick, Collectors on roof, 62% design solar fraction, No solar DHW
Bond Construction, Gladstone, Missouri	Flat-plate, 43 m² (465 ft²), Ethylene glycol, Water, steel tank, 3032 l (800 gal.)	Residence, single, 179 m² (1932 ft²), Unfinished basement	Natural-gas furnace, Fireplace, Furnace coil	DHW system, 75% solar contribution
Contemporary Systems, Inc. Walpole, New Hampshire	Flat-plate, 74 m² (800 ft²), Air, Rock, 55 m³ (1940 ft³)	Commercial, manufacturing, 344 m² (3700 ft²), Beneath building	No auxiliary, Forced air from collectors or storage	Includes large passive system
Cushing Home, Murray, Utah	Flat-plate, elastromeric mat, 42 m² (457 ft²), Water, drainback, Water, 1706 l (450 gal.)	Residence, single, 228 m² (2450 ft²), First floor	Electric resistance heater, Mats embedded in floor slab	DHW preheat
First Manufactured Homes #9, Lubbock, Texas	Flat-plate, 27 m² (288 ft²), Air, Rock, 6 m³ (216 ft³)	Residence, single, 119 m² (1280 ft²), Within building	Air-to-air heat pump, Forced air from collectors or storage	DHW preheat, Direct collector-to-load option

Table 14.1 (continued)

Site Location	Collector type Collector size Collector fluid, Freeze protection Storage type Storage capacity	Building type Heated floorspace Storage location	Auxiliary 1 Auxiliary 2 Solar delivery	Comments
First Manufactured Home # 10 Lubbock, Texas	Flat-plate 27 m² (288 ft²) Air Rock 6 m³ (216 ft³)	Residence, single 119 m² (1280 ft²) Within building	Electric strip heater Forced air from collectors or storage	DHW preheat Direct collector-to-load option
GSA/Federal Youth Center Bastrop, Texas	Flat-plate 2022 m² (21,760 ft²) Water, draindown Water 56,850 l (15,000 gal.)	Commercial, institution 13,904 m² (149,671 ft²) Outside (2 tanks)	Natural-gas boilers Water-to-air heat exchanger	Several individual buildings Solar space cooling and DHW
Helio Thermics 8 Greenville, South Carolina	Flat-plate 39 m² (416 ft²) Air Rock 25 m³ (870 ft³)	Residence, single 101 m² (1086 ft²) Under house	Electric strip heater Forced air from collectors or storage	Glazed roof and attic space serve as collectors Direct collector-to-load option DHW preheat
Karasek Home Blackstone, Massachusetts	Trickle-down 59 m² (640 ft²) Water Water 8717 l (2300 gal.)	Residence, single 100 m² (1075 ft²) Basement	Oil water heater Forced air through rock heat exchanger	DHW preheat
La Quinta Motor Inn Las Vegas, Nevada	Flat-plate 111 m² (1200 ft²) Ethylene glycol (18%) Water 9475 l (2500 gal.)	Commercial, motel 114 rooms Outside	Electric boiler Water-to-air heat pump	114 heat pumps on a distributed parallel circuit DHW preheat

	Collector and storage	Building	Auxiliary system	Comments
M. F. Smith Jamestown, Rhode Island	Flat-plate 48 m² (512 ft²) Water, draindown Water, concrete tank 11,939 l (3150 gal.)	Residence, single 163 m² (1752 ft²) Basement	Heat pump—solar Electric element Heat exchanger	Overheating, 1/2 losses to structure DHW system
Montecito Pines Santa Rosa, California	Flat-plate 88 m² (950 ft²) Water, draindown Water 7580 l (2000 gal.)	Residence, multi 642 m² (6912 ft²) Underground	Natural-gas boiler Water-to-air duct coil	DHW preheat 8 apartments
Oakmead Industries Santa Clara, California	Flat-plate 244 m² (2622 ft²) Propylene glycol (10%) Water 24,635 l (6500 gal.)	Commercial 5582 m² (60,085 ft²) Within building	Natural-gas furnace Water-to-air duct coils	Large passive Trombe wall DHW preheat
RHRU Clemson Clemson, South Carolina	Flat-plate 31 m² (338 ft²) Air Rock 33 m³ (1161 ft³)	Residence, single 138 m² (1484 ft²) Basement	Electric strip heater Forced air from collectors or storage	Air entering collector may be preheated in greenhouse DHW preheat Direct collector-to-load option
Saddle Hill Trust Lot 36 Medway, Massachusetts	Flat-plate 29 m² (315 ft²) Glycerol (60%) Water 2843 l (750 gal.)	Residence, single 178 m² (1916 ft²) Basement	Oil furnace Wood stove Water-to-air duct coil	DHW preheat
Scattergood School West Branch, Iowa	Flat-plate 232 m² (2496 ft²) Air Rock, 59,020 kg (130,000 lb)	Commercial, school 740 m² (7966 ft²) Adjacent shed	Propane-gas furnace	75% design solar fraction Grain drying DHW preheat

Table 14.1 (continued)
Characteristics of selected systems and sites in the NSDN

Site Location	Collector type Collector size Collector fluid, Freeze protection Storage type Storage capacity	Building type Heated floorspace Storage location	Auxiliary 1 Auxiliary 2 Solar delivery	Comments
Spearfish High School Spearfish, South Dakota	Flat-plate 746 m² (8034 ft²) Air Rock 117 m³ (4150 ft³)	Commercial, school 3995 m² (43,000 ft²) Basement mechanical room	Natural-gas boilers to water-to-air coils Forced air from storage or collectors	DHW preheat coil in collector duct Direct collector-to-load option
Summerwood M Old Saybrook, Connecticut	Flat-plate 32 m² (340 ft²) Water, draindown Water 2274 l (600 gal.)	Residence, row house 128 m² (1375 ft²) Crawl space	Heat pump—solar and conventional Electric strip heater Water-to-air duct coil	Poured-concrete, insulated storage tank Solar DHW Heat pump uses solar at >4°C (40°F) Air-to-air mode <4°C (40°F)
Telex Communications Blue Earth, Minnesota	Flat-plate 1070 m² (11,520 ft²) Water, draindown Water, steel tank 75,800 l (20,000 gal.)	Larger commercial office 9011 m² (97,000 ft²) Machinery room	Electric strip heater Electric unit heaters Multizone AHU	70% design solar fraction No solar DHW
Terrell E. Moseley Lynchburg, Virginia	Flat-plate 37 m² (400 ft²) Water, drainback Water 7580 l (2,000 gal.)	Commercial, office 165 m² (1780 ft²) Unheated attached building	Solar-assisted heat pump Natural-gas boiler Water-to-air duct coil >29°C (85°F) Solar-assisted heat pump <29°C (85°F)	Site-built collectors DHW preheat, low DHW load

Table 14.2
System performance

	Collector efficiency %	Collector COP	Incident energy delivered to load %	Solar fraction %	Energy savings[a]
Bell Telephone of Pennsylvania	26	43	19	47	2.25 (198)
Bond Construction	25	29	19	29	0.99 (86.8)
Contemporary Systems, Inc.	36	34	36	63	4.22 (372)
Cushing Home	23	34	20	49	2.38 (210)
First Manufactured Homes #9	43	30	29	67	4.01 (353)
First Manufactured Homes #10	29	21	18	86	2.26 (199)
GSA/Federal Youth Center	21	21	7	8	1.62 (143)
Helio Thermics 8	18	13	15	52	2.34 (206)
Karasek Home	23	51	23	35	2.27 (200)
La Quinta Motor Inn	32	32	16	—	— (—)
M. F. Smith	30	34	31	77	— (—)
Montecito Pines	28	14	21	28	3.61 (318)
Oakmead Industries	43	30	35	53	4.43 (390)
RHRU Clemson	22	16	14	77	2.29 (202)
Saddle Hill Trust Lot 36	33	44	31	25	4.44 (391)
Scattergood School	25	30	18	93	— (—)
Spearfish High School	25	12	17	52	3.23 (285)
Summerwood M	35	18	22	56	1.48 (130)
Telex Communications	28	32	28	82	2.28 (201)
Terrell E. Moseley	28	13	24	80	5.53 (487)

a. MJ/m^2 day (Btu/ft^2 day).

14.2.2 Collector and Storage Subsystem Factors

14.2.2.1 Total and Operational Collector Array Efficiencies

Table 14.2 presents the total collector array efficiencies for the selected systems. Heating subsystem performance is indirectly related to the efficiency of the collectors. Solar energy is provided to the heating subsystem components either directly from the collectors or indirectly through storage. The temperature requirements of the heating subsystem dictate the temperature at which the collector/storage subsystem operates. In general, a heating subsystem that can use heat at low temperature operates at

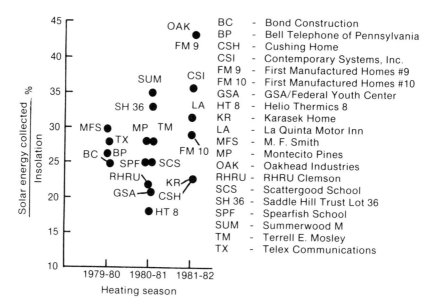

Figure 14.1
Collector efficiency.

higher collection efficiencies by lowering the collector operating temperature. Figure 14.1 shows the total collector array efficiency for all sites. The total efficiency is the ratio of solar energy collected to the total available insolation. Factors affecting total efficiency include control system strategy, collector type, collector heat-loss coefficient, collector and storage operating temperatures, weather conditions, and system application.

Operational efficiency is calculated as the ratio of solar energy collected to the solar insolation available while the collector is operating. Operational efficiency is higher than total efficiency because radiation at an intensity too low for successful collection is not included in the ratio.

The solar collector coefficient of performance (COP) is the ratio of solar energy collected to the energy required by the collector blowers or pumps. This value is a measure of the effectiveness of electricity use in operating the collector subsystem.

The best performing collector arrays used only for space heating systems were the Telex Communications and Bell Telephone of Pennsylvania systems. Telex Communications and Bell Telephone of Pennsylvania systems had overall collector efficiencies of 28% and 26%, respectively. For

Bell Telephone of Pennsylvania, the collector efficiencies would have been higher for the season if the automatic collector controller had been repaired earlier in the heating season. Solar energy was collected manually for three months of the heating season. Their storage temperatures averaged 48° and 44°C (118° and 111°F), respectively. At these moderate temperatures, the collectors could operate at satisfactory efficiencies.

The three top performers for space heating and hot water applications were First Manufactured Homes #9, Oakmead Industries, and Contemporary Systems, Inc. Their overall collector efficiencies were 43%, 43%, and 36%, respectively. Their average storage temperatures were 39° and 33°C (102° and 91°F), respectively.

14.2.2.2 Telex Communications

This site is a single-story manufacturing plant (9,011 m^2 [96,994 ft^2]) in Blue Earth, Minnesota. The solar energy system is a retrofit using 360 (1,070 m^2 [11,517 ft^2]) flat-plate collectors. Water is the medium for delivering solar energy from the collectors to the 75,800 l (20,024 gal.) storage tank. Gravity draindown is used for collector freeze-protection.

14.2.2.3 Bell Telephone of Pennsylvania

This site is a single-story office building located in West Chester, Pennsylvania, using draindown-water-type collectors (196 m^2 (2,110 ft^2)). Solar energy is stored in a 22,740 l (6,007 gal.) tank, from which both heat pumps and direct heat exchangers deliver solar energy to the building load. Auxiliary thermal energy is delivered by circulating liquid through an electric boiler.

14.2.2.4 First Manufactured Homes #9

This site is a single-family residence in Lubbock, Texas. The solar system consists of 27 m^2 (291 ft^2) flat-plate air collectors providing solar energy for space heating and domestic hot water (DHW). Heat is stored in a 9,297 kg (20,500 lb) rock storage system. The excellent performance of this system was due to good control strategy (i.e., activating the collection subsystem early in the morning and turning it off late in the afternoon) and the use of a temperature-stratified pebble bed for heat storage that continuously supplied the collector with air at approximately room temperature. The air flow rate was about 23 m^3/min (13.5 ft^3/s), somewhat higher than the rule-of-thumb value of 0.6 m^3/min (0.4 ft^3/s) per square meter of collector.

14.2.2.5 Oakmead Industries

This site is a 5,582 m² (60,084 ft²) office and manufacturing facility in Santa Clara, California. The building is normally occupied six days a week. This retrofit solar system has a 244 m² (2,626 ft²) flat-plate collector array and a Trombe-type air collector on the south wall. Only the liquid system was monitored. The system has 24,635 l (6,508 gal.) storage and a 455 l (120 gal.) DHW preheat tank. The solar storage tank was slightly oversized, which allowed the collectors to run at higher efficiencies because of the lower collector inlet temperatures. The collector controls were designed to activate the collector when the collector plate stagnation temperature was greater than 49°C (120°F). An adjustable time delay, nominally five minutes, maintained pump operation. At the end of the delay period, the collector pump would continue to operate if the plate temperature was 3°C (5°F) greater than the thermal control loop. Otherwise, the pump would deactivate and the cycle would begin again. This control device functioned extremely well.

14.2.2.6 Contemporary Systems, Inc. (CSI)

This site is a 344 m² (3,703 ft²) office and manufacturing facility in Walpole, New Hampshire. The active solar energy system consists of 74 m² (797 ft²) of flat-plate air collectors. These collectors serve the space heating and DHW load of the building. The storage subsystem consists of two 27.5 m³ (971 ft³) rock bins. The DHW is preheated with an air-to-water heat exchanger in the collector outlet duct.

14.2.3 System Problems

Common problems that reduced collector array performance in many systems were the following:

1. Low flow rates. If the flow rate is lower than designed, the collectors operate at a higher temperature and consequently at a reduced efficiency.

2. Control misadjustment or failure. Some sites had early "on" and late "off" cycles, which rejected energy from storage. One site rejected solar heat because of a low set point on the heat rejection coil loop. At another site, an improper set point for the freeze-protection mode activated the collector pump at 9°C (48°F).

3. Control sensor failure. This failure was caused by improper mounting.

4. Tracking problems with concentrators.

5. High inlet temperatures to collectors. Small storage-to-collector ratios or auxiliary heat added to storage caused the high inlet temperatures.

6. Degradation of components as a result of excessive temperatures during summer application when demand was very low.

7. Poor storage utilization (storage was totally supplied by auxiliary energy, energy removal rate, etc.).

8. Poor construction techniques.

9. Excessive complexity of systems.

14.2.4 Overall System Savings

The "bottom-line" comparison of the solar systems is the overall net energy savings from the solar systems, expressed in megajoules per square meter of collector per day. Based on experiences of operating solar heating systems, it can be concluded that substantial savings in nonrenewable energy resources can be achieved by using solar energy. Figure 14.2 shows overall normalized system net energy savings, which are the quantities of solar energy used, minus the energy required for operation of the solar system.

Telex Communications and Bell Telephone of Pennsylvania had similar savings, around 2.27 MJ/m^2-day (18.39 BTU/ft^2-day). The straightforward design of Telex Communications contributed to its large energy savings. Solar energy temperatures as low as 18°C (64°F) could be used to heat the building, a warehouse; the storage temperature never got much lower than that, so solar energy could be, and was, used continuously to heat the building. The sizing of the storage tanks was also within the rule-of-thumb range (40–100 l/m^2 [0.4–1.1 gal/ft^2]) for solar collector area. For Telex Communications, the ratio was 71 l/m^2 (0.8 gal/ft^2). The solar system was operational for the entire heating season, and this reliability contributed to the large energy savings. The harsh climate and low solar resource in the area should be considered when examining the savings figures.

At Bell Telephone of Pennsylvania, a large percentage of the thermal losses contributed heat to the building.

For the space heating and DHW sites, the best system, when comparing space heating savings, was Oakmead Industries, which had a thermal savings of 4.4 MJ/m^2-day (387.6 BTU/ft^2-day). This liquid-based system also

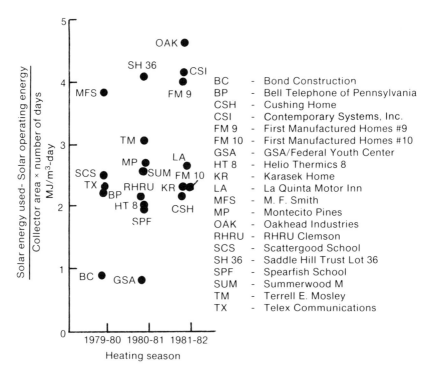

Figure 14.2
Normalized savings.

had the best overall collector array efficiency and the highest load-to-collector ratio of 101 l/m² (1.1 gal/ft²).

The second highest space heating savings were obtained by Contemporary Systems, Inc. This air system had a space heating thermal savings of 4.2 MJ/m² day (370.0 BTU/ft²-day). This system had the second highest load-to-collector ratio and the highest percentage of incident and collected solar energy delivered to the load. The storage-to-collector area ratio was 0.74 m³/m² (2.43 ft³/ft²).

14.3   Performance Summary

Table 14.3 and figure 14.3 present the average performance of the NSDN sites discussed in the previous sections. The percent of incident solar energy delivered to the total load is a measure of the overall system effi-

Table 14.3
Summary of performance

System type	System efficiency %	Average daily savings MJ/m²-day (Btu/ft²-day)	Maximum system efficiency %	Maximum daily savings MJ/m²-day (Btu/ft²-day)
Heating	22	4 (352.4)	24	6 (528.6)
Heating + hot water	22	3 (264.3)	36	4 (352.4)

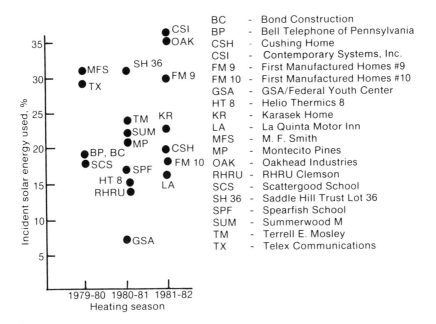

BC	– Bond Construction
BP	– Bell Telephone of Pennsylvania
CSH	– Cushing Home
CSI	– Contemporary Systems, Inc.
FM 9	– First Manufactured Homes #9
FM 10	– First Manufactured Homes #10
GSA	– GSA/Federal Youth Center
HT 8	– Helio Thermics 8
KR	– Karasek Home
LA	– La Quinta Motor Inn
MFS	– M. F. Smith
MP	– Montecito Pines
OAK	– Oakhead Industries
RHRU	– RHRU Clemson
SCS	– Scattergood School
SH 36	– Saddle Hill Trust Lot 36
SPF	– Spearfish School
SUM	– Summerwood M
TM	– Terrell E. Mosley
TX	– Telex Communications

Figure 14.3
Solar system efficiency.

ciency. Energy losses from the collector, transport, and storage subsystems and distribution equipment are included in this overall efficiency.

Combined heating and hot water system performances were similar to heating-only systems. Using the heating system to provide another load appears to increase average efficiencies, but this could not be determined from the sample sites. The average efficiency was 22%, with a net savings of 3 MJ/m^2 day (264.3 BTU/ft^2-day), similar to the heating-only site results. Maximum performance was better than heating-only systems, at 36% efficiency, and poorer at 4 MJ/m^2-day (352.4 BTU/ft^2-day) savings. Performance was limited because arrays oversized for summer application increased the proportion of operating energy available for hot water use (heating-only sites must be dormant during the summer), excessive thermal losses in the summer could increase air conditioning loads, and control problems occurred in collection and distribution systems.

14.4   Energy Flow Diagram

The energy flow diagram illustrates the energy flow relationships among the solar energy system components and subsystems at a particular site. It presents the energy flows through the system for the month, from solar energy collection and storage, through distribution of solar and auxiliary energy to building thermal energy load requirements, including losses. The flow diagram consists of collector performance, storage performance, load subsystems, and losses. Figure 14.4 shows a typical flow diagram.

14.4.1   Collector Performance

Total solar energy incident on the gross upper surface area of the modules composing the collector array is illustrated, as well as energy "loss" from the array caused by reflection, solar energy level too low to activate collectors, back-radiation, and resultant energy available from the array and supplied either to storage or directly to building thermal energy load subsystems. Operating energy of pumps or blowers for circulating liquid or air for heat transfer from the collector array to storage is contributed either to the block for the collector array or to the energy flow between the two components. The transfer of energy from the piping or duct work through the insulation is likewise subtracted from the energy flow from the collector array to storage.

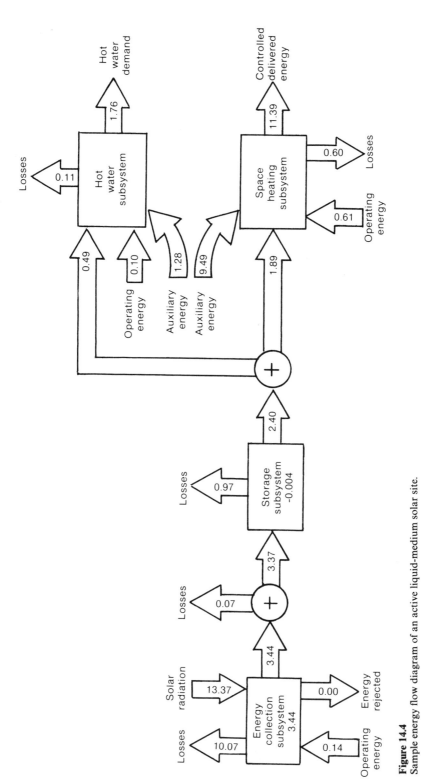

Figure 14.4
Sample energy flow diagram of an active liquid-medium solar site.

14.4.2 Storage Performance

Energy to and from storage is shown as well as energy losses through the surface of the storage vessel. If there is a change in stored energy from the prior month, it is also indicated in this block.

14.4.3 Load Subsystems

Energy flows in and out of building thermal energy load subsystems (hot water supply, space heating, space cooling, etc.) are shown as well as those for the storage. Usually, these subsystems have auxiliary energy sources (electric power, fossil fuels), and the thermal energy from these sources is shown as a contribution to the blocks for the subsystems. The arrows from the subsystems show the total energy, from both solar and auxiliary sources, which the subsystems contribute to their respective loads. In active solar energy systems, however, these contributions may not represent the total loads because electric lights, fireplaces, and portable heaters that may be used to supplement energy for space heating or hot water are not accounted for in the contributions from the load subsystems to the loads. The foregoing does not generally apply for sites with passive solar designs. Passive sites are analyzed by considering the building structure as a thermal energy storage subsystem, and all contributions of energy to the building thermal energy loads, including fireplaces and, of course, passive solar heating of the structure, are generally accounted for in the analysis process for these sites.

14.4.4 Losses

Figure 14.4 shows energy losses for subsystems, heat exchangers, and piping runs. Losses are inferred from energy flow balances computed for system components. Thus, for example, the reflection and back-radiation loss from the collector array is calculated as the difference of the solar energy incident on the array and the collected solar energy. Any losses that result in space heating will be indicated in parentheses with an asterisk next to the appropriate loss arrow.

14.5 Performance Limits

Each solar system operates at a characteristic efficiency level resulting from the interaction of the subsystems, environmental conditions, and system configurations. The net savings per square meter of solar collector indicate the relative performance of each of these systems. In light of these differences between systems, this section examines the theoretical and practical limits of the performance of solar energy equipment.

Table 14.4, the basis for the following discussion, presents five categories of system-level design parameters that limit solar system performance. These are:

1. *Solar resource assessment.* This category represents the reference weather data values used by the solar design community.

2. *Collection subsystem.* This category represents the solar collection subsystem, including devices used to capture incoming solar radiation.

3. *Storage subsystem.* This category deals with all aspects of the system effects caused by storage components.

4. *Controls.* This category refers to equipment and methods for controlling solar components within the solar system.

5. *Load.* This category deals with the types and magnitude of the heat requirements in the buildings.

14.5.1 Solar Resource Assessment

Designers of solar energy systems for buildings use a variety of reference weather data values to assess the quality of the solar radiation at a particular location. All design programs used today rely on the long-term estimates of environmental data available in the literature (Howard 1981). The SOLMET data base is the long-term source of insolation for selected locations in the United States. These data had been collected since the 1950s by the U.S. Department of Commerce and recently had instrumentation at 38 sites around the United States administered by the National Oceanic and Atmospheric Administration (NOAA) and the National Weather Service (NWS). Insolation was monitored in the horizontal plane by pyranometers, typically at airports. Howard (1981) discusses data gaps and adjustments to the data base resulting from variability in sensor type, maintenance problems, and calibration compensation.

The NSDN insolation data are measured in the plane of the collector array, eliminating the need to convert horizontal radiation values into tilted values. These data were compared by Howard (1981) to long-term insolation data similar to those available from common design programs.

Table 14.4
Limitations to solar system performance

Solar resource assessment	Collection subsystem	Storage subsystem	Controls	Load
• Correlation between long-term and modeled data	• Air locks	• Auxiliary mixing	• Component failure	• Design using constant temperature requirement
• Data requirements for design	• ASHRAE 93–77 design errors	• Baffles	– electrical	• Heat load vs. solar availability
• Lack of data for some region of United States	• Collection freeze protection	• Charge/discharge rate	– mechanical	• Hot water use
• Measured SOLMET data-irregular, accuracy corrections in refurbishment of data base	• Collector control	• Clogging (dust)	• Design	• Hybrid (combined) loads
• Modeled data may be inaccurate	– differential	• Cold storage	– components	– DHW/SH
• Rules of thumb may be inaccurate	– light level	• Collector interaction	– mode logic	– passive/active
• Variable resource for continual demand	– operational incident	• Construction	– modification	– SH/SC
	• Collector type-mismatching of use	• Corrosion	• Sensors	• Load determination
	• Control sensor location	• Excessive losses	– installation	– new system
	• Corrosion	• Insulation	– location	– retrofit
	• Flow distribution	– degradation	• Set points and calibration	• Short cycles
	• Insulation/glazing	– thickness		– oversized auxiliary
	– broken	• Leaks		• Storage discharge rate
	– outgassing	• Location		– high
	• Leaks	– above ground		– low
	• Sizing	– buried		• Timing
	• Snow	– inside building		
	• Stagnation	• Media		
	– boiling	• Short circuiting		
	– expansion	• Storage energy change		
	– peeling	• Stratification		
	• Storage temperature effect			
	• Tracking			
	• Transport mechanisms			
	– fans			
	– fluids			
	– losses			
	– pumps			
	• Weathering			

Nineteen NSDN sites in four geographic regions were used to compare actual measured solar data with reference values generated using a modified TRNSYS radiation processor. Overestimation of available insolation was shown to occur in the four regions studied. Suspected reasons for this included microclimatological effects, regional variability, sensor uncertainties, and cloud index variability.

Figure 14.5 shows a U.S. map indicating the locations of the sites used in the study, plus a short table of values as measured at the NSDN sites. The highest percentage of difference between reference and actual data was for region 7, while the lowest difference was for region 8. In all cases, the measured radiation values were at least 8% lower than expected.

The implication is obvious; if the solar energy input is overestimated, then the expected levels of performance will be achieved, even if 8% sensor bias is added to the results.

As Howard (1981) pointed out, there is a significant research need in the area of insolation assessment as applied to solar design tools. Among the unknowns at present are the relationships between direct and diffuse components of the solar radiation, the relationship between tilt angle and radiation variability, and the reliability of the SOLMET data base. Improving the quality of these data will improve the outputs of the various solar design and simulation programs.

14.5.2   Collection Subsystem

The active solar approach to system design is characterized in most cases by using solar collector panels to capture incoming solar radiation. Collection subsystem performance is related to system performance, storage temperature, and system output.

The NSDN results show several key limitations to system performance in the collection subsystem.

The American Society of Heating, Refrigeration, and Air Conditioning Engineers (ASHRAE) *Standard 93–77* collector performance graphs are commonly used as a design tool; however, ASHRAE test results only indicate collection subsystem output. Generally used are arrays consisting of many single panels in some arrangement. Eck (1981) reports that field data collected from NSDN differ from ASHRAE single-panel test results in several important ways.

First, a test of an array of single panels produces a different performance curve than a single-panel test would produce. Second, controlled tests,

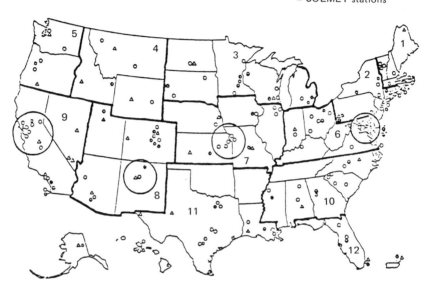

Figure 14.5
SOLMET versus NSDN measured data.

Geographic dispersion of
SOLMET stations and NSDN sites

Legend
o NSDN sites
△ SOLMET stations

Region	Mean reference insolation (MJ/m²-day)	Mean measured at site (MJ/m²-day)	Percentage difference (%)	Mean reference temperature (°C)	Mean measured temperature (°C)
6	147	135	9.4	12	13
7	235	190	23.7	14	14
8	168	155	8.5	12	12
9	194	168	15.0	18	16

(Monthly averages)

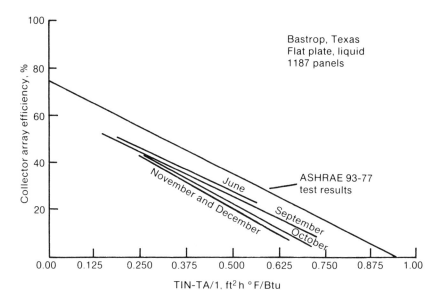

Figure 14.6
Typical ASHRAE test results versus actual performance.

such as the procedures used in ASHRAE *Standard 93–77*, do not account for seasonal weather variability, which subjects panels to a wider variety of incidence angles, solar radiation levels, wind speeds, differing percentages of direct and diffuse components, inlet fluid temperature variability, and transient system effects. Other effects not accounted for include fluid flow rate; flow balancing effects; blockage; corrosion; stagnation, wind or storm damage; broken glazing; wet or degraded insulation; outgassing; air locking; leakage; differential thermal expansion; snow loads; dirt; and weathering of components.

One prediction method (f-chart) addresses some of these effects, including incidence-angle modifier and system effects. Figure 14.6 presents a sample plot of actual field data compared to ASHRAE *Standard 93–77* test results to illustrate the performance of actual large-scale arrays. Two large collector arrays were studied. In both cases, all-day efficiency was lower than ASHRAE *Standard 93–77* single-panel test results, which appeared to be the upper limit of performance of the arrays.

Figure 14.7 plots the least-squares curve fits of collection efficiencies with operating points at Terrell E. Moseley in the form of linear equations

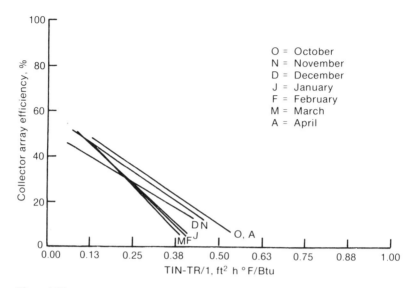

Figure 14.7
Collector efficiency results, Terrell E. Moseley, October 1980 through April 1981

with the collector transmission/absorption products $(F_R \tau \alpha)$ indicated by the y-intercept, and the effective collector heat loss coefficients $(F_R U_L)$ indicated by the slope of the curve-fit lines. The variability of intercept and slope illustrates the substantial dependence of collection efficiency on system interactions (Vitro 1981a) and on other operating conditions cited previously.

Figure 14.7 shows that the y-intercept ranged from 0.62 to 0.56 during this heating season. The highest value was in February, the lowest in December. Part of the difference is probably due to variation in incidence angle of the sun with respect to the 50-degree collector plane. The intercept is the combined $F_R \tau \alpha$ value for these single glazed, nonselective collectors.

Collector controllers, such as the differential type, are another major area of performance limitation in solar energy systems. Consider the following examples from the NSDN.

Waterman (1981), prompted by several performance reductions at NSDN sites, examined cause-effect relationships in control systems. He divided problems into two major areas—catastrophic failure and improper collection pump operation. His study narrowed the focus to the latter problem, because catastrophic failures are more easily found and rectified than more subtle sensor effects. Waterman discussed two collector control scenarios—start up cycling and delayed shutoff of collector pumps. In the first scenario, storage temperature and collector sensors were not sensing correct temperatures, causing startup cycling. In the second scenario, the placement of the tank temperature sensor delayed shutoff of the collector pump at the end of the day. The sensor was mounted on a space heating pipe that was subject to different temperature regimes from inside the tank. This placement caused delayed pump shutoff and wasted energy.

Another example of poor placement of a collector control sensor occurred at Brookhaven National Laboratory when the control thermistor on the collector plate was shaded frequently during the day (Vitro 1981b). The poor location of this thermistor was responsible for poor array efficiency at this site. After the thermistor was moved, the collector operated normally. Many other sites experienced problems with collection control. Waterman (1981) describes more of these problems in greater detail.

Collectors also need to be matched to the system in which they are used. Several designs encountered in the NSDN have had shortcomings in collection system selection. A site in Idaho used linear Fresnel concentrating collectors at a single-family residence where flat-plate collectors would have been adequate and more reliable. The collectors required tracking controls and mechanisms (which failed) and excessive movable pipe connections (which leaked). Moreover, other problems related to storage location and configuration occurred. The implication here is clear; the least complex components that will perform the task should be used; using tracking collectors for a single-family dwelling increases potential reliability problems.

Collection subsystem flow rates also limit system performance and deserve comment. An air system, RHRU Clemson, in South Carolina (Vitro 1981c) was upgraded from a single blower/damper system to a dual-fan air handler, increasing airflow in the system. The collection subsystem output increased from 29 KJ (27.5 BTU) in 1979–1980 to 48.1 KJ (45.6 BTU) in 1980–1981 with roughly similar conditions. Storage use was also increased (see section 14.5.3). Solar collection COP (ratio of energy collected to collection subsystem operating energy) decreased from 20.80 to 16.11, but the overall collection subsystem output offset the increased power consumption.

Installation problems encountered with collection subsystems are another area of concern. For example, one solar system at Lawrence Berkeley Laboratory (Vitro 1981d) rejected energy from the collector array because of a thermosyphon loop that began only after a short pumping cycle for freeze-protection. During a sample month, a total of 0.83 KJ (0.79 BTU) was lost through this mechanism. The solution appeared to be the addition of check valves to prevent backflow. However, when this correction was implemented along with control set point adjustments, the thermosyphon problem continued. The final solution to the problem came with the installation of solenoid valves in the supply and return pipes to the collector. These valves prevented the thermosyphoning, which occurred when the pumps would start and stop momentarily for freeze-protection or otherwise.

Related to installation problems are flow distribution limitations caused by less-than-adequate piping networks, air locking caused by improperly installed purge valves (or lack of air purge valves), support structure failure caused by snow or wind loads, and other basic problems caused by lack of concern for sound engineering practices.

14.5.3 Storage Subsystem

Storage subsystems affect solar system performance in several ways. Combined systems serving hot water loads generally use a preheat type system for energy storage. Preheat tanks store solar energy at a lower temperature than the auxiliary tank, which is generally thermostatically controlled. Kennedy, Sersen, and Rossi (1980) discuss system losses.

Typical solar space heating systems (figures 14.8a and 14.8b) normally include a collector array with a pump or fan to circulate the liquid or air collection fluid to storage and, in some system designs, also directly to the heated space. Heat exchangers are used when subsystem isolation is required or when energy conversion from one medium to another (liquid to air, etc.) is used.

Variations in design of storage type, auxiliary energy source, other loads (DHW, cooling), valving, piping, and system controls may be expected to affect subsystem and overall performance.

Other factors affect storage systems at solar sites; these include mixing, stratification, baffling, and internal flow patterns. Kennedy, Sersen, and Rossi (1980) studied the performance of three storage subsystems for commercial/institutional hot water solar systems. Their results showed that

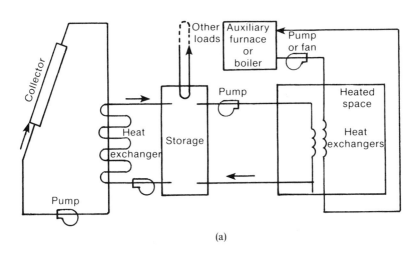

(a)

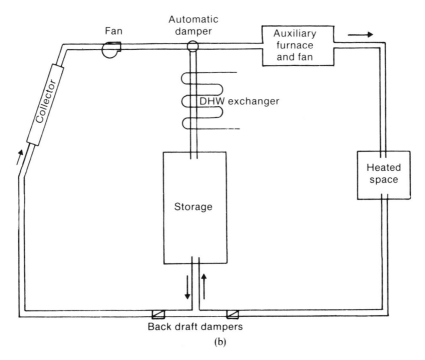

(b)

Figure 14.8
(a) Typical solar space-heating system—liquid type; (b) typical solar space-heating system—air type.

direct-feed tanks are well mixed during collection and at the end of the charge cycle despite initial stratification. In immersed heat exchanger tanks, stratification was a function of load pattern. The effect of tank orientation (vertical or horizontal cylinder types) on stratification of temperature regions was discussed. Vertical tanks held stratified for a longer period of mixing.

Many systems also have unique storage problems that limit maximum solar performance. Vitro (1981e) discusses storage systems and highlights the following from the NSDN:

• Designers often do not consider losses that will occur within conditioned spaces. Standby losses from storage tanks in solar energy systems are generally higher than in conventional systems because of the use of large tanks with resulting larger surface areas. These tanks should be better insulated than the small conventional hot water tanks have been. In addition, solar storage tanks located in conditioned spaces should be well insulated to minimize losses in the summer, which may cause increased cooling loads.

• Care should be taken to examine seasonal effects on solar performance. Solar energy systems with both space heating and hot water will often meet 100% of the summer hot water load, but solar operating costs may exceed energy savings. This occurs when the collector array is oversized for the small summer hot-water-only load. In addition, some sites require operation during the summer simply to protect the collectors from overheating, regardless of any potential contribution to load.

Unless properly controlled, a combined hot water and space heating system may develop excessive temperatures in storage during the summer, which may damage or degrade the tank and cause larger cooling loads. If the savings for hot water do not exceed the cost of operating pumps and the extra costs for cooling, and if the storage tank cannot sustain high temperatures, then part or all of the solar system may have to be deactivated on a seasonal basis.

The magnitude of storage losses is a strong function of storage location and type. Figure 14.9 identifies storage losses (as percentages of the solar heat delivered to storage from the collectors) for liquid tanks in buildings, outside buildings, or in unheated buildings or sheds, and underground. Also shown are losses from rock bins. The percent loss is smallest for liquid storage inside buildings and largest for rock bins.

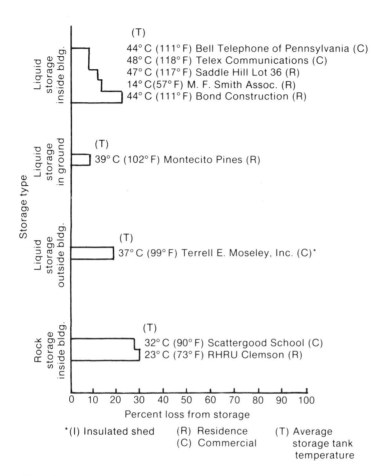

Figure 14.9
Percent energy loss from storage for various storage types.

With storage losses ranging from 10% to 30%, the losses can have a significant impact on system performance. If storage is located in the building, the losses can contribute to interior heating requirements; however, these losses can also add to the cooling load or cause human discomfort if they are not adequately controlled when heat is not needed. If storage is located outside the building, the losses will not affect space heating or cooling loads, but they will affect energy cost because they result in higher fuel use. Additionally, the proper installation of underground (buried) storage has proven to be very difficult. The effectiveness of some types of tank insulation is often severely degraded because of the permeation and retention of groundwater. Even closed-cell foams will eventually be permeated under conditions of a high water table or poor drainage.

The actual overall heat losses from storage tanks are often considerably larger than indicated by design calculations. Small amounts of uninsulated tank surface area and heat loss through connecting pipes and tank supports can cause increases in the heat loss coefficient (UA) of 20% to 50%. An example of poor installation and/or wet insulation was the Bell Telephone of Pennsylvania site. The ratio of measured heat loss to design-predicted heat loss was 2.3.

14.5.4   Controls

Control systems may impose an effective limit on solar thermal performance. Vitro (1981e) discusses four major control subsystem problem categories that were studied at 40 NSDN sites. The four categories were:

1. *Design.* This category refers to the matching of the control components and logic with the intended system operational sequence. It also refers to postinstallation modification of the control systems or logic, intentional or unintentional.

2. *Sensor installation and location.* This category covers control system problems resulting from processing of incorrect information by the controller. The frequent cause of these problems is incorrect location or installation of control sensors.

3. *Set points and calibration.* This category deals with the selection of set points on adjustable controllers, nonadjustable controls, or improper control calibration.

4. *Component failure.* This broad category covers electrical and mechanical failure of control system components, including linkages, circuits, relays, valves, solenoids, activators, and other elements.

Figure 14.10 shows the relative percentages of the causes of control failures at the 40 study sites. It is interesting that nearly half of the problems were directly related to set points and calibration of the control system. These problems may often be solved at minimum cost, as many controls allow adjustments to be made.

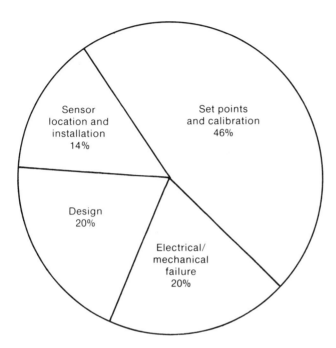

Figure 14.10
Distribution of NSDN site solar system control problems.

Design problems are generally a result of logic errors by system designers and poor adherence to engineering principles. The following examples from the NSDN illustrate typical control design problems:

• At a site in Maryland (J. D. Evans A), the occupant added a timer to the normal control system. The timer conflicted with the timing of available solar radiation for the system and resulted in a major decline in the energy savings attributable to solar contribution. Waterman (1981) describes the timer's effect and the resulting increase in auxiliary power consumption. The timer was added to force the system to use stored solar energy during evening hours rather than on demand. As temperature levels in storage decreased during stored energy use, the heat output from the storage heat exchanger also decreased. In the early morning, the tank temperature was often above the 32°C (90°F) required to activate the heat pump auxiliary; however, the heat output dropped below that needed to satisfy the heating load of the home. To satisfy the heating demand, the electric resistance elements then activated (when the heat pump could otherwise have been used) and seasonal power consumption increased by 23%.

• A site in Texas (College Houses) used a minicomputer-based control system to operate a solar cooling/heating/DHW site of moderate size. The site computer software was constantly revised by site personnel, to the detriment of system operation. Incorrect modes of operation (such as simultaneous cooling and heating) resulted from this inconsistent control logic. This example of complicating the solar system illustrates the principle of simplicity discussed in Vitro (1981f).

• A site in Kansas (Kansas City Fire Station) did not operate as well as possible because of control problems. The central control panel was located in an area subjected to high levels of employee traffic, resulting in tampering with control set points by untrained personnel. The solar system operated erratically. This example illustrates that control units having variable set points should be located in secure areas or be "factory-sealed" or key operated to prevent tampering and resultant degraded performance.

• At one site in Florida (Matt Cannon), a hot water system pump was replaced with a nonvariable flow pump without modification to the control system. This replacement resulted in excessive storage losses and reduced energy available for the preheat tank. Corrective action to provide control system matching was required to ensure proper operation of other systems.

• At a site in California (San Anselmo School), improper control of an automatic valve on the collector-to-storage piping loop prevented collection of solar energy . The set point on the valve did not allow the valve to open until the collector water reached 104°C (219°F). The scale of the thermostat adjustment was incorrect. When the grantee was able to get this thermostat adjusted to the desired level of 79°C (174°F), there was a dramatic increase in the amount of solar energy collected.

14.5.4.1 Location of Sensors
Control sensors must be placed where they will be subjected to the correct environmental conditions. Examples of site-specific problems follow:

• At a site in Maryland (J. D. Evans B), temperature sensors were improperly located for sensing storage tank temperature. The resulting problems included pump cycling, late collector pump turnoff, and excessive use of auxiliary energy for space heating. The improper location of the sensor did not merely result in an incorrect temperature reading of a few degrees. Rather, the effect was a pronounced hysteresis phenomenon, equivalent to having controller set points that depended on the state of the system.

• At a site in Virginia (Loudoun County School), a temperature sensor, used to determine when the collector pump should activate, was located next to the cold water inlet (city water supply) to the solar preheat tank. This sensor then monitored a temperature that depended on the load rather than on actual storage temperature.

14.5.4.2   Control Set Points

The matching of control set points to optimize system performance is a major criterion for solar performance maximization. The overall efficiency of conversion of solar energy to thermal energy in buildings is a function of the set points throughout the system.

Collection control set points are related to late onset of collection and early shutoff, thereby reducing the potential for energy collection.

Many systems use a heat rejection strategy for freeze-protection. The control set points for circulating warm storage water through a collector array are also important. Consider the following examples:

• At a church in South Dakota (First Baptist Church), there was an improper set point in the freeze protection circuit of the DHW subsystem. Freeze protection was activated at about 10°C (50°F) instead of 2°C (35.6°F) to 3°C (37.4°F), causing excessive heat rejection. During severe winter weather, the preheat tank temperature was reduced to about 16°C (60.8°F). There was also a design problem with a damper motor. This damper controlled the airflow between the solar system and the building system. However, when the air-handling fan would come on, the negative pressure on the damper pulled it open. The problem caused auxiliary energy to be added to the return air to charge storage. A larger damper motor was installed and the problem was solved.

• A site in California (Lawrence Berkeley Laboratory) had a similar control problem with the collector array freeze-protection set point. Initially, the thermostat called for freeze-protection when the collector plate temperature was about 9°C (48°F). The grantee was able to calibrate the freeze-protection thermostat with the help of NSDN data. It was recalibrated to about 2°C (35.6°F) with a substantial reduction in freeze-protection losses.

Other set points affect the utilization of solar energy in conjunction with auxiliary energy. The following examples illustrate these effects:

• At a site in Hawaii (Honolulu Ramada Inn), high rates of natural gas consumption were experienced during the evening and early morning, at the same time that solar energy was available from storage. The boiler set point was 60°C (140°F), and average supply to the boiler was 45°C (113°F). The water from the boiler was then tempered to within a few degrees of the storage temperature (49°C [120°F]). Lowering the boiler output temperature decreased the consumption of auxiliary fuel by 74% during periods when solar energy was available, increasing the total solar contribution from an average of 42% to 52%.

• At one site in South Dakota (Mount Rushmore Visitor's Center), the heating subsystem used stored solar energy at various threshold levels. A series of tests conducted during 1979–1980 determined the optimal temperatures for onset of stored energy use and termination of use. Set points of 49°, 43°, 38°, and 32°C (120°, 109°, 100°, and 90°F) were used as upper limits (onset of energy use) versus termination temperatures of 42°, 38°, 32°, and 27°C (108°, 100°, 90°, and 81°F), respectively. The optimal set points were based on an examination of storage energy discharge versus recovery (energy charge) time and were 38° to 32°C (100° to 90°F). This study indicates that actual measurements may be required to ascertain the effects of changes in set points.

Component failures often go unnoticed by system operators. The NSDN data analysis resulted in the discovery of many component-related failures. One example is a defective modulating valve transformer at a Wyoming site (Wyoming Rural), which redirected heated water from storage back to the heat exchanger coil. This flow prohibited heat from being withdrawn from storage, and maintained storage temperatures near 82°C (180°F), thus adversely affecting collector operation. Following valve repair, storage temperature and system operation returned to normal. Another example occurred at the Pennsylvania site. Bell Telephone of Pennsylvania experienced a failure of the collector differential control unit, which was not repaired for several months. The collector control device failed to activate the collector pump when proper conditions were met, so site personnel operated the collection subsystem manually. The collection subsystem efficiency increased from 16% to 30% when the control circuits were replaced, indicating that well-operating differential control should be more effective than manual controls based on operator judgment.

14.5.5 Load

This section discusses the magnitude and timing of energy demands on solar energy systems.

Heating systems and combined heating/DHW systems generally have higher loads during evening hours, coinciding with greater temperature difference between interior and exterior. This peak-loading necessitates greater storage capacity for heating systems than for DHW alone and increases the importance of storage discharge rates.

An NSDN site in Utah (Cushing Home) using a radiant floor heating distribution system experienced a problem with energy removal from storage. The collector performance was quite adequate during the 1981–1982 heating season, collecting an average of 116 MJ (109,968 BTU) per day, or 2.7 MJ/m^2 day (237.9 BTU/ft^2). The overall system load was 200 MJ (189,600 BTU) per day, or 4.7 MJ/m^2 day (414.1 BTU/ft^2) of collector area. Measured energy removal from storage was 53 MJ (50,244 BTU) per day, of which 42 MJ (39,816 BTU) were used for space heating. This removal represents a system efficiency of 8%, without considering the effect of storage energy loss.

The actual heat removal rate from storage was lower than the collection rate of charging. If the storage heat exchanger capacity were doubled, for example, this change would improve energy removal from storage and increase overall system efficiency. If a system cannot use stored energy, it must increase auxiliary consumption to make up the difference. System efficiency at this site would increase dramatically when a portion of storage losses entered the living space.

14.5.6 Corrosion

Corrosion often causes component failures. Vitro (1981g) discusses much of what has been learned concerning corrosion in solar systems.

Corrosion, if not checked, can destroy components in liquid solar system. Extent of corrosion depends on many physical and chemical factors, including water properties (hardware, ion content, pH, buffering capacity, etc.), temperature and pressure in the system, admission of air (oxygen), and system design (dissimilar metals).

Corrosion is the chemical reaction between a metal and its environment. Scaling is the deposition of a solid material on a metal surface, usually with no chemical reaction. Erosion is the physical removal of surface metal as a result of turbulence or particulate activity.

A number of systems failed because of component corrosion, scaling, and erosion. Corrosion-inhibiting practices are discussed in the literature; the following do's and don'ts from Vitro (1981) presented some of these practices. Do:

1. Use deionized water whenever possible;

2. Maintain the pH slightly on the alkaline side (i.e., between 7.5 and 9);

3. Exclude oxygen and carbon dioxide from the system whenever possible;

4. Use electrically insulated unions (or flanges) whenever connecting dissimilar metals;

5. Use a rust inhibitor containing a surface-coating agent, such as sodium silicate, in systems containing aluminum;

6. Use an inhibited glycol-base antifreeze where necessary and follow the manufacturer's instructions.

Do not:

1. Allow pH to fall much below seven, particularly if there is any iron in the system;

2. Allow pH to rise above nine if the system contains aluminum;

3. Use water from a water softener, particularly if the system contains copper;

4. Allow concentrations of heavy metal ions, such as mercury, copper, lead, tin, etc., to persist in aluminum systems;

5. Allow sodium silicate rust inhibitors to stay in a system for more than two years;

6. Use rust inhibitors with sodium silicate in systems that do not contain aluminum.

14.6 Conclusions

There are some significant observations to be made from the data presented in this chapter. The most striking one is that no specific type of solar system has proved superior for space heating applications. An examination of the top three solar systems in each of the five primary performance factors—collector efficiency, collector COP, solar fraction, percent of incident solar radiation delivered to load, and savings—shows that a variety of system types and components are represented. The list includes both air and liquid systems; flat-plate and trickle-down collectors, some with selective coatings and some without; and rock (both lateral and vertical flow types) and water storage, even one with both rock and water storage. Delivery of the solar energy to the load is accomplished by forced air, hot water baseboard, and solar-assisted heat pumps. The auxiliary systems include oil, natural gas, heat pump, and electric resistance heat.

Based on the previous observations, it can be concluded that good solar performance does not depend as much on the type of system or subsystems used as it does on the skill of the designer in selecting, sizing, and combining solar system components for a specific application. Second, using top-quality hardware, especially controls and sensors, is mandatory for good system performance. A review of the systems that showed the lowest efficiencies and energy savings indicates that the most common cause of poor performance was a controller that failed or malfunctioned. The third important requirement for good system performance is proper installation and adjustment of the system. Finally, the frequency of noncatastrophic failures, such as early/late startup and shutdown of the system and sticking or leaking valves and dampers, suggests the need for routine system maintenance. This maintenance should not be limited to a cursory check but should include instrumented measurements of temperatures, flow rates, and pressure drops.

The data also suggest the following to be true (however, a broader data base is required to confirm these statements):

1. System performance can be improved by keeping the collector area-to-load ratio low.

2. Collector efficiency is improved by maintaining low collector inlet temperatures. The data also suggest that low collector inlet temperatures produce better overall system performance; however, the resulting lower outlet temperatures may actually reduce overall performance if the system is not designed to use heat effectively at the lower delivery temperatures.

3. A significant improvement in space heating efficiency can be achieved by providing for space heating directly from the collectors when solar energy is available.

4. A solar system that is oversized in relation to the heating load is less cost-effective than a smaller system that operates more nearly at full capacity for a large fraction of the year. Equipment use is thus maximized, and the value of energy saved in relation to system cost is maximized.

5. Solar systems should be designed to prevent excessive heat losses and to use any heat losses that do occur to meet heating loads.

6. Space heating systems that also provide heat for DHW should have a summer mode that bypasses the space heating storage, since losses from storage can cause an added cooling load in the summer.

7. Performance of air systems was as good or better as liquid system performance.

   In conclusion:

1. Systems that were based on long-standing and recognized design and installation practices, that used proven system design, and that were properly controlled performed well.

2. Many of the problems associated with operating solar systems resulted from failure to follow conventional HVAC (heating, ventilating, and air-conditioning) practices. Other problems resulted from:

- Use of inadequate or nonexistent specifications
- Use of inadequate or erroneous design tools or methods
- Unacceptable cost reduction attempts
- Lack of detailed design and/or planning
- Improper or nonexistent maintenance
- Lack of availability of maintenance or operating manuals

3. Solar-related problems that reduced the performance of solar heating systems included the following:

- Inadequate or erroneous application of solar design methods

- Faulty selection and integration of system components, specifically, the poor matching of components with load and with other components within the solar system

• Inappropriate or unacceptable components (e.g., collectors) that contain design flaws

• Design and installation of innovative, unique, or "experimental" systems without adequate testing, control, and/or instrumentation to ensure proper operation

• Faulty installation procedures

• Poor selection of operating modes

• Insufficient analysis of system hydraulics

4. The major factors that resulted in reduced or unacceptably low performance included the following:

• Oversizing of solar system in relation to heating load

• Excessive thermal losses from the system to the interior and exterior of the conditioned space

• Unacceptable and excessive electric energy consumption in operating the solar system

• Lack of proper controls in operating the solar system and specific solar components

• Lack of adherence to architectural constraints

5. Several active space heating systems that received careful attention to detail in their design, installation, and operation performed well.

6. Active space heating systems using air-heating solar collectors with pebble-bed storage, and water-heating solar collectors with water storage performed with equivalent efficiencies and savings in nonrenewable energy resources. Neither of these two major types of solar heating systems was shown to operate at significantly higher performance levels.

References

Eck, T. 1981. The all-day efficiency of collector arrays in the National Solar Data Network. *National Solar Data Program performance results, vol. 3*. SOLAR/0005–81/81. Silver Spring, MD: Vitro Laboratories.

Howard, B. D. 1981. Microclimate of solar collection installations. *National Solar Data Program performance results, vol. 1*. SOLAR/005–81/36. Silver Spring, MD: Vitro Laboratories.

Kennedy, M. J., S. J. Sersen, and S. M. Rossi. 1980. *Comparison of liquid solar thermal storage subsystems in the National Solar Data Network*. ASME 80–WA/Sol–36. Silver Spring, MD: Vitro Laboratories.

Vitro Laboratories. 1980. *The National Solar Data Network.* SOLAR/0003–80/17, prepared under DOE contract no. DE–AC01–79CS30027. Silver Spring, MD: Vitro Laboratories.

————. 1981a. *Solar energy system performance evaluation, Terrell E. Moseley, Inc., October 1979 through April 1980.* SOLAR/2011–81/14. Silver Spring, MD: Vitro Laboratories.

————. 1981b. *Solar energy system performance evaluation, Brookhaven National Laboratories, October 1980 through August 1981.* SOLAR/2046–81/14. Silver Spring, MD: Vitro Laboratories.

————. 1981c. *Solar energy system performance evaluation, RHRU, Clemson, November 1979 through April 1980.* SOLAR/2086–81/14. Silver Spring, MD: Vitro Laboratories.

————. 1981d. *Solar energy system performance evaluation, Lawrence Berkeley Laboratory, June 1980 through July 1981.* SOLAR 2050–81/14. Silver Spring, MD: Vitro Laboratories.

————. 1981e. *Solar design and installation experience: An overview of results from the National Solar Data Network.* SOLAR/0009–81/37. Silver Spring, MD: Vitro Laboratories.

————. 1981f. *Guidelines for selecting a solar heating, cooling or hot water design.* SOLAR/0091–81/13. Silver Spring, MD: Vitro Laboratories.

————. 1981g. *Corrosion and scaling in solar heating systems.* SOLAR/0909–81/70. Silver Spring, MD: Vitro Laboratories.

Waterman, R. E. 1981. Solar energy control systems installation and modification, and effects on system performance. *National Solar Data Program performance results, vol. 1.* SOLAR/0005–81/36. Silver Spring, MD: Vitro Laboratories.

15 Costs of Solar Space Heating Systems

Thomas King and Jeff Shingleton

15.1 Introduction

This chapter addresses the construction cost of solar space heating systems, primarily in nonresidential applications, and is a companion to chapter 10, which focuses on the cost of solar water heating systems but also includes ' information on a broad range of other systems applications and subsystems. The reader is referred to chapter 10 for information and discussion on:

• the general topic of construction cost and its relation to system cost effectiveness

• comparisons between the costs of different system applications (space heating, water heating, cooling)

• data on subsystems that are essentially common to all system applications (e.g., collectors, support structure, storage systems)

• cost estimating guidelines

Unfortunately, the construction costs of solar space heating systems have been defined accurately for relatively few installed systems. Data are limited to several commercial systems from the National Solar Heating and Cooling Demonstration Program (King et al. 1979). These projects from the demonstration program were installed between 1976 and 1979. Comparable detailed cost data are not available for residential applications. Thus, this chapter presents data almost exclusively for nonresidential applications.

The cost analysis procedures included site visits, discussions with construction contractors and subcontractors who actually installed the systems, and detailed review of their records. The primary aim of the cost analysis procedures was determination of accurate total system costs. However, in all cases, subsystem level data have been obtained for the following subsystems:

• collector array

• support structure

• energy transport (piping and pipe insulation)

• storage

• electrical and controls
• general construction

The "demonstration" or research effect on costs was minimized in the analysis through the following methods:

1. No costs for instrumentation other than for normal control were included. Many of the early systems were instrumented extensively through the National Solar Data Network (DOE 1979).

2. Many different types of organizations were involved in the construction of the systems. To permit better comparisons, the effect of the widely varying overhead of these organizations was eliminated by identifying "bare" construction costs and multiplying the bare costs by an overhead and profit margin typical for the construction industry.

3. The costs presented do not include design costs.

4. Auxiliary energy system costs are not included.

All the costs are presented in terms of 1981 dollar values by utilizing the construction cost index to account for inflation during the period. In addition to the detailed construction cost data obtained for individual installations, this chapter also discusses several other projects that involved cost analysis. For instance, the DOE 1983 Active Program Research Requirements program (APRR) included estimates of the cost of then-current and projected solar energy systems (Science Applications 1983).

The 1981 Active Solar Installation Survey also collected data on solar space heating systems as well as other types, and the data, although collected through less rigorous procedures, are compared here to the data from the demonstration program (Applied Management Science 1983).

15.2 Presentation and Analyses of Cost Data

This section provides a description of the characteristics of the systems for which cost data have been collected, the detailed cost data for each system in the data set, and various analyses of system and subsystem cost trends. A more detailed analysis of costs for subsystems that are "generic" to all commercial solar thermal systems can be found in chapter 10 and are not repeated here. The subsystems that have been considered "generic" and not significantly different between heating and hot water systems are collector array, support structure, and storage.

The data are presented here in two forms. Tables are provided that contain all the detailed cost data for each system organized according to system application and collector type. Trends identified in these detailed data are the basis for the discussions provided. In addition, the availability of these detailed data makes it possible for the reader to perform independent analyses. The data are presented in the form of bar graphs, which illustrate the range of costs associated with various systems, subsystems, and applications. The bar graphs are intended to facilitate the visualization of trends and are not particularly precise. The reader is cautioned to rely only on the detailed data tables for accurate individual cost data.

15.2.1   Presentation of Data Tables

Table 15.1, Summary of System Characteristics, and table 15.2, Summary of System Costs, provide descriptive information and basic cost data for

Table 15.1
Summary of system characteristics

System name	Solar application	New/retro	Area ft^2	Collector type
Moseley	SH, HW (LIQ)	R	376	FPL
Telex	SH, HW (LIQ)	R	10700	FPL
Billings	SH, HW (LIQ)	R	1660	FPL
Charlotte	SH, HW (LIQ)	R	3950	FPL
Blakedale	SH, HW (LIQ)	N	928	FPL
Averages	SH, HW (LIQ)		3523	
Howard's Grove	SH (AIR)	R	2357	FPA
DuCat	SH (AIR)	R	7000	FPA
Aberdeen	SH, HW (AIR)	R	1260	FPA
Scattergood	SH, HW (AIR)	R	2240	FPA
Concord	SH (AIR)	R	1737	FPA
Averages	SH (AIR)		2919	
SH averages			3221	

Source: King et al. (1979).
Notes:

Funding Program: SHAC, Solar Heating and Cooling Demonstration Program.

Solar Application: HW—Service hot water
                                  SH—Space heating
                                  C—Space cooling
Collector type: FPL—Flat-plate, liquid
                              FPA—Flat-plate, air

Table 15.2
Summary of system costs, 1981 dollars

Project number	System name	Cost $/ft²	Collector $/ft²	Array %	Support structure $/ft²	%	Energy $/ft²	Transport %	Thermal $/ft²	Storage %	Electric & controls $/ft²	%	General $/ft²	Construction %
1	Moseley	47.40	13.80	29.11	4.60	9.70	16.00	33.76	5.10	10.76	6.40	13.50	1.10	2.32
2	Telex	53.30	17.30	32.46	9.20	17.26	20.70	38.84	2.10	3.94	2.80	5.25	1.30	2.44
3	Billings	56.80	20.40	35.92	10.80	19.01	19.80	34.86	2.50	4.40	3.50	6.16	0.00	0.00
4	Charlotte	77.50	18.00	23.23	8.20	10.58	35.10	45.29	4.60	5.94	4.60	5.94	7.10	9.16
5	Blakedale	86.00	12.00	13.95	21.40	24.88	24.20	28.14	11.80	13.72	11.80	13.72	0.10	0.12
	Averages	64.20	16.30	29.93	10.84	16.29	23.16	36.18	5.22	7.75	5.82	8.91	1.92	2.81
6	Howard's Grove	26.20	13.40	51.15	3.30	12.60	4.40	16.79	3.80	14.50	1.30	4.96	0.00	0.00
7	DuCat	41.60	21.90	52.64	12.90	31.01	4.90	11.78	0.00	0.00	1.90	4.57	0.00	0.00
8	Aberdeen	59.30	24.80	41.82	1.90	3.20	18.40	31.03	7.80	13.15	4.70	7.93	1.30	2.19
9	Scattergood	61.00	24.50	40.16	16.00	26.23	11.20	18.36	5.30	8.69	3.80	6.23	0.30	0.49
10	Concord	87.80	18.60	21.18	35.70	40.66	17.10	19.48	7.20	8.20	8.60	9.79	0.60	0.68
	Averages	55.18	20.64	41.39	13.96	22.74	11.20	19.49	4.82	8.91	4.06	6.70	0.44	0.67
	SH averages	59.69	18.47	34.16	12.40	19.51	17.18	27.83	5.02	8.33	4.94	7.81	1.18	1.74

Source: King et al. (1979).

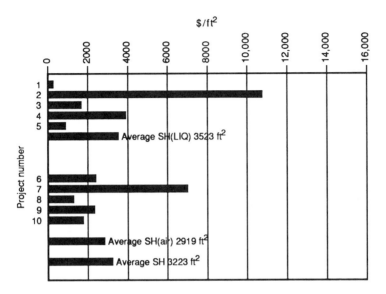

Figure 15.1
Array size vs. application.

each solar space heating system that has been used in the analysis (King et al. 1979). The systems are listed generally in order of increasing total system cost.

As described in section 15.1, every system for which detailed construction cost data are available has been funded for construction by the National Solar Heating and Cooling Demonstration Program (SHAC). Only those data that are specifically related to solar heating systems are discussed here.

In table 15.1 and figure 15.1, the systems in the data set for solar heating exhibit a broad range of array sizes, from 350 to 10,700 square feet (King et al. 1979). In addition, within both liquid-based SH systems (SH-LIQ) and air-based SH systems (SH-AIR), there is a broad range of system sizes. This is illustrated by figure 15.1, a bar chart of array size versus system application. SH-LIQ systems use flat-plate liquid (FLP) collectors and evacuated tubular (ET) collectors, and SH-AIR systems all use flat-plate air (FPA) collectors. The SH(AIR) systems, with an average size of 2919 square feet are slightly smaller than the SH(LIQ), which average 3523 square feet.

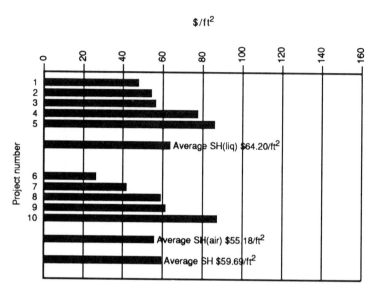

Figure 15.2
Total system cost vs. application.

15.2.2 Total System Costs

Figure 15.2 is a bar graph of the total system cost data from table 15.2 for SH systems. SH-LIQ systems averaged $64.20/ft^2, SH-AIR systems averaged $55.18/ft^2. The reader is cautioned, however, to consider not only the average but also the range of system costs within each group. Chapter 10 contains a discussion of economy-of-scale effects in the cost of solar thermal systems as well as other trends. The balance of this section presents analyses of those cost trends that are identifiable for SH systems. These include trends in the cost of the collector array, support structure, and storage subsystems.

15.2.3 Energy Transport Costs

The data in table 15.2 and in figure 15.3 suggest that energy transport costs are strongly related to the system application. Energy transport costs average $23.16/ft^2 for SH-LIQ systems (which is close to the $17.13/ft^2 average for HW systems in chapter 10), and $11.20/ft^2 or SH-AIR systems. Energy transport represents about 28 percent of the total system costs for space heating systems in general, though this percentage is much lower for

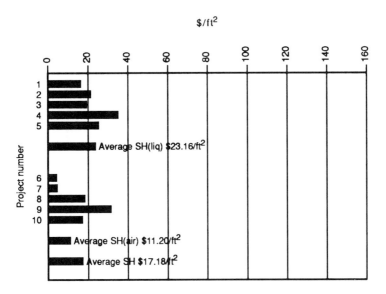

Figure 15.3
Energy transport cost vs. application.

SH-AIR than SH-LIQ systems, with averages of about 19 percent and 36 percent, respectively.

15.2.4   Electric and Control Costs

Electric and control costs tend to be primarily dependent on the system application, not the collector type, and a discussion of these costs for HW and STEAM systems can be found in chapter 10. Figure 15.4 presents a bar graph of electric & control costs. These costs for the SH-LIQ and SH-AIR systems are very similar with averages of $5.82/ft^2 and $4.06/ft^2.

Figure 15.5 is a bar graph of the percentage of total system cost that the electric and control costs represent for SH systems. Percent of total may provide a better measure of electric and control costs than system size, as evidenced by the tight ranges on figure 15.5. SH-LIQ and SH-AIR electric and control costs accounted for averages of 8.9 and 6.7 percent of total system costs.

15.2.5   General Construction Costs

The category of general construction costs includes any construction that is not directly related to a specific subsystem in a particular solar thermal

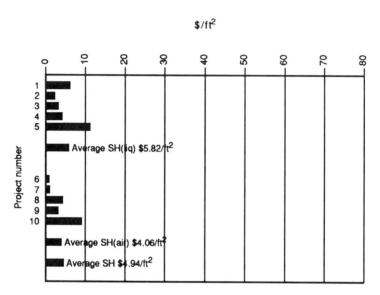

Figure 15.4
Electric and controls cost vs. application.

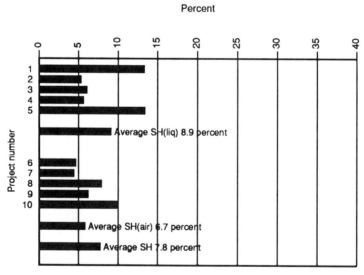

Figure 15.5
Electric and controls cost percentage vs. application.

system but which is necessary for the installation of that solar thermal system at a particular site. These costs are therefore very site-specific. Many SH systems had no general construction costs.

For those SH systems with general construction costs, these costs were generally less than $2.00/ft^2, and less than 3 percent of total system construction costs (which is consistent with those costs for HW systems which had general construction costs).

15.3   Other Detailed Cost Data

15.3.1   Background

During 1982, the DOE Office of Solar Heat Technologies sponsored a project to identify the research priorities in active solar thermal energy systems (Science Applications 1983). That project, known as the APRR project, took a "top-down" approach and centered around system concepts in order to identify areas of potential cost or performance improvement in solar thermal energy systems. Research activities were defined and evaluated with respect to their potential for contributing to the development of one or more cost-competitive active solar thermal systems.

The APRR current system cost estimates were based primarily on manufacturer and installer surveys and conventional construction cost-estimating manuals rather than on actual construction cost data from completed systems.

The APRR cost estimates compare favorably with total system cost data from actual systems that are presented in section 15.0. This is illustrated in table 15.3, which contrasts summary cost estimates from the APRR project (organized according to the cost categories previously used) with average cost data from similar system types, developed from actual construction cost data from section 15.2. The APRR total system cost estimates for combined space heating and hot water systems, ranging from $49.54/ft^2 to $66.94/ft^2, compare well to the average of total system cost for systems, $59.69/ft^2, presented earlier. Table 15.3 shows similar good agreement between estimated and actual subsystem costs.

15.3.2   Active Solar Installation Survey

The Office of Solar Heat Technology and the Energy Information Administration sponsored the EIA-701 program, known as the Active Solar

Table 15.3
Summary of selected APRR project results vs. actual construction cost data

Application (1981 dollars)	APPR case #	Area ft²	Total $	Cost $/ft²	Collector type	Collector and support structure		Thermal $/ft²	Storage %	Energy $/ft²	Transport %	Electric & controls	
						$/ft²	%					$/ft²	%
Residential SH & HW	1.00	250.00	13295.00	53.18	FPA	26.25	49.36	7.74	14.55	17.08	32.12	2.10	3.95
Residential SH & HW	2.00	250.00	16736.00	66.94	ET	38.25	57.14	8.29	12.38	16.60	24.80	3.80	5.68
Residential SH & HW	3.00	250.00	14236.00	56.94	FPL	28.25	49.61	8.28	14.54	16.60	29.15	3.80	6.67
Commercial SH & HW		800.00	39634.00	49.54	FPL	33.23	67.07	4.90	9.89	10.22	20.63	1.19	2.40
Construction cost averages for 10 commercial SH (& HW) systems in table 15.1		3221		59.69		30.87		5.02		17.18		4.94	

Source: Science Applications (1983).

Installation Survey (Applied Management Sciences 1983). The EIA and contractors performed surveys of the activities of private solar thermal system installation contractors during 1980 and 1981. In addition to publishing the results of the individual year surveys, the EIA published results of a detailed comparison made of both years' survey data. The discussion here is concerned primarily with the results of the 1981 survey.

Detailed questionnaires were mailed to every known solar thermal system installation contractor in the United States. The questionnaires solicited information concerning level of business activity with several categories of solar thermal systems. The questions were structured so that the EIA could perform statistical analyses of the results to determine state, regional, and national patterns in system-installation size and cost, and in numbers of installations for residential and commercial pool heating, hot water, space heating, and cooling systems. These categories were further subdivided to show trends in both single-family and multifamily hot water residential applications.

Some of the survey cost data are shown in table 15.4 for space heating systems and combined water and space heating systems for single-family, multifamily, and commercial applications. Review of these data reveals several features:

1. The costs presented for each application varied greatly between 1980 and 1982, as much as 112% in the case of commercial space heating systems.

2. The large differences between space heating and combined (water and space heating) system costs do not seem reasonable. For instance, multi-

Table 15.4
Shift in the average cost of active solar installations between 1980 and 1981 in dollars per square foot (based on *Survey Data*)

| Applications | Market sector | | | | | | | | |
| | Single-family | | | multifamily | | | Commercial | | |
	1980	1981	%	1980	1981	%	1980	1981	%
Space heaters	24.60	32.71	+33	16.68	20.95	+26	15.57	33.01	+112
Combined (water & space)	22.57	43.14	+91	30.83	36.65	+19	40.44	29.85	−26

Source: Applied Management Science (1983).

family space heating systems in 1980 are listed at $16.68, while combined systems are listed at $30.83.

3. Some extremely low costs are presented. As an example, commercial space heating systems in 1980 are listed at $15.57. What is known about the costs of collectors and other subsystem costs as previously presented in this chapter and in chapter 10 makes it appear that such low total system costs are unreasonable.

4. The reasonable agreement of costs reported in the National Solar Data Network program (table 15.1 and table 15.2) and in the APRR studies (table 15.3), showing average solar space heating system costs approximating $60 per square foot of collector, lends confidence to that general level of costs and places the lower costs reported in the Active Solar Installation Survey (table 15.4) in considerable doubt.

The survey approach relies on the accuracy and motives behind the individual survey responses, and there is a lack of connection between reported costs and system performance. The degree to which the systems reported on reflected the state-of-the-art is unknown. On the other hand, the data collection in the National Solar Data Network continues to be of value for estimating the cost of commercial systems.

References

Applied Management Sciences. 1983. *Key trends in active solar installation activities between 1980 and 1981*. Final report, DOE/CE/30692–1.

Dodge manual for building construction pricing and scheduling. 1987. New York: McGraw-Hill.

DOE (U.S. Department of Energy). 1979. *Instrumented solar demonstration systems project description summaries*. DOE Solar/0017–79/34.

King, T. A., et al. 1979. Cost effectiveness—an assessment based on commercial demonstration projects. *Proc. 2d Annual Solar Heating and Cooling System Operational Results Conference*, Colorado Springs, CO, November 1979.

R. S. Means Company, Inc. 1981. *1981 Means mechanical and electrical cost data*. Kingston, MA: R. S. Means.

R. S. Means Company, Inc. 1987. *1987 Means mechanical cost data*. Kingston, MA: R. S. Means.

National construction estimator. 1987. Carlsbad, CA: Craftsman Book Company.

Science Applications, Inc. 1983. *Active program research requirements*. Final report. Washington, DC: U.S. Department of Energy.

16 Community Scale Systems: Seasonal Thermal Energy Storage and Central Solar Heating Plants with Seasonal Storage

Dwayne S. Breger

16.1 Introduction

This chapter documents research projects and technological results directly concerned with central solar heating plants utilizing seasonal storage designed to meet loads of relatively large building complexes or to supply district heating systems. Material for this documentation is derived from experimental studies on storage subsystems, analytical studies on the design, optimization, and operation of systems, and the experience of design, construction, and operation of projects. The latest work in this field has largely been carried out as part of or as input to the International Energy Agency (IEA) Solar Heating and Cooling Program, Task VII on Central Solar Heating Plants with Seasonal Storage (CSHPSS). Only a portion of this task effort has been carried out by the United States as part of the Solar Program; the methods, results, and experience of the Task VII work comes from all participating countries. The most significant project construction and operational experience comes from a number of northern European countries. The United States has made significant experimental contribution on seasonal energy storage on a subsystem level, and these test facilities and studies will be discussed along with the system analytical work performed within the IEA context. The scope of this chapter does not include discussion of community energy systems designed to meet cooling loads with seasonal chill or ice storage, solar total energy, or other cogeneration systems. Also excluded from this discussion is solar pond technology using salt gradients for long-term thermal energy storage.

This introductory section provides a basis for the interest and applicability of CSHPSS, and provides a background and summary of the work performed and its current status. Section 16.2 provides greater detail of the major projects and results of CSHPSS and related research in the United States and in other participating IEA countries.

16.1.1 CSHPSS—Concept and Applicability

Meeting low-temperature building space heat and hot water loads with solar energy is of interest in any energy planning scenario determined to reduce the use of nonrenewable energy resources. Over an annual meteorological cycle, however, the solar availability and space heat load demand are, of course, out of phase. In many geographical regions (particularly northern latitudes), poor winter insolation makes a significant solar contribution to space heating demand particularly difficult. The seasonal storage concept, in contrast to conventional diurnal systems, is shown in figure 16.1. The concept is based on the storage of heat collected during the sunny, warm months for use in meeting winter heating demands. The heat stored can come from a solar array or another source, such as waste heat, an off-peak low-cost heat source, or heat rejected from summer cooling systems. The heat store is gradually charged throughout the spring, summer, and fall months so that a large heat source is available to meet the winter load, independently of short-term meteorological conditions, and is normally designed to meet essentially the entire load (including the constant year-round hot water load).

The collector subsystem is the dominant cost in most solar heating systems. For diurnal storage systems designed to provide a significant portion of a winter space heating load, a large collector array is required, which then remains idle (except to meet the relatively small hot water demand) during the summer months of high insolation. A seasonal storage system allows more or less unconstrained collector operation throughout the year and thus greater utilization of the capital intensive subsystem and a reduction in the required array area. For effective collection during the summer months, numerous collector designs are feasible including low-cost, light-weight unglazed flat-plate collectors. Collector tilt is optimized for maximum annual collection and will thus differ from diurnal system tilt angle. Temperature stratification is normally maintained in the storage to provide a low collector inlet temperature. A heat pump can be a valuable design option to extend the effective storage capacity and reduce the operating temperatures.

Seasonal heat storage can be accomplished in several ways, as illustrated in figure 16.2. Heat can be stored in water in a constructed above- or below-ground tank, or in an excavated or natural cavern or abandoned mine, or in an excavated earth pit that is lined, covered, and insulated. For very large-scale systems, storage may be uninsulated or partially insulated. Solar salt-gradient ponds can also be used but are not included in this discussion. Heated water can be stored in aquifers of suitable characteristics by pumping through wells and displacement of ambient aquifer water. Heat can also be stored directly in earth or rock of suitable characteristics

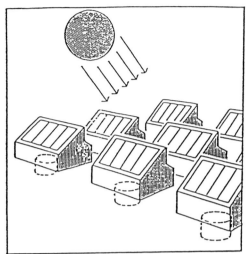

(a) Distributed Solar
and Storage

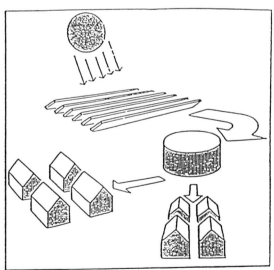

(b) Central Solar Heating Plant
with Seasonal Storage

Figure 16.1
Conceptual comparison: CSHPSS versus distributed solar approach. Source: Breger, Sunderland, and Bankston 1989.

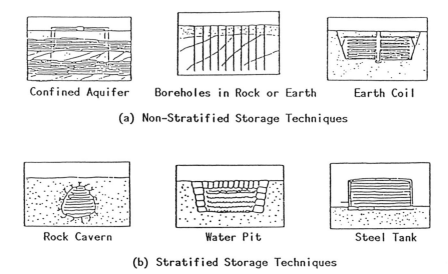

Confined Aquifer   Boreholes in Rock or Earth   Earth Coil

(a) Non-Stratified Storage Techniques

Rock Cavern   Water Pit   Steel Tank

(b) Stratified Storage Techniques

Figure 16.2
Options for large scale seasonal thermal energy storage. Source: Breger, Sunderland, and
Bankston 1989.

with heat exchange accomplished by an array of horizontal pipes (for
earth storage) or vertical boreholes (for earth or rock storage) through
which a heat exchange fluid (water) is circulated. These methods have all
been topics for analytical and experimental efforts which will be detailed
in section 16.2.

Efficient and economically effective seasonal heat storage can be accomplished only on a rather large scale. This characteristic is due to the economies-of-scale in preparing a storage facility (in terms of cost per unit
volume of effective storage) and storage heat loss, which is a function of
storage surface area that decreases rapidly in proportion to the storage
volume with increasing storage size. For these reasons, seasonal storage
systems are designed for large loads, particularly a building or group of
buildings served by a district heating system. A large central collector
array or distributed collectors provide heat to the central storage facility.

The seasonal storage concept is attractive because it provides, with solar energy, major portions of large space heat and hot water loads in
climates with poor winter insolation. Limitations to the CSHPSS system applicability include the need to have a large, dense, low-temperature

heating load (in terms of energy use per land area) and available unobstructed area (on nearby land or roofs) for a large collector array. Storage design, cost, and performance are heavily dependent on local geological conditions, which normally are not known in sufficient detail at a specific site without initial geological testing and evaluation. The hydrogeological conditions are particularly important in terms of storage heat loss to the surrounding earth, which may be prohibitively high if local ground water movement is substantial. Large systems may take several years before the storage heat loss reaches a steady-state condition as the surrounding earth is gradually heated. A CSHPSS system requires a great deal more planning and capital expenditure than conventional heating methods routinely pursued by large-scale commercial and residential developers.

16.1.2 Status of Seasonal Storage and CSHPSS Technology before the IEA Task VII

The U.S. Department of Energy (formerly ERDA) involvement in seasonal thermal energy storage began in 1975. Aquifer thermal energy storage (ATES) was identified as having significant near-term potential and thus has been the focus of major study, experimentation, modelling, and demonstration. This work was originally managed by Oak Ridge National Laboratory and later (1979) by Pacific Northwest Laboratory.

A supplementary seasonal storage program, aimed at solar applications only, was also begun in early 1975, under the direction of Argonne National Laboratory. Projects were sponsored in seasonal storage using underground storage tanks (University of Toronto), earth pit (University of Virginia), and saturated earth (Colorado State University). These projects are discussed in section 16.2.

Early work in aquifer thermal energy storage included a research project begun in 1975 by Auburn University at an ATES test site near Mobile, Alabama. A cycle of heated water injection, storage, and recovery was completed in 1977 and, despite some difficult problems with injection-well plugging and fracturing of the impermeable caprock overlayer, served to demonstrate that ATES could be achieved at low temperatures. At Lawrence Berkeley Laboratory, a numerical hydrodynamic model (CCC —Conduction, Convection, Compaction; later renamed P-T) was adapted to use for simulation and design of ATES in 1977. A survey of ATES applicability in the United States was performed by the Tennessee Valley

Authority in 1978, and several feasibility and design studies were undertaken for heat and chill ATES projects. Major results of this ATES work in the United States will be discussed in section 16.2.

At the Solar Energy Research Institute, CSHPSS system analysis work was performed from 1979 through 1981. A computer simulation model was developed and used to analyze design trade-offs between collector area and storage volume and provide economic and sensitivity analyses. This system approach provided insight to the unique design considerations and operation of seasonal storage systems and was used as initial U.S. input to the IEA Task VII work. Parallel work on system analysis and modelling was performed at the same time by Arthur McGarity at Swarthmore College, Pennsylvania, and was also contributed to the IEA Task effort. Significant results of these works will be summarized in section 16.2.

Prior to the IEA Task VII, Sweden had developed several CSHPSS demonstration projects along with conceptual and feasibility studies. Three projects were constructed and operating in 1980, using thermal storage in constructed water volumes—an earth pit, an excavated (blasted rock) tank, and an above-ground tank. Different collector types and storage and distribution temperature ranges provided a range of design and operational experience. These projects and their operation are briefly described in section 16.2.

16.1.3   IEA Task VII—Description and Summary of Results

Since 1980, a major focus of work on seasonal solar thermal energy storage has been within the International Energy Agency Solar Heating and Cooling Program, Task VII on Central Solar Heating Plants with Seasonal Storage. The objective of this task is to determine the technical feasibility and cost-effectiveness of solar seasonal storage energy systems for large-scale district heating. The participants in the task are Austria, Canada, the Commission of European Communities (CEC), Denmark, Finland, Germany, Italy, the Netherlands, Sweden, Switzerland, the United Kingdom, and the United States (Austria and the United Kingdom participated only in Phase I, the CEC only in Phases I and II, and Finland and Italy only in Phase III). The task work was divided into three phases—Phase I was on preliminary design methods and assembly, and evaluation of subsystem data (costs, performance, etc.); Phase II was on detail design, evaluation of system design concepts, and on system monitoring and evaluation methods. Phase III involved an exchange of design, construction, and monitoring information on existing systems and analysis of improved designs. Below is a summary of the task structure in the three phases.

Phase I

Subtask I(a) System Studies and Optimization

Subtask I(b) Solar Collector Subsystem

Subtask I(c) Heat Storage Subsystem

Subtask I(d) Heat Distribution System

Subtask I(e) Preliminary Site Specific System Design

Phase II

Subtask II(a) MINSUN (Simulation Program) Support

Subtask II(b) Evaluation of System Concepts

Subtask II(c) System Monitoring and Evaluation Methods

Phase III

Assessment of General Conditions for CSHPSS

Analysis and Evaluation of Existing Projects

Analysis and Evaluation of Planned Projects

Generic Systems Analysis

Phase I was completed at the end of 1983, Phase II at the end of 1985, and Phase III (and Task VII) at the end of 1989. The results of each subtask are published in official IEA reports, and other associated work is found in task working documents and published national reports. A new IEA Task XV on advanced CSHPSS systems is planned to follow Task VII and will continue the international collaboration on CSHPSS.

The overall results and significant system studies of the task are summarized in section 16.2. Phase I results were mainly a compilation of cost, performance, and design information for the major system components or subsystems, which could then be used as a data base for system evaluation and analysis. The task developed the MINSUN computer simulation model specifically for CSHPSS systems and has actively enhanced its capabilities. The program has been used extensively by the Task VII in all phases for system design, analysis, and optimization, and evaluation and comparison of system concepts. In Phase I, each country contributed a National Application Case using MINSUN, which provided a range of system design studies and tested the application of MINSUN and the Phase I subsystem data base.

In Phase II, system concept evaluation was performed for systems based on water, aquifer, and duct (earth or rock) storage systems. Other design parameters analyzed included use of a heat pump, meteorological conditions, and load distribution temperature. This work is fully summarized in section 16.2. Within the context of Phase II, several countries proceeded to perform National Assessments of CSHPSS, using specific national costs, climate, and design parameters. The U.S. participants performed an assessment of systems specifically in the New England region, using duct storage in rock and a heat pump. The U.S. national study is also summarized in section 16.2. A result of Phase II was also the development of system monitoring and evaluation requirements for R&D purposes and specific reporting formats to be used in Phase III CSHPSS projects. These requirements and formats will be used to assure that a data exchange can continue between participants in a consistent, comparable, and detailed manner.

The results of Phase III included a detailed analysis and evaluation of over ten operating systems and the design analysis of new projects. The Phase II system concept evaluation was extended into a generic analysis of the most appealing CSHPSS designs and the redesign of some existing projects under new conditions. During Phase III, two CSHPSS design studies were initiated in Massachusetts with the objective of developing a first U.S. project. These studies are summarized in section 16.2.

16.1.4 Status of Other U.S. Seasonal Storage Studies

Separate from, or in association with, the IEA task work, several seasonal storage projects have been undertaken in the United States. These include system analytical and design studies, numerical modelling, and simulation of storage systems, and experimental aquifer and other storage test facilities. The description and results of these projects are provided in section 16.2.

Since 1979, Pacific Northwest Laboratory has acted as the lead laboratory for the U.S. DOE program in underground thermal energy storage. This effort includes the Seasonal Thermal Energy Storage (STES) program. Federal funding for this program was substantially reduced during the early 1980s. The STES effort emphasized Aquifer Thermal Energy Storage (ATES) systems, and most recent efforts have concentrated on field and laboratory research on high-temperature ATES systems and on the technology assessment and development of nonaquifer STES concepts.

The goals of the U.S. effort in the seasonal storage program are to reduce the technical and economic uncertainties that inhibit development and implementation of the technology in the private sector. The results of the IEA Task VII Phase II work have provided reason for optimism regarding the economic competitiveness of CSHPSS systems, though an effort to expose these results to private developers is needed. A demonstration or field experiment system is of great interest for the purpose of meeting these goals, though suitable funding and arrangements have not yet been found.

16.2 Description of Seasonal Storage and CSHPSS Research Activity

In this section, the major research activity relating to seasonal storage and CSHPSS is documented. Activities are described and major results discussed, though detailed information must be sought from the reference list at the end of the chapter.

The organization of the section is as follows. First, section 16.2.1 documents research on the seasonal storage subsystem, particularly on Aquifer Thermal Energy Storage (ATES), which has been the concept emphasized in the DOE program. The numerical methods and simulation procedures developed to analyze the storage system will first be discussed, followed by the analytical and experimental studies. Section 16.2.2 will then document CSHPSS system studies within the United States and as part of the IEA Task VII work. Again, numerical/simulation methods developed will be discussed first, followed by system analytical and design studies and assessments. Section 16.2.3 provides brief documentation on experimental/demonstration projects undertaken by the DOE program and within the participating IEA countries.

16.2.1 Seasonal Storage Subsystem Studies

16.2.1.1 Numerical Models and Simulation Programs

P-T (PRESSURE-TEMPERATURE) MODEL   The P-T model (formerly CCC) was developed by Lawrence Berkeley Laboratory (LBL) for the ATES program (Mercer et al. 1981; Tsang et al. 1978). The model is an integrated finite-difference model in one, two, or three dimensions, which simulates transient pressure and temperature fields. It is the most thoroughly tested multidimensional ATES model and has been used extensively to model activity at the Mobile Aquifer Field Test Facility (FTF) and the St. Paul, Minnesota, ATES FTF. In addition, a simplified version was used to generically examine the energy recovery factor as a function of aquifer properties and storage parameters (Tsang, Buscheck, and Doughty n.d.). In this study, graphical presentation showed the influence and sensitivity of these factors and indicated the optimized combinations of geological and operating conditions.

The P-T model was used to predict results of temperature and total energy recovery factor for the first and second Mobile FTF test cycles and showed excellent agreement (Tsang, Buscheck, and Doughty 1980). This success demonstrated that the main physical properties occurring at this FTF are probably well understood. The model was then used to study alternative injection and production schemes (accounting for the strong buoyancy flow) to maximize the recovery factor for a third test cycle (see discussion in section 16.2.1.2).

CFEST MODEL   The CFEST (Coupled Fluid, Energy, and Solute Transport) code was developed at Pacific Northwest Laboratory (PNL) (Gupta et al. 1982). It is a multidimensional finite-element model of heat, fluid, and solute flow in a confined aquifer. The code has been used extensively for simulation of chemical transport characteristics in an aquifer. The code development and testing was undertaken to provide capability for the evaluation of experimental designs and field data. It can be used to gain quantitative understanding of the aquifer environment and the complex physical and chemical interactions within the system.

AFM (AREAL FLOW MODEL)   The AFM is a simple numerical model developed at PNL, capable of analyzing injection pumping policies in a multi-well ATES system. The model couples basic aquifer performance characteristics and economic considerations in a user-oriented code. Aquifer system operation (temperatures and flow rates) are controlled by the user to analyze trade-offs and simulate management decisions.

AQUASTOR   The AQUASTOR code was developed at PNL as a model for cost analysis of ATES systems coupled with a district heating (or cooling) system (Brown, Huber, and Reilly 1982). The model is not so much a model of the aquifer storage as a tool to analyze the overall system cost broken down into its basic subsystems—energy supply, well-field and aquifer store, distribution, and heat exchangers. Economic and financing data are input and can cover a wide range of financial and tax structures and economic scenarios. The cost of energy supplied to the aquifer storage facility is input (be it from solar, waste heat, cogeneration, etc.), though no explicit energy supply system is modeled. Climate data are input to calculate annual residential district or single point loads based on degree-days. The distribution system is modeled as the combination of various district elements, each characterized by its load and geometric layout. The aquifer is not modeled by detailed geohydrological simulation but through representation by three parameters: fluid recovery fraction, thermal storage efficiency, and natural (ambient) aquifer temperature. The model is particularly useful in analyzing the cost effect of varying a single parameter or group of parameters and has been so used, as described in the economic assessment summarized in section 16.2.1.2.

LUND UNIVERSITY CODES   A set of codes developed at Lund University (Lund, Sweden) are AST (aquifer storage model), DST (duct storage model), and SST (stratified storage model) (Hellström and Eftring n.d.; Hellström, Bennet, and Claesson 1982; Hellström 1982). The three codes each have versions that can run independently or in a subroutine form for use with TRNSYS or MINSUN.

AST simulates thermal energy storage in a confined aquifer by dealing with the thermal processes in the aquifer and surrounding ground. A single well is modeled for injection and extraction of water. An implicit second well can be used conceptually as a water supply and dump to and from the storage well. AST assumes negligible ground water flow, negligible buoyancy flow, and negligible disturbance from other wells.

The DST model simulates heat storage in a region of rock or soil. Heat is injected and extracted through a duct system in which a heat carrier fluid is circulated. The model accounts for the "global" temperature variation from the store center to its boundary and into the surrounding ground. The "local" heat transfer to the store and temperature distribution around and along each duct is also modeled, and all these processes are fitted together by superposition. The storage region is described by user input for number and depth of boreholes and total store volume.

Detailed parameters are input to define the regional ambient characteristics and heat transfer process between the ducts and ground.

The SST model simulates heat storage in a water-filled tank, cavern, or pond. Water is taken from the bottom of the tank, heated, and returned to the top. Heat extraction is accomplished by reversing this circulation. The temperatures in the storage volume are stratified in horizontal layers. The thermal process in the storage volume and in the surrounding ground are coupled by the heat flow through the store boundary. The model assumes a storage shape of a vertical cylinder and, for stratification, is divided into layers of equal thickness. Heat loss to the surrounding ground decreases over the first few annual cycles until a steady-state condition is approached. A simplified preheating simulation can be specified to provide a steady-state simulation.

16.2.1.2   Analytical and Experimental Studies

ATES THERMAL BEHAVIOR PARAMETRIC STUDY   Study of ATES thermal behavior was performed at LBL during the 1970s and early 1980s. The main purpose and results of this work were the development of characterization techniques to describe the hydrothermal behavior of ATES systems. This work allows graphical representation for a range of ATES cases of system production temperatures and energy recovery factors in terms of a few dimensionless parameters. This procedure readily allows prediction of basic ATES performance for a specific site, given some general site-characteristic parameters. These studies were extended to consider site-specific analysis, particularly of the Mobile FTF test cycles.

The main study (Doughty et al. 1981) assumes a single-well ATES system with negligible regional ground water and buoyancy flow, and a cycle of injection, storage, production, and rest. The calculations provide the time-varying temperature of the water extracted during the production period. The energy recovery factor for each cycle is determined from the injected and produced energies in reference to ambient aquifer conditions. The calculations were made using a steady-flow simulation model for a large number of ATES systems. Graphical results of thermal behavior are presented in terms of the dimensionless groups and some individual parameters. In addition, other effects including unequal cycle period lengths, finite caprock thickness, and long-term effects of multiyear cycles are determined.

Conclusions from the study include the following. Storage volume is a fundamental system parameter and must exceed a minimum value to attain a good recovery factor; thus, for a seasonal cycle, ATES must be applied on a large scale. The relative duration of injection, storage, production, and rest periods of a cycle has a relatively small influence on the energy recovery factor. The recovery factor increases with the number of cycles due to the increase of residual heat in the surrounding ground. The heated volume within the aquifer should have as compact shape as possible to reduce heat loss. The recovery factor reaches a flat maximum when the radius of the heated volume is about 2/3 of the aquifer thickness.

ATES REGIONAL ASSESSMENT   An assessment of the suitability of aquifers for ATES systems was performed by Century West Engineering Corporation for PNL (CWEC 1981). The assessment was done on a regional basis for each of the twelve identified major ground-water regions in the United States (10 continental, Alaska, and Hawaii). For each region, the major water-bearing rock formations are identified and defined, selected aquifers are evaluated for their individual ATES potential, and hydrological data are tabulated and referenced. It is noted that small local aquifers are not covered by the study but may be suited for ATES and that the major aquifers identified may have local characteristics that are inconsistent with the general ATES potential indicated. The goal of this assessment was a qualitative description of ATES potential of major aquifers as a precursor to site-specific studies.

The initial task in this assessment involved a literature search and development of a bibliography. Information was obtained from each State and the U.S. Geological Survey, and data were formatted and tabulated. Limiting guidelines were established with assistance from PNL for evaluating ATES potential from aquifer parameters and properties. These guidelines were used to determine the ATES suitability of each major aquifer in a region and were established for small systems and large systems with respect to three storage scenarios: low temperature (60–95°C), high temperature (95–150°C), and chill (0–10°C). The guidelines were separated into the following four categories describing the aquifer's properties: (1) physical properties; (2) hydrodynamic properties; (3) thermodynamic properties; and (4) geochemical properties.

For each region, a summary is provided that discusses and tabulates the suitable major aquifers for ATES applications. The region's general potential is summarized in a qualitative manner. Overall results provided indication of widespread applicability of ATES systems in the United States and thus provided the information needed for further interest in DOE-sponsored ATES development.

ATES ECONOMIC ASSESSMENT   An economic assessment of ATES systems was performed by PNL using the AQUASTOR computer model (Reilly, Brown, and Huber 1981). The assessment systematically considered a range of system parameters and economic conditions. The factors that proved to have a substantial influence on system cost included the cost of purchased thermal energy, source temperature, cost of capital, system size, transmission distance, and aquifer storage efficiency. Load scenarios considered were point-demand, residential development, and multidistrict citywide systems. Each scenario was analyzed separately in terms of base-case conditions and parameteric sensitivity analyses. ATES-system-delivered energy costs are computed on a levelized basis and compared to conventional technologies.

The assessment provides quantitative measures of effect on system cost of many system parameters. The purchase cost of thermal energy was determined to be a major cost component of ATES systems at prices as low as \$2/MBtu (\$1.9/GJ), and this cost is magnified by price escalation or by system thermal losses. Cost of financing is also an important consideration, and thus financing possibilities or assistance by municipalities at lower interest rates are attractive. The source temperature of heat delivered to the ATES is economically important. Higher temperatures would have higher thermal losses but would allow smaller piping requirements and lower pumping costs. At a constant energy price, the higher source temperature lowered the system delivered energy cost. It was found that fairly large-scale systems (serving an 8 to 10 MW, and larger, peak load at a 25% load factor) are required for cost-effectiveness. This scale is increased substantially (to over 20 MW) for systems requiring extensive distribution networks to nonpoint loads (single-family residential housing). Aquifer thermal efficiency is an important cost parameter, particularly with high purchased thermal energy cost. Generally, the assessment indicates that ATES-delivered energy can be cost-competitive with conventional sources under a number of economic and technical conditions.

MOBILE, ALABAMA, ATES FIELD TEST FACILITY   Major field experimentation of ATES has been carried out by Auburn University under contract to

DOE at the Mobile Field Test Facility (FTF) (Molz et al. 1979, 1983). A suitable aquifer formation was determined, and wells for injection, production, and observation were drilled. Initial work involved injection of heated water at 37°C from a power-plant discharge into the confined aquifer. Volumes injected and recovered were about 7570 m³. Injection well plugging caused a major problem due to the high level of suspended solids in the injection water. Backflushing was required to clear the well, which subsequently led to the fracturing of the caprock overlayer. Recovery continued until the production temperature fell to 21°C, and about 67% of the injected energy was recovered. Injection, storage, and recovery periods lasted 420, 1416, and 2042 hours, respectively.

A second set of experiments involved injection of water pumped from an unconfined aquifer and heated with a boiler to about 55°C. Injection, storage, and recovery periods were 1900, 1213, and 987 hours, respectively, and recovery was terminated at a temperature of 33°C. A 65% energy recovery factor was determined, and a second test provided an increase in recovery to 76% due to the preheating effect of the first test cycle. Plugging problems were reduced and buoyancy effects were not substantial.

Starting in 1980, a set of three cycles was performed after revising the FTF to a doublet configuration in a different region of the aquifer with supply, injection, and production coming from the same aquifer. A boiler was again used to heat the water before injection.

For the first two cycles, injection volumes were 25,402 m³ and 58,063 m³, at average temperatures of 58.5°C and 81°C, respectively. Recovery factors were determined with a production volume equal to the injection volume. The recovery factor for the first cycle was 56%, and production temperatures were lower than expected, declining linearly over time from 55°C to 35°C. At the higher temperatures in the second cycle, free thermal convection was more pronounced due to buoyancy. Initial recovery temperature was only 55.1°C. The recovery well was modified to produce only from the top half of the aquifer, and production temperature was increased by 2.5°C but continued to decrease as before well-modification. Final recovery factor was 45.2%. During these cycles, well-clogging problems were not encountered.

Production strategy for the third cycle was significantly changed, determined with the assistance of numerical/simulation analysis by LBL of various alternatives. A modified injection/production well was used to pump water into, and out of, only the top half of the aquifer. A second well was drilled nearby and used simultaneously during production to pump out cool water from the bottom part of the aquifer. The rationale was to keep the heated water more compact and avoid dilution with the cooler water. Injected and produced volumes were similar to the second cycle, and the average injection temperature was 79°C. The scheme proved unsuccessful, and a recovery rate of only 42% was achieved. Mainly, it was felt that substantial lateral distribution of heat occurred through the high permeability horizontal layer through the center of the aquifer, and high conduction losses occurred to the upper confining layer of the aquifer.

ST. PAUL, MINNESOTA, ATES FIELD TEST FACILITY   A high-temperature (150°C) FTF was developed in St. Paul as part of the University of Minnesota ATES Demonstration Project (Kannberg 1988). The facility is designed to store 5 MW (thermal) using a doublet well configuration and a maximum water temperature of 150°C. A series of tests have been made to determine the aquifer characteristics and well development. Laboratory test on aquifer core samples indicated a storage temperature limit of 125°C, above which severe dissolution and chemical reactions occurred.

The first short-cycle injection/storage/recovery tests were completed in 1982 after encountering serious problems of calcite precipitation from the heated water. The problem was solved by adding precipitator/filter elements to the system after the heat exchanger. Heat injection was completed in five increments. Total volume injected was 8328 m^3 at an average temperature of 90.9°C. After 11 days of storage, water was pumped from the injection/production well into the second well. The average production temperature was 60.2°C and the recovery factor was about 59%. Data indicated very little buoyancy flow during storage and general confinement of heat near the injection well. Three other short-cycle tests using increasingly higher injection and recovery temperatures were completed by the end of 1983 with recovery factors between 46% and 62%.

Two longer-cycle tests (180 days) have been completed through 1988. The cycles included approximately equal periods of injection, storage, and recovery, and operated at temperatures of approximately 110°C (injection) and 80°C (recovery). Recovery factors of 62% were reported for both tests. Careful attention and evaluation were made of the geochemical problems and of solutions during the long-term tests. Injected water was effectively softened using a simple ion-exchange (sodium for calcium) treatment to avoid scaling; however, the environmental acceptability and cost of the water treatment solution was of concern.

16.2.2   CSHPSS System Studies

16.2.2.1   Simulation Programs and Numerical Studies

UNIVERSITY OF TORONTO   A simulation program was developed at the University of Toronto to study CSHPSS, particularly the storage heat loss. This effort was funded by the U.S. DOE Solar Heating and Cooling Program and the Canadian government. The NTACT (Numerical Thermal Analysis for Cylindrical Tanks) simulation code (Hooper and McClenahan 1982) was designed to calculate heat losses to the soil from cylindrical storage tanks. The model is a two-dimensional, explicit-form transient heat transfer program written in PL/1. The model is developed to simulate the operation of a CSHPSS system, though the subsystem development is relatively simple except for the storage. The detailed storage heat loss simulation is useful for storage design in terms of required sizing and associated trade-offs with insulation thickness and distribution of insulation around the tank. Optimum insulation-thickness distribution on the top, side, and bottom of the tank can be determined and analyzed.

NTACT requires a relatively large amount of computer time and memory, and thus a greatly simplified heat loss procedure (SIMLOS) was developed for use by groups unaccustomed to these requirements (Hooper and McClenahan 1982). SIMLOS utilizes tabulated lumped heat loss parameters based on detailed NTACT computations performed for a wide range of configurations of below-ground thermal storage tanks. Both codes were validated against experimental data from a solar seasonal storage project in Ontario (see section 16.2.3). A procedure was also developed for analysis of buried rectangular tanks. The simplified procedure was determined to offer excellent accuracy for seasonal storage simulations, where storage temperature approaches a sinusoidal cycle.

MCGARITY SYSTEM OPTIMIZATION ANALYSIS   An analytical study of the optimum storage volume and collector area design parameters for seasonal storage solar systems was performed by Arthur E. McGarity at Swarthmore College, Pennsylvania, in 1978 and 1979 (McGarity 1979). McGarity developed a computer-implemented optimization model and determined cost minimizing combinations of collector area and storage volume for a water-based solar system. Optimum systems were determined for a wide range of solar fraction from 0.1 to 1.0 and for different relative collector and storage costs. Expansion paths were derived to map the cost-minimizing collector area/storage volume combinations as solar fraction is increased.

The study showed that as solar fraction is increased beyond 0.9, two relative cost minima exist. One involves storage volumes representative of a short-term storage system, and the other involves seasonal energy storage.

The analysis shows that many different combinations of collector area and storage volume could be selected to achieve a desired solar fraction. The optimum combination is determined by an economic trade-off dependent on the relative costs of these design parameters. McGarity shows that at low solar fractions, the optimum ratio of collector area and storage volume is rather insensitive to their relative costs and is associated with short-term storage. However, this short-term storage ratio (which has often been assumed to be universally applicable for solar system design) is misleading when a solar fraction above 0.7 is specified. At high solar fractions, the optimum area/volume combination becomes very sensitive to the relative cost assumptions, as the system design can either operate on a short-term or annual-cycle storage basis.

McGarity's analytical procedure and results were initial contributions to the IEA task effort and clearly demonstrated the importance of optimization and proper design analysis for seasonal storage systems. This work also provided input to the SERI analytical effort described next.

SERI SIMULATION CODE AND ANALYTICAL STUDY   Work performed at the Solar Energy Research Institute provided a broad and generically based analytical investigation of seasonal storage solar systems (Baylin, Monte, and Sillman 1980; Baylin et al. 1981; Sillman 1981). This system analysis work along with McGarity's was an initial contribution for the U.S. participation in the IEA Task VII. The system analysis performed was based on the SASS (SERI Annual Storage Simulation) code developed for this purpose. SASS is a simple interactive code using a daily time step and is easy to modify for different circumstances. It is limited to systems with cylindrical storage tanks with insulation and flat-plate and evacuated-tube collectors, though preprocessing programs can be used to develop other collector capabilities or storage strategies. The simulation can be run for a

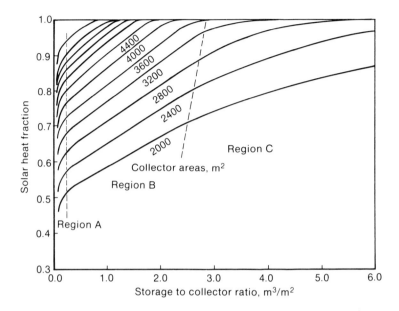

Figure 16.3
System performance curves for a typical annual storage system. System is for a Boston,
Massachusetts fifty-house space heat load, using flat plate collectors. Source: Sillman 1981.

system with both a long-term (seasonal) storage tank and a small short-
term (diurnal) buffer tank.

System analysis work was carried out at SERI to study system performance, economics, and sensitivities (Sillman 1981). System designs were
compared over a range of locations, collector types, and building types
that determine load size and distribution. The basic system performance
results are discussed here.

System performance is indicated by the solar fraction and is presented
in figure 16.3 from the SERI analysis. These results show the increase in
performance due to increase of storage volume with a constant collector
area. As storage volume increases, three ranges are described. Region A
corresponds to diurnal or weekly storage; region B is an intermediate
storage range; and region C provides annual storage. The border between
regions B and C is referred to as the point of "unconstrained operation,"
where storage is just large enough to store all the heat collected during the
summer without exceeding a maximum design temperature. Sillman con-

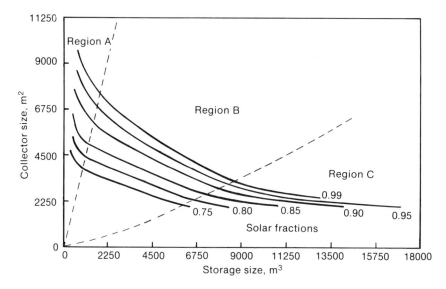

Figure 16.4
Collector–storage trade-off curves for a typical annual storage system. Source: Sillman
1981.

cludes that this point is the only likely economic optimum (Sillman 1981,
p. 9). Figure 16.4 provides a presentation of isoperformance curves on a
graph of collector area versus storage volume. This trade-off plot is essentially the standard economic tool used to plot performance isoquants in a
resource space. An economic optimum can be determined by considering
the relative costs of collector area and storage volume.

Annual-cycle space heating systems were determined to have a typical
storage volume/collector area ratio of 3–5 m^3/m^2, though systems with
large portions of domestic hot water load (constant throughout the year)
could have a ratio as low as 1 m^3/m^2 and still operate on an annual
storage basis (Sillman 1981, p. 9). A key parameter used for system evaluation is the "net energy added by storage" or the increase in energy supplied
due to an increase in volume. It is the slope of the performance curves
in figure 16.3. The two-tank system (a diurnal tank used for a buffer storage) was determined to provide substantial performance benefits and
its effect is shown in figure 16.5. The buffer tank is used to collect low-
temperature heat when the annual storage is fully charged and to produce
high-temperature heat for domestic hot water when unavailable from

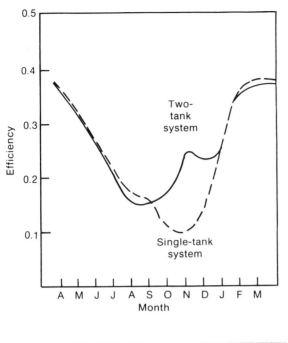

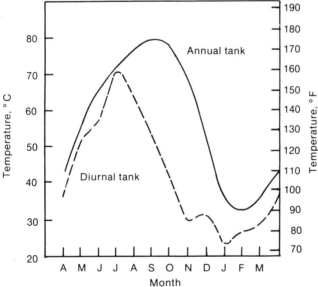

Figure 16.5
Monthly efficiency and end-of-month storage temperature for two-tank annual storage system. System is for flat-plate collectors, standard load, in Boston, Massachusetts. Source: Sillman 1981.

the annual storage. The benefit of the two-tank system is reduced if the seasonal storage tank is well stratified.

Sillman continues to show that while diurnal solar systems tend to reach a clear economic optimum between solar fraction of 40–70%, annual-cycle system cost either decreases or remains constant as solar fraction increases to 100%. Therefore, optimum designs would provide solar fractions near 100% when fuel costs are high enough to favor active solar heating.

Economic analysis was performed using simple assumptions for component costs. The analysis indicated that annual storage systems are economically favored in northern climates relative to diurnal systems. Economies-of-scale were evident and clearly showed annual storage systems to be most advantageous for large apartment complexes or districts. Generally, results indicated that annual storage systems may be cost-effective in large parts of the United States if high fuel costs escalation or the solar tax credits (available at the time of the study) are assumed.

MINSUN: IEA TASK VII COMPUTER SIMULATION   The MINSUN code was originally written at Studsvik Energiteknik in Sweden and has been greatly developed, extended, and utilized by the IEA Task VII studying central solar heating plants with seasonal storage (CSHPSS). The model is applicable specifically for seasonal storage simulations and is useful for solar applications research and as a preliminary system design, analysis, and optimization tool (Chant and Håkansson 1983; Chant and Biggs 1983).

As shown in figure 16.6, the MINSUN routine begins with a collector preprocessing program that takes hourly weather data and prepares an input collector file for the MINSUN simulation. MINSUN performs an annual system thermal simulation on a daily time-step and a detailed subsystem costing and system economic analysis. The program is built around the subsystems, and all options available are shown in figure 16.7. The detailed storage models used with MINSUN for aquifer (AST), duct (DST), and stratified tank (SST) storage were previously described in section 16.2.1.1. A range of collector types with user-specified performance parameters is available, and systems with or without heat pumps can be simulated.

Analyses can be performed with MINSUN in a number of ways. A simulation of a single system can be performed, and detailed output is provided describing the subsystem and overall annual and daily thermal

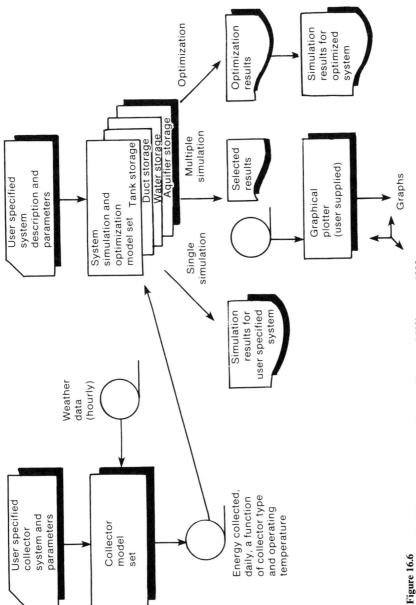

Figure 16.6
Overview of MINSUN set of programs. Source: Chant and Håkansson 1983.

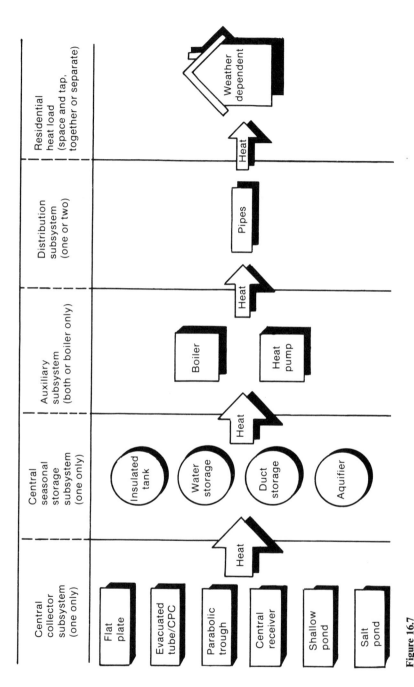

Figure 16.7
Basic options within MINSUN programs. Source: Chant and Håkansson 1983.

performance and a detailed cost breakdown and annualized economic results. A computer-optimization algorithm can also be used to automatically determine the system configuration with minimum annualized cost. The variables to be optimized are specified by the user. Task VII work initially placed a great deal of importance on this optimization capability, though problems became apparent that were inherent with the algorithm (based on steepest descent) and the rather flat cost function of CSHPSS systems over a range of parameter values. More importantly, this optimization procedure failed to reveal a great deal about CSHPSS system design, operation, and sensitivity to parameters. A multirun driving function was developed and incorporated into MINSUN to provide an interactive optimization procedure and output conducive for graphical analysis. This graphical optimization procedure has been developed by Task VII and is documented in some of the system studies presented in section 16.2.2.2.

Work within the IEA Task VII has involved a preliminary effort to validate the MINSUN program against TRNSYS. Difficulty arises in describing the TRNSYS model with the same configuration and operating strategy that is used in MINSUN. Nonetheless, simulation results proved similar between the two models, though no formal validation effort was undertaken. A number of CSHPSS operating systems do exist and offer a means of model validation of MINSUN and its storage subroutine models. Such comparisons have supported the confidence in MINSUN but have not been formally used for model validation. One difficulty is the limitations in the MINSUN design and system control configuration compared with the flexible control strategies used in the actual projects. For many of the IEA projects, additional computer codes were developed within the respective countries to model specific site and operational conditions.

16.2.2.2 System Design and Analytical Studies

UNIVERSITY OF ARIZONA   A study was performed by the University of Arizona with funding from the Arizona Solar Energy Research Commission on potential of annual storage solar heating systems (Cluff, Kinney, and Eskandani n.d.). The study focused on the design and optimization of systems using an azimuth tracking parabolic collector array that floated on a water reservoir used for annual heat storage.

A computer code was developed to simulate the system, and design was carried out for systems serving subdivisions of 248, 50, and 10 homes.

Systems were designed to serve space heating and domestic hot water loads, and the alternative of including a central cooling load (served by the heat source) was also considered.

The designs showed a significant reduction in the collector area required with an annual storage system (relative to a short-term storage system). Collector area required was increased substantially for loads including cooling, but still an overall reduction was indicated. Storage efficiencies for the systems ranged from 82–90% and distribution losses of 10% were assumed. Storage temperatures cycled between approximately 120°F (140–150°F with cooling) and 200°F (about 50°C and 93°C). Systems were designed to supply 100% of the heating requirements in an average year. A detailed description of the collector, storage, and distribution subsystem designs is provided.

An economic analysis is provided for the system serving 248 housing units. Collector costs are assumed to be $10/ft² and $4/ft² for the high and low cost scenarios, respectively. Total annualized cost (in 1979) was determined to be between $352 and $220 per house. These values are approximately the same as local conventional costs using electricity or propane gas, and about twice as much as natural gas heating costs at the time of the study.

FOX RIVER VALLEY PROJECT   The Fox River Valley Project was a CSHPSS system design and optimization effort based on a hypothetical community. This work was carried out by SERI and Argonne National Laboratory (ANL) for the U.S. contribution to the IEA Task VII (Michaels et al. 1983). This study, along with similar contributions from the other participants during Phase I of the task, served to provide initial CSHPSS design experience with various national computer codes and MINSUN.

The project considered a total annual load of 55,000 MWh with a peak demand of 17 MW located in Madison, Wisconsin. The load is a mix of residential and commercial buildings and includes a domestic hot water load of 24 MWh/day. The original study performed by SERI considered heat storage in an aquifer with several alternative operating strategies including an intermediate storage tank. The analysis was carried out originally with the SASS simulation code and later with MINSUN. Aquifer storage behavior was based on simulations performed by LBL. Five collector types were modeled using the MINSUN presimulation routine— flat plate, evacuated tube, CPC, parabolic trough, and a central receiver.

Simulations were performed for a range of collector areas, and each of the alternative operating strategies was modeled for each collector type. Component-cost assumptions were made based on previous SERI and IEA work. The economic analysis was based on the system break-even cost, defined as the conventional heating cost at which the solar option would become cost-effective. Generally, it was found that the evacuated-tube collectors provided no significant improvement over flat-plate collectors, though CPC collectors offered benefits in performance and reduced area. The range of break-even costs was about $0.05/kWh to $0.09/kWh, considered encouraging compared to conventional costs. The sensitivity analysis concluded that the distribution system and aquifer heat exchanger efficiency were two design parameters with major impact on overall system performance and economics.

This analysis was followed by work at ANL using MINSUN to simulate the Fox River Valley system using storage in a buried concrete tank (SST storage subroutine) (Chant and Biggs 1983, pp. 107–19). This work became the official U.S. MINSUN Application Case required by participants in the IEA Task VII, Phase I. The analysis approach was based on system optimization over a wide range of simulated areas and volumes. Optimization was based on minimum total annual cost to meet the load. Simulations were run with auxiliary fuel costs of $0.05/kWh and $0.09/kWh, and only the $0.09/kWh scenario provided an optimum system that included solar. The approach, however, pointed out that the minimum cost criterion should not be applied strictly. Consideration of the sensitivity of cost to solar fraction is necessary in order to assess whether the system may be designed for a greater solar fraction at a relatively small additional cost. The expansion path was developed to plot minimum total annual cost versus solar fraction to depict this sensitivity. The expansion path has been used extensively in the IEA task analytical work and will be illustrated in the more recent and detailed system studies.

CHARLESTOWN NAVY YARD PROJECT   The Charlestown Navy Yard (CNY) is a decommissioned Navy yard in Boston, Massachusetts, part of which is a National Park Service (NPS) historic park, and the other section is the site of a major redevelopment project directed by the Boston Redevelopment Authority (BRA). The availability of two large underground concrete storage tanks (totalling 5700 m^3) stimulated the study of a CSHPSS system for this area. This study was sponsored by ANL as part of the U.S.

participation in the IEA Task VII. The project provided the first detailed, site-specific design analysis and optimization work in the United States within the IEA context using the MINSUN program. The study was performed in two phases—the first carried out at the Massachusetts Institute of Technology Energy Laboratory (Breger 1982), and the follow-up work at ANL considered the addition of a heat pump into the system design (Breger and Michaels 1983).

A unique and simplifying aspect of this project was the fixed storage volume. This constraint limited the amount of energy that could be supplied to the selected load of 2200 MWh, and thus solar fractions are low compared to the normal design of CSHPSS systems. Product and cost information for storage renovation, distribution piping, collectors, and small components were obtained from previous IEA work and through direct contact with appropriate manufacturers and distributors. Design analysis concentrated on variation in collector type and area and the use of a heat pump.

System performance determined with MINSUN is shown in figure 16.8 for all options analyzed. The results show that the systems with a heat pump provide substantially greater solar fraction due to the greater effective storage capacity of the fixed-volume storage. Also, at the lower operating temperatures of the heat pump systems, the flat-plate collectors provide performance equal to the more expensive designs. System economics were analyzed in terms of the unit solar cost derived to account only for the solar (and heat pump) portion of the system (both in terms of cost and energy delivered). This method allows comparison of all system designs on merit of their solar contribution alone, independent of conventional energy costs. Solar cost (in $/MWh) is defined as the annualized solar and heat pump capital cost plus the heat pump operational cost, divided by the solar and heat pump energy delivered. Economic results are shown in figure 16.9. Clearly, the heat pump systems provide energy at a lower unit cost, and flat-plate collectors would be the design choice. The solar cost of approximately $66/MWh is similar to electric heating costs and somewhat higher than oil and gas heating costs in the Boston area at the time of the study.

U.S. AIR FORCE ACADEMY PROJECT   The U.S. Air Force Academy (AFA) in Colorado Springs, Colorado, was the site of another CSHPSS design analysis by the U.S. participants in the IEA Task VII. This effort was

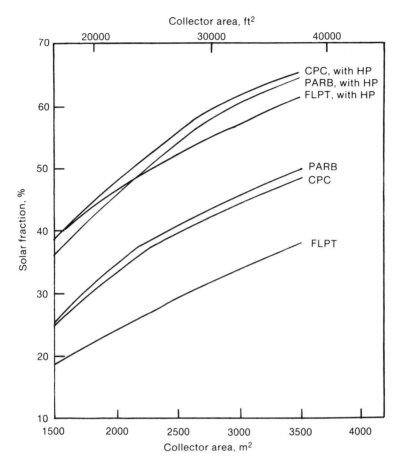

Figure 16.8
Charlestown Navy Yard CSHPSS study: System performance for all options. Source:
Breger and Michaels 1983.

performed at ANL at the request of the AFA (Breger, Michaels, and
Bankston 1983). The system was designed to substantially displace the use
of heating plant #2 that serves the service and supply area of the campus.
The present system is subject to high fuel-escalation rates and large in-
efficiencies due to operation at a small fraction of its capacity and to heat
distribution at a high temperature.

System design was based on information provided by the AFA Civil
Engineering group regarding heating loads, distribution network, and the

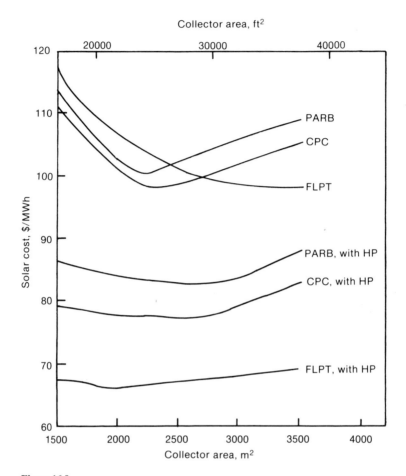

Figure 16.9
Charlestown Navy Yard CSHPSS study: Solar cost for all options. Source: Breger and
Michaels 1983.

operation and fuel cost history of heating plant #2. Though conventional fuel costs were low, the escalation rate and price uncertainty were sources of major concern. Heat load for the system was determined from previous year records to be 9000 MWh. Heat was distributed from the heating plant at 157°C, a temperature inappropriate for the load or for efficient operation of a CSHPSS system. The study assumed the distribution piping was adequate for a CSHPSS system operating between 55–80°C, otherwise small storage tanks could be installed at each load site for peak conditions. Storage was designed as an underground tank or water pit, though geological conditions might warrant aquifer or duct storage.

The analysis was executed with MINSUN and considered variation in collector type and area, storage volume, and use of a heat pump. A sensitivity study investigated the influence of higher distribution temperatures. System analysis was performed in a manner similar to the CNY study. A reference (or optimized) system was selected that used flat-plate collectors and a heat pump, providing a solar fraction of 85% and a delivered energy unit cost of $51/MWh. This cost was slightly less than the annualized conventional heating costs assumed from AFA data—$0.03/kWh (delivered heat) at 6% real escalation over 20-year lifetime.

NEW ENGLAND ASSESSMENT STUDY   Following the IEA Task VII effort (Subtask II[b]) to evaluate CSHPSS system concepts over a wide range of designs and scenarios (see section 16.2.2.3), some countries concentrated on a National Evaluation to assess CSHPSS technology under their more specific circumstances. This effort by the United States looked specifically at the New England region and the concept of duct storage in the bedrock and use of a heat pump (Breger and Michaels 1985). The New England region had been previously cited as a promising region for CSHPSS technology due to the climatic conditions and high energy costs. Prevalent bedrock (mostly granite) is available throughout most of the region, though the extent of fracturing and ground water movement, which may limit the storage effectiveness, must be determined on a site-specific basis.

The study assumed an annual heating load of 10,000 MWh (with 20% domestic hot water) and considered a high and low temperature distribution requirement. The distribution network cost and heat losses were not included due to their dependence on the site layout and assumption that a conventional district heating system would have similar cost and losses; thus results are correctly understood as heat delivered to the distribution

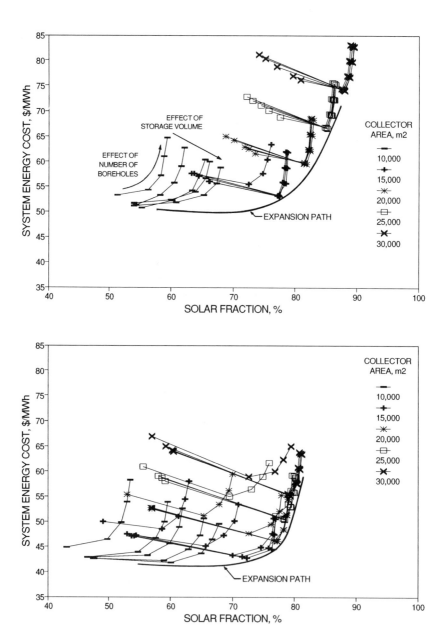

Figure 16.10
New England CSHPSS assessment: Economic results shown as expansion paths for base
case with flat-plate (top) and unglazed (bottom) collectors. Source: Breger and Michaels 1985.

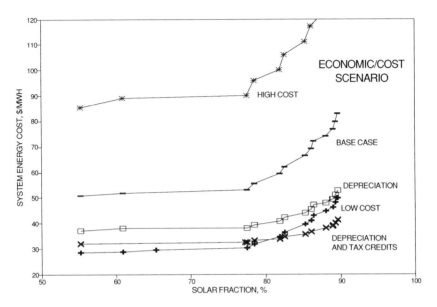

Figure 16.11
New England CSHPSS assessment: Results shown as expansion paths for all economic scenarios with flat-plate collectors. Source: Breger and Michaels 1985.

network. Economics was derived for a base-case economic scenario of best available costs and economic data, and for variations including the effect of financing options and tax incentives. System analysis was performed using MINSUN to simulate a range of systems with varying collector type, collector area, storage volume, and number of storage boreholes. Analysis was based on the system energy cost ($/MWh) defined as the annualized solar and heat pump system capital and heat pump operating costs, divided by the solar and heat pump energy delivered to the load. Expansion paths were derived as the minimum cost envelope for increasing solar fractions. This result is shown for both flat-plate and unglazed flat-plate collectors in figure 16.10 for the base-case scenario. Extending this analytical methodology, system optimization is derived by selecting the system along the expansion path that provides a marginal system energy cost equal to the conventional heat cost. Figure 16.11 shows the expansion path summary (for flat plate collectors) for the other economic scenarios, and figure 16.12 shows the expansion path summary for both collector types for the high and low distribution temperatures.

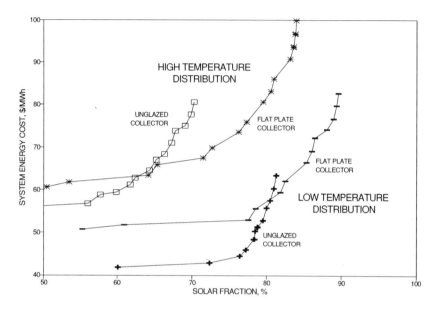

Figure 16.12
New England CSHPSS assessment: Expansion path summary for both collectors and
distribution temperatures. Source: Breger and Michaels 1985.

The general results of the analysis indicated that the optimized systems
provide 75–80% of the low temperature demand and 60–65% of the high
temperature demand with solar energy, with the remaining portion coming primarily from the heat pump electrical input energy. The annualized
cost of the energy supplied to the distribution network using the base-case
economic scenario is \$44–52/MWh and \$58–62/MWh for the low and
high temperature loads, respectively. Significant reductions in cost are
found when systems are financed with incentives, provided by the federal
government at the time of the study. These results are encouraging in
terms of system cost-effectiveness relative to the energy costs in New England, which are comparable to these values.

TRICOUNTY REGIONAL VOCATIONAL TECHNICAL SCHOOL—FRANKLIN, MASSA-
CHUSETTS   A preliminary CSHPSS design was completed in 1988 for the
Tricounty Regional Vocational Technical School in Franklin, Massachusetts. The CSHPSS design study was supported by the U.S. DOE under an
agreement with the Massachusetts Executive Office of Energy Resources

(MEOER) (Breger, Sunderland, and Bankston 1989). The school is a 280,000 ft^2 (26,000 m^2) facility with 2600 MWh heating and hot water load, which is currently met by an electric boiler and combined forced hot air and hydronic heat distribution system. The school has received a capital grant from the MEOER to design and construct an alternative energy project to reduce the electric heating demand.

The MEOER and the school are also considering a solar-assisted ground coupled heat pump (SA-GCHP) system for the site. The design of the SA-GCHP system and a final CSHPSS design are to be technically and economically compared, and a decision made by the school board regarding which project to implement.

The preliminary CSHPSS design calls for a 15,000 m^3, 10-meter deep excavated water-filled earth pit, a 3500 m^2 flat-plate collector array, and a heat pump to allow low storage temperatures. A solar fraction of 73% is simulated. Supplementing the solar source by operating the existing electric boiler off-peak and using the top of the store for diurnal storage during the heating season is being considered. With the heat pump operation, the bottom of the storage may be useful for spring cooling loads.

UNIVERSITY OF MASSACHUSETTS, AMHERST   In 1988, DOE supported efforts at the University of Massachusetts (UMass) to identify, evaluate, and design a major CSHPSS to serve as a first U.S. project (Breger, Sunderland, and Bankston). The site selected is on the UMass campus in Amherst, and preliminary design work has been completed. Project engineering and final design has received additional DOE support and is schedule to be completed by January 1993. Funding to construct the project is being pursued through state, federal, and industry sources.

The site geology is of particular interest; a 100-foot-thick soft saturated clay formation is to be used for the seasonal store through the installation of vertical heat exchange boreholes. Land is available for a collector array, and the heat load is a new Arena/Convocation Center to be constructed near the site and an existing large gymnasium building. The Arena HVAC system is being designed to accommodate the low temperature heat source delivered at 130°F (55°C). The gymnasium will require retrofit of the steam-supplied airhandling units and hot water generators.

Preliminary system analysis and optimization has been completed (Breger, Sunderland, Elhasnaoui 1991). Simulation results show that 89% of the 3500 MWh load can be met by the CSHPSS—designed with 11,000 m^2 flat-plate collectors and a 75,000 m^2 clay storage with 600 boreholes. The clay temperature cycles between 52–73°C and, after attaining a steady state, an 81% efficiency is expected. Cost and economic analysis has shown that the CSHPSS can provide an annualized delivered heat cost of $70/ MWh. This cost exceeds the predicted regional heating costs for the first year of operation, but anticipated increases in fuel cost are expected to exceed this value and to result in substantial savings over the system economic life.

16.2.2.3   IEA Task VII—Evaluation of System Concepts

A major accomplishment of the IEA Task VII was the effort within Sub-task II(b), which provided an evaluation of CSHPSS system concepts (Bankston 1983). The participants examined a wide range of CSHPSS configurations, operational strategies, and load/climate conditions in order to compare and rank CSHPSS designs and applications on a thermal and economic basis. Much of the previous work of the task was used to define subsystem parameters and component cost data, and to provide the necessary modeling tools and analysis approach for this effort.

The conditions assumed for analysis of CSHPSS design included two climates (Madison, Wisconsin, and Copenhagen, Denmark), three load sizes (3.6, 36, and 360 TJ; or approximately 50, 500, and 5000 housing units), three load patterns (0, 20, and 50% domestic hot water, or constant load), and a high and low distribution demand temperature scenario (115/80/50 and 60/50/30°C; demand temperature at −20°C ambient, 0°C ambient, and constant return temperature). The distribution network cost and heat loss were not considered due to their site-specific nature, so results are understood to be for heat delivered to the distribution network. System concepts were analyzed for the following variations in subsystem components and system operation. Collector types examined included unglazed flat-plate, flat-plate, and evacuated CPC collectors. Storage was examined for aquifer, duct in rock or earth, and water storage in a cavern, earth pit, or tank. Systems operating with and without a heat pump were examined.

A common set of parameters characteristic of 1984 technology was used (except for storage description). These characteristics were based on sub-system studies performed in Phase I of the IEA Task (Bankston 1983; Hadorn and Chuard 1983; Bruce and Lindeberg 1982). For each storage type, hundreds of MINSUN simulations were run to span all design variations over a wide range of collector area and storage capacity or temperature (Chant and Breger 1984; Hadorn, Havinga, and van Hattem 1984; Zinko et al. 1984). System economic results were based on the annualized unit solar cost (annualized solar system component costs, divided by the solar fraction times the load). All relevant simulations were then plotted on one graph of unit solar cost versus solar fraction, so that the expansion path of least-cost systems versus solar fraction could be identified. This procedure allows for direct comparison of the expansion paths of various system concepts (e.g., comparison of storage or collector alternatives, heat pump option, or determination of effect of climate or distribution temperature).

The expansion-path methodology proceeds into a straightforward optimization procedure. Provided conventional heat cost is greater than the minimum unit solar cost on the expansion path, the overall economic optimum is reached by proceeding along the expansion path to higher solar fractions until the solar marginal cost (derivative of curve) is equal to the conventional cost.

For each load temperature condition and storage type, a composite expansion path was derived from the individual paths representing each collector type and heat pump option simulated. An example is shown in

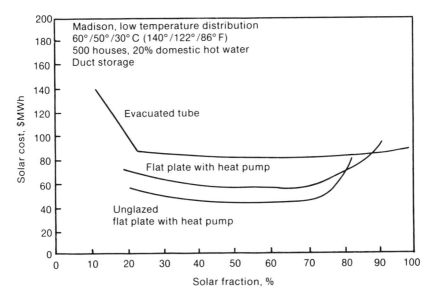

Figure 16.13
IEA CSHPSS evaluation: Duct storage system expansion path diagram. Source: Bankston 1984.

figure 16.13 for duct storage in Madison with a low temperature load. The expansion diagram for all system concepts evaluated are summarized in figure 16.14 for Madison, with high and low temperature distribution. The general results of this evaluation are as follows. For most storage types (except constructed water pits and caverns), unglazed collectors operating with a heat pump offer the best economics. However, to increase solar fraction, design would need to change to flat-plate collectors and, for near-100% solar systems, systems without heat pumps are required. (The analysis assumed that heat pump electrical unit cost and auxiliary heat cost were equivalent, which would provide relatively optimistic solutions for heat pump systems in most locations where electricity costs are higher.)

The aquifer and duct storage systems offer the lowest unit solar cost, though solar fractions are limited to about 75–80% and 60–70% for the low and high distribution temperature systems, respectively. For solar fractions greater than these values, systems without heat pumps and with water pit or cavern storage appear most economical. From figure 16.14, it is apparent that CSHPSS systems are competitive with conventional heat cost (to the distribution network) ranging from \$30–80/MWh and \$80–130/MWh for the low and high temperature distribution demands, respectively. Results for Copenhagen were similar. Sensitivity studies were performed to analyze variation in load size and percent domestic hot water. Results lacked the extensive analysis of the base-case conditions but clearly indicated system economic improvement with larger loads and greater domestic hot water demand.

General economic conclusions regarding CSHPSS cost-effectiveness relative to competing technology are difficult due to the international basis of this study. However, indications from this work show that the solar costs of the best CSHPSS systems are low enough to be competitive in areas where conventional energy costs are high—about \$50–100/MWh. Systems with seasonal storage do seem to offer clear cost advantages over diurnal storage systems designed for substantial solar fractions. The low temperature distribution scenario may not be applicable for a retrofit site, and many areas are not familiar with this technology or comfortable with its reliability. These results do make clear, though, the economic importance of reducing demand temperatures for effective solar system design and operation. In addition, distribution piping material costs and heat losses are reduced with the low temperature systems, although heat exchangers and pumping power requirements would partially offset this ad-

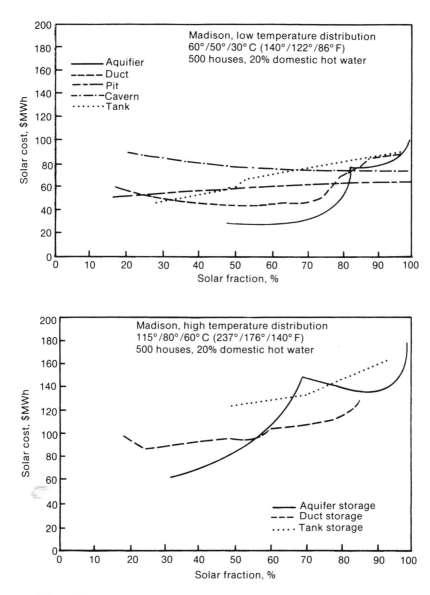

Figure 16.14
IEA CSHPSS evaluation: Composite expansion path diagrams. Source: Bankston 1984.

vantage. Clearly, for point-source or high-density loads, the distribution network cost and heat losses would be relatively small.

In Phase III of the IEA task, a generic systems analysis was performed (Dalenbäck 1989) to update and supplement the comparative systems analysis discussed in this section. The analysis included improvements in the cost and performance of collector technology and other system components to reflect current experience and future expectations. The analysis concluded that all systems for large loads (over 200 houses) showed heat costs of about \$50/MWh using anticipated future costs assumptions; although under current costs, site- and system-specific CSHPSS configurations already do exist. For large loads, high temperature flat-plate collectors and system operation without a heat pump were found most suitable. Smaller systems were not economical and required additional R&D, particularly regarding storage design and materials.

16.2.3   CSHPSS Systems—Experimentation and Demonstration

Operational experience with CSHPSS systems in the United States has been limited to a few experimental projects. Through the IEA Task VII participation, the experiences of a number of demonstration systems in other countries have provided insight to the problems and successes encountered in the design, construction, and operation of these systems. This section will summarize the U.S.-sponsored experimental work and briefly describe some of the operating systems in other countries.

UNIVERSITY OF VIRGINIA   A seasonal storage project was performed at the University of Virginia with DOE funding through ANL (Beard et al. 1977). The experimental system consisted of an excavated pool or earth pit (27,400 gallons) and corrugated glazed trickle collectors. The original pool encountered soil stability problems as the bermed sides caved in. A smaller pool was constructed with four-foot wooden bermed sides, a liner and no insulation on the sides or bottom, and 30-centimeter floating loose styrofoam bead insulation to cover the pool. The trickle collectors were floated on top and water circulated directly from the storage through the collectors and trickled back into the top of the pool.

Heat loss was carefully monitored and determined to be extremely high. The bermed sides were uninsulated and exposed to the highest temperatures (due to the natural storage stratification) and experienced tremendous heat loss. The storage reached its maximum temperature in July, when heat loss began to equal the heat added to the storage. Heat extraction was accomplished with a heat pump to meet the demands of a 140 m² house. Only one cycle of heat storage was completed. The experiment determined that on this small scale, uninsulated earth storage is not feasible. Additional modeling effort, following the experiment, indicated system improvements by insulating the berm and using higher efficiency collectors. More favorable results were indicated for a system scaled up to serve a much larger load.

COLORADO STATE UNIVERSITY   Work at the Colorado State University under funding from ANL concentrated on the study of soil water migration and thermal properties (Walker, Sabey, and Hampton 1981). As part of this work, an experimental seasonal storage system was designed and implemented. Soil was excavated in the shape of an inverted truncated cone (50 feet in diameter at top, 20 feet in diameter at bottom, and about 20 feet deep). The excavated soil was spread out to dry. A double-wall cylindrical plastic "bag" was filled with saturated wet soil and placed in the center of the excavated pit and then surrounded by the dry soil. The saturated soil storage measured 12 feet in diameter and was 9 feet high. The dry soil was used to insulate the store and reduce heat loss due to conductivity and water migration. Within the saturated soil, polyflow plastic tubing, 1/2 inch in diameter, was placed in several horizontal spiral layers to create a heat exchanger in the soil storage.

An experiment was conducted by initially charging the saturated soil store until heat losses equaled heat gains, followed by a period of steady state and then discharge. The store and surrounding earth was monitored for heat loss as the store was left undisturbed. Results indicated that the presence of the dry soil, assisted by layers of coarse sand, was effective in preventing capillary flow back into the dry area and thus served as an effective insulation method. Modeling work on this project indicated low-cost applications of this technology for systems scaled up to serve one home and that annual heat storage efficiency could be as high as 70%.

UNIVERSITY OF TORONTO   Two CSHPSS experimental systems were built in Canada with funds from the U.S. DOE and Canadian government in the late 1970s (Hooper, McClenahan, and Williams 1980). The work was carried out at the University of Toronto. The Provident House system, located 20 miles north of Toronto, serves a single family house and uses an insulated seasonal storage tank in the basement that was excavated deeper than usual to allow sufficient storage volume and useable basement area. Flat-plate collectors were located on the roof of the house. The heat losses from the tank were approximately 20% higher than expected due to water leakage into the insulation. Otherwise the system operated well and provided experimental data for model validation.

A second system, located in Aylmer, Ontario, serves a senior citizen apartment complex of about 30 one-story units. Flat-plate collectors are located on the roofs of the units, and a seasonal storage tank or pit was constructed underground with concrete walls and sprayed-foam insulation. The top of the storage is insulated and supports a patio. The storage was initially subject to a serious leak problem but was repaired with the installation of a rubber liner, and subsequently operated without major problems.

IEA COUNTRIES   Over 30 CSHPSS projects are in operation around the world, predominantly in the countries participating in the IEA Task VII. In this section, a few of the more significant and well-documented projects are summarized.

SWEDEN   Sweden has been the forerunner of CSHPSS and low temperature distribution technology. Investigators in that country have made substantial contribution to the IEA Task VII and offer the most experience with relatively large-scale CSHPSS system construction and operation (Dalenbaĉk 1988). The four major Swedish projects are summarized in table 16.1, which provides the main system design specifications. A brief description of the operating experience with each system follows.

The system at Studsvik (Dalenbäck, Gabrielson, and Ludvigsson 1981; Roseen and Perers 1980; Dalenbäck 1988, 1989) has operated more or less as intended, though several problems were encountered. The collector reflecting surface has become faded, and draining the collectors during times of frost-risk did not work satisfactorily. The heat store experienced some leakage, and some rainwater leaked into the tank-cover insulation. This small system experienced large storage heat losses, but this was not unexpected for this small-scale system.

The Lambohov project (Dalenbäck, Gabrielsson, and Ludvigsson 1981; Norback and Hallenberg 1980; Dalenbäck 1988, 1989) has been in operation since 1980. There were considerable problems throughout the first two years concerning construction, operation, and control. In time for the summer of 1982, the project was substantially rebuilt and since then has operated as intended.

The Ingelstad plant (Dalenbäck, Gabrielsson, and Ludvigsson 1981; Finn 1979; Dalenbäck 1988, 1989) has been in operation since 1979. The system has operated as intended in terms of the technical and control aspects. However, low collector efficiency due to failures with the collector tracking system have been major problems. The collector subsystem was replaced with flat-plate collectors, which have so far operated reliably.

The Lyckebo system (Bjurstrom 1983; Dalenbäck 1988, 1989) was constructed in 1983 and includes a huge uninsulated, water-filled annular rock cavern. System operation has begun with only a fraction of the required solar array area in place, with the remainder simulated by an electric boiler. Depth-adjustable inlet and outlet pipes are used effectively to maintain a high degree of temperature stratification in the 30-meter deep cavern. As expected, several years were necessary before the surrounding rock was pre-heated so that a steady-state heat loss was reached. Unexpected heat losses were attributed to convection losses through an old construction access tunnel.

THE NETHERLANDS   The Groningen project (Wijsman 1983; Dalenbäck 1989) is based on heat storage in soil, exchanged through vertical plastic pipes connected to a collector array and a low-temperature distribution system to 96 housing units. The vertical pipes were installed with a machine designed to press the pipes into the clay soil, and thus it could be done quite quickly and inexpensively. The seasonal store contains a volume of 23,000 m^3 of soil that is designed to reach a maximum temperature of 60°C and have an expected efficiency of 68%. An uninsulated short-term storage tank (100 m^3) is buried in the seasonal store ground to allow heat exchange to and from the seasonal store to continue over a full daily cycle, therefore reducing the required heat exchange capacity and increasing its effectiveness. The collector subsystem includes 2500 m^2 of high-performance evacuated-tube collectors distributed on the multiunit houses being served. The houses are equipped with radiators with large heating surface, which require a supply temperature of only 42.5°C. System operation was begun in 1984. After two years, the store had reached a steady-state heat loss condition. The third year performance, in terms of heat delivery, was approximately 14% lower than design expectation. Reductions in performance were attributed to lower collector efficiency and

Table 16.1
Technical comparison (as initially designed) of four major Swedish CSHPSS projects

Project		Studsvik	Lamvohov	Ingelstad	Lyckebo
Nearest town		Nyköping	Linköping	Växjö	Uppsala
Volume of heat store	m^3	640	10000	5000	100000
Storage capacity	MWh/year	19	750[a]	300	5500
Store temperature	°C	70/30	70/5	95/40	90/40
Type of store		Ground excavation	Excavated (blasted) cylindrical tank	Above-ground cylindrical concrete tank	Rock cavern annular form
Dimensions	m	Depth 6 Diameter (surface) 16 Diameter (bottom) 6	Depth 12 Diameter 32	Depth 8 Diameter 28	Depth 30 Diameter (outside) 75 Diameter (inside) 35
Thermal insulation:	mm				
walls		400 mineral wool	250 lightweight concrete wall 750–1200 cementbound lightweight granules	100 glass fibre 920 mineral wool	—
bottom		400 mineral wool	Approx 1200 lightweight concrete	320 glass fibre	—
cover		400 polyurethane	400 polyurethane	100 glass fibre 900 mineral wool	—
Waterproof layer		Butyl rubber sheet	Butyl rubber sheet	Concrete	
Heat supplied to		Office building, 500 m^2	55 terraced houses	55 detached houses	350 detached houses, 200 apartments
Annual coverage	%	100	100[a]	50	100[b]
Distribution system design temperature	°C	30	55/25	80/50	80/50
Heat distribution medium in buildings		Air	Air	Water	Water

Solar collectors		Concentrating (CPC) on the tank cover, solar tracking	Flat plate, on building roofs	Parabolic concentrating, ground mounted, solar tracking	Flat plate, ground mounted
Surface	m²	120	2900	1300	4300
Inclination		25°	55°	35°	42°
Heat pumps, electrical drive power	kW	—	156 for space heating: 29 for domestic hot water	—	—
Building and installation time	months	6	18	13	20
Support from BFR	Million SEK	1	13	8	17
Owner		Studsvik Energiteknik AB	AB Östgötabyggen, Linköping	Växjö local authority	Uppsala Power and Heating Co

Source: Dalenbäck, Gabrielsson, and Ludvigsson (1981).
Reference: Dalenbäck, J. O., and T. Jilar. 1983. Two Swedish solar heating plants with seasonal storage: Evaluation program and experiences from three years of operation. ANL Seminar Report, September 1983.
a. Including heat pumps.
b. 85% simulated by an electric boiler.

higher storage heat loss than expected. The higher storage losses were due to higher U-value in the top insulation and greater than anticipated ground water movement through the store region.

CANADA   A seasonal storage aquifer system has been in operation since 1985 for a newly designed energy efficient Government of Canada building in Scarborough, Ontario (Morofsky 1983; Dalenbäck 1989). An upper and a lower aquifer are used for heat and chill storage. A heat pump is used and energy is exchanged between the two storage aquifers. A small collector array provides additional thermal energy to the hot aquifer storage. The aquifer storage has operated without significant hydrogeological problems. Problems have resulted from incorrect assumptions on building occupancy and heat pump sizing; data acquisition for system monitoring has not been fully operational, and less data than desired has been available.

16.3   Summary and Conclusions

The DOE solar and conservation programs have laid substantial groundwork in the development of seasonal storage thermal energy systems. Storage in aquifers has been strongly emphasized in terms of experimental work, though assessment of other storage means has also been performed. The participation of the United States in the International Energy Agency Solar Heating and Cooling Program, Task VII on Central Solar Heating Plants with Seasonal Storage, has provided direct access and active contribution to a collaborative effort and expertise from other countries and demonstration CSHPSS projects.

A great deal has been learned regarding the performance of seasonal storage systems and the methods needed for their design, analysis, and optimization. System economics has been addressed with promising results but remain very uncertain and untested for private-market development. This conclusion is particularly important for these systems, which inherently require a large-scale application to properly demonstrate their performance and economics. Implementation of seasonal thermal storage and CSHPSS systems in the private sector will probably require the awareness and interest of real-estate commercial and residential developers and investors as sources of third-party financing, or the development and ownership by thermal-energy utilities. The large capital requirement makes private development of the new technology particularly difficult.

Also of great importance for the utilization of seasonal storage technology is the development and more widespread use of community district heating networks. Once a distribution network is in place for a community of residential or commercial buildings served by a single-point heating source, the integration of a seasonal thermal storage with a solar (or waste or other low-cost heat) source can readily be connected and displace the conventional energy supply. Low-temperature heating distribution networks should be cost-effective for high load density areas, and the solar source can then be phased in when its economics is clearly more favorable than the conventional heat source.

The overall conclusions that can be drawn from the research work performed provide substantial demonstration of the technological readiness of solar seasonal storage systems and all subsystem component technology. Analytical work has provided tools for system analysis and design, and assessment studies have indicated the appropriateness and widespread applicability of seasonal storage systems. Economic evaluations have concluded that life-cycle system costs are on the order of conventional costs particularly in regions with high energy prices. The economics of CSHPSS systems seems to favor this solar approach compared to systems based on diurnal storage when designing for large winter loads and substantial solar fractions. The advantage of seasonal storage is especially clear in cold-winter climates characterized by low solar availability.

References

Bankston, C. A. 1986. *CSHPSS: evaluation of concepts.* IEA Task VII, Subtask II(b). Report # T.7.2.B. U.S. Government Printing Office.

Bankston, C. A. 1983. *CSHPSS: Basic performance, cost, and operation of solar collectors for heating plants with seasonal storage.* IEA Task VII, Subtask I(b). Argonne National Laboratory.

Baylin, F., R. Monte, and S. Sillman. 1980. *Annual-cycle thermal energy storage for a community solar system: Details of a sensitivity analysis.* SERI/TR–721–575.

Baylin, F., et. al. 1981. *Economic analysis of community solar heating systems that use annual cycle thermal energy storage.* SERI/TR–721–898.

Beard, J. T., et. al. 1977. Heat transfer analysis of a system for annual collection and storage of solar energy (system at the University of Virginia). *Heat Transfer in Solar Energy Systems.* ASME.

Bjurstrom, S. 1983. Heat storage in rock caverns in Sweden. *Proc. International Conference on Subsurface Heat Storage*, Stockholm, June 1983.

Breger, D. S., J. E. Sunderland and C. A. Bankston 1989. CSHPSS: The development of two projects in Massachusetts. *Proc. 1989 Annual Conference, American Solar Energy Society*, Denver, CO.

Breger, D. S., J. E Sunderland, and H. Elhasnaoui. 1991. Preliminary Design Development of a Central Solar Heating Plant with Seasonal Storage at the University of Massachusetts, Amherst, *Proc. International Solar Energy Society 1991 Solar World Congress*, Denver, CO, August.

Breger, D. S. 1982. *A solar district heating system using seasonal storage for the Charlestown, Boston Navy Yard Redevelopment Project*. ANL–82–90.

Breger, D. S., and A. I. Michaels. 1985. *An assessment of solar heating systems with seasonal storage in New England: Systems using duct storage in rock and a heat pump. ANL/EES–TM–279.*

Breger, D. S., and A. I. Michaels. 1983. *A seasonal storage solar energy heating system for the Charlestown, Boston Navy Yard National Historic Park: Phase II. Analysis with heat pump.* ANL–83–58.

Breger, D. S., A. I. Michaels, and C. A. Bankston. 1983. *A solar heating system using seasonal storage for the U.S. Air Force Academy, Colorado Springs, Colorado—A preliminary study.* Argonne National Laboratory (unpublished).

Brown, D. R., H. D. Huber, and R. W. Reilly. 1982. Aquastor: A computer model for cost analysis of aquifer thermal energy storage coupled with district heating or cooling systems. *Proc. 17th IECEC*, 2012–2018.

Bruce, T., and L. Lindeberg. 1982. *CSHPSS: Basic design data for the heat distribution system.* IEA Task VII, Subtask I(d). Swedish Council for Building Research.

CWEC (Century West Engineering Corporation) 1981. *Regional assessment of aquifers for thermal energy storage*, vol. 1, 2, and 3. PNL–3995–1,2,3.

Chant, V. G., and D. S. Breger. 1984. *CSHPSS: Evaluation of systems concepts based on heat storage in aquifers.* IEA Task VII, Subtask II(b), working document. Canada.

Chant, V. G., and R. Håkansson, eds. 1983. *CSHPSS: The MINSUN simulation and optimization program—Application and user's guide.* IEA Task VII, Subtask I(a). National Research Council of Canada.

Chant, V. G., and R. C. Biggs. 1983. *CSHPSS: Tools for design and analysis.* IEA Task VII, Subtask I(a). National Research Council of Canada.

Cluff, C. B., R. B. Kinney, and F. Eskandani. N.d. *Final report on an evaluation of annual heat storage of solar energy for Arizona subdivisions using an azimuth tracking floating collector.* University of Arizona for the Arizona Solar Energy Research Commission.

Dalenbäck, J. O. 1988. *Large-scale Swedish solar heating technology—system design and rating.* D6:1988. Swedish Council for Building Research, Stockholm.

Dalenbäck, J. O. 1989. *CSHPSS: A status report.* IEA Task VII, Phase III, final draft. Chalmers University of Technology, Sweden.

Dalenbäck, J. O., E. Gabrielsson, and B. Ludvigsson. 1981. *Three Swedish Group solar heating plants with seasonal storage: A summary of experience from the Studsvik, Lambohov and Ingelstad plants up to the end of 1980.* D5:1981. Swedish Council for Building Research, Stockholm.

Doughty, C., G. Hellström, C. F. Tsang, and J. Claesson. *A study of ATES thermal behavior using a steady flow model.* LBL–11029.

Finn, L. 1979. *A Swedish solar heating plant with seasonal storage: The Inglestad project design and construction stage.* D14:1979. Swedish Council for Building Research, Stockholm.

Gupta, S. K., et. al. 1982. *A multi-dimensional finite element code for the analysis of coupled fluid, energy and solute transport (CFEST).* PNL-4260.

Hadorn, J. C., and P. Chuard. 1983. *CSHPSS: Heat storage models—evaluation and selection, CSHPSS: Heat storage systems: Concepts, engineering data and compilation of projects, CSHPSS: Cost data and cost equations for heat storage concepts.* IEA Task VII, Subtask I(c). Eidgenossiche Drucksachen und Material Zentrale, Switzerland.

Hadorn, J. C., J. Havinga, and D. van Hattem. 1984. *CSHPSS: Evaluation of systems concepts based on duct storage.* IEA Task VII, Subtask II(b), working document. Switzerland.

Hellström, G., and B. Eftring. N.d. *Description of storage subroutines in TRNSYS form—aquifer storage (AST), duct storage (DST), stratified storage (SST).* Lund University, Department of Mathematical Physics, Sweden.

Hellström, G., J. Bennet, and J. Claesson. 1982. *Model of aquifer storage system—manual for computer code.* Lund University, Department of Mathematical Physics, Sweden.

Hellström, G. 1982. *Model of duct storage system—manual for computer code.* Lund University, Department of Mathematical Physics, Sweden.

Hooper, F. C., J. D. McClenahan, and G. T. Williams. 1980. *Solar space heating systems using annual heat storage.* Final report. DOE–CS–32939–12.

Hooper, F. C., and J. D. McClenahan. 1982. *User manual for a simplified heat loss program based on the extended NTACT model.* Final report. University of Toronto, Department of Mechanical Engineering, Canada.

Kannberg, L. D. 1988. *Seasonal Thermal Energy Storage Program—progress summary for the period April 1986, through March 1988.* PNL–6705.

McGarity, A. E. 1979. Optimum collector-storage combinations involving annual cycle storage. *Proc. Solar Energy Storage Options,* San Antonio, Texas, CONF–790328.

Mercer, J. W., C. R. Faust, W. J. Miller, and F. J. Pearson. 1981. *Review of simulation techniques for aquifer thermal energy storage.* PNL–3769.

Michaels, A. I., S. Sillman, F. Baylin, and C. A. Bankston. 1983. *Simulation and optimization study of a solar seasonal storage district heating system: The Fox River Valley case study.* ANL–83–47.

Molz, F. J., et. al. 1983. *Design, performance and analysis of an aquifer thermal energy storage experiment using the doublet well configuration.* PNL–4849.

Molz, F.J., et. al. 1979. *Thermal energy storage in confined aquifers.* Water Resources Research Institute, Auburn University.

Morofsky, E. L. 1983. Overview of Canadian aquifer thermal energy storage field trials. *Proc. International Conference on Subsurface Heat Storage,* Stockholm, June 1983.

Norback, K., and J. Hallenberg. 1980. *A Swedish Group solar heating plant with seasonal storage: Technical-economic description of the Lambohov project.* D36:1980. Swedish Council for Building Research, Stockholm.

Reilly, R. W., D. R. Brown, and H. D. Huber. *Aquifer thermal energy storage costs with a seasonal heat source.* PNL–4135.

Roseen, R., and B. Perers. 1980. *A solar heating plant in Studsvik: Design and first-year operational performance.* D21:1980. Swedish Council for Building Research, Stockholm.

Sillman, S. 1981. *The trade-off between collector area, storage volume and building conservation in annual storage solar heating systems.* SERI/TR–721–907.

Tsang, C. F., T. A. Buscheck, and C. Doughty. N.d. Aquifer thermal energy storage—recent parameter and site-specific studies. Lawrence Berkeley Laboratory.

Tsang, C. F., T. Buscheck, D. Mangold, and M. Lippmann. 1978. *Mathematical modeling of thermal energy storage in aquifers.* LBL–9970.

Tsang, C. F., T. Buscheck, and C. Doughty. 1980. Aquifer thermal energy storage: A numerical simulation of Auburn University field experiment. LBL-10210. *Water Resources Research* 17(3): 647–58.

Walker, W. R., J. D. Sabey, and D. R. Hampton. 1981. *Studies of heat transfer and water migration in soils.* Final report, DOE/CS/30139.

Wijsman, A. J. Th. M. 1983. The Groningen project: 100 houses with seasonal solar heat storage in the soil using a vertical heat exchanger. *Proc. International Conference on Subsurface Heat Storage*, Stockholm, June 1983.

Zinko, H., S. Rolandsson, K. K. Hansen, and D. Krischel. 1984. *CSHPSS: Evaluation of water storage systems.* IEA Task VII, Subtask II(b), working document. Sweden.

IV SOLAR COOLING

17 Research and New Concepts

Noam Lior

17.1 Integral and Direct Heating Systems

17.1.1 Space Heating Systems with Integrated Collectors

Included in this category are active solar heating systems in which solar collectors are an integral part of the building structure, that is, they replace an external facade or roof material. In this way the collectors serve a dual function, possibly leading to improved economy.

An early design was the roof-integrated Solaris trickle collector (Thomason 1960; Thomason and Thomason 1975). Since water condensation on the interior of the window panes constitutes an effective heat transfer process, the associated heat losses from such collectors were found to be higher than those from other flat-plate solar collectors—by almost a factor of two (Beard et al. 1976, 1978; Beard 1978)—and the layer of condensate on the glass also tended to reduce its transmittance, thus also reducing the solar energy absorbed by the collector. To minimize evaporation and avoid freezing and corrosion, Scientific Atlanta, Inc. (Beard 1978), manufactured a collector that used a film of silicone oil instead of water in a collector of similar design. Tests with this collector indicated that the overall efficiency was still quite low.

Higher efficiency is obtained when the water passes through tubes, as practiced in detached conventional solar collectors. For example, the Colorado State University Solar House I used a roof-integrated solar collector consisting of a roll-bond absorber laid on thermal insulation, which was mounted on the roof sheathing, with two glass plates above, and showed a collector efficiency slope of about 4.5 W/m²C, twice better than that of the Solaris collector (Löf and Ward 1976; Karaki, Duff, and Löf 1978). The design, construction, and testing of a solar collector combined into a structural unit so that it would double as the building roof was successfully performed by the Los Alamos Scientific Laboratory (Moore, Balcomb, and Hedstrom 1974).

A significant effort to develop building-integrated air heating collectors was undertaken by Total Environmental Action, Inc. (Kohler et al. 1978; Moore, Temple and Adams, 1980). The collector design chosen is shown in figure 17.1. It was claimed that the installed costs are about half of those

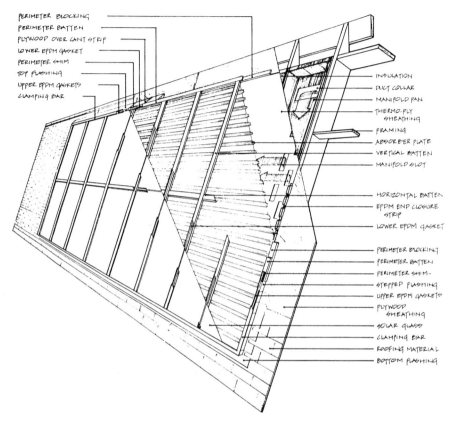

PERIMETER BLOCKING
PERIMETER BATTEN
PLYWOOD OVER CANT STRIP
LOWER EPDM GASKET
PERIMETER SHIM
TOP FLASHING
UPPER EPDM GASKETS
CLAMPING BAR

INSULATION
DUCT COLLAR
MANIFOLD PAN
THERMO-PLY SHEATHING
FRAMING
ABSORBER PLATE
VERTICAL BATTEN
MANIFOLD SLOT

HORIZONTAL BATTEN
EPDM END CLOSURE STRIP
LOWER EPDM GASKET

PERIMETER BLOCKING
PERIMETER BATTEN
PERIMETER SHIM
STEPPED FLASHING
UPPER EPDM GASKETS
PLYWOOD SHEATHING
SOLAR GLASS
CLAMPING BAR
ROOFING MATERIAL
BOTTOM FLASHING

Figure 17.1
MODEL-TEA roof collector. Source: Temple and Adams (1980).

for commercially available air heating solar collectors and that the testing
has shown that the performance is at least as good.

The study has confirmed the fact that eliminating leaks is perhaps the
most critical aspect of site building an air collector. A smoke test was
specified as an integral part of the construction process, and it has been
reported that actual installations demonstrated that these collectors can
be readily made air-tight. A low-cost air handling system and rock-bin
thermal storage that could be easily incorporated into buildings were also
developed. A comprehensive construction manual was prepared and made
available (Temple and Adams 1980).

Another design of a collector integrated into the building roof was developed by Contemporary Systems, Inc. (Christopher 1979–1985). Manufactured collector modules, 2 feet (0.61 m) wide and 10 to 16 feet (3.05 to 4.88 m) long, are mounted between the rafters of a sloped roof or the studs of a vertical wall. The collectors form a maintenance-free durable outer shell for the building and thus replace any external sheathing or roofing materials. Rock-bin storage may be used with these collectors in one of the offered designs as well as in a hybrid active/passive system. In a performance monitoring and evaluation project of such a building by DOE in 1981–82, it was found that the system performed very well, delivered a solar fraction of 82%, and had a 10-year payback (DOE 1982). It should be noted that although a 15 to 20-year care-free lifetime was projected by the manufacturer for the plastic collector windows, there is no experimental evidence that such plastic covers would last or perform effectively for that long. Plastic windows of this and other types have failed after a relatively shorter time in many collector applications, and a useful life of about 5 years seems to be the state of the art. The short life will have negative impact on the economic outlook of such systems.

Forbes and co-workers from Mississippi State University have developed and tested solar air heaters integrated with preengineered metal buildings, primarily for agricultural applications, such as poultry broilers, farm shops, etc. (Forbes and McLendon 1977, 1979; McLendon, Forbes, and Hanks 1979). One of the walls of the building is built as a collector where the air flow channel is formed by horizontally corrugated steel sheeting on one side and vertically corrugated steel sheeting on the side exposed to the sun, with a 3-cm air gap between the sheets. Rock-bin storage was used. Efficiencies of up to 70% were observed but did not correlate with the $\Delta T/I$ ratio. The cost for the collector (in 1979) was stated to be less then $10/ft^2 (about $100/m^2). That design was adapted for commercial manufacturing and sale by Gulf State Manufacturing, Inc. (Starkville, MS).

Payne and Doyle (1978) calculated the energy needed for the production (including processing of the raw materials), transportation, installation, and maintenance of a typical solar collector with an aluminum or copper absorber and glass window, and found it to be about 10^6 Btu/ft^2 (11.34 GJ/m^2). Assuming that the energy collected over a heating season is about 100,000 Btu/ft^2 (1.13 GJ/m^2), they indicated that it would therefore take about 10 years of solar energy collection just to recoup this

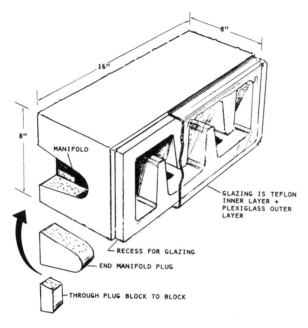

a. The standard size (8″ × 8″ × 16″) air-heating concrete block

b. A prototype of the standard size open-face air-heating concrete block with its outer glazing panel. This is one of several glazing options.

Figure 17.2
The concrete-block solar collector (Ketron, Inc.). (a) The standard size (8″ × 8″ × 16″) air-heating concrete block. (b) A prototype of the standard size open-face air-heating concrete block with its outer glazing panel. This is one of several glazing options. Source: Payne (1978).

energy investment. Year-round use was predicted by them to recoup the energy in 3 to 5 years. A more recent estimate (considering higher quality commercial systems that collect 200,000 Btu/ft^2/year, or 2.26 GJ/m^2/year) indicates values of 1.6 to 1.9 years for solar space heat with fossil fuel backup (Cleveland et al. 1984). Payne and Doyle have consequently proposed that the use of collector materials which require substantially less energy should be considered, and have designed various collectors made of concrete and fired clay. The energy use for such collectors is almost three orders of magnitude lower, about 4,000 Btu/ft^2 (45.4 MJ/m^2). They have also built and tested a few units including concrete blocks that were designed to serve as solar collectors and, at the same time, replace structural blocks in south-facing walls. One of the designs considered is shown in figure 17.2. With a block-to-ambient temperature difference of about 5°C, efficiencies of about 70% were obtained. The support for this promising work stopped before the R&D was completed.

17.1.2   Direct Heating of Makeup Air, and Space Heating Assistance from Large Solar Water Heating Systems

Often the simplest and most cost-effective way to use solar energy for heating is by ducting the makeup air used in a heating system through air-heating solar collectors. In the simplest scheme, no storage or controls are needed, and the existing air handler is used to draw the air through the collectors, as shown in figure 17.3. Shown also is a collector bypass loop that can be controlled automatically to allow better temperature control of the makeup air (Solaron 1978).

When a large water heating load coexists with a space heating load, a solar heating system that supplies both loads in an effective combination would increase the annual solar contribution and most likely reduce costs. A combined system like this is shown in figure 17.4 (Solaron 1979), where the heat exchanger coil is used for water heating.

17.1.3   Direct Room Heating by Wall-Mounted Collectors

An integrated wall-mounted collector, such as those described in 17.1.1 above, or a factory-built collector mounted on the roof or south wall can be used in the simplest way by directing its heat to the space that it borders, without thermal storage. In that case, heat is supplied from the collector whenever needed and available, and an auxiliary heater can be

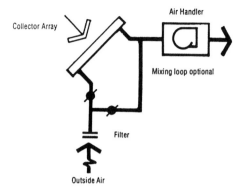

Heating outside air is often the simplest, most cost-effective use of the
Solaron system. Outside air is drawn or blown through the collector array
where the air is heated. BTU delivery is maximized in this system since the
inlet air is at ambient temperature, thereby reducing losses from the collectors.

APPLICATIONS
• Make-up Air
• Process Hot Air for Industrial Drying
• Agricultural Air Drying

Figure 17.3
Makeup air and process hot air heating. Source: Solaron (1978).

used for periods when heat is needed but not available from the collector.
A simple automatic controller can regulate this entire operation.

Reif (1981) describes the construction details of an air collector that is
mounted on the exterior sheathing of a building wall. Air is driven in a
horizontal direction by a thermostatically controlled fan through a channel formed between a corrugated sheet aluminum absorber (painted black)
in front and an aluminum-foil-faced cardboard sheet in the back. The
collector is insulated by a double-glazed window in front and by the wall
insulation in the back. It was suggested that storage is typically not necessary in a regular-size house if the collector area does not exceed 200 ft^2
(18.6 m^2). The cost estimate (in 1981) for a 110 ft^2 (10.2 m^2) collector
was about \$9/ft^2 in materials and about six days of work for two people.

Extensive residential use of factory-built daytime solar air heaters
commenced in the mid-1980s, primarily in the western and south-central
United States.

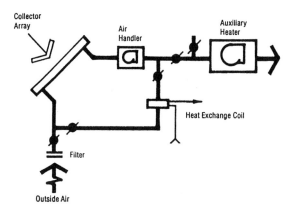

This system combines outside air and water heating. The combined system is often preferable in order to utilize the solar system throughout the year. In cases where hot air demands are intermittent, the system stores energy for hot water requirements.

APPLICATIONS:
• This system combines the applications for make-up air/process hot air heating and process hot water heating listed above.

Figure 17.4
Combined makeup air/process hot air heating and process water heating. Source: Solaron (1979).

17.1.4 Space Heating by Window Concentrators

One way to increase the solar energy collection area without increasing the size of the absorber is to add flat mirrors that reflect solar energy to the absorber. Mirrors are much lighter and less expensive than absorbers, and thus cost reduction of the solar heat may be feasible. At the same time, their angle must be changed with time to obtain maximal capture of the insolation, and they must be kept relatively clean. A system like this was developed by Wormser Scientific Corporation (Pyramidal Optical Collector 1977), and several buildings were equipped with solar heating systems based on this type of concentrating collector.

Figure 17.5 shows this system installed in the attic, using a skylight as its aperture. The size of the aperture is changed with a movable reflective surface within the attic to follow the sun angle changes during the year. Stationary and movable reflective surfaces form a pyramid shape

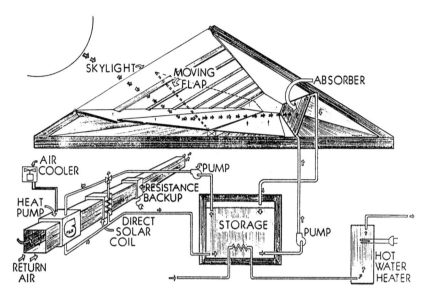

Figure 17.5
Pyramidal optics solar system. Source: Pyramidal Optical Collector (1977).

that directs the sunlight onto the absorber plate. The maximum theoretical concentration factor is 4. Although the system developers predicted a three-year payback, a number of reservations about this design, including the need to move the mirrors periodically, the occupation of potentially valuable attic space, and the actual operating experience, have stopped its further development.

17.2   Solar Space Heating Retrofit: Research and General Methodology

17.2.1   Introduction

Approximately 26% of the energy consumed in the United States is used for comfort heating and cooling, and service hot water heating in buildings. A significant satisfaction of this energy demand by the sun would be feasible over a reasonable time period only if a major fraction of the existing building stock (about 100 million in 1982) is retrofitted to solar energy use. Most of the effort in both active and passive solar heating and cooling of buildings has been so far oriented to components and new construction, and primarily to retrofit water heaters.

Having to be performed on an existing building that was not originally designed to include a solar system, retrofit is usually more costly and complex than the inclusion of a solar heating and cooling system into a new building that was designed to accommodate and integrate it with optimal installation procedures, performance, economics, and esthetics. Furthermore, a major barrier to the implementation of a countrywide solar retrofit program is the nonuniformity of existing buildings and, in many cases, the lack of adequate information about the buildings' structure and present condition, facts that tend to necessitate custom design and installation for each building.

The retrofit of existing buildings to solar heating and cooling has, nevertheless, an important national potential for the conservation of depletable energy resources and for the reduction of pollution, and in the longer term also for the reduction of comfort conditioning costs for the individual citizen. The review in this section is a very brief summary of work that has originated, in part, from the deliberations and conclusions of the Solar Retrofit Review Meeting held by the USDOE Solar Heating and Cooling R&D Branch in 1978 (unpublished; see acknowledgment). The review also includes the analysis of more recent information to present some of the main technical research aspects of active solar heating and hot water retrofit.

17.2.2 Potential for Saving Depletable Energy and Reducing Pollution

Out of the 26 quads of energy used annually by the residential and commercial sector, 18.5 are used for building service hot water and comfort heating and cooling (EIA 1981). The cost of this energy in 1981 was 89.7 billion dollars. DOE estimates that about 3.9 quads can be saved by energy conservation measures (DOE 1984). If the full energy conservation measures would indeed be implemented (a rather unlikely proposition), that would leave an annual consumption of 14.6 quads. Based on current economic constraints, solar systems are typically designed to supply around 50% of the total heating, service hot water, and air-conditioning load (although the latter is not a solar market yet). Consequently, if all buildings in the United States were retrofitted to solar heating and cooling (which is impossible in practice, but provides an upper limit), the potential annual saving in depletable fuels would be 7.3 quads, which represents an approximate cost of 35.4 billion dollars or 58% of the U.S. annual oil import. Furthermore, it would eliminate the generation of about 825 million tons (based on oil as fuel, or about 980 million tons for coal as fuel) of pollutants (including CO_2) annually.

Since only a fraction of the buildings can and will be retrofitted, the actual fuel replacement and savings would be lower, but even if, say, only one tenth of the current consumption would be supplied by solar energy by means of solar retrofit, the impact would be highly significant.

17.2.3 Technical Aspects

The general design principles specific to solar retrofit are: (1) minimal changes should be made to the building and existing heating/cooling system, (2) careful matching (interfacing) of the solar system with the existing heating/cooling system is necessary, and (3) adequate provisions for maintenance and repair should be ensured.

Even more than other solar energy applications, economical retrofit is favored by using lightweight, highly efficient solar collectors. Optimization of the spacing between collectors, which may allow some mutual shading to increase the overall amount of solar energy collected, is likely to be necessary (Lior, O'Leary, and Edelman 1977). Compact lightweight thermal storage is desirable.

Despite the fact that solar air heating systems are not subject to freezing problems and not prone to corrosion as compared to water heating types, the equipment is much more bulky and somewhat less efficient.

It is noteworthy that the existing heating system has most likely been designed for operating temperatures that are typically higher than those optimal in solar systems. Consequently, the interfacing of the solar and the existing system must be done carefully and with consideration of this difference (Dubin 1975).

The economics of solar energy applications in general has been discussed in many publications (cf. Kreith and West 1980; Ruegg and Sav 1980) and are a subject of continuous update. A survey by the U.S. Energy Information Administration conducted in 1981 indicated an *average* installed cost of $55/ft^2 (single-family) and $27/ft^2 (multifamily) for service hot water installations, and $43/ft^2 (single-family) and $37/ft^2 (multifamily) for combined water and space heating installations. At these prices and at present and near-future costs of fuel, the prospect for widespread use of solar energy is dim. Cost reductions will be attained by improved system design and manufacturing, and through transition to mass production and assembly. It should be noted that many manufacturers of solar systems and components find that the ultimate cost to the customers is overwhelmingly dominated by marketing costs (Löf, pers. com. 1987).

One way to improve the economics of solar retrofit is to design it in a way that would allow more than a single use, for example, generating additional sheltered space. Solar collectors may also alleviate maintenance costs: if a building needs a new roof or refacing with a new facade, the collectors can double up as a reliable new roof or facing material, and credit may be taken for the material that it displaces (cf. Balcomb et al. 1975).

In view of the need to reduce installation costs, and of many bad experiences with installed systems (cf. Jorgensen 1984; ESG 1984) that generated poor customer confidence in performance and reliability, the development and training of a ready work force of craftsmen skilled in solar retrofit is of great importance.

17.2.4 Some Key Research and Development Needs

CREATION OF A DATA BASE TO CLASSIFY GROUPS OF TYPICAL BUILDING/LOAD COMBINATIONS AMENABLE TO RETROFIT. A primary need for the implementation of a major solar retrofit program and the better definition of R&D requirements for that program is the conduct of a nationwide survey that would classify groups of typical building/load combinations amenable to retrofit. One of the amenability criteria should be the magnitude of energy conservation impact.

SYSTEM PRIORITIES:

1. Building service hot water

2. Space heating and low-to-medium temperature process heat

3. Heat for agricultural needs

4. Cooling

SOLAR COLLECTORS.   The principal needs are the development of: (1) economical and light-weight methods and equipment for collector support and interface to the building; (2) durable, reliable, efficient, easy-to-install, relatively lightweight, and attractive collectors; and (3) dual-purpose collector and support-structure applications.

THERMAL STORAGE.   The principle needs are: (1) the development of modular or collapsible tanks, bins, and associate materials, for easy access into existing buildings; (2) expansion of the existing R&D effort on low volume/mass storage systems (such as those using phase-change materials); and (3) study of the implementation of the thermal mass of the building for storage.

SPACE HEATING.   Since solar heat is collected more efficiently at temperatures lower than those used in conventional space heating systems, it is necessary, for buildings that presently use hot water heating, to develop lower temperature convectors heated by liquid. R&D is also needed on system integration, interfacing with auxiliary backup, whole-system design (including control), and on economical and efficient, packaged solar space heating units. A space-heating retrofit handbook needs to be developed, which would include detailed design, assembly and maintenance instructions, and detailed case studies, past experience, and costs. The format and content of *Solar* for *Your Present Home* (Barnaby et al. 1978) and SERI's *Specification and Cost Manual for Energy Retrofits* (1984) may serve as a good beginning for this effort.

SYSTEM ASSEMBLY.   The main R&D needs are: (1) development of design information about the spacing and piping of collector arrays, and its provision at the contractor level, and (2) development of the solar-related assembly and work methods, and on-site construction procedures for solar retrofit systems that represent minimal cost and disruption to the building's function. Related to the above, training of installation personnel at the vocational-school level is needed.

INSTRUMENTATION.   Development of: (1) lower-cost, simple, and reliable instrumentation packages; (2) a clamp-on Btu meter that does not require the disassembly of existing pipes; and (3) human comfort-level measurement instruments, especially for the experimental evaluation of passive systems, is needed (cf. Boehm 1978, Lior 1979, Ferraro, Godoy, and Turrent 1982).

17.3   Solar Space Heating in Urban Environment: Research and General Methodology

17.3.1   Urban Characteristics Relative to Solar Heating

Despite the high potential and challenge, relatively little work has been done so far on the overall question of massive use of solar energy in cities.

Apart from futuristic architectural projects for the design of radically new cities, such as that by Soleri (1973), which are not in the scope of this chapter, the only studies found are the work at the University of Pennsylvania (Lior, Lepore, and Shore 1976; Lior, O'Leary, and Edelman 1977; Lior et al. 1978; Smith et al. 1976; Shore, Lepore, and Lior 1977; Lepore, Shore, and Lior 1978; Lior 1980), the ambitious Solar Cities and Towns program supported by the DOE and the National Endowment for the Arts, a brief summary of which is contained in IUD (1982), and a fair number of publications that describe specific solar energy projects in cities, such as the technical description of solar retrofits of individual buildings, partially referenced in section 17.2 above.

The Solar Cities and Towns program was initiated by the Department of Energy to provide models of energy conservation and solar design at urban scale. The Design Arts Program of the National Endowment for the Arts awarded related grants emphasizing the architectural and urban design implications. Some of the projects in this program are described below:

• In the effort to clarify the question of solar access and the legislation needed to ensure it in cities, the "solar envelope" approach to design and zoning was developed (Knowles 1982). A solar envelope is a volumetric set of limits in which development can occur without shadowing the natural or built surround at specified times of day and season.

• To develop a neighborhood energy strategy for the low-income neighborhood of Roxbury, Massachusetts (about 63,000 people), a study was made of building typologies and their amenability to solar energy use. Twelve housing types were identified, and recommendations for the best conservation and solar options were made (Schnee 1982).

• In combining the concepts of urban agriculture and solar energy, a schematic design, performance simulation, and cost-benefit analysis of an 1120 ft^2 (104 m^2) rooftop greenhouse were developed (Weinstein and Smith 1982). The greenhouse was designed for construction by local volunteer labor on top of a renovated six-story tenement building in the Bronx. The greenhouse is based on concepts of cooperative ownership and provides an opportunity for vegetable crop production at little cost to the tenants, as well as potential heating benefits for the apartment building. The estimated annual crop yield was $4,252 (in 1980 dollars), but the amount of solar heat available for heating was too small. At the same time, the greenhouse provided better insulation for the roof.

17.3.2   Solar Philadelphia

Since the early 1970s, Philadelphia served as the cradle of some of the major ideas, studies, and projects for the introduction of solar energy into cities. The University of Pennsylvania (cf. Lior et al. 1976, 1977, 1978, Smith et al. 1976, Shore et al. 1977, Lepore et al. 1978, Lior 1980) initiated an effort in the early 1970s that was focused on the technical, economical, and social aspects of retrofit of row homes. This type of building constitutes more than three-quarters of the residential housing in that city and is also the dominant fraction of residential housing in many other major cities in the country. These houses are inherently well insulated, with rows of them built in a single structure often a street-block long. They usually have flat roofs, and the solar collectors may thus be placed on the roof and possibly shared by the entire block. All these attributes lead to the fact that row-homes are suitable for mass production and installation techniques of solar equipment, and thus have improved solar energy economics.

To demonstrate this concept and provide information for builders and homeowners, a row home near the campus ("SolaRow") was successfully retrofitted for space heating and domestic water, using about 500 ft^2 (47 m^2) of double-glazed flat-plate collectors on the roof and about 1,000 gallons (3.7 m^3) of hot water storage in the basement (cf. Lior et al. 1978). The system, shown in figure 17.6, was well instrumented, integrated with an automatic data acquisition system, and is still in operation. At about the same time, Drexel University (also in Philadelphia) has retrofitted one of its smaller dormitories to solar heating of domestic water, emphasizing the use of double-exposure flat-plate collectors installed on the roof, with flat mirrors used to reflect solar radiation to the back of the collector (Larson, Narayanan, and Savery 1976).

The Philadelphia Solar Planning Project (cf. IUD 1982, Burnette & Assoc. 1980, Prowler and Legerton 1980a, 1980b, Coughlin and Ervolini 1981, Coughlin et al. 1980, Miller et al. 1980, Sims and Gilmore 1980, Wallenrod 1980, Flanagan et al. 1980, Burnette and Pauman 1981) represented the first comprehensive, citywide attempt to assess the potential for solar and energy conservation applications in a major Northeastern city and to implement them through policy changes within city departments. Some of the highlights are briefly described below.

Figure 17.6
SolaRow, in a row-home neighborhood. Source: Lior (1980).

A residential housing survey indicated that four most numerous housing types accounted for 76% of the 484,000 residential buildings in the city, that over 80% are mid-row houses having two shared walls, and that over 90% have flat roofs with ample solar access (Prowler and Legerton 1980a, 1980b). An economic analysis of conservation and solar energy investments has indicated that all the conventional conservation measures considered (such as storm windows, insulation, etc.) have a payback period of less then 10 years, and all solar applications a payback of more then 10 years (at 1981 prices: $4/1000 ft^3 gas, $1/gallon oil, $0.06/kWh electricity). The Trombe wall was determined to have the shortest payback among the solar installations (Coughlin and Ervolini 1981).

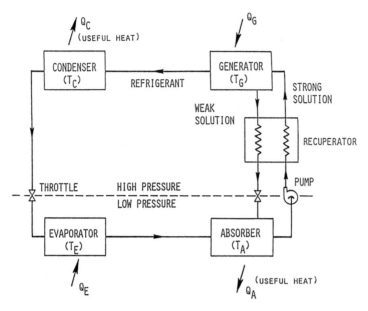

Figure 17.7
Absorption cycle heat pumps. (a) Standard absorption heat pump: $T_E < (T_A, T_G) < T_G$. Q_A and Q_C are useful heat outputs, Q_G is heat input at the highest temperature, and Q_E is "free" heat input. (b) "Reverse" absorption cycle heat pump: $T_C < (T_G, T_E) < T_A$. Q_A is the useful heat output, Q_E and Q_G are heat inputs from intermediate-temperature source, and Q_C is heat rejection to sink.

An economic input-output model for the Philadelphia Metropolitan Region was used to determine the impact of the implementation of conservation and solar measures on the economy, employment, and energy consumption in the region (Coughlin, Michel, and Cohen 1980). One of the findings was that the value of the sum of the direct and indirect benefits of these conservation and solar measures is approximately double the financial investment. Another was that the accompanying losses to the economy due to reduced demand for fossil energy are small relative to the economic gains, typically on the order of 10% or less.

17.4 Novel Solar-Driven Heat Pump Systems

17.4.1 Scope

The objective of this section is to briefly describe various novel solar-driven heat pump systems that include not only the heat pump, but also

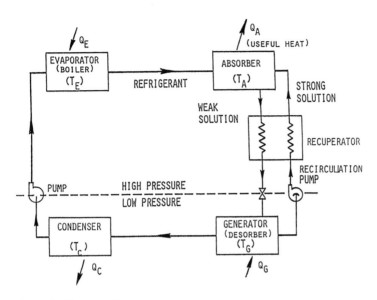

Figure 17.7 (continued)

the energy collection, storage, and demand elements. The heat pumps themselves are described in more detail in other chapters in this book. Electrically or engine-driven solar-assisted vapor-compression heat pumps are also not included in this chapter.

17.4.2 Absorption Heat Pump

Two different absorption cycles operate as heat pumps. One is the commonly used absorption cooling cycle, shown in figure 17.7a, in which there are (a) two heat inputs: one at a high temperature in the generator (Q_G at T_G) and one at a low temperature in the evaporator (Q_E at T_E), and (b) two useful heat outputs obtained from the absorber and condenser, at temperature levels (Q_A at T_A, and Q_C at T_C, respectively) between those of the two heat inputs. The temperature boosting is from the low-temperature source (applied to the evaporator) to the intermediate-temperature output (typically from the absorber and condenser), where that source could be the ambient, or low-temperature solar or waste heat. This type of system would be of benefit for heating only when the COP (based on the high-temperature heat quantity as the input) is sufficiently larger than 1, so that the trade-off of the high-temperature heat input for the lower temperature heat output is justified. From the systems standpoint, one advantage of this cycle is that it can be used year-round, for cooling in summer and for heating in winter, with minimal or no change. More detailed descriptions of this cycle are given in Niebergall (1959), Baughn and Jackman (1974), Schrenk and Lior (1975), Schwartz and Shitzer (1977), Harris and Shen (1977), Baughn and McDonald (1977), Knoche and Stehmeier (1979), and Lazzarin (1981).

The other absorption heat pump cycle, shown in figure 7b, has a heat input at an intermediate temperature, into the evaporator (which is sometimes called the "boiler" in this cycle; Q_E at T_E) and into the generator (here called the "desorber"; Q_G at T_G), it has a useful heat output from the absorber at a higher temperature (Q_A at T_A), and rejects heat at the lowest temperature from the condenser (Q_C at T_C). Compared to the previously described cycle, here the lower temperature heat input is boosted to the higher temperature without need for a high-temperature heat input. The same components and flow direction exist in both cycles, with one marked difference: by appropriate valving and pumps, the evaporator and absorber are on the high-pressure side, while the condenser and generator are on the lower-pressure side, the reverse of the pressure distribution maintained in the previous cycle. This cycle, sometimes called the "reversed absorption cycle", is described in more detail in Schwartz and Shitzer (1977), Lazzarin (1981), Isshiki (1977), Cohen, Salvat, and Rojey (1979), Perez-Blanco and Grossman (1982), and Kouremenos (1985).

The first system, *the absorption cooling cycle used as a heat pump*, has been successfully built and tested. LiBr absorption cooling units of 4.5 kW and 25 kW cooling capacity, manufactured by the Yazaki Corporation, have been tested in that mode of heating operation at the University of Padova (Lazzarin 1981). For a heat output at 35°C and generator temperatures of 80°C to 95°C, the COP of the smaller unit was between about 1.3 and 1.4 for evaporator temperatures above about 15°C, and has declined rapidly as the evaporator temperature fell below that level. For the larger unit, the COP was higher, reaching values of 1.6 to 1.7 for evaporator temperatures above 22°C. An ARKLA air-source gas-fired 7 kW ammonia-water cooling unit was built and tested as a heat pump to provide 14.65 kW heating (Kuhlenschmidt and Merrick 1982). For a delivered water temperature of 51°C to 52°C, the COP for an outdoor ambient temperature of 8.3°C was 1.26, and it declined to 1.12 for an outdoor temperature of −8.3°C. The electrical power requirement was small, 0.5

kW. For both types of heat pumps, an improvement in COP was recommended through further R&D.

Several conceptual solar-assisted heating systems based on this cycle have been proposed and analyzed (cf. Lauck et al. 1965, Schrenk and Lior 1975, Schwartz and Shitzer 1977, Harris and Shen 1977, Baugh and McDonald 1977, Cocchi et al. 1979, Lazzarin 1981, McLinden and Klein 1982). The most recommended system configuration is the one where the solar collectors supply the heat directly to the load when the temperature is high enough, and to the heat pump evaporator when it is not. With water as refrigerant (in a LiBr-water system, for example), the temperature of the evaporator must be kept, with possible assistance from solar heat, above 0°C. The fluid used for supplying heat to the load is preferably circulated in parallel through the absorber and condenser (Cocchi, Raffellini, and Stopponi 1979), since series circulation through the condenser and absorber provided a somewhat lower COP. The generator in this configuration is heated by a fossil-fuel source, since the efficiency of collectors for such high temperatures is relatively low and their cost relatively high.

Several studies have been made to examine the effects of using tanks of refrigerant and solution, integral to the absorption loop, as thermal storage (cf. Baugh and Jackman 1974, Semmens et al. 1974, Schrenu and Lior 1975, Harris and Shen 1977, McLinden and Klein 1982). The thermal storage is charged by supplying the energy required to separate the refrigerant from the absorbent solution (in the generator) and is discharged by releasing latent heat in the condenser and by the exothermic absorption of the refrigerant in the weak absorbent solution in the absorber. The condensed refrigerant (which is on the high-pressure side of the system) can also be stored and expanded into the evaporator whenever cooling is needed. The advantages of such thermochemical storage are the ability to store more energy per unit volume or weight than that of sensible heat storage systems for the same purpose, the almost isothermal nature of the storage system, and the typically ambient temperature of the storage system that minimizes heat losses and need for thermal insulation. The studies by McLinden and Klein (1982) and Semmens, Wilbur, and Duff (1974) propose only two storage tanks: one for the condensate and one for the absorbent solution. In intermittent and transient operation, characteristic for systems with thermal storage, this system configuration would, however, lead to changes in solution concentration that may either lead to crystallization or to higher-than-optimal generator temperatures. The ad-

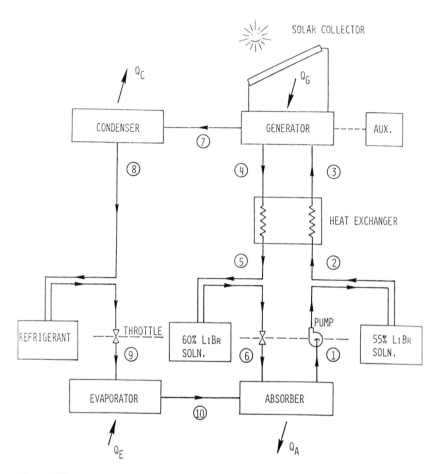

Figure 17.8
Schematic of solar-assisted absorption heat pump with integral refrigerant and solution storage. Source: Schrenk and Lior (1975).

dition of one more tank, as proposed by Baughn and Jackman (1974), Schrenk and Lior (1975), Harris and Shen (1977), and Baughn and McDonald (1977), and shown in figure 17.8, allows the maintenance of constant concentrations in the system throughout its operation, within the limits of stored mass.

The second system, *the reversed absorption cycle*, was tested at the Oak Ridge National Laboratory with the primary objective of making use of industrial waste heat in the 60°C to 90°C range (Huntley 1982). The test unit was constructed by ARKLA Industries by using several modified components of their standard 88 kW (25 ton) LiBr-water chiller, to result in a heating capacity of 42 kW. For hot water input temperatures of 60°C to 80°C, and condenser cooling water inlet of 14.7°C to 34.6°C, the COP was 0.38 to 0.5, and the temperature boost 15.1°C to 32.1°C. The parasitic power was relatively small, with electric COPs of 50 to 70. One may conclude that the relatively small temperature boost and COP at these conditions do not make this heat pump commercially attractive. A two-stage heat pump of this type was proposed and analyzed, indicating potentially larger temperature boosts, at the expense of a more complex system using somewhat more parasitic power (Perez-Blanco and Grossman 1982).

17.4.3   Chemical Heat Pump

The absorption heat pumps discussed above are part of the broader family of chemical heat pumps. All of the chemical heat pumps described here share the same principle: heat is stored through two reversible thermochemical reactions, one that releases the more volatile component and the other that absorbs it (sometimes the second reaction is simply a phasechange process). Due to the difference in concentration in the two reactions, they occur at two different temperatures. Going in one direction, the process transfers heat from the high temperature region to the low. Reversal of the reaction allows the transfer of heat from the lower temperature to the higher, which thus produces the heat pump effect, i.e., the boosting of a low temperature energy source to a higher level. Another attractive feature is the high energy storage density involved with such reactions, because it includes here not only sensible heat but primarily heat of reaction and/or phase change. The reactants may be either a liquid and a gas (the gas being the volatile component), or a solid and a gas. More details about the fundamental processes involved and their thermodynamic characteristics can be found in Offenhartz (1976) and Raldow and Wentworth (1979). Some of the key system-related aspects are described below.

Apart from the related work on absorption cooling, very little work has been done on chemical heat pumps until the 1970s. The energy crisis has prompted the study of a number of such concepts. The DOE Office of Energy Systems Research supported five studies for systems based on calcium chloride/methanol, magnesium chloride/water, sulfuric acid/water, ammoniated salts, and paired metal hydrides. Other systems were under development privately, including a zeolite/water cooling system, and a sodium hydroxide/water chiller to operate at temperatures of about 160 F

(71°C); a Na_2S/H_2O system is under development in Sweden; and an ispopropanol/hydrogen/acetone system in France (Mezzina 1982).

The system based on the *reaction of* $CaCl_2$ *and* CH_3OH vapor was investigated theoretically and experimentally by the EIC Corporation (Offenhartz 1978; Offenhartz and Brown 1979). The reactions are:

$$CaCl_2 \cdot 2CH_3OH(\text{solid}) \rightarrow CaCl_2(\text{solid}) + 2CH_3OH(\text{gas}) \quad (\text{at } T_2),$$

$$CH_3OH(\text{gas}) \rightarrow CH_3OH(\text{liquid}) \quad\quad\quad\quad\quad\quad\quad (\text{at } T_1).$$

If a temperature T_1 of 40°C is wanted for heating or for heat rejection, T_2 must be about 130°C for regeneration in the charging mode. These conditions allow a pressure drop from the solid $CaCl_2 \cdot CH_3OH$ to the condensing methanol, and a respective transport of the methanol vapor in that direction.

Experiments (Offenhartz and Brown 1979) have proven the concept in principle and also identified some of the problems: the difficulty in producing good rates of mass and heat transfer with this solid inorganic salt without appreciable pressure drop in the methanol vapor, the need to keep the system sealed against air in-leakage, the doubling of the volume of $CaCl_2$ during the transformation to $CaCl_2 \cdot CH_3OH$, and corrosivity. It was also found that a temperature driving force of 14°C to 20°C was needed to drive the reaction at the required rates.

The *sulfuric acid/water* system was studied and tested by Rocket Research Company (Clark and Hiller 1978; Hiller and Clark 1979a, 1979b; Clark and Carlson 1980; RRC 1982). In the charging mode, a mixture of H_2SO_4 and H_2O is heated, and the water is removed by boiling at a pressure maintained by a cooled condenser. The boiling point keeps rising with the concentration of the acid in this process, going from about 60°C at a concentration of 70%, to 170°C at 95%. Thus the first reaction is that of separation of water from the acid, and the second is condensation. When discharging, the concentrated acid solution is cooled to a point where its vapor pressure becomes lower than that of the water in the condenser, and the process reverses: heat supplied at the condenser temperature causes the water to evaporate and flow to the acid solution, with which it recombines. This reaction is exothermic and raises the temperature of the solution. Heat can thus be withdrawn from the solution for some useful purpose, at a temperature higher than that of the condenser, resulting in the heat pump or temperature boosting action. This process is

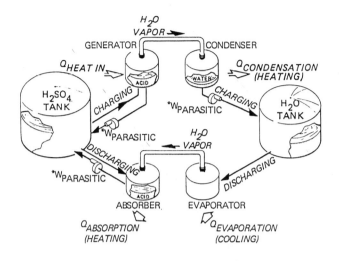

*$W_{PARASITIC}$ = 3 TIMES THE ELECTRICAL WORK

Figure 17.9
Chemical heat pump/chemical energy storage system schematic. Source: Clark and Carlson (1980).

rather similar in principle to that in a conventional absorption heat pump with internal energy storage. The heating COP was found in the laboratory and in theoretical predictions to be about 1.6 for heating and about 0.6 for cooling.

A system flow diagram for both heating and cooling is shown in figure 17.9. The test facility operated well with charging temperatures between about 100°C and 200°C. A major effort was made to develop an industrial heat pump to boost the temperature of waste heat sources at up to about 250°F (121°C) to temperatures of up to about 380°F (193.3°C). Temperature boosts of up to about 113°F (62.8°C) were obtained at a thermal COP of 0.2–0.4.

Metal hydrides are typically metal alloys of AB_5 composition, which were found to reversibly absorb and desorb large amounts of hydrogen at a relatively constant pressure with excellent kinetics. For example, stoichiometric $LaNi_5$ can absorb over six hydrogen atoms while undergoing a 25% lattice expansion, with over 95% of the hydrogen equilibrium pressure attained in a few minutes at room temperature. The absorption reaction is exothermic and the desorption endothermic. These properties, combined with the fact that different hydrides have different temperatures for the same hydrogen pressure plateau, indicate that a heat pump can be constructed by using the hydriding reaction with two metals, one at the higher temperature (T_H) and one at the lower (T_L) (cf. Gruen, Mendelsohn, and Sheft 1978; Argabright 1982a, 1982b).

The process is in principle similar to that of the previously described chemical heat pumps and is depicted in figures 17.10a (heat amplification) and 17.10b (temperature boosting). Four tanks are used: 1 and 2 for the warm-side alloy, and 3 and 4 for the cold-side alloy. One can see from figure 17.10a that approximately two units of heat are produced at T_M for each unit of heat supplied at T_H. The temperature-boosting cycle shown in figure 17.10b uses an intermediate temperature heat source (at T_M) and a low temperature (T_L) source (say, the ambient) to produce heat at the highest of the three temperatures, T_H. Here the COP is lower than 1 (about 0.5), but temperature-boosting is achieved.

As shown in figure 17.10, the cycles must operate intermittently, requiring charge cycles between cycles of useful output. Because of the high cost of the alloys (about \$20/1b for LaNi$_5$ in 1982), which constitutes the major cost item in the proposed heat pump, it is necessary to minimize the quantity of the alloy per unit energy output. This can be accomplished by reducing the cycle time in this intermittent cycle and by minimizing the thermal losses due to the intermittent heating and cooling of the heat pump structure itself. Thus it is desirable to provide high rates of heat transfer with a low thermal mass heat exchanger. The DOE-supported effort by the team of Southern California Gas Company and Solar Turbines, Inc. has indeed been focused in this direction. Using LaNi$_5$ and MmNi$_{4.15}$Fe$_{0.45}$ (Mm is Mischmetal) as the high and low temperature alloys, respectively, they developed heat exchangers consisting of finned copper tubes in a staggered tube bundle arrangement in which the hydride powder is stored in the annular spaces between the fins. Water is the heat transfer medium that flows through the tubes. Tests were performed (for refrigeration only) with the approximate temperatures being: $T_H = 200°F$ (93.3°C), $T_M = 85°F$ (29.4°C), and $T_L = 45°F$ (7.2°C). Due to the general problem of inadequate cycling rate and change in the federal research program directions, the funding of this project was discontinued before the necessary R&D for the solution of these problems was completed, and thus before a complete operating heat pump system could be constructed.

Possible environmental impact and hazards due to the use of the hydrides and hydrogen were not discussed or explored.

An *open cycle desiccant* heat pump was developed by Robison and co-workers (Robison 1982; Robison and Griffiths 1984), analysis on such concepts was also made by others (cf. Schlepp and Collier 1981), based on the reaction between the humidity in the air and a desiccant. The desiccants tried by Robison were triethylene glycol, calcium chloride, and mixtures of calcium chloride and lithium chloride. A heat pump based on this principle was installed in a 2800 ft^2 (260 m^2), house in South Carolina and provided all the heating and cooling for a few years. The authors estimate a cost of around $2,000 (in 1984) for a 3-ton heat pump of this kind, but more detailed work and economic analysis need yet to be performed to draw reliable conclusions on the commercial viability of this concept.

A comparative study of the energy performance and economics of the sulfuric acid/water and methanolated salt chemical heat pumps discussed above and of conventional and emerging-technology HVAC systems was conducted by TRW, Inc. (1981). It was concluded that solar-driven chemical heat pumps consume less resource energy than either baseline or emerging systems, that chemical heat pumps can be cost-competitive in specific residential applications and locations (for example, in Albuquerque but not in Boston), and that their application in commercial buildings is not attractive either from the energy-conservation or from the cost-saving standpoints.

17.4.4 Rankine Cycle Heat Pump

This chapter focuses on overall *heat pump system* innovations; more information on mechanical cooling systems is given in chapter 19 of this volume. Solar-assisted heat pump systems that are not driven by solar engines are not treated here; information can be found in chapter 12 of this volume.

The major characteristics of five of the main Rankine cycle heat pumps studied are described in table 17.1 (cf. Koai, Lior, and Yeh 1984). Four of the systems use organic fluids in the power cycle, and so do most of the Rankine cooling systems. The advantages of using organic fluids are their high molecular weight (typically 100 or more), which provides improved cycle efficiency in a less costly single stage expander, and a positive slope $\partial T/\partial S$ vapor saturation limit curve, which produces a dry (superheated)

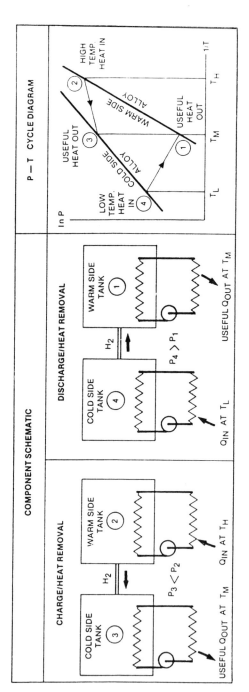

(a) Heat amplification

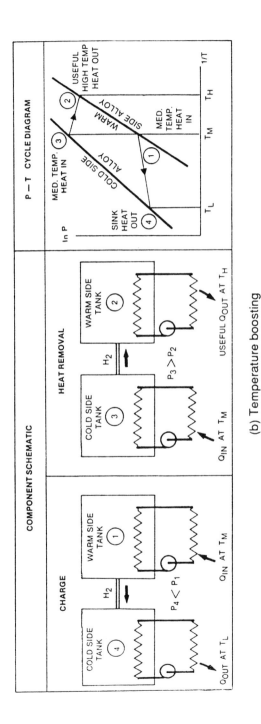

Figure 17.10
Metal hybride heat pump cycles. Source: Argabright (1982a). (a) Heat amplification. (b) Temperature boosting.

Table 17.1
Rankine-cycle solar heat pump projects

Investigator	Power cycle fluid	Engine	RPM	Compressor	Cooling cycle fluid	Solar supply temp., °C	Cooling capacity, tons	Comments	
United Technologies Research Center	R-11	Turbine	45,000	Centrifugal	R-11	143	18	same-shaft turbo compressor	a
Garrett-Air Research	R-11	Turbine	24,800–82,600	Centrifugal	R-11	93	3, 25, 75	same-shaft turbo compressor	b
Battelle Laboratories	R-12	Pivoting-tip rotary-vane	3,600	Pivoting-tip rotary-vane	R-12	100	3		c
General Electric Corp.	FC88 (fluorinated hydrocarbon)	Rotary vane	1,200	Reciprocating	R-22	100–140	3, 10		d
University of Pennsylvania	steam	Turbine	15,300	1750 RPM Commercial (Trane) open, reciprocating, or other	R-22	100	25	Hybrid cycle: 20% of the energy supplied by fuel to super-heat steam to 600°C	e

References: a. Biancardi, Sitler, and Melikian (1982), Melikian et al. (1982).
b. Rousseau and Noe (1976), McDonald (1978).
c. Fischer (1978).
d. Eckard (1976), Graf (1980).
e. Lior (1980), Lior, Yeh, and Zinnes (1980), Lior and Koai (1984a, 1984b), Sherburne and Lior (1986), Koai, Lior, and Yeh (1984).

vapor upon expansion from the saturated state, thus avoiding problems associated with droplet formation in the turbine. In general, organic fluids also have disadvantages, such as possible toxicity, flammability, corrosivity, instability at high temperatures, susceptibility to oxygen and contaminants, and mutual degradation when in contact with lubricants used in the engine/compressor. Several have performed well, however, for the duration of the tests (up to several thousand hours with several fluids, and several million hours with tri-chloro-benzene at temperatures up to 200°C); for further references see chapter 19 of this volume. Thermal stability limitations also do not permit superheating by the addition of heat, an important detriment to the potential improvement of cycle efficiency.

Steam is used as the working fluid in one of the systems described below (Koai, Lior, and Yeh 1984; Lior 1977; Lior, Yeh, and Zinnes 1980; Lior and Koai 1984a, 1984b, Sherburne and Lior 1986; Curran and Miller 1975), and although it has the disadvantage of a negative $\partial T/\partial S$ curve and a molecular weight at least 6 times lower than most organic fluids used in Rankine cycles, thereby requiring a more complex and multi-stage expander, it is stable, nontoxic, nonflammable, and relatively noncorrosive. It has a much lower back-power ratio (requires less pumping power), its properties are well known, and it can be easily superheated by the addition of heat to increase cycle efficiency. Its cost is relatively negligible, a fact that also allows economical integration of water-based thermal storage with the steam and power generation components (cf. Lior and Koai 1984a).

The solar Rankine heat pump development program, which was probably the most successful in meeting design and performance goals, was conducted by the United Technologies Research Center (UTRC) with its subsidiary, Hamilton Standard Division (Biancardi, Sitler, and Melikian 1982; Melikian et al. 1982). Primarily supported by DOE, the most advanced model developed and tested was the "MOD-2", with a capacity of 18 tons (63.3 kW) cooling and 110–170 kW heating (figure 17.11). The turbocompressor used was developed specifically for this application, consisting of a single-stage fixed-geometry turbine and centrifugal compressor mounted on a common shaft and operating at a design speed of about 42,000 RPM.

Cooling tests in the laboratory, using a water-cooled condenser (but with the water discarding heat to ambient air in an outdoor heat exchanger) at the ASHRAE rated conditions of 95°F (35°C) ambient air and

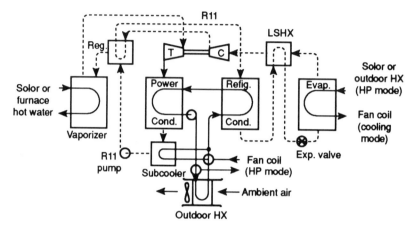

Figure 17.11
UTRC "MOD-2" heat pump module schematic. Source: Melikian et al. (1982).

45°F (7.2°C) chilled water output, with a vapor generator inlet water temperature (equivalent to about 10°C less than the solar collector outlet
temperature) of 295°F (146°C), have indicated a cooling COP of about
0.64, turbine efficiency of 76.5%, and compressor efficiency of 78%. Extra-
polating from data obtained through about 120 hours of testing of the
MOD-2 and from the seasonal COP of 0.39 for a UTRC earlier model
heat pump adapted to cooling only, which operated for two years at a site
in Phoenix, Arizona, a seasonal cooling COP for the MOD-2 may conservatively be estimated to be about 0.44. At present and near-future costs of
conventional energy and solar system components, neither this nor the
other solar Rankine heat pumps described below were found to have significant market penetration potential.

UTRC also presented a conceptual design of a more advanced solar
Rankine heat pump, the MOD-3, emphasizing improvements of the parallel solar-assisted heat pump concept as opposed to the MOD-2 series
solar assisted concept. As shown in figure 17.12, vapor compression loop
improvements considered include two-stage compressors with inter-stage
flash economizer, condenser subcooler, liquid suction heat exchanger, and
compressor exhaust recuperator. Power loop improvements include two-
stage expansion with regenerative feed heating, and turbine and exhaust
recuperation. Additionally, better matching of the turbine and compressor
designs is proposed by use of multi-staging or intermediate gearboxes, to

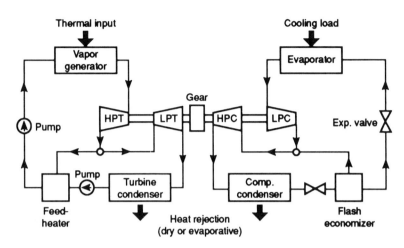

Figure 17.12
UTRC-proposed advanced-design Rankine cycle heat pump. Source: Melikian et al. (1982).

ensure that both turbine and compressor will be running near their best efficiency points over a wide range of operating conditions. It was predicted that such a unit would double the COP of the MOD-2 at a relatively-low increase in cost and may thus achieve commercial viability.

The only solar Rankine heat pump system using steam was proposed and developed by the University of Pennsylvania (Lior 1977; Lior, Yeh, and Zinnes 1980; Lior and Koai 1984a, 1984b; Sherburne and Lior 1985, 1986; Curran and Miller 1975). Depicted in figures 17.13 and 17.14, and named SSPRE ("solar steam powered Rankine engine"), this is a hybrid solar-powered/fuel-assisted power cycle that is used to drive a conventional vapor compression heat pump. The underlying principle of this cycle is the use of energy from two different temperature levels to arrive at (1) a better thermodynamic matching with the energy sinks in the power cycle, and (2) improved system economics. The design conditions for the system that was developed under DOE contracts specified the supply of solar energy at about 100°C to convert the water into steam, and the supply of heat from a fuel-fired, or from point-focusing solar concentrators—Sherburne and Lior (1985), superheater to superheat the steam to 600°C, the top temperature compatible with engineering materials used in conventional power plant technology. Exhaustive analyses both by the University of Pennsylvania researchers and others (Koai, Lior, and Yeh 1984; Lior

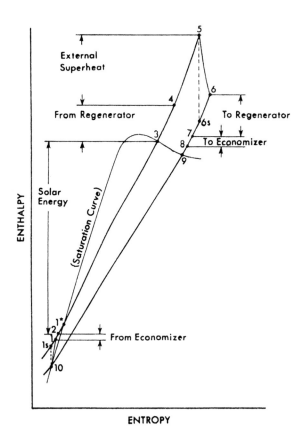

Figure 17.13
The solar-powered/fuel-assisted hybrid Rankine cycle (SSPRE) Mollier diagram. Source:
Lior (1977).

1977; Lior and Koai 1984a, 1984b; Curran and Miller 1975) have shown
that when about 20% to 26% of the total energy input is supplied by fuel
(in the superheater), the power cycle efficiency is essentially doubled above
that of organic fluid Rankine cycles, which operate at similar solar collector temperatures, to values of about 15% (seasonal) to 18% (at design,
with condensation at 46°C). This doubling of the efficiency halves the
required area of solar collectors, which is the primary cost-component of
all solar power cycles. Furthermore, solar energy at 100°C can be obtained
from flat-plate or evacuated collectors, a major advantage in cost and
reliability when compared to concentrating collectors with tracking. It is

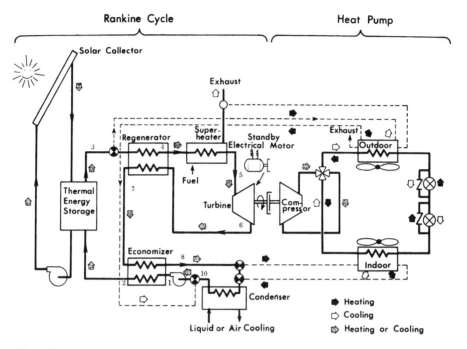

Figure 17.14
The University of Pennsylvania solar-powered/fuel-assisted hybrid Rankine cycle driven heat pump: system diagram. Source: Lior (1977).

also noteworthy that since steam generation occurs at 100°C, the highest pressure in the system is atmospheric.

Computer simulations (Koai, Lior, and Yeh 1984; Lior 1977; Lior and Koai 1984a, 1984b) have indicated that this system should have a cooling COP of about 0.62 under ASHRAE standard test conditions while operating with solar collector water output of 98°C. This is about the same COP as obtained in tests of the UTRC system described above. However, the UTRC system was operating at the much higher collector water temperature of about 156°C, which necessitates tracking solar concentrators and more expensive thermal storage and insulation. Using water-cooling, instead of air-cooling and making two other relatively simple changes, it was predicted that a cooling COP of about 1.35 could be attained by the SSPRE system.

Operating in the heating mode, the SSPRE heat pump system was compared (through computer simulation, Lior [1977]) with ten other heating

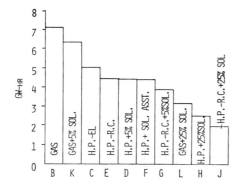

Figure 17.15
Annual resource-energy consumption for heating with use of SSPRE heat pump—30-story
office building, New York. Source: Lior (1977).

systems, nine of which are shown in figure 17.15. Applied to a New York
skyscraper with a design heat load of 3MW, figure 17.15 shows that the
SSPRE system has the lowest resource energy consumption of all systems
compared; for example, a 3.5-fold decrease in annual resource energy consumption for heating occurs when the SSPRE system with collector area
of about 27,000 ft² (about 2,900 m²) is used instead of a gas-fired heating
system.

Since low-horsepower commercial steam turbines that operate under
the conditions of the SSPRE cycle have an efficiency typically below 50%,
a novel 30-HP radial-flow 10-stage turbine with 25-cm-diameter counter-
rotating rotors, which uses reaction blading, was designed at the University of Pennsylvania and built for operation in the system. The design
predicts an efficiency of 75% and excellent off-design performance. Preliminary tests, which had to be interrupted due to a leakage problem, have
indicated efficiencies up to 79.5%. The design speed is 15,300 RPM, significantly lower than that of the organic fluid Rankine cycle turbines used,
and thus with a potentially higher reliability. The speed-reduction gear
box incorporates an over-running clutch system to allow the automatic
engagement of a backup electric motor.

Another novel feature, made possible by the use of steam as the working fluid, is the incorporation of a combined thermal-storage/steam-
generation system. The heat from the collectors is stored in a hot water
tank that releases steam to the turbine by flashing when it is exposed
to the lower pressures governed by the power-cycle condenser. Con-

sequently, the same water is used both as the thermal storage medium and the power-cycle working fluid, with resulting economy in equipment.

The entire 25-ton system was constructed, and both component and system testing was started; the work was interrupted, however, by major reductions in DOE solar energy budgets.

The Rankine-driven heat pump work may be summarized by saying that the organic fluid cycles have been demonstrated, improved, and made more reliable primarily by using good engineering. The SSPRE hybrid steam cycle was the least conventional and had the highest predicted performance and economic potential but has not been advanced yet through the stages of shakedown and full testing. The discontinuation of these programs under DOE support are a consequence primarily of poor economic prospects at present fuel prices (for more details on the economic status and commercialization prospects, see also chapters 19 and 22 in this volume).

17.4.5 Note on Performance Criteria

The definition of indices of performance of energy systems that have more than one type of useful output (say, heating *and* cooling) and/or input (say, fuel *and* electricity) poses a problem, because the unit value and cost of these energy quantities may be different (cf. Bonne [1978]). Simply added up in energy units, the efficiency or COP may be misleading: for example, a higher efficiency or COP may actually result in a higher cost-per-unit desired energy output because of a shift to using a larger amount of the more expensive energy type.

The definition of energy-performance indices for solar heating and cooling systems poses even greater difficulties, primarily because a larger number of variables must be specified to describe the conditions under which the performance criterion is to be determined.

Taking these considerations into account, Lior, Yeh and Zinnes (1980) proposed a new performance criterion, the "economic COP" (with the symbol COP$):

$$\text{COP\$} = \frac{H_0}{AH_s + BH_f + CE},$$

where

H_0	is total useful energy for heating and cooling;
H_s	is solar energy input;

H_f is fuel energy input;

E is electric energy input;

A, B, and C individual energy source costs of solar heat, fossil fuel
                         energy, and electricity inputs, respectively, in \$/energy unit.

A, B, and C may be the present-value life-cycle costs, or calculated by any
other method that the customer wishes to use in the evaluation of heating/
cooling options. This method complicates somewhat the evaluation of the
coefficient of performance (relative to a ratio based on energy), but it still
is a relatively small penalty in view of this straightforward resolution of a
complex problem. It is worth noting again that C is also a function of the
overall system configuration.

    This COP\$ must be determined for the same cooled/heated space and
ambient conditions, possibly as recommended by ASHRAE and ARI
standards, but it should also be determined for a "typical" cooling and
heating season for several loads and geographic locations.

    To those interested in the overall supply, demand, consumption, and
conservation of fuel resources, a COP based on resource energy, COP_r,
should be calculated as follows:

$$COP_r = \frac{H_0}{H_f + E/\eta},$$

where η is the efficiency of conversion of heat to electricity, usually taken
at 0.25 to 0.3, which would allow comparing the solar-powered unit to
those that are entirely electric-powered:

$$COP_{r,e} = \frac{H_0}{E/\eta},$$

or fossil-fuel fired:

$$COP_{r,f} = \frac{H_0}{H_f}.$$

17.5 Large-Scale Concepts

17.5.1 Space Heating with Annual Energy Storage

It is an appealing idea to collect solar heat during the summer, when it is
more abundant and less needed, store it, and then use it in winter (along-

side with the heat collected in winter), when it is more needed and less available. Several studies were conducted to determine the effect of storage size on the solar fraction gained, on required collector area for a desired solar fraction (Speyer 1959; Hooper and Cook 1980; Drew and Selvage 1980; Braun, Klein, and Mitchell 1981; Sillman 1981), and on the heat losses incurred from the storage device (Hooper and Attwater 1977). The common conclusion was that for a desired solar fraction, the use of very large storage reduces the required solar collector area significantly. For the climatic conditions of the northern states, it was found that for buildings of more than 10,000 ft^2 (955 m^2) floor area, annual storage systems providing 100% of the space heating requirement from the solar source were more cost effective than shorter term storage systems supplying their optimum proportion of the load (usually about 60%). A consensus conclusion from these studies is that small district heating systems in which a single storage is shared by the community appear to offer the most cost-effective means of solar heating for smaller buildings in the same region. It is generally felt, however, that clearer evidence of the cost advantages of systems with long-term energy storage vs. those with short-term energy storage still needs to be generated.

Quantitative guidelines for the effect, optimization, and design of solar heating systems with seasonal storage were obtained from a computer simulation for a district of 50 family houses in Boston, Medford, Oregon, Bismarck, Nevada, and Albuquerque, New Mexico, (Sillman 1981). Several types of building loads and energy conservation measures were considered. The results are summarized in figure 17.16 and 17.17.

It is noteworthy that the solar fraction in any region could actually decrease with increased storage volume if the collector area is below a certain size, because of the accompanying drop in temperature below useful levels. At the other limit, if the solar fraction needs to be raised in systems with diurnal-type storage, the larger collector area required for that purpose produces temperatures above allowed limits for a sizable fraction of the time (especially in summer), and thus cost-optimization does not favor high solar fractions, typically showing a minimal cost at a solar fraction of about 40–70%. A number of easy-to-use design tools have been developed for the sizing of such systems (Baylin and Sillman 1980; Braun, Klein, and Mitchell 1981; Drew and Selvage 1980).

In the three northern cities, the optimal design required about 1 to 2.5 m^3 of water storage per m^2 collector. Due to the much higher insolation,

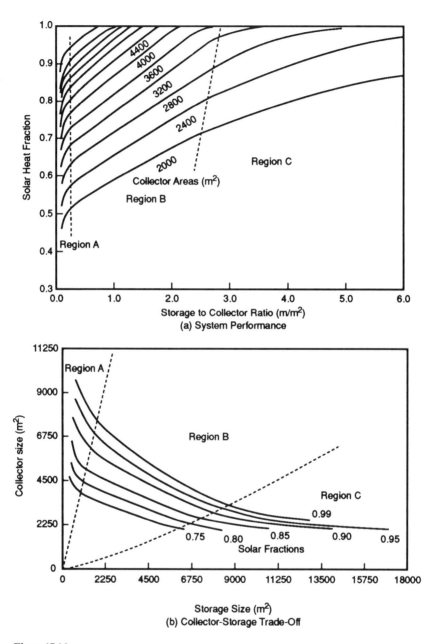

Figure 17.16
System performance and collector storage tradeoff curve for a typical annual storage
system. Source: Sillman (1981).

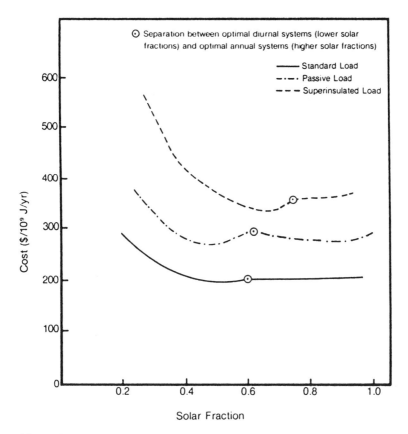

Figure 17.17
System cost per unit heat delivered vs. solar fraction for combined space heat and hot water systems. Source: Sillman (1981).

5 m^3 of storage were needed per m^2 of collector in Albuquerque, New Mexico. Depending on the particular system and location chosen, using the economic parameters listed under figure 17.17 and a capital discount rate of 10%, the seasonal storage system was reported to break even against other heat sources costing 4–9¢/kWh.

A more formal optimization conducted by Lund (1984) for a 60° latitude location (Helsinki), with a system that includes a heat pump using the lower-temperature stored heat as the source, indicated, contrary to what was predicted in the previous analyses, that the present-value life-cycle cost of a system with 50% solar fraction would be lower than half that of one with a 90% solar fraction. For a partially passive heated system, a 2% annual fuel-escalation rate and a real 4% discount rate, Lund predicted a life-cycle energy cost of 8.9¢/kWh for a typical system. It should be noted that the studies from Sweden and Finland use electric power costs that are significantly lower than those in the United States and thus their final cost conclusions are not applicable here.

In view of the traditional use of district heating in Europe, significant interest exists there to develop central solar heating plants with seasonal storage (Bankston 1982; Lund et al. 1982). Six such systems were built in Sweden and one in Finland, described briefly in table 17.2. These systems have in general operated well. It would be of value to the program to try and validate the theoretical design and prediction methods (quoted above) with the performance results from these operating plants. Much of the activity is indeed integrated through the International Energy Agency Solar Heating and Cooling Program, Task VII on Central Solar Heating Plants with Seasonal Storage, and the Energy Conservation through Energy Storage Program (a recent review is given in Bankston [1986]).

A 27,400-gallon (103.7 m^3), 12-ft (3.66 m) deep experimental system for annual collection and storage of solar energy was constructed in 1976 at the University of Virginia (Beard et al. 1979). It was operated and evaluated for one collection-heating cycle from February 1977 through January 1978, and it was found that the heat losses were excessive. At the end, the collector suffered wind and snow damage and the experiment was terminated.

An active project in Hatfield, Massachusetts, heated a 5,000-m^2 school building with a system that includes a 2,000-m^3 insulated earth-storage unit charged by 170 m^2 of solar collectors and discharged through 27 water source heat pumps to different building zones. The system has been monitored by the University of Massachusetts since 1982 (Krupezak et al. 1985).

17.5.2 Space Heating by Solar Ponds

17.5.2.1 Shallow Ponds
The shallow solar pond (SSP) is a large-area horizontal solar collector that consists of a layer of water, a few centimeters deep, resting on a thermally-insulating base material, with one or two sheets of glazing placed over the top. A transparent (typically plastic) evaporation-suppres-

Table 17.2
Solar central heating projects with seasonal storage in Sweden and Finland

Location	Collector Type	Area (m²)	Heat Pump	Storage Type	Volume (m³)	Planned Solar Contribution (%)	Application
Studsvik, Sweden	CPCᵃ on rotating lid	120	No	Water in earth pit	640	100	Office building
Ingelstad, Sweden	Line-focus PTCᵇ	1,250	No	Water in concrete tank	5,000	50	52 single-family houses
Lambohov, Sweden	Single-glazed, selective FPCᶜ	2,900	Yes (storage to consumer)	Water in rock	10,000	85	55 single-family houses
Sigtuna (Sunstore project), Sweden	Black FPC with one glass; black FPC uncovered	36 126	No	Rock, 40 bore holes	10,000	80	One large single family house
Kingsbacka (Sunclay project), Sweden	Black FPC, uncovered	1,600	Yes (storage to consumer)	Clay, submerged U-tube	85,000	60	School for 800 students
Lyckebo, Sweden	Selective FPC, covered by one glass and two Teflon sheets	4,320	Not yet decided	Rock cavern	1000,000	15	500 dwellings
Kerava, Finland	Single-glazed FPC	1,100	Yes	Water in rock + rock with 54 bore holes	1,500 + 300 equivalent in rock	45	44 apartments

Source: Bannston 1982, Lund et al. 1982.
a. CPC: compound parabolic concentrator
b. PTC: parabolic-trough concentrator
c. FPC: Flat-plate collector

sion layer is in contact with the top water surface, and the water rests on a black radiation-absorbing bottom surface. Several designs have been developed and tested since the early work by Willsie and Boyle was published in 1909, and a detailed description of the state of the art, especially of the more recent work at the Lawrence Livermore Laboratory, was given by Clark and Dickinson (1980), and co-workers (Casamajor and Parsons 1979). The purpose of the ponds is to heat water to about 40–70°C (100–160°F), at which the ponds operate at a typical average daily efficiency of about 30–50%.

Intended ultimately for providing process heat for the Sohio uranium mining and milling complex (near Grants, New Mexico), an SSP prototype test facility, consisting of three 3.5m by 60m modules and hot and cold storage reservoirs, was constructed by researchers at Lawrence Livermore Laboratory. Temperatures of up to the 60°C range were obtained in summer, up to the 40°C range in fall, and up to 19.5°C in winter. The all-inclusive installed costs were $60.20/m^2 ($5.50/ft^2) in 1975 dollars. At the economic conditions of that time, it was estimated that the price should come down to about $40/m^2 to be competitive with oil at $15/barrel.

An SSP system was designed and constructed by the same group to provide heating for army barracks at Fort Benning, Georgia (LLNL 1985). The overall pond area was 26,500 m^2 (composed of 80 pond modules and covering 11 acres of land) to supply 2,000 m^3 of hot water per day for the barracks and laundry. The project was constructed by the Army Corps of Engineers for an approximate cost of $4.5 million and planned to save more than 11,000 barrels of oil per year.

17.5.2.2 Salt-Gradient Ponds

This section will describe only the application of salt-gradient ponds to the heating of buildings; other details about solar pond principles, design, construction, and operation can be found in a number of references, such as chapter 11 of *Economic Analysis* of *Solar Thermal Energy Systems*, volume 3 of this series, Tabor (1981), Tabor and Weinberger (1981), Nielsen (1980, 1986), and a brief manual on this topic was prepared by Fynn and Short (1982). Table 17.3, taken from that manual, shows the estimated solar pond areas needed to meet an annual average load at a given latitude, average annual insolation, and temperature difference.

Produced typically at temperatures of 40°C to 90°C, the heat from salt-gradient solar ponds is very well suited for heating service water and

Table 17.3
Initial estimate for salt-gradient solar pond areas (in m^2) need to meet an annual average load at a given latitude, average annual insolation, and temperature difference

Tropical and subtropical climate

Latitude	0–29°N			
Insolation	500 Langleys			
	77 Btu/hr sq ft			
	242 Watts/sq m			

Temperature difference °C	Load (kilowatts) 30	60	120	300
33	692	1359	2683	6632
44	889	1730	3396	8347
55	1229	2364	4600	11224

Mediterranean to northern U.S. climate

Latitude	30–43°N			
Insolation	400 Langleys			
	61 Btu/hr sq ft			
	193 Watts/sq m			

Temperature difference °C	Load (kilowatts) 30	60	120	300
33	1056	2065	4066	10025
44	1572	3036	5922	14482
55	2952	5572	10695	25781

Intermediate climate

Latitude	44–29°N			
Insolation	300 Langleys			
	46 Btu/hr sq ft			
	145 Watts/sq m			

Temperature difference °C	Load (kilowatts) 30	60	120	300
33	2115	4102	8027	19682
44	5619	10549	20173	48463

Northern European climate

Latitude	50–53°N			
Insolation	200 Langleys			
	31 Btu/hr sq ft			
	97 Watts/sq m			

Temperature difference °C	Load (kilowatts) 30	60	120	300
33	38872	69410	128052	297918

Source: Fynn and Short (1982).

buildings. Methods of heat extraction from the ponds are discussed in Wittenberg and Etter (1982). An early study of the applicability of solar ponds to space heating (Rabl and Nielsen 1975) proposed the possibility of using a heat pump during periods in which the pond temperature falls below values useful for direct heating of space and supplying then the pond heat to the heat pump evaporator. This concept was successfully tried by Shah, Short, and Fynn (1981a). Rabl and Nielsen's calculations (for a system without a heat pump) for heating a home that has a heat load of 25,000 Btu/degree(°F)/day (47.475 MJ/degree(°C)/day) indicated that about 130–140 m^2 of pond area would be needed in Boston, Seattle, and Columbus, Ohio, 60 m^2 in Albuquerque, New Mexico, and 240 m^2 in Fairbanks, Alaska. The pond depths were 3 to 5 m, and the average temperatures about 60–70°C. The produced heat cost (including the costs of salt) was reported to be 1.1 ¢/kWh ($3.22 per million Btu; $3.05/GJ). Because of the lower relative costs of pond construction, and lower edge heat losses, this cost was reduced to 0.41 ¢/kWh ($1.20 per million Btu; $1.14/GJ) when a larger pond to supply heat to twenty houses was considered. Lebeouf (1980, 1981) studied the application of such ponds to district heating and cooling for the climates of Fort Worth, Texas, and Washington, D.C.

A number of salt-gradient solar ponds have been built in this and other countries (principally in Israel for power production). For space heating, a 155 m^2 (1668 ft^2), 3 m deep pond was built in 1975 by the Ohio Agricultural Research and Development Center at Wooster, Ohio, to supply heat to a greenhouse (Badger et al. 1977; Shah, Short, and Fynn 1981b, 1982). The maximum efficiency achieved was 12%. As indicated above, the system was converted in 1979 to a solar pond heat pump system (Leboeuf 1981). A 2,000 m^2 (about half acre) pond, about 3 m deep, was built in Miamisburg, Ohio, for heating a community recreational building and outdoor swimming pool (cf. Wittenberg and Harris [1979, 1980, 1981]). The estimated cost of the heat delivered by the pond at that time was $8.95/GJ ($9.45/million Btu), competing favorably with the cost of fuel oil.

The largest salt-gradient solar pond in the United States was constructed in Chattanooga, Tennessee, by the Tennessee Valley Authority (Chinery and Siegel 1982) and has an area of 4,000 m^2 (one acre). The largest salt-gradient solar pond in the world was constructed in Israel for producing 5 MW peak electric power; it has an area of 250,000 m^2 (68.6 acres).

A comprehensive study of the potential of salt-gradient solar ponds in the United States (Lin 1982) has concluded that conventional salt-gradient solar ponds can provide heat at sufficiently high temperatures for space heating and service water heating in all of the regions except Alaska. The minimal economical size is a half-acre pond (about 2,000 m^2), to supply heat to a number of houses or a large commercial or apartment building. The availability of low-cost land in the proximity of the heated buildings is a limiting factor, since vacant land in most developed areas is scarce and costly. The total U.S. pond potential for heating buildings and service water was estimated to be 3.27 quads/year. The capital costs were estimated (in 1981 dollars) to range from $31/m^2 to $87/m^2 ($2.90/ft^2 to $8.10/ft^2). Assuming a discount rate variation from 11% to 20%, the cost of delivered heat ranged from $6 to $54.7 per million Btu ($5.69 to $51.84 per MJ).

In comparison with flat-plate solar collectors ($400/m^2 to $800/m^2 installed, 30–35% efficiency) and concentrating solar collectors ($750/m^2 to $1,200/m^2 installed, 50–60% efficiency), salt-gradient solar ponds ($30/m^2 to $90/m^2, 15–20% efficiency) appear to be the least expensive solar collector/thermal storage systems available. It should, however, be noted that the economic analyses of solar ponds have not taken marketing costs into account; that the mere potential for construction and its subsequent inception drive land costs up, resulting possibly in land costs that are much higher than estimated by Lin (1982) and other economic studies; and that the experience with solar ponds and with their economic analysis is not yet sufficient to decisively substantiate the present claims.

U.S. government funding for salt-gradient pond research and development ended in fiscal year 1984.

17.6   Photovoltaic/Thermal (PV/T) Hybrids

Much synergism exists between photovoltaic and thermal collectors of solar energy. They can share common components, such as the transparent cover, frame, absorber, and supports. Since the efficiency of the photovoltaic cells increases as the temperature decreases, it is desirable to cool them and use the discarded heat. The total energy conversion efficiency of a combined system increases and the total collector area needed for a given energy load decreases. Concentrating PV/T collectors in particular are inherently hybrid photovoltaic-thermal collectors because of the requirement to cool the photovoltaic cells. Since excess electric energy could conceivably be sold back to the utility, while excess heat is of no use at present, the combined unit is bound to have a better efficiency than any single unit. If the incremental cost of a hybrid system, above that of a single-purpose system, is lower than the value of the additional energy obtained, such hybrid systems could also have an economic advantage.

Early studies (Wolf 1976) showed through computer analysis that a liquid heating solar collector that contained photovoltaic (PV) cells can supply the diversified electric, domestic hot water, and space heating loads of a home in the northeastern United States. The state of the art was advanced through a number of theoretical and experimental studies (Boër, Higgins, and O'Connor 1975; Kern 1979; Russell 1979; Loferski 1982; Raghuraman 1979, 1980; Hendrie 1982). It became clear that in cases where the heat load was high and when the solar thermal collector contribution was significant, the combined photovoltaic/thermal collector system was more cost effective than systems containing PV modules only or side-by-side PV modules and solar thermal collectors.

Another conclusion derived from these studies was that the PV/T collector needs to be designed so that it can attain the maximal combined efficiency. For example, the first PV/T panels that were constructed under DOE sponsorship by ARCO-Solar and Spectrolab, and tested by the MIT Lincoln Laboratories showed disappointing results (Hendrie 1982). The designs included both air-cooled and liquid-cooled collectors, and were simply extensions of existing, commercially available solar thermal collectors, with only minor modifications for conversion to PV/T units. Neither has attained the required performance of at least 6.5% maximum-point electrical efficiency at 15V and thermal efficiency of 40% and a maximum cell-to-average fluid temperature difference of 15°C at the test conditions of 35°C difference between outlet fluid and ambient temperatures, 45°C cell temperature, and 1 kw/m^2 insolation level. The test results were also well below typical solar thermal collector efficiency values at the same conditions. Analysis of the results indicated several avenues of improvement, including the improvement of the thermal conductance between the absorber and the heat transfer fluid (by better conducting bonds and by improved heat convection through jet impingement or increased surface area), improvement of the thermal insulation of the window, improved cell absorptance by texturing on both sides (because the cells are partially transparent to solar radiation), addition of a second absorber to capture radiation that is transferred through the cell-mounting absorber, and maximal cell packing. Tests have shown that the electrical efficiency at the

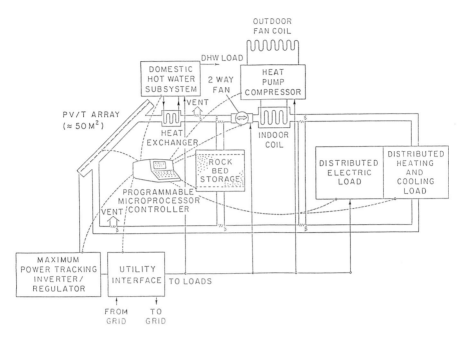

Figure 17.18
PV/T system schematic. Source: Loferski (1982).

above conditions reached 9.8%, and the thermal efficiency 42%, a marked improvement over the first-generation units. Further improvements were recommended but not carried through due to termination of funding.

The work at Brown University (Loferski 1982) resulted in the development of an advanced PV/T air-heating collector that incorporated a number of the above-recommended improvements. A 360 ft² (33.5 m²) collector system of this type was installed as a part of the roof of a prefabricated building having a 1,200 ft² (111.6 m²) floor area. All electric loads in the house were connected to the AC electric service, which, in turn, was connected to a synchronous inverter. Excess electric power was sold to the utility. As shown in figure 17.18, the solar heat is first used to heat domestic water and then, according to demand, is sent either directly to heat the house or to the rock thermal storage. The parallel air-to-air heat pump is used when the heat supplied from the collectors is insufficient. The system capacity is 2.86 peak kW, and it can displace about 6000 kWh per annum from the 11,000 kWh total needed for the building. About half of the

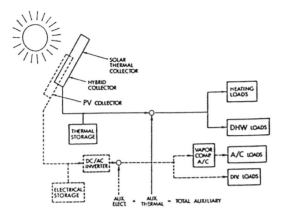

(a)   Baseline Hybrid Solar
      Power System

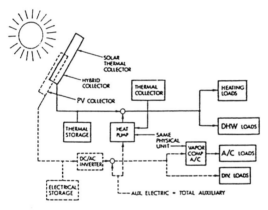

(b) Parallel Heat Pump Hybrid
     Solar Power System

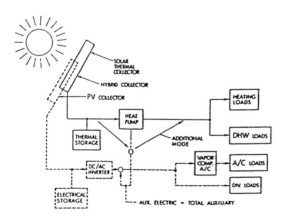

(c) Series Heat Pump Hybrid
     Solar Power System

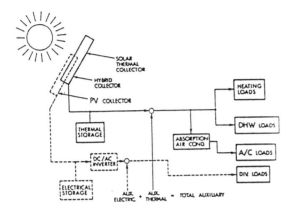

(d) Absorption Cooling Hybrid
     Solar Power System

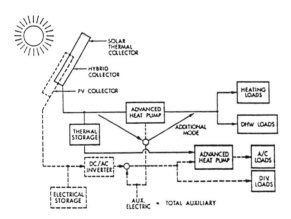

(e) Series Advanced Heat Pump
     Hybrid Solar Power System

Figure 17.19
Five hybride PV/T systems. Source: Kern and Russell (1978).

energy supplied is thermal. The add-on cost of the PV/T was $54,570, resulting in a payback period of 91 years (the PV part alone has a payback of 95 years). Assuming that DOE cost goals for PV cells of $1/peak Watt and for other components were met, the PV/T add-on costs would be reduced to $13,700, resulting in a payback of 23 years. Although the system was built, it was not run due to termination of funding.

Computer simulation studies were conducted to compare five different hybrid PV/T systems (Kern and Russell 1978), shown in figure 17.19, with a conventional system that provides heating from an oil or gas-burning furnace and cooling from a central vapor-compression electric-powered air conditioner. In addition to thermal storage, batteries were used in the simulation for electric energy storage. All of the systems were applied to a 10,000 ft² (930 m²) office building and a 1200 ft² (111.6 m²) single-family residence at four climatic regions: Phoenix, Arizona, Miami, Florida, Boston, and Ft. Worth, Texas.

The calculations have shown that the series advanced heat pump provides the largest energy savings at all locations. At the same time, the direct solar heat system with vapor compression air-conditioning is the most economical due to its lowest first cost. The relative need of thermal vs. electric energy depends on the nature of the load: the fraction of PV to thermal energy should increase as the fraction of cooling-load to heating-load increases.

A prototype PV/T concentrating collector, an E-Systems cylinder-shaped line focusing Fresnel lens system with a concentration ratio of 25:1, aperture of 0.914×2.44 m, and 46 silicon cells 2.3×2.3 cm each, mounted on the focal line, was tested by the University of Arizona (Wood et al. 1982). A cost analysis, using the actual cost of this PV/T system ($300/m²) indicated that it is not economically attractive and that the dominant factor is the cost of the PV cells.

In addition to the analyses and design methods developed by Kern (1979), Russell (1979), and Kern and Russell (1978) discussed above, several other contributions have been made to the design methodology of hybrid PV/T systems (Wood et al. 1982; Florschuetz 1979; Venkateswaran and Anand 1980; Herman and Schwinkendorf 1982; Schwinkendorf 1982, 1984). A weathertight and simple method for mounting solar thermal or photovoltaic collectors as an integral part of a structure is described by Rost, Amerduri, and Groves (1981).

An analysis for the optimization of a spherical-reflector/tracking-absorber (SRTA) concentrator for combined hot water and PV electricity generation resulted in recommendations for optimization (Bar-Lev, Waks, and Grossman 1982). A summary of the technical viability of flat-plate PV/T solar collectors is given by Andrews (1981) and of experimental results with such collectors by Kern and Pope (1982).

Acknowledgment

Section 17.2 (Solar Space Heating Retrofit) draws, in part, on some of the conclusions of the Solar Retrofit Review Meeting organized by the USDOE Solar Heating and Cooling R&D Branch, held in March 1978 at Columbia, Maryland. The author, who served as the chairman of the session on R&D and systems development, is grateful to the meeting's general chairman, Mr. Herbert Yim, and to the meeting participants for their contributions, particularly to Mr. J. M. Davis and Dr. F. H. Morse from the USDOE, who initiated and sponsored the meeting.

References

Aitken, D. W., Rozwell, D. K. 1976. Project sunshower: San Jose State University dormitory retrofit to solar-assisted water heating. *Proc. Joint Conf. AS/ISES and SESC*, Winnipeg, Canada, August 1976, pp. 105–228.

Andrews, J. W. 1981. *Evaluation of flat-plate photovoltaic hybrid systems for solar energy utilization*. Brookhaven National Laboratory report BNL 51435.

Argabright, T. A. 1982a. *Metal hydride/chemical heat pump development project phase I*. Final report BNL–51539, Brookhaven National Laboratory

Argabright, T. A. 1982b. Heat/mass flow enhancement design for a metal hydride assembly phase II. Final report BNL–51727, Brookhaven National Laboratory.

Badger, P. C., Short, T. H., Roller, W. L., and Elwell, D. L. 1977. a prototype solar pond for heating greenhouses and rural residences. *Proc. Annual Meeting AS/ISES*, pp. 33/19 33/22.

Balcomb, J. D., Hedstrom, J. C., Moore, S.W., and Herr, K. C.. 1975. Research on integrated solar collector roof structures. Paper presented at 1975 ISES Congress and Exposition, Los Angeles, CA. Also Los Alamos Scientific Laboratory Report LA–UR–75–1335.

Bankston, C. A. 1982. *Central solar heating plants with seasonal storage*. Report on the International Energy Agency Solar Heating and Cooling Program Task VII, document ANL/ES-139, Energy and Environmental Systems Division, Argonne National Laboratory.

Bankston, C. A. 1986. The status and potential of central solar heating plants with seasonal storage. *Proc. ASES 86*, Boulder, CO, pp. 79–91.

Bar-Lev, A., Waks, S., and Grossman, G. 1982. Analysis of a combined thermal-photovoltaic solar system based on the spherical reflector/tracking absorber concentrator. *J. Solar Energy Eng.*. 195 (3): 322–28.

Barnaby, C. S., Caesar, P., Wilcox, B. and Nelson, L. 1978. *Solar for your present home*, Prepared for the California Energy Commission, Berkeley Solar Group, Berkeley, CA.

Baughn, J. W., and Jackman, A. 1974. *Solar energy storage within the absorption cycle.* ASME paper 74-WA/HT-18.

Baughn, J. W., and McDonald, M. J. 1977. Theoretical modeling of an ammonia/water absorption cycle with solar energy storage. *Proc. AS/ISES*, vol. 1, pp. 7-24 to 7-28.

Baylin, F, and Sillman, S. 1980. *Systems analysis techniques for annual cycle thermal energy storage systems.* Report SERI/RR-721-676. Golden, CO: Solar Energy Research Institute.

Beard, J. T., Dirhan, L. A. Jr., Iachetta, F. A., Lilleleht, L. U., and Duvall, M. D. 1979. Analysis and experimental evaluation of a system for the annual collection and storage of solar energy. *Trans. ASHRAE*, 85 (1): 502-11.

Beard, J. T. 1978. *Engineering analysis and testing of water trickle solar collector* Report no. UVA/527121/MAE78/107 to the Energy Research and Development Administration, University of Virginia, Charlottesville, VA.

Beard, J. T., Iachetta, F. A., Lilleleht, L. U., Huckstep, F. L., and May, W. B., Jr. 1978. Design and operational influences on thermal performance of "Solaris" solar collector. *J. Eng. for Power* 100 (4): 497-502.

Beard, J. T., F. L. Huckstep, W. B. May, F. A. Iachetta, and L. U. Lilleleht. 1976. Analysis of performance of "Solaris" water trickle solar collector. ASME paper 76-WA/Sol-21.

Biancardi, F. R., Sitler, J. W., and Melikian, G. 1982. Development and test of solar Rankine cycle heating and cooling systems. *Int. J. Refrigeration*, 5 (6): 351-60.

Boehm, R., ed. 1978. *Proc. Conf. on Performance Monitoring Techniques for Evaluation of Solar Heating and Cooling Systems*, sponsored by the Solar R&D Branch, Office of Conservation and Solar Applications, USDOE, April 1978, Washington, DC.

Boër, W. K., Higgins, J. H., and O'Connor, J. K. 1975. Solar One, two years experience. *Proc. 10th IECEC*, pp. 7-13.

Bonne, U. 1978. Definition of efficiency for hybrid heating systems. *ASHRAE J.* (March 1978): 31-33.

Braun, J. E., Klein, S. A., and Mitchell, J. W. 1981. Seasonal storage of energy in solar heating. *Solar Energy*, 26 (5): 403-11.

Burnette, C. W., & Associates 1980. *Philadelphia solar planning: Project Summary.* Philadelphia: Charles Burnette & Associates.

Burnette, C., and Pauman, M. 1981. *Solar access protection for Philadelphia.* A "Philadelphia Solar Planning" report. Philadelphia: Charles Burnette & Associates.

Casamajor, A. B., and Parsons, R. E. 1979. *Design guide for shallow solar ponds.* Report UCRL-52386 rev. 1, Lawrence Livermore Laboratory, Livermore, CA.

Chinery, G. T., and Siegel, G. R. 1982. Design, construction, and cost of TVA's 4000 m2 (1 acre) nonconvecting salt gradient solar pond. *Solar Engineering* 150-56.

Christopher, J. 1979-1985. *Sales catalogs, brochures, and installation, manual.* Contemporary Systems, Inc., Rt. 12, Walpole, NH 03608.

Clark, C. E., and Hiller, C. C. 1978. Sulfuric acid-water chemical heat pump/energy storage system demonstration. *Proc. ASME Winter Annual Meeting*, pp. 45-50.

Clark, A. F., and Dickinson, W. C. 1980. Shallow solar ponds. Ch. 12 *Solar Energy Technology Handbook*, edited by W. C. Dickinson and P. N. Cheremisinoff. New York: Marcel Dekker. pp. 377-402.

Clark, E. C., and Carlson, D. K. 1980. Development status and utility of the sulfuric acid chemical heat pump/chemical energy storage system. *Proc. 15th IECEC*, vol. 2, pp. 926–31.

Cleveland, C. J., Costanza, R., C. A. S. Hall, and Kauffman, R. 1984. Energy and U.S. economy: A biophysical perspective. *Science*, (225): 890–97.

Cocchi, A., Raffellini, G., and Stopponi, V. 1979. A solar operated water-LiBr absorption refrigeration machine used also as a heat pump: Technical ane economic analysis. *Proc. ISES Congress*, vol. 1, Atlanta, GA, pp. 777–81.

Cohen, G., Salvat, J. and Rojey, A. 1979. A new absorption cycle process for upgrading waste heat. *Proc. 14th IECEC*, vol. 2, pp. 1720–1724.

Cornelius, K. A., and Benson, J. D. 1981. The performance of the USAF Academy retrofit solar test house—a summary report. *Proc. Annual Meeting AS/ISES*, Philadelphia, PA pp. 466–70.

Copeland, C. C. 1975. Manhattan tenement adds solar heater. *Building Systems Design*, 72(6): 15–17.

Coughlin, R. E., Michel, and Cohen, P. F. 1980. *Economic impacts of conservation and solar programs in Philadelphia*. A "Philadelphia Solar Planning" report. Philadelphia: Charles Burnette & Associates.

Coughlin, R. E., and Ervolini, M. 1981. *The decision to invest in conservation and solar applications*. A "Philadelphia Solar Planning" report. Philadelphia: Charles Burnette & Associates.

Curran, H., and Miller, M. 1975. Evaluation of solar-assisted Rankine cycle concepts for the cooling of buildings. *Proc. 10th IECEC*, paper 759203, pp. 1391–98.

Daystar Corp. 1977. *Solar energy in high-rise buildings*, Report no. DSE–1963 to the Energy Research and Development Administration, Daystar Corp., Burlington, MA.

Dean, T. 1975. Design constraints for retrofit solar systems. Paper presented at ISES, International Solar Energy Congress, University of California, Los Angeles, CA.

DOE (U.S. Department of Energy) National Solar Data program. 1982. *Solar energy system performance evaluation: Contemporary Systems, Inc.* Report Solar/2116–82/14.

DOE Office of the Assistant Secretary for Conservation. 1984.

Drew, M. S., and Selvage, R. L. G. 1980. Sizing procedure and economic optimization methodology for seasonal storage systems. *Solar Energy* 25 (1): 79–82.

Dubin, F. S. 1975. Solar energy design for existing buildings. *ASHRAE J.* (November 1975): 53–55.

Eckard, S. E. 1976. Test results for a Rankine engine powered vapor compression air-conditioner for 366 K heat source. *Proc. IECEC*, paper 769205, pp. 1169–73.

Edgel, W. R. 1978. *Development of retrofit energy conservation and solar heating systems (phase I)*, New Mexico Energy Institute, report no. NMEI 20.

ESG, Inc. 1984. *Survey of system operational failure modes from 122 residential solar water heater systems over a period of approximately two years*, Report ESG-R107-82, Atlanta, GA.

EIA (U.S. Energy Information Administration). 1981. Annual report to Congress.

Ferraro, R., Godoy, R., and Turrent, D., eds. 1982. *Monitoring solar heating systems: A practical handbook*, Oxford: Pergamon Press.

Fischer, R. D. 1978. Feasibility study and hardware design of a pivoting-tip rotary-vane compressor and expander applied to a solar-driven heat pump. *Proc. 1978 Purdue Compressor Technology Conf.*, Edited by J. F. Hamilton, pp. 233–40.

Flanagan, J. R., Schultz, B., Burnette, C., Peretta, S., and De Rosa, E. 1980. *Financial incentives.* A "Philadelphia Solar Planning" report. Philadelphia: Charles Burnette & Associates.

Florschuetz, L. 1979. Extension of the Hottel-Whillier model to the analysis of combined photovoltaic/thermal flat plate collectors. *Solar Energy* 22 (4): 361–66.

Forbes, R. E., and McLendon, R. W. 1977. The heating of a poultry broiler house using solar energy. Paper no. 77–3005, *Proc. 1977 Annual Meeting of the American Society of Agricultural Engineering*, North Carolina State University, Raleigh, NC.

Forbes, R. E., and McLendon, R. W. 1979. Addition of solar air heaters to a pre-engineering metal building. ASME Paper 79–WA/Sol–33.

Fynn, R. P., and Short, T. H. 1982. *The salt-stabilized solar pond for space heating-a practical manual.* Special circular 106, The Ohio State University, Ohio Agricultural Research and Development Center, Wooster, OH.

General Electric Company. 1974. *Solar heating and cooling of buildings study conducted for department of the army.* GE doc. no. 74SD4226, 1, Executive Summary and Implementation Plans, June 1974.

Gottschalk, W. L. 1977. System performance of first commercial solar installation in Orange, Virginia (retrofitted office building). *Proc. 1977 Annual Meeting of AS/ISES*, Orlando, FL June 1977.

Graf, J. C. 1980. Rankine cycle solar driven heat pump development. *Proc. Annual DOE Active Heating and Cooling Contractors' Review Meeting*, NTIS CONF 800340, pp. 1–13 to 1–18.

Gruen, D. M., Mendelsohn, M. H., and Sheft, I. 1978. Metal hydrides as chemical heat pumps *Solar Energy.* vol. 21 pp. 153–56.

Harris, A. W., and Shen, C. N. 1977. Operational dynamics of coupled flat plat solar collector and absorption cycle heat pump system with energy storage. In *Heat transfer in solar energy systems.* Atlanta, GA: ASME–WAM, pp. 103–11.

Herman, R. W., and Schwinkendorf, W. E. 1982. Evaluation of actively cooled concentrating photovoltaic systems for combined photovoltaic-thermal applications. *Proc. ASES Annual Meeting*, pp. 199–204.

Hill, J. E., and Richtmyer, T. E. 1975. *Retrofitting a residence for solar heating and cooling: The design and construction of the system*, U.S. Department of Commerce, NBS technical note 892, report no. PB–247 482, November.

Hudson, W. T., and Williams, J. R. 1977. Installation and performance of a retrofit home solar energy system with reflector augmented array. *Proc. 1977 Annual Meeting of AS/ISES*, Orlando, FL, June 1977, pp. 3–20 and 3–21.

Huntley, W. R. 1982. Performance test results of an absorption heat pump that uses low temperature (60°C[140°F]) industrial waste heat. *Proc. 18th IECEC*, vol. 4, pp. 1921–26.

IUD (Institute for Urban Design) 1982. *Solar cities and towns.* Information kit. Purchase NY: IUD.

InterTechnology Corp. 1974. *Solar energy school heating augmentation experiment—design, construction and initial operation*, NSF/RANN, contract no. NSF–C868, 4 December.

Hendrie, S. D. 1982. *Photovoltaic/thermal collector development program.* Report DOE/ET/20279–168.

Hiller, C. C., and Clark, E. C. 1979a. Development and testing of the sulfuric acid/water chemical heat pump/chemical energy storage system. *Proc. 14th IECEC*, Boston, pp. 51–515.

Hiller, C. C., and Clark, E. C. 1979b. Development of the sulfuric acid-water chemical heat pump/chemical energy storage system for solar heating and cooling. *Proc. ISES Congress*, vol. 1, pp. 149–53.

Hooper, F. C., and Attwater, C. R. 1977. A design method for heat loss calculation for in-ground heat storage tanks. In *Heat transfer in solar energy systems*. New York: ASME, pp. 39–43.

Hooper, F. C., and Cook, J. D. 1980. Design of annual storage space heating systems. Fundamentals and Applications of Solar Energy, A.I.Ch.E. Symp. Ser. 198, vol. 76, pp. 80–91.

Ishiiki, N. 1977. Study on the concentration difference energy system. *J. Nonequilibr. thermodynamics*, 2, 85–107.

Jorgensen, G. J. 1984. *A summary and assessment of historical reliability and maintainability data for active solar hot water and space conditioning systems*, Report SERI/TR–253–2120.

Karaki, S., Duf, W. S. and Löf, G. O. G. 1978. *A performance comparison between air and liquid residential solar heating systems*. Report COO–2868–4 to USDOE.

Kern, E. C., and Pope, M. D. 1982. *Development and evaluation of solar photovoltaic systems: Final report*. MIT Lincoln Laboratory DOE contract DE–AC12–76ET20279.

Kern, E. C. 1979. *Phase-one experiment test plan: Solar photovoltaic/thermal laboratory*. MIT Lincoln Laboratory report on DOE Contract EG–77–S–02–4577, Lexington, MA.

Kern, E., Jr., and Russell, M. C. 1978. Hybrid photovoltaic/-thermal solar energy systems. *Proc. DOE 3rd Workshop on the Use of Solar Energy for the Cooling of Buildings*, pp. 260–79.

Klebba, J. M. 1980. Insuring solar access on retrofits: The problem and some solutions. *Solar engineering*, (January 1980): 16–19.

Koai, K, Lior, N. and Yeh, H. 1984. Performance analysis of a solar-powered/fuel-assisted Rankine cycle with a novel 30 hp turbine. *Solar Energy*, vol. 32, pp. 753–64.

Kohler, J. T., Temple, P. O., Coleman, J. J., and Sullivan, P. W. 1978. Evaluation of six designs for a site-fabricated, building-integrated air-heater, parts I and II. *Proc. 1978 Annual Meeting AS/ISES, 2.1*, pp. 239–249.

Kouremenos, D. A. 1985. A tutorial on reversed NH_3/H_2O absorption cycles for solar application. *Solar Energy* 34, 101–15.

Knoche, K. F., and Stehmeier, D. 1979. Absorption heat pumps for solar space heating systems. In *Studies in heat transfer*, edited by J. P. Hartnett, T. F. Irvine, Jr., E. Pfender, and E. M. Sparrow. Washington, DC: Hemisphere.

Knowles, R. L. 1982. *Sun rhythm form*. Cambridge, MA: MIT Press.

Kreith, F., and West, R. E., eds. 1980. *The economics of solar energy and conservation systems* West Palm Beach, FL: CRC Press.

Krupczak, J., Jr., Skillman, P., Brancic, A., and Sunderland, J. E. 1985. Seasonal storage of solar energy using insulated earth. *Proc. 9th Biennial ISES Congress*, vol. 2, pp. 806–10.

Kuhlenschmidt, D., and Merrick, R. H. 1982. An ammonia-water absorption heat pump cycle. *Trans. ASHRAE*, vol. 89, pt. 1B, pp. 215–19.

Lauck, F., Myers, P. S., Uyehara, O. A., and Glander, H. 1965. Mathematical model of a house and solar-gas absorption cooling and heating system. *Trans. ASHRAE*, vol. 71, pt. 1, pp. 273–85.

Larson, D. C., Narayanan, R. and Savery, C. W. 1976. Retrofit solar water-heating system for urban buildings. *Proc. 2d Southeastern Conf. on Applications of Solar Energy*, Baton Rouge, LA, April, 1976.

LLNL (Lawrence Livermore National Laboratory) 1985. Shallow solar ponds: Economical water heaters for industry. *Energy Technology and Review*, Lawrence Livermore National Laboratory, April 1985, pp. 11–26.

Lazzarin, R. M. 1981. Solar assisted heat pumps feasibility. *Solar Energy*, 26, 223–30.

Leboeuf, C. M., et al. 1980. Solar ponds for district heating and electricity generation. *Proc. 15th Intersociety energy Conversion Engineering Conference*, pp. 1453–58.

Leboeuf, C. M. 1981. Solar ponds applied to district heating and cooling. *Proc. AS/ISES Annual Meeting*, 4.1, pp. 772–76.

Lepore, J. A., Shore, S. and Lior, N. 1978. Retrofit of urban housing for solar energy conversion. *Housing Science*, 2(6): 482–98.

Lin, E. I. H. 1982. Regional applicability and potential of salt-gradient solar ponds in the United States. vol. 1 and 2, report DOE/JPL–1060–50. Prepared for the USDOE and the USDOD by the Jet Propulsion Laboratory.

Lior, N., Lepore, J. A., and Shore, S. 1976. Residential solar retrofit in the urban environment. *Proc. Joint Conf. AS/ISES and SESC*, Winnipeg, Canada, August, pp. 36–53.

Lior, N. 1977. Solar energy and the steam Rankine cycle for driving and assisting heat pumps in heating and coolingmodes. *Energy Conversion*, vol. 16 (3): pp. 111–23.

Lior, N., O'Leary, J. and Edelman, D. 1977. Optimized spacing between rows of solar collectors. *Proc. 1977 Annual AS/ISES Meeting*, Orlando, FL pp. 3–15 to 3–20.

Lior, N., Shore, S. Lepore, J. A., and Jones, G. 1978. Solar heating retrofit of an urban row house: Construction and start up. *Proc. Annual Meeting AS/ISES*, Denver, CO August 1978, pp. 425–31.

Lior, N. 1979. Instrumentation principles for performance measurement of solar heating systems, *Energy* (4): 561–73.

Lior, N. 1980. First year performance monitoring of an urban row house retrofitted to solar heating. *Proc. AS/ISES Annual Meeting*, vol. 3. i, pp. 261–65.

Lior, N., Yeh, H. and Zinnes, I. 1980. Solar cooling via vapor compression cycles *Proc. AS/ISES Annual Meeting*, vol. 3.1, pp. 210–13.

Lior, N., and Koai, K. 1984a. Solar-powered/fuel-assisted Rankine cycle power and cooling system: Simulation method and seasonal performance. *J. Solar Energy Eng.*, 106 142–52.

Lior, N., and Koai, K. 1984b. Solar-powered/fuel-assisted Rankine cycle power and cooling system: Sensitivity analysis. *J. Solar Energy Eng.*, vol. 106, pp. 447–56.

Löf, G. O. G. 1986. Personal communication with author.

Löf, G. O. G., and Ward, D. S. 1976. *Design, construction and testing of the Colorado State University Solar House 1 heating and cooling system*. Report COO/2577–76/1 to NATO Committee on Challenges of Modern Society.

Lund, P. D., Makinen, R., Routti, J. T., and Vuorelma, H. 1982. Simulation studies of the expected performance of Kerava solar village. *Int. J. Ener. Res.* 7 347–57.

Lund, P. D. 1984. Optimization of a community solar heating system with a heat pump and seasonal storage. *Solar Energy*. 33 353–61.

Marshall, M. M. 1977. 1976 ERDA/DoD 50-unit residential solar heating demonstration program. DOE report CONF-771229-P2. *Proc. Solar Heating and Cooling Demonstration Program Contractor's Review Meeting*, New Orleans, LA December 1977, pp. 438–44.

McCartney, K. E. 1977. Retrofitting potential in the London housing stock. Paper presented at International Conf. and Exhibition on Solar Bldg. Tech., London, July 1977.

McDonald, G. 1978. An advanced integrated Rankine/Rankine solar heating and cooling system. *Proc. 3rd Workshop on Use of Solar Energy for the Cooling of Buildings*, USDOE, pp. 186–91.

McLendon, R. W., Forbes, R. E., and Hanks, J. E. 1979. Efficiency of a pre-engineering metal building solar collector. Paper no. 79–4553, *Proc. 1979 Annual Meeting American Society Agricultural Engineering*, New Orleans, LA.

McLinden, M. O., and Klein, S. A. 1982. Simulation of an absorption heat pump solar heating and cooling system. *Solar Energy*, 31, 473–82.

Melikian, G., Biancardi, F. R., Landerman, A. M., Meader, M. D., and Anderson, T. 1982. Development, test, and evaluation of an improved solar heat pump Rankine cycle. *Proc. 18th IECEC*, paper 829303, pp. 1856–61.

Mezzina, A. An overview of the U.S. chemical heat pump program at Brookhaven National Laboratory. *Int. J. Ambient Energy*, 3 (3): 137–40.

Miller, M., Krauss, R. I., Burnette, C., Schultz, R., Allen, P., and Pauman, M. 1980. *Neighborhood energy planning kit.* A "Philadelphia Solar Planning" report. Philadelphia: Charles Burnette & Associates.

Moore, S. W., Balcomb, J. D., Hedstrom, J. C. 1974. Design and testing of a structurally integrated stell solar collector unit based on expanded flat metal plates. Paper presented at the AS/ISES Meeting, Ft. Collins, CO, August 1974. Also Report LA–UR–74–1093, Los Alamos Scientific Laboratory, Los Alamos, NM.

Mumma, S. A., and Dzioba, J. 1976. Energy conservation through residential solar retrofit. *Proc. Joint Conf. AS/ISES and SESC*, Winnipeg, Canada, August 1976, pp. 67–92.

Niebergall, W. 1959. *Sorptions-Kaltemaschinen*. Vol. 7 of *Handbuch der Kaltetechnik*, edited by, R. Planck. Berlin: Springer-Verlag.

Nielsen, C. E. 1986. The latest on solar ponds. *Proc. ASES 86*, technical papers, pp. 134–43.

Nielsen, C. E. 1980. Nonconvecting salt gradient solar ponds. In *Solar Energy Technology Handbook*, edited by W. Dickinson and N. Cheremisinoff. New York: Marcel Dekker, pp. 345–76.

Offenhartz, P. O'D. 1976. Chemical methods of storing thermal energy. *Proc. Joint Conf. of AS/ISES and the CSES*, vol. 8, Winnipeg, Canada, pp. 48–72.

Offenhartz, P. O'D. 1978. Chemically driven heat pumps for solar thermal storage. *Proc. ISES Congress*, New Delhi, India, vol. 1, New York: Pergamon Press. pp. 488–89b.

Offenhartz, P. O'D. and Brown, F. C. 1979. Methanol based heat pumps for storage of solar thermal energy. *Proc. 14th IECEC*, vol. 1, pp. 507–509.

Othmer, P., and Scott, J. 1977. Solar heating and cooling of the El Camino Real elementary school in Irvine, California (retrofit ERDA demonstration project). *Proc. 1977 Annual Meeting of AAS/ISES*, Orlando, FL June 1977.

Payne, P. R. (presently Ketron, Inc., Annapolis, MD) 1978. Air heating structural blocks. *Proc. 3d Annual Solar Heating and Cooling R&D Branch Contractors' Meeting*, September 1978, Washington, DC, Report CONF-780982, USDOE, Assistant Secretary for Conservation and Solar Applications, Division of Solar Applications, March 1979, pp. 120–21.

Payne, P. R., and Doyle, D. W. 1978. The fossil fuel cost of solar heating. Paper no. 789504, *Proc. 13th IECEC*, San Diego, CA, pp. 1650–1656.

Perez-Blanco, H. and Grossman, G. 1982. Open cycle absorption heat pumps for low grade heat utilization. *Trans. ASHRAE*, 88 (1): 825–43.

Prowler, D., and Legerton, J. 1980a. *Solar and conservation applications analysis.* A "Philadelphia, Solar Planning" report. Philadelphia: Charles Burnette & Associates.

Prowler, D., and Legerton, J. 1980b. *The Philadelphia solar audit*. A "Philadelphia Solar Planning" report. Philadelphia: Charles Burnette & Associates.

Pyramidal optical collector system operates in condominium project. *Solar Engineering*, (July 1977): 30–32.

Rabl, A., and Nielsen, C. E. 1975. Solar ponds for space heating. *Solar Energy* 17 1–12.

Raghuraman, P. 1980. Analytical prediction of the performance of an air photovoltaic/thermal flat-plate collector. MIT LL/DOE report DOE/ET/20279–93, MIT Lincoln Laboratory, Lexington, MA.

Raghuraman, P. 1979. *Analytical prediction of liquid photovoltaic/thermal flat plate collector performance*. MIT LL/DOE report COO–4094–66, MIT Lincoln Laboratory, Lexington, MA.

Raldow, W. M., and Wentworth, W. E. 1979. Chemical heat pumps—a basic thermodynamic analysis. *Solar Energy*, 23, 75–79.

Reif, D. K. 1981. *Solar retrofit*, Andover, MA: Brick House Publishing.

Robison, H. I. 1982. Operational experience with a liquid desiccant heating and cooling system *Proc. 18th IECEC*, vol. 4, pp. 1952–57.

Robison, H. I., and Griffiths, W. 1984. Energy storage and heat pumping with liquid desiccants. *Proc. 19th IECEC*, vol. 3, pp. 1344–49.

RRC (Rocket Research Company) 1982. *Sulfuric acid/water chemical heat pump/chemical energy storage*. Final report, BNL 51540, Brookhaven National Laboratory.

Rost, D. F., Amerduri, G., and Groves, L. 1981. Installation system for integral mounting of thermal or photovoltaic panels. *Proc. AS/ISES Annual Meeting*, vol. 4.1, pp. 136–40.

Rousseau, J., and Noe, J. 1976. Solar powered Rankine-cycle heat pump system. *Proc. IECEC*, paper 769202, pp. 1163–68.

Ruegg, R. T., and Sav, G. T. 1980. *Microeconomics of solar energy*. NBS report.

Russell, M. C. 1979. *Solar photovoltaic thermal residential systems demonstration—phases I and II*. Final technical progress report on USDOE Grant no. DE-ACO30205, Brown University, Providence, R. I.

Schlepp, D. R., and Collier, R. K. 1981. The use of an open-cycle absorption system for heating and cooling. *Proc. AS/ISES Annual Meeting*, vol. 4.1, pp. 651–55.

Schnee, G. 1982. *Executive summary, neighborhood-scale building energy typologies*. Roxbury, MA: Greater Roxbury Development Corporation.

Schrenk, G. L., and Lior, N. 1975. The absorption cycle heat pump and the potential role of thermochemical storage. *Proc. ERDA Workshop on Solar Energy for the Cooling of Buildings*, Los Angeles, CA August 1975, pp. 207–26.

Schreyer, J. M. 1977. *Economy of a retrofit solar system*, U.S. Department of Commerce, report no. Y-2098, Oak Ridge Y-12 Plant, Oak Ridge, TN.

Schwartz, I., and Shitzer, A. 1977. Thermodynamic feasibility study: Solar absorption system for space cooling and heating. *ASHRAE J.* 11 51–54.

Schwinkendorf, W. E. 1984. Photovoltaic/thermal system sizing. *Proc. 6th Annual Technical Conference, ASME Solar Energy Division*.

Schwinkendorf, W. E. 1982. Technique for preliminary analysis and sizing of photovoltaic-thermal (PV-T) systems and associated components. *Proc. ASES Annual Meeting* 265–70.

Semmens, M. G., Wilbur, P. J. and Duff, W. S. 1974. *Liquid refrigeration storage in absorption air conditioners*, ASME paper 74–WA–HT–19.

SERI (Solar Energy Research Institute) and Oak Ridge National Laboratory. 1984. *Specification and cost manual for energy retrofits on small commercial and multi-family buildings.* Report SERI/SP–253–2373 (USDOE DE84012753).

Sherburne, D. C., and Lior, N. 1985. *The solar-powered/fuel assisted hybrid Rankine cycle ('SSPRE'): First test results, control optimization, and evaluation of the use of a solar superheater.* USDOE report SAN/ET/20110–15.

Shah, S. A., Short, T. H., and Fynn, R. P. 1981a. A solar pond assisted heat pump for greenhouses. *Solar Energy*, 26 491–96.

Shah, A. A., Short, T. H., and Fynn, R. P. 1981b. Modeling and testing a salt gradient solar pond in northern Ohio. *Solar Energy* 27 393–401.

Sherburne, D., and Lior, N. 1986. Evaluation of minimum fuel consumption control strategies in the solar-powered/fuel-assisted hybrid Rankine cycle. *Proc. ASES Annual Meeting*, pp. 300–303.

Sillman, S. 1981. Performance and economics of annual storage solar heating systems. *Solar Energy* 27 (6): 513–28.

Sims, D. H., and Gilmore, S. 1980. *Solar and the fossil fuel industry.* A "Philadelphia Solar Planning" report. Philadelphia: Charles Burnette & Associates.

Shore, S., Lepore, J. A., and Lior, N. 1977. Solar system retrofit of row houses. *Proc. International Conf. on Solar Technology of Buildings*, London, England, July 1977, London: RIBA Publications 2, pp. 692–99.

Smith, A. E., Lior, N., Klausner, S. Z. and Stanic, S. I. 1976. Solar energy application considerations in depressed communities. *Proc. Joint AS/ISES and SESC Conf.*, Winnipeg, Canada pp. 137–54.

Smith, R. O., and Meeker, J. F. 1978. One hundred solar water heaters: Some lessons learned about installation and performance at New England Electric. *Proc. Annual Meeting AS/ISES*, Denver, CO, August 1978, pp. 578–85.

Solaron Corporation. 1978. *Industrial and agricultural process heating systems*, Bulletin 13.11a/sol. Denver, CO: Solaron.

Solaron Corporation. 1979. *The solar air heating system*, Bulletin 13.11/sol. Denver, CO: Solaron.

Soleri, P. 1973. *Arcology: The city in the image of man*, Cambridge, MA: MIT Press.

Speyer, E. 1959. Optimum storage of heat with a solar house. *Solar Energy* 3 24–48.

Tabor, H. 1981. Review article-solar ponds. *Solar Energy*, 27 181–94.

Tabor, H., and Weinberger, Z. 1981. Nonconvecting solar ponds. In *Solar Energy Handbook*, edited by J. Kreider and F. Kreith. New York: McGraw Hill, pp. 10–1 to 10–29.

Temple, P. L., and Adams, J. A. 1980. *Construction manual: Model TEA Solar Heating System*, Total Environment Action, Inc., Harrisville, NH.

Thomason, H. E. 1960. Solar space heating and air conditioning in the Thomason House. *Solar Energy* 4 (4): 11–19.

Thomason, H. E., and H. J. L. Thomason, Jr. 1975. *Solar House Plans II-A*. Barrington, NJ: Edmund Scientific Co.

Thomsen, C. L. 1977. First Year's operation-residential solar retrofit. *Proc. 1977 Annual Meeting of AS/ISES*, Orlando, FL, June 1977.

TRW, Inc. 1981. *Chemical heat pump cost effectiveness evaluation*, Report BNL-51484, Brookhaven National Laboratory.

Venkateswaran, S. R., and Anand, K. K. 1980. A design procedure for combined photovoltaic/thermal solar collector-heat pump systems. *Proc. AS/ISES Annual Meeting*, vol. 3.1, pp. 281–85.

Wallenrod, M. H. 1980. *Solar Analysis for the electric utility*. A "Philadelphia Solar Planning" report. Philadelphia Charles Burnette & Associates.

Weinstein, A., Duncan, R. T., Jr. and Sherbin, W. C. 1976. Lessons learned from Atlanta (Towns) solar experiment, *Proc. Joint Conf. AS/ISES and SESC*, Winnipeg, Canada, August 1976, pp. 153–68.

Weinstein, S., and Smith, S. *Urban rooftop greenhouses: Schematic design, performance analysis, and cost benefit analysis*. Boston: Ehrenkrantz Group.

Williams, J. R., Hudson, W. T. and Swoffort, A. L. 1977. analysis of a 35,000 GPD solar domestic hot water system for Camp LeJeune, North Carolina. *Proc. 1977 Annual Meeting of AS/ISES*, Orlando, FL June 1977.

Willsie, H. E. 1909. *Engineering News*, 61 511.

Wittenberg, L. J., and Etter, D. F. 1982. Heat extraction from a large solar pond. ASME paper 82–WA/Sol–31.

Wittenberg, L. J., and Harris, M. J. 1981. Construction and performance of the Miamisburg salt-gradient solar pond. *J. Solar Energy Eng.*, 103 11–16.

Wittenberg, L. J., and Harris, M. J. 1980. Management of a large operational solar pond. *Proc. 1th Intersociety Energy Conversion Engineering Conference*, pp. 1435–37.

Wittenberg, L. J., and Harris, M. J. 1979. Performance of a large salt-gradient solar pond. *Proc. 1979 Intersociety Energy Conversion Engineering Conference*, pp. 49–52.

Wolf, M. 1976. Performance analysis of combined heating and photovoltaic power systems for residence. *Energy Conversion* 16 79–90.

Wood, B. D., Evans, D. L., Rule, T. T., and McNeil, B. W. 1982. *Analytical and experimental studies of combined thermal and photovoltaic systems*. Arizona State University report no. ERC-82027.

Wood, B. D., and Warnock, J. F., Jr. 1976. Solar retrofit applications for public buildings. *Proc. Joint Conf. AS/ISES and SESC*, Winnipeg, Canada, August 1976, pp. 105–28.

18 Solar Cooling—Introduction and Summary

Michael Wahlig

18.1 Introduction

Three main approaches to solar cooling have been pursued actively over the past ten years: mechanical, absorption, and desiccant cooling. Each of these is the subject of a succeeding chapter in this section (schematic diagrams of the three approaches are given in figures 18.1, 18.2, and 18.3).

In both the mechanical and absorption approaches, the cooling effect is produced by the evaporation of a pure refrigerant fluid in a closed container (an "evaporator"), a cooling process identical to that occurring in a conventional electric-driven vapor compression air conditioner. In fact, a solar-driven mechanical cooling unit contains a vapor compression chiller as a subcomponent. The cooling effect in a desiccant cooling system is generated by the evaporation of water into an air stream that has been dried by the cycle, using essentially the same cooling mechanism as that produced by a conventional evaporative cooler in a dry climate.

The details of the thermodynamic cycles for the three approaches are given in their respective chapters. This introductory chapter will present general concepts applicable to all these cooling approaches, relative advantages among them, and a brief picture of how their developments have fared over the past ten years. The latter part of this chapter will summarize the status and outlook for each approach.

18.1.1 General Perspective for Solar Cooling

All of the solar cooling technologies considered here are solar thermal ones; i.e., the solar energy is converted to thermal energy (i.e., heat) via a solar collector, and then the heat is used to drive a cooling cycle.

In this respect, there is a basic difference between solar heating and solar cooling. For heating, the collected solar heat may be applied directly to a building heating load. For cooling, the solar-collected heat must be used to drive an energy conversion device, using a thermodynamic cycle to produce cooling. The output cooling effect produced (i.e., the amount of heat removed from the building) divided by the input heat used to drive the cycle is the efficiency of the cooling cycle. Traditionally, this efficiency is called the thermodynamic coefficient of performance, or COP, of the cooling cycle. Typically, the COP value and the required driving temperature vary with the load and ambient temperatures.

676   Michael Wahlig

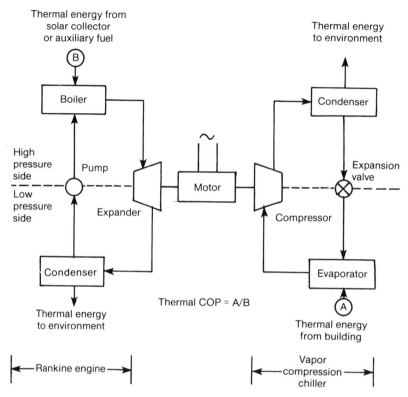

Figure 18.1
Schematic of a mechanical cooling unit.

The challenge for solar cooling is to collect the solar energy efficiently and inexpensively—and to convert it to cooling efficiently and inexpensively—so that the cooling delivered to the load is competitive with nonsolar alternatives on a cost and reliability basis.

However, it must be remembered that solar energy is just one of a number of thermal sources and usually not the most attractive one because of its variability in supply temperature and availability. Improvements in the efficiency of converting thermal energy to cooling are available to nonsolar (e.g., gas, heating oil) as well as solar thermal sources. For simplicity, we will use gas energy to represent all of the fossil fuel alternative sources. Therefore, the crux of solar cooling competitiveness is often reduced to a simple consideration: can a solar collector subsystem deliver thermal

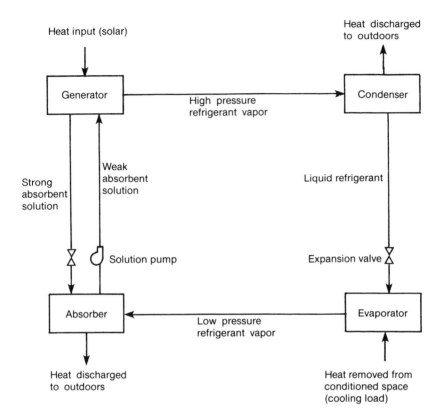

Figure 18.2
Schematic of an absorption cooling unit.

energy to a cooling conversion unit at the same or lower cost than a gas source can?

Improvements in the efficiency of the cooling conversion device are always worthwhile from an energy conservation viewpoint, independent of which thermal source is used. But, from the solar cooling perspective, the conversion device must also be amenable to solar energy input. That is, it must be efficient over a range of input temperatures characteristic of solar collector outputs and must accept hot water or hot air inputs.

A further consideration is meeting the heating load. As the most expensive element of the solar system, the collectors should be used year-round to displace as much conventional energy as possible. Therefore, it is usually taken for granted that any solar cooling system will also be used to

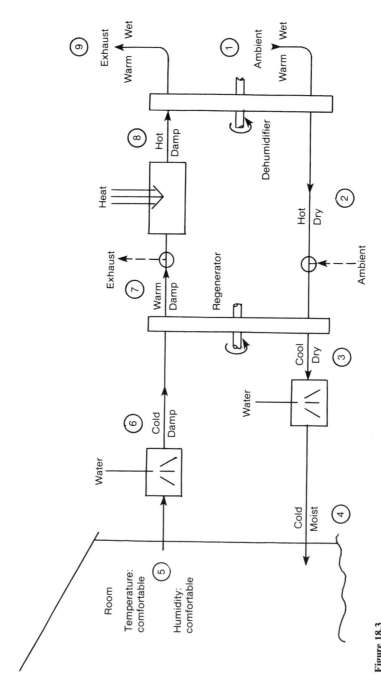

Figure 18.3
Schematic of a desiccant cooling system.

satisfy heating loads as far as practical. For mechanical and absorption cooling systems, in which the energy conversion device is essentially a heat pump, it may be more cost-effective to meet the heating load by operating such units in a heat-pump mode rather than applying the solar-collected heat directly to the load. In general, different methods for meeting the heating loads will be optimum for the different solar cooling system approaches, which brings us to the next topic.

18.1.2 Relative Advantages and Disadvantages among Solar Cooling Approaches

All of the approaches to solar cooling are bound by the same laws of thermodynamics and by the same limiting conditions: the thermal source (i.e., collector outlet temperature) and sinks (i.e., ambient dry bulb temperature and wet bulb temperature or humidity).

In principle, only one solar cooling approach is needed—the "best" one. But which one is that? And is the same approach best for all applications (e.g., in arid vs. humid climates)? The answer is far from obvious from a technical point of view, even after ten years of research and development (R&D). The advantages and disadvantages of the three approaches are given in considerable detail in the following three chapters; some highlights are given here.

Mechanical systems represent the least departure, technically, from currently accepted and widely used hardware. Most mechanical systems consist of a Rankine power subcycle (the left-hand side of figure 18.1) coupled to a Rankine cooling subcycle (the right-hand side of figure 18.1). The cooling subcycle is the commonly used vapor compression cycle. The power subcycle consists of a heat engine very similar to those in common use in electric power plants and in cogeneration systems. The common shaft that couples the heat engine output to the compressor input can be and has been integrated with a motor-generator unit, thereby allowing use of auxiliary electric energy for driving the compressor when solar energy is not available and allowing generation of electricity when cooling is not needed. If coupled to the electric grid, the electricity so generated can be considered to be an efficient energy storage mechanism, and credit for this electricity is an important element of the solar cooling economics.

However, the commonality of the mechanical solar cooling approach to conventional Rankine-cycle electricity-generation technology represents a disadvantage as well as an advantage. Invariably the question arises: why not replace the common mechanical shaft with an electrical coupling between the heat engine subcycle (which then becomes an electric power generation module) and the cooling subcycle (which then becomes completely indistinguishable from a conventional vapor compression air conditioner)? The electric coupling option results in a small efficiency penalty (a few percent) but gains a great deal in flexibility by allowing the power generation module to be located remotely from the vapor compression cooler. Adoption of the electric coupling option logically leads to a couple of follow-up questions: (1) Should the solar cooling development program work on the vapor compression cooling unit, which is now purely conventional, electric-driven technology? (2) Should the power generation module be operated at inlet solar collector temperatures associated with the DOE solar cooling program (up to 300°F) rather than be operated much more efficiently at the higher temperatures (up to 750°F) associated with the DOE solar thermal electric program? If the responses to these questions are negative, selection of the attractive, electric coupling option may result in marching the mechanical approach right out of the DOE solar cooling program.

To be fair, there are two advantages associated with maintaining the direct mechanical shaft coupling rather than the electric coupling. First, the direct coupling is somewhat more efficient because it avoids the mechanical to electric conversion step and the subsequent electric to mechanical reconversion step; typical efficiencies of 95% each are normally assigned to these two steps. Second, for heating applications where it is advantageous to operate in the heat-pump rather than the direct-heating mode, the heat discharged from the power loop condenser can be added to the heat discharged from the vapor compression loop condenser, more efficiently satisfying the heating load.

Absorption cooling has benefited from being the conventional cooling approach most readily adaptable to being driven by hot water from solar collectors. Conventional, thermally driven (steam, hot water, or gas) absorption chillers have been marketed for decades. Driving temperatures have been approximately 240°F. It was fairly straightforward to develop solar-driven versions of these chillers. The major modification was enlargement of the heat exchanger areas so that the chillers could be driven effectively by 185°F solar-heated hot water.

Not surprisingly, nearly all the solar cooling systems installed over the past decade have used absorption chillers as the solar cooling units. Several industrial firms in the United States and Japan have marketed absorption chillers for solar cooling applications.

Thermodynamically, absorption machines have an efficiency advantage over mechanical cooling units, for any given set of source and sink temperatures, because the mechanical units require two extra energy-conversion steps: thermal to mechanical, then mechanical back to thermal, each step entailing an efficiency loss of about 10 to 20% from the theoretically maximum conversion efficiencies.

However, absorption cooling suffers from several disadvantages. One is that it is not practical to use electric auxiliary energy to drive the absorption chiller; another thermal energy source such as gas must be used. (Electric resistance heating could be used to drive an absorption chiller, but the higher cost, compared to the use of gas auxiliary energy, makes it impractical.) The use of conventional absorption cooling machines has declined significantly over the last ten years, with absorption machines being displaced by electric vapor compression units. One reason for this decline in popularity was concern over future availability of natural gas, aggravated by the natural gas shortages of the mid-1970s. In principle, absorption chillers can be modified to generate electricity when there is no cooling load, but in practice no machines with this feature have yet been built.

Desiccant cooling systems have not been used to any appreciable extent as a conventional cooling alternative, although the two principal elements have been widely used as separate devices: desiccant materials for drying air and evaporative coolers for space cooling.

The desiccant cooling approach appears very attractive because of its promise of higher efficiencies at lower regeneration temperatures (140°F to 160°F) than those possible with the mechanical or absorption approaches. These lower driving temperatures mean that the solar collectors will operate more efficiently as they drive the cooling system. On the other hand, the source of the high efficiency values of the desiccant part of the cooling system is undoubtedly the judicious use of the evaporative cooling potential of ambient air, based on the difference between the dry bulb and wet bulb temperatures. This ambient air potential may also be tapped by incorporating evaporatively cooled condensers into mechanical and absorption cooling systems, thereby narrowing or eliminating any relative efficiency advantage of desiccant cooling systems.

Many of the earlier desiccant cooling system configurations suffered from relatively high parasitic electric power requirements to transport the air streams. This problem has been largely overcome with more recent desiccant system designs, and the desiccant approach may now be on a par with other cooling options with respect to parasitic power usage.

Thermodynamically, the desiccant approach is quite distinct in that it meets the latent part of the cooling load by chemical adsorption (or absorption) of the water vapor in the air directly onto (or into) the desiccant material. Mechanical and absorption cooling systems must remove the water vapor by condensing it on a cold evaporator coil, even though such a low temperature at the evaporator may not be necessary to meet the sensible portion of the cooling load. This unique characteristic of the desiccant approach has several consequences: (1) The relative advantage of desiccant cooling may be significantly greater for applications with a larger ratio of latent-to-sensible load; and (2) A hybrid cooling system consisting of a desiccant component to satisfy the latent load and a refrigeration-cycle (vapor compression or absorption) unit to meet the sensible cooling load may outperform all the other options.

On the other hand, higher efficiency multiple-stage designs and conversion devices have been developed for absorption-cycle and Rankine-cycle machines (more closely approximating the ideal Carnot efficiency), whereas all the desiccant cooling cycles developed to date are essentially single-stage devices.

With all these relative advantages and disadvantages of the mechanical, absorption, and desiccant approaches to solar cooling, how can one hope to compare them on a common basis? The answer lies in the common economic criteria that must be met by all the solar cooling technologies.

18.1.3 Economic Commonality

All the solar cooling approaches must measure up to the same economic yardstick. This is treated in considerable detail in the final chapter of this section on solar cooling. (As was stated in section 18.1.1, it is always assumed that, whenever practical, a solar cooling system will also supply space heating and hot water heating so that maximum year-round use can be made of the solar collectors.)

In concept, the economics issue is simple. A determination is made of how much conventional energy (gas and electricity) is saved on an annual basis by operating the solar cooling system rather than a conventional (typically an electrically driven vapor-compression cycle) cooling system. The "value" of the saved energy is calculated. To be cost-effective, the incremental cost of the solar cooling system (i.e., the extra cost over that of the conventional cooling system, taking into account first cost, maintenance cost, replacement cost, etc.) must be equal to or less than the "value" of the energy saved. When this economic criterion is satisfied, the consumer saves money by installing the solar cooling system. Of course, criteria other than purely economic ones would be involved in any such decision-making process (aesthetics, etc.).

Complexities arise in the calculation of the "value" of the saved energy. This value contains uncertainties and subjectivity. It is uncertain because the energy saved is future energy, and the cost of that energy depends upon its cost-escalation rate over future years. It is subjective because the value of that future money saved will be different for different consumers; i.e., an acceptable discount rate or payback period is an individual (or corporate) decision.

Further complicating the picture are the numerous factors affecting the incremental cost of the solar cooling system. The cost to the consumer (the one who decides whether or not to purchase and install the solar cooling system) depends upon tax credits and other incentives (or sometimes disincentives) offered by governmental and utility organizations.

In practice, reasonable values are assigned to the uncertain parameters, and economic analyses are performed, as shown in chapter 22. Importantly, whatever set of economic parameters is selected, that set should apply uniformly to all the solar cooling approaches. Thus, comparative economic assessments of the solar cooling options are meaningful.

18.2 Summary

A brief summary of the research and development activities conducted over the past ten years is given here, followed by a statement on the current status and a forecast of the future outlook for each of the solar cooling technologies.

18.2.1 Ten-Year Historical Overview

As evidenced just by the sizes of the following three chapters, a substantial research effort has been expended during the past ten years on all three of the solar cooling approaches. It would be far too much duplication of information to mention here even briefly all the companies and research projects involved. Therefore an attempt will be made in this section to provide a flavor of the research and development performed over this period, and the reader is referred to the individual chapters for more details.

In the early 1970s, a modest research effort was under way on all three approaches: mechanical, absorption, and desiccant. The absorption research was most significant during this period (largely privately financed by Arkla Industries) and led to the marketing by Arkla of 3-ton and 25-ton, single-effect (COP up to 0.7) absorption chillers for solar cooling systems. These chillers have been used in most of the solar cooling installations in the United States over the past ten years. However, Arkla discontinued the manufacture of these chillers in the early 1980s due to the lack of a significant market, leaving the small-capacity absorption chiller field to several Japanese firms currently marketing their products in the United States.

Meanwhile, much of the absorption cooling research in the late 1970s was directed towards the development of air-cooled absorption chillers (eliminating the separate wet cooling tower) in sizes appropriate for residential and small commercial applications, and the development of large-capacity, production-quality, water-cooled absorption chillers for commercial applications. These were all single-effect absorption machines with typical COPs of 0.7. These developments (by the Carrier Corporation and Arkla Industries) were successful technically, but the market for them never developed; solar cooling system costs were too high. The main cost component was still the solar collector.

Several of the newly developed absorption chillers were tested in a system context at the Colorado State University Solar Houses, generating valuable practical information on the operating characteristics and idiosyncrasies of these machines.

A portion of the absorption cooling research was directed toward reducing the collector array area needed (and thus its cost) through the development of higher-efficiency absorption chillers. The first of these advanced-cycle absorption chillers was built and tested successfully (by Lawrence Berkeley Laboratory—LBL) in the early 1980s. The development of evacuated-tube collectors appropriate for driving these high-efficiency absorption chillers has proceeded (by University of Chicago, Argonne National Laboratory—ANL, and several commercial manufacturers) in parallel with the chiller development over the past ten years. Steady improvements in the performance of these collectors have been achieved over this period.

Open-cycle absorption systems, in which the solar collector is simply a rooftop over which an absorption solution flows, offered an alternative very-low-cost-collector method of reducing solar cooling system costs. Arizona State University, Colorado State University, and other organizations have investigated various open-cycle system alternatives analytically and experimentally, addressing the maintenance issues as well as the thermal performance.

In 1976, R&D on mechanical, Rankine-cycle cooling systems erupted with the advent of the NASA-managed, DOE-supported "404" program, one element of the Development in Support of the Demonstration Program. On the basis of a competitive solicitation, NASA selected three projects for multiyear, multimillion dollar R&D support. All three were Rankine-cycle projects. One of these projects was dropped after several years of research due to persistent technical problems, mainly with the foil bearings for the high-speed rotating shaft. A second project developed several small Rankine cooling units, but they suffered breakdowns soon after being installed in the field; the units were removed and that project ended. Only one of the three projects (a collaboration between the Honeywell, Lennox, and Barber-Nichols corporations) resulted in a significant number of successful field installations of Rankine cooling systems. About a dozen units, of 3-ton and 25-ton capacities, were fabricated and installed, with generally successful field operation.

Several other mechanical, Rankine-cycle cooling R&D projects were initiated throughout the 1970s as the result of DOE support of solicited and unsolicited proposals. Bearings and seals for the high-speed rotating components constituted serious technical problems and led to the termination of at least one of the major projects. The most successful of these projects, by United Technologies Research Center (UTRC), developed a fairly reliable 18-ton Rankine-cycle heat pump that passed successful test runs in the laboratory and in a field installation. However, since a sizable solar cooling market has yet to develop, these Rankine cooling units remain in the category of expensive, made-to-order, individual cooling packages.

In the early 1980s it became clear that the solar cooling equipment developed up to that time, with typical COP values of 0.6 to 0.7, was unlikely to become cost-competitive with electric-driven vapor compression machines. Accordingly, UTRC and Barber-Nichols designed advanced, Rankine-cycle chillers with higher efficiencies (COP $\gtrsim$ 1.0). However, the combination of a decreasing solar cooling research budget and the lack of a clear-cut advantage (technical or economic) of solar Rankine systems (over absorption or desiccant cooling systems) has led to a cessation of almost all solar cooling research based on the mechanical approach.

The desiccant cooling research that started in the early 1970s was typified by packed-desiccant-bed designs, characterized by mediocre efficiencies (COP about 0.5) and high parasitic power requirements. Then, in the later 1970s and early 1980s, this field was revitalized by theoretical predictions by the Illinois Institute of Technology (IIT) that substantially higher COP values could be achieved, and by new dehumidifier designs that broke through the previous efficiency and parasitic power barriers.

Research by the Exxon corporation, partially supported by the Gas Research Institute, achieved a desiccant cooling COP of 1.1 and offered promise of eventual improvement to COP values of 1.5. The parallel-passage dehumidifier design developed first by IIT and then by the Solar Energy Research Institute (SERI) overcame most of the parasitic power concerns and also led SERI to predict desiccant cooling COPs of 1.1.

The concept of hybrid desiccant-vapor compression cooling systems was also introduced, in which the desiccant component is used to meet the latent load and the vapor compression component is used to meet the sensible cooling load. Many of the variations and advantages of such hybrid systems have been explored analytically by the University of Wisconsin. In many respects, these hybrid systems will outperform separate dessicant or vapor compression cooling systems. However, the role of (i.e., need for) solar energy as a heat source to regenerate the desiccant in these hybrid systems is not clear.

Taken as a whole, substantial technical progress has been made and reported for all three approaches to solar cooling over the ten-year period from 1974 to 1984. It is also true that much remains to be done before the performance, economics, and reliability are such that solar cooling systems will achieve appreciable market penetration.

18.2.2   Current Status

Technically aggressive research activities are continuing on the absorption and desiccant approaches to solar cooling, although the overall magnitude of the solar cooling research effort continues to decline in concert with decreases in the DOE solar heating and cooling budget. Mechanical, Rankine-cycle solar cooling is essentially in a dormant state. Generally successful field tests of solar Rankine cooling systems in Phoenix were recently concluded by UTRC and by Barber-Nichols as part of the U.S.-Saudi Arabian SOLERAS Program. These two companies also completed preliminary designs of advanced, higher-efficiency versions of their respective Rankine-cycle equipment as the final phase of their DOE-supported projects. The only DOE-funded solar Rankine activity as of March 1985 is the final, test phase of the fossil-boosted, solar-evaporated, superheated steam turbine concept being performed by the University of Pennsylvania.

Work on closed-cycle absorption chillers is continuing at LBL. Testing has concluded on a double-effect regenerative absorption chiller (COP about 1.2 at 280°F inlet temperature), and the design phase is starting on a single-effect regenerative absorption chiller (design COP of about 1.5 at 280°F inlet temperature). Analytic calculations are also in progress at LBL to predict the performance of advanced-cycle absorption cooling systems, the control and storage interactions of such systems, and comparisons between the performance of absorption and desiccant cooling systems.

System development and testing of solar heating and cooling systems containing single-effect absorption-cycle chillers continues at Colorado State University.

Experimental research on solid desiccant cooling components and systems is proceeding at SERI. Single-pass tests of a parallel-passage dehumidifier wheel containing small silica-gel particles adhered to spirally wound Mylar tape have been completed. Construction is nearing completion on a complete two-pass test loop that will enable realistic testing of this type of desiccant wheel in a solar cooling system context. Parallel research at SERI is determining desiccant material and component characteristics, such as isotherm hysteresis under cyclic conditions.

Related to and coordinated with this SERI work is research on new desiccant materials by ANL and analytical performance calculations by the University of Wisconsin. The Wisconsin analysis encompasses a variety of desiccant cooling configurations and control strategies, extending to hybrid desiccant-vapor compression cooling systems.

Finally, an appreciable amount of work is now under way on open-cycle cooling systems using liquid working fluids, based on both the open-cycle-absorption and the liquid-desiccant approaches. These two technologies share a common collector-regenerator subsystem, which is the solar-unique part of these cooling systems. Arizona State University is pursuing an open-rooftop-regenerator technique, whereas Colorado State University is investigating a packed-tower-regenerator method, coupled to air collectors. Both projects include analytical and experimental activities, and most of the work to date has concentrated on open-cycle absorption configurations. A small effort has been started specifically studying a particular liquid desiccant cooling system recently installed at the Virginia Science Museum, which has solar potential (although no solar collectors have yet been installed); participants in this project include Meckler Engineering, the University of Wisconsin, and the Tennessee Valley Authority.

18.2.3 Future Outlook

Not surprisingly, the future outlook for solar cooling is clouded. Solar cooling is a technical success. But, in general, solar cooling is not yet economically justifiable for widespread applications.

Assuming continued DOE support, significant technical improvements should continue well into the future, based upon research now under way. If DOE support were to cease, it is questionable whether any appreciable new work on solar cooling would be undertaken by private firms in the near future. A few private firms might continue to market solar cooling systems based on current technology. Several Japanese firms (including Yazaki, Hitachi, and Sanyo) are selling solar absorption cooling systems. American Solar King is just beginning to market a solar desiccant cooling system.

The big uncertainties are associated with the economic issues, not the technical ones. It appears unlikely that significant market penetration (say, greater than 10%) will occur until solar cooling system costs (i.e., combination of first costs, operating costs, and maintenance costs) are approximately on a par with the cost of conventional cooling systems (typified by an electrically driven vapor compression air conditioner). In other words, a solar cooling consumer would have to realize a fairly short payback period (say, 5 years or less) for the investment in a capital-intensive solar cooling system; alternatively, the solar cooling system financing would have to be structured in such a way that the consumer experiences no real increase in the cooling expenses associated with the solar cooling system installation.

More optimistic projections for solar cooling utilization have been made in prior years, based partially on the supposition that conventional fuel costs would continue to escalate at significant rates. These escalation rates have slowed considerably over the past few years, dampening the outlook for early competitiveness of solar cooling alternatives.

Meanwhile, the competition is on the move. As an example, the Electric Power Research Institute (EPRI) has given Carrier a multimillion dollar contract to improve the efficiency of electric-driven vapor compression air conditioners.

To make an informed prediction on the future outlook for solar cooling, we must have a clear picture of how solar cooling compares with the conventional cooling alternatives. This comparison can be made in two steps.

The first step involves the recognition that all three of the solar cooling approaches are *heat-driven* cooling technologies (with the heat produced by various types of solar collectors). As such, the cost of the solar-produced heat must compete with the cost of conventionally produced heat (e.g., by a natural gas flame). Thus the solar collector subsystem is the key to the solar system economics; it must deliver the heat to the energy conversion device (the chiller, or heat pump) at the required temperature at the same or lower cost than that of natural gas or fuel oil.

Of course, the corollary to this requirement is also true. The advanced, high-efficiency cooling technology being developed through solar cooling research efforts is quite amenable to spin-off applications driven by fossil-fuel sources. For example, the Gas Research Institute is supporting gas-driven desiccant cooling system development, using desiccant technology originally researched with solar-driven cooling applications in mind. These nonsolar spin-off applications can constitute an important benefit resulting from solar cooling technology development, especially in the near term.

The comparative economics of solar-driven vs. gas-driven systems must be calculated comprehensively, taking into account heating as well as cooling (and also hot water heating, if appropriate) and the parasitic electric power required. However, a simplified example will serve to establish an approximate target cost for the solar collector subsystem. In 1985, typical natural gas prices were about $7 or $8 per million Btu. An efficient solar collector system, operating year-round for heating and cooling, can collect about 0.25×10^6 Btu/yr ft^2 of collector area. Thus a square foot of collector area might save about \$2 worth of gas per year. To be an acceptable investment by the consumer (say, a 5 or 6-year payback), the installed solar collector subsystem (collector, storage, piping, controls) would have to cost the consumer no more than \$10 or \$12/ft^2. This is a difficult target for a solar collector subsystem that has to be of adequate quality to drive a cooling system.

Therefore, the future outlook for solar cooling is partially tied to the requirement for renewed substantial escalation of natural gas prices or substantial reduction in the installed cost of solar collector subsystems, or a combination of both. Of course, solar cooling tax credits or utility rebates, if available, would directly contribute to making the solar alternative more competitive.

The second step in the process for assessing the future outlook for solar cooling entails the recognition that the primary competition is the *electric-driven* vapor-compression air conditioner or chiller, the efficiency of which is being substantially increased by manufacturers' development plans. These electric-driven chillers have the lion's share of the cooling market, and this share has grown in recent years at the expense of heat-driven cooling alternatives. This trend would have to be reversed for solar cooling systems to capture a significant fraction of the market. The question must be asked: If conventional heat-driven (e.g., gas-driven) cooling equipment has not penetrated the market, what is so different about solar-collector-heat-driven cooling systems that would enable them to capture a significant share of the electric-dominated cooling market? After all, the auxiliary, backup energy source for a solar absorption or desiccant cooling system will likely be natural gas.

The answer to this question is not obvious. Solar energy is still popular in concept, even though it is more expensive than conventional energy sources. People would likely use solar cooling if it were reliable and cost-effective. But who can foresee what economic conditions (based on fossil-fuel availability) or national security issues (based on energy independence) will again become paramount and render important the step of weaning ourselves from dependence on conventional energy supplies?

Meanwhile, the usual "chicken and egg" situation prevails: (1) a market will materialize for solar cooling systems when costs are reduced to competitive levels; and (2) costs will be reduced significantly when a large enough market exists to justify volume production and sale of solar cooling systems. One avenue to solve this dilemma is closely related to the second step in the process described just above. Penetration of the conventional cooling market by fuel-driven cooling equipment should be assisted by the development of advanced, high-efficiency, heat-driven chillers that are more efficient energywise than the electric-driven competition. This is the type of chiller hardware being developed in the solar cooling program. Market acceptance of this fuel-driven cooling equipment should greatly facilitate subsequent acceptance of solar-heat-driven cooling systems in which the same or similar-type chillers are used. This realization provides a justification and a requirement for continued research on solar chiller technology. And the outlook for technical success in developing this high-efficiency cooling technology is excellent.

The solar-unique part of solar cooling systems, as stated above, is comprised of the solar collector subsystem: collectors, storage, piping, and controls. There is potential for continued efficiency improvements and cost reduction in this subsystem over the next few years. Methods for accomplishing these improvements have already been identified and the necessary research is under way.

A number of system considerations are also very important for eventual commercialization of solar cooling systems, including effective integration of heating, cooling, and hot water functions, development of standardized designs, reliability improvements, and major reductions in marketing costs.

As a final consideration, the relative attractiveness of the solar cooling approaches has long been a topic of discussion. It becomes relatively more important as solar cooling research budgets decrease and choices must be made concerning which solar cooling options to continue supporting. Reviewing the past ten-year period, the relative attractiveness of each of the three solar cooling approaches—mechanical, absorption, and desiccant—has peaked and ebbed at different times, so if a decision had been made to select a "best" option, that decision would have depended strongly upon when it was made, and it probably would not have been the best option. This situation still holds true today.

Overall, the future outlook for ample benefits ensuing from the solar cooling research program is quite favorable. However, the eventual outcome in terms of market adoption and use of a specific solar cooling technology is still quite uncertain.

19 Mechanical Systems and Components

Henry M. Curran

19.1 Summary of State of the Art

This chapter covers the mechanical approaches to solar cooling, as distinguished from the absorption and desiccant approaches. This first section summarizes the state of the art. Subsequent sections describe major components and systems, performance data, economic considerations, and research on advanced concepts.

19.1.1 Introduction

In the mechanical approach to solar cooling, the cooling effect is produced by the evaporation of a refrigerant. All mechanical systems operate on the basis of a closed refrigeration cycle. That is, a fixed quantity of refrigerant is continuously cycled through evaporation and condensation processes. Since the evaporation process requires a lower pressure than the condensation process, a compressor is required to maintain continuity of operation. The motive power needed to operate the refrigerant compressor is provided by an energy conversion device. Applicable energy conversion devices are either heat engines or vapor-jet units. These convert a thermal energy input into a mechanical energy output.

19.1.1.1 Heat Engines
There are two basic categories of heat engines: (a) working-fluid engines that operate on the basis of thermodynamic cycles, and (b) solid-state engines based on the enthalpy of transformation corresponding to temperature-related, diffusionless phase transformations in certain metallic alloys, particularly Nitinol. Research on various Nitinol engine configurations has been conducted at the Lawrence Berkeley Laboratory (Hollander and Banks 1975; Banks and Wahlig 1976). Because of the very low conversion efficiencies and other limitations, engines of this type have not been found suitable for use in solar cooling units.

Two types of working-fluid engines are technically feasible for use in solar cooling units. In one type, the working fluid cyclically changes phase from liquid to gas and from gas back to liquid. The most widely used thermodynamic cycle of this type is the Rankine cycle. Rankine cycle engines have been developed in various forms and for diverse applications for many years. Reports of work on their application to solar cooling began to appear in the U.S. literature in the mid-1970s (Sargent and Teagan 1973; Prigmore and Barber 1974; Curran et al. 1974; Curran, Miller, and Alereza 1975; Biancardi et al. 1975). The principles of operation of the Rankine cycle are discussed in section 19.2.

In the other type of working-fluid heat engine, the working fluid remains a gas. Engines of this type operate on various thermodynamic cycles, including the Stirling, Brayton, and Vuilleumier cycles. For relatively low thermal energy input temperatures (less than about 392°F [200°C]), Rankine cycle engines are superior in energy conversion efficiency to gas cycle engines, whereas at high temperatures gas cycle engines have efficiencies equal to, or better than, Rankine cycle engines. Since, historically, research on solar cooling has been associated with relatively low temperature solar collectors, there has been very little research on gas cycle engines for this application (Shelpuk 1975).

19.1.1.2 Vapor-Jet Units

An alternative to a heat engine is a vapor-jet unit. Consider first that the component of a heat engine that provides motive power to drive a refrigeration compressor is a rotating or reciprocating expander. This component converts an enthalpy drop in the working fluid, corresponding to a pressure drop, into kinetic energy. When an expander is coupled to a compressor, the kinetic energy is converted to an enthalpy increase in the refrigerant fluid, corresponding to an increase in pressure. In a Rankine unit the same working fluid can be used in both the power cycle and the refrigeration cycle.

A vapor-jet unit functions equally as an expander and a compressor in a single component without moving parts. This unit comprises a convergent nozzle and a divergent nozzle separated by a mixing chamber. The convergent nozzle is analogous to a mechanical expander and converts an enthalpy drop in the working fluid into kinetic energy. Refrigerant vapor entering the mixing chamber at subsonic velocity is entrained, with momentum exchange, by the supersonic flow from the convergent nozzle. The mixture, still at supersonic velocity, enters the divergent nozzle, or diffuser, where the kinetic energy is converted into an enthalpy increase, with an increase in pressure. Thus the diffuser is analogous to a compressor. The vapor is condensed to liquid in a condenser, and the flow is then divided into two streams, one going to the thermal energy input loop and the other to the refrigeration loop.

When compared with heat engine-compressor units, vapor-jet units have the advantages of being less complex and less costly. Theoretical analysis and hardware research have shown, however, that vapor-jet units are less efficient than heat engine-vapor compressor units (Zeren, Holmes, and Jenkins 1978; Zeren and Holmes 1981; Lansing and Chai 1979). Consequently, there has been little development and application of this technology to the solar cooling field.

19.1.1.3 Research and Development Focus

With minor exceptions, the U.S. and foreign research and development (R&D) activities related to mechanical concepts for solar cooling have concentrated primarily on the Rankine cycle technology. This is the focus of the remainder of this chapter.

19.1.2 Solar Cooling Using Rankine Engines

Several basic configurations are possible in which a Rankine expander can be used to provide the energy input to a cooling unit. Within each of these configurations, there may exist many variations depending on the design approach. Figure 19.1 illustrates the basic configurations.

19.1.2.1 Mechanical Coupling

Figure 19.1a shows the mechanical coupling of a Rankine expander to a vapor compressor. In the simplest variation, a rotary expander and a rotary compressor are mounted on a common shaft. Alternatively, a high-speed rotary expander may be coupled to a low-speed rotary or reciprocating compressor by a gearbox.

19.1.2.2 Mechanical Coupling with Electric Unit

Figure 19.1b illustrates a configuration in which an electric unit, either a motor or a motor-generator, is mechanically coupled with a Rankine expander and a vapor compressor. Depending on the relative speeds of the three components, the coupling may be by in-line shafts or a gearbox. Including an electric motor provides for operation of the compressor when solar energy is not available to operate the Rankine expander. Adding the generator provides for generation of electricity when solar energy is available, but cooling is not required. When the motor is used, a clutch may be included to permit decoupling of the expander to eliminate frictional and windage losses. Similarly, when the motor-generator is used, two clutches may be included so that either the expander or the compressor may be decoupled.

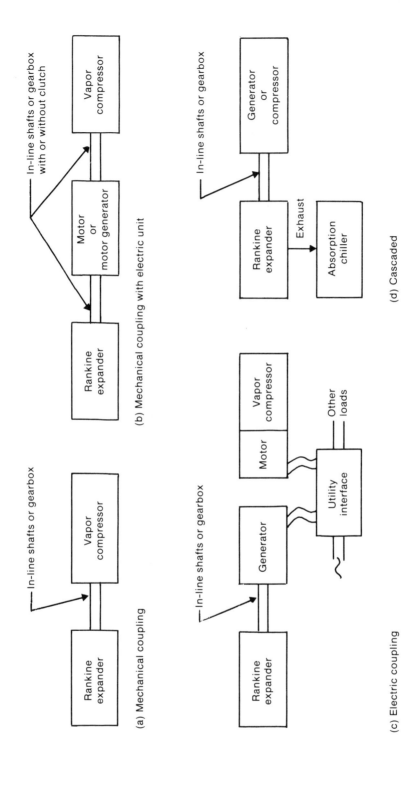

(a) Mechanical coupling

(b) Mechanical coupling with electric unit

(c) Electric coupling

(d) Cascaded

Figure 19.1
Configurations of subsystems using Rankine expander for space cooling.

19.1.2.3 Electric Coupling

Figure 19.1c illustrates an electric coupling configuration. Here the Rankine expander drives an electric generator, either through an in-line shaft or a gearbox. The electric output from the generator is used to operate an electrically driven vapor compressor. When solar energy is not available, electric energy from the utility is used. When cooling is not required, any electric output from the generator may be fed back to the utility or used for other electric loads in the building.

19.1.2.4 Cascaded Configuration

Figure 19.1d shows a cascaded configuration. In this design, the hot exhaust from a high-temperature Rankine expander is used to operate an absorption chiller. The shaft power output of the expander may drive either an electric generator or a vapor compressor. Theoretically, for equivalent temperature conditions, cascaded and noncascaded configurations are thermodynamically equivalent (Curran 1978).

19.1.2.5 Provision for Heating

The configurations shown in figure 19.1 can be used for heating as well as for cooling if the refrigerant circuits connected to the compressors have reversing valves that provide for interchanging the evaporator and condenser functions. That is, in the reversed condition, the circuit operates as a heat pump, receiving thermal energy from the environment into the evaporator and discharging thermal energy at a higher temperature to the building space from the condenser. Thermal energy discharged from the Rankine cycle condenser may also be used for heating.

The alternative to designing the system to operate as a thermally driven heat pump is to bypass the mechanical system and to use thermal energy from the solar collector directly for heating. The heat pump approach has been shown to be theoretically preferable for specified steady-state conditions (Curran and Miller 1976). However, in practical, nonsteady-state applications, system designers must make many tradeoffs in deciding which approach to use, and both approaches are represented in the state of the art.

19.1.3 U.S. and Foreign Solar Cooling Projects

Rankine engines that could be used for solar cooling applications have been developed in the United States, Germany, France, Italy, Israel, and Japan. Many of these engines were developed for nonsolar applications, and some for solar applications other than cooling, such as water pumping and electric power generation. Most of the research on mechanical systems for solar cooling has been done in the United States. Other countries involved to a small extent are Saudi Arabia and Japan.

Practically all of the research in the United States has been funded by federal agencies, primarily the National Science Foundation (NSF), the National Aeronautics and Space Administration (NASA), the Energy Research and Development Administration (ERDA), and its successor, the Department of Energy (DOE).

19.1.3.1 NSF, ERDA, and DOE Projects

Research and development on mechanical systems for solar cooling were initiated in 1973 by the Division of Advanced Research and Technology of NSF. In 1975, this activity was transferred to the Division of Solar Energy of the newly formed ERDA. In 1978, the ERDA activities were incorporated into its successor agency, DOE. From 1973 to 1983, the major contractors funded by these agencies for development of Rankine solar cooling were the Energy Resources Center of Honeywell Inc., Barber-Nichols Engineering Co., the Energy Systems Division of Carrier Corp., Energy Technology Inc., the Mechanical Engineering Department of the University of Pennsylvania, and Sandia Laboratory.

19.1.3.2 NASA 404 Program

In 1976, NASA initiated the 404 Program at the George C. Marshall Space Flight Center to develop prototype solar heating and cooling systems for application in single-family residences, multifamily residences, and small commercial buildings. Funding was awarded to three contractors: AiResearch Manufacturing Co., in association with Dunham-Bush Corp.; the Space Division of General Electric Co.; and the Energy Resources Center of Honeywell Inc., in association with Barber-Nichols Engineering Co. and Lennox Industries. The projects included development, manufacture, testing, installation at selected sites, maintenance, problem resolution, and performance evaluation.

19.1.3.3 SOLERAS Program

The United States-Saudi Arabian Joint Commission on Economic Cooperation established the SOLERAS Program for solar energy research and development in October 1977 (Martin, Corcoleotes, and Williamson 1981). The solar cooling hardware aspects involved the design and construction, by U.S. manufacturers, of absorption and Rankine chillers suitable for use under Saudi Arabian weather and building cooling conditions. The program provided for two years of testing of each chiller in commercial buildings in the United States at a site having climate characteristics similar to those of Saudi Arabia. Phoenix, Arizona, was selected for this purpose. The two Rankine projects are those of Honeywell Inc., in collaboration with Barber-Nichols Engineering Co.; and United Technologies Research Center (UTRC).

19.1.3.4 Japan

The Mitsubishi Co. has reported R&D on a small Rankine solar cooling system installed in an experimental house (Nishiyama et al. 1979).

19.2 Rankine Solar Cooling Projects

Rankine solar cooling projects have proceeded along two paths: (1) the development of Rankine engines to drive conventional vapor compression cooling units, and (2) the development of turbocompressor units in which both the power and cooling components are integrated into a single machine. Following a review of the operating principles of the Rankine power cycle and the vapor compression cooling cycle in subsection 19.2.1, the major Rankine solar cooling projects in the two categories mentioned above are described in subsections 19.2.2 and 19.2.3. Some of these projects included operational test sites, and some performance results are given for those for which data have been published. The limited space available here permits only very limited summaries, of course. The reference documents should be consulted if more detailed information is desired.

19.2.1 The Rankine Power and Refrigeration Cycles

The components of a Rankine chiller are shown in figure 19.2, the configuration being of the general type shown in figure 19.1b.

The Rankine power cycle is a closed sequential series of thermodynamic processes that convert thermal energy into mechanical energy. This conversion is effected by cyclic changes in the state conditions of a circulating working fluid. The fluid in the liquid state is pumped at a relatively high pressure into a boiler in which it is evaporated by a thermal energy input. The high-pressure vapor generated in the boiler is then expanded through an expander to a condition of lower pressure and temperature. This expansion process converts the thermal energy input to the boiler

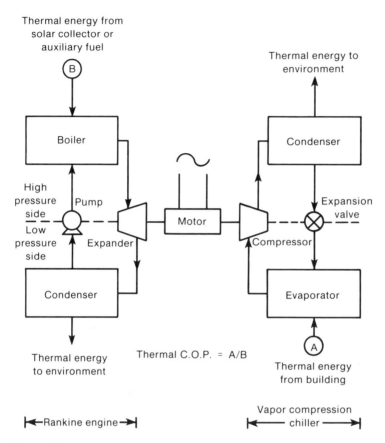

Figure 19.2
Solar Rankine-vapor compression cooling unit.

into mechanical energy at the expander shaft. The vapor exhaust from the expander flows into a condenser where it returns to the liquid state by giving up the enthalpy of condensation to cooling water or ambient air. This liquid is then pumped back to the boiler to continue the cyclic operation.

The Rankine refrigeration cycle also operates on a closed sequential series of thermodynamic processes and is essentially a reversed Rankine power cycle; i.e., a kinetic energy input is converted into an enthalpy increase in a working fluid. Liquid working fluid at relatively high pressure flows from a condenser through an expansion valve into an evaporator.

The decrease in pressure and the low temperature thermal energy flowing from the cooled space into the evaporator cause the working fluid to evaporate. The vapor is then compressed to the higher cycle pressure in a compressor. The high pressure vapor is then condensed in the condenser, and the cyclic operation continues.

In some systems, the same working fluid is used in both the power cycle and the refrigeration cycle. A common condenser can then be used and special seals to maintain fluid separation in turbocompressor units eliminated.

In considering the application of the Rankine cycle to the solar cooling of buildings, the thermal coefficient of performance (COP) is an important parameter. The COP of a heat-engine-driven chiller is the product of the cycle thermal efficiency of the heat engine and the thermal energy efficiency ratio (TEER) of the vapor compression chiller. Of these two, only the heat engine cycle thermal efficiency is a function of the maximum system temperature, but both are functions of the condensing temperature. The maximum possible cycle thermal efficiency for a heat engine is the reversible Carnot cycle efficiency:

$$E_c = \frac{T_s - T_0}{T_s},$$
(19.1)

where T_s is the thermal energy source absolute temperature and T_0 is the thermal sink absolute temperature. The actual cycle efficiency for heat engines is always much less than the Carnot efficiency because of deviations of real heat engine cycles from the ideal reversible Carnot cycle and because of thermodynamic irreversibilities in real cycles.

Figure 19.3 shows the cycle thermal efficiency as a function of maximum cycle temperature for the Carnot cycle and for small engines of up to 134 HP (100 kW) operating on the Rankine (Curran et al. 1974), Stirling, and Brayton (Fujita et al. 1980) cycles (Curran 1982). The Rankine envelope curve is for engines condensing at 122°F (50°C), corresponding to the use of air-cooled condensers. For maximum cycle temperatures below 302°F (150°C), the thermal efficiency for small, state-of-the-art Rankine engines is about 0.5 Carnot, whereas at 932°F (500°C), the relative efficiency is somewhat less, about 0.38 Carnot. This decrease in efficiency relative to Carnot with increasing temperature is due to the fact that, because of the thermodynamic properties of real working fluids, the basic Rankine cycle used in small engines departs more from the Carnot cycle at high tempera-

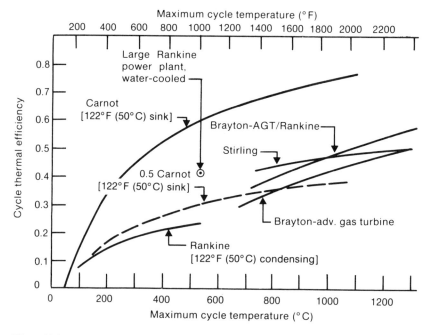

Figure 19.3
Heat engine design point thermal efficiency versus maximum cycle temperature (power output is 100 kW or less, except large power plant). Source: Curran (1982).

tures than at low temperatures. As indicated in figure 19.3, the state of the art for large, water-cooled Rankine power plants corresponds to a cycle thermal efficiency of about 0.42 at 1004°F (540°C) maximum cycle temperature. This is about two-thirds of the Carnot cycle value and is attained by using interstage reheaters and other special components to obtain a close match between the real cycle and the Carnot cycle. This is economically feasible because of the large size and base-load operation of such power plants. The 0.5 Carnot curve is an indication of the approximate efficiency limit that might be approached in the 392° to 932°F (200° to 500°C) range with advanced design of small Rankine engines.

As also indicated in figure 19.3, for maximum cycle temperatures above 1112°F (600°C), small engines using the Stirling cycle, the Brayton cycle, and combinations of these can provide efficiencies above 0.5 Carnot (Fujita et al. 1980). This is because the working fluids in the Stirling and Brayton cycles remain in the gaseous state, so that they are not subject to the

limitations imposed by the thermodynamic properties of fluids undergoing change-of-phase cycles, such as in the Rankine cycle.

The maximum possible thermal energy efficiency ratio (TEER) for refrigeration is that of the reversed Carnot cycle,

$$\text{TEER}_c = \frac{T_e}{T_0 - T_e}, \tag{19.2}$$

where T_0 is the thermal sink absolute temperature and T_e is the refrigeration absolute temperature.

The thermal COP of a Carnot engine driving a Carnot refrigerator is
the product of equations (19.1) and (19.2),

$$\text{COP}_c = E_c \times \text{TEER}_c = \frac{T_s - T_0}{T_s} \times \frac{T_e}{T_0 - T_e}. \tag{19.3}$$

The actual design point cycle thermal efficiency for small Rankine engines
typically falls in the range 0.08 to 0.20, depending primarily on the input
and condensing temperatures. The refrigeration cycle TEER is a measure
of the number of units of thermal energy the machine removes from the
cooled space, via the evaporator, for each unit of shaft energy supplied to
the machine. Average actual steady-state values of this ratio are about 5
for water-cooled machines and about 3 for air-cooled machines.

As noted in figure 19.2, the thermal COP of an actual Rankine engine-
vapor compression unit is the ratio of the cooling effect to the thermal
energy input rate to the boiler. This is the product of the actual power
cycle thermal efficiency and the actual refrigeration cycle TEER. Thus,
taking the product of the approximate ranges of these performance measures given above, a typical range of values for the thermal COP of actual
Rankine chillers is about 0.24 to 1.0. This is, of course, only a measure of
the energy performance of a chiller subsystem and not of an entire solar
cooling system.

Note that, even when operating in the solar mode, solar heating and
cooling systems normally require an electric power input to operate pumps,
fans, and controls. This is commonly referred to as parasitic power. In the
case of a Rankine engine, the expander power developed from a solar
input should substantially exceed the parasitic power, otherwise there
would be no reason to use the Rankine engine instead of an electric motor
to drive the compressor. This consideration applies not only to overall
system design, but also to control design for specific modes of opera-

tion. For example, one possible mode of operation may be to use thermal energy from storage to operate the Rankine engine when solar radiation is not being received by the collectors. As the storage temperature decreases, the electric power required for pumps, etc., may exceed the Rankine expander power, requiring the controls to discontinue this mode of operation.

When a Rankine chiller is installed in a complete solar cooling system with various electrically driven components, the energy performance of the solar system can be compared to that of a conventional electrically powered chiller system by computing the ratio, in equivalent units, of the seasonal cooling effect to the seasonal electric energy inputs required for the two systems.

19.2.2   Rankine Engine Driving Conventional Vapor Compressor

Much of the Rankine solar cooling R&D activity has been on the type of subsystem in which a conventional vapor compression chiller is driven by a Rankine engine, usually with an electric motor or motor-generator included. Projects involving the design and construction of complete subsystems of this type include those of Barber-Nichols Engineering Co., both independent and subcontracted by Honeywell Inc.; General Electric Co.; and Mitsubishi. Projects limited only to the development of Rankine engines for this type of subsystem are those of Energy Technology Inc. and the University of Pennsylvania.

19.2.2.1   Honeywell Inc./Barber-Nichols Engineering Co.

The initial Honeywell/Barber-Nichols project in 1973 included the development and demonstration of a 3-ton (10.5-kW) solar cooling system with an electric power generation capability of 1.3 HP (1 kW), using a configuration of the basic type shown in figure 19.1b. The Rankine subsystem was designed and constructed by Barber-Nichols in support of Honeywell's mobile solar energy research laboratory (Prigmore and Barber 1974; Barber 1978). The design point turbine inlet temperature was 200°F (93°C). Subsequently, the Honeywell/Barber-Nichols team participated in the NASA 404 Program and in the SOLERAS program.

19.2.2.1.1   HONEYWELL NASA 404 PROJECTS   Over a period of about eight years, Honeywell, in collaboration with Barber-Nichols and Lennox Industries, designed, constructed, installed, and operated ten solar Rankine systems. The operations statistics include 30 system years, with 20,000

Rankine engine hours and 60,000 air-conditioner hours. Two basic Rankine chiller sizes were developed: a nominal 3-ton (10.5-kW) for single-family residential application, and a nominal 25-ton (87.8-kW) for multifamily residential and small commercial application (Honeywell 1977–1978; Honeywell 1978a, 1978b; Batton and Barber 1978; Technical Support Package 1983). The Rankine engine design power values were 2.4 and 20 HP (1.8 and 15 kW), respectively, at 195°F (90.6°C) input temperature. The corresponding cycle thermal efficiencies were 0.08 and 0.086, at turbine speeds of 35,000 and 24,000 RPM, respectively.

Compressors from two manufacturers, York and Trane, were used, both having a TEER of about 5.4. Using this value and the power cycle thermal efficiency of 0.086, the nominal thermal COP for the 25-ton (87.8-kW) unit was 0.46.

Figure 19.4 shows a general system schematic for the Rankine chiller subsystem. This is a Barber-Nichols design, using R-113 fluid.

Under the NASA program, Honeywell installed and operated solar heating and cooling systems at eight sites: four single-family residential, one student residence, and three commercial. Summary details for two representative sites, one single-family and one commercial, are presented below.

19.2.2.1.2 HONEYWELL OTS 41, SHENANDOAH (NEWNAN), GEORGIA Installation of the Honeywell system (Honeywell 1982b) at this single-family residence began in mid-July 1980 and was completed in mid-November. Operational data were collected between January and August 1981. The solar energy system is a hydronic heating and cooling system consisting of 702 ft^2 (65.2 m^2) of liquid-cooled, flat-plate collectors; a 1000-gal. (3785 l) thermal storage tank; a 3-ton (10.5-kW) capacity Barber-Nichols Rankine chiller; and associated equipment for space heating and hot water heating. The system operation is controlled automatically by a Honeywell microprocessor-based control system that also provides diagnostics. During the 7 months of the operational test, the solar system collected 53 MBtu (55.8 GJ) of thermal energy from the total incident solar energy of 219 MBtu (230.7 GJ) and provided 11.4 MBtu (12 GJ) for cooling, 8.6 MBtu (9.1 GJ) for heating, and 8.1 MBtu (8.5 GJ) for hot water. The projected net annual energy resource effects of the use of the solar system were a saving of approximately 49,300 ft^3 (1395 m^3) of natural gas and a small increase, 0.9 MBtu (280 kWh), in the use of electrical energy.

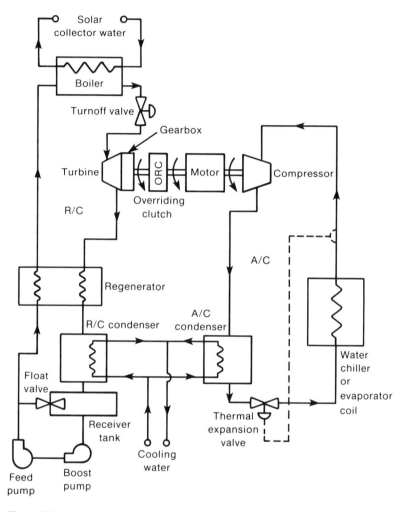

Figure 19.4
General system schematic for Honeywell Rankine cycle air-conditioning system. Source:
Honeywell (1978b).

During the cooling season, 30% of the cooling load was met while the Rankine engine was operating. Of this, 83% was during direct cooling and the remainder during cooling from storage. Since the Rankine engine was not designed to produce enough shaft power to provide all the power required by the compressor, some auxiliary electric input was always needed to meet the cooling load. Because of this requirement, the solar cooling fraction was 15%. The Rankine engine usually did not operate with a thermal efficiency as high as predicted during most of the test period. The precise cause of the degraded performance was not isolated but appeared to be related to air entrapment caused by leaks. The operating mode of cooling from storage was found to be uneconomical, largely because of the parasitic energy required.

19.2.2.1.3 HONEYWELL OTS 45, SALT RIVER PROJECT, PHOENIX, ARIZONA
This solar heating and cooling project (Honeywell 1982a) is in the Salt River Crosscut Operation and Maintenance Building, which contains offices, repair shops, and warehouse facilities. The solar system was installed in the spring of 1981 and provides solar heating only for the repair shops and solar cooling only for the offices. It was designed to provide 89% of the required space heating and a maximum of 15.9% of the required space cooling. Additional cooling is provided by a 228-ton (800-kW) Westinghouse centrifugal chiller.

The hydronic solar system consists of 8208 ft^2 (762.8 m^2) of liquid-cooled, flat-plate collectors; a 2500-gal. (9464 l) thermal storage tank; two 25-ton (87.8-kW) capacity Barber-Nichols Rankine chillers; and associated equipment. The system operation is controlled automatically by a Honeywell microprocessor-based control system. During the 8 months of the test, the solar system collected 1143 MBtu (1204 GJ) of thermal energy from the total incident solar radiation of 3440 MBtu (3623 GJ) and provided 241 MBtu (254 GJ) for cooling and 64 MBtu (67.4 GJ) for heating. The projected net annual energy savings due to the use of the solar system were approximately 136.5 MBtu (40,000 kWh) of electric energy. The solar contribution expressed as a percentage of the building load was 6.6%. Performance of the Rankine engine and the refrigeration compressor in unit no. 1 generally agreed with results obtained during laboratory testing. In unit no. 2, both the compressor and the Rankine engine performed substantially below the laboratory measured values. The precise causes for this difference were not determined.

19.2.2.1.4 HONEYWELL SOLERAS TEST SITE In 1981, Honeywell completed installing a 14-ton (49.1-kW) air-cooled Rankine cycle unit with 2351 ft² (218.5 m²) of evacuated-tube collectors on an Arizona Public Service building in Phoenix (Martin, Corcoleotes, and Williamson 1981). The system was designed to provide approximately one-half of the cooling needed for the one-story office and laboratory building. The Rankine chiller unit, designed and constructed by Barber-Nichols, contains a radial inflow 39,500-RPM turbine driving a 49-kW (13.1-ton), 5-cylinder R-12 compressor. The Rankine engine is designed for a 300°F (149°C) inlet temperature, a 129°F (54°C) condensing temperature, and a 61°F (16°C) evaporator temperature. The configuration is of the basic type shown in figure 19.1b with a motor-generator.

The system does not include thermal energy storage, but it does have a 200-gal. (757-l) hot expansion tank with a nitrogen gas blanket to stabilize system pressure and allow operation up to 325°F (163°C).

Operational data indicated poor initial performance of the Rankine turbine. Analysis by Barber-Nichols indicated that a faulty blade design caused excess turbulence or blockage and, therefore, an excess drag on the turbine wheel. The turbine wheel was replaced, and the performance was subsequently improved to a slightly greater than predicted value. Another major problem was a high rate of collector tube breakage, apparently caused by power outages and surges.

19.2.2.1.5 HONEYWELL/BARBER-NICHOLS POWER GENERATION UNITS The Honeywell/Barber-Nichols team has also investigated the feasibility of power generation configurations of the basic type shown in figure 19.1 (Honeywell 1980; Gossler and Orrock 1981; Batton and Barber 1981). A power generation unit was installed at the New Mexico State University in Albuquerque. This unit is coupled to evacuated-tube collectors and had operated for 1500 hours as of July 1984.

19.2.2.1.6 HONEYWELL OFFICE BUILDING PROJECT Figure 19.5 is a schematic diagram of a solar Rankine system designed and constructed to supply shaft power to conventional water chillers in a Honeywell Inc. office building in Minneapolis, Minnesota (Warner 1978). The system includes two identical Barber-Nichols 85-HP (63.4-kW) Rankine engines, each driving a 100-ton (351-kW) York centrifugal water chiller. The chillers can also be operated using the conventional electric motors.

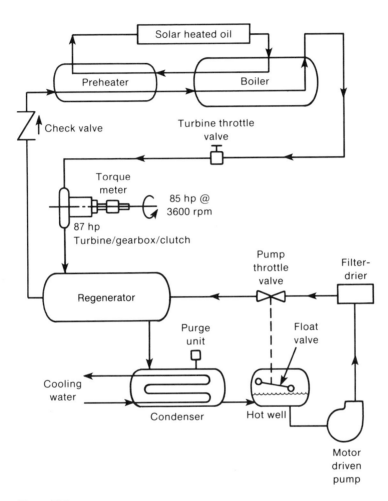

Figure 19.5
Schematic diagram of Barber-Nichols solar Rankine system for driving conventional vapor compression chiller. Source: Werner (1978).

The energy source to the Rankine engines is solar-heated Caloria HT-43 heat transfer oil, and the Rankine working fluid is R-113. At the design point, the Caloria temperature is 300°F (148.9°C), the cooling water temperature is 85°F (29.4°C), and the cycle thermal efficiency is 0.156. The solar input is provided by 20,250 ft^2 (1881.3 m^2) of advanced concentrating collectors.

The turbine operates at 22,400 RPM and the output shaft of the two-stage gearbox at a nominal 3540 RPM. The Rankine subsystem includes a one-way mechanical clutch that allows the turbine to provide shaft power to the compressor and prevents the compressor motor from back-driving the gearbox and turbine.

19.2.2.2 Barber-Nichols Engineering Co.

Independently of Honeywell, Barber-Nichols also provided Rankine chillers for other projects. Examples are described below.

19.2.2.2.1 LOS ALAMOS PROJECT In 1977, Barber-Nichols constructed a 77-ton (270-kW) solar-powered Rankine water chiller for the National Security and Resources Study Center at Los Alamos, New Mexico (Barber-Nichols Engineering 1977; Barber et al. 1978; Hedstrom et al. 1978). This was installed in parallel with a solar-powered absorption chiller for comparison studies. The design-point turbine inlet temperature was 200°F (93°C). The turbine and compressor were directly coupled and operated at 11,950 RPM. A motor-generator was not included. The working fluid in both the power and refrigeration cycles was R-11, and the condensers were water-cooled. Performance results reported for four months of operation in the summer of 1978 were as follows:

Average chiller thermal COP = 0.72;

$$\frac{\text{Total cooling effect}}{\text{Electric energy input to solar system}} = 6.71;$$

$$\frac{\text{Total cooling effect}}{\text{Electric energy input to building system}} = 3.87.$$

In the last of these ratios, the building system includes the solar system, the chilled water pump, the condenser pump, and the cooling tower fan, which permits comparison with a conventional electrically driven vapor compression chiller. The value of 3.87 is approximately the same as would be obtained with a conventional system, indicating that the solar system did not provide any energy advantage.

19.2.2.2.2 UNIVERSITY OF PETROLEUM AND MINERALS PROJECT Barber-Nichols supplied a 10.4-HP (7.8-kW) Rankine engine to the University of Petroleum and Minerals at Dhahran, Saudi Arabia. This unit receives thermal energy from a combination of parabolic trough and evacuated-tube solar collectors. A gas-fired heater provides thermal energy when the collectors cannot supply the design energy load. The turbine is loaded with an electric generator or an air conditioning compressor. The collectors provide pressurized water at 250 psi (1724 kPa) and 350°F (176.7°C) to the Rankine engine boiler. The Rankine cycle uses R-113 fluid with a turbine inlet temperature of 300°F (48.9°C) and a pressure of 175 psi (1207 kPa). The condensers are air cooled. The unit was started up in February 1984.

19.2.2.3 General Electric Co. (Space Division)

General Electric (GE) was also a participant in the NASA 404 Program. The GE project involved the design and development of solar heating and cooling systems for single-family and small commercial applications, and the delivery, installation, and monitoring of the prototype systems (Eckard 1976; General Electric 1978, 1979). Two basic Rankine chiller sizes were developed: a nominal 3-ton (10.5-kW) size for single-family residential applications and a nominal 10-ton (35-kW) size for multifamily residential and small commercial applications.

Figure 19.6 shows a cross-section of the expander/motor/compressor unit. In contrast to the open-shaft type Honeywell/Barber-Nichols subsystems described above, the GE units were hermetically sealed, and a magnetic clutch was used to separate the Rankine expander compartment from the motor/compressor compartment. An overrunning clutch located inside the magnetic clutch self-engaged when the Rankine expander started to rotate. The Rankine engine cycle used FC-88 working fluid and had a two-stage multivane rotary expander with a speed of 1725 RPM. The expander design-point inlet temperature was 285°F (140.6°C), and the design cycle thermal efficiency was 0.13 for the 3-ton (10.5-kW) unit and 0.15 for the 10-ton (35-kW) unit, based on the use of air-cooled condensers. For each size, the refrigeration unit was a modified GE Weathertron reciprocating unit using R-22 working fluid. Test values of the thermal COP for the 10-ton (35-kW) subsystem for 285°F (140.6°C) collector fluid fell in the

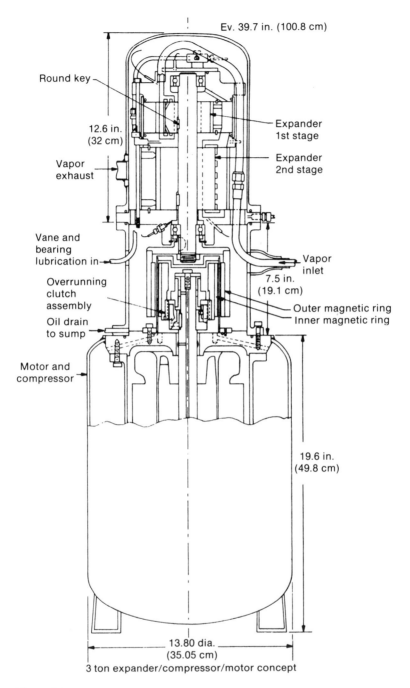

Ev. 39.7 in. (100.8 cm)

Round key

12.6 in.
(32 cm)

Vapor
exhaust

Expander
1st stage

Expander
2nd stage

Vane and
bearing
lubrication in

Vapor
inlet

Overrunning
clutch
assembly

7.5 in.
(19.1 cm)

Outer magnetic ring

Oil drain
to sump

Inner magnetic ring

Motor and
compressor

19.6 in.
(49.8 cm)

13.80 dia.
(35.05 cm)

3 ton expander/compressor/motor concept

Figure 19.6
General Electric three-ton expander/compressor/motor unit. Source: General Electric
(1978).

range of 0.66 to 1.0, depending on the condenser and evaporator temperatures (General Electric 1979).

Figure 19.7 shows a schematic diagram of the GE cooling system. All of the components, except the indoor coil, were assembled as a packaged outdoor unit. The Rankine and refrigeration condensing coils were integrated into a single fin-tube assembly to obtain equal air flow and temperature distribution, and to achieve "fin-sharing" on the refrigeration side when the Rankine engine was inoperative. Cooling air was pulled through the coil by a single two-speed fan that operated at high speed when the Rankine engine was operating and at low speed when the electric motor was driving the compressor.

Under the NASA 404 Program, General Electric installed solar heating and cooling systems at three sites: a 3-ton unit in a single-family residence in Dallas, Texas; a 10-ton (35-kW) unit in an office building in Muscle Shoals, Alabama; and a 10-ton (35-kW) unit in a school in Murphy, North Carolina. During the test period, the Rankine chillers in these systems developed serious failures and the contract was terminated.

19.2.2.4 Mitsubishi

Mitsubishi has reported the design, construction, and installation of a small Rankine solar cooling unit in an experimental solar house in Japan in 1977 (Nishiyama et al. 1979). Flat-plate collectors were used, with an area of 516 ft^2 (48 m^2). The Rankine expander/motor/compressor unit is shown in figure 19.8. This was similar in configuration to the GE units described above and also incorporated a sliding vane expander. The working fluid in the Rankine power cycle was R-114 and the expander speed was 2800 RPM. The power output was 1 HP (0.75 kW) at an input temperature of 194°F (90°C).

The compressor was a rolling piston type, and the refrigeration capacity was 1 ton (3100 kcal/h) at an evaporator temperature of 41°F (5°C) and a condenser temperature of 100.4°F (38°C). The working fluid was R-22. For the temperatures given above, the following performance values were determined experimentally: Rankine cycle thermal efficiency, 0.082; refrigeration cycle TEER, 4.6; and thermal COP, 0.38.

19.2.2.5 Energy Technology Inc.

One of the problems associated with using solar collectors having operating temperatures below 300°F (148.9°C) for solar Rankine cooling systems is the resulting low cycle thermal efficiency. A concept for obtaining

714   Henry M. Curran

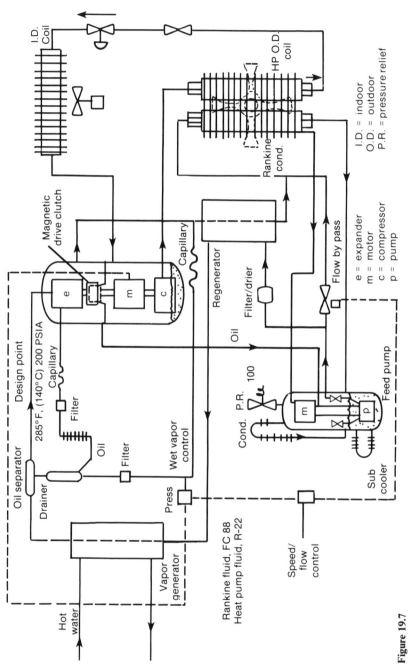

Figure 19.7
General Electric Rankine heat pump. Source: General Electric (1979).

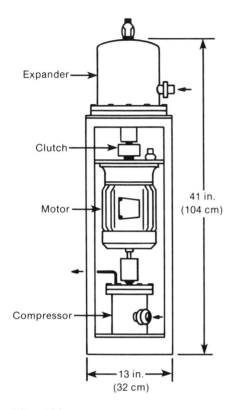

Figure 19.8
Mitsublshi Rankine cycle expander/motor/compressor unit. Source: Nishiyama et al. (1979).

higher efficiency is to use solar collectors to evaporate the working fluid, and to use the combustion of fossil fuel to superheat the working fluid to a higher temperature (Martin and Swenson 1976). Energy Technology Inc. (ETI) was awarded a DOE contract in 1977 to develop a 31-HP (23-kW) steam turbine that would operate in accordance with this concept (Energy Technology 1980). The evaporation temperature was selected at 300°F (148.9°C) and the superheat temperature at 1000°F (537.8°C). This high temperature required the use of water as the working fluid. The hybrid arrangement substantially increases the cycle thermal efficiency by the use of about a 20% to 25% input of a nonsolar energy resource. This higher efficiency requires correspondingly less solar collector area, and therefore

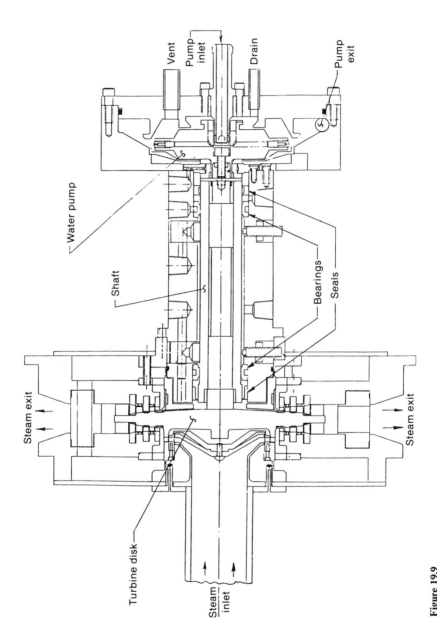

Figure 19.9
Cross-section of ETI steam turbine. Source: Energy Technology (1980).

lower first cost, for a solar cooling system using this concept, compared with a system using only the solar collector. (The eventual marketability of solar cooling systems requires that the unit cost of solar collectors be substantially reduced and/or that the unit costs of fossil fuels increase substantially. Both of these requirements, if met, would tend to cancel the present life cycle cost advantage of the solar/fossil concept [Curran, 1981b].)

Figure 19.9 shows a cross-section of the ETI turbine. This machine was designed for an operating speed of 75,000 RPM and a cycle thermal efficiency of 0.20. It is of the multistage radial outflow type. The design provides for steam flow radially outward through four stages of concentric alternating stator and rotor blade rows. The first-stage nozzle operates at Mach 1.15, and the remaining seven blade rows operate at subsonic velocities. This design choice yields high performance at the off-design conditions that would be frequently encountered in the intended application.

The turbine was completed in 1980 and tested for several months at speeds up to 60,000 RPM and a power level of 24 HP (18 kW). The testing was discontinued with the unit in good condition due to limitations of contract funding. Extrapolation of the measured data indicated that at the design speed of 75,000 RPM, a power output of 35 hp (26.1 kW) and a turbine efficiency of at least 75% could have been achieved.

19.2.2.6 University of Pennsylvania (Mechanical Engineering Dept.)

In 1978, DOE awarded a contract to the Mechanical Engineering Department of the University of Pennsylvania for the development of a 30-HP (22-kW) solar powered/fuel-assisted Rankine cooling subsystem (Lior 1977; Lior and Yeh, 1978; Lior, Koai, and Yeh 1985; Lior and Koai 1982, 1982b; Liu et al. 1982; Subbiah and Lior 1983). The contract initially included design, construction, and testing of a steam turbine coupled through a gearbox to a conventional vapor compressor and an auxiliary electric motor. Due to cost limitations, the contract scope was later revised to eliminate the compressor.

This steam turbine is intended to operate with a solar/fossil hybrid energy input, as in the case of the ETI turbine described above. The thermodynamic and aerodynamic (blading) design was carried out at the University of Pennsylvania. Philadelphia Gear Co. and its subcontractors were responsible for the structural design, manufacturing, assembling, and spin testing of the turbine/gearbox combination. A cross-section of the

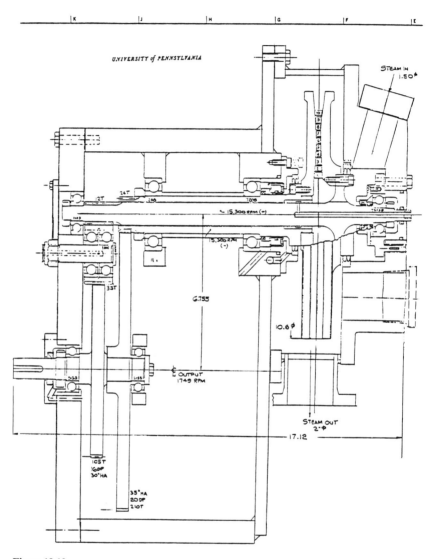

Figure 19.10
Cross-section of University of Pennsylvania steam turbine and gearbox. Source: Kroon et al. (1981).

unit is shown in figure 19.10 (Kroon et al., 1981). The turbine is of the counter-rotating radial outflow type. Each of the two discs carries five rows of blades, and each shaft rotates at 15,300 RPM. Each of the counter-rotating shafts drives a gear through a pinion directly attached to the shaft, with one idler interposed between one of the gears and one pinion to reverse the rotation. This design allows both gears to be mounted on the same output shaft, which rotates at 1749 RPM.

The design evaporation temperature is 212°F (100°C) and the superheat temperature is 1112°F (600°C). As shown in figure 19.11, when this turbine is used in a solar cooling system, the evaporation will take place in a flash storage tank instead of a boiler. This arrangement is included in the laboratory test facility.

The turbine and gearbox were completed and spin-tested by Philadelphia Gear in 1981. Assembly of the laboratory test facility with the turbine/gearbox installed was completed at the University of Pennsylvania, and testing with steam was in progress as of December 1983.

19.2.3   Turbocompressor Units

In the projects described above, the principal objective was the development of Rankine engines that could be used to drive existing conventional vapor compressors. In the projects described below, the principal objective was the development of matched turbines and compressors, configured as hermetically sealed common-shaft turbocompressors. Projects of this type were undertaken by AiResearch Manufacturing Co., Energy Systems Division of Carrier Corporation, and United Technologies Research Center.

19.2.3.1   AiResearch Manufacturing Co.

AiResearch, in association with Dunham-Bush Corp., was a participant in the NASA 404 Program (Rousseau and Noe 1976; AiResearch Manufacturing 1978). The contract initially provided for the development of 12 turbocompressor heat pumps in three sizes: 3 ton (10.5 kW) for single-family residential applications; 10 ton (35 kW) for commercial applications; and 25 ton (87.8 kW) for multifamily residential applications. The contract was later modified to delete the 10-ton (35-kW) size and substitute a 75-ton (263-kW) size.

The basic design of all sizes was similar, size being the major difference between them. Each turbocompressor unit was hermetically sealed and included a high-speed, permanent-magnet electric motor located between

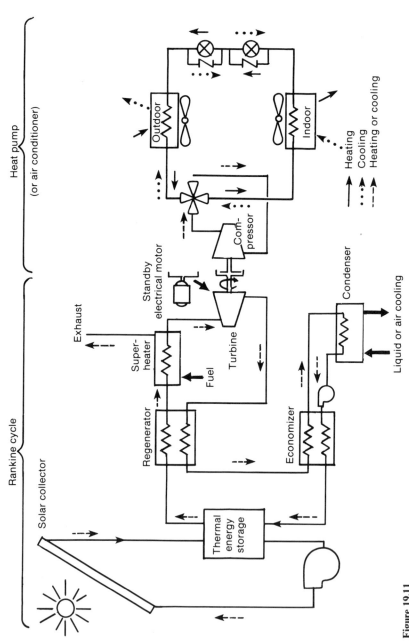

Figure 19.11
University of Pennsylvania solar-powered, fuel-superheated steam Rankine-cycle-driven vapor-compression heat pump (or air conditioner).
Source: Lior (1977).

the turbine and the compressor, as shown in figure 19.12. The working fluid was R-11 in both the power and the refrigeration circuits, and the design input temperature for the cooling mode was 190°F (87.8°C).

As shown in figure 19.13, a reversing valve was included in the refrigeration circuit so that the unit could be used for either heating or cooling. In the cooling mode, hot water from storage was directed to the boiler to provide energy input to the power cycle. In the heating mode, the boiler was isolated and low-temperature warm water from storage was directed to the outdoor coil to evaporate R-11 at low pressure. The electric motor drove the compressor, raising the R-11 pressure and temperature. Condensation of the R-11 in the indoor coil then provided thermal energy for heating the building air.

Several turbocompressor units of the various sizes were completed and tested in laboratory facilities. During the testing, several failures occurred, primarily of the bearings, and the contract was terminated in 1980.

19.2.3.2   Carrier Corporation (Energy Systems Division)

The Energy Systems Division (ESD) of Carrier Corp. was awarded a DOE contract in 1977 for the development of a solar-powered, air-cooled Rankine water chiller in the 15 to 25-ton (52.7 to 87.8-kW) range, using 250° to 300°F (121.1° to 148.9°C) input from solar collectors (English 1978, 1979a, 1979b). ESD subcontracted to Mechanical Technology Inc. (MTI) the design and construction of the turbocompressor unit. As shown in figure 19.14, the configuration was similar to that of the AiResearch design: a hermetic unit with a variable-speed, permanent-magnet motor located between the turbine and the compressor. The working fluid was R-113 in both the power and the refrigeration loops. At the design speed of 24,000 RPM and design inlet temperature of 290°F (143.3°C), the design output was 25 tons (87.8 kW) and the COP was 0.70. No oil was used, lubrication of the bearings being provided by the working fluid.

As designed, this unit was intended to operate only in the cooling mode. Figure 19.15 shows the arrangement of the subsystem components. After the turbocompressor had been completed by MTI, it was installed in the Carrier test loop. During one of the initial test runs, a serious bearing failure occurred, and the contract was subsequently terminated.

19.2.3.3   United Technologies Research Center

Since 1973, UTRC, in collaboration with Hamilton Standard Division (HSD), has been involved in the development of a series of solar-powered

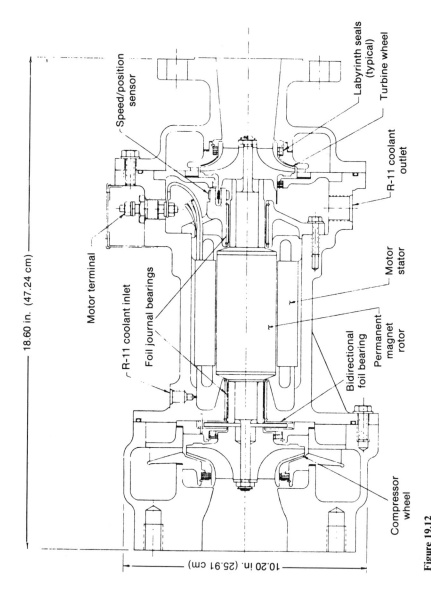

Figure 19.12
Cross-section of AiResearch 25-ton turbocompressor. Source: AiResearch (1978).

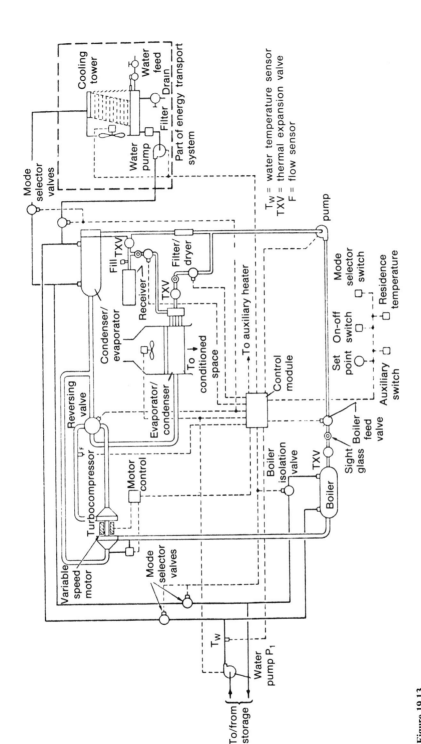

Figure 19.13
AiResearch 3-ton/60 kBtu/h cooling and heating subsystem. Source: AiResearch (1978).

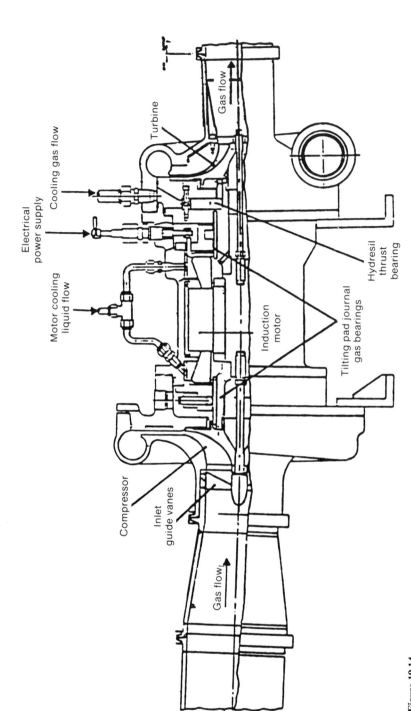

Figure 19.14
Carrier/MTI solar-powered water chiller turbocompressor subsystem. Source: English (1979b).

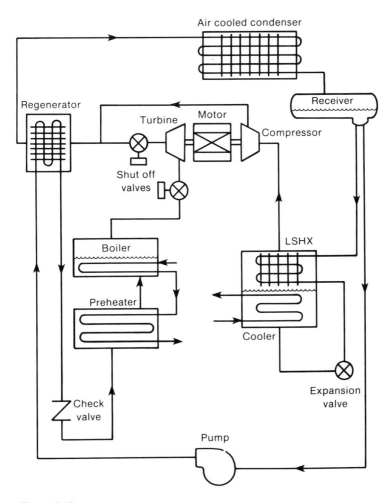

Figure 19.15
Carrier air-cooled dual-loop Rankine driven vapor compression cycle. Source: English
(1979b).

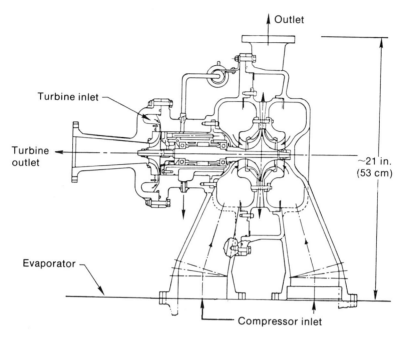

Figure 19.16
Cross-section of United Technologies 18-ton turbocompressor. Source: United
Technologies Research Center (1979).

turbocompressor heat pumps, ranging from a 5-ton (17.6-kW) laboratory
proof-of-concept unit to an advanced 18-ton (63.2-kW) site-installed operational unit (Biancardi et al. 1975; United Technologies Research Center,
1977, 1979, 1982; Biancardi and Melikian, 1978; Melikian et al., 1980;
Melikian, Biancardi, and Meader, 1980). The basic concept is a hermetically sealed turbocompressor that does not have an auxiliary electric
motor. Auxiliary thermal energy is furnished by a gas furnace. The 18-ton
unit was specifically designed for maximum performance over the range
of operating temperatures available from medium-concentration, high-
efficiency collectors. The turbine is a single-stage, single-entry radial inflow type, and the compressor is a single-stage, double-entry type. A
cross-section of the turbocompressor is shown in figure 19.16. Five operational modes are possible: solar cooling, furnace cooling, solar assisted
heat pump, direct furnace heating, and direct solar heating.

Figure 19.17 illustrates the cooling modes—thermal energy from either
the solar collector subsystem or the furnace operates the Rankine power

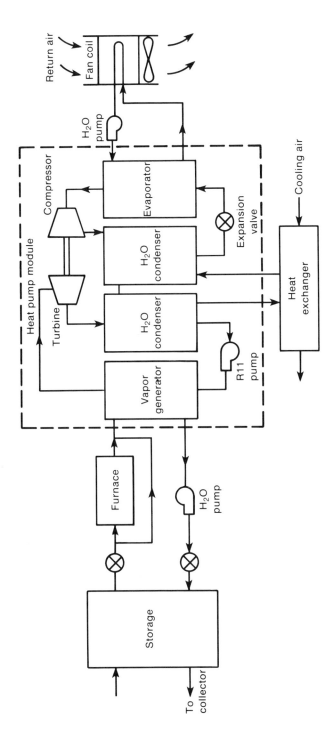

Figure 19.17
United Technologies system configuration—cooling mode. Source: United Technologies Research Center (1979).

cycle. Figure 19.18 shows the solar-assisted heat pump heating mode—low temperature water from storage evaporates the working fluid in the evaporator, and thermal energy from the furnace operates the Rankine power cycle. Thermal energy rejected from the power and refrigeration cycles at the condensers heats the building air. Changeover from cooling to heating involves using flow control valves in the water circuits, instead of a reversing valve in the refrigeration circuit.

The turbocompressor heat pump module is designed to provide 18 tons (63.2 kW) of cooling at ambient air temperatures of 95°F (35°C) dry bulb and 75°F (23.9°C) wet bulb, while providing an evaporator-chilled water outlet temperature of 44°F (6.7°C). The design-point inlet temperature to the boiler is 305°F (151.7°C), and the corresponding design COP is 0.6. The working fluid in both the power and the refrigeration cycles is R-11, and the design speed is 42,000 RPM.

As of 1983, UTRC and HSD were developing an advanced solar-powered heat pump module with higher cooling mode performance capability and lower cost features that were identified as a result of laboratory and field testing of the earlier prototypes.

19.2.3.3.1 UNITED TECHNOLOGIES SOLERAS TEST SITE In 1981, UTRC, as a participant in the SOLERAS Program (Martin, Corcoleotes, and Williamson 1981), completed installing an 18-ton (63.2-kW) air-cooled Rankine cycle unit with 1316 ft² (122.3 m²) of tracking parabolic trough collectors at their Hamilton Test Systems subsidiary in Phoenix, Arizona. This solar system provides approximately 75% of the cooling for a single-story office building. Backup is provided by a parallel electrically driven chiller. The design inlet temperature is 302°F (150°C). The storage subsystem consists of two buried tanks and two expansion tanks. The largest tank, about 2000 gal. (7570 l), is for cold storage for system operation during periods when the Rankine chiller cannot be operated. A 1500-gal. (5678-l) hot storage tank extends the system's daily operating period.

Highly successful checkout tests were begun in March 1981, and the system was delivering up to 20 tons (70 kW) of cooling in April 1981. During a typical day, with hot storage at about 300°F (148.8°C), the system provided 18 to 21 tons (63.2 to 73.7 kW) of cooling, about half of which was used to cool the cold tank and the remainder to cool the building. The system COP varied from 0.55 to 0.75.

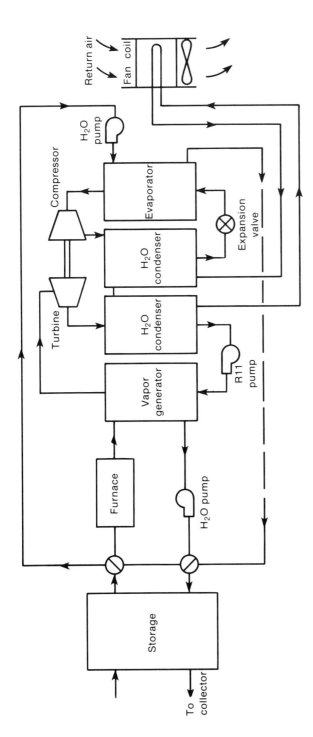

Figure 19.18
United Technologies system configuration—heating mode. Source: United Technologies Research Center (1979).

19.2.4   Sandia Laboratories

Under an ERDA contract, Sandia Laboratories began to develop a Solar
Total Energy System Test Facility (Veneruso 1976; Abbin 1978, 1979) in
1975. This facility utilizes 8759 ft^2 (814 m^2) of solar tracking, east-west
oriented, parabolic trough collectors to heat Therminol 66 heat transfer
fluid to 600°F (316°C). The energy operates a Sundstrand Rankine cycle
engine using toluene as the working fluid. A cascaded configuration, as
shown in figure 19.1 (d), is used with an electric output of 43 HP (32 kW)
for 11 hours per day, and a maximum of 645 kBtu/h (679 MJ/h) for heating or a maximum of 30 tons (380 MJ/h) for cooling. This test facility, with
programmable inputs and loads, provides a flexible experimental tool for
evaluating components and subsystems.

19.2.5   Other Rankine Engine Projects

Two smaller hardware projects were also funded by DOE. The contractors were Biphase Engines, Inc., and Scientific-Atlanta, Inc. The Biphase
project involved the development of a two-phase turbine in which two
different fluids, one in the liquid state and one in the gaseous state, flowed
through the turbine (Amend 1977–1978). The objective of the Scientific-
Atlanta project was to develop a Rankine free-piston expander-compressor
unit (Shelton 1977–1978).

19.2.6   Configuration and Design Point Summaries

Table 19.1 contains a summary of typical equipment configurations as
developed in some of the major solar cooling projects discussed above.
Table 19.2 is a summary of design-point data for a representative set of
Rankine engines developed in these projects.

19.2.7   Condensers

If a solar Rankine chiller operating in the solar mode is considered in
the "black-box" sense, the major thermal energy flows are seen to be the
energy input from the solar collector subsystem, the energy input from
the cooled space, and the energy rejected by the power and refrigeration
cycle condensers. Thus, from a simple energy balance, the rejected thermal
energy is equal to the sum of the two thermal energy inputs, neglecting
minor thermal energy exchanges between the chiller equipment and the
environment, and neglecting parasitic energy flows. Because of the need to

Table 19.1
Equipment configurations of representative solar Rankine projects

Developer	Expander		Coupling	Compressor		Thermal energy rejected to	No. of condensers	Electric unit
	Type	Fluid		Type	Fluid			
AiResearch	Turbine	R-11	In-line, clutch	Centrifugal	R-11	Water	1	Motor
Barber-N.	Turbine	R-113	Gearbox, clutch	Centrifugal	R-11	Water	2	Motor-generator
Carrier	Turbine	R-113	Common shaft	Centrifugal	R-113	Air	1	Motor
Energy Tech.	Turbine	Water	NIP[a]	NIP[a]		Water	2[b]	NIP[a]
General El.	Rot. vane	FC-88	In-line, clutch	Recip.	R-22	Air	2	Motor
Honeywell/BN	Turbine	R-113	Gearbox, clutch	Recip.	R-12	Water	2	Motor-generator
Mitsubishi	Rot. vane	R-114	In-line, clutch	Roll. pist.	R-22	Water	2	Motor
Sandia	Turbine	Toluene	Gearbox	None		Absorption unit	—	Generator
Un. Penn.	Turbine	Water	Gearbox, clutch	NIP[a]		Water	2[b]	Motor
United Tech.	Turbine	R-11	Common shaft	Centrifugal	R-11	Air	2	None

a. NIP—Not in project as funded.
b. Two required if the expander is coupled to a compressor.

Table 19.2
Summary of design-point data for representative solar Rankine expanders

Contractor Nominal capacity (tons)	AiResearch			Barber-Nichols	Carrier	Energy Tech.
	3	25	75	77	25	25
Rankine fluid	R-11	R-11	R-11	R-11	R-113	Water
Expander type	Turbine	Turbine	Turbine	Turbine	Turbine	Turbine
Design point data:						
Boiler input, °F	200	200	200	200	300	320
°C	93	93	93	93	149	160
Expander inlet, °F/psia	190/86	190/85	190/86	186/85	273/132	1000/67
°C/kPa	88/593	88/586	88/593	86/586	134/910	538/462
Condenser coolant/°F	Wat/84	Wat/84	Wat/84	Wat/65	Air/95	Air/95
°C	29	29	29	18	35	35
Condenser, °F/psia	96/22	96/22	96/22	78/16	120/15	115/1.5
°C/kPa	36/152	36/152	36/152	26/110	49/103	46/10
Expander power, HP	2.3	16	50	46	27	29
kW	1.7	12	37	34	20	22
Expander speed, kRPM	60	30	18	12	20	75
Expander efficiency	0.81	0.85	0.85	0.75	0.80	0.78
Cycle efficiency	0.08	0.10	0.09	0.10	0.14	0.20
Pump power, HP	0.4	0.9	5.4	1.5	1.3	0.13
kW	0.3	0.7	4.0	1.1	1.0	0.1
Mass flow, lbm/h	660	4450	13240	12200	4780	250
kg/h	299	2019	6007	5535	2169	113

reject all of the thermal energy received from the solar collector subsystem in addition to the thermal energy from the cooled space, substantially more condenser surface is required than that for an electrically driven chiller.

In the case of Rankine engines driving conventional vapor compression chillers, two condensers are used, one for the Rankine engine and one for the refrigeration unit, since they use different working fluids. In the case of the turbocompressor designs discussed above, a common working fluid was used in the two cycles, which makes it possible to use a common condenser. The AiResearch and Carrier designs used a common condenser, whereas UTRC used a separate condenser for each cycle. Numerous trade-offs are made in establishing this design decision. One of the major considerations is whether or not lubricating oil is used only in the refrigeration cycle and is excluded from the power cycle. If this is the case, oil is prevented from entering into the power cycle by using appropriate seals between the expander and compressor and using separate condensers. The

Table 19.2 (continued)

General Electric		Honeywell/ Barber-Nichols		Univ. Penn.	UTRC
3	10	3	25	18	18
FC-88	FC-88	R-113	R-113	Water	R-11
Rot. vane	Rot. vane	Turbine	Turbine	Turbine	Turbine
300	300	195	195	225	300
149	149	91	91	107	149
285/200	285/200	176/38	176/38	1100/10	285/246
141/1379	141/1379	80/262	60/262	593/69	141/1697
Air/95	Air/95	Wat/85	Wat/85	Wat/85	Air/95
35	35	29	29	29	35
109/22	109/22	95/10	95/10	115/1.5	114/30
43/152	43/152	35/69	35/69	46/10	26/207
3.1	10.5	2.4	20	29	21
2.3	7.9	1.8	15	22	16
1.625	1.625	35	24	15	42
0.83	0.83	0.67	0.80	0.72	0.76
0.13	0.15	0.08	0.086	0.183	0.13
0.5	1.1	0.13	0.5	0.13	1.6
0.4	0.8	0.1	0.4	0.1	1.2
1100	3400	1080	8490	300	4140
499	1543	490	3852	136	1878

AiResearch and Carrier machines did not use oil, so this was not a consideration. The UTRC machines do use oil in the refrigeration cycle only, hence the choice of two condensers. Another factor to be considered in making this design decision is whether or not the refrigeration cycle is to be used to provide heating by operating in the heat pump mode and whether the Rankine engine or an electric motor is to provide power in this mode. The UTRC designers found two condensers to be preferable in this respect, the Rankine engine being used in the heat pump mode; the AiResearch designers found one condenser to be preferable, an electric motor being used in the heat pump mode.

Condensers may be air or water type. In the air type, the rejected thermal energy is transferred to the ambient air by a fan blowing air through the condenser coil. In the water type, the transfer is to a stream of water. This may be a once-through flow of water, a recirculated flow from a cooling tower, or a closed circuit flow in which the water is used only as a heat transfer medium to an air heat exchanger. (See, for example, the UTRC subsystem in figure 19.17.) Water may also be used in evaporative condensers in which water is evaporated from the condenser surface instead of in a cooling tower. Some of the Barber-Nichols 3-ton (10.5-kW) Rankine chillers discussed above used evaporative condensers, for example.

Water cooling is preferable to air cooling because it permits a lower condensing temperature, and therefore a higher Rankine cycle thermal efficiency and a higher refrigeration cycle TEER. However, in many places, using water is impractical because scarcity or local ordinances prohibit its use for condensing. In such places air cooling must be used.

19.2.8 Working Fluids

The following subsections provide background information on the selection and testing of working fluids for use in Rankine cooling units, particularly with respect to the Rankine power cycle.

19.2.8.1 Selection of Working Fluids

The original working fluid for Rankine cycle engines was water, and this is still used in power plants and other applications of high-temperature Rankine engines. However, the low molecular weight of water requires the use of multistage expanders to obtain high cycle efficiency. Organic fluids with high molecular weights, on the other hand, generally require only single-stage expanders but have the disadvantage of being limited by temperature (Curran 1981a).

Considerable care is required in the selection of an organic working fluid for a Rankine engine in order to minimize the possibility of temperature-related decomposition of the fluid. Chemical decomposition of an organic fluid can produce noncondensable gases that degrade the heat transfer rate in condensers, as well as liquid and gaseous compounds that have a corrosive effect on the system materials. The decomposition may be caused by reactions of the organic fluid and lubricating oil in the presence of metals and contaminants, or it may be caused by pyrolytic instability of the organic fluid or the oil, or some combination of these factors. Chemical reactions are temperature-dependent and usually increase in rate by a factor of two or more for every 18°F (10°C) increase in temperature (ASHRAE 1976).

With the exception of a few oil-free engines, organic fluids must coexist with lubricating oil in Rankine engines. Various research studies have shown that some oils are more stable toward organic fluids than others, with increasing temperature accelerating the organic fluid-oil reaction. The reaction rate also depends on the kinds of metal in contact with the oil and the organic fluid, the amount of air and moisture present, and the additives present in the oil.

The choice of an organic working fluid for a Rankine engine is governed primarily by the physical and thermodynamic properties of the fluid. A candidate fluid should be chemically stable at the maximum expected cycle temperature and in contact with the engine materials, lubricating oil, and unavoidable contaminants. Preferably, it should also be noncorrosive, nonflammable, nontoxic, and low in cost.

A partial list of organic fluids that have been used or proposed for use in Rankine engines is given in table 19.3. The table shows the molecular weight, critical point, freezing point at atmospheric pressure, approximate thermal stability limit, vapor pressure at 104°F (40°C) or saturation temperature at 1.45 psia (10 kPa), and whether the fluid is "wetting" or "drying." Wetting fluids become saturated or partially condensed upon expansion; drying fluids become superheated upon expansion.

The thermal-stability limits given in table 19.3 are approximate. The results of tests of thermal stability for a given fluid, as reported in the literature, vary considerably, depending on the test equipment, procedure, test duration, and materials with which the fluid was in contact. The thermal-stability limit of an organic fluid cannot be defined in a precise manner that would be applicable to all possible uses. Test results generally indicate a chemical decomposition rate as a function of temperature and the test conditions. Thus the thermal stability limit values in table 19.3 are representative of temperatures at which the chemical decomposition rate could be unacceptably high for use in Rankine engines.

19.2.8.2 Testing of Organic Working Fluids

Although much of the information developed in the refrigeration industry on the use of organic fluids in thermodynamic cycle equipment is directly applicable to Rankine engine design, additional information specific to the design is usually required. Because of the higher temperatures, the thermal stability may be as important a factor in the selection of a working fluid as the thermodynamic properties. Extensive tests in capsules are commonly performed to determine the chemical and thermal stability of candidate fluids near the maximum expected engine temperature and in contact with the engine materials and probable contaminants. Less common are long-

Table 19.3
Organic fluids data

Fluid	Chemical formula[a]	Molecular weight	Critical point, C/kPa	Freezing point, C	Approx. thermal stability limit, C	Vapor pressure at 40°C, kPa	Saturation temperature at 10 kPa, C	Wetting (W); drying (D); neither (N).
R-11	CCl_3F	137	198/4413	−111	120	159		N
R-21	$CHCl_2F$	103	178/5102	−135		276		W
R-22	$CHClF_2$	86	96/4978	−160	200	1455		W
R-113	$C_2Cl_3F_3$	187	214/3441	−35	175	76		D
R-114	$C_2Cl_2F_4$	171	146/3261	−94	175	317		D
R-133a	$C_2H_2ClF_3$	118	152/4068	−106	200	310		N
CP-9[b]	$C_{15}H_{16}$	196	522/2448	−55	370		200	D
CP-25[c]	C_7H_8	92	321/4254	−95	480	7		D
CP-27[d]	C_6H_5Cl	113	359/4523	−55	320	7		D
CP-32[e]	C_5H_5N	79	347/5633	−42	370	7		D
CP-32/ Water[f]	$\{0.23\,C_5H_5N / 0.77\,H_2O\}$	33	366/8964	−18	400	14		D
CP-34[g]	C_4H_4S	84	307/5461	−40	290	21		<66°C: W / >66°C: D
FC-75[h]	$C_8F_{16}O$	416	227/1607	−62	320	7		D
FC-88[i]	C_5F_{12}	288	150/2131	−115	200	140		D
F-85[j]	$\{0.85\,CF_3CH_2OH / 0.15\,H_2O\}$	88	240/6412		290	21		W
Dowtherm A[k]	$\{0.265\,(C_6H_5)_2 / 0.735\,(C_6H_5)_2O\}$	166	499/3241	−48	370		~140	D
Biphenyl Allied	$(C_6H_5)_2$	154	498/3289	69	370		~140	D
P-1 D	$C_{10}F_{22}O_2$	570	243/1186	−85	370	2	~75	D

Source: Curran (1981a).
a. Mixture fractions on mole basis.
b. Monoisopropyl biphenyl $(C_6H_5)(C_6H_4) - CH(CH_3)_2$.
c. Toluene.
d. Monochlorobenzene.
e. Pyridine.
f. Azeotrope of pyridine and water.
g. Thiophene.
h. Perfluoro-2-butyltetrahydrofuran.
i. Perfluoropentene.
j. Fluorinol (trifluoroethanol/water mixture).
k. Eutectic of biphenyl and phenyl ether.

duration dynamic tests in which a candidate fluid is circulated in a closed loop to simulate engine temperature, pressure, and material conditions.

In 1980, DOE awarded a contract to Thermo Electron Corporation to determine the thermal stability of organic working fluids by using dynamic test loops simulating the conditions in Rankine engines (Thermo Electron 1982). The fluids tested and the ranges of boiler outlet test temperatures were as follows:

Test fluid	Temperature test range
R-11/Zephron Oil	250° to 330°F (121.1° to 165.6°C)
R-11	290° to 325°F (143.3° to 162.8°C)
R-113	315° to 350°F (157.2° to 176.7°C)
R-113/Zephron Oil	300° to 330°F (148.9° to 165.6°C)

The test results indicated that R-113 has a significantly higher temperature capability than R-11, with or without the Zephron oil. Acceptable maximum boiler outlet temperatures for long-duration operation were found to be as follows:

Test fluid	Maximum acceptable temperature
R-11/Zephron Oil	below 250°F (121.1°C)
R-11	300°F (148.9°C)
R-113	above 350°F (176.7°C)
R-113/Zephron Oil	above 330°F (165.6°C)

In three of these cases the acceptable values were determined to lie outside the test range, so precise values could not be reported.

19.2.9 Thermal Storage

The uses and types of thermal storage in solar systems are covered in depth in another chapter of this book. The following is a brief general discussion of the use of thermal storage in solar systems that incorporate Rankine chillers.

Although thermal energy storage is not essential for the operation of an active solar cooling system, it is commonly used to provide continuity of operation when solar energy is not directly available. Thermal storage may be provided on the hot side and/or on the cold side of the cooling unit. In particular, if the system is to provide heating and hot water, hot-side storage is normally required independent of the cooling function. The type and amount of storage to be used in solar cooling systems are strongly dependent on site and system considerations. Design decisions are usually guided by empirical data from operating systems and/or computer analyses. On the basis of the results from the Honeywell NASA 404 test sites, eliminating thermal storage for the cooling function was recommended because of the excess operating energy required and the reduced Rankine cycle efficiency caused by the low inlet temperatures available from storage (Technical Support Package 1983).

Using electric coupling (see figure 19.1c) would avoid the need for more than a small amount of thermal storage for buffering, since this configuration permits using the electric utility as the equivalent of a very large storage unit.

19.3 Economic Considerations

Solar cooling and heating systems using Rankine technology are substantially more expensive in first cost than the conventional alternatives. This difference in first cost is partially offset on a life-cycle cost basis by the lower expenses for nonsolar energy resources required by a solar system. For solar systems to become economically competitive with conventional systems, the first-cost differential must be reduced or the unit costs of nonsolar energy resources must increase.

In 1980, DOE initiated the Active Program Research Requirements (APRR) project (Scholten and Morehouse 1983). The focus of this project was identifying long-term, high-risk research and development activities for the Active Solar Program, specifically, activities that would contribute toward active solar heating and cooling systems becoming cost competitive in the year 2000. A cost catalog was developed as an element of the project. This provided installed cost data for all of the components required for solar and conventional systems.

Another element of the project was the development and application of a methodology to establish the cost and performance goals required for cost competitiveness in the year 2000 (Warren and Wahlig 1982). The principal concept is that the goal for the incremental first cost is the present value of the system savings in nonsolar energy resources over a specified payback period.

The following table gives the values developed for systems using the 1982 Rankine technology, assuming the solar and conventional systems to be located in Phoenix, using 1982 dollars, and payback periods of 7 years and 5 years for residential and commercial systems, respectively:

	Residential	Commercial
Solar system cost	$16,920	$68,700
Conventional system cost	$3,000	$19,700
Incremental cost	$13,920	$49,000
Solar cost goal, year 2000	$807	$14,275
Incremental cost/cost goal	17.3	3.4

These results indicate that, using the predicted higher unit costs of nonsolar energy resources in the year 2000, solar cooling systems based on 1982 Rankine technology cannot be expected to be cost-effective in the year 2000. Evidently, additional research and development will be required to reduce the incremental cost and to increase the nonsolar energy resource savings. Some possible approaches are discussed below.

19.4 Future Research and Development

For solar cooling systems to approach cost-effectiveness, the major directions indicated for future R&D are these:

1. Increase the annual savings in nonsolar energy resources by increasing system efficiency and reducing parasitic energy requirements, using a combination of developing more efficient components and developing system designs that reduce the number of energy-related components.

2. Reduce the system first cost by a combination of developing lower cost components and developing system designs that reduce the number of components.

3. Improve the system annual operating efficiency by using more sophisticated control subsystems.

In the case of solar systems using Rankine technology, a study of future research approaches was prepared by the team of Barber-Nichols, Honeywell, and Science Applications Inc. (Barber-Nichols Engineering 1982; Honeywell 1982c; Choi, Stokley, and Morehouse 1982). The objective was to define an advanced solar system and a Rankine subsystem that would have significantly higher performance than present solar cooling technologies. The approach used was to evaluate existing data and results of Honeywell/Barber-Nichols solar cooling projects, to develop and analytically compare by simulations promising future technical approaches, and to select the technically best system. Annual, rather than design-point, performance and energy resource savings were used as final selection criteria. System cost was factored into the final analysis to weight appropriately the optimized technical comparison and to provide a means for a quantified comparison.

Barber-Nichols's task was to simulate various alternative Rankine engine configurations and to produce the corresponding probable performance maps. A companion solar system (collectors, pumps, controls, etc.) was devised by Honeywell, and then the annual performance and system costs were simulated by Science Applications. Ten different configurations of solar-driven Rankine chillers were defined and modeled. The principal conclusions drawn from this study follow:

1. The advanced solar Rankine systems should perform significantly better and have lower installed costs than present-day solar cooling technologies. These characteristics are reflected in the projected reduced total annual operating energy resources and the reduced life-cycle cost.

2. The second-generation (2G) performance and cost savings are due primarily to optimized (simplified) system configurations and control strategy, improved Rankine chiller performance, and electrical generation, rather than thermal storage, during periods of excess solar energy.

3. The 2G Rankine chiller and system performance could be improved further by the use of such characteristics as variable-speed pumps, advanced system and Rankine chiller controls, variable-flow turbine nozzles, improved turbine-motor-generator-compressor unloading concepts, and high-speed motor-generators.

4. The major parametric studies resulted in the following advanced system or Rankine chiller characteristics:

a. Electrical generation is preferred over thermal storage.

b. Rankine engine thermal energy rejection to the building (for space heating) during electricity generation is not an effective mode of operation.

c. The boil-in collector (use of solar collector as boiler for Rankine engine) performs better than the 2G system with concentrator collectors, but the life-cycle costs are nearly the same.

d. Energy resource savings and life-cycle costs decrease as collector area decreases. The higher-temperature, 300°F (148.9°C), systems are more cost effective than lower-temperature systems.

e. Some preferred Rankine chiller machine design characteristics are a 20-HP (15-kW) turbine for a 25-ton (87.8-kW) compressor; an evaporative condenser; site-specific thermal energy rejection control; and a 55°F (12.8°C) chilled water temperature.

f. Matching (undersizing is optimal) the solar system to the load is important.

Although the study results define advanced solar Rankine cooling systems that could be significantly better than the 1982 state of the art, even these second-generation systems have projected life-cycle costs that are three times as large as those for conventional air-conditioning systems. Additional research along the lines indicated above could possibly be effective in reducing the life-cycle cost differential.

References

Abbin, J. P., Jr. 1978. *Solar total energy test facility project test summary report: Rankine cycle energy conversion subsystem.* Sandia Laboratories Report SAND 78–0396.

Abbin, J. P., Jr. 1979. Sandia Laboratories operational experience with small heat engines in solar thermal power systems. *Proc. of 14th IECEC*, no. 799025, Boston, MA.

AiResearch Manufacturing Co. 1978. *Status report on preliminary design activities for solar heating and cooling systems.* DOE/NASA report CR–150673.

Amend, W. 1977–1978. *Study of the two-phase engine for solar space cooling.* Progress reports to U. S. Department of Energy on contract EG-77-C-03–1542.

ASHRAE. 1976. *ASHRAE handbook and product directory, 1976 systems volume,* chapter 30. New York: American Society of Heating, Refrigerating and Air Conditioning Engineers.

Banks, R., and M. Wahlig. 1976. *Nitinol engine development.* Lawrence Berkeley Laboratory report LBL-5293, Berkeley, CA.

Barber, R. E. 1978. Performance tests of an improved three ton air conditioner for the Honeywell Mobile Solar Research Facility. *Solar Cooling and Heating,* vol. 2. Washington, DC: Hemisphere Publishing.

Barber, R. E., et al. 1978. Design, fabrication and testing of a 77-ton solar powered water chiller for the National Security and Resources Study Center at Los Alamos, New Mexico. *Proc. 3d Workshop on the Use of Solar Energy for the Cooling of Buildings,* AS/ISES, University of Delaware.

Barber-Nichols Engineering Co. 1977. *Design, fabrication and testing of a 77 ton solar pow-ered water chiller for the National Security and Resources Study Center at Los Alamos, New Mexico*. Final report.

———. 1982. *Modeling of a second-generation solar-driven rankine air conditioner*. Report to U.S. Department of Energy on contract no. DE-AC03–81CS30576.

Batton, W. D., and R. E. Barber. 1978. A solar engine used to power a 23-ton water chiller. *Proc. 1978 AS/ISES Meeting*, Denver, CO, pp. 455–59.

Batton, W. D., and R. E. Barber. 1981. Rankine engine solar power generation: part II—the power generation module. ASME paper 81-WA/Sol-23. New York: American Society of Mechanical Engineers.

Biancardi, F. R., et al. 1975. Design and operation of a solar-powered turbocompressor air-conditioner and heating system. *Proc. of 10th IECEC*, 759031.

Biancardi, F. R., and G. Melikian. 1978. Analysis and design of an 18-ton solar-powered heating and cooling system. *Proc. of 13th IECEC*, no. 78924, San Diego, CA.

Choi, M. K., J. R. Stokley, and J. H. Morehouse. 1982. *Analysis of commercial and residential solar absorption and Rankine cooling systems*. Final report to U.S. Department of Energy on contract no. DE-AC03–81SF11573.

Curran, H. M. 1978. Thermodynamic equivalence of thermally driven refrigeration systems. *Solar Energy* 21(5):451–52.

———. 1981a. The use of organic working fluids in Rankine engines. *Journal of Energy* 5(4):218–23.

———. 1981b. Solar/fossil Rankine cooling. *Proc. of 16th IECEC*, no. 819713, Atlanta, GA.

———. 1982. High temperature solar cooling systems. *ASHRAE Trans.*, vol. 88, part 1.

Curran, H. M., et al. 1974. *Assessment of the Rankine cycle for potential application to the solar-powered cooling of buildings*. National Science Foundation report NSF/RA/N-74–108.

Curran, H. M., and M. Miller. 1976. Comparative evaluation of solar heating alternatives. *Proc. 2d Southeastern Conference on Application of Solar Energy*. Baton Rouge, LA: Louisiana State University, pp. 404–11.

Curran, H. M., M. Miller, and T. Alereza. 1975. *Assessment of solar powered cooling of buildings*. National Science Foundation report NSF-RA-N-75–012.

Eckard, S. E. 1976. Test results for a Rankine engine powered vapor compression air conditioner for 366 K heat source. *Proc. of 11th IECEC*, no. 769205.

Energy Technology, Inc. 1980. *Design and demonstration tests of a steam turbine for solar cooling*. Final report to DOE on contracts DE-AC03–77CS34543 and DE-AC03–80CS30214.

English, R. A. 1978. *Development of a high temperature solar powered water chiller*. Phase I technical progress report for September 1977 to June 1978, vol. 1–4, SAN-1590–1/1–4, U.S. Department of Energy.

———. 1979a. *Development of a high temperature solar powered water chiller*. Phase II technical progress report for June 1978 to March 1979, vol. 1 and 2 (draft), U.S. Department of Energy.

———. 1979b. *Development of a high temperature solar powered water chiller*. Phase III technical progress report for April 1979 to September 1979, vol. 1 (draft), U.S. Department of Energy.

Fujita, T., et al. 1980. Comparison of advanced engines for parabolic dish solar thermal power plants. *Proc. of 15th IECEC*, no. 809347, Seattle, WA.

General Electric Co. 1977–1978. *Prototype solar heating and combined solar heating and cooling systems.* DOE/NASA quarterly reports: CR-150686, October 1976; CR-150687, January 1977; CR-150691, April 1977; CR-150692, July 1977; CR-150693, October 1977: CR-150694, January 1978; CR-150705, April 1978.

General Electric Co. 1978. *Solar heating and cooling system design and development.* DOE/NASA status summaries: CR-150735, April 1978; CR-150803, July 1978.

General Electric Co. 1979. *Solar heating and cooling system design and development.* Final report. DOE/NASA CR-161422.

Gossler, A. A., and J. E. Orrock. 1981. Rankine engine solar power generation: part I— performance and economic analysis. ASME paper 81-WA/Sol-22. New York: American Society of Mechanical Engineers.

Hedstrom, J. C., et al. 1978. Initial operation and performance of a Rankine chiller and an absorption chiller in the National Security and Resources Study Center. *Proc. 3d Workshop on the Use of Solar Energy for the Cooling of Buildings.* AS/ISES, University of Delaware.

Hollander, J. M., and R. Banks. 1975. *Nitinol project test bed.* National Science Foundation report NSF/RANN/SE/AG-550/FR 75/2.

Honeywell, Inc. 1977–1978. *Design and development of single-family, multi-family, light commercial and commercial size solar heating, cooling, and hot water systems.* DOE/NASA quarterly reports: CR-150770, June 1977; CR-150542, September 1977; CR-150640, December 1977; CR-150732, March 1978.

———. 1978a. *Preliminary design package for prototype solar heating and cooling systems.* DOE/NASA Report CR-150853.

———. 1978b. *Preliminary design package for residential heating/cooling system—Rankine air conditioner redesign.* DOE/NASA Report CR-150871.

———. 1980. *Solar power generation study.* Final report, NASA contract NAS8-32093.

———. 1982a. *Solar energy system performance evaluation, final report for Honeywell OTS 45, Salt River Project, Phoenix, Arizona.* DOE/NASA report.

———. 1982b. *Solar energy system performance evaluation, final report for Honeywell OTS 41, Shenandoah (Newnan), Georgia.* DOE/NASA report.

———. 1982c. *Second generation solar Rankine cycle air conditioning system study.* Final report to U.S. Department of Energy on contract no. DE-AC03–81CS0574.

Kroon, R. P., et al. 1981. *Design of a 30 HP solar steam powered turbine.* Report no. DOE/ET/20110-3, U.S. Department of Energy.

Lansing, F. L., and V. M. Chai. 1979. *A thermodynamic analysis of a solar-powered jet refrigeration system.* Jet Propulsion Laboratory invention report NPO-14550/30–4168.

Lior, N. 1977. Solar energy and the steam Rankine cycle for driving and assisting heat pumps in heating and cooling modes. *Energy Conversion,* vol. 16, pp. 111–23.

Lior, N., and K. Koai. 1982a. Solar powered/fuel assisted Rankine cycle power and cooling system: sensitivity analysis. ASME Paper No. 82-WA/Sol-7. (Also published as U.S. DOE Report SAN/20110-8, August 1982).

———. 1982b. Solar powered/fuel assisted Rankine cycle power and cooling system: simulation method and seasonal performance. ASME paper no. 82-WA/Sol-8. Also published as U.S. DOE report SAN/20110–7, August 1982.

Lior, N., K. Koai, and H. Yeh. 1985. *Analysis of the solar powered/fuel assisted rankine cycle cooling system.* Phase I topical report no. SAN/20110–5, U.S. Department of Energy.

Lior, N., and H. Yeh. 1978. *Chiller driven by a solar steam-powered Rankine engine.* Report no. DOE/ET/20110–1, U.S. Department of Energy.

Liu, J., et al. 1982. *Detail design, construction and spin-testing of the 30 HP solar steam powered turbine.* Report no. DOE/ET/20110–10, U.S. Department of Energy.

Martin, C. G., and P. F. Swenson. 1976. Method of converting low grade heat energy to useful mechanical power. U.S. patent no. 3,950,949.

Martin, R., G. Corcoleotes, and J. Williamson. 1981. SOLERAS—solar cooling engineering field tests. Paper presented at DOE Solar Heating and Cooling Contractors Review Meeting.

Melikian, G., et al. 1980. Results of systems simulation and economic analysis of a solar-powered turbocompressor heat pump. *Proc. of DOE 2d Annual Systems Simulation and Economic Analysis Conference,* San Diego, CA.

Melikian, G., F. R. Biancardi, and M. D. Meader. 1980. Test evaluation of a prototype 18-ton solar powered heating and cooling system. *Proc. 15th IECEC,* no. 809430, Seattle, WA.

Nishiyama, E., et al. 1979. Test results of the solar powered Rankine cycle refrigerator installed in the experimental house. *Proc. ISES Congress.* Pergamon Press, vol. 1, pp. 691–95.

Prigmore, D. R., and R. E. Barber. 1974. A prototype solar powered Rankine cycle system providing residential air conditioning and electricity. *Proc. 9th IECEC,* San Francisco, CA.

Rousseau, J., and J. C. Noe. 1976. Solar-powered Rankine cycle heat pump system. *Proc. 11th IECEC,* no. 769202.

Sargent, S. L., and W. P. Teagan. 1973. Compression refrigeration from a solar-powered organic Rankine-cycle engine. ASME paper 73-WA/Sol-8. New York: American Society of Mechanical Engineers.

Scholten, W. B., and J. H. Morehouse. 1983. *Active Program Research Requirements.* Science Applications, Inc., draft final report to U.S. Department of Energy on contract DE-AC03–81SF11573.

Shelpuk, B. 1975. *Regenerative gas cycle air-conditioning using solar energy.* National Science Foundation report NSF/RANN/SE/GI44098/FR/75/2.

Shelton, S. V. 1977–1978. *Development of a solar driven free piston dual loop heat pump.* Progress reports to U.S. Department of Energy on contract EG-77-C-02–4539.

Subbiah, S., and N. Lior. 1983. *The test facility for the solar-powered/fuel assisted hybrid Rankine cycle (SSPRE).* Phase III report. SAN/ET/20110–11, U.S. Department of Energy.

Technical support package—solar heating and cooling development program. 1983. NASA Tech Briefs, MFS-27015, vol. 8, no. 1.

Thermo Electron Corp. 1982. *Determination of the thermal stability of organic working fluids used in Rankine-cycle power systems for solar cooling.* Final report to U.S. Department of Energy on contract DE-AC03–80CS30220.

United Technologies Research Center. 1977. *Test and evaluation of a solar-powered laboratory turbocompressor system for building heating and cooling.* Final report R79–952529–1, ERDA.

———. 1979. *Design, development and testing of a solar-powered multi-family residential prototype turbocompressor heat pump.* Final technical report, R79–953050–1, U.S. Department of Energy.

———. 1982. *Design, development and testing of a solar-powered multi-family residential prototype turbocompressor heat pump.* Extended test program interim report, R82–953050–1, U.S. Department of Energy.

Veneruso, A. F. 1976. Simulation and operation of a solar powered organic Rankine cycle turbine. ASME paper 76-WA/Sol-18. New York: American Society of Mechanical Engineers.

Warren, M. L., and M. Wahlig. 1982. *Cost and performance goal methodology for active solar cooling systems.* Lawrence Berkeley Laboratory report LBL-12753.

Werner, D. K. 1978. Design and test results of a 63.4 kW (85 hp) solar powered Rankine cycle. *Proc. 1978 AS/ISES Meeting,* Denver, CO, pp. 805–10.

Zeren, F., and R. E. Holmes. 1981. Performance evaluation for a jet pump solar cooling system. ASME paper 81-WA/Sol-30. New York: American Society of Mechanical Engineers.

Zeren, F., R. E. Holmes, and P. E. Jenkins. 1978. Design of a freon jet pump for use in a solar cooling system. ASME paper 78-WA/Sol-15. New York: American Society of Mechanical Engineers.

20 Absorption Systems and Components

Michael Wahlig

20.1 Introduction

Absorption cycles have been used for cooling applications for over one hundred years. Being heat-driven cooling devices and requiring only moderately high inlet temperatures ($\geq 180°F$ [$82°C$]), it was natural for absorption machines to be coupled with solar collectors to form solar cooling systems. The modifications required to adapt existing, marketed absorption chillers to solar-driven applications have been fairly modest in general. In fact, some larger solar cooling installations have used commercially available absorption chillers as is, although usually with a substantial capacity de-rating due to operation at below-design-point inlet temperatures.

The specific needs of solar energy systems are sufficiently different from the traditional absorption chiller applications, however, that they have spurred a resurgence in absorption technology development over the past decade. The driving element, the solar collectors, is expensive, and maximum use must be made of the hot water collectors produce. This places a premium on the development of absorption equipment that can operate at low inlet temperature (i.e., higher solar collector efficiency) and high efficiency (i.e., smaller solar collector area needed to meet a given cooling load).

A further goal has been the development of absorption machines that can operate air-cooled (rather than water-cooled, as most large absorption machines have been) to address solar-driven residential and small commercial cooling needs. The use and maintenance of a cooling tower for these applications is considered to be unlikely. This market is currently served almost exclusively by electric-driven, air-cooled air conditioners. Since air-cooled absorption chillers require higher driving temperatures than do water-cooled units, higher performance solar collectors are needed to maintain high system efficiency.

The consequences of these solar cooling system requirements led to a two-tiered R&D program. For the short term, the emphasis was on development of relatively efficient, single-stage, water-cooled absorption chillers that could operate on fairly low temperature ($180°F$, $82°C$) solar-heated hot water. This equipment could be (and was) used in first-generation solar cooling systems, such as solar cooling demonstration program installations. For the longer term, research aimed at substantial advances in absorption-cycle technology has been undertaken with the objective of eventual attainment of technically mature, cost-effective solar cooling systems. This research has involved investigation into new absorption cycles, new working fluids and fluid additives, better system control, and heating as well as cooling applications.

This chapter summarizes much of the R&D that has taken place over this period, starting around 1974. Section 20.2 contains a review of absorption-cycle technology: types of cycles, fluids used, and options for external heat exchange couplings. The heart of the chapter is in sections 20.3 and 20.4, describing first the absorption chiller development itself and then its incorporation into solar cooling systems. The mechanism for presenting this information is by description of individual projects that have been performed, grouping the projects that have had similar objectives. Section 20.3 on absorption chiller R&D includes descriptions of absorption technology R&D that was occurring during the same period for nonsolar applications, since the overlap among all these projects (in objectives, technical issues and advances, end-use applications) is substantial.

A brief statement on the economic situation in section 20.5 is followed by an outlook for the future of absorption cooling in section 20.6.

Much of the work done and progress made has been reported in proceedings of solar cooling and solar building workshops and conferences, project summaries reports, and similar documents. These are listed at the front of the reference list, and then several of the subsequent references refer back to these major documents (Proceedings 1–14; Grossman and Johannsen 1981; Eisa et al. 1986).

20.2   Fundamentals of Absorption Cycles

In absorption-cycle cooling, just as is the case for mechanical cooling cycles, the cooling effect is produced by the evaporation of a refrigerant fluid. The heat required to evaporate the liquid refrigerant is taken from the conditioned space, thus cooling the space. The various types of absorption cycles represent various methods of pressurizing and condensing the refrigerant fluid so that the cooling effect may be repeated in a cyclic process.

20.2.1 Cycle Descriptions: Closed Cycles

In closed absorption cycles, the refrigerant is conserved and reused repeatedly in successive cycles. Heat exchange, but not fluid exchange, takes place between the refrigerant and the atmosphere.

20.2.1.1 Single-Effect Cycle

20.2.1.1.1 BASIC CYCLE   The basic single-effect absorption cycle is depicted in figure 20.1. The four basic components are heat exchangers: the generator, condenser, evaporator, and absorber. The generator contains a mixture of two fluids: the absorbent and the refrigerant. As external heat (e.g., from a solar collector) is added to the generator, refrigerant is boiled off from the absorbent/refrigerant mixture. This refrigerant vapor flows to the condenser where it is liquefied by releasing heat to the ambient air, either directly or through a cooling water loop (cooling tower). The condensed refrigerant passes through an expansion valve where its pressure is reduced and it is partially vaporized. After the expansion, the refrigerant flows to the evaporator, where heat is introduced from the conditioned space, completely evaporating the low pressure refrigerant. The conditioned space is thereby cooled, as the evaporator temperature is typically 45°F (7°C).

Referring again to the generator, the process of boiling off the refrigerant produces a refrigerant-deficient solution. This solution is directed to the absorber through a pressure-reducing valve. Being refrigerant-deficient, this is a strong absorbent solution and vigorously absorbs the low pressure refrigerant vapor coming from the evaporator. This absorption process produces heat that must be discharged to the ambient, in a similar manner to the heat discharged by the condenser.

The output of the absorption process is a refrigerant-rich solution (i.e., a weak absorbent solution). This solution is pumped to the generator where heat is added to drive off the refrigerant, repeating the cycle.

The thermal performance of the absorption cycle is usually expressed in terms of its coefficient of performance (COP). This is defined as the dimensionless ratio of the heat removed from the conditioned space (i.e., the heat input to the evaporator) to the heat input to the generator over the same time period. The cooling rate, the energy per unit time removed from the conditioned space, is usually expressed in kW or Btu/hr or tons of refrigeration, where one ton equals 12,000 Btu/hr.

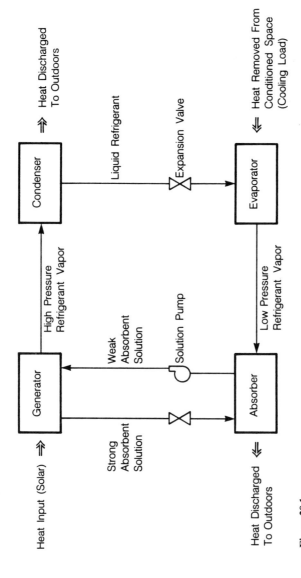

Figure 20.1
Schematic diagram of basic single-effect absorption cycle.

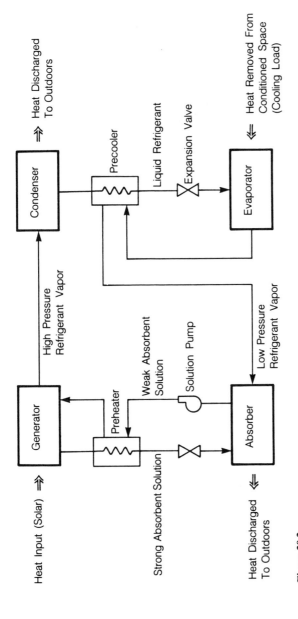

Figure 20.2
Schematic diagram of single-effect absorption cycle with additional heat exchangers for higher efficiency.

A second measure of performance is the electric power required to operate the pumps and fans relative to the output cooling rate of the cycle.

20.2.1.1.2 CYCLE WITH ADDITIONAL HEAT EXCHANGERS In practice, one or more additional heat exchangers must be added to reclaim sensible heat between the basic four components in order to attain acceptable COP values. The most critical is the preheater, shown in figure 20.2, which reclaims the sensible heat from the hot strong absorbent solution exiting the generator to preheat the weak absorbent solution being pumped from the absorber. The effectiveness of this preheater is often more critical than that of the four basic components in terms of its effect on the overall cycle COP.

For some absorbent/refrigerant fluid pairs (e.g., water/ammonia), it is also advantageous to include a precooler heat exchanger. This element precools the liquid refrigerant entering the evaporator, using the cool refrigerant vapor exiting the evaporator.

Depending upon the application and the fluid pair, it may also make sense to include other minor heat exchangers in the cycle. For example, a heat exchanger may be used to reclaim the sensible heat contained in the hot refrigerant vapor exiting the generator; this heat may be used to further preheat the weak absorbent solution before it enters the generator (Dao, Rasson, and Wahlig 1983). Additional heat exchangers such as this one can only be justified if the incremental improvement in cycle COP outweighs the incremental cost of adding the component to the chiller.

20.2.1.1.3 THERMODYNAMIC DESCRIPTION OF THE CYCLE A pressure-concentration-temperature (P-X-T) plot, such as shown in figure 20.3, is a convenient method of describing the absorption cycle thermodynamically. Two pressure levels and three temperatures are involved. The generator (G) and the condenser (C) are at the higher pressure, and the absorber (A) and evaporator (E) are at the lower pressure. The highest temperature (from the solar collectors) is used to produce the input heat (Q_{in}) to the generator. The condenser and absorber are close to the ambient (T_{sink}) temperature. The evaporator sits at the lowest (T_{cold}) temperature (typically 40°F [4°C]), providing the cold reservoir for cooling the conditioned space.

The cycle is illustrated as follows, referring to figure 20.3. As heat, Q_{in} is added to the generator, the absorbent/refrigerant solution mixture releases refrigerant vapor (shown as the dashed line going to the condenser), and

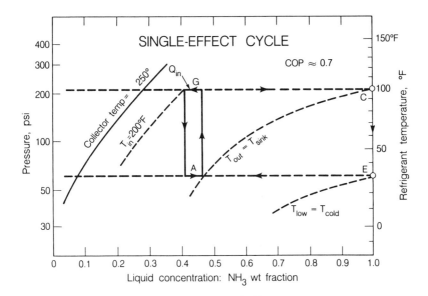

Figure 20.3
Pressure-concentration-temperature (P-X-T) diagram of single-effect absorption cycle, using ammonia/water as the refrigerant/absorbent fluid pair.

the solution concentration (percent refrigerant) decreases. This reduced-concentration solution (solid line in figure 20.3) undergoes a pressure drop en route to the absorber. There it absorbs the low-pressure refrigerant vapor coming from the evaporator, increasing the solution concentration to its initial value. The heat produced by this absorption process is discharged to the ambient, T_{sink}, temperature. This high-concentration solution is then pumped back up to the generator pressure, completing the cycle. Note that the concentration difference between the high- and low-concentration solutions may be only a few percent.

The points C and E for the condenser and evaporator states appear at the right-hand edge of the figure 20.3 diagram, corresponding to pure refrigerant. As the vapor is condensed to liquid refrigerant in the condenser, heat is discharged to ambient temperature. The refrigerant liquid pressure is reduced between the condenser and evaporator. The heat required to evaporate the refrigerant at point E is taken from the conditioned space being cooled, at temperature T_{cold}.

Referring to figures 20.2 and 20.3, the amount of refrigerant vapor boiled off from the solution in the generator is the same quantity of refrigerant vapor boiled in the evaporator. Thus the amount of heat exchanged with the external sources and sinks is approximately the same (within about 30%) for each of the four major components. The heat supplied to the generator must exceed that supplied to the evaporator because of the heat of solution and heat losses in the cycle. Therefore, the COP of the single-effect absorption cooling cycle is always less than 1. Thermodynamic second-law calculations (Anand et al. 1984) have shown that the maximum theoretical COP at standard ARI conditions (see section 20.2.4.1) is about 0.8, but in practice maximum achievable COP values are about 0.7. These efficiency limits may be exceeded, however, by using more complex cycles, as described in the next section.

20.2.1.2 Double-Effect Cycle

20.2.1.2.1 TYPICAL DESIGN   A typical double-effect absorption cooling cycle is diagrammed in figure 20.4. There are two generator components. The first generator operates just as does the generator in a single-effect cycle: it uses the heat from the solar collectors to drive off refrigerant vapor from the weak absorbent solution. The partially concentrated solution exiting from this generator is reduced in temperature (by sensible heat exchange with the weak absorbent solution en route to the first generator) and enters a second generator component. Here heat from the condensation of refrigerant vapor (coming from the first generator) is used to drive additional refrigerant vapor from the partially concentrated solution. Both the liquid and vapor refrigerant streams pass into the condenser component.

The condenser, evaporator, and absorber components are essentially the same as those used in a single-effect cycle. The fully concentrated solution (i.e., the strong absorbent solution) exiting the second generator is first used to partially preheat the weak absorbent solution and then flows into the absorber where it absorbs the low-pressure refrigerant vapor. Finally, the weak absorbent solution exiting the absorber is pumped back to the first generator, completing the cycle.

The thermodynamic state points and processes for this double-effect cycle are shown in figure 20.5. In addition to the three temperatures characteristic of a single-effect cycle ($T_{high} = T_{in}$, $T_{sink} = T_{ambient}$, and $T_{low} = T_{cold}$), there is a fourth, intermediate temperature, T_{inter}. This is the temperature at which the high pressure vapor (driven off from generator G_1 at

DOUBLE-EFFECT LITHIUM BROMIDE CYCLE

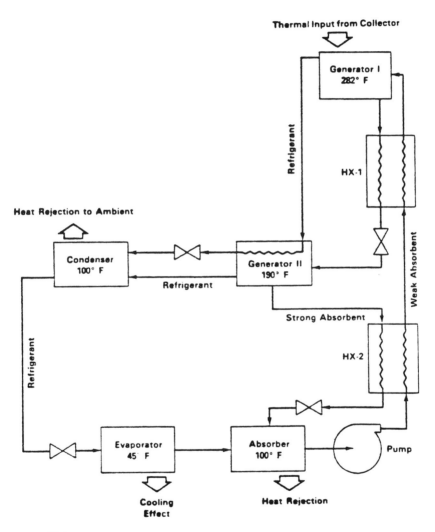

Figure 20.4
Schematic diagram of a typical double-effect absorption cycle.

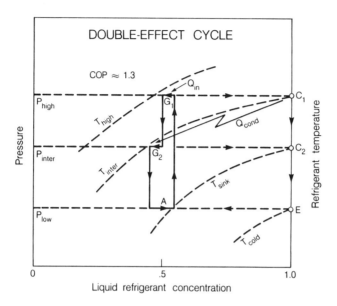

Figure 20.5
P-X-T diagram of a typical double-effect absorption cycle: a condenser-coupled design.

pressure P_{high}) condenses (at point C, in figure 20.5). The heat of condensation Q_{cond} at temperature T_{inter} is used to drive off the additional refrigerant vapor from the second generator G_2. Since the heat of condensation is the heat transfer mechanism between the high-temperature stage and low-temperature stage of this double-effect cycle, it is often referred to as a condenser-coupled design. This is the type of cycle used for most LiBr/H$_2$O double-effect absorption machines.

20.2.1.2.2 ALTERNATIVE CONFIGURATION For some fluid pairs (e.g., H$_2$O/NH$_3$), the high pressure level required for the cycle shown in figure 20.5 is impractical. For such cases, the alternative cycle configuration shown in figure 20.6 may be appropriate. In this design, there are two absorption components A_1 and A_2, in contrast with the two condensers C_1 and C_2 in figure 20.5. In the figure 20.6 configuration, the heat of absorption released in the intermediate temperature, T_{inter}, absorber A_1 is used to drive off the additional refrigerant vapor from generator G_2. Therefore, this is usually termed an absorber-coupled design. By operating the higher-temperature stage at low concentration ratios, only two pres-

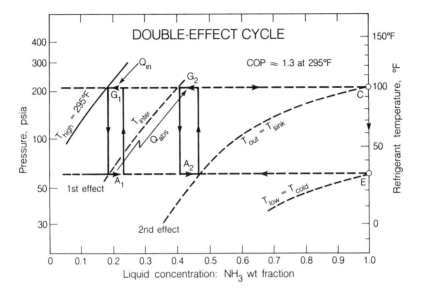

Figure 20.6
P-X-T diagram of an alternative configuration for a double-effect absorption cycle: an absorber-coupled design.

sure levels are required, avoiding the high pressure level of the figure 20.5 configuration.

Both these designs exhibit the primary important characteristic of double-effect cycles; namely, higher COP values than those of single-effect cycles. By driving off refrigerant vapor from both generators G_1 and G_2, the amount of cooling produced per unit of heat Q_{in} to the cycle is nearly doubled. Thus cooling cycle COPs of about 1.3 are achievable.

The price one pays for the higher COP values is the higher input temperatures required: typically 280°F (138°C) or higher. This requires the use of higher-quality solar collectors (i.e., evacuated-tube collectors or tracking collectors) than are needed to drive single-effect cycles (typically 180°F [82°C] or higher).

20.2.2 Cycle Description: Open Cycle

In open absorption cycles, the refrigerant, after its use for cooling, is released to the atmosphere, and replacement refrigerant fluid is introduced for each cycle. Economically, water is the only fluid that is expendable and available to this extent, so water is the refrigerant for all open absorption

Michael Wahlig

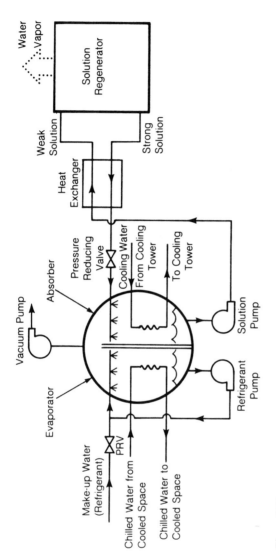

Figure 20.7
Schematic of a typical open-cycle absorption diagram. Source: Leboeuf and Löf (1980).

cycles. Fortunately, water is an excellent refrigerant, so the requirement for its use is not a serious constraint.

A schematic diagram for a typical open absorption cycle is given in figure 20.7. The evaporator and absorber components are essentially identical to those found in a closed cycle machine. These two components operate at a common low pressure, the evaporator operates at a cold temperature (typically 40°F [4°C]), removing heat from the space being cooled, and the absorber operates at near-ambient temperature, discharging its heat to the environment.

Referring back to the closed-cycle diagrams in figures 20.1 and 20.3, the absorber and evaporator functions and thermodynamic conditions still hold for open-cycle configurations. However, the generator and condenser operate at atmospheric pressure in open cycles. The water vapor driven off (from the absorbent/water solution) in the generator is released to the atmosphere. Its condensation into water is left to the big condenser in the clouds. Makeup water is supplied to the evaporator through a pressure reduction valve (PRV) as shown in figure 20.7. Thus the condenser component is completely missing in an open-cycle machine, reducing its cost.

The open-cycle design must have a generator-type component in which the weak absorbent is regenerated; i.e., is made more concentrated in absorbent by giving up some water vapor to the atmosphere. This is shown as the "solution regenerator" component in figure 20.7. The open nature of this cycle provides various options for accomplishing this; in particular, the regenerator can be combined with the solar collector subsystem in ways not possible with closed-cycle systems.

For example, a solar collector open to the atmosphere can itself serve as the regenerator. The weak absorbent solution can be passed over a very low-cost open collector surface (glazed or unglazed, but open to the atmosphere), using direct absorption of the sunlight to drive off water vapor from the solution as it flows down over the roof surface. Ordinary roof shingles have served as the collector/regenerator surface in some experimental installations. Another regenerator option is to drip the weak solution through a packed column, through which solar-heated air is passed to drive off vapor from the solution.

The main advantages of the open-cycle system, compared to closed-cycle systems, are therefore the reduced cost of the cooling equipment (no condenser) and the possibility of using a low-cost collector subsystem for the solution regeneration. The cycle as shown in figure 20.7 is essentially a single-effect cycle and therefore has a practical COP limit of about 0.7, similar to that of closed single-effect absorption cycles.

Potential disadvantages arise from the exposure of the absorbent solution to the atmosphere. Contaminants are picked up in the regeneration stage—from the atmosphere and from surfaces over which the solution is passed. Nondissolved solid contaminants can be filtered out. Noncondensible gases can be purged using a small vacuum pump intermittently, as shown in figure 20.7. In the reverse direction, there is a mass transfer from the absorbent solution to the atmosphere. Small amounts of solution will be lost in this way, requiring periodic makeup. The degree to which any fluid additives (discussed in the next section), which may have a higher volatility, are lost to the atmosphere must be considered. The environmental effects of transferring small amounts of the absorbent or additives to the atmosphere must be determined. None of these disadvantages appears insurmountable, and many of these issues are topics of current and recent research, some of which will be covered in later sections of this chapter.

20.2.3   Fluids for Absorption Cycles

20.2.3.1   Lithium Bromide/Water Fluid Pair

By far the most widely used absorbent/refrigerant fluid pair for solar absorption cooling equipment has been lithium bromide/water. Water is an excellent refrigerant because of its high latent heat of vaporization, its chemical stability, and its low cost. Its high freezing point (32°F, 0°C) represents a problem if it is desired to operate the absorption machine in a heating as well as a cooling mode, because the heating mode requires that the evaporator exchange heat with outdoor winter conditions.

Lithium bromide has a high affinity for water vapor. This high solubility of water in the lithium bromide absorbent reduces the amount of solution that has to be pumped per unit of refrigerant (water) produced. Moreover, the high latent heat of water results in low parasitic pumping requirements for lithium bromide/water systems. Also important is the low vapor pressure, or nonvolatility, of the lithium bromide at the generator temperatures used to drive off the water vapor, thus minimizing the carryover of absorbent into the condenser and evaporator.

A disadvantage of the lithium bromide/water fluid pair is that it is subject to crystallization at high concentrations, as a function of the solution temperature and pressure. Normal absorption-machine operating conditions are designed to avoid the crystallization region. However, experience to date is that many of the LiBr/H$_2$0 absorption chillers used in solar applications have undergone crystallization at some time in their operation, due to inadvertent operation at off-design conditions. Fortunately, decrystallization procedures, although a nuisance, are not difficult. Fluid additives, as discussed below, can assist in avoiding crystallization conditions.

20.2.3.1.1 FLUID ADDITIVES In practice, the fundamental absorption process that takes place within the absorber component has proved to be the limiting heat and mass transfer process in terms of cycle performance. The boiling process that occurs in the generator, as well as the refrigerant condensation and evaporation, are all processes that can be modeled adequately theoretically and verified experimentally. But this is not true yet for the absorption process.

The experimentally observed performance of the absorber component invariably falls far short of the performance predicted on the basis of heat and mass transfer coefficients. Empirical correction factors (i.e., oversizing the heat exchanger) have been employed to achieve desired machine capacity. The fundamental problem is the difficulty in achieving adequate distribution of the refrigerant vapor within the absorbent solution so that efficient absorption can occur.

One method of promoting vapor/solution mixing is through the use of chemical additives. Manufacturers of LiBr/H$_2$O machines have used the compound 2-ethyl hexanol-1 to produce a turbulent mixing effect (Marangoni effect) in the absorber, considerably improving its performance.

Another use of fluid additives, as mentioned in the previous section, is to assist in preventing crystallization of the absorbent/refrigerant solution. Carrier investigated (Biermann 1978) the addition of ethylene glycol to the lithium bromide solution, which was known to "move" the crystallization line away from the normal pressure-concentration-temperature operating regime. However, the standard heat transfer additive is not operative in the lithium bromide/ethylene glycol/water ternary fluid. Carrier's contributions were to optimize the ratio of LiBr to glycol and, more importantly, to develop a heat transfer additive (1-nonylamine) that would enable the equilibrium properties to be realized in a dynamic system (Reimann and Biermann 1984).

Subsequently, 1-nonylamine was discovered to have an undesirable side effect: when heated in the presence of copper oxide (formed from oxygen leaking into the machine), it forms chemically refractory copper soaps. A different additive with similar mixing properties, phenylmethylcarbinol (PMC), was then substituted for the 1-nonylamine.

Use of the resulting lithium bromide/ethylene glycol/water ternary fluid (called "Carrol") allows air cooling of the condenser, i.e., condenser operation in the 110° to 120°F (43° to 49°C) range. This modification overcomes one of the limitations of the use of the LiBr/H$_2$O fluid pair; namely, that the condenser has to be operated at temperatures requiring water cooling towers (typically 85° to 95°F [29° to 35°C]), to avoid the crystallization region.

Aqueous lithium bromide solutions are corrosive, especially if any air is introduced into the solution. Besides the need to use containment materials that are resistant to attack by lithium bromide, this property introduces the need for other types of fluid additives: ones that serve as corrosion inhibitors by forming protective films.

The resulting mix of additives to absorption solutions has developed partially as an "art" and partially as a result of years of experience by absorption equipment manufacturers. Part of this information remains proprietary.

20.2.3.2. Water/Ammonia Fluid Pair

Water/ammonia was the original absorption fluid pair used when absorption refrigeration was discovered over one hundred years ago and is still widely used for refrigeration equipment. All the old Servel gas-driven refrigerators in this country used water/ammonia solutions.

Although this pair has excellent thermodynamic properties, its applicability for space cooling is diminished by the toxic and flammable properties of ammonia. In the United States, the ammonia-containing equipment must be either outdoors or in a separate equipment room. In some other countries, as in Japan, its use in space cooling equipment is more restricted than in the United States.

The low boiling temperature of ammonia makes it possible to use the water/ammonia fluid pair for heating-mode operation as well as for cooling. In addition, there is no possibility of crystallization of the solution, and the condenser and absorber may be air-cooled as well as water-cooled.

The vapor pressure of the water in the generator is not negligible with respect to that of the ammonia, so a small amount of water vapor is boiled off along with the ammonia vapor. This necessitates the use of a rectifier component between the generator and the condenser; the rectifier recaptures and condenses the water vapor, returning it to the generator. Designs have been developed to accomplish this step very efficiently (Dao, Rasson, and Wahlig 1983).

20.2.3.3 Other Fluid Pairs

Many other absorbents, refrigerants, and pairs of these fluids have been proposed, investigated, and had their properties measured. Perhaps ten to twenty pairs have actually been tested experimentally in working models of absorption-cycle machines. However, essentially none except the LiBr/ H_2O and H_2O/NH_3 pairs (including their additives) has been used to date in commercially marketed absorption machines.

An April 1982, a workshop on R&D needs and priorities for new absorption working pairs was held in Berlin (Proceedings 8). The four background papers presented at that workshop provide a good review of work on absorption heat pumps and absorption fluid pairs in Europe up to 1974 (Stephan 1982), in Europe since 1974 (Hodgett 1982), in the United States up to 1974 (Macriss 1982), and in the United States since 1974 (Hanna and Wilkinson 1982).

Some investigations into new candidate fluid pairs were supported by the DOE Solar Program in the 1970s, with the work performed by the Institute for Gas Technology (IGT) (Macriss et al. 1979) and by the Southern Research Institute (Lacey 1978), in addition to the previously mentioned study by Carrier (Biermann 1978). More recently, measurements were made by SRI International (Podoll et al. 1982) of the properties of candidate fluids for higher temperature ($\geq 280°F$, 138°C) absorption-cycle applications.

Meanwhile, a significant amount of research into absorption fluid pairs has been supported by the DOE Conservation Program, which cosponsored the 1982 Berlin workshop on absorption fluids. An organic absorbent was identified and tested for a freon refrigerant by Allied Corporation, as discussed in section 20.3.3.2. IGT recently completed a comprehensive data survey of absorption fluids (Macriss and Zawacki 1984). The required characteristics and selection methods for new absorption heat pump fluids have been summarized by Ryan (1983).

The March 1985 Paris Congress on Absorption Heat Pumps (Proceedings 9) included a session on absorption fluid pairs and represented a follow-up to the Berlin workshop held three years earlier.

Major work now under way on new, higher efficiency (higher than single-effect) absorption cycles for heating and cooling, supported by DOE/Conservation and being performed by Trane, Carrier, and Phillips Engineering, is placing a significant emphasis on using nonconventional fluids. Ternary fluid combinations are being proposed and tested for these developmental absorption-cycle machines.

20.2.4   Coupling to the Heat Sink

The basic absorption cycle of figure 20.1 can be interpreted as consisting of two subcycles: a driving cycle (the left-hand side of the diagram) and a driven, or refrigeration, subcycle (the right-hand side of the diagram). The closer the discharge temperature's approach to the atmospheric heat sink, the larger the temperature difference ΔT (and the higher the efficiency) for the driving subcycle and the smaller the ΔT (and the higher the efficiency) for the driven subcycle. Therefore, a close temperature approach of the condenser and absorber to the atmospheric temperature is critical for high-performance operation of absorption cycles. This is more important for solar-driven than for gas-driven operation because of the lower solar driving temperatures.

In general, there are two methods of coupling the condenser/absorber to ambient: water-cooling and air-cooling.

20.2.4.1   Water-Cooling the Condenser/Absorber

In water-cooled operation, a cooled-water loop couples the condenser/absorber to the atmosphere via a water cooling tower. Through the process of water evaporation to the atmosphere, this system is essentially coupled to the ambient wet-bulb temperature. For standard ARI test conditions (95°F [35°C] dry bulb and 75°F [24°C] wet bulb), the return cooled water temperature from the cooling tower is normally 85°F (29°C) and the internal condenser and absorber temperatures are typically 95°F (35°C). If series rather than parallel cooling water flow is used, the condenser temperature is usually somewhat higher.

The advantage of water cooling (compared to air cooling), as mentioned above, is the higher cycle efficiency associated with the lower condenser/absorber temperature. The disadvantages include the initial cost and maintenance cost of the cooling tower and the usage of the water that evaporates from the cooling tower.

Open absorption cycles already require use of makeup water as the refrigerant, so the use of water for a wet cooling tower represents only an incremental use of water. Moreover, the low solar collector temperature ($< 180°F$, $82°C$) normally associated with open-cycle absorption systems makes it more important that the absorber operate at (low) temperatures attainable only by water-cooled methods.

20.2.4.2 Air-Cooling the Condenser/Absorber

Closed absorption cycles have the option for water-cooled or air-cooled operation. For the latter, ambient air is usually blown directly across condenser and absorber coils that contain the fluids being cooled to near ambient temperatures. "Near ambient" in this case means the $15°$ to $20°F$ ($8°$ to $11°C$) approach characteristic of good liquid-to-air heat exchangers. For standard $95°F$ ($35°C$) outdoor dry-bulb temperature, the condenser/ absorber fluids would operate in the $110°$ to $115°F$ ($43°$ to $46°C$) range. These conditions require that the inlet generator temperature be at least $200°F$ ($93°C$), necessitating good quality collectors. In practice, the absorber and condenser coils are usually in series with respect to the cooling air stream, resulting in a higher temperature for one of the coils.

The advantages and disadvantages of air cooling are just the reverse of those given above for water cooling. Air cooling results in lower cycle efficiencies but avoids the need for makeup water or a wet cooling tower. The fan power required for air cooling usually exceeds that required for the liquid pumping and fan power of a wet cooling tower, so the parasitic power is greater.

Air cooling is not practical for the $LiBr/H_2O$ fluid pair because of the crystallization problem. This problem can be circumvented to some extent by the use of fluid additives, as is the case for Carrier's "Carrol" working fluid (Reimann and Biermann 1984).

20.2.4.3 Evaporatively Cooling the Condenser/Absorber

Evaporatively cooled operation is a variant of water-cooled operation, in which the cooling by evaporation of water into the atmosphere takes place right at the condenser/absorber coils rather than remotely in a wet cooling tower. This introduces constraints on the absorption machine design and placement (i.e., outdoors) but has advantages of reduced cost (no cooling tower) and reduced parasitic pumping power compared to wet cooling tower operation.

20.2.5 Coupling to the Load

The heat removed from the conditioned space (i.e., the cooling load) is coupled to the absorption cycle through the mechanism of vaporizing the refrigerant in the evaporator component. This heat transfer can occur directly, or indirectly via a chilled-water loop.

20.2.5.1 Chilled-Water Loop

In this configuration, a closed water loop is used to transfer heat from the conditioned space to the absorption-cycle evaporator. The water is chilled by having its coils in contact with the evaporating refrigerant. At the other end of the loop, the water is warmed by sensible heat exchange with air from the conditioned space.

Most absorption air conditioners have operated using chilled water loops; hence, the commonly used descriptive term: absorption "chillers". The advantage is in flexibility of placement of the machine. It can be located remotely in an equipment room or outdoors, with the chilled-water loop piped throughout a building. Multiple machines may be plumbed into the same chilled-water distribution system.

Disadvantages are mainly twofold. First, the extra heat exchange step requires lower temperature operation of the evaporator. To produce 45°F (7°C) chilled water (standard ARI conditions), the evaporator must operate at about 40°F (4°C). Lowering the evaporator temperature from 45°F to 40°F (7°C to 4°C) requires a corresponding increase in generator temperature or decrease in condenser/absorber temperature to maintain cycle efficiency. The second disadvantage is the danger of freezing the chilled water in the loop, since normal operation is only 8°F (4°C) above the freezing point; this danger can be mitigated by using antifreeze in the chilled-water circuit.

20.2.5.2 Direct Expansion Evaporator

In a direct-expansion configuration, the building air is blown directly across the coils in which the refrigerant evaporation is occurring. This arrangement has the advantage of a higher evaporator temperature (45°F [7°C] instead of 40°F [4°C]) and eliminates the cost of the chilled-water loop. On the negative side, either large ducts must be used to transport the building air to a centralized absorption machine, or a collection of smaller machines must be distributed around the building, giving up some of the economy of scale associated with a larger, centralized chiller.

For residential applications, direct-expansion cooling coils are normally the preferred type of installation but have not been used for absorption equipment because of limitations of the fluid pairs, specifically the refrigerant. Water has too large a specific volume in the vapor phase, and ammonia cannot be used because of its toxic and flammable nature.

20.3   Absorption Chillers and Heat Pumps

20.3.1   Cooling and Heating Modes of Operation

An absorption chiller, as described in figures 20.1 and 20.3, is also a heat pump. Using the high temperature heat from the solar collectors, heat is pumped "uphill" from the cooler conditioned space to the hotter outdoor ambient temperature sink.

Similarly, this type of absorption cycle can also be used to heat a building in the wintertime. To accomplish this, the role of the evaporator and the condenser/absorber are reversed: the evaporator coils are coupled to the cooler out-door air, and the condenser and absorber coils are coupled to the warm indoor space being heated. To accomplish this in practice, reversing valves are used to redirect the fluid flow paths, or two sets of heat exchangers are used.

The only difference in the P-X-T diagram (figure 20.3) for heating-mode operation, as contrasted with cooling-mode operation, is in the temperature values used. The condenser and evaporator must run at temperatures high enough ($\geq$ 120°F, 49°C) to heat the building air. The evaporator is heated by the cool ambient air and so operates typically at 15°F (8°C) below the ambient temperature.

The definition of thermal efficiency, or COP, remains the same for heating-mode as for cooling-mode operation; namely, the COP equals the energy delivered to the load divided by the heat input to the generator. The heat removed from both the condenser and the absorber is used to heat the conditioned space (the load). Looking at figure 20.1, it is intuitive that the heat input to the evaporator (Q_E) is essentially the same as the heat removed from the condenser (Q_C). Also, the heat input to the generator (Q_G) must be very close to the amount of heat removed from the absorber (Q_A). That is, $Q_E \approx Q_C$ and $Q_G \approx Q_A$. The definitions of the COPs are:

$$\text{COP}_C = Q_E/Q_G \qquad \text{(for cooling)}, \tag{20.1}$$

$$\text{COP}_H = (Q_A + Q_C)/Q_G \qquad \text{(for heating)}. \tag{20.2}$$

By substitution, equation (20.2) becomes:

$$\text{COP}_H = (Q_G + Q_E)/Q_G,$$

$$\text{COP}_H = 1 + \text{COP}_C. \tag{20.3}$$

Thus the heating COP equals one plus the cooling COP, taken at the same operating temperatures. Since typical cooling-mode operating temperatures are normally quite different from typical heating-mode operating temperatures, care must be used in applying equation (20.3).

There is a third mode of operation of absorption cycles: the temperature-boosting mode. Such devices are also called heat transformers. This operating mode is described in section 20.3.6, together with specific research projects on temperature-boosting machines.

20.3.2 Single-Effect Absorption Chiller R&D within the DOE Solar Cooling Program

Most of the research and development (R&D) on absorption cooling supported by DOE/Solar over the past ten years has focused on single-effect machines. The main objectives of this R&D have been to improve the efficiency and capacity under realistic operating conditions, to achieve reliable operation in the field, and to attain air-cooled operation.

20.3.2.1 Arkla Projects: Chiller Development

Arkla Industries has been by far the primary supplier of absorption chillers for solar cooling applications in the United States. In the early and mid-1970s, using internal company funds, Arkla developed a basic, single-effect LiBr/H_2O absorption chiller, designed for inlet operating temperatures (195°F, 91°C) achievable in warm weather with good quality flat-plate collectors. Arkla manufactured two models, a 3-ton residential-size model, and a small commercial-size model with 25-ton capacity. Most solar cooling demonstration projects during the 1970s and early 1980s used one of these Arkla chillers. These machines required the use of wet cooling towers (85°F [29°C] returning cooled water from the cooling tower), produced 45°F (7°C) chilled water out of the evaporator, and had a design COP value of about 0.7. Figures 20.8 and 20.9 show the schematic

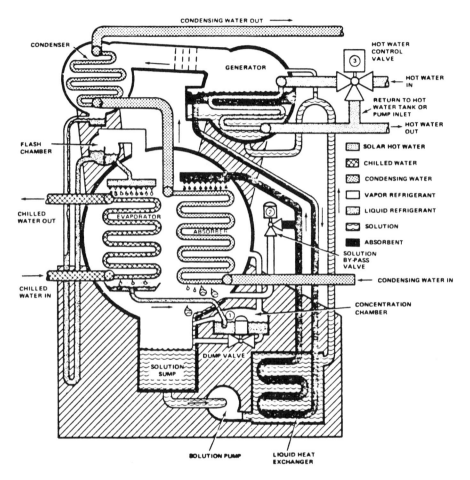

Figure 20.8
Schematic diagram of Arkla 25-ton water-cooled absorption chiller (SOLAIRE WFB–300).
Source: Arkla commercial literature.

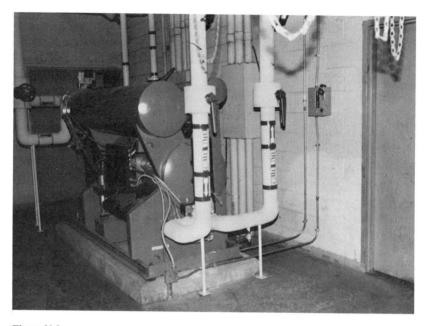

Figure 20.9
Arkla WFB-300 water chiller installed at the National Agricultural Library.

diagram and a typical installation photograph of a 25-ton chiller. While marketing these chillers, Arkla pursued their further development with support from the DOE/Solar program.

20.3.2.1.1 WATER-COOLED CHILLERS To reduce installation costs and improve reliability, Arkla developed unitary packages for the 3-ton and 25-ton water-cooled chillers. In addition to the chiller itself, the skid-mounted, 3-ton unitary package included a systems tank (a small storage tank), all valves, pumps, and controls, and an auxiliary heater. All the installer had to do was hook up the lines from the collector, the main storage tank, the cooling tower, and the fan coil unit.

Arkla built several of these 3-ton unitary packages, and field-tested three of them in residential installations in the Evansville, Indiana, area. They worked quite satisfactorily. However, they never became a part of the regular Arkla manufacturing product line. There were two main reasons for this. First, the cost was high. Arkla was selling the 3-ton chiller for about $3,000. The company's estimated selling price for the 3-ton unitary package was about $11,000. The second reason was small market demand. The market was falling off for the 3-ton chillers, making it impossible for Arkla to justify the tooling-up costs involved in marketing the unitary packages.

Arkla also developed a unitary package for the 25-ton chiller. Because of the large tank size, it was decided to omit a systems tank from the 25-ton package; otherwise, it was very similar to the 3-ton unitary package. Only one of these was built. It was shipped to Lawrence Berkeley Laboratory (LBL) for inclusion in a Solar Federal Buildings Program cooling system at that laboratory.

In about 1982, in light of the very small demand for these 3-ton and 25-ton chillers, Arkla made a corporate decision to discontinue their manufacture. The company announced its intention about 6 to 9 months before the cessation, saying it would accept new orders for a final manufacturing run before termination. After supplying those final units, Arkla shut down the absorption chiller operations. This left the United States without a regular domestic manufacturer of small-capacity, water-cooled absorption machines suitable for solar cooling applications. Japanese imports are available, however, through U.S. distributors.

20.3.2.1.2 EVAPORATIVELY COOLED CHILLER The requirement for a separate wet cooling tower represented one impediment to widespread residential use of absorption chillers, mainly because of the maintenance needed for the cooling tower. Recognizing this, Arkla undertook the development (with DOE/Solar support) of an evaporatively cooled, 3-ton absorption chiller (Merrick 1982; Merrick and Murray 1984). A schematic of this unit is shown in figure 20.10.

In this design, the condenser and absorber are fabricated as concentric shells, with large surface areas. Water at atmospheric pressure is distributed to flow down the outside of these shells. A fan is used to force air upward between the shells, promoting the evaporation of water on the shell surfaces. Thus the design essentially "builds in" a wet cooling tower directly into the absorption chiller. Advantages, compared to use of a separate cooling tower, include lower cost, smaller size, and reduced parasitic power, in addition to the reduced maintenance requirement.

With this design, it is critical that the scale buildup on the absorber and condenser shells be held to an acceptable level over the lifetime of the machine (say, 20 years). Otherwise, the heat exchange rate from the

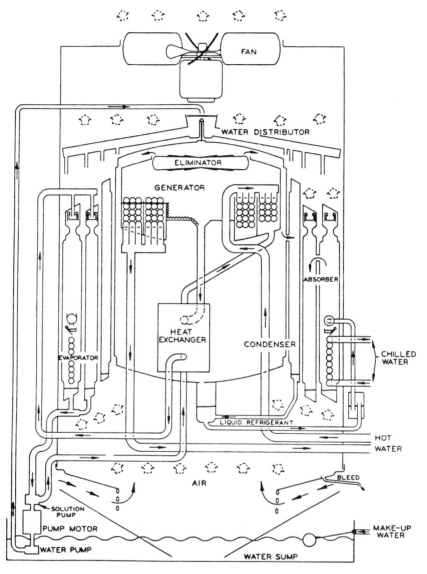

Figure 20.10
Schematic diagram of Arkla 3-ton evaporatively cooled absorption chiller (model XWF-36). Source: Merrick and Murray (1984).

absorber and condenser to the cooling water would diminish, impacting chiller performance. The Arkla design featured large surface area and generous cooling water recirculation rate to reduce the required heat flux across the shell walls. It was determined that an 80-mil buildup of scale would result in a 1°F (0.6°C) increase in temperature across the shell walls. A laboratory chiller test determined experimentally a scale buildup of only 6 mils after 9,000 hours of operation, using city water of average hardness. Thus this design goal was met.

Another design goal proved more difficult. As these chillers normally operate out-doors, a reliable protective coating for the mild steel shells is mandatory. A number of plastics-type coatings were tested over extended periods, and all failed. The final prototype units used hot-dip galvanized shells and thin-walled stainless steel walls. The galvanized unit operated successfully in the laboratory and in the field. The stainless steel unit was never run because of other problems with that unit.

The final prototype design essentially met the machine design performance: COP of 0.75 at 195°F (91°C) hot water supply, 85°F (29°C) condensing water, and 45°F (7°C) leaving chilled water (Merrick and Murray 1984). After testing at Arkla, this unit (with the galvanized shells) was shipped to Colorado State University, where it was operated to provide cooling for Solar House III starting in 1982 (see section 20.4.6.1.1).

20.3.2.1.3 EVAPORATIVELY COOLED CHILLER WITH GAS-DRIVEN SECOND STAGE   Arkla recognized that a serious economic drawback to the single-effect absorption chiller was the poor performance in the backup gas-driven mode (when solar energy was not available). With a gas-driven boiler efficiency of 0.8 and a chiller COP of 0.75, the gas-driven COP is only 0.6. This could not compete economically with a reasonably efficient electric-driven air conditioner. Accordingly, Arkla undertook the development of a solar/gas absorption chiller that would operate in a single-effect mode using solar-heated hot water and in a double-effect mode using gas-heated hot water.

A schematic diagram of this unit is shown in figure 20.11. The design COP for double-effect operation was 1.1, including the gas burner efficiency. The unit is quite similar to the Arkla single-effect evaporatively cooled chiller, with the addition of a gas-fired high temperature generator and a second set of hot water coils in the lower temperature generator.

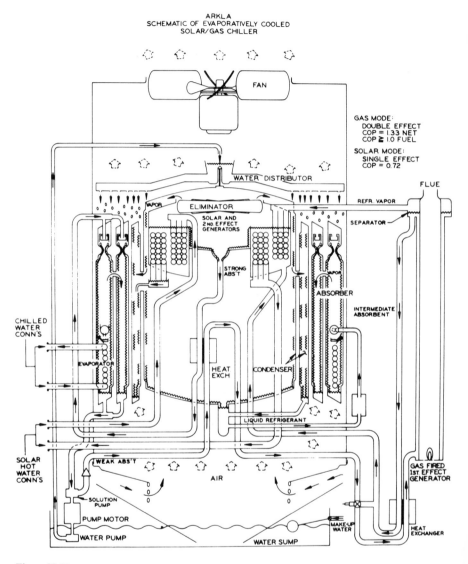

Figure 20.11
Schematic diagram of experimental model of Arkla 3-ton evaporatively cooled chiller with double-effect operation in the gas-driven backup energy mode.

Arkla fabricated and conducted preliminary tests on two of these solar/gas chillers. However, development was discontinued when Arkla decided to terminate its absorption chiller R&D activities, as part of its (previously mentioned) decision to stop manufacturing the 3-ton and 25-ton LiBr/H_2O machines.

20.3.2.2 Carrier Projects: Chiller Development

The Carrier Corporation has long been a manufacturer of absorption machines for nonresidential applications, using LiBr/H_2O as the working fluids, and for residential applications (in its Bryant Division), using H_2O/NH_3 as the working fluids. Under the auspices of the DOE/Solar Program, Carrier undertook two parallel development efforts: (1) an air-cooled absorption chiller for residential applications, and (2) a water-cooled absorption chiller for nonresidential applications, redesigned specifically for solar-driven operation.

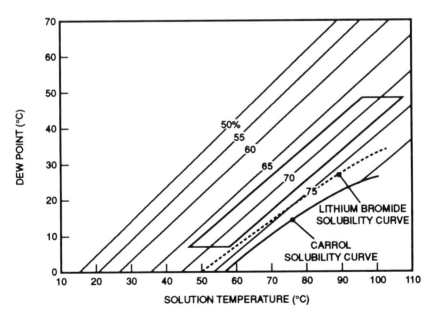

Figure 20.12
Equilibrium chart for aqueous solutions of Carrol. For reference purposes, the crystallization line for lithium bromide-water has been superimposed as well as a typical cooling cycle path.

20.3.2.2.1 AIR-COOLED CHILLERS   As discussed in section 20.2.3.1.1, Carrier successfully utilized fluid additives to lithium bromide that permitted air-cooled (i.e., higher temperature) operation of the condenser and
absorber without crystallization of the absorbent/refrigerant solution
(Biermann 1978; Reimann and Biermann 1984). The resulting solution
("Carrol") consisted of lithium bromide and ethylene glycol in a weight
ratio of 4.5:1, with water as the refrigerant and 1-nonylamine as an additive for enhanced absorption characteristics. Phenylmethylcarbinol was
later used to replace the 1-nonylamine. The effect of the ethylene glycol in
moving the crystallization line (the solubility curve) away from the region
of cycle operation is shown graphically in the equilibrium chart of figure
20.12.

A schematic diagram of the air-cooled chiller (Reimann and Biermann
1984) is given in figure 20.13. The major operational consequence of air-
cooled operation is the requirement of higher inlet temperature to the
generator. With 95°F (35°C) ambient air and a typical temperature
approach of 15°F (8°C) for air-cooled operation of the condenser and
absorber, these components operate near 120°F (49°C). This sets a design
requirement of 230°F (110°C) entering hot water to drive the generator.
These design conditions call for a COP of 0.75 at 45°F (7°C) chilled water
temperature.

A falling-film generator is used in place of the more commonly used
submerged tube bundle. The falling-film design reduces the temperature
approach by permitting counterflow heat transfer, minimizes superheating
losses associated with nucleate boiling and hydrostatic head, and reduces
cycling losses by eliminating storage of the large quantity of hot solution
that would be required to cover a generator tube bundle.

Prototype machines were made in three sizes. Following extensive laboratory testing on the first 10 kW (3-ton) unit, a 35 kW (10-ton) machine
(shown in figure 20.14) was built as part of the SOLERAS program (see
section 20.4.7.4) and installed in a solar cooling system in Phoenix. In
retrospect, it was probably a mistake to use such an early prototype
machine in a field installation; it suffered from repeated leaks and only
operated a small fraction of the time.

A set of three 10 kW second prototype chillers was built next. Two
of these machines were earmarked for testing at other laboratories: one
at Solar House I at Colorado State University (see section 20.4.6) and the
other at Arizona State University (see section 20.3.2.6). The third 10 kW

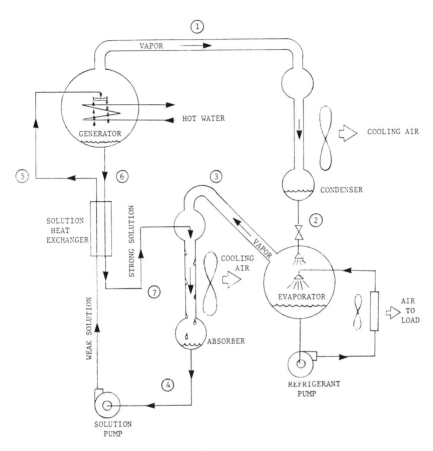

Figure 20.13
Schematic diagram of Carrier air-cooled absorption chiller. Source: Reimann and Biermann (1984).

machine remained at Carrier for testing. The Colorado State University unit worked well; the other two suffered from repeated leaks, as had the previous air-cooled machines.

All these air-cooled chillers essentially achieved design efficiency (COP ≈ 0.75) and were successful in avoiding crystallization when operated as designed. The measured capacities were typically 10% to 20% less than design values. This was originally thought to be a generator problem but then (after generator improvements were made) was identified as an absorber problem.

Figure 20.14
Carrier prototype 10-ton air-cooled absorption chiller.

Finally, a 70 kW (20-ton) machine was fabricated and tested briefly at Carrier at the end of the project. This provided a range of sizes (3-ton to 20-ton) as a potential market line for this new technology.

Overall, the development was quite successful. Air-cooled operation was achieved, design COP was attained, and the design proved scalable to a range of sizes. The below-design capacity values are not a problem for solar-driven operation. This problem would likely be solved with additional research. But even if it weren't, it could be overcome by oversizing the heat exchangers. This would add to chiller cost, but that is a minor part of solar system cost. Attaining design COP is a more important consideration. If the COP value were low, it would require oversizing the collectors to compensate, and the collectors constitute a large part of system cost.

An important lesson learned is the need for more extensive laboratory testing of newly developed equipment before engaging in field tests. The series of leaks that developed in these chillers could have and should have been discovered (and fixed) in the laboratory, not in field installations.

Upon completion of the DOE/Solar-supported development, Carrier opted not to proceed with commercializing the air-cooled chillers because of the lack of market demand for them.

20.3.2.2.2 WATER-COOLED CHILLERS   Carrier developed a series of water-cooled, lithium bromide/water absorption chillers (Biermann and Reimann 1981) for solar-driven applications, based on modifications of existing Carrier designs. The absorber, condenser, evaporator, and solution heat exchanger closely resembled those in Carrier's 16JB product line but with increased heat exchange area to reduce the temperature approach. The generator was a falling-film type, similar to that used in Carrier's air-cooled chiller, described in the previous section.

Chiller design conditions were: COP of 0.74, inlet hot water at 180°F (82°C), inlet cooling water at 85°F (29°C), and leaving chilled water at 45°F (7°C). A range of chiller capacities was designed and built, from 50 kW (15 tons) to 450 kW (130 tons). The first three of these absorption machines were 15-ton units, jointly sponsored by Bonneville Power Administration (BPA) and DOE. Two were installed in the BPA service area, at the Ross Control House in Vancouver, Washington, and at the Big Eddy Control House in The Dalles, Oregon. The third unit was installed at the Carrier factory in Tyler, Texas. All of these machines operated

Figure 20.15
Carrier 130-ton water-cooled absorption chiller (SAM-130) for the Frenchman's Reef,
Virgin Islands, installation. Source: Biermann and Reimann (1981).

satisfactorily, although field operating data information is minimal because of absence of instrumentation. A fourth 15-ton machine was built
and installed at the BDP Distribution Center in Phoenix as part of the
U.S.-Saudi Arabian SOLERAS program (see section 20.4.7.4).

The largest machine of this design was built for retrofit installation at
the Frenchman's Reef Holiday Inn, U.S. Virgin Islands demonstration
project. A picture of this SAM-130 chiller (Solar Absorption Machine—
130 tons) is shown in figure 20.15. This machine was highly successful. It
operated reliably from the minute it was turned on. The hotel owners later
purchased a second similar machine with their own funds to replace the
second (older) chiller at that installation.

The final two water-cooled chillers of this design built using DOE/Solar
support were 75-ton machines. These were constructed for installations in
Houston and Las Vegas, at sites previously earmarked as solar cooling
field installations for Rankine-cycle chillers as part of the earlier NASA/
DOE Development in Support of Demonstration Program. The solar col-

lectors and balance of system had been installed at these sites, but the Rankine chiller development (by AiResearch) was discontinued before the chillers were finished. It was subsequently decided to substitute the two Carrier 75-ton absorption machines to complete these solar cooling system installations. However, negotiations on the retrofit and operational details between DOE and the site owners eventually stalled, and the Carrier chillers were never installed at these sites. Later plans called for the potential use of at least one of these 75-ton machines in one of the Solar in Federal Buildings Program cooling systems.

This solar-driven water-cooled chiller development constitutes an example of successful DOE/Solar support of new technology development leading to a new marketable product by a major manufacturer. These Carrier chillers, although of new design, are manufactured using Carrier normal production line facilities. Carrier announced that they would make to order any customer-requested machines of this design in the 15-ton to 150-ton capacity range.

20.3.2.2.3 WATER-COOLED CHILLER FOR REFRIGERATION PURPOSES Carrier conducted a small experimental and analytical research effort to investigate possible use of the water-cooled chiller for refrigeration applications down to $23°F$ ($-5°C$) refrigerant temperatures (Best, Biermann, and Reimann 1985). Freezing of the refrigerant (water) would be prevented by bleeding absorbent into the evaporator to an absorbent concentration of about 10%. With lithium bromide absorbent, this would prevent freezing to about $20°F$ ($-7°C$); using Carrol as the absorbent would allow even lower evaporator temperatures.

To test the concept, one of the existing 15-ton chillers (the one that had been installed in Tyler, Texas) was used. Test conditions included raising the inlet hot water temperature to $195°F$ ($91°C$), keeping the entering condenser cooling water at $85°F$ ($29°C$). The expected $23°F$ ($-5°C$) evaporator temperature was achieved, and ice formation was observed in the evaporator. However, increased viscosity of the absorbent solution partially clogged the spray nozzles in the absorber, severely limiting machine capacity. This problem, which became even worse using the higher viscosity Carrol solution, could be solved by changing from a spray to a dripper-type absorber configuration. This type of hardware modification, although beyond the scope of the DOE/Solar-supported project, was successfully carried out with corporate funding.

This project served as part of a joint Mexico-U.S. solar refrigeration collaboration, including participation by a researcher from UNAM (the National Autonomous University of Mexico).

20.3.2.3 Lawrence Berkeley Laboratory Project: Chiller Research

The Lawrence Berkeley Laboratory (LBL) performed research on single-effect absorption-cycle machines that could operate in an air-cooled mode. Ammonia/water was chosen as the refrigerant/absorbent fluid pair because this pair allows operation at low evaporator temperatures and the high condenser/absorber temperatures needed for air-cooled operation, without any danger of crystallization.

As a first step, LBL modified an Arkla gas-fired, 5-ton ammonia/water absorption chiller to operate with solar-heated inlet hot water at about 200°F (93°C). This required rebuilding the generator, rectifier, and preheater components. Under air-cooled operation at 110°F (43°C) condenser/absorber temperature and 45°F (7°C) evaporator temperature, the chiller would be operating very close to cutoff conditions and with a small concentration difference between the weak and strong absorbent solutions. The experimental results (Dao et al. 1977) confirmed the stability of operation under these conditions and confirmed the accuracy of the predicted values of operating parameters and coefficient of performance.

LBL next proceeded to design, build, and test an entirely new single-effect, ammonia/water, air-cooled absorption chiller for solar cooling applications. Design operating conditions for the 3-ton machine were 220°F (104°C) hot water inlet to the generator, 112°F (44°C) condenser temperature, 45°F (7°C) leaving chilled water temperature, and 0.7 COP. Special features of the machine included tube-in-tube heat exchanger for high effectiveness at low cost, an internally powered solution pump that recoups the normally wasted mechanical energy in the strong absorbent stream to reduce the parasitic power required, and a three-channel preheater that recoups the sensible heat in the ammonia vapor leaving the generator in addition to the sensible heat in the strong absorbent stream to increase the COP. A photograph of this chiller, before the external connections, instrumentation, and insulation were added, is shown in figure 20.16.

Experimental test results generally confirmed the predicted operating performance of the chiller. As with most absorption machines, the absorber component did not produce its calculated performance. A detailed

Figure 20.16
Lawrence Berkeley Laboratory experimental single-effect, ammonia-water absorption
chiller.

report (Dao, Rasson, and Wahlig 1983) includes the design details for all
the components, the testing results, and a comprehensive computer code
that accurately models the chiller performance.

This single-effect chiller design subsequently served as the lower-
temperature stage of the LBL double-effect regenerative absorption
chiller, described in section 20.3.4.2.

20.3.2.4   University of Maryland Project: Chiller Modeling
The University of Maryland group developed the SHASP (Solar Heating
and Air-conditioning Simulation Programs) computer model (Anand et

al. 1978) and then applied it to steady-state and dynamic simulations of absorption cycle and system performance (Anand, Allen, and Kumar 1981). An important outgrowth of this work was the evolution of a simplified design method for solar cooling system performance based on regional design charts (Anand, Kumar, and Allen 1983). By subdividing the country into four climatic regions, it was possible to overcome previous barriers to the development of an f-chart-like simplified performance model for solar cooling systems.

Another direction taken was the development of a closed-form representation for modeling cooling system performance (Anand, Kumar, and Hall 1982). A summary document on absorption fluid studies (Anand, Allen, and Venkateswaran 1980) was also produced as part of this project. Finally, the computer simulation models developed under this project made possible the subsequent work on second law analysis described in section 20.3.4.3.

20.3.2.5 Brookhaven National Laboratory Project: Chiller Testing

Brookhaven National Laboratory (BNL) constructed a hardware test facility capable of measuring in detail the actual performance of absorption chillers. This test facility was first used to determine the steady-state performance of an Arkla 3-ton, model WF36 absorption chiller. The laboratory test data (Auh 1979b) showed a strong dependence of capacity and COP on the evaporator chilled-water outlet temperature and the condenser cooling-water inlet temperature.

The second phase of this project involved measurements of transient and cycling performance of the same Arkla chiller. These tests showed that, during the ten minutes following shutoff from steady-state operation, significant residual cooling could be obtained from the chiller, equivalent to about four minutes of running at steady-state capacity (Auh 1979a). These results indicated that the losses in average values of capacity and COP, which normally occurred during on/off cycling operation of a chiller, could be almost completely eliminated by continuing to run the chiller for 10 or 15 minutes after the inlet hot water source has been removed. This performance is shown in figure 20.17.

This demonstration of the "spin-down" chiller operating strategy by Paul Auh of BNL, by which realistic on/off operation of an absorption chiller could attain well over 90% of the steady-state capacity and COP values, was an important contribution to the solar cooling program during the late 1970s.

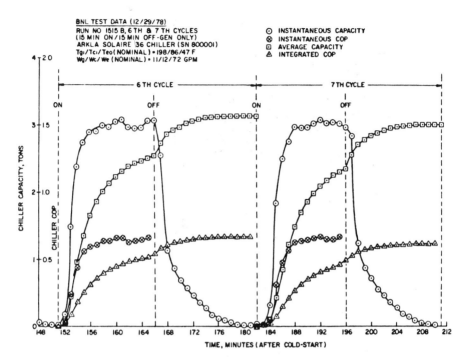

Figure 20.17
Results of dynamic chiller tests at Brookhaven National Library, using an Arkla 3-ton, water-cooled chiller. The capacity and COP curves show how the average chiller performance can be restored almost to its steady-state value by "spin-down" control during 15-minutes-on/15-minutes-off cycling. Source: Auh (1979a).

20.3.2.6   Arizona State University Project: Chiller Testing and Modeling

The work at Arizona State University (ASU) paralleled, to a significant extent, that of BNL, described in the preceding section. ASU constructed a laboratory chiller test facility and used it to test steady-state and dynamic performance of the Arkla 3-ton absorption chiller (Rauch and Wood 1976; Froemming et al. 1979; Guertin and Wood 1980; Guertin, Wood, and McNeill 1981). The experimental test results, such as those shown in figure 20.18, revealed correlations between chiller cycling and the effects on capacity and COP similar to those observed by BNL. The ASU work also included development of dynamic simulation models for comparison with and interpretation of the experimental test results.

ASU subsequently modified its chiller test facility to enable performance testing of the Carrier 3-ton air-cooled absorption chiller, as mentioned in

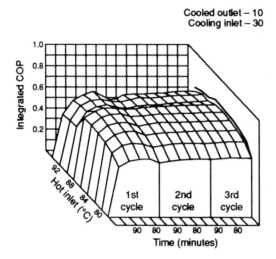

Cooled outlet – 10
Cooling inlet – 30

Figure 20.18
Results of dynamic chiller tests at Arizona State University, using an Arkla 3-ton, water-cooled chiller. This figure shows the effect on integrated COP of time, cycling, and hot water inlet temperature. Source: Guertin and Wood (1980).

section 20.3.2.2.1. The requirement of air cooling necessitated that a room-size enclosure be built to contain the chiller under test. Unfortunately, the chiller was damaged during shipment from Carrier to ASU, and numerous repair attempts never succeeded in producing satisfactory machine performance. The project final report (Borst and Wood 1985) documents the test facility parameters and the extensive efforts to achieve satisfactory machine operation, along with a small amount of test-run results. The report also contains a detailed description of the Carrier chiller and a summary of performance test results on absorption chillers conducted at other test facilities.

20.3.3 Single-Effect Absorption Heat Pump R&D within the DOE Conservation Program
As was the case with the DOE/Solar Absorption Cooling Program, the emphasis in the early years of the DOE/Conservation Absorption Heat Pump Program was on single-effect machines.

20.3.3.1 Arkla Project: Heat Pump Development
Drawing on its extensive experience in the manufacture of gas-driven, ammonia/water absorption chillers, Arkla undertook the development

of a similar machine for heating-only applications (Merrick 1981; Kuhlenschmidt and Merrick 1983). The air-source, gas-fired heat pump was designed to supply output hot water at 125°F (52°C) and sized for an input of 50,000 Btu/h (14.7 kW).

Three prototype machines were built and tested in the course of finalizing the design. COP values of 1.25 at 47°F (8°C) ambient and 1.12 at 17°F (−8°C) ambient were attained. These were close to the initial target COP values of 1.3 and 1.2, respectively. These COP values, and those of other gas-fired machines described in this chapter, include combustion losses unless explicitly mentioned otherwise.

Following the prototype development, six additional machines were built to the final design, three for life testing at Arkla and three for field testing. One of these heat pumps was sent to NBS for laboratory testing of its steady-state and transient performance, as described in section 20.3.3.4.

20.3.3.2   Allied Corporation Project: Residential Heat Pump Development for Heating and Cooling, Using New Organic Fluids

The centerpiece of this project (Murphy 1981; Phillips 1982; Murphy and Phillips 1983; Allied Corporation 1985) is the use of organic working fluids. A promising new fluid pair was identified, consisting of ETFE (ethyltetrahydrofurfurylether) as the absorbent and R133a (chlorotrifluoroethane) as the refrigerant. R133a is suitable for both heating- and cooling-mode operations.

A major impediment to previous use of organic refrigerants—the high solution circulation rates (and therefore high pumping power) required because of the low heat of vaporization—was overcome by the successful development of a reliable, high efficiency, magnetically driven turbine pump.

A second potential disadvantage—relatively low heat and mass transfer rates of organic working fluids compared to commonly used inorganic fluids—was overcome by use of extended heat and mass-transfer areas made of formed aluminum. Compatibility of these organic fluids with aluminum is a definite advantage in reducing machine fabrication costs. Using test units constructed essentially 100% of aluminum, the organic working fluids have proven stable over operating temperatures between 0°F and 390°F (−18°C and 200°C).

The schematic diagram for the heat pump is shown in figure 20.19. The mode valve allows switching between heating and cooling modes of opera-

Absorption Heat Pump Circuit

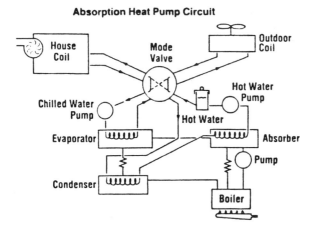

Figure 20.19
Schematic diagram of the Allied Corporation-Phillips Engineering gas-fired single-effect
absorption heat pump. Source: Murphy (1981).

tion. Prototype 72,000 Btu/h (21 kW) units have essentially achieved target performance of a heating COP of 1.25, with 36°F (2°C) evaporator
temperature and 130°F (54°C) delivered hot water temperature to the load
coil. The cooling COP value was measured to be 0.5, slightly less than the
0.6 target value.

   Of the six prototype units constructed initially, two were field-installed
in residences in the St. Joseph, Michigan, area for two heating and cooling
seasons. The overall field-test experience was quite successful, with the
units providing 100% of winter heating and summer cooling requirements.
One of these installations is shown in figure 20.20. One unit was sent to
NBS for laboratory testing of its steady-state and transient performance,
as described in section 20.3.3.4. In all, six field-test units were operated
outdoors for two years.

   Major subcontractors for this project were Phillips Engineering (which
built and tested the heat pumps), Arthur D. Little, and United Technologies Research Center. Funding was supplied partially by DOE and
partially by the Gas Research Institute (GRI).

   After much of the development had been completed, R133a was discovered (unrelated to this project) to be possibly very slightly toxic. To
avoid a negative impact on this project, nontoxic R123a (dichlorotri-

Figure 20.20
Field-test installation of the Allied-Phillips gas-fired absorption heat pump for both heating and cooling at a residence in St. Joseph, Michigan. Source: Murphy (1981).

fluoroethane) was substituted for the R133a refrigerant, without substantially affecting heat pump performance.

20.3.3.3 Battelle Columbus Laboratories Project: Development of Compound Absorption Chiller for Waste Heat Recovery

The Compound Absorption Chiller (CAC) uses a proprietary cycle to produce $45°F$ ($7°C$) chilled water for cooling applications, while being driven by a $140°F$ ($60°C$) waste (or solar or geothermal or other) heat source (Hanna and Wilkinson 1981). The design COP value is 0.42 at $85°F$ ($29°C$) condenser temperature. A proof-of-concept CAC unit was fabricated and tested using $LiBr/H_2O$ as the fluid pair, although the CAC cycle can be used with other working fluids. This 35-ton unit attained a 0.35 COP value.

The project also included the identification and testing of alternative advanced desorber designs. Three candidate configurations were built and tested. The best of these, a counterflow design, exceeded test goals for heat and mass transfer.

20.3.3.4   National Bureau of Standards Project: Heat Pump Testing

The National Bureau of Standards (NBS) has conducted independent measurements of the performance of gas-driven absorption heat pumps developed under DOE-GRI sponsorship (McLinden, Radermacher, and Didion 1983; Radermacher 1984). Controlled performance measurements in the NBS laboratory were made on the Arkla and the Allied/Phillips machines described in sections 20.3.3.1 and 20.3.3.2. The testing program included determination of transient startup, shutdown, and cyclic modes, in addition to steady-state operation. Typical results for the Arkla heat pump at part load under cycling conditions are shown in figure 20.21.

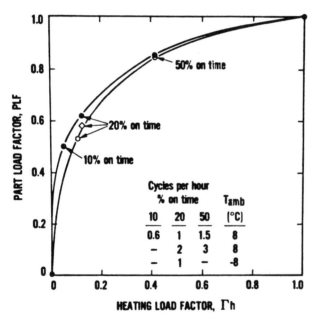

Figure 20.21
Measurements at National Bureau of Standards of part-load performance of the Arkla absorption heat pump under various cycling conditions. The part-load factor is the ratio of the cycling COP to the steady-state COP. The heating load factor is the ratio of the cycling capacity to the steady-state capacity. Source: Radermacher (1984).

NBS also worked on the development of procedures for testing and rating absorption heat pumps for heating- and cooling-mode operation (Weber, Radermacher, and Didion 1984).

20.3.4 Double-Effect Absorption Chiller R&D within the DOE Solar Cooling Program

Although the development of single-effect absorption chillers for solar cooling applications was quite successful technically (as evidenced by the projects described in sections 20.3.2.1, 20.3.2.2, and 20.3.2.3), it became evident that the economics could not be easily satisfied. The main problem has been the cost of the solar collector array to drive the chiller; the development of collectors (that can produce the 180° to 210°F [82° to 100°C] output fluid temperatures required to drive the single-effect absorption chillers) at a cost that can compete with conventional cooling systems has proven to be an elusive target. Moreover, the preferred backup operating mode is to drive the chiller with natural gas when solar energy is not available, to avoid the need for also purchasing an electric-driven chiller as backup. But, in most circumstances, the efficiency of a gas-driven single-effect absorption chiller is too low to compete economically with electric-driven machines. For these reasons, the solar cooling program began moving in the late 1970s toward higher efficiency, multiple-effect absorption cooling technology.

20.3.4.1 University of Texas Project: Chiller Modeling

The University of Texas at Austin (UTA) had developed a computer model capable of detailed simulation of a double-effect, lithium bromide/water absorption chiller. With DOE/Solar support, UTA modified the simulation model for solar-driven hot water operation, and then used the model to perform sensitivity analysis of the effect on chiller performance of various changes in design and operating parameters (Vliet and Kim 1983).

The schematic diagram of the double-effect chiller that was modeled is given in figure 20.22. It contains seven heat exchangers: a condenser, evaporator, absorber, two generators, and two solution heat exchangers. The major issue addressed was how best to allocate the heat exchanger area among the seven heat exchangers, for any fixed value of overall total chiller heat exchange area. The most difficult part of the investigation was identifying what parameter to optimize in order to determine the "best" relative areas of the seven heat exchangers.

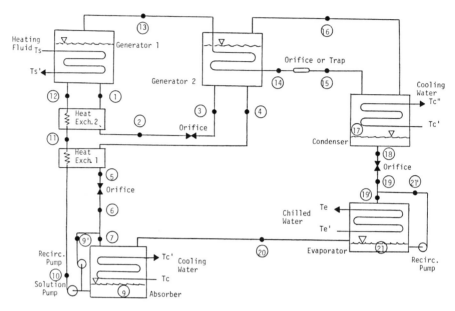

Figure 20.22
Schematic diagram of double-effect LiBr/H_2O absorption chiller modeled and analyzed by
the University of Texas. Source: Vliet and Kim (1983).

Initially, it was decided to optimize the chiller COP, but this resulted in
a drastic reduction in chiller capacity. Similarly, basing the optimization
on maximum capacity produced unsatisfactory results. Finally, it was decided to optimize the present value of the chiller (including initial cost,
depreciation, and fuel costs over the life of the chiller) per unit of capacity.
This proved quite satisfactory. Compared to initial heat exchange areas
typical of current practice, the optimized reallocation of component relative heat-exchanger areas produced a 25% increase in COP (from 1.21
to 1.52) and a 16% decrease in present value per unit capacity. Unex-
pectedly, this optimized allocation of heat-exchanger area assigned a much
larger than anticipated (by a factor of 7) area to the two solution heat
exchangers.

The UTA study also explored the effects on calculated chiller performance of different control strategies and a range of source, cooling, and
chilled water temperatures.

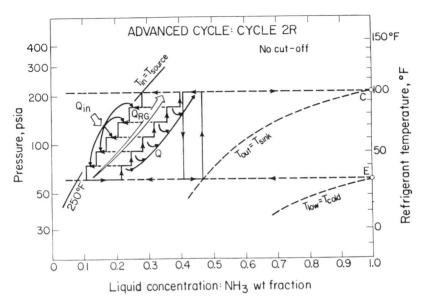

Figure 20.23
P-X-T diagram of the Lawrence Berkeley Laboratory double-effect regenerative (2R) absorption cycle, using ammonia/water as the fluid pair.

20.3.4.2 Lawrence Berkeley Laboratory Project: Chiller Research

LBL generated the concept of regenerative absorption cycles and worked out preliminary designs for a double-effect regenerative (2R) cycle (Dao 1978b) and a single-effect regenerative (1R) cycle (Dao 1978a). Both of these cycles have the characteristic of a COP value that increases with inlet temperature, approximating a constant percentage of the theoretical maximum Carnot cycle efficiency. The 2R cycle is predicted to operate at about 60% of Carnot efficiency, and the 1R cycle at about 70% of Carnot. In addition, these cycles eliminate the sharp cutoff minimum inlet temperature characteristic of normal single- and double-effect absorption cycles. Estimated cooling COP values for 280°F (138°C) inlet hot water temperature, 95°F (35°C) absorber/condenser temperature, and 40°F (4°C) evaporator temperature were 1.25 for the 2R cycle and 1.55 for the 1R cycle.

   LBL decided to design, build, and test a 2R cycle chiller first, because it represented a smaller departure from conventional absorption technology than did the 1R cycle. Figure 20.23 shows a pressure-concentration-temperature diagram for the 2R cycle. Ammonia/water was selected as the fluid pair because it allowed heating- as well as cooling-mode operation and permitted air-cooling as well as water-cooling of the absorber and condenser.

Comparing the P-X-T diagram of figure 20.23 with those in figures 20.3 and 20.6, it is seen that the lower temperature stage of the 2R cycle is a conventional single-effect cycle. However, the high temperature stage is a constant-inlet-temperature, multiple-pressure-step subcycle.

A photograph of the 2R chiller being assembled for testing in the LBL laboratory test facility is shown in figure 20.24. Test results successfully proved the regenerative absorption cycle concept. The 2R chiller achieved stable operation over a range of operating conditions and essentially at-

Figure 20.24
Lawrence Berkeley Laboratory experimental double-effect regenerative absorption chiller.

tained predicted COP values. Publication of design and test results is planned for 1986 (Rasson, Dao, and Wahlig 1986).

20.3.4.3   TPI Project: Second Law Analysis of Chiller Performance

Using the second law of thermodynamics, TPI, Inc. calculated the sources of irreversibility associated with absorption cycles and their components (Anand et al. 1984). This technique is valuable in establishing the theoretically maximum possible cycle COP and in identifying where improvements might be made in a real cycle to increase its efficiency.

For the single-effect lithium bromide/water cycle, TPI calculated the cycle COP to be 0.83 under ideal assumptions and 0.72 using more realistic assumptions for heat and mass transfer. The calculated COP for the double-effect lithium bromide/water absorption cycle was 1.52. In studying sensitivities of cycle COP to the effectiveness of all the components, TPI determined that the biggest improvements in COP were obtained by improvements in the effectiveness of the solution and evaporator heat exchangers. Finally, TPI applied second-law analysis to complete solar cooling systems in two case studies. The energy and exergy flows throughout the system were calculated, identifying which components were the major sources of irreversibility and causes of decreased system performance.

20.3.5   Double-Effect Absorption Heat Pump R&D within the DOE Conservation and GRI Programs

Gas-driven absorption heat-pump development within the DOE/Conservation Program paralleled that of solar-driven absorption chiller development within the DOE/Solar Program. Single-effect gas-driven heat pumps were developed first, successfully in a technical sense. But the economics proved unfavorable: the efficiency of a gas-driven single-effect heat pump is too low to compete with a high efficiency gas furnace in the heating mode or with a good electric-driven air conditioner in the cooling mode. Accordingly, DOE/Conservation shifted program emphasis to double-effect cycles, issuing a solicitation that resulted in three new projects starting in late 1982 by Carrier, Trane, and Phillips Engineering. Meanwhile, GRI continued support of an ongoing double-effect project by Columbia Gas. As mentioned previously, the COPs of these gas-driven machines include the combustion losses unless explicitly stated otherwise. It should also be mentioned that much of this advanced-cycle work is proprietary.

20.3.5.1   Columbia Gas Systems Service Corporation Project: Advanced-Cycle Heat Pump

Columbia Gas initiated the development of a high efficiency, residential-sized absorption heat pump for heating and cooling, and then continued the development with partial support from GRI (Reid 1981; Marat 1984). The gas-fired, air-to-air unit uses a proprietary working fluid. The performance targets established for this proprietary cycle design are a heating COP of 1.5 and a cooling COP of 0.8, at normal rating conditions, including parasitic electric power consumption.

A first, nongas-fired breadboard unit met the performance goals, achieving a heating COP of 1.55 and a cooling COP of 0.80. This was followed by the construction and testing of an advanced, gas-fired breadboard unit using near-commercial components. This heat pump achieved stable operation at heating COP of 1.54 and cooling COP of 0.82. In addition, acceptable corrosion rates were determined for candidate construction materials, and a 5000-hour lifetime test was run successfully for the solution pump. For the next phase, a packaged prototype heat pump was designed and built, and will be tested and evaluated.

20.3.5.2   Carrier Project: New Cycle Analysis and Heat Pump Development

As the first step in development of a new, high-efficiency gas-fired absorption heat pump for heating and cooling (Reimann 1984; Biermann and Reimann 1984), an extensive screening of candidate cycle configurations and fluid pairs was conducted. This selection process led to the choice of the novel "dual-loop" cycle shown in figure 20.25. This design incorporates heat exchange but not fluid exchange between two independent single-stage machines; thus different absorbent-refrigerant fluid pairs can be used in each loop.

Lithium bromide/water was selected as the fluid pair for the high temperature loop. For the low temperature loop, a 60% lithium bromide/water solution was chosen as the absorbent, with methylamine as the refrigerant. Calculated heating mode performance gave an expected COP, including combustion, of 1.7 for a high-side generator solution temperature of 295°F (146°C), ambient of 17°F (-8°C), and 90°F (32°C) supply air to the load. The predicted cooling mode COP, including combustion, was 1.2 for a generator solution temperature of 280°F (138°C), 85°F (29°C) cooling tower water, and 45°F (7°C) chilled water to the load.

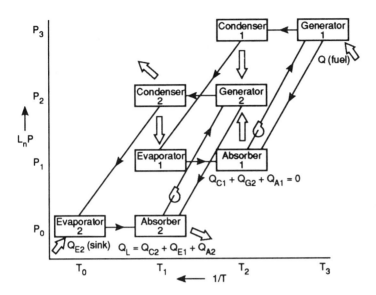

Figure 20.25
The dual-loop cycle selected for the Carrier advanced absorption heat pump. Source:
Reimann (1984).

The physical layout of the initial design for the dual loop heat pump,
consisting of five basic shells, is shown in figure 20.26. This is a water-to-
water design, sized for 17 tons (58.5 kW) heating at the 17°F (−8°C)
ambient air rating point. It is planned to perform fluid stability tests and
heat exchange design testing, possibly involving the construction and testing of an experimental single-stage machine to test the concepts and then
proceed to build and test the dual-loop machine.

20.3.5.3 Phillips Engineering Project: New Cycle Analysis and Heat Pump Development

The objective of this project was the design and development of a gas-
fired, air-cooled, high-efficiency absorption heat pump for heating and
cooling in residential and small commercial applications (Phillips 1984).
The first phase concentrated on the identification and detailed analysis
and comparison of candidate cycle configurations. An initial screening of
candidate fluid pairs with sufficiently known properties to allow the cycle
calculations and having properties compatible with the 400°F (204°C)
driving temperature and evaporator temperatures down to −10° or −20°F
(−23° or −29°C) identified water/ammonia as the pair of choice.

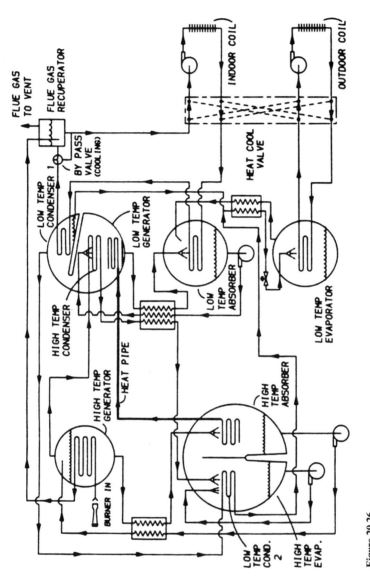

Figure 20.26
Schematic diagram of the Carrier dual-loop absorption heat pump. Source: Reimann (1984).

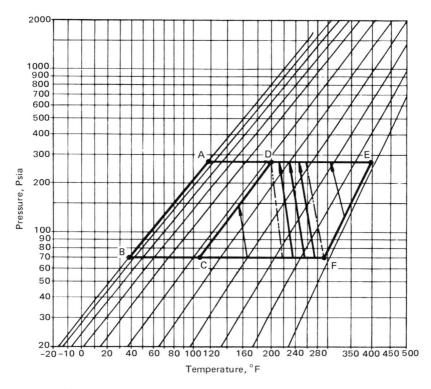

Figure 20.27
P-X-T diagram of the Phillips Engineering generator-absorber heat exchange (GAX) cycle, with water/ammonia as the fluid pair. Source: Phillips (1984).

Six candidate high-efficiency absorption cycles were identified for evaluation: double-effect cycle, resorber-augmented cycle, generator-absorber heat exchange (GAX) cycle, two-stage GAX cycle, double-effect regenerative (2R) cycle, and variable-effect cycle. Although high cooling and heating COP values were the primary consideration, other evaluation criteria included parasitic pumping power, total heat exchanger area, operating temperature range, complexity, and probable cost.

The GAX cycle emerged as the preferred choice for the second phase development. The pressure-concentration-temperature diagram for this cycle is shown in figure 20.27. Calculated heating and cooling COP values and their dependence on ambient temperature are given in figure 20.28.

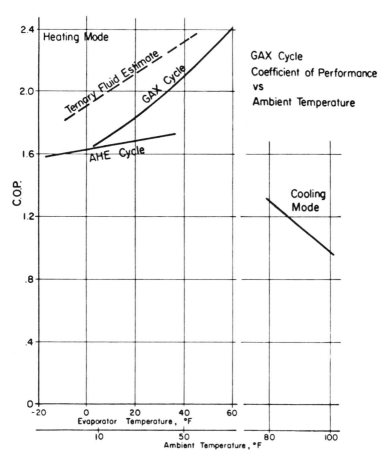

Figure 20.28
Calculated heating-mode and cooling-mode COP values (solid lines) for the GAX cycle
using water/ammonia as a function of the ambient temperature. The heating-mode
performance of the absorber heat-exchange (AHE) cycle is shown for comparison. Also
shown is the estimated heating-mode performance improvement for the GAX cycle through
use of the lithium bromide/water/ammonia ternary fluid combination. Source: Phillips
(1984).

Replacing water (as the absorbent) with a 60% lithium bromide/water solution should allow higher input boiler temperatures and higher COP values. A rough estimate of the heating COP gain by using this ternary fluid combination is indicated in figure 20.28.

In the second phase of this project, a 3-ton, GAX-cycle breadboard test unit, using water/ammonia as the fluid pair, has been constructed and is being tested. Preliminary test results produced COP values very close to the predicted values of 2.0 for heating and 1.0 for cooling, not including flue losses.

20.3.5.4   Trane Project: New Cycle Analysis and Heat Pump Development

Trane chose to concentrate on the study and development of high-efficiency, air-cooled, absorption heat pump systems for heating and cooling for commercial applications (Hayes and Modahl 1984). In the first phase of the project, five efficiency-enhancing types of cycle modifications were considered: double effect, separate generator and absorber heat exchange, direct generator-absorber heat exchange, regeneration, and resorption-desorption. These are not exclusive; many configurations are possible that combine many of these features.

The initial set of candidate fluids was limited to four binary mixtures whose properties were reasonably well known and acceptable. Subsequently, three ternary fluid mixtures were added to the list. The 24 cycle/fluid combinations shown in table 20.1 were selected for detailed computer modeling, which produced cycle operating parameters including COP, parasitic pump power, and amount of internal heat exchange. A ranking scheme was established and applied, resulting in selection of five systems for economic analysis. Three of these systems appeared acceptable economically and thus met all of the criteria; these are combinations ABS4, ABS6B, and ABS20 in table 20.1.

The preferred absorption cycle contained four of the previously-mentioned five efficiency-enhancement features (all except double effect). The preferred fluid was the ternary mixture lithium bromide/water/ammonia, with the binary mixture water/ammonia close behind. One of these top three combinations has to be selected for the second-phase breadboard unit fabrication and testing.

Table 20.1
Advanced cycle-fluid mixture combinations selected by Trane

Cycle/fluid designation	Fluid	Cycle
ABS1	NH_3/H_2O	Absorber heat exchanger/generator heat exchange
ABS2	R22/DMETEG	"
ABS3	R123/ETFE	"
ABS4	NH_3/H_2O	Absorber/generator heat exchange, Absorber heat exchange/ Generator heat exchange
ABS5	R22/DMETEG	"
ABS6A	R22/DMETEG	Regeneration, abs./gen. heat exchanger Abs. heat exch./gen. heat exch.
ABS6B	NH_3/H_2O	"
ABS7A	R22/DMETEG	Two-stage absorption, partial flash regeneration, resorp./desorp. cycle Abs/gen. heat exch., abs. heat exch./gen. heat exch.
ABS7B	NH_3/H_2O	"
ABS8	$LiBr/H_2O$	Double-effect generator, cascaded evap.
ABS9	NH_3/H_2O	Use of resorp./desorp. cycle to achieve double-effect evaporation
ABS10A	R123A/ETFE	Double-effect generator with resorp./desorp. cycle
ABS10B	R123A/ETFE	"
ABS11	NH_3/H_2O	Double-effect generator, common condenser, abs. heat exch., gen. heat exch.
ABS12	$NH_3/H_2O/LiBr$	Abs./gen. heat exch., abs. heat exch./ gen. heat exch.
ABS13	$NH_3/H_2O/LiBr$	Same as ABS12 plus absorbent recirculation
ABS14	$NH_3/H_2O/LiBr$	Same as ABS13 plus regeneration
ABS15	$CH_3NH_2/H_2O/LiBr$	Same as ABS12
ABS16	"	Same as ABS13
ABS17	"	Same as ABS14
ABS18	$NH_3/H_2O/LiBr$	Same as ABS12 except resorp./desorp. cycle substituted for conventional evap./cond.
ABS19	"	Same as ABS13 except resorp./desorp. cycle substituted for conventional evap./cond.
ABS20	"	Same as ABS14 except resorp./desorp. cycle substituted for conventional evap./cond.
ABS21	$CH_3OH/LiBr/ZnBr$	Abs. heat exch./gen. heat exch.

Source: Hayes and Modahl (1984).

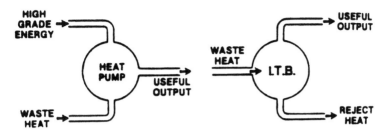

Figure 20.29
Simplified schematic illustrating the difference between a conventional heat pump and a temperature-booster (ITB). Source: Hanna, Lane, and Whitney (1983).

20.3.6   Temperature-Boosting Absorption Heat Pump R&D within the DOE Conservation Program

An alternative application of absorption cycles is for the upgrading of heat sources to produce an output temperature higher than that of the driving source. Typically, this technique would be used to boost the temperature of a waste heat stream to a temperature high enough to be useful for driving an industrial process.

The difference between a temperature booster (also called a heat transformer) and a conventional heat pump is illustrated (Hanna, Lane, and Whitney 1983) in a simple way in figure 20.29. Thermodynamically, this can be seen by comparing the pressure-concentration-temperature diagram for a temperature booster (figure 20.30) with that of an ordinary heat pump (figure 20.3). The two obvious differences are that for the temperature-booster: (1) the input heat source is at the intermediate temperature, with the output at the high temperature, and (2) the generator and condenser are at the lower pressure. The second difference requires that a second pump be used, to pump the refrigerant from the condenser to the evaporator.

Multiple stages may be used to increase the temperature boost or to increase the COP, with these two parameters being traded off against each other, depending on the application. However, since the input heat source is used to power the cycle as well as to provide the source of the heat boosted, the COP must always be less than 1.

20.3.6.1   Oak Ridge National Laboratory Project: Single-Effect and Multiple-Effect Cycles for Temperature Boosting

ORNL explored the potential of using an absorption cycle for recovering waste heat by boosting its temperature. The conceptual design and perfor-

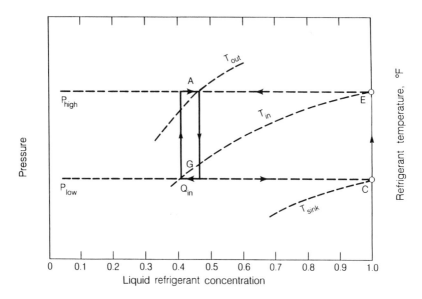

Figure 20.30
P-X-T diagram of a single-effect absorption cycle for temperature boosting (heat transformer).

mance analysis phase looked promising enough (Grossman and Perez-Blanco 1982) to proceed to experimental verification. A lithium bromide/water absorption heat pump for this purpose was fabricated by Arkla to ORNL's specifications. This machine represented a modification of the Arkla 25-ton, single-effect, water-cooled chiller described in section 20.3.2.1.1.

The schematic of the temperature-boost unit as built by Arkla is shown in figure 20.31. It was designed to operate over a range of inlet waste-heat temperatures from 140° to 180°F (60° to 82°C), with condenser cooling water temperatures from 59° to 95°F (15° to 35°C). The expected temperature boost ranged from 30° to 60°F (17° to 33°C), with a COP value between 0.4 and 0.5.

The unit was tested extensively at ORNL (Huntley 1983). Measured performance data agreed well with theoretical values predicted by a detailed computer simulation model of the machine (Grossman and Childs 1983). Parasitic electric power requirements were very low, corresponding to electrical COP values between 50 and 70.

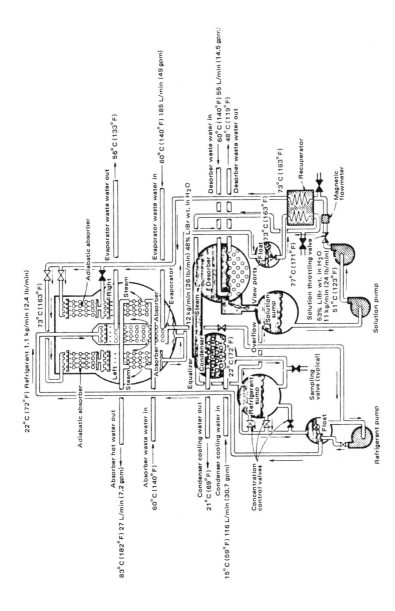

Figure 20.31
Schematic of single-effect absorption heat pump for temperature boosting, built by Arkla to Oak Ridge National Laboratory's specifications. Source: Huntley (1983).

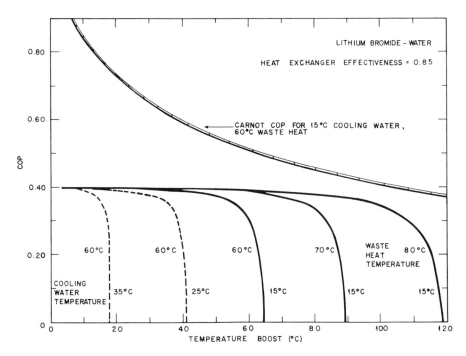

Figure 20.32
Predicted performance characteristics—COP and temperature boost—for a two-stage
system for different heat source and sink temperatures, for the Oak Ridge National
Laboratory design. Source: Grossman and Perez-Blanco (1982).

The ORNL group also extensively explored multiple-stage options for
temperature boosting. Figure 20.32 shows predicted performance characteristics for a two-stage heat pump (Grossman and Perez-Blanco 1982).
Later work (Grossman 1985a) investigated a multitude of multi-stage configurations in a systematic approach to identify systems with improved
COP, greater temperature boost, and combined cooling/heating effects.

20.3.6.2 Battelle Columbus Laboratory Project: Heat Pump for
Temperature Boosting
Battelle developed and tested an industrial temperature booster (ITB), a
waste-heat-powered heat pump for industrial applications (Hanna, Lane,
and Whitney 1983). Formerly called the Battelle temperature booster, this
is a single-stage, lithium bromide/water design, as shown in figure 20.33.
This work has been entirely privately funded.

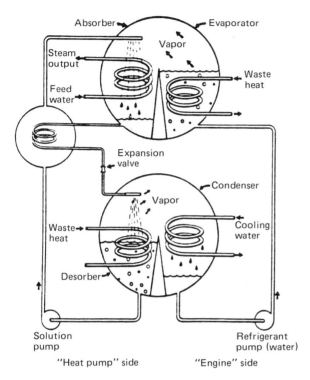

Figure 20.33
Schematic of single-effect industrial temperature booster developed by Battelle Columbus Laboratories and licensed to Adolph Coors Company. Source: Hanna, Lane, and Whitney (1983).

A 180-ton absorption chiller was modified to fabricate a test unit with a capacity of two million Btu/h (586 kW). At an input waste heat temperature of 235°F (113°C) and 80°F (27°C) cooling water temperature, the unit delivered 346°F (174°C) output water at a COP value between 0.4 and 0.5. The measured performance was in good agreement with that predicted by a computer simulation model.

Following this successful development, the technology was licensed by Battelle to the Adolph Coors Company, so that Coors could proceed to design, fabricate, and market these machines.

20.3.6.3 Another Temperature-Boost Heat Pump Project
Mitsubishi Electric Corporation in Japan has recently developed and tested a two-stage lithium bromide/water absorption heat pump for large

temperature-difference boosting (Ikeuchi et al. 1985). A prototype 100 kW unit achieved its predicted performance. With a source temperature of 192°F (89°C) and cooling water temperature of 73°F (23°C), a boost of 112°F (62°C) to an output temperature of 304°F (151°C) was measured. The coefficient of performance was about 0.3.

20.3.7 Chemical Heat Pump R&D within the DOE Storage Program

A chemical heat pump is a device or a cycle that uses reversible chemical reactions to achieve heating or cooling or temperature boosting. Thus all absorption heat pumps are a subset of chemical heat pumps. In practice, the chemical heat pump R&D supported by the DOE/Storage Program has emphasized different features than have the absorption heating and cooling programs supported by DOE/Solar and DOE/Conservation. The central feature for chemical heat pumps has been internal energy storage in the form of the chemical potential energy of stored, separated chemical compounds. Other features in some of the projects include intermittent (rather than continuous) operation and use of solid (rather than liquid) absorbents.

20.3.7.1 EIC Laboratories Project: Chemical Heat Pump Development for Solar Heating and Cooling

The design criteria for this project were that the chemical heat pump be capable of providing both heating and cooling, have sufficient storage for 24-hour-a-day operation, be able to operate at 45°F (7°C) evaporator temperature with air cooling of the condenser/absorber in the cooling mode, and be able to deliver hot air above 100°F (38°C) in the heating mode when outdoor temperatures were below freezing. Calcium chloride/methanol was selected as the absorbent-refrigerant pair to meet these criteria (Offenhartz 1981). (EIC had done earlier development work on a calcium chloride/water chemical heat pump for solar cooling-only applications [Offenhartz 1978].)

A generalized schematic of the cycle is shown in figure 20.34. Since calcium chloride is a solid absorbent, it cannot be circulated, so the cycle works in a batch fashion, with solar energy being used to drive off refrigerant from one salt bed, then from the other.

Cycle analysis predicted a COP of 0.6 in the cooling mode and 1.6 in the heating mode. A 1/3-scale test unit was constructed, with 100,000 Btu (29 kWh) storage capacity, to test the design. The measured COP values of

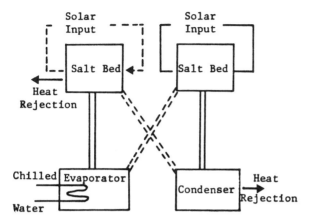

Figure 20.34
Generalized scheme for a chemical heat pump design by EIC based on solid-gas reactions.
A solid absorbent like calcium chloride is used with a refrigerant such as methanol or
water. The functions of the two salt beds (absorber and generator) are periodically
switched. Source: Offenhartz (1978).

0.52 for cooling and 1.5 for heating came acceptably close to the design
values for a first prototype unit.

20.3.7.2 Rocket Research Company Project: Sulfuric Acid/Water
Chemical Heat Pump

A chemical heat pump using sulfuric acid and water as the absorbent-
refrigerant pair was designed for space-conditioning and waste heat pumping applications, including the option of internal energy storage (Clark
1981). A simplified schematic diagram is shown in figure 20.35.

A test unit with nominal output of 150,000 Btu/h (44 kW) and 1 million
Btu (293 kWh) storage was constructed to verify the design. An important concern was the long-term integrity of the materials used to contain
the sulfuric acid solution; namely, porcelain-lined steel shells and teflon-
lined plumbing and valves.

20.3.8 Open-Cycle Absorption Chiller R&D

As discussed in section 20.2.2, the chiller hardware for open cycles represents a subset of that used for closed cycles. As was shown in figure 20.7,
the open-cycle system requires an absorber, an evaporator, and a solution
heat exchanger but eliminates the need for a condenser or a conventional-
type generator. Accordingly, there has been little independent develop-

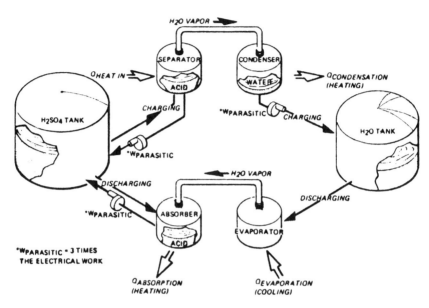

CHP/CES SYSTEM SCHEMATIC

Figure 20.35
System schematic of the Rocket Research sulfuric acid/water chemical heat pump, chemical
energy storage verification test unit. Source: Clark (1981).

ment of open-cycle chillers (i.e., absorber/evaporator units) other than the
modification of chiller components originally developed for closed-cycle
absorption systems.

20.3.8.1   Colorado State University Project: An Evaporator-Absorber
Unit for Open-Cycle Absorption Cooling

The diagram used in figure 20.7 is taken from a schematic of the Colorado
State University (CSU) open-cycle absorption system. The collector-
regenerator component and system performance will be discussed in following sections; the evaporator-absorber subsystem is treated here.

   Preliminary testing at CSU made use of evaporator-absorber components removed from an early model Arkla absorption chiller. More extensive later experimentation and analysis were based on use of a modified
Arkla WF-36 absorption chiller (Lenz et al. 1984). Lithium chloride was
used as the absorbent fluid. Detailed analytical calculations of heat and

mass transfer in the absorber were compared with the experimental results. The limiting rate process in the absorber was identified as the surface resistance to the transfer of the absorbed water molecules through the surface film of lithium chloride solution. Recommendations were formulated for overcoming this limitation.

It had been determined in earlier experimental testing (Lenz et al. 1982) that the problem of buildup of noncondensible gases in the evaporator-absorber unit could be solved in a practical manner. Maintenance of a steady vacuum in the unit required only 15 watts of vacuum pumping power.

For the next phase (Löf and Lenz 1985), CSU obtained a modified Carrier 15-ton absorption chiller, one of the units described in section 20.3.2.2.2. The generator and condenser were removed, and approximately 80% of the heat transfer surfaces in the absorber and evaporator were disabled by plugging the tubes. This resulted in a 3-ton evaporator-absorber unit, matched in capacity to the rest of the CSU open-cycle system. Initial tests used lithium chloride solution as the absorbent, but then lithium bromide was selected for final use because of its superior properties in this application. Preliminary tests showed satisfactory performance of the unit, at slightly reduced capacity. Addition to the solution of a heat transfer additive, as is commonly used in closed-cycle absorption chillers, was expected to increase the capacity to the 3-ton design level. However, the volatility of effective additives was expected to limit, and perhaps to prevent, their use in an open cycle.

20.3.9   Heat and Mass Transfer in the Absorption Process

The ability to predict accurately the performance of absorption chillers and heat pumps and to execute efficient designs depends upon the degree of understanding of the heat and mass transfer mechanisms in the various components. In general, the condensing and evaporating processes are well understood. The boiling processes that take place in the generator can usually be adequately modeled. However, the heat and mass transfer processes that occur in the absorption process, i.e., in the absorber component, are still inadequately described. As a result, the measured performance (COP and capacity) of many of the chillers and heat pumps developed under DOE sponsorship has been less than anticipated. In practice, this deficiency has often been overcome by using an empirical experience factor to simply oversize the absorber component.

During many of the projects described in this chapter, extensive modeling, testing, or design of absorber components have been performed; e.g., see descriptions in Dao, Rasson, and Wahlig (1983), Reimann and Biermann (1984), and Lenz et al. (1984). In addition, an extensive theoretical analysis effort has been launched by Grossman at ORNL (Grossman 1983; Grossman and Heath 1984) and the Technion (Grossman 1985b) to gain the necessary fundamental comprehension of the absorption process.

20.3.10 Commercially Available Absorption Chillers in the United States

20.3.10.1 U.S.-Made Machines

At one time, the small-capacity, gas-driven, air-cooled water/ammonia chiller was a large-volume product, with hundreds of thousands of units built and sold by Whirlpool, Arkla, and Carrier (Bryant Division). Current availability, however, is limited to the chillers periodically manufactured by Arkla when accumulated orders justify a production run.

The three largest U.S. manufacturers of air conditioning equipment—Carrier, Trane, and York—have long manufactured large tonnage machines for absorption cooling applications and continue to market these machines (note, however, the recent announcement by Carrier mentioned below). With the exception noted below, these are all single-effect, lithium bromide/water machines. The nominal capacities range from 100 to 1600 tons, at rating conditions of 85°F (29°C) entering condenser cooling water and 44°F (7°C) leaving chilled water. The machines may be driven by steam or hot water, with typical hot water temperatures of 240° to 270°F (116° to 132°C) required for capacity operation.

With an eye toward solar-driven cooling applications, Trane and York literature has pointed out that these chillers may be driven with hot water temperatures as low as 170° to 180°F (77° to 82°C) with accompanying derating of the machine capacity. For example, a Trane chiller operating at 170°F (77°C) will produce about 20% of its nominal capacity. A number of these machines, with varying degrees of derating, have been used for large solar cooling system installations.

The exception mentioned above is the Trane double-effect absorption chiller, which has been manufactured by that company for many years. It is available in nominal capacities from 385 to 1060 tons. Typical hot water driving temperature is 370°F (188°C). It achieves a COP of approximately 1.0, compared to a COP of about 0.7 for the single-effect machines.

For about 10 years, from the early 1970s to the early 1980s, Arkla was the major supplier of absorption machines for solar cooling applications. More than one hundred Arkla 3-ton and 25-ton water-cooled absorption chillers were installed, many of them in government-supported demonstration projects. However, as mentioned in section 20.3.2.2.1, these machines are no longer made, and the United States is without a domestic manufacturer of small-capacity absorption chillers.

As described in section 20.3.2.2.2, Carrier used DOE support to develop a line of medium- to large-capacity (15 to 150 ton) water-cooled absorption chillers with enhanced heat transfer area and other modifications so that the machines could be used for solar applications without need for derating. Although not actively marketed, Carrier made it known that it would build these chillers to order, using regular production facilities. More recently, however, Carrier has publicly announced the termination of all its U.S. manufacturing of absorption machinery.

Thus, none of the absorption chillers or heat pumps described in sections 20.3.2 through 20.3.8 has led to a currently marketed product. The present near-term outlook for relative costs of natural gas and solar collectors indicates that market demand for solar-driven absorption machines will remain essentially nonexistent in the near term. The DOE-supported advanced absorption heat pumps by Carrier, Phillips Engineering, and Trane, described in section 20.3.5, appear to have a reasonable chance of leading to commercializable products for space heating and cooling and for temperature-boosting in the near future.

20.3.10.2 Japanese-Made Machines

Yazaki, Hitachi, and Sanyo are all manufacturing absorption machines for solar cooling applications and marketing them in the United States.

Yazaki sells both single-effect and double-effect lithium bromide/water absorption machines. The single-effect chiller is available in residential and commercial sizes. The residential-sized chiller models have capacities of 1.3, 2, 3, and 5 tons. At rating conditions of 190°F (88°C) inlet hot water temperature, 85°F (29°C) cooling water to the condenser/absorber, and 48°F (9°C) leaving chilled water temperature, the COP value is 0.60. The commercial-size units are available in 7.5-ton and 10-ton modules, which can be joined in combinations to provide incremental capacities up to 50 tons. At the same rating conditions, these units have COP values of 0.70.

Yazaki has two types of double-effect machine. The plain gas-fired double-effect chiller operates with a COP of 0.95, including burner efficiency. The basic modules with capacities of 7.5, 10, 20, and 30 tons can be combined to produce incremental sizes up to 150 tons. The second type has the capability of operating in a double-effect mode with gas input (at a COP of 0.95) or operating in a single-effect mode with hot water input (at a COP of 0.75), or both. The basic 20 and 30 ton modules can be combined to provide up to 150 ton capacity. Therefore, this unit can operate using solar-heated hot water at 190°F (88°C) in a single-effect mode, with backup double-effect operation using a gas flame. Both types of double-effect machines can also provide heating. However, the 130°F (54°C) outlet hot water for heating purposes is generated directly by the gas flame, with an efficiency of 83%; it is not provided through absorption heat pump operation. Yazaki also markets the additional hardware to make complete solar cooling systems: collectors, cooling towers, and control units.

Sanyo markets an absorption chiller line for solar applications that parallels that described above for Yazaki. Included are a hot water-driven single-effect chiller, a gas-driven double-effect chiller, and a combined single/double-effect machine, run at single-effect on hot water and at double-effect on high temperature steam. The Sanyo single-effect capacities range from 10 to 250 tons, the double-effect from 10 to 1150 tons, and the combined single/double-effect from 30 to 360 tons. The double-effect machine operates at a cooling COP of 1.0, including combustion efficiency, and has a direct heating-mode efficiency of about 83%. Sanyo also markets a high performance evacuated-tube solar collector, well capable of driving the solar cooling systems.

Hitachi similarly manufactures and markets a line of single- and double-effect absorption chillers for solar-driven and gas-driven applications.

Although not oriented toward operation with solar heat supply, several Japanese manufacturers are producing limited numbers of absorption chillers and absorption heat pumps for cogeneration installations and waste heat supply.

20.4 Absorption Systems

20.4.1 Typical System Configurations

A generalized solar cooling system schematic is shown in figure 20.36. Besides the chiller, the key component in terms of both performance and

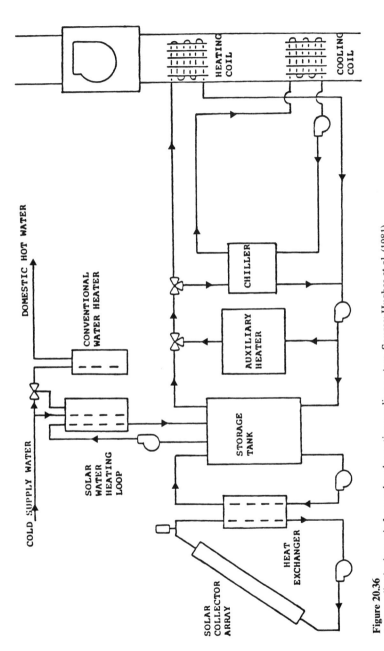

Figure 20.36
A generalized schematic for a solar absorption cooling system. Source: Hughes et al. (1981).

cost is the collector. The storage component is also important in terms of system performance, chiller sizing, and system operating strategy. Figure 20.36 shows only hot-side storage. Cold-side storage (storage of the "coolth" output of the chiller) may also be used instead of or in addition to hot-side storage.

Figure 20.36 includes the features of direct heating from the hot storage to the building load and preheating of the domestic or service hot water. In general, these features would always be included as long as a heating load exists, as they allow for a fuller use of the collectors.

The auxiliary heater is shown in a configuration in which it can provide heat to drive the chiller for cooling backup and provide heat directly to the building load for heating backup. For advanced, high-efficiency absorption heat pumps that can supply both cooling and heating, with a heating COP in the region of 1.5 to 2.0, the auxiliary heater would likely provide heat input to the chiller/heat pump for both heating and cooling backup.

In the backup cooling mode, an auxiliary gas heater driving a single-effect absorption chiller only achieves an overall COP of about 0.5. At this low an efficiency, it has often been more economically sensible to use a separate electric-driven air conditioner as backup. But this entails a substantial additional first cost penalty for the system. This has been the driving force for development of chillers that operate as double-effect chillers in the gas backup mode.

Not shown in Figure 20.36 is the heat exchange loop from the condenser/absorber coils of the chiller to the ambient air. For a water-cooled chiller, a wet cooling tower is required.

20.4.2 Collector and Regenerator Requirements

The primary function of the solar collector subsystem in closed-cycle absorption systems is for sensible heating of the collector fluid, which transfers the heat to the sealed generator element of the absorption chiller. In open-cycle absorption systems, the solar energy is used primarily to regenerate the absorbent solution, concentrating it by driving off water vapor into the atmosphere.

20.4.2.1 Collectors for Closed-Cycle Systems

Single-effect chillers require input temperatures in the 180° to 200°F (82° to 93°C) range, which translates into a need for a good-quality flat-plate collector. Double-effect or other types of higher-efficiency chillers generally require temperatures above 200°F (93°C), necessitating use of evacuated tube or tracking collectors. The higher the efficiency, the higher the allowed cost per unit collector area in terms of achieving cost-competitiveness.

Flat-plate collector development has generally been associated with solar heating, not solar cooling systems. For solar cooling applications, collector R&D has concentrated on the development of high performance, moderate-cost collectors, capable of high efficiency when driving high-performance absorption chillers, such as the LBL 2R and 1R cycle machines.

Tracking trough collectors have been developed with extensive support by the DOE/Solar Thermal Technology Program since the early 1970s. Tracking troughs have been used to drive chillers in many solar cooling applications and, if cost goals for these collectors can be met, they may eventually be the most economical method of driving solar cooling systems in many parts of the country.

High-quality evacuated-tube collectors have been developed and marketed by a number of commercial firms, and have proved quite capable of driving solar cooling systems in many installations. The DOE/Solar Buildings Program has supported research and development on moderately-concentrating, nontracking, evacuated-tube collectors for the above-200°F temperature regime needed for solar cooling applications.

In the mid-1970s, the CPC (compound parabolic concentrator) collector was proposed by the University of Chicago (UC), and early research on the CPC concept and similar collectors using nonimaging optical reflector shapes was conducted by UC and Argonne National Laboratory (ANL). This led to commercial collectors manufactured by several manufacturers who added features of their own. Collector arrays using nonimaging reflectors have been used to drive absorption chillers in a number of solar cooling installations.

More recently, UC has evolved the concept of the ISEC (integrated stationary evacuated concentrator) collector (O'Gallagher, Snail, and Winston 1982; Snail, O'Gallagher, and Winston 1984). In this design, as shown in figure 20.37, the nonimaging reflector is integrated within the evacuated tube itself. The result is a very high-efficiency collector with excellent durability, since the reflective surface is protected within the vacuum envelope. The predicted high performance of this collector has been confirmed experimentally (Snail, O'Gallagher, and Winston 1984).

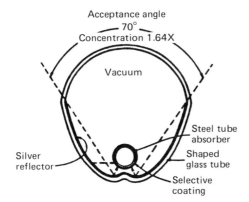

Figure 20.37
Cross-sectional view of University of Chicago integrated stationary evacuated concentrator (ISEC) collector design. Source: Schertz et al. (1985).

As shown in figure 20.38, its efficiency is expected to exceed or match that of a tracking trough collector at temperatures up to 400°F (200°C).

GTE Laboratories worked closely with UC in the early development of the ISEC collector. UC and ANL have worked together on its continued development and testing (Schertz et al. 1985). In terms of performance, the ISEC collector is an excellent candidate to drive high-efficiency absorption chillers/heat pumps at any temperature up to 400°F (200°C). The critical and difficult issue remains the collector cost. Further R&D is concentrating on design modifications that can dramatically reduce collector cost while preserving the high performance. One such option is to use a thin, shaped-metal insert for the reflector rather than shaping the glass envelope.

20.4.2.2 Regenerators for Open-Cycle Systems
In open-cycle systems, the solar energy may be used directly to regenerate the absorbent solution, as on a rooftop, or an air stream heated by solar energy may, in turn, be used to regenerate the absorbent solution. The following two subsections present examples of these two different types of regeneration.

20.4.2.2.1 OPEN ROOFTOP REGENERATOR   In a system similar to one previously operated in the USSR, Arizona State University (ASU) designed, constructed, and tested a 2500 ft² open rooftop collector-regenerator to

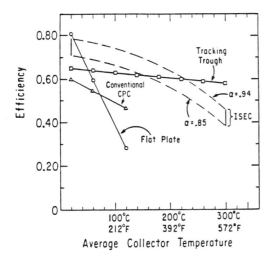

Figure 20.38
Calculated instantaneous efficiency curves normalized to beam insolation for the ISEC and
three other collector types. The lower ISEC curve (for $\alpha = 0.85$) is based on a fit to the
actual measured prototype performance, while the upper curve ($\alpha = 0.94$) represents what
can be expected with a higher absorptance coating. Source: Schertz et al. (1985).

study the practical feasibility of this approach (ASU 1981). A lithium chlo-
ride/water solution flowed down a sloped roof, as shown schematically
in figure 20.39, consisting of black asphalt shingles mounted on a standard roof structure. The collector-regenerator performed well, as expected.
After four years of operation there was no abnormal degradation of the
asphalt shingles nor other parts of the collector loop in contact with the
lithium chloride solution. ASU accompanied this experimental testing
with a detailed study of the heat and mass transfer characteristics of the
collector-regenerator subsystem (Wood et al. 1983).

A parallel experimental study of an open sloped roof collector-
regenerator was conducted by Lockheed-Huntsville Research and Engineering Center (Lockheed 1981) with similar positive results.

One problem area was experienced by both ASU and Lockheed: rain
water would wash away residual lithium chloride salt that had dried on
the shingles. Either the initial rain water had to be collected and processed
to recover the lithium chloride, or lost lithium chloride salt had to be
made up. This problem could be overcome by installing a low-cost glazing
over the roof. Lockheed tested this modification by installing a glazing

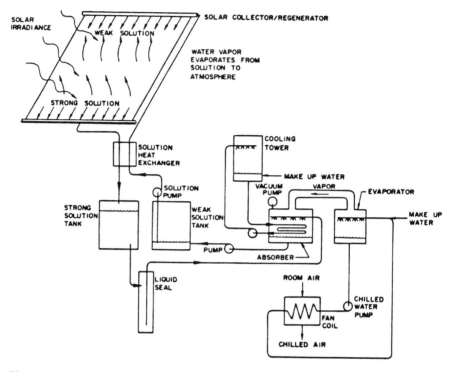

Figure 20.39
Schematic diagram of Arizona State University open-cycle absorption system using an
open rooftop collector-regenerator. Source: Wood et al. (1983).

over their roof and repeating the collector-regenerator test. The glazed
unit performed essentially the same as the unglazed one, so the only disadvantage is cost, not performance. It appears likely that future installations
will use this glazing feature in which the rooftop is open to the ambient air
but covered. It not only eliminates the rainfall problem but also greatly
reduces contamination of the lithium chloride solution by foreign material
such as dust and insects.

20.4.2.2.2 PACKED TOWER REGENERATOR   Colorado State University
(CSU) has chosen to use a packed bed as the solution regenerator, in the
configuration shown in figure 20.7. A schematic of the packed-bed regenerator is given in figure 20.40. Solar air collectors are used to generate hot
air that is blown vertically upward through the packed bed, driving off
water vapor and concentrating the lithium chloride or lithium bromide

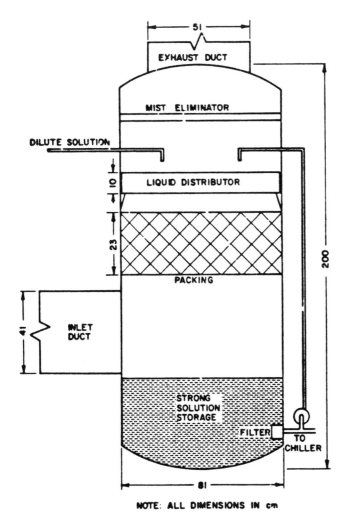

Figure 20.40
Schematic diagram of packed-bed regenerator used in Colorado State University open-cycle absorption cooling system. Source: Löf and Lenz (1985).

solution that drips downward through the bed. The packed-bed regenerator was tested extensively separately (Löf, Lenz, and Rao 1984) and then
operated as part of the CSU Solar House II open-cycle absorption cooling
system (Löf and Lenz 1985). The regenerator worked satisfactorily in the
cooling system, but mass transfer (diffusion of water) was identified as one
of the cooling rate limiting effects, and improvements were planned.

The packed-bed regenerator has the advantage (compared to the open
rooftop collector-regenerator) of a more controlled environmental interaction with outdoor air (the absorbent solution is confined to the packed-
bed component, not trickled down the roof) but has disadvantages of
higher relative cost (the cost of the air collectors plus packed-bed regenerator would likely exceed the solar-related cost of the rooftop collector-
regenerator) and higher parasitic electric power requirements (to blow the
air through the packed bed). Another cost-benefit consideration must be,
however, the space heating capability of a conventional air collector during cold weather.

20.4.3 Storage Requirements

The storage requirements for solar cooling systems are similar qualita-
tively to those for solar heating: solar energy is converted to thermal
energy and stored for use when sunlight is not available. Quantitatively,
however, hot thermal storage for cooling applications requires higher temperature, dictated by the inlet conditions for driving the chiller. Alternatively, solar cooling systems may use lower temperature storage, storing
the output "coolth" produced by the solar cooling system, using the stored
"coolth" to cool the building directly at a later time.

20.4.3.1 Sensible Heat Storage

Most solar cooling systems that have included storage have simply used a
tank of hot water. The storage capacity is proportional to ΔT, the temperature difference between the temperature of the stored fluid and that required by the chiller inlet. Since absorption chillers normally employ inlet
temperatures of 180°F (82°C) or higher, a storage temperature capability
above 212°F (100°C) may be necessary, requiring that the hot water tank
be pressure-rated.

Storage of solar heat is discussed in part 3 of this book, Solar Space
Heating. Specific for cooling, however, is the storage of chilled water being
delivered from the chiller when the supply exceeds cooling demand. This

design may be advantageous in systems not employed for heating and therefore having no heat storage capability. Cooling demand is usually fairly well correlated with solar availability. However, the small ΔT between the chilled water output from the chiller (typically 40° to 45°F [4° to 7°C]) and the 32°F (0°C) freezing point of water necessitates the use of a large volume of chilled water to achieve an appreciable storage capacity, and this technique has not been commonly used.

Most solar cooling systems have also been used for heating (space heating or hot water heating or both), and the hot water storage tank has been used for all these end uses.

20.4.3.2 Latent Heat Storage Using Phase-Change Materials

Just as for solar heating systems, the storage requirements for solar cooling systems can be met by phase-change materials (PCMs) rather than sensible (ΔT) heat storage. Most of the developments in PCMs have been associated with solar heating rather than cooling, but limited work has been directed toward cold storage. The latter application is included here, whereas solar heat storage is discussed in association with space heating systems.

The advantages (compared to sensible heat storage) are that phase-change materials have a higher storage energy density, require less volume, and avoid limitations set by the boiling or freezing point of water. Disadvantages include high material cost, the additional cost and ΔT losses associated with the required heat exchangers, and the finite lifetime (i.e., number of charge-discharge cycles) of many candidate materials.

Partially under DOE support, CALMAC developed low temperature PCMs for solar cooling applications (MacCracken 1980). The 45°F (7°C) PCM was an eutectic combination of seven different materials, including a clay-type thickener to prevent stratification, plus sodium sulfate, ammonium chloride, and potassium chloride. A measured charge-discharge cycle for the 45°F (7°C) PCM is shown in figure 20.41. Stirring of the mixture was not required, and more than 1000 cycles confirmed its stability.

One of the CALMAC low temperature PCM units was tested at CSU as part of a Solar House I solar cooling system (Karaki et al. 1984). It had a phase-change temperature of 55°F (13°C). It was similar to the 45°F (7°C) module mentioned previously but with sodium chloride substituted for potassium chloride as one of the mixture components. It contained 4000 lbs (1800 kg) of PCM, with a latent heat storage capacity of 250,000

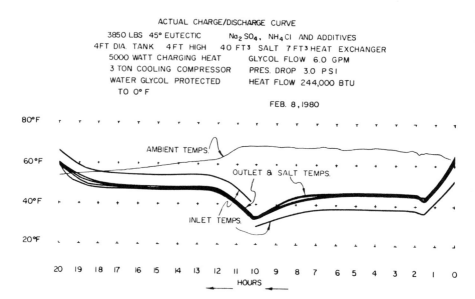

Figure 20.41
Charge-discharge cycle performance for the CALMAC 45°F phase-change storage unit.
Source: MacCracken (1980).

Btu (264 MJ). During very limited testing in the summer of 1983, heat transfer rates to and from the storage unit were lower than expected. During charging, the temperature of the water returning from the storage coils to the chiller gradually decreased to about 42°F (6°C) instead of leveling off near the 55°F (13°C) phase-change temperature; these low returning temperatures produced an adverse effect on chiller efficiency, degrading the overall cooling system performance. These results indicated serious decreases in heat transfer rates as the PCM material solidified around the heat exchange tubing.

There has been no commercial application of phase-change cold storage in solar cooling systems.

Ice is an obvious PCM material for cooling systems in general and has been used in some solar cooling applications. It is ideal in all respects (stable, low cost) except that its melting temperature is too low. The efficiency of most solar cooling systems would be substantially degraded by operating the evaporator of the chiller at a low enough temperature (about 27°F [−3°C]) to freeze water at 32°F (0°C).

Cross-linked high-density polyethylene (HDPE), which was developed by Monsanto (Botham et al. 1978), is a candidate high temperature PCM for solar cooling applications in which the thermal storage is on the input side rather than the output side of the chiller. HDPE undergoes a crystalline-to-amorphous phase change at 266°F (130°C), retaining its shape without appreciable flow during the phase-change process.

A latent heat storage device using HDPE was developed and tested at Argonne National Laboratory (Cole 1981, 1983). Shortly afterwards, a 4000-liter storage tank containing HDPE pellets was constructed and incorporated into the CSU Solar House I solar cooling system (Karaki et al. 1984) for test as a hot-side PCM storage unit. The storage unit was charged as expected during the first two days of operation, but then it was discovered that the pellets had fused into a solid block that could not even be broken apart by an air-powered jack hammer. Testing of the device was discontinued. It appears that further development is in order before it is ready for practical use.

Finally, a summary of much of the early work on thermal storage for solar cooling applications can be found in the proceedings of the 1979 storage workshop (DOE 1979b), including article by Curran on storage for solar cooling systems (Curran and Heibein 1979) and a panel report (Wahlig 1979) on the same topic.

20.4.3.3 Electricity Generation as Storage Option

Mechanical Rankine solar cooling systems have the capability of generating electricity rather than using thermal storage when there is a solar energy source but no cooling load. The electricity can be used in the building or fed back into the utility grid. Thus, no thermal storage is required in the solar cooling system. Instead, it is replaced by the energy in the fossil fuel (or other energy source) that is not consumed by the utility company to generate the equivalent amount of electricity.

It has not been widely discussed, but solar absorption systems have a similar option for generating electricity. An expander can be installed in parallel with the condenser and absorber components in such away that high pressure refrigerant vapor from the generator serves as the input to the expander, and the low pressure output vapor from the expander passes to the absorber. Suitable valves and controls would direct the refrigerant vapor to the condenser when cooling is desired and to the expander when electricity generation is desired.

Although no development has taken place on this concept for solar cooling applications, expanders for the refrigerants commonly used in absorption chillers (ammonia, water, freons) have been developed and used for many other applications. The thermodynamics are also favorable. A solar absorption system would generate electricity at a higher efficiency than would a solar Rankine (mechanical) system at the same driving temperature (Dao 1979).

The feasibility of this option would depend on a number of considerations, such as the relative cost of the expander component compared to that of the storage device, the recovery efficiency for the thermal energy put into storage, and the value received for the electricity generated.

20.4.3.4 Comparisons of Storage Options for Solar Cooling and Heating Systems

Either prior to or accompanying the development of thermal storage devices for solar cooling systems, it would be prudent to conduct analytic studies to predict which storage options make the most sense from performance and economic points of view. To a large extent, however, these analytic studies were not completed until after much of the storage component development had been undertaken.

Hittman Associates made a qualitative study of thermal storage options for solar cooling systems (Curran and DeVries 1981) in a project that did not include computer modeling. The major options considered were hot-side vs. cold-side and sensible vs. phase-change storage. The only specific conclusion made was that hot-side phase-change storage was less favorable than the other three options. The study pointed out the lack of quantitative analysis of the subject and the need for better information as a result of computer modeling and experimental measurements. Two interesting observations were: (1) the storage subsystem only accounts for about 6% of solar cooling system cost, so cost reductions in this component are of minor importance; and (2) an improvement in thermal storage subsystem efficiency to 100% would increase the overall solar fraction only a few percent in a well-designed system.

A more detailed and quantitative investigation of thermal storage options for solar cooling systems was performed by Science Applications, Inc. (SAI) (Hughes et al. 1981). Options studied included hot-side vs. cold-side, sensible vs. phase change, and single vs. multiple tank. Computer models were run separately for residential-size (heating and cooling) and commercial-size (cooling only) systems in four climate locations. Cold-side storage systems were modeled with storage also on the hot side, as is seen in figure 20.42. Figure 20.36 showed a system with hot-side storage only.

Some of the major findings of the SAI analysis were: (1) none of the solar cooling systems investigated, on a life-cycle cost basis, was economically competitive with conventional cooling systems, regardless of the storage option used; (2) of all the storage options investigated, cold-side phase-change storage appeared the most promising in nearly all cases; (3) multiple tank sensible storage should be dropped from further consideration.

20.4.4   Control of Absorption Cooling Systems

In a real system, the absorption chiller will encounter part-load operation most of the time, and ambient conditions will vary considerably from the design point. The capacity and efficiency of the chiller will depend strongly on these external parameters and, in a coupled way, on the inlet driving temperature of the chiller. The efficiency of the solar collectors is also a function of ambient conditions and of the outlet temperature used to drive the chiller. Accordingly, the solar absorption cooling system control strategy is crucial to the attainment of good, overall system efficiency over a cooling season.

Alwin Newton has been for decades one of the foremost proponents of employing proper control strategies to achieve good performance for absorption cooling systems in general and solar absorption cooling systems in particular. In one of his publications (Newton 1978), he shows how use of appropriate controls can increase the amount of useful solar heating or cooling that can be obtained from a given collector array by as much as 35 or 40%.

20.4.4.1   Drexel University Project: Control of Solar Cooling Systems

Drexel developed detailed time-dependent computer models of solar collector loops. These models tracked the fluid flow in time increments, enabling accurate analysis of the effects of various control strategies and control sensor locations. This analytical technique was then applied to solar cooling system control (Herczfeld, Fischl, and Anand 1984).

The Drexel project sought to include experimental verification in addition to the development of the analytic control model itself. For this reason, it was decided to model the solar cooling and heating system in CSU Solar House III as a test case for the analytic technique. With the co-

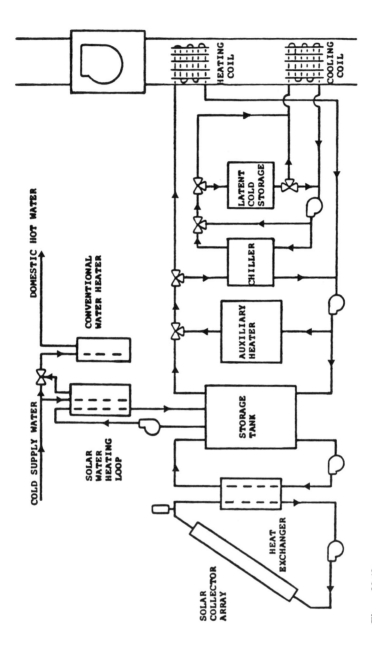

Figure 20.42
Schematic of a solar absorption cooling system using cold-side latent (phase-change) thermal storage. Source: Hughes et al. (1981).

operation of CSU, test runs of the Solar House III system were made for purposes of identifying component and system parameters. Necessary supplementary instrumentation was installed.

The dynamic model development occupied the first phase of the project, including the development of alternate control strategies. The CSU solar cooling and heating system was operated during the second phase of the project in such a way that the effects of the control strategies on actual system performance could be observed and measured. Drexel plans to conclude the project with comparison of the analytically predicted and experimentally measured results of the control strategies. TPI is a major subcontractor in this project.

20.4.5   Analysis of Absorption Cooling System Performance

A substantial amount of analysis of absorption cooling system performance has been performed since the early 1970s. These calculations serve to predict expected system performance, to assist in deciding optimal design parameters for system components, to allow comparison with and understanding of experimental results, and to provide input to economic analyses of system cost-effectiveness.

20.4.5.1   Closed-Cycle Absorption Systems Analysis
Most analysis work has dealt with closed-cycle systems. They have been in the development phase for a longer period than have open-cycle systems. Also, the analysis of solar cooling system performance for closed-cycle cooling systems is very similar to that for heating systems—i.e., distinct components linked together into a system—except that the additional cooling components increase the system complexity. Almost every solar cooling project listed in this chapter has involved systems analysis to a degree, as a tool to assist in component design or project justification or applications understanding. The projects listed in this section are those in which systems analysis was a primary objective of the work.

20.4.5.1.1   UNIVERSITY OF MARYLAND PROJECT: SHASP PROGRAM   As mentioned in section 20.3.2.4 on chiller modeling, the University of Maryland (UM) developed the SHASP (solar heating and air-conditioning simulation programs) computer model (Anand et al. 1978) for analysis of complete solar cooling systems (Anand, Kumar, and Allen 1983; Anand, Kumar, and Hall 1982) as well as for detailed chiller component analysis (Anand, Allen, and Kumar 1981; Anand, Allen, and Venkateswaran 1980). Since SHASP

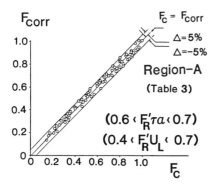

Figure 20.43
Comparison of solar cooling fraction calculated using the University of Maryland simplified correlation model (F_{corr}) with the solar cooling fraction calculated using the University of Maryland detailed SHASP simulation model (F_c), showing agreement within $\pm 5\%$. Source: Anand, Kumar, and Allen (1983).

was not intended to be made available and supported as a general purpose computer model for general use by other groups, UM undertook the development of a simplified model that could serve that purpose.

The simplified model (Anand, Kumar and Allen 1983) was designed to predict the average long-term solar cooling system performance. Using a simple procedure and input requirements, it would be suitable for widespread use for system design and long-term performance predictions. It appeared that the subdivision of the United States into four regional design charts was sufficient to apply the model to all U.S. climate regions. The simplified model was verified by comparison of its results with the more detailed predictions of the SHASP model, in much the same way that the simplified f-chart model (for heating systems) was verified using TRNSYS detailed calculations. As shown in figure 20.43, the output of the simplified model agreed with the SHASP predictions to within ± 5 percent of the cooling load.

20.4.5.1.2 UNIVERSITY OF WISCONSIN PROJECT: TRNSYS PROGRAM Undoubtedly, most of the calculations of solar absorption cooling system performance have been made using the widely available TRNSYS computer model developed and supported by the University of Wisconsin (UW) under DOE/Solar sponsorship. TRNSYS is a modular program, so any user has a choice of using a chiller module supplied by UW with the TRNSYS package or of substituting the user's own chiller module. Since many of the solar projects described in this chapter are concerned with the development of new chiller equipment, those groups have been able to make cooling system calculations using TRNSYS with their own model of the newly-developed chiller, without need for developing a new separate model of the rest of the solar cooling system.

20.4.5.1.3 SCIENCE APPLICATIONS, INC., PROJECT: ANALYSIS OF SOLAR COOLING SYSTEMS   Under DOE support, Science Applications, Inc. (SAI), carried out the most comprehensive analysis to date of solar cooling system performance (Choi, Stokely, and Morehouse 1982). SAI analyzed the thermal and economic performance of solar absorption and Rankine cooling systems for commercial and residential buildings. For commercial applications, the systems analyzed included the Carrier water-cooled absorption chiller (section 20.3.2.2.2), the Carrier air-cooled absorption chiller (section 20.3.2.2.1), and the Rankine chillers developed by Barber-Nichols and by United Technologies Research Center. The residential systems included the Arkla evaporatively cooled absorption chillers (sections 20.3.2.1.2 and 20.3.2.1.3), the Carrier air-cooled absorption chiller (section 20.3.2.2.1), and the Barber-Nichols evaporatively cooled Rankine chiller.

The cooling systems were simulated in four climate areas: Miami, Phoenix, Fort Worth, and Washington, D.C. TRNSYS-compatible models were developed by SAI for each of the separate chillers, with the cooperation and assistance of the chiller developers. Standardized assumptions (Leboeuf 1979) were made for the building loads, weather data, and other parameters. The thermal performance of each of the solar cooling systems was calculated via a series of TRNSYS simulation runs.

Conventional HVAC systems were also simulated, to allow calculation of the energy savings of the solar systems. Information on the costs of electricity and gas was used to determine the energy cost savings of the solar cooling systems, which, in turn, established cost goals (Warren and Wahlig 1982) for the solar systems. Estimates of equipment costs were used to compare solar cooling system incremental capital costs with cost goals for systems installed in the year 2000.

The study concluded that even the best solar cooling systems had incremental costs that exceeded the cost goals by a factor of 3 or 4. This wide difference between the cost-effectiveness of solar and conventional cooling systems masked the differences between the relative economics of the various solar cooling systems. To meet the cost goals, SAI stated that the efficiences of the solar cooling systems must be increased, the parasitic power consumption must be reduced, and solar collector and chiller costs must be significantly reduced by volume production.

20.4.5.1.4 TPI PROJECT: SECOND LAW ANALYSIS OF SYSTEM PERFORMANCE
As mentioned in section 20.3.4.3, TPI applied second law analysis to complete solar cooling systems (Anand et al. 1984) as well as to chillers. For solar cooling systems, which require higher driving temperatures than solar heating or hot water systems, exergy (also called available energy or equivalent work) is a more meaningful parameter than energy in the analysis of system performance. Figure 20.44 shows a comparison between

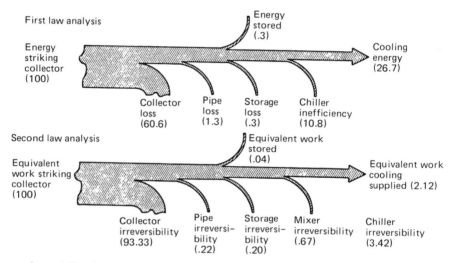

Legend: The theoretically maximum portion of the thermal energy which can be converted to work is dependent upon the temperature at which the thermal energy is available. This theoretically maximum amount of work is known as the "equivalent work" or "exergy" associated with the thermal energy. These two figures demonstrate an energy and equivalent work of exergy balance of a solar absorption cooling system. The energy balance shows that 39.4% of the energy striking the collector is captured, i.e., the collector has a 39.4% First Law efficiency. The 39.4 units of energy collected have an equivalent work value of 6.67 units, i.e., the Second Law efficiency of the collector is only 6.67%. Similarly the efficiencies of other components are indicated The First Law efficiency of the entire system is shown to be 26.7% while the Second Law efficiency is 2.12%.

Figure 20.44
Comparison of first and second law analysis of a solar absorption system. Source: Anand et al. (1984).

a first law (energy) analysis and a second law (exergy) analysis for a typical solar absorption cooling system. The exergy analysis determines the irreversibility of each component in the system, identifying targets of opportunity for increasing overall system performance.

20.4.5.2   Open-Cycle Absorption Systems Analysis

In open-cycle absorption systems, there is a stronger coupling between the collector and chiller components than there is in closed-cycle systems, because of the presence of mass (water) as well as heat transfer between components. Serious work on open-cycle systems started more recently than that on closed-cycle systems. For these reasons, open-cycle systems analysis to date is less developed and more project-specific than is the analysis of closed-cycle solar absorption cooling systems.

20.4.5.2.1   COLORADO STATE UNIVERSITY PROJECT   The open-cycle absorption cooling project at CSU, which has been mentioned previously in sections 20.3.8.1 and 20.4.2.2.2, was begun with an analysis that demonstrated the technical feasibility of this approach to solar cooling (Leboeuf and Löf 1980). The analysis showed that regular solar air collectors with an area of about 700 ft^2 (65 m^2) could drive a 3-ton capacity open-cycle cooling system using a packed-bed regenerator during a clear summer day in Fort Collins and St. Louis.

Systems analysis calculations by CSU have continued in conjunction with the experimental study of the open-cycle absorption system in CSU Solar House II. The results have been reported in the annual reports and other publications (Lenz et al. 1983) of the CSU group.

20.4.5.2.2   SOLAR ENERGY RESEARCH INSTITUTE PROJECT   Analytical studies that showed the promise of open-cycle absorption systems using open rooftop collector-regenerators were performed by the Solar Energy Research Institute (SERI). The first analysis (Collier 1979) showed the technical feasibility of solar cooling using open-cycle absorption systems in five U.S. cities: Phoenix, Albuquerque, Dallas, New York, and Miami.

A subsequent analytical investigation (Schlepp and Collier 1981) expanded the scope of this technology by showing how open-cycle systems could be used for heating as well as cooling. As shown in figure 20.45, the system would operate as a heat pump in the heating mode, with the absorber output providing hot water at 104°F (40°C) to the heating load and with the solar collector output providing thermal input to the evapora-

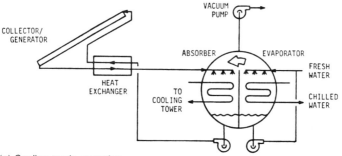

(a) Cooling-mode operation

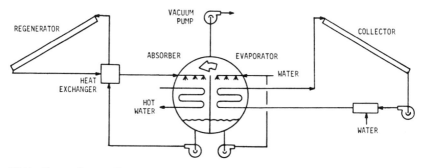

(b) Heating-mode operation

Figure 20.45
Schematic diagram of open-cycle absorption system using rooftop regeneration for both heating and cooling operation. Source: Schlepp and Collier (1981).

tor. The solar collector surface would be divided into two sections, one part operating as a regular collector to drive the evaporator thermally, and the other part serving as the solution regenerator for the absorber solution loop. The simulations for heating-mode operation were run for the same five cities previously used for the cooling-mode calculations. These analyses showed that solar open-cycle absorption systems could supply full heating and cooling needs for office buildings in all these locations.

20.4.5.2.3 ARIZONA STATE UNIVERSITY PROJECT   The open cycle absorption cooling project at ASU, which has been mentioned previously in section 20.4.2.2.1, has also involved a substantial amount of systems analysis.

A detailed computer simulation of the open-cycle system shown in figure 20.39 predicted that it could supply over 99% of the cooling load for a

2000 ft² (186 m²) structure in Phoenix (Wood, Siebe, and Collier 1985). The calculations revealed that a mixture of 3 parts calcium chloride and 1 part lithium chloride would be as effective as 100% lithium chloride. Economically, this is an important result, because the high cost of lithium chloride posed a serious barrier to the application of these open-cycle systems. The use of mostly calcium chloride, which has 1/30 the cost of lithium chloride, largely removes this cost barrier. The analytical study also investigated the effects of a number of control strategies by use of a series of TRNSYS runs with ASU-developed component models.

20.4.6 Testing of Absorption Cooling System Performance by Use of Systems-Level Test Facility

For solar cooling and heating, the primary performance goal is the attainment of good *system* efficiency. Therefore, it is as important that the system components (chiller, collector, etc.) work well together as it is to develop components with high individual efficiencies. A systems-level test facility is essential to learn how to achieve this degree of integrated system performance, so that initial mistakes and understanding can take place in a controlled, instrumented environment, not in a poorly characterized (and often highly visible) field installation. The Solar Houses at CSU have served as the systems-level test facility for the DOE Solar Heating and Cooling Program.

20.4.6.1 Colorado State University Solar Houses

Four solar houses have been constructed at the CSU Solar Energy Applications Laboratory. These can be seen in figure 20.46. Three of these (Solar Houses I, II, and III) are essentially identical structures and have been used for studying active solar systems. The fourth house has been used to study passive solar systems.

Although the houses themselves have been built using standard construction practice for this type of structure, many of the solar heating and cooling components have been installed in such a way that system configurations and components can be changed every few years. Thus a sequential series of systems tests has been cycled through each of the solar houses since they were built in the mid-1970s.

Since the early days of the tests, solar cooling operation has been incorporated along with the solar heating and hot water system investigations. Even an abbreviated account of all the cooling testing and reporting that

Figure 20.46
The four solar houses at the Solar Energy Applications Laboratory at Colorado State
University.

have transpired over the past ten years is beyond the scope of this section.
Several aspects of the work (chiller testing, open-cycle chiller and regenerator development, storage and controls studies, and systems analysis) have
been mentioned in previous sections of this chapter. A brief overview of
the cooling systems investigations will be given here.

20.4.6.1.1 CLOSED-CYCLE ABSORPTION COOLING SYSTEM TESTS   One of the
early models of the Arkla 3-ton water-cooled absorption chiller was tested
in Solar House I. This was later replaced with an early model of the
Yazaki absorption chiller and with a second-generation Arkla chiller.

Several important lessons were learned from these early cooling systems
studies. The average absorption chiller efficiency during system operation
was observed to be about half the rated chiller efficiency. This reduction in
efficiency was found to be partially caused by cycling the chiller on and off
during normal system operation. (A solution to this problem was found
later by BNL: the spin-down technique described in section 20.3.2.5.) Another problem that surfaced was that a large fraction of the cooling load
was being caused by thermal losses from the hot storage tank. Increased
insulation could not solve the problem. The hot storage tank must be
located outside of the conditioned space. This is an example of one of the
rules of practice that have been established over the years as a result of the
CSU work.

Solar House I became the testing ground for high-performance evacuated-tube solar collectors. Therefore, it was a natural choice for the location to system-test the Carrier air-cooled absorption chiller (see section 20.3.2.2.1) that was designed for a driving temperature of about 230°F (110°C). The most complete testing of this 3-ton Carrier chiller, as a component as well as part of a solar cooling system, was a result of this work at Solar House I (Karaki, Vatano, and Brothers 1984). During the summer of 1982, the monthly average thermal COP of the chiller ranged from 0.71 to 0.76. It was also found that the chiller would operate at driving temperatures as low as 190°F (88°C), although only at about 60% of design capacity.

Solar House I was also used to investigate hot-side and cold-side storage options for solar cooling systems (Karaki et al. 1984), using the Carrier air-cooled chiller, as described in section 20.4.3.2.

Solar House III was used for several years for solar cooling system tests incorporating the Arkla 3-ton evaporatively cooled absorption chiller described in section 20.3.2.1.2 (Löf, Westoff, and Karaki 1984). A schematic diagram of this cooling system, as configured during the 1983 summer season, is shown in figure 20.47. An active interchange took place between CSU and Arkla during the first two summers of chiller operation at CSU, resulting in improved understanding of the operating characteristics of this chiller.

During the 1983 summer season (Karaki, Brisbane, and Löf 1984), the Arkla chiller achieved an excellent seasonal average COP of 0.69, with daily COP values between 0.65 and 0.75. The overall cooling system performance was lower than anticipated, however, due to a number of factors, such as losses from storage and nonoptimum collector flow rate strategy. These factors are all correctable. The knowledge gained during this season of cooling system operation can be directly applied to improving solar cooling system performance in the future, at CSU or other installations using similar equipment. This is representative of the value to the DOE/ Solar Program of operating a systems-level test facility over many seasons with many cooling and heating system configurations.

Solar House III was also used as the cooling and heating system for testing and validating the system control strategies developed by Drexel University, as discussed in section 20.4.4.1.

20.4.6.1.2 OPEN-CYCLE ABSORPTION COOLING SYSTEM TESTS   CSU Solar House II incorporates an air collector array and air distribution system,

838   Michael Wahlig

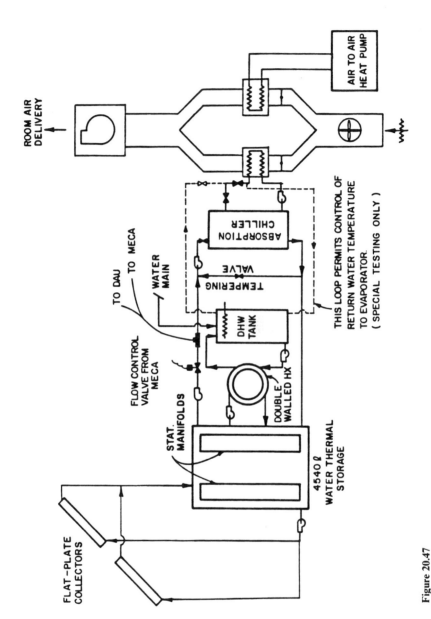

Figure 20.47
Solar cooling system for Colorado State University Solar House III during the summer of 1983, including the Arkla evaporatively cooled absorption chiller. Source: Karaki, Brisbane, and Löf (1984).

in contrast to Solar Houses I and III that use liquid collection and distribution systems. CSU pioneered the development of open-cycle absorption systems in this country as a method of including cooling as part of a solar air system. The solar-heated air is used to regenerate the absorbent solution in a packed-bed regenerator, as discussed in section 20.4.2.2.2 and shown in figures 20.7 and 20.40.

CSU has been using the Solar House II system as a test bed for testing and evaluating the individual chiller and packed-bed components during their development phase. Operation of the complete system on a continuing basis over one or more cooling seasons is planned, so that operational data and understanding comparable to those gathered for closed-cycle absorption cooling systems can be obtained.

20.4.7 Field Operation of Absorption Cooling Systems

Numerous solar cooling systems have been built and operated for building space conditioning, with private as well as governmental support. More information, in terms of reporting and data collection, is available for those cooling systems built under government programs, so this discussion is mostly limited to that subset.

20.4.7.1 Demonstration Projects

In 1974, Congress passed the Solar Heating and Cooling Demonstration Act that declared, among other things, that solar cooling technology would be developed and demonstrated within five years. Although the majority of the solar demonstration projects were hot water systems and space heating systems, there were a number of solar cooling projects from the outset of the demonstration program (Doering 1978). Most of these projects involved Arkla 3-ton or 25-ton absorption chillers. A few of the largest installations used Trane, York, or Carrier absorption chillers, usually derated for the lower solar driving temperatures.

The field performance of the solar cooling demonstration projects ranged from very good to very poor. Most of the problems were related to the conventional HVAC aspects of the systems, not specifically to the solar collectors or to the chiller itself. Improper control of the systems was a major problem with many solar heating and solar cooling projects. The additional complexity and operating modes of the solar cooling (compared to solar heating-only) systems exacerbated the controls problems.

A DOE-sponsored *ASHRAE Handbook* (Ward and Oberoi 1980) of experiences in the design and installation of solar heating and cooling systems contains a detailed analysis of problems encountered and lessons learned from a variety of federally funded demonstration projects. Nine solar cooling systems were included in the study. The majority of these cooling systems were net energy losers; i.e., they used more (electric) energy to operate the system than they supplied (as thermal energy) from the solar components. One residential and one commercial solar cooling system, however, performed well and achieved significant energy savings. Major problems found, in general, were improper integration of components with system and load requirements, improper controls, excessive heat losses, and excessive electrical energy operating requirements.

A great deal of documentation exists on the physical descriptions of demonstration projects. Periodic summary publications have included brief descriptions, photographs, and schematic diagrams of these systems (DOE 1979a).

20.4.7.2 Bonneville Power Administration Sites

As was reported in section 20.3.2.2.2, three Carrier 15-ton, water-cooled absorption chillers were developed and then field-installed and operated as part of joint BPA-DOE projects (Biermann and Reimann 1981; Biermann 1980). These were well-controlled field tests, with BPA and/or Carrier closely monitoring field operation for at least a year to uncover potential problem areas. The chillers operated well, and the field tests concluded their development process, with Carrier going on to manufacture these machines for other installations. The overall solar cooling systems experienced some problems characteristic of early system field tests: high parasitic electric power usage due to oversized pumps (recognized in advance, but it was the only equipment available at the time) and some control and collector problems. It was planned to gather a small amount of thermal performance data for these systems, but very little quantitative data were actually obtained because of unsatisfactory performance of the Btu-meter instrumentation used.

20.4.7.3 National Solar Data Program: Data Acquisition and Analysis

A subset of solar heating and cooling demonstration projects and other field installations were selected for inclusion in the National Solar Data Network (NSDN) program. Extensive instrumentation was installed at the sites and transmitted over phone lines to a central location for data collection, checking, storage, analysis, and publication. For several years, the NSDN program was run for DOE by IBM, and then by Vitro Laboratories.

A specific solar system would typically be included in NSDN for a few heating or cooling seasons, to gather data on its performance, and then removed from the network as a new site was instrumented to replace it. Approximately thirty solar cooling sites were included in the NSDN program over the years. A large number of reports—system descriptions, system performance, system costs, reliability and material assessment, comparative assessments, etc.—were issued as part of the NSDN Program (DOE 1984).

Perhaps the most informative of all these reports, as far as solar cooling system performance is concerned, are the comparative assessment reports on the performance of active solar space cooling systems. Separate reports were issued for the 1980 cooling season (Blum et al. 1981), the 1981 cooling season (Wetzel and Pakkala 1982), and the 1982 cooling season (Logee and Kendall 1983).

The 1980 cooling season comparative report (Blum et al. 1981) covered eight solar cooling systems. All but one of these used Arkla absorption chillers; the exception was a 228-ton Carrier 16JB absorption chiller. Six of the eight sites were net energy losers, requiring more energy to operate the solar cooling system than if an auxiliary chiller had been used. Design values generally exceeded the solar cooling fractions actually achieved. Improper control and ineffective use of storage were major problems.

Five solar cooling systems were included in the 1981 cooling season comparative report (Wetzel and Pakkala 1982). One system used Rankine chillers. The other four used absorption chillers, including three with Arkla 25-ton chillers and one with a 148-ton Trane chiller. One of the sites had also been included in the 1980 cooling season report. Three of the systems, including the one with Rankine chillers, provided significant energy savings. The other two systems were net energy losers by slight amounts. None of the systems was cost-effective in terms of a reasonable payback period. Those systems that were built with a storage bypass option had better solar utilization. For all the systems, the actual solar fractions fell far short of the design values. Energy losses throughout the systems were excessive.

The comparative report for the 1982 cooling season (Logee and Kendall 1983) included five solar cooling systems. Four of these were repeats from the previous report. The fifth used another 25-ton Arkla chiller, replacing a similar system monitored the previous summer. These were all commercial-building-size systems, three using evacuated-tube, one flat-plate, and one linear concentrating tracking collectors. Four of the five systems were net energy losers, with most systems performing somewhat poorer than they did during the previous cooling season. High parasitic power usage was a major problem. Most of the cooling systems were over-sized for the average load by a factor of three, reducing chiller performance.

Overall, looking at the results of the three cooling seasons, it was premature to include these cooling systems in a demonstration program. The adage applies, "the systems should be developed before they are demonstrated."

20.4.7.4   United States-Saudi Arabian SOLERAS Projects

As part of the Saudi Arabian-United States Program for Cooperation in the Field of Solar Energy (SOLERAS), four solar cooling systems were designed, built, installed, and operated for several cooling seasons in a U.S. climate area that approximated climate conditions found in Saudi Arabia (Davis and Williamson 1980). Phoenix was selected as the field test location for all four cooling systems. Two systems used absorption chillers and two used Rankine chillers.

Carrier Corporation was the prime contractor for both of the absorption systems. One of these systems (Biermann and Orbesen 1982b) used the 35 kW (10 ton) air-cooled absorption chiller mentioned in section 20.3.2.2.1, driven by linear Fresnel-lens tracking collectors. Pressurized hot water in the collector loop drove the chiller at temperatures between 210° and 250°F (100° and 120°C). There was no hot-side storage used, only cold-side storage, which consisted of a 4000 gallon (15 m³) tank of chilled water. The chiller schematic was shown in figure 20.13.

Several major problems plagued the operation of this system. The most serious was the succession of air leaks occurring in the air-cooled absorption chiller, starting with a leak that developed during shipment to the installation site, and continuing throughout the period of the field test. A large amount of time was spent locating and repairing the leaks, only to have a new leak surface shortly after chiller operation restarted. The oxygen that entered the chiller produced an adverse reaction with the heat transfer additive, 1-nonylamine. A more suitable additive, phenylmethyl-carbinol, was identified and substituted, as has been mentioned in sections

20.2.3.1.1 and 20.3.2.2.1. The consequence of these problems was that the chiller generally operated far below its design capacity and COP values.

Collector problems served to worsen the situation. They operated at below-design capacity. In addition, the tracking control system suffered several failures. To compound the situation, one control failure allowed the collectors to rotate beyond their stop limits, causing the main support frame to be bent. This reduced collector output from that time onward. Finally, a failure in the data acquisition unit resulted in an absence of data while it was being repaired.

The second Carrier project within the SOLERAS program (Biermann and Orbesen 1982a) used a 53 kW (15 ton) water-cooled absorption chiller, as mentioned in section 20.3.2.2.2. Tracking parabolic-reflector trough collectors were used to drive the chiller. A 2500-gallon (9.5 m^3) hot-side water tank provided storage.

This system also experienced major operational problems. The most serious was the absorption chiller's inability to maintain its vacuum. The first air leak occurred during shipment of the unit to Phoenix, and an inability to find and fix a series of air leaks resulted in substandard chiller performance throughout the system's operation. This type of vacuum leak was unexpected in this equipment; three similar 15-ton chillers (for the BPA-DOE projects in section 20.4.7.2) had been installed in field sites and operated for 3 to 5 years with no history of leaks.

This project also experienced collector tracking problems, cooling tower problems, controls problems, and chiller crystallization. These two Carrier solar cooling SOLERAS projects offer further evidence of the need for additional chiller and cooling system testing in controlled situations (such as at CSU or near a manufacturer's facility) before these systems are ready for successful operation at field sites.

20.4.7.5 Solar in Federal Buildings Program Projects

Of the 720 solar demonstration projects within the DOE Solar in Federal Buildings Program (SFBP), eight included solar cooling (Rockwell 1983). Although a small percentage of the projects numberwise, the solar cooling projects were among the largest and most expensive, accounting for about 15% of the overall SFBP cost. All eight sites used absorption chillers: 25-ton Arkla units (at four sites), 3-ton Arkla units (at one site), a conventional Carrier 100-ton chiller, a conventional Trane 130-ton chiller, and one of the Carrier 75-ton chillers developed for solar applications (see section 20.3.2.2.2). Five of the systems were driven by evacuated-tube collectors, two by parabolic trough tracking collectors, and one by flat-plate collectors.

As SFBP is a demonstration program, most sites included minimal or no data acquisition equipment. However, eight sites were selected for inclusion in the NSDN program (section 20.4.7.3), including two solar cooling systems; it is expected that detailed performance information will be gathered for these sites. In addition, one cooling system was included in the set of ten SFBP sites that were being closely monitored using Btu-meter instrumentation. Reliability and maintenance data will also be collected and reported (Billings 1985).

20.5   Economic Considerations

As mentioned numerous times in this chapter, economic considerations must be foremost if solar cooling systems are ever to achieve widespread applications. Analytic and economic analysis, such as that of SAI in section 20.4.5.1.3, has shown that first-generation and even some second-generation solar cooling systems fall far short of meeting cost goals that would make them economically competitive with conventional nonsolar cooling systems. This economics topic is important enough to merit being the subject of a separate chapter (chapter 22) in this book, so it will not be treated further here.

20.6   Outlook—Future R&D Directions

Technically, the outlook is bright. There is a good match between solar collector capabilities and absorption chiller requirements. Collectors under development can operate efficiently up to 400°F (200°C), which is near the upper temperature limit of absorption chiller operation. At the lower temperature end, collectors can efficiently regenerate absorbent solutions for open-cycle absorption chiller requirements. Absorption chillers designed specifically for solar applications have been developed and operated successfully, some just to the prototype stage, others all the way to commercially available products.

Economically, the outlook is quite another story. To be economically competitive with nonsolar cooling alternatives, solar cooling systems must achieve lower costs or higher efficiencies or both. Higher efficiencies include reduced parasitic power requirements as well as higher thermal efficiency. The main factor in the economics equation is the cost of conventional fuels (gas/oil and electricity). As these fuel costs level off in many parts of the country, the horizon for solar cooling cost-effectiveness stretches out further into the future.

For single-effect closed-cycle absorption, the chiller development phase has been completed. The equipment developed under DOE/Solar support is close enough to commercializable products that private manufacturers could readily market this technology if a demand existed for it. However, the relatively low efficiency of single-effect chillers presents a serious economic impediment because the cooling produced (and conventional fuel displaced) per unit collector area is directly proportional to the chiller efficiency. Thus the single-effect chiller efficiency directly limits the acceptable (i.e., cost-effective) cost of collectors. For any specific case, an accurate economic evaluation requires consideration of the heating as well as the cooling provided by the solar system.

Low chiller efficiency may also present a practical problem in terms of the manufactured cost of the chiller itself, as production volume is usually an important factor in unit cost. In solar cooling systems, the chillers are normally gas-driven in the backup mode when solar energy is not available. But gas-driven single-effect chillers are relatively inefficient and usually not cost-competitive compared with good electric-driven chillers. (To compare gas-driven and electric-driven efficiencies, it is assumed that the electricity is generated at a 1/3 efficiency by a power plant.) As a result, gas-driven single-effect absorption machines are no longer being developed or commercialized by the private sector. It may prove very difficult to achieve, for solar applications alone, sufficient market volume to drive down the chiller manufacturing costs to a viable level.

The development of double-effect and other high-efficiency closed-cycle absorption machines for cooling and heating continues to be a promising direction, for both solar and nonsolar applications. It seems likely at this point that nonsolar applications (e.g., gas-driven) will become cost-effective first; i.e., they will compete successfully with electric-driven heat pumps for cooling and heating, and with gas furnace/electric air conditioner combinations. When this occurs, the competition for solar absorption cooling will be gas-driven versions of the same high-efficiency absorption machines. Then the economics picture for solar cooling and heating will become simple: the solar collector must be capable of delivering heat to the absorption machine at the same cost as natural gas can deliver that heat. Moreover, the absorption equipment must be capable of operating with solar-heated hot water as well as with a gas-flame input.

The implications for the solar cooling R&D program are twofold. First, the solar program must develop collectors capable of driving the high-efficiency chillers. The University of Chicago ISEC collector meets the performance requirements, but practical modifications to lower the costs are still needed. Secondly, the solar program must develop, either on its own or in collaboration with the conservation program, versions of the advanced-cycle absorption chillers/heat pumps that can operate on variable temperature hot water from solar collectors, in addition to the constant temperature input provided by a gas flame.

The other future solar cooling direction that still offers promise is at the other extreme: the very low-cost, low-efficiency (i.e., single-effect) open-cycle absorption systems. Since backup for these systems will likely require purchase of conventional cooling equipment, no economic credit can be gained for displacing such costs. Therefore, the solar cooling system incremental cost includes the open-cycle chiller as well as the collector/regenerator subsystem. This further reduces the allowable costs for the collectors. Costwise, it appears that the open (but glazed) rooftop collector/regenerator has the best chance among the open-cycle options of meeting these cost goals for solar cooling applications. Since these systems are generally inappropriate for providing heating, their use may be limited to installations having year-round cooling loads.

The R&D needs for the open-cycle absorption systems center on the environmental issues: the effects on chiller and regenerator performance of exposure of the working fluids to the environment (e.g., absorbent contamination by sulfur compounds in the air, loss of fluid additives by evaporation, etc.) and the effects on the environment of release to the air of small quantities of the working fluids and additives. No insurmountable obstacles have surfaced to date, but an extensive and comprehensive research effort is required before the marketplace would accept the fact that an absorbent solution can flow for 10 or 20 years over a rooftop with acceptable reliability, maintenance level, and environmental consequences.

For both approaches (high-efficiency chillers with high-efficiency collectors and open-cycle systems), the economic outlook is fairly pessimistic in the near term and clouded in the far term.

Before development of any of these approaches is finished, however, it is clear that a period of system integration and test is required before the systems will be ready for unattended field installations. Matching component characteristics, proper control (especially for part-load and off-design conditions), component reliability in a systems context, and maintenance guidelines must all be resolved before a further attempt is made to "demonstrate" solar cooling systems. Further, since successful solar cooling systems will likely be based on future chiller components, not those available today, it would be of limited usefulness to engage in extensive system integration testing for closed-cycle systems until the chiller development is closer to completion, and potentially marketable products can be identified.

Acknowledgments

The assistance of several colleagues in locating, providing, and reviewing material used in this chapter is greatly appreciated, especially the contributions of Al Heitz, Joseph Rasson, Kim Dao, Mashuri Warren, and Robert DeVault. Also of considerable benefit to the finished product were the comprehensive reviews of the draft chapter by George Löf, Wendell Biermann, Dave Anand, and Phil Anderson.

References

Proceedings

1. Premeeting Proceedings and Project Summaries for the Annual DOE Active Solar Heating and Cooling Contractors' Review Meeting. Washington, DC, September 1981, DOE report CONF.810912-PRELIM.

2. Proceedings of the Annual DOE Active Solar Heating and Cooling Contractors' Review Meeting. Incline Village, NV, March 1980, DOE report CONF 800340.

3. Active Solar Heating and Cooling System Development Projects. DOE report DOE/CS/30211–01, October 1980.

4. Solar Heating and Cooling Research and Development Project Summaries. DOE report DOE/CS-0010, May 1978.

5. Proceedings of the DOE Heat Pump Contractors' Program Integration Meeting. McLean, VA, June 1981, DOE report CONF-810672, published March 1982.

6. U.S. Heat Pump Research and Development Projects. DOE report DOE/CE-0035, August 1982.

7. IEA Heat Pump Center Report on R, D&D Project Descriptions. Vol. 1–4, 1st ed., December 1984. Volume 1: Basic Research in Heat Pump Technology, Report HPC-R1–1; volume 2: Heat Pump Applications in Residential and Commercial Buildings, Report HPC-R1–2; volume 3: Application of Heat Pumps in Industry, District Heating, and in Integrated Community Energy Systems, Report HPC-R1–3; and volume 4: Indexes, Report HPC-R1–4.

8. Proceedings of the Workshop on New Working Pairs for Absorption Processes. Berlin, April 1982. Edited by Wiktor Raldow. Stockholm, Sweden: Swedish Council for Building Research, publication D23:1982.

9. Proceedings of Absorption Experts 85 at the Absorption Heat Pumps Congress. Paris, March 1985.

10. Proceedings of the DOE/ORNL Heat Pump Conference: Research and Development on Heat Pumps for Space Conditioning Applications. Washington, DC, December 1984, DOE report CONF-841231, published August 1985.

11. Proceedings of the Solar Buildings Conference. Washington, DC, March 1985, DOE report DOE/CONF-850388, published July 1985.

12. Proceedings of the Third Solar Cooling Workshop, "The Use of Solar Energy for the Cooling of Buildings." San Francisco, CA, February 1978, DOE report CONF-780249, published August 1978. This document also contains a comprehensive solar cooling bibliography compiled by Francis de Winter et al., p. 368.

13. Proceedings of the Second Solar Cooling Workshop, "The Use of Solar Energy for the Cooling of Buildings." Los Angeles, CA, August 1975, DOE report SAN/1122–76/2, published July 1976.

14. Workshop Proceedings, "Solar Cooling for Buildings." Vol. 1, report NSF-RA-N-74–063, U.S. Government Printing Office stock no. 3800–00189, February 1974.

Other References

Allied Corporation. 1985. *Development of a residential gas fired absorption heat pump.* Final report under ORNL contract no. 86X-24610C and GRI contract no. 5014–341–0113. Vol. 1, *Component development and field trial program*; vol. 2, *Physical and thermodynamic properties of R123a/ETFE, system development and testing, and economic analysis.* Oak Ridge National Laboratory reports ORNL/Sub/79–24610/3 and ORNL/Sub/79–24610/4.

Anand, D. K. et al. 1978. SHASP, solar heating and air-conditioning simulation programs. Solar Energy Projects Report, University of Maryland.

Anand, D.K., R. W. Allen, and S. R. Venkateswaran. 1980. *Review of absorbent/refrigerant cycle fluid studies.* Solar Energy Projects Report, University of Maryland.

Anand, D. K., R. W. Allen, and B. Kumar. 1981. Transient simulation of absorption machines. *Proc. ASME 3d Annual Technical Conference on Solar Engineering,* Reno, NV.

Anand, D. K., B. Kumar, and N. Hall. 1982. *Estimation and validation of cooling system performance using closed form representation.* Solar Energy Projects Report, University of Maryland.

Anand, D. K., B. Kumar, and R. W. Allen. 1983. Regional simplified solar cooling design charts. *Proc. 16th Intersociety Energy Conversion Engineering Conference,* Atlanta, GA, August 1981, p. 1668. Also, University of Maryland report under DOE contract DE-AC03–79CS30204.

Anand, D. K., K. W. Lindler, S. Schweitzer, and W. J. Kennish. 1984. Second law analysis of solar powered absorption cooling cycles and systems. *J. Solar Energy Eng.* 106:291.

ASU. 1981. *Heat and mass transfer characteristics of liquid desiccant open-flow collectors.* Final report by Arizona State University under SERI subcontract XE-0–9179–2.

Auh, P. C. 1979a. *Development of hardware simulators for tests of solar cooling/heating subsystems and systems; phase 2. Unsteady state hardware simulation of residential absorption chiller.* Brookhaven National Laboratory report BNL-51121, September 1979. Also, Evaluation of performance enhancement of solar powered absorption chiller with an improved control strategy using the BNL-built hardware simulator. *Proc. International Solar Energy Society Silver Jubilee Congress,* Atlanta, GA, p. 715.

————. 1979b. *Development of hardware simulators for tests of solar cooling/heating subsystems and systems; phase 1. Residential subsystem hardware simulator and steady state simulation.* Brookhaven National Laboratory report BNL-51120.

Best, R., W. Biermann, and R. Reimann. 1985. *Operation of a low temperature absorption chiller at rating point and at reduced evaporator temperature.* Final report by Carrier Corporation under DOE contract DE-AC03–83SF11905, available from NTIS as report no. DOE/SF/11905-T1 (DE85006308).

Biermann, W. J. 1978. *Candidate chemical systems for air cooled solar powered absorption air conditioner design: part I—Organic absorbent systems; part II—Solid absorbents, high latent heat refrigerants; part III—Lithium salts with antifreeze additives.* Report by Carrier Corporation under DOE contract EG-77-C-03–1587.

Biermann, W. J. 1980. Prototype modular 15 ton solar absorption air conditioning system. *Proc. Annual DOE Active Solar Heating and Cooling Contractors' Review Meeting,* Incline Village, NV, p. 2–13.

Biermann, W. J., and R. C. Reimann, 1981. Water cooled absorption chillers for solar cooling applications. *Premeeting Proc. Annual DOE Active Solar Heating and Cooling Contractors' Review Meeting,* Washington, DC, p. 2–15.

Biermann, W. J., and G. B. Orbesen. 1982a. *53 kW water cooled solar cooling package for BDP distribution center in Phoenix, Arizona.* Final report by Carrier Corporation under DOE Contract DE-FC02–80ET20643.

————. 1982b. *35 kW air cooled solar cooling package for Sigler-Faigen parts and supply building computer room in Phoenix, Arizona.* Final report by Carrier Corporation under DOE Contract DE-FC02–80ET20643.

Biermann, W. J., and R. C. Reimann. 1984. *Advanced absorption heat pump cycles.* Phase 1 final report under DOE contract no. DE-AC05–84OR21400.

Billings, G. J., III. 1985. Solar in Federal Buildings Program. *Proc. Solar Buildings Conference,* Washington, DC, p. 269.

Blum, D., S. Frock, T. Logee, D. Missal, and P. Wetzel. 1981. *Comparative report: performance of active solar space cooling systems, 1980 cooling season.* DOE report SOLAR/0023–81/40, prepared by Vitro Laboratories under DOE contract DE-AC01–79CS30027.

Borst, R. R., and B. D. Wood. 1985. Dynamic performance testing of prototype 3-Ton air-cooled Carrier absorption chiller. Report ERC-R-85013, College of Engineering and Applied Sciences, ASU.

Botham, R. A., G. L. Ball, III, G. H. Jenkins, and I. O. Salyer. 1978. *Form-stable crystalline polymer pellets for thermal energy storage—high density polyethylene intermediate products.* Final report ORNL/Sub-7398/4 by Monsanto Research Corporation under subcontract to Oak Ridge National Laboratory.

Choi, M. K., J. R. Stokely, and J. H. Morehouse. 1982. *Analysis of commercial and residential solar absorption and Rankine cooling systems.* Final report by Science Applications, Inc., under DOE contract DE-AC03–81SF11573.

Clark, E. C. 1981. Sulfuric acid-water chemical heat pump. *Proc. DOE Heat Pump Contractors' Program Integration Meeting*, McLean, VA, p. 55.

Cole, R. L. 1981. Development of TES device using cross-linked HDPE. *Premeeting Proc. Annual DOE Active Solar Heating and Cooling Contractors' Review Meeting*, Washington, DC, p. 10–1.

Cole, R. L. 1983. *Thermal storage device based on high-density polyethylene.* Argonne National Laboratory report ANL-83–52.

Collier, R. K. 1979. The analysis and simulation of an open cycle absorption refrigeration system. *Solar Energy* 23:357.

Curran, H. M., and J. DeVries. 1981. *Options for thermal energy storage in solar cooling systems.* Final report by Hittman Associates, Inc., under DOE contract DE-AC03–79CS30202.

Curran, H. M., and S. Heibein. 1979. Thermal storage for solar Rankine and absorption cooling systems. Proceedings of Solar Energy Storage Options, an Intensive Workshop on Thermal Energy Storage for Solar Heating and Cooling. San Antonio, TX. DOE report CONF-790328, vol. 1, p. 97.

Dao, K., M. Simmons, R. Wolgast, and M. Wahlig. 1977. *Performance of an experimental solar-driven absorption air conditioner.* Lawrence Berkeley Laboratory report LBL-5911.

Dao, K. 1978a. *A new absorption cycle: The single-effect regenerative absorption refrigeration cycle.* Lawrence Berkeley Laboratory report LBL-6879.

———. 1978b. *Conceptual design of an advanced absorption cycle: The double-effect regenerative absorption refrigeration cycle.* Lawrence Berkeley Laboratory report LBL-8405.

Dao, K., J. Rasson, and M. Wahlig. 1983. *Testing results of an ammonia-water absorption air conditioner designed for solar cooling applications.* Lawrence Berkeley Laboratory report LBL-16929, draft.

Dao, K. 1979. Personal communication with author.

Davis, R. E., and J. S. Williamson. 1980. Five solar cooling projects. *Proc. 1980 Annual Meeting of the American Section of ISES*, Phoenix, AZ, p. 196.

DOE. 1979a. *Solar heating and cooling demonstration project summaries.* DOE report DOE/CS-0038–2.

———. 1979b. *Proceedings of Solar Energy Storage Options, an Intensive Workshop on Thermal Energy Storage for Solar Heating and Cooling.* San Antonio, TX. DOE report CONF-790328-P1, P2, P3.

———. 1984. *Availability of Solar Energy Reports from the National Solar Data Program.* DOE report SOLAR/0020–84/23.

Doering, E. R. 1978. Solar cooling projects in the Commercial Demonstration Program—summary. *Proc. 3d Solar Cooling Workshop*, San Francisco, CA, p. 55.

Eisa, M. A. R., S. K. Chandhari, D. V. Paranjapa, and F. A. Holland. 1986. Classified references for absorption heat pump systems from 1975 to May 1985. *Heat Recovery Systems* 6:47.

Froemming, J. M., B. D. Wood, and J. M. Guertin. 1979. Dynamic test results of an absorption chiller for residential solar applications. *ASHRAE Trans.* 85 pt. 2, p. 777.

Grossman, G., and A. Johannsen. 1981. Solar cooling and air conditioning. *Progress in Energy and Combustion Science* 7:185.

Grossman, G., and H. Perez-Blanco. 1982. Conceptual design and performance analysis of absorption heat pumps for waste heat utilization. *International Journal of Refrigeration*, vol. 5, p. 361; also, *ASHRAE Trans.* 88 pt. 1, p. 451.

Grossman, G., and K. W. Childs. 1983. Computer simulation of a lithium bromide-water absorption heat pump for temperature boosting. *ASHRAE Trans.* 89 pt. 1B, p. 240.

Grossman, G. 1983. Simultaneous heat and mass transfer in film absorption under laminar flow. *Int. J. Heat and Mass Transfer* 26:357.

Grossman, G., and M. Heath. 1984. Simultaneous heat and mass transfer in absorption of gases in turbulent liquid films. *Int. J. Heat and Mass Transfer* 27:2365.

Grossman, G. 1985a. Multistage absorption heat transformers for industrial applications. *ASHRAE Trans.* 91 pt. 2B, p. 2047.

Grossman, G. 1985b. Heat and mass transfer in film absorption. Chapter 6 in *Handbook for heat and mass transfer operations*, edited by, N. P. Cheremissinoff. Gulf Publishing.

Guertin J. M., and B. D. Wood. 1980. Transient effects on the performance of a residential solar absorption chiller. *Proc. American Section of ISES Conference*, Phoenix, AZ, p. 186.

Guertin, J. M., B. D. Wood, and B. W. McNeill. 1981. Residential solar absorption chiller thermal dynamics. Report ERC-R-81013, College of Engineering and Applied Sciences, ASU.

Hanna W. T., and W. H. Wilkinson. 1981. Compound absorption cooling from low-grade heat using advanced desorbers. *Proc. DOE Heat Pump Contractors' Program Integration Meeting*, McLean, VA, p. 43.

Hanna, W. T., and W. H. Wilkinson. 1982. Absorption heat pumps and working pair developments in the United States since 1974. *Proc. Workshop on new Working Pairs for Absorption Processes*, Berlin, p. 71.

Hanna, W. T., M. L. Lane, and L. T. Whitney. 1983. Industrial applications for waste heat powered temperature booster. *Proc. 18th Intersociety Energy Conversion Engineering Conference*, Orlando, FL, p. 1906.

Hayes, F. C., and R. J. Modahl. 1984. Evaluation of advanced design concepts for absorption heat pumps. *Proc. DOE/ORNL Heat Pump Conference*, Washington, DC, p. 291.

Herczfeld, P. R., R. Fischl, and D. K. Anand. 1984. *Research on the control of active solar space conditioning systems*. First annual report by Drexel University and TPI, Inc., under DOE contract DE-AC03–83SF11950.

Hodgett, D. L. 1982. Absorption heat pumps and working pair developments in Europe since 1974. *Proc. Workshop on New Working Pairs for Absorption Processes*, Berlin, p. 57.

Hughes, P. J., J. H. Morehouse, M. K. Choi, N. M. White, and W. B. Scholten. 1981. *Evaluation of thermal storage concepts for solar cooling applications*. Report SERI/TR-09083–1 by Science Applications, Inc., under SERI subcontract XI-0–9083–1.

Huntley, W. R. 1983. Performance test results of an absorption heat pump that uses low temperature (140°F) industrial waste heat. *Proc. 18th Intersociety Energy Conversion Engineering Conference*, Orlando, FL, p. 1921.

Ikeuchi, M., T. Yumikura, E. Ozaki, and G. Yamanaka. 1985. Design and performance of a high-temperature-boost absorption heat pump. *ASHRAE Trans.* 91 pt. 2B, p. 2081.

Karaki, S., P. Vatano, and P. Brothers. 1984. *Performance evaluation of a solar heating and cooling system consisting of an evacuated-tube heat pipe collector and an air-cooled lithium bromide chiller*. Report SAN-30569-32, prepared by Colorado State University under DOE contract DE-AC03–81SF30569.

Karaki, S., D. Wiersma, T. E. Brisbane, and G. O. G. Löf. 1984. *Performance of a solar heating and cooling system with phase change storage in CSU Solar House I*. Report SAN-11927–16, prepared by Colorado State University under DOE contract DE-AC03–83SF11927.

Karaki, S., T. E. Brisbane, and G. O. G. Löf. 1984. *Performance of the solar heating and cooling system for Solar House III for the summer season 1983 and winter season 1983–84*. Report SAN-11927–15, prepared by Colorado State University under DOE contract DE-AC03–83SF11927.

Kuhlenschmidt, D., and R. H. Merrick. 1983. An ammonia-water absorption heat pump cycle. *ASHRAE Trans.* 89 pt. 1B, p. 215.

Lacey, R. E. 1978. *Analysis of advanced conceptual designs for single-family-sized absorption chillers*. Final report by Southern Research Institute under DOE contract DE-AC03–77CS31586.

Leboeuf, C. 1979. *Standard assumptions manual for solar system simulation*. Solar Energy Research Institute report.

Leboeuf, C. M., and G. O. G. Löf. 1980. Open-cycle absorption cooling using packed-bed absorbent reconcentration. *Proc. 1980 Annual Meeting of the American Section of ISES*, Phoenix, AZ, p. 205.

Lenz, T. G., G. O. G. Löf, R. Iyer, and J. Wenger. 1983. Residential cooling with solar heated air and open-cycle lithium chloride absorption chiller. *Proc. 8th Biennial Congress of ISES: Solar World Congress*, Perth, vol. 1, p. 125.

Lenz, T. G., G. O. G. Löf, J. Wenger, R. Iyer, V. Kaushik, and G. Sandell. 1984. Open cycle lithium chloride cooling system. Report SAN-11927–17, prepared by Colorado State University under DOE contract DE-AC03–83SF11927.

Lenz, T. G., et al. 1982. *Removal of non-condensable gases in an open cycle lithium chloride absorption air conditioner*. Report SAN-30569–16, prepared by Colorado State University under DOE contract DE-AC03–81SF30569.

Lockheed 1981. *Heat and mass transfer characteristics of liquid desiccant open-flow solar collectors*. Final report by Lockheed-Huntsville Research and Engineering Center under SERI subcontract XE-0–9179–1.

Löf, G. O. G., T. G. Lenz, and S. Rao. 1984. Coefficients of heat and mass transfer in a packed bed suitable for solar regeneration of aqueous lithium chloride solutions. *J. Solar Energy Eng.*, vol. 106, p. 387.

Löf, G. O. G., M. A. Westoff, and S. Karaki. 1984. *Performance evaluation of an active solar cooling system utilizing low cost plastic collectors and an evaporatively-cooled absorption chiller*. Report SAN-30569–30, prepared by Colorado State University under DOE contract DE-AC03–81SF30569.

Löf, G. O. G., and T. Lenz. 1985. Open cycle absorption cooling. *Proc. Solar Buildings Conference*, Washington, DC, p. 87.

Logee, T., and P. Kendall. 1983. *Comparative report: Performance of active solar space cooling systems, 1982 cooling season*. DOE report SOLAR/0023–83/40 (DE83016710), prepared by Vitro Laboratories under DOE contract DE-AC01–79CS30027.

MacCracken, C. D. 1980. Salt hydrate thermal energy storage system for space heating and air conditioning. *Proc. Annual DOE Active Solar Heating and Cooling Contractors' Review Meeting*, Incline Village, NV, p. 11–15.

Macriss, R. A., J. T. Cole, M. T. Kouo, and T. S. Zawacki. 1979. *Analysis of advanced conceptual designs for single-family-sized absorption chillers*. Final report by Institute of Gas Technology under DOE contract DE-AC03–77CS31439, available from NTIS as report SAN-1439–2.

Macriss, R. A. 1982. Overview and history of absorption fluids developments in the U.S.A. (1927–1974). *Proc. Workshop on New Working Pairs for Absorption Processes*, Berlin, p. 37.

Macriss, R. A., and T. S. Zawacki. 1984. *Absorption fluids data survey*. Final report by Institute of Gas Technology under DOE contract AC05–84OR21400; also, *Proc. DOE/ORNL Heat Pump Conference*, Washington, DC, p. 235.

Marat, A. R. 1984. The GRI gas heat pump program—an update. *Proc. DOE/ORNL Heat Pump Conference*, Washington, DC, p. 247.

McLinden, M. O., R. Radermacher, and D. A. Didion. 1983. A laboratory investigation of the steady-state and cyclic performance of an air-to-water absorption heat pump. *Proc. Congress of the International Institute of Refrigeration*, Paris, session E2, paper 472.

Merrick, R. H. 1981. Development and demonstration of an ammonia-water absorption heat pump. *Proc. DOE Heat Pump Contractors' Program Integration Meeting*, McLean, VA, p. 41.

Merrick, R. H. 1982. A direct, evaporatively cooled, three-ton lithium bromide-water absorption chiller for solar application. *ASHRAE Trans.* 88 pt. 1, p. 797.

Merrick, R. H., and J. G. Murray. 1984. *Evaporatively cooled chiller for solar air conditioning systems design and field test.* Final report by Arkla Industries under DOE contract AC03-77CS-34593.

Murphy, K. P. 1981. Organic absorption gas-fired residential heat pump. *Proc. DOE Heat Pump Contractors' Program Integration Meeting*, McLean, VA, p. 38.

Murphy, K. P., and B. A. Phillips. 1983. Development of a residential gas absorption heat pump. *Proc. 18th Intersociety Energy Conversion Engineering Conference*, Orlando, FL, p. 1911.

Newton, A. B. 1978. Using controls to reduce component size and energy needs for solar HVAC. *ASHRAE Trans.* 84 pt. 2, p. 387.

O'Gallagher, J. J., K. A. Snail, and R. Winston. 1982. A new evacuated CPC collector tube. *Solar Energy* 29:575.

Offenhartz, P. 1981. Solar heating and cooling with the calcium chloride-methanol chemical heat pump. *Proc. DOE Heat Pump Contractors' Program Integration Meeting*, McLean, VA, p. 75.

Offenhartz, P. 1978. Development of a solid-phase absorption air conditioner with built-in storage. Proc. 3d Solar Cooling Workshop, San Francisco, CA, p. 35.

Phillips, B. A. 1982. Development of absorption heat pumps using organic fluid pairs. *Proc. Workshop on New Working Pairs for Absorption Processes*, Berlin, p. 137.

Phillips, B. A. 1984. Analyses of advanced residential absorption heat pump cycles. *Proc. DOE/ORNL Heat Pump Coference, Washington*, DC, p. 265.

Podoll, R. T., T. E. Gray, C. E. Rolon, and K. A. Sabo. 1982. *Determination of properties of fluids for solar cooling applications.* Final report by SRI International under DOE contract DE-AC03-80CS30221.

Radermacher, R. 1984. Laboratory measurements on absorption heat pumps. *Proc. DOE/ORNL Heat Pump Conference*, Washington, DC, p. 227.

Rasson, J., K. Dao, and M. Wahlig. 1986. The double-effect regenerative absorption heat pump: Cycle description and experimental test results. Paper submitted to 1986 IIR Conference on Progress in the Design and Construction of Refrigeration Systems, Purdue Univ..

Rauch J. S., and B. D. Wood. 1976. Steady state and transient performance limitations of the Arkla Solaire absorption cooling system. *Proc. Joint Conference of the American Section of ISES and the Solar Energy Society of Canada*, Winnipeg, Canada, vol. 3.

Reid, E. A. Jr. 1981. High efficiency advanced absorption heat pump. *Proc. DOE Heat Pump Contractors' Program Integration Meeting*, McLean, VA, p. 33.

Reimann, R. C. 1984. Advanced absorption heat pump cycles. *Proc. DOE/ORNL Heat Pump Conference*, Washington, DC, p. 259.

Reimann, R. C., and W. J. Biermann. 1984. Development of a single-family absorption chiller for use in a solar heating and cooling system. Phase III final report by Carrier Corporation under DOE contract EG-77-C-03–1587.

Rockwell International Corporation. 1983. Unpublished documents on the Solar in Federal Buildings Program by Paul J. Pekrul and Oscar R. Hillig of the Energy Technology Engineering Center, Rockwell International Corporation, Canoga Park, CA.

Ryan, J. D. 1983. Fluid pairs for advanced absorption cycles: Selection methods and data requirements. *Proc. 18th Intersociety Energy Conversion Engineering Conference*, Orlando, FL, p. 1915.

Schertz, W. W., J. R. Hall, R. Winston, and J. O'Gallagher. 1985. Advanced evacuated tube collectors. *Proc. Solar Buildings Conference*, Washington, DC, p. 59.

Schlepp, D. R., and R. K. Collier. 1981. The use of an open-cycle absorption system for heating and cooling. *Proc. 1981 Annual Meeting of the American Section of ISES*, Philadelphia, p. 651.

Snail, K. A., J. J. O'Gallagher, and R. Winston. 1984. A stationary evacuated collector with integrated concentrator. *Solar Energy* 33:441.

Stephan, K. 1982. Absorption heat pumps and working pair developments in Europe until 1974. *Proc. Workshop on New Working Pairs for Absorption Processes*, Berlin, p. 19.

Vliet, G. C., and J. Kim. 1983. Modeling of double effect water-lithium bromide absorption cycle. *Proc. 18th Intersociety Energy Conversion Engineering Conference*, Orlando, FL, p. 1900.

Wahlig, M. 1979 Central storage for building cooling. Panel report, *Proc. Solar Energy Storage Options, an Intensive Workshop on Thermal Energy Storage for Solar Heating and Cooling, San Antonio, TX.* DOE report CONF-790328, vol. 3, p. 139.

Ward, D. S., and H. S. Oberoi. 1980. *ASHRAE handbook of experiences in the design and installation of solar heating and cooling systems.* DOE report DOE/CS/32224-T1, prepared by ASHRAE and Colorado State University under DOE contract AC01–76CS32224.

Warren, M. L., and M. Wahlig. 1982. *Cost and performance goal methodology for active solar cooling systems.* Lawrence Berkeley Laboratory report LBL-12753.

Weber, B., R. Radermacher, and D. Didion. 1984. *Test procedures for rating residential heating and cooling absorption equipment.* National Bureau of Standards report NBSIR 84–2867.

Wetzel, P., and P. Pakkala. 1982. *Comparative report: Performance of active solar space cooling systems, 1981 cooling season.* DOE report SOLAR/0023–82/40 (DE82016025), prepared by Vitro Laboratories under DOE contract DE-AC01–79CS30027.

Wood, B. D., D. A. Siebe, M. A. Applebaum, K. S. Novak, and L. M. Ballew. 1983. *Performance characteristics of open-flow liquid desiccant solar collector/regenerator for solar cooling applications and system simulation and performance measurements.* Final report by Arizona State University under DOE contract DE-AC03–82SF11691.

Wood, B. D., D. A. Siebe, and R. K. Collier. 1985. Open cycle absorption system: A low cost solar heating and cooling option. *Proc. Solar Buildings Conference*, Washington, DC, p. 94.

21 Desiccant Systems and Components

John Mitchell

21.1 Introduction

Solar desiccant air-conditioning systems use a desiccant to remove moisture from an air stream. The dried air is processed by evaporative coolers to produce a relatively dry, cool air stream, which then cools the building space. Solar energy can be used to regenerate the desiccant. These systems offer the promise of a high thermal coefficient of performance using the moderately low temperatures compatible with solar collectors. The system configurations can be relatively simple and often use low-cost components. The technology required appears to be possible. As a result, increasing DOE support has gone into developing these systems.

In this chapter, the two major types of desiccant systems are described: solid desiccant systems, in which the desiccant is a stationary bed or rotating matrix, and liquid desiccant systems, in which the desiccant is pumped around between air streams. The solid desiccant development has received considerable DOE support and has been extensively studied, whereas liquid desiccant systems have been investigated to a much lesser extent.

Desiccant air-conditioning systems are classified as either open- or closed-cycle systems. Air flows through the components in open-cycle systems, and moisture and heat are transferred between the air stream, the atmosphere, and heat sources. The ambient serves as the sink for the discharged moisture. Closed cycles only exchange heat with the surrounding atmosphere.

Open-cycle systems are characterized by whether the desiccant is a liquid or a solid. In a liquid system, the desiccant is circulated, and water vapor in air streams is alternatively absorbed by and desorbed from the desiccant. In absorption, the vapor is condensed and goes into solution. The saturation vapor pressure above the liquid desiccant is less than that for pure water. The liquid is regenerated by heating to a relatively high temperature to increase the saturation vapor pressure and evaporate the water.

An analogous set of processes occurs for solid systems. Water vapor in an air stream is adsorbed by a solid desiccant. In adsorption, water vapor is condensed on the desiccant surface at a pressure lower than the normal saturation pressure of the vapor. The common mechanisms are condensation in small pores in the solid and chemical attraction between molecules. In the regeneration of a solid desiccant, the adsorbed water is evaporated into a high temperature air stream passing over the desiccant.

Many materials can serve as solid or liquid desiccants. Calcium chloride, lithium chloride, and glycols are common liquids; silica gel, zeolite, molecular sieves, calcium chloride, and lithium chloride may be solid desiccants. Often an inert substrate is required to support the desiccant.

The ability of a given desiccant to remove water vapor is characterized by the relationship between the equilibrium water content of the desiccant and the relative humidity of the air-water vapor mixture in contact with it. The relationship, known as an isotherm, is given in figure 21.1 for typical materials. Desiccants with high water content at low relative humidity are favorable for open-cycle systems.

In dehumidification, air at a given humidity and temperature passes over the desiccant, which eventually comes to its equilibrium water content. To regenerate the desiccant, air at a lower relative humidity must be used. Heated air, which at the same humidity ratio is at a lower relative humidity, passes over the desiccant to remove moisture.

The energy associated with the sorption and desorption processes is also important. This energy is equal to the latent heat of condensation plus a differential heat of sorption. It is beneficial to have a low total heat of sorption. For most desiccants, the values are within 20% of the heat of condensation of pure water.

There are several measures of the thermal performance of desiccant systems. These ratios characterize the cooling effect relative to the energy inputs. Desiccant systems as conceived at present use combinations of solar energy, electricity, and auxiliary energy for the processes. The four relevant coefficients of performance (COPs) are

$$COP_{thermal} = \frac{Cooling}{Thermal\ input} = COP_t, \qquad (21.1)$$

where the cooling is the total cooling provided, and the thermal input is the sum of all thermal energies supplied;

$$COP_{solar} = \frac{Cooling}{Solar\ input} = COP_s, \qquad (21.2)$$

where the solar input is the solar energy incident on the collector array

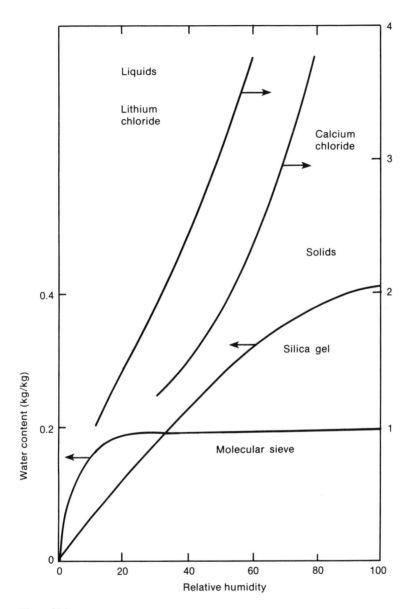

Figure 21.1
Isotherms for typical desiccants.

(this term has not been used for desiccant systems but becomes an important measure for comparison to other systems);

$$COP_{electric} = \frac{Cooling}{Electricity\ input} = COP_e, \tag{21.3}$$

where the electricity input is the sum of the fan, pump, etc. energies used to provide the air and liquid flows (this COP is often reported as an EER with units of Btu per watt-hour); and

$$COP_{resource} = \frac{Cooling}{Resource\ energy\ input} = COP_r, \tag{21.4}$$

where the resource energy input is the sum of all the resource energy required to produce cooling (auxiliary and electricity).

These various measures are helpful in comparing desiccant systems among themselves and with other cooling options.

21.2   Solid Desiccant Systems

21.2.1   Dehumidifier Development

The dehumidifier is the essential component of all desiccant systems, and considerable work has gone into its development. Prior to the interest in solar air-conditioning, industrial desiccant dryers with rotary beds were available, marketed by several manufacturers. These rotary beds are shown schematically in figure 21.2. The Bry-Air Company produced a unit with silica-gel particles packed between metal screens; Cargocaire, a unit with a fiberglass matrix impregnated with lithium chloride; and Munters, a unit with paper impregnated with a molecular sieve. The latter two units have parallel flow passages that are triangular in flow cross-section with a corresponding low pressure drop. In industrial applications, regeneration is usually accomplished through combustion of natural gas at high temperatures.

The application to solar air-conditioning placed new constraints on bed design. Regeneration must be achieved with the relatively low temperatures of 60°–80°C (140°–176°F) that can be obtained from solar collectors. High heat and mass transfer coefficients coupled with low diffusion resistance inside the desiccant are needed to produce adequate dehumidification at these low regeneration temperatures. Pressure drop has to be

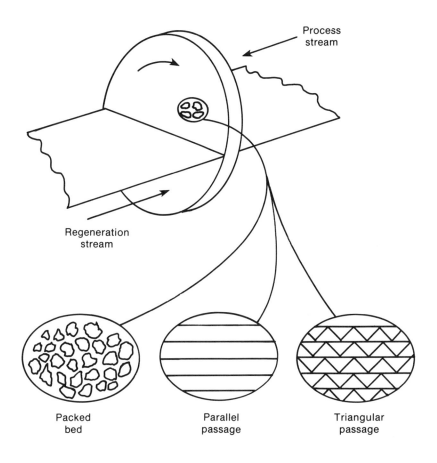

Matrix configuration

Figure 21.2
Schematic of adiabatic dehumidifier design.

minimal. Material selection and the configuration of the matrix are important. The generally accepted goal has become a rotary bed composed of parallel flow passages that are rectangular to allow laminar flow and a high ratio of transfer coefficient to friction factor. The desiccant will probably be fabricated in a thin layer to minimize diffusion resistance. In the solar desiccant program, both laboratory and model development have become geared to these goals.

21.2.1.1 Experimental Tests on Dehumidifiers

A variety of bed designs have been experimentally evaluated during the course of the solar-desiccant program. Tests on rotating packed beds made of silica-gel particles were carried out by investigators at the University of Wisconsin and the Commonwealth Scientific and Industrial Research Organization (CSIRO) in Australia (van Leersum and Close 1982). The gel particles were approximately 3 mm (0.12 in.) in diameter. The bed was found to be unable to adsorb sufficient moisture at the low regeneration temperatures employed, and the parasitic power was high. The Lewis number (ratio of mass transfer resistance to that for heat transfer) was in the range of 15 to 20; it was desired to have Lewis numbers of 1 to 2. These tests verified that beds of packed particles were not suitable.

A fixed, cooled bed was developed by Illinois Institute of Technology (IIT) (Worek and Lavan 1980, 1982). Cooling reduced regeneration temperatures. Paperlike silica-gel sheets 1.5 mm (0.06 in.) thick were constructed from micron-size silica-gel particles held in a teflon web. These sheets were then bonded to 0.2 mm (0.008 in.) thick aluminum sheets. An air-to-air cross-flow heat exchanger consisting of 2 mm (0.08 in.) wide rectangular flow channels was then constructed from these sheets (figure 21.3). Tests showed that thin silica gel sheets could be obtained. However, the resistance associated with diffusion through the teflon significantly reduced performance. The high thermal capacitance aluminum sheets made the approach unsatisfactory for a rotary matrix.

Experimental work at UCLA and the Solar Energy Research Institute (SERI) was aimed at producing a rotary bed composed of parallel flow passages (Barlow 1981). The process eventually developed consisted of coating both sides of a Mylar® sheet with a temperature-setting adhesive and then sprinkling approximately 50 μm diameter gel particles on the sheet. The bed was formed by winding the sheet spirally around a hub with spacers to form parallel passages. This wheel was tested extensively.

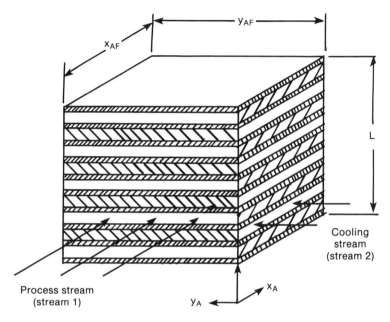

Figure 21.3
Schematic of a cooled desiccant dehumidifier. Source: Worek and Lauan 1980.

Proprietary work was conducted at Exxon on a concept similar to that at SERI (Epstein et al. 1983). Although many details of this work were never made public, the wheel construction and performance were similar to those at SERI. This study initiated a significant effort by the Gas Research Institute (GRI) in gas-fired desiccant cooling systems in 1982. GRI is cosponsoring a hybrid desiccant air-conditioning application for supermarkets. The dehumidifier is made by Cargocaire, with lithium chloride as the desiccant. Regeneration energy is supplied by natural gas. The application appears promising.

21.2.1.2   Model Development
The models used to analyze and design rotary desiccant dehumidifiers are based on the governing equations for heat and mass transfer. It is assumed that the dehumidifier is composed of a matrix that rotates between the two air streams. The air flow can be either counter- or parallel-flow. The following assumptions are usually made: the air flow in each period (process or regeneration) is at constant pressure and velocity; the interstitial air

thermal and moisture capacities are negligible compared to the matrix capacities; axial heat and moisture diffusion in air and matrix are negligibly small; the combined diffusion-convective film resistances to heat and mass transfer are equal (Lewis number of unity) and can be described by composite coefficients; and there is no transfer flux coupling. Under these constraints, the equations of mass and energy conservation for a coordinate system moving with respect to the matrix are:

$$Conservation\ of\ mass \quad \mu\frac{\partial W}{\partial \theta} + v\frac{\partial \omega}{\partial x} = 0; \tag{21.5}$$

$$Conservation\ of\ energy \quad \mu\frac{\partial H}{\partial \theta} + v\frac{\partial h}{\partial x} = 0, \tag{21.6}$$

where θ is time, x is the axial coordinate, μ is the ratio of matrix mass to that for the air, W and H are the matrix water content and enthalpy, v is the air velocity, and ω and h are air water vapor content (humidity ratio) and enthalpy.

The transfer rate equations for mass and energy are:

$$Transfer\ of\ water\ vapor \quad v\frac{\partial \omega}{\partial x} = J(\omega - \omega_f); \tag{21.7}$$

$$Transfer\ of\ heat \quad v\frac{\partial h}{\partial x} = J(h - h_f), \tag{21.8}$$

where J is the transfer coefficient, and f stands for properties measured at the matrix surface.

Since the coordinate system is fixed with respect to the matrix, the rotary problem is handled using periodic boundary conditions with a flow reversal at the beginning of each new period. Most of the analyses conducted have assumed that the flow velocity, temperature, and moisture content are uniform over each inlet face. The inlet conditions are assumed either to be constant with time or to vary slowly compared to the rotation time of the matrix.

The equations are commonly solved by use of a finite-difference approach. They are written in difference form for both time and space coordinates. The wave nature of the equations requires care in the numerical solution, and a variety of methods are employed. Solutions can be obtained for different time and grid sizes and then extrapolated to zero grid size (Jurinak 1982; Epstein et al. 1983; Pla-barby and Vliet 1978). The main disadvantage of using finite differences is the extensive computer time required to solve the coupled energy and mass conservation equations for both periods and to continue the solution until a cyclic steady state is reached. As a result of this disadvantage, research effort has been directed toward simplified solutions.

If the heat and mass transfer coefficients are infinite, then the equations describe equilibrium changes in the local air and matrix moisture and enthalpy. Banks and Close (1972) and Banks (1972) have shown that by proper definition of combined heat and mass transfer potentials and capacitance rates, the equations can be written in terms of two independent potentials, F_i. The solution to the equations produces the process lines for an equilibrium sorption process. On a psychrometric chart, the process lines are similar to enthalpy and relative humidity lines. The intersection points represent the outlet states for the equilibrium dehumidifier.

The equilibrium solution has been extended to account for finite transfer coefficients. In a series of papers, Banks (1972), Banks and Close (1972), and Maclaine-Cross (1972) have shown that the performance of a dehumidifier can be determined by analogy to that for a sensible heat exchanger. This analogy allows two effectivenesses similar to those for sensible heat exchangers to be defined, and facilitates a rapid calculation of dehumidifier performance. The values of effectiveness are either calculated from the finite-difference solution or obtained from measurements. The analogy approach has been extensively used in system simulations (Jurinak, Mitchell, and Beckman 1984; Nelson et al. 1978; Schultz, Mitchell, and Beckman 1982; Howe, Beckman, and Mitchell 1983a, 1983b; Sheridan and Mitchell 1985) for both adiabatic and cooled dehumidifiers.

A different method has been developed by Barlow (1981) for computing dehumidifier performance. The approach is based on an analogy between the flow of air over the desiccant to that between a flow stream of air and a stream of desiccant. The performance is then calculated from counterflow heat exchanger effectiveness relations. The method has been compared favorably with experiments. However, it is a heuristic approach with no basic theory underlying the method.

The main limitation of these models is the lack of treatment of adsorption phenomena in detail. In many desiccants, hysteresis is present and sorption and desorption isotherms differ. Diffusion resistance within the solid is generally not directly accounted for. Inclusion of the transient diffusion term inside the solid is prohibitively expensive in computer time. For system simulations, the effectiveness approach is useful, while for design purposes, a more fundamental approach must be employed.

21.2.1.3 Performance Predictions

The performance of a rotary dehumidifier can be predicted by use of the models discussed in section 21.2.1.2. The outlet state of the process air is conveniently shown on psychrometric coordinates as in figure 21.4. The outlet state lies between the inlet states of the process and regenerating streams, and depends on the rotational speed of the dehumidifier.

The outlet states for an ideal dehumidifier follow the characteristic lines labeled F_1 and F_2 as shown in figure 21.4. These characteristics are similar in shape to enthalpy and relative humidity, and represent the outlet state of the process air in equilibrium with the desiccant throughout the dehumidifier.

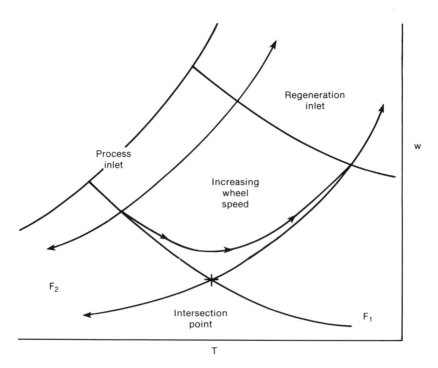

Figure 21.4
Schematic of adiabatic dehumidifier operation.

The theoretical performance of a rotary dehumidifier has been studied extensively, and much is known about its operation (Jurinak, Mitchell, and Beckman 1984; Nelson et al. 1978; Pla-barby and Vliet 1978, 1979; Jurinak 1982; Van den Bulck 1983). For an ideal dehumidifier, the processes follow the F_1 and F_2 lines. For a dehumidifier with finite heat and mass transfer coefficients, the process states follow the F_1 and F_2 lines in shape. At very low rotation speed, the process air experiences little dehumidification and exits essentially at the process inlet state. As the rotational speed increases, the outlet humidity first decreases. There is a point of minimum outlet humidity near the intersection of the characteristic F_1 and F_2 lines. Further increases in rotational speed allow the process stream to come to equilibrium with the desiccant at the regeneration stream inlet state. For optimal dehumidification, the humidifier should be tuned so that the outlet state is near the intersection point.

The minimum outlet humidity occurs at an optimum nondimensional rotational speed (mass of desiccant divided by rotational speed times mass flow of air) of 0.2. The performance of the dehumidifier is not very sensitive to speeds greater than 0.2, but performance decreases sharply as the speed drops below 0.2. The other important parameter is the heat and mass exchange size (number of transfer units, N_{tu}). Satisfactory dehumidification is achieved if the N_{tu} is greater than 15 (Jurinak, Mitchell, and Beckman 1984).

The sorption properties of desiccants are strongly dependent on temperature. This dependence allows the regeneration flow rate to be lower than the process flow. In cycles in which the regeneration temperature is relatively low, the regeneration flow rate can be about 60% of the process flow rate. For cycles with a higher regeneration temperature, the regeneration flow rate can be 80% of the process stream rate. Unbalancing the flow rates in this manner requires less regeneration energy (Jurinak, Mitchell, and Beckman 1984).

Diffusion resistance in the desiccant or the matrix reduces the dehumidification performance considerably. The Lewis number is the ratio of mass transfer resistance to that for heat transfer. This ratio should be below 2 for the wheel to operate satisfactorily.

The portion of the process stream in contact with the matrix that has just rotated from the regeneration side is hot and wet. This process air may be recirculated back to the regeneration side to lower the bulk process stream humidity. Such operation produces a drier average process stream, but the system flow process and capacity are lowered (Jurinak and Mitchell 1984b).

The choice of desiccant and its sorption properties can influence performance. Jurinak and Mitchell (1984a) studied generic desiccants to determine the influence of isotherm shape, maximum water content, heat of sorption, and isotherm hysteresis. The range of property variations was selected to represent available solid adsorbents, such as silica gel and molecular sieve. The characteristic isotherm shapes are depicted in figure 21.5. The study was aimed at evaluating the influence of property variation on system performance. Other studies (Collier, Barlow, and Arnold 1982; Barlow and Collier 1981) were aimed at staging desiccant or selecting properties for optimal operation.

Several rough guidelines emerged from the studies. For maximum dehumidifier performance, an absorbent such as molecular sieve with a type 1 isotherm yields best dehumidification performance. The linear isotherm, which is characteristic of silica gel, yields about 10% lower performance. Type 3 and 5 isotherms are very unfavorable for dehumidifiers and produce little drying in system configurations. Maximum water content of the desiccant should be high, and heat of sorption, low. Hysteresis loops dras-

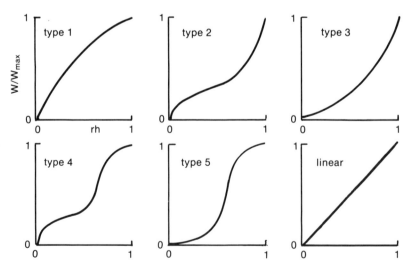

Figure 21.5
Characteristic isotherm types. Source: Jurinak and Mitchell 1984a.

tically reduce dehumidifier performance. Existing silica gel was found to be satisfactory for solar applications.

21.2.1.4 Potential and Directions for Further Development

The analyses to date produce a relatively clear picture of the qualities of a satisfactory dehumidifier. There should be laminar flow passages to reduce friction power. The overall N_{tu} should be 15 to 20, and Lewis number should be less than 2. Diffusion resistance and hysteresis should be minimal. The rotational speed should be related to process stream flow, and the flow rate of the regeneration stream should be 60% to 80% of the process stream flow.

The major developmental difficulty is to produce a desiccant fabricated with a parallel passage geometry. Molecular sieve wheels are constructed with triangular passages and are acceptable, but parallel passages are preferable in terms of pressure drop. Bonding silica to plastic films such as Mylar® to form parallel passages is in the experimental stages and represents the construction desired for solar desiccant systems.

21.2.2 System Description

21.2.2.1 Adiabatic Dehumidification Systems

Systems with adiabatic dehumidifiers have been studied analytically, and some experimental prototypes have been built. There are two basic modes of operation (Rush and Macriss 1969; Rush et al. 1975; Nelson et al. 1978). In the ventilation mode, fresh ambient air is continuously introduced, processed to provide cooling for the building, and then discharged. In the recirculation mode, the building air is continuously reprocessed and reintroduced into the building with a minimum amount of fresh air for makeup. Regeneration is accomplished by heating either a discharge or ambient air stream and passing it through the desiccant.

The main developmental emphasis has been on residential and light commercial applications, but the principles apply to industrial and multizone commercial situations. The ventilation mode is shown schematically in figure 21.6 along with a psychrometric representation of the states of the air as it passes through the system. Ambient air (state 1) is dried in the dehumidifier and simultaneously heated by the energy released during the sorption process. The air at state 2 is regeneratively cooled by exhaust air to state 3, then evaporatively cooled to state 4, and introduced into the building at a lower temperature and humidity than that inside. The use of

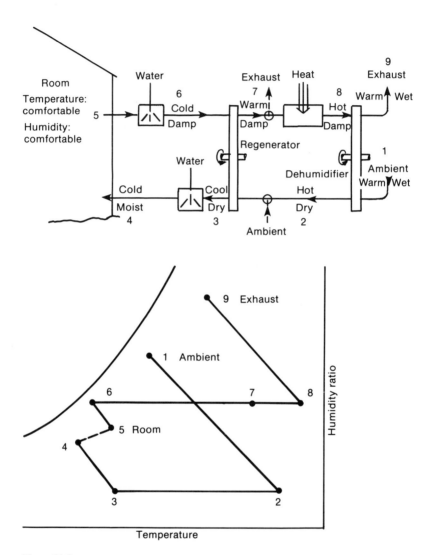

Figure 21.6
Schematic of ventilation cycle desiccant air conditioner and the corresponding psychrometric representation. Source: Jurinak, Mitchell, and Beckman 1984.

evaporative cooling at this stage allows control of the supply air temperature and humidity so that both the sensible and the latent loads may be met. The building exhaust air (state 5) is first evaporatively cooled (state 6) to provide a low temperature for the regenerative heat exchanger. It is first heated in the exchanger and then additionally heated by solar energy or a conventional fuel, either together or separately, to state 8 to allow regeneration of the desiccant. Under some operating conditions, a heat exchanger may be used to heat the regeneration stream using the exhaust air from the dehumidifier. The psychrometric diagram is for typical operating conditions. Ambient conditions change over the course of the day, with resulting changes in the operating states and loads.

The recirculation mode shown in figure 21.7 employs the same components as the ventilation mode. However, the building air is recirculated and ambient air is used only for regeneration. The building air (state 6) is dehumidified (state 7), cooled (state 8), and then evaporatively cooled (state 9) prior to reentering the building. Ambient air (state 1) is evaporatively cooled (state 2), regeneratively heated (state 3), and then further heated (state 4) before passing through the dehumidifier to regenerate the desiccant. The recirculation mode has higher thermal COP in cooler climates, whereas the thermal COP of the ventilation system performance is higher in hotter climates.

The prototype systems developed by IGT (Macriss and Zawacki 1982) and AiResearch (Rousseau 1982) have used rotary dehumidifiers, which have the advantage of compactness and simplicity. The disadvantage of an adiabatic humidifier is that a relatively high temperature ($60°-80°C$ [$140°-176°F$]) is required for regeneration. A dehumidifier that is cooled by ambient air during the sorption process has been developed by IIT (Worek and Lavan 1980). Cooling during adsorption removes the heat of sorption and maintains the desiccant at a low temperature, which allows a lower regeneration temperature ($40°-50°C$ [$104°-122°F$]). Alternatively, the process stream can be dried to a lower humidity for the same regeneration temperature as an uncooled bed. The schematic and process diagrams are similar to those of figures 21.6 and 21.7, except that two fixed desiccant beds are required. One bed is used to dry the supply air and is cooled during this process, while the other is regenerated by heated air. This system is more complex than an adiabatic one, and an additional cooling stream is required. Rejecting the heat of sorption to the ambient reduces the COP.

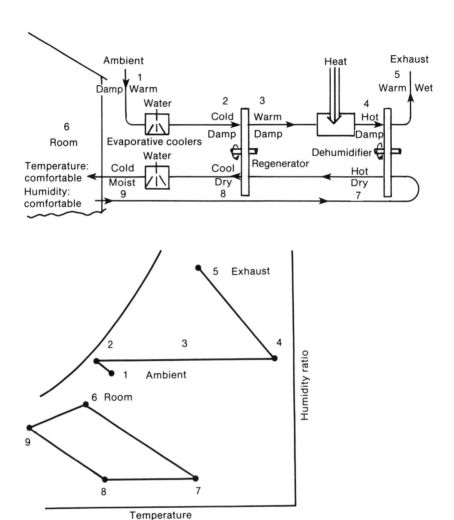

Figure 21.7
Schematic of the recirculation cycle desiccant air conditioner and the corresponding psychrometric representation. Source: Jurinak, Mitchell, and Beckman 1984.

The regenerative heat exchanger is commonly a rotary unit, although a direct transfer unit could also be employed. Satisfactory rotary exchangers are made of plastic films wound spirally to produce a series of parallel passages. In the ventilation mode, a very high heat exchanger effectiveness, greater than 90%, is required to obtain a high thermal COP. The temperature at state 2 is high, and a high effectiveness allows reclaiming significant energy and reduces the solar or auxiliary energy requirements. In addition, a high effectiveness is needed to reduce the temperature at state 3 and allow the sensible and latent loads to be met. The heat exchanger in the recirculation mode is not as critical since the temperature differences between state 2 and 7 are lower and less heat is reclaimed.

High-effectiveness direct or indirect evaporative coolers are required in the regenerating air stream to provide low temperatures for cooling the process air. The building supply streams require temperature and humidity control by the evaporative cooler at the building entrance. This control is accomplished through varying the cooler effectiveness.

System control is complicated by the ability of the system to meet both sensible and latent components of the load through varying mass flow rates, rotational speed, regeneration temperature, and effectiveness of both evaporative coolers. A controller that senses both room temperature and relative humidity and delivers two functions to indicate the need for sensible cooling, dehumidification, or both is required. The systems operate most efficiently at the highest supply temperature and humidity (states 4 and 9 in figures 21.2 and 21.3, respectively) that allow the total load to be met. The maximum flow rate is dictated by parasitic power requirements, and a flow rate that results in an enthalpy difference between supply and room conditions of about 15 kJ/kg (6.5 Btu/lbm) is acceptable. The rotating speed of the dehumidifier is tuned to this flow rate to yield maximum dehumidification, while the regeneration flow is the minimum necessary to regenerate the dehumidifier. The control of these parameters must occur dynamically to yield a high COP.

The optimal control strategy maximizes cooling capacity while minimizing auxiliary thermal energy input. The hierarchy is first to use regenerative evaporative cooling and bypass the dehumidifier. If the cooling in this mode is insufficient to meet the load, dehumidification with regeneration by solar is employed. Further cooling is provided by solar and auxiliary energy for regeneration. The energy input is always controlled to provide the lowest temperature at which the desiccant can be regenerated.

The sensible-to-latent load ratio is controlled by the effectiveness of the evaporative cooler at the room inlet.

These cooling systems can also provide heating through increasing the rotational speed of the dehumidifier and ceasing rotation of the heat exchange wheel. The evaporative coolers are turned off. However, this mode is less efficient than either direct solar or auxiliary heating.

21.2.2.2 Hybrid Solar Desiccant Systems

Hybrid systems use a dehumidifier to remove moisture and a standard vapor compressor air conditioner to provide sensible cooling. This combination employs the best attributes of both devices to advantage. Figure 21.8 is a schematic of a hybrid system and the associated psychrometric diagram. The ventilation air for the building (state 2) is first dehumidified to state 3 and then sensibly cooled to state 4 in an indirect evaporative cooler. This heat exchanger uses evaporatively cooled ambient air as the heat sink. The supply air is then sensibly cooled by the air conditioner prior to entering the building at state 5. Regeneration is accomplished through use of heat from a solar system, from the air-conditioner condenser, from an auxiliary supply, or from a combination of these units.

As with solar desiccant systems, hybrid systems can be controlled to meet both sensible and latent loads. Dehumidification is controlled by varying the regeneration temperature to produce a desired humidity level at state 3. The supply air temperature is controlled by the air conditioner. These cycles require lower regeneration temperatures ($40°$–$50°$C ($104°$–$122°$F]) than do the solar-only cycles (Howe, Beckman, and Mitchell 1983a, 1983b; Sheridan and Mitchell 1985).

21.2.2.3 Closed-Cycle Systems

A closed-cycle desiccant system has been developed in which the refrigerant (water) is cycled (Tchernev 1981). During the cooling process, water previously condensed is injected into a flash evaporator and evaporates to provide cooling. The vapor is adsorbed in a zeolite desiccant, which allows a low ($10°$–$15°$C [$50°$–$59°$F]) temperature in the evaporator. The desiccant is built into a solar panel and regenerated by solar heat during the day. The water evaporated from the zeolite during regeneration is condensed in a sensible exchanger by heat rejection to the surroundings. The condensed water is then recycled to the evaporator.

The system is simple. However, it provides cooling only at night. It does not employ heat reclaim, and thus more energy is required to regenerate

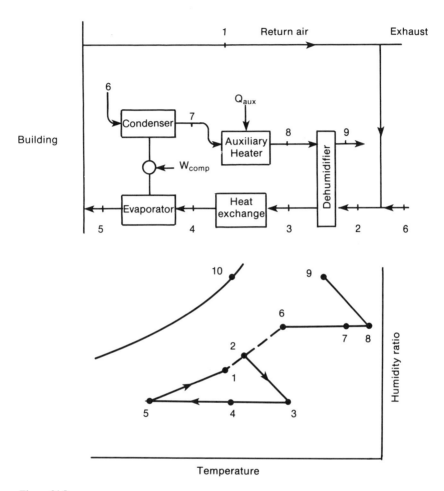

Figure 21.8
Schematic of a hybrid system and the corresponding psychrometric representation. Source:
Howe, Beckman, and Mitchell 1983a.

the desiccant than in some of the more complex cycles. The prototypes built to date have very low cooling capacities.

21.2.3 System Development

Progress has been made in two areas of system development. First, there have been several prototype systems built and tested under both laboratory and field conditions. One system is nearing the market stage. These experiments demonstrate that the system concepts are valid and that the system can provide air-conditioning. Second, system analyses have been conducted for residential and commercial applications. These studies have further established the potential of the systems and identified areas crucial to further improvement.

21.2.3.1 Testing on Full-Scale Systems
The first system tests of residential-size desiccant systems were conducted by IGT in the early 1970s in Los Angeles (Rush and Macriss 1969; Rush et al. 1975). The system could be operated in either the ventilation or recirculation mode (figures 21.6 and 21.7). Initially, regeneration was with natural gas as the energy source, and later solar energy was used. The rotary desiccant was a molecular sieve manufactured by Munters. In the recirculation mode, between 3.5 and 7 kW (1 and 2 tons) of cooling were produced at a thermal COP between 0.26 and 0.51. The COP was highest at low ambient humidity and at a low temperature difference between the room and outdoor air. These tests proved that a desiccant system could be built and would produce cooling.

Laboratory tests were conducted by IGT on improved versions of the prototype system (Macriss and Zawacki 1982). By use of a simulated solar input and a regeneration temperature of 80°C (180°F), a COP of 0.5 at ARI standard conditions was obtained. The unit developed 9.4 kW (2.7 tons) of cooling. It had an electric COP of 8 (EER of 27.5 Btu/Wh), which is an acceptable value. The sensitivity of the performance to regeneration temperature was studied. It was found that COP decreased 12% and capacity rose 60% as the regeneration temperature was increased from 80° to 110°C (180° to 230°F). These results were in agreement with concurrent system analyses.

Numerous other tests were conducted to map out the performance over a range of inlet (ambient) temperatures and humidities. It was found that unbalance of the regeneration flow improved the COP and that a heat exchanger effectiveness of 95% significantly increased COP. The measured performance was found to agree with model results. These results established the usefulness of modeling techniques and allowed exploration of the methods required to improve performance.

Laboratory tests were also conducted at AiResearch during the 1970s (Rousseau 1982) on a system similar to that of IGT. The dehumidifier was drum-shaped and consisted of silica-gel particles packed between perforated steel sheets. At a regeneration temperature of 95°C (200°F), a COP of 0.6 for the recirculation mode and a COP of 0.48 for the ventilation mode at ARI conditions was obtained. The capacity was 4.9 kW (1.35 tons), and the electric COP was 5.8 (EER of 20 Btu/Wh). Various control options were explored; varying process flow rate directly with load was best. A map of the performance over other ambient conditions was developed.

A recirculation-cycle system using a cooled desiccant bed was constructed and tested by IIT in the late 1970s (Monnier, Worek, and Lavan 1981; Majumdar, Worek, and Lavan 1982; Worek and Lavan 1982). It had a 3.5 kW (1 ton) capacity and operated in the recirculation mode. At a regeneration temperature of 63°C (145°F), the COP was 0.63. The performance was measured over a wide range of system operating parameters and ambient conditions. A model was developed to explain and extrapolate system performance. Another approach to a cooled bed used a staged, cooled rotary dehumidifier (Lunde 1975), but it proved impractical.

These system tests demonstrated that solid desiccant systems were a promising concept for solar cooling. However, further development of the components was needed, particularly the desiccant and the controls. Thus, system construction and testing awaited a next generation of advanced components.

American Solar King, of Waco, Texas, has developed a system that promises to enter the consumer market in 1986. The details of construction and operation of the unit are proprietary, but the design appears to be based on the earlier IGT concepts. It uses a ventilation cycle system such as that depicted in figure 21.6. Solar energy supplies some of the regeneration energy. The private-sector developments in this area are promising.

21.2.3.2 System Analysis
System analyses have been conducted for a wide variety of configurations and applications. Ventilation and recirculation cycles with both adiabatic

and cooled beds have been studied. The applications cover residential and, to a limited extent, commercial situations. Regeneration energy sources included solar, natural gas, condenser heat rejection, and a combination of these. System parameters have been varied over a wide range of present and expected values. Design point studies as well as seasonal simulations have been conducted.

The general goals of these studies were to evaluate the best system configuration for a given application and location and to determine the optimal control strategies for minimum auxiliary heating requirements and parasitic power. The operation is sensitive to design parameters such as wheel thickness, rotational speed, flow rate, and regeneration temperature. The performance of the system is a function of ambient temperature and humidity, and the building latent and sensible loads. The evaluation of a control strategy to accommodate these factors became a major goal.

The performance of a given configuration at a design point is important but does not represent seasonal operation. Both COP and capacity vary with temperature and humidity of ambient and regeneration streams in a nonlinear fashion. Locations with only slightly different design conditions often have significantly different ambient variations over the course of the cooling season. Further, systems with solar auxiliary depend on weather patterns that affect storage. As a result, seasonal simulations are required.

System models that interact dynamically with weather and loads have been developed. In general, these have been TRNSYS-compatible component models that may be connected together and driven by weather records. The dehumidifier component is commonly modeled by the analogy relations together with realistic values for component effectivenesses. Heat exchangers and evaporative cooler models are based on heat and mass transfer mechanisms. Load models reflect buildings with both latent and sensible load components.

Seasonal evaluations have been conducted of residential systems with adiabatic dehumidifiers (Oonk et al. 1978; Jurinak, Mitchell, and Beckman 1984; Nelson et al. 1978). The performance of the recirculation mode has been compared to that of the ventilation mode. The heat exchanger effectiveness is critical, and at low effectiveness, the recirculation system has a higher COP, while for effectiveness values greater than about 0.90, the ventilation system COP is higher. For both systems, COP decreases as regeneration temperature increases. Increased regeneration temperature and energy supply do not produce a proportionate decrease in process

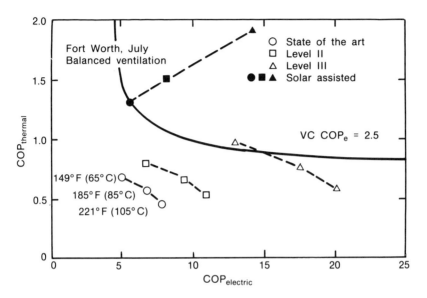

Figure 21.9
Performance of ventilation cycle. Source: Jurinak, Mitchell, and Beckman 1984.

stream humidity and consequent ability to meet the load. However, cooling capacity increases as regeneration temperature increases. Overall, it appears that regeneration temperatures between 60° and 80°C (140° and 176°F) produce most acceptable performance.

Desiccant systems use thermal energy for regeneration and electric energy for fans to create air flow. A common basis must be established to compare the total purchased energy consumption for different levels of components to that for vapor compression machines. In figure 21.9, the thermal COP is plotted as a function of the electric COP. The reference curve is the resource energy for a vapor compression machine with a COP of 2.5. For a desiccant machine to compete with a standard air conditioner on a resource basis, the values of purchased $COP_{thermal}$ and $COP_{electric}$ must be above this curve.

Three sets of desiccant component parameters were chosen to reflect the current prototypes and two levels of improved design. The IGT ventilation and recirculation cycles provide a baseline for current systems (Macriss and Zawacki 1982). The model parameters were roughly matched to those of these systems to provide level I systems. The level II system parameters mirror the changes proposed by IGT. The level III parameters are representative of carefully engineered components that appear to be technically feasible.

The open symbols represent gas-fired operation at regeneration temperatures ranging from 65° to 105°C (149° to 221°F). For gas-fired operation, all three levels are able to maintain comfort 90% of the time. Below 65°C (149°F), the $COP_{electric}$ is high, but the capacity is too low to meet the load. At high regeneration temperature, the capacity is greater, and less process air flow, and consequently less parasitic power, is needed. Only the level III system with a 65°C (149°F) regeneration temperature uses less resource energy than a standard air conditioner. Since current systems designs are at about level II, considerable engineering development is required to produce a competitive gas-fired desiccant system.

The resource energy consumption is considerably reduced by adding a 40 m² (432 ft²) flat-plate collector array in series with the auxiliary supply. The solar system provides at least 50% of the thermal energy required for regeneration, and thus values of purchased $COP_{thermal}$ are about doubled. Because the temperature of the solar supply is often low, a greater running time is required than with gas, which reduces $COP_{electric}$. For level II and III systems, solar-assisted desiccant air conditioners use less resource energy than standard air conditioners. However, if solar-only operation is attempted, the 40 m² (432 ft²) of collector area is insufficient to provide enough energy to meet the load, and the area must be about 60 m² (648 ft²) to provide comfort 90% of the time.

Figure 21.10 shows the results for a recirculation cycle with the same conditions as the ventilation cycle. The purchased $COP_{thermal}$ is 40% lower for level I, even though the cooling capacities are about equal. At the high level, the recirculation $COP_{thermal}$ is about 70% lower than that for the ventilation cycle. With the addition of a solar energy system, the purchased energy is reduced but is still 35% greater than that of the ventilation cycle.

Unbalancing the flow through the system improves performance significantly, as shown in figure 21.11. The $COP_{thermal}$ increases 10% to 15% with only a slight drop in capacity. The increase in $COP_{thermal}$ for the recirculation cycle is about 50%. This greater increase results from the relative insensitivity of the recirculation cycle to regeneration temperature.

The gas-fired results show that for desiccant systems to become more competitive with vapor compression machinery, the thermal COP rather

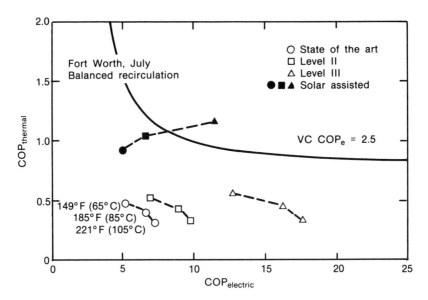

Figure 21.10
Performance of recirculation cycle. Source: Jurinak, Mitchell, and Beckman 1984.

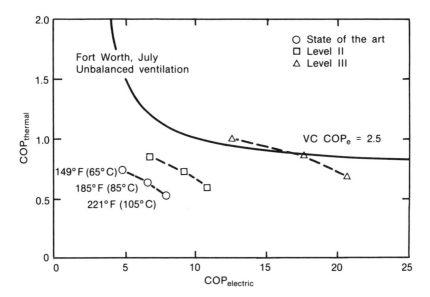

Figure 21.11
Performance of ventilation cycle with unbalanced flow. Source: Jurinak, Mitchell, and Beckman 1984.

than the electrical COP must be increased. Additional improvements can be made by increasing the dehumidifier N_{tu}, purging the dehumidifier, and increasing the dehumidifier thermal capacitance. Doubling the N_{tu} improves the $COP_{thermal}$ by 20%, although capacity drops 10%. More important, dehumidifier thickness is doubled with a consequent increase in parasitic power. Purging the first portion of the regeneration air from the process stream improves $COP_{thermal}$ by about 10%. An increase in thermal capacity also produces an approximately 10% higher $COP_{thermal}$. Thus, it can be expected that with these improvements, level III ventilation systems may approach a seasonal thermal COP of 1.1 to 1.3.

Figure 21.12 shows the performance of solar-assisted desiccant-cooling systems with unbalanced flow and an 85°C (185°F) regeneration temperature. There is marked improvement in $COP_{thermal}$, and even the level I systems consume less energy than conventional vapor compression units. In contrast to gas-fired units, the solar-fired recirculation system has better performance than the ventilation cycle. The amount of energy collected

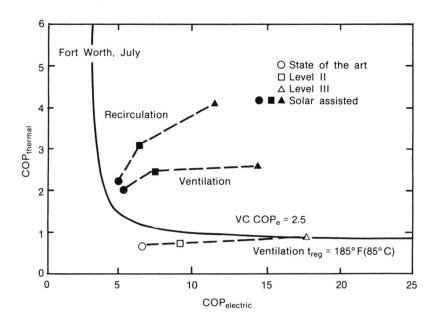

Figure 21.12
Performance of solar and natural gas-fired cycles. Source: Jurinak, Mitchell, and Becktran 1984.

in the recirculation cycle is relatively insensitive both to the level of performance and flow unbalance, since regeneration temperatures are always
low, while in the ventilation cycle the amount of solar collected decreases
as the performance and regeneration temperature increase. If the same
benefits of increased N_{tu}, wheel purging, and increased thermal capacitance occur in the solar-fired machine as in gas fired units, the $COP_{thermal}$
would be an additional 20% to 30% higher than shown. The energy
consumption is significantly lower than that of the vapor compression
machine.

The results presented in figures 21.9 through 21.12 are for July in Fort
Worth, Texas. The seasonal $COP_{thermal}$ is somewhat higher than for July
due to lower average cooling requirements. The performance of systems in
Miami, Florida, is only slightly lower than for Fort Worth, while in a
climate such as in Columbia, Missouri, the systems perform even better.
Desiccant systems appear best suited to moderate (low temperature and
humidity) climates.

The performance of desiccant air-conditioning systems with cooled desiccant beds and with adiabatic dehumidifiers is compared in figure 21.13

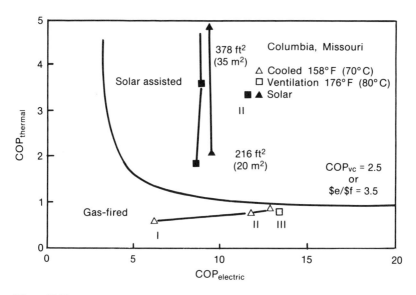

Figure 21.13
Performance of cooled and adiabatic dehumidifier systems. Source: Schultz, Mitchell, and
Beckman 1982.

for Columbia, Missouri (Schultz, Mitchell, and Beckman 1982). The lower open symbols are for gas-fired units with the three levels of component performance comparable to adiabatic units. The cooled system with a 70°C (158°F) regeneration temperature is comparable to that of the ventilation system with a 80°C (176°F) regeneration temperature.

The use of solar energy to meet the regeneration energy requirements significantly reduces fuel use. For both cooled and adiabatic systems, the thermal COP increases dramatically. The electrical COP drops slightly due to increased parasitic power with increased collector area. The cooled-bed system is able to use the collectors much more efficiently since it operates at a lower regeneration temperature. The performance in a climate such as in Fort Worth is comparable to that in Columbia, although the difference between the solar-heated ventilation and cooled-bed systems is less. For year-round operation, a climate such as in Columbia is probably the most appropriate for a desiccant system. The collector area is not excessively large since cooling requirements are relatively low. The collectors may be used year-round to provide space heating, water heating, and cooling.

Hybrid systems for commercial applications that combine solar, desiccant, and vapor compression components have been studied (Howe, Beckman, and Mitchell 1983a, 1983b; Sheridan and Mitchell 1985). Figure 21.8 shows a promising configuration. There is a variety of options for regeneration and condenser heat rejection, solar heat, and auxiliary that may be used in various combinations. Preliminary studies demonstrate that significant electrical savings are possible by employing a desiccant to meet the latent portion of the load and a conventional vapor compression machine to provide sensible cooling.

These studies are not clear, though, on the role of solar collectors to provide regeneration energy. A system that uses only condenser heat for regeneration produces nearly the same energy savings as a system in which the condenser rejects heat to the ambient and solar provides regeneration. It appears that heat recovery from the vapor compression machine is a valuable source of energy.

These results appear to depend heavily on several factors. First, in these studies, the sensible load was large enough to provide sufficient condenser heat rejection. A larger latent-to-sensible load ratio would require more heat for regeneration than available from the condenser and make solar more valuable. Second, the increase in parasitic power for a solar array tends to offset the reduction in vapor compression COP because of an elevated condenser temperature. However, it may not be possible to provide duct work to collect condenser heat rejection, and solar may be the only source. Third, in drier climates, the indirect evaporative cooler may eliminate the need for sensible cooling by the chiller, and solar would be the only source of regeneration energy. The only established conclusion is that hybrid systems offer a large energy-saving potential over conventional vapor compression machines.

21.2.3.3 Directions and Potential for Further Improvements

The system analyses establish that desiccant systems may save operating-resource and end-use energy over conventional vapor compression machines. The successful development of a low-pressure-drop dehumidifier would allow system tests of either solar-fired or solar-hybrid systems. New configurations or devices such as staged indirect evaporative coolers may provide incremental improvement. Fine tuning of the control strategies will also increase performance. Overall, a breakthrough does not appear possible; rather, a continued development of the system will yield a viable option.

21.3 Liquid Desiccant Systems

The use of liquid desiccants to provide air-conditioning has received a limited amount of study under DOE sponsorship. Preliminary assessments show that liquid systems promise lower cost components than do solid systems, but there are additional complexities. The work to date has mainly focused on developing a low-cost regenerator that can also be used during the heating season.

21.3.1 Major Components

Liquid desiccant systems may be configured similarly to those for solid systems, with the liquid desiccant absorbing and desorbing water. The desiccants suggested are lithium chloride, calcium chloride, and glycol solutions. In one type of liquid desiccant system, a major component is an open-flow collector. It differs from a conventional collector in that it is a heat- and mass-transfer device with evaporation of water as an output. Solar energy is absorbed by the flowing liquid. The fluid temperature increases and water evaporates. This moisture is removed by a flow of air over the liquid. The collector may or may not have a transparent cover plate, but in either case the desiccant is exposed to the atmosphere. Recent studies have shown that the cover is virtually a necessity in most U.S. climates.

The collector performance has been modeled (Collier 1979; Peng and Howell 1981). Energy and mass balances are performed for differential elements. Heat and mass transport relations are incorporated to relate concentration and temperatures of the air and the desiccant. These differential equations are integrated over the length of the collector to obtain the outlet values of desiccant concentration and temperature. Two critical parameters in these models are the heat and mass transfer coefficients. Predictions have been made for the performance of the collector.

The open-flow collector concept has been successfully tested (Wood et al. 1983). A thin film of desiccant has been made to flow successfully down the roof (collector surface) and acceptable moisture removal obtained. Contamination by dust and rain was found not to be detrimental. The experimental results have been compared favorably to model predictions of concentration. However, there was not good agreement for prediction of temperature. A major uncertainty was the correct relationships for heat and mass transfer in the natural wind.

An alternative to regeneration in an open-flow collector is to accomplish regeneration in a packed bed (Leboeuf and Löf 1980; Lenz and Löf 1983). In this system, solar-heated air passes up through a packed bed through which the desiccant trickles. The air removes moisture from the desiccant and exhausts it to the environment. The remaining components of the cycle are the same. An advantage of this approach over the open-flow collector is the ability to use the solar system for winter heating.

The packed bed is a heat and mass exchanger. The physical difference between the bed and collector is that the desiccant is directly heated by solar energy in the collector, whereas the desiccant is heated by warm air in the bed configuration. In both devices, a counterflow of air removes moisture.

The bed has been analyzed in a manner similar to that for the open-flow collector (Lebouef and Löf 1980; Schlepp and Collier 1981). Fundamental heat and mass transfer equations are used to predict the outlet states. As with the collector analysis, heat and mass transfer coefficients are critical parameters. The packed-bed system has also been tested (Lenz and Löf 1983), and performance found to be lower than expected. This difference is suspected to be due to concentration gradients in the desiccant films that lower the overall mass transfer.

21.3.2   System Studies

There is a limited amount of information available relative to system performance. Satisfactory models are not available for all components, and no system analyses have been conducted. Limited tests have been carried out on an open-cycle absorption system using a packed bed (Lenz and Löf 1983), and a COP of 0.58 was obtained. Regeneration was accomplished with air at temperatures between 37° and 70°C (99° and 158°F). The performance would probably improve with further development.

An entire system using lithium chloride as the desiccant has been built and used to air-condition a house (Robinson 1979–1983). Regeneration is accomplished in an open-flow collector and conditioning of the air by a spray chamber. This system has operated successfully for several summers in the extreme high humidity of the South Carolina coast. No quantitative performance data are available.

The successful development of liquid desiccant systems hinges on the advantages they may have over other solar-driven air conditioners. From a thermodynamic viewpoint, the potential COP is equal to that for a single-stage absorption unit. Their advantage is either simplicity, as in the low-cost, open-flow collector, or their ability to be used for winter heating, as in the packed-bed concept. System analyses, component development, and full-scale experiments are necessary in order to adequately evaluate system performance.

References

Banks, P. J. 1972. Coupled equilibrium heat and single adsorbate transfer in fluid flow through a porous medium. 1. Characteristic potentials and specific capacity ratios. *Chem. Engr. Science* 27:1143–55.

Banks, P. J., and D. J. Close. 1972. Coupled equilibrium heat and single absorbate transfer in fluid flow through a porous medium. 2. Predictions for a silica-gel air-dryer using characteristic charts. *Chem Engr. Science* 27:1157–69.

Barlow, R. S. 1981. A simple predictive model for performance of desiccant beds for solar dehumidification. *Proc. International 1 Passive and Hybrid Cooling Conference*, Miami Beach, FL.

Barlow, R. S., and R. K. Collier. 1981. Optimizing the performance of desiccant beds for solar regenerated cooling. *Proc. AS/ISES* Philadelphia, PA, pp. 496–500.

Collier, R. K. 1979. The analysis and simulation of an open cycle absorption refrigeration system. *Solar Energy Journal* 23:357–66.

Collier, R. K., R. S. Barlow, and F. H. Arnold. 1982. An overview of open cycle desiccant—cooling systems and materials. *Trans. ASME* 104:28–34.

Epstein, M., M. Groimes, K. Davidson, and D. Kosar. 1983. Advanced desiccant dehumidification/cooling technology. *Proc. ASES*, Minneapolis, MN pp. 398–407.

Howe, R. R., W. A. Beckman, and J. W. Mitchell. 1983a. Commercial applications for solar hybrid desiccant systems. *Proc. ASES*, Minneapolis, MN, pp. 253–58.

————. 1983b. Factors influencing the performance of commercial hybrid desiccant air conditioning systems. Paper presented at the 8th Annual IECEC conference in Orlando, FL, 1983.

Jurinak, J. J. 1982. Open-cycle solid desiccant cooling—component models and system simulation. Ph.D. diss., University of Wisconsin-Madison.

Jurinak, J. J., and J. W. Mitchell. 1984a. Effect of matrix properties on the performance of a counterflow rotary dehumidifier. *J. Heat Transfer* 106:369–75.

————. 1984b. Recirculation of purged flow in an adiabatic counterflow rotary dehumidifier. *J. Heat Transfer* 106:252–260.

Jurinak, J. J., J. W. Mitchell, and W. A. Beckman. 1984. Open cycle desiccant air conditioning as an alternative to vapor compression cooling in residential applications. *J. Solar Energy Eng.* 106:252–60.

Leboeuf, C. M., and G. O. G. Löf. 1980. Open cycle absorption cooling using packed bed absorption reconcentration. *Proc. AS/ISES*, Phoenix, AZ, pp. 205–9.

Lenz, T., and G. O. G. Löf 1983. Residential cooling with solar heated air and open-cycle, lithium chloride absorption chiller. *Proc. ASES*, Minneapolis, MN.

Lunde, P. J. 1975. Solar desiccant cooling with silica gel. In *Use of solar energy for the cooling of buildings*, edited by F. de Winter and J. W. de Winter. Los Angeles: UCLA, pp. 280–93.

Maclaine-Cross, I. L. 1972. Coupled heat and mass transfer in regenerators—predictions using an analogy with heat transfer. *Int. J. Heat and Mass Transfer* 15:1225–42.

Macriss, R. A., and T. S. Zawacki. 1982. High COP rotating wheel solid desiccant system. Paper presented at 9th Energy Technology Conference, Washington, DC.

Majumdar, P., W. M. Worek, and Z. Lavan. 1982. Simulation of a cooled bed solar powered desiccant air-conditioning system. *Proc. ASES*, Houston, TX, pp. 627–32.

Monnier, J. B., W. Worek, and Z. Lavan. 1981. Testing of a solar-powered cooling system using cross-cooled desiccant dehumidifier. *Proc. AS/ISES*, Philadelphia, PA, pp. 505–9.

Nelson, J. S., W. A. Beckman, J. W. Mitchell, and D. J. Close. 1978. Simulations of the performance of open cycle desiccant systems using solar energy. *Solar Energy* 21:273–78.

Oonk, R. L., L. E. Shaw, M. Cash, P. J. Hughes, and J. W. Mitchell. 1978. Performance of solar—desiccant cooling systems in New York City. *Proc. AS/ISES*, Denver, CO, pp. 460–67.

Peng, C. S. P., and J. R. Howell. 1981. Analysis and design of efficient absorbers for low-temperature desiccant air conditioners. *Trans. ASME* 103:67–74.

Pla-barby, F. E., and G. C. Vliet. 1978. Performance of rotary bed silica gel solid desiccant dryers. ASME paper 78-HT-36. New York: ASME.

————. 1979. Rotary bed solid desiccant drying: An analytical and experimental investigation. ASME paper 79-HT-19. New York: ASME.

Robinson, H. I. 1979–1983. Internal reports on open cycle desiccant air conditioning. University of South Carolina, Physics Department.

Rousseau, J. 1982. *Development of a solar desiccant dehumidifier.* Phase II summary report, DOE/CS/31591-T8, AiResearch Manufacturing Co.

Rush, W. F., and R. A. Macriss. 1969. Munters environmental control system. *Appliance Engineer* 3:23–28.

Rush, W. F., J. Wurm, L. Wright, and R. Ashworth. 1975. A description of the solar—MEC field test installation. Paper presented at 1975 International Solar Energy Congress, Los Angeles, CA, July 28, 1975. Summarized in *Extended Abstracts ISES 1975 Congress*, p. 405.

Schlepp, D. R., and R. K. Collier. 1981. The use of an open cycle absorption system for heating and cooling. *Proc. AS/ISES*, pp. 651–55.

Schultz, K. J., J. W. Mitchell, and W. A. Beckman. 1982. The performance of desiccant dehumidifier air-conditioning systems using cooled dehumidifiers. *Proc. ASES*, Houston, TX, pp. 1–7.

Sheridan, J. C., and J. W. Mitchell. 1985. Hybrid solar desiccant cooling systems. *Solar Energy* 34:187–93.

Tchernev, D. 1981. Integrated solar zeolite collector for heating and cooling. *Proc. AS/ISES*, Philadelphia, PA pp. 520–24.

Van den Bulck, E. 1983. Analysis of solid desiccant rotary dehumidifiers. M.S. thesis, University of Wisconsin-Madison.

van Leersum, J. G., and D. J. Close. 1982. Experimental Verification of open-cycle cooling system component models. DET 24, Highett, Victoria, Australia: Division of Energy Technology, CSIRO.

Wood, B. D., G. A. Buck, D. A. Siebe, and M. Breslauer. 1983. Performance characteristics of an open-flow liquid desiccant solar collector/regenerator in a hot arid climate. Perth, Australia: Solar World Congress.

Worek, W. M., and Z. Lavan. 1980. Cross-cooled desiccant dehumidifier for solar powered cooling systems. *Proc. AS/ISES*, Phoenix, AZ, pp. 224–27.

———. 1982. Performance of a cross-cooled desiccant dehumidifier prototype. *J. Solar Energy Eng.* 104:187–96.

22 Comparative Performance and Costs of Solar Cooling Systems

Mashuri Warren

22.1 Introduction

In this chapter, the work done to evaluate and compare the performance and costs of absorption, Rankine, and desiccant solar cooling technologies will be reviewed and compared. Absorption and Rankine technologies are well understood, and several large systems have been built and evaluated. The performance of these systems must be improved and their cost reduced through research and development before they become attractive in the market place. Solar desiccant cooling technologies have undergone rapid development in the laboratory and show considerable promise.

To assess the economic viability of future active solar energy systems requires establishing the solar technology costs compared to the conventional technology; establishing the system performance to meet the specified heating, cooling and domestic hot water load; and making assumptions about the expected future costs of fuel, inflation, interest rates, taxes, etc. The potential buyer can then establish economic criteria or figures of merit that form the basis of decision, such as simple payback, life-cycle cost, discounted present value, return on investment analysis, and levelized annual cost.

A comprehensive assessment of the energy savings potential of both residential and commercial active solar cooling systems was made by use of TRNSYS simulation analysis (Choi and Morehouse 1982; Hughes et al. 1981). To determine cost and performance requirements for future active solar cooling systems, a cost-goal methodology was developed (Warren and Wahlig 1983) by estimating the present value of future energy savings of the solar system, discounted at a rate sufficient to ensure market acceptance. The electrical and thermal performance improvements that can and must be made were projected for both absorption and Rankine cooling systems to establish future cost goals (Warren 1982a, 1982b, 1982c; Warren and Wahlig 1982) and to define research needs (Warren and Liers 1983).

Beginning in the fall of 1981, a systematic evaluation of the economic potential was undertaken for various active solar space conditioning technologies (both heating and cooling) that were still in the research phase. In the Active Program Research Requirements (APRR) assessment (Scholten

1983) that was conducted intermittently between January 1982 and November 1983; the performance and economics of different solar space heating and cooling technologies were compared. The solar cooling technologies included absorption heat pumps, desiccant cooling systems, and heat-engine-driven heat pumps still in the research and development phase. Bankston (1986) has evaluated effective cost of delivered solar energy with several different technologies and estimated the potential for energy use displacement under different future cost scenarios.

22.1.1   Evaluating Cost Effectiveness

To evaluate the cost effectiveness of systems that have not yet been built requires: (1) estimating the load to be served and calculating the contribution of the active solar system to meet that load; (2) calculating the contribution of an efficient conventional system to meet the same load and thereby establish the energy and fuel cost savings attributable to the solar system; (3) estimating the cost of the active solar system with full backup and the cost of the conventional system; and (4) comparing the energy cost savings to the incremental solar system cost with suitable economic assumptions.

In the design of a solar system, the performance of a specific application is calculated by use of simplified design methods, or it is modeled in detail. For the purposes of evaluating different technologies, standardized residential and light commercial loads have been used (Leboeuf 1980). A detailed discussion of these different calculation tools is beyond the scope of the present chapter.

The performance of future conventional space-conditioning systems affects the economic potential of active solar systems. The performance and cost of today's conventional heating, cooling, and domestic hot water systems can be readily determined, but conventional heating and cooling technology is constantly improving. To evaluate the cost-effectiveness of active solar systems to be built 5 or 10 years from now, the anticipated efficiency of new pulsed combustion gas furnaces, electric heat pumps, and chillers should be used in the comparison.

22.1.2   Performance Indices

Performance indices for thermal and electrical operation are useful in comparing both chiller performance and overall solar cooling system performance.

Collection efficiency, $\eta_{collector}$, is the ratio of the useful thermal energy collected to the solar energy incident on the collector. The collection efficiency gives a measure of how well the solar array operates.

Storage efficiency, η_{store}, is the ratio of the useful heat taken from the storage tank to the heat put into storage.

Chiller thermal performance, $COP_{Chiller}$, is the ratio of the useful cooling effect produced to the heat put into the generator of the absorption chiller or put into the boiler of a Rankine turbine.

Chiller electrical performance, $ECOP_{Chiller}$, is the ratio of cooling effect produced at the cooling coil to the electric energy required to operate the chiller. The electric operating energy of the chiller consists of energy to run the pumps for supplying hot water to the generator, the absorbent solution, and the chilled water in the absorption chiller; or to run pumps to supply the boiler and chilled water in the Rankine system. In addition, energy is required for the condenser loop pumps and fans necessary for heat rejection. In general, absorption and Rankine systems must reject more heat than vapor-compression systems. The additional fan and pumping energy for heat rejection penalizes the solar system. This electric energy consumption can be reduced by proper design and sizing of pumps, piping, and heat exchangers in the chiller.

System electrical performance, $ECOP_{system}$, is the ratio of cooling effect produced at the cooling coil to the electric energy required to operate the system. The system electrical performance gives a measure of the overall system effectiveness including all of the operating energy required to collect, store, and move solar energy, to operate the chiller, and to reject heat to the environment.

System thermal performance, COP_{system}, is the ratio of cooling effect delivered at the cooling coil to the solar energy incident on the collectors. The system thermal efficiency gives a measure of overall effectiveness in driving a cooling process with solar energy.

Solar cooling performance factor, *SCPF*, is the ratio of the daily cooling delivered by the system to the daily total insolation incident at the collector and adjusted for daily change in thermal storage. The solar cooling performance factor can broadly be used to determine the effectiveness of a system to convert solar energy to useful cooling.

System performance indices for solid and liquid desiccant systems that do not use cooling coils must be based on cooling effect delivered. In these systems, thermal energy is used to regenerate the desiccant. Electric energy is used for fans to move air through the desiccant and other components. These systems may also have significant water consumption.

Developing system performance indicators for systems with multiple end-uses is more difficult. The utilization of collected solar energy can be improved if the system is designed for multiple uses in addition to cooling, such as space heating and domestic hot water. The domestic hot water load is more or less constant throughout the year. The space heating load and the summer cooling load cover the winter and summer. In Rankine systems that develop shaft power, electric power generation can improve solar utilization. Thermally driven heat pumps have the potential to operate with high efficiencies in the heating mode and provide cooling at a small additional cost. The ability to meet both heating and cooling loads makes the system more cost-effective. Ultimately, the value of solar energy collected is the value of the energy end-use served by conventional technology.

22.2   Economic Criteria and Assumptions

The economic viability of a solar energy system depends on two factors: (1) the value of conventional energy displaced (fuel cost savings); and (2) the capital cost of hardware necessary to achieve those savings. There are two principal approaches to economic analysis that are related: payback analysis and cost of delivered energy.

There are many factors that affect the decision to adopt a new technology, including the real or perceived risks, the economic attractiveness as expressed by some figure of merit, and the status as an innovator or trendsetter. The economic evaluation of a solar application includes factors such as the capacity cost of delivering solar energy, the optimum sizing of collectors and other equipment, the cost of competing technologies, and financial analyses. There figures of merit that are used to accept or reject a particular solar application, including simple payback, cash flow, capital cost per unit of energy saved, life-cycle cost, net present value, and levelized energy cost.

22.2.1   Cost Goals

To establish cost goals for a future technology, the analysis of the economics of active solar systems can be turned around to ask the question: Based on the value of energy savings, how much can be spent on the system? This value then becomes the cost goal for the system. For a range of economic assumptions, an analysis considering a 5-year or 7-year simple payback is assumed to be an adequate figure of merit to establish cost goals for active solar heating and cooling technologies (Scholten 1983). The value of energy saved in the first year of system operation is multiplied by the simple payback period to establish the cost goal. For commercial applications, a payback of less than 5 years is necessary for market acceptance. For residential applications, a payback of 7 years may be acceptable.

For the relatively short paybacks required for market acceptance of a new technology, fuel-escalation rates do not greatly change the required cost-goal multiplier. Certainly, for estimating the approximate cost goals for future technologies, the uncertainties generated by assuming a simple payback are smaller than the uncertainties of future costs of collectors and other components. The actual cost of energy at the time when future systems are proposed and built has a significant impact on the future economic attractiveness of solar options. Fuel-escalation rates are projections that are used to estimate future first-year energy cost savings.

22.2.2 Fuel Cost Savings

Any specific application of solar energy must be evaluate by use of energy prices that apply to the specific location. There are also some differences in the cost of energy for residential and for commercial customers as well as different utility rate structures that may depend on both the quantity of energy used and, for electricity, the peak rate of usage. There is considerable variation in the cost of different fuels from locality to locality.

There will also be an escalation of energy prices over the next 20 years. For the purpose of comparison of different technologies, the national average energy prices as given by the National Energy Policy Plan (NEPP 1983) will be used. The residential and commercial national average energy prices in 1982 dollars for the years 1982 and 2000, and the average fuel-price escalation rates are shown in table 22.1.

There is considerable uncertainty in the price and supply of natural gas. In a recent study (OTA 1983), the Office of Technology Assessment found

"insufficient evidence on which to base an optimistic or a pessimistic outlook for conventional domestic gas production. Given market conditions favorable to gas exploration and production, the production of natural gas from conventional sources within the lower 48 states could range from 9 to 19 TCF/yr (10^{12} ft^3/yr) by the year 2000."

Table 22.1
NEPP-IV national average energy prices in 1982 dollars

	Residential price estimates		
	1982	escalation	2000
Natural gas	$5.39/MBtu	3.1%	$9.28/MBtu
Heating oil	$8.47/MBtu	2.4%	$12.99/MBtu
Electricty	$20.11/MBtu	1.0%	$24.07/MBtu
	Commercial price estimates		
	1982	escalation	2000
Natural gas	$5.00/MBtu	3.3%	$8.97/MBtu
Heating oil	$7.80/MBtu	2.4%	$12.00/MBtu
Electricty	$20.11/MBtu	1.3%	$25.56/MBtu

Source: NEPP 1983.

Because of the large uncertainty in predicted supply and demand for natural gas, there is also a large uncertainty in the future price and availability of natural gas. OTA reports that if the recent surge in the rate of additions to the proven natural gas reserves is not sustained, there will be strong upward pressures on natural-gas prices in particular and energy prices in general. This could significantly change the conclusions about the attractiveness of solar cooling technologies.

If a solar system is built to deliver energy over a 20-year period, the value of the energy produced by that system must be compared with the cost of alternative fuel over a similar period, in which fuel costs rise because of inflation and fuel-price escalation. The levelized energy cost is a constant cost that, when discounted and summed over the given period, gives the same present value as the escalated fuel cost over the same period. Table 22.2 shows the levelized energy-cost factor that gives the effect of discount rate, fuel-escalation rate, and time horizon on the energy cost. The energy cost for the first year of operation must be multiplied by the energy cost factor to give the levelized energy cost.

If a solar system is evaluated over a long period of time, say, 20 years, then the levelized price will be about twice the first-year price. If, however, we look at fuel prices for only the next 5 years, then the average price is only 20% higher than first-year prices. For smaller discount rates, the importance of escalation of fuel prices is larger than for large discount rates. The annualized cost of delivered solar energy should be compared

Table 22.2
Levelized energy-cost factor assuming an 8% annual inflation rate

Discount	Escalation	Levelized energy cost factor		
		20 yr.	10 yr.	5 yr.
18%	2%	1.762	1.403	1.183
18%	4%	2.017	1.508	1.224
12%	2%	2.025	1.459	1.195
12%	4%	2.391	1.580	1.238

with the levelized cost of conventional energy, for energy costs will probably rise faster than inflation.

For the commercial owner, there are loan-interest deductions and the disincentives of income tax to be paid on energy cost savings. For the residential owner, there are loan-interest-expense tax deductions. The effects of tax rate, depreciation, operation and maintenance, and leveraging the investment with borrowed money can be shown to be a multiplier of the simple cost goal (Scholten 1983). For residential systems, this multiplier is typically 1.03 to 1.10 without energy or investment tax credits. For commercial systems, the negative impact of income tax paid on energy savings makes the multiples 0.5 to 0.7 without energy and investment tax credits.

22.2.3. Capital Costs

The payback period for a solar system or the annualized cost of solar energy delivered depends on the capital cost of installing the system. Estimating costs of active solar systems requires a detailed design including the size of the collector array and storage. As part of the Active Program Research Requirements (APRR) assessment, a unified cost catalog (SAI 1983) was developed, showing present and projected future costs of different solar components. This catalog is not sufficiently accurate or timely to be used for cost estimation of actual construction projects, but it does establish a uniform basis for comparing the costs of different technologies. With the experience developed in the many active solar installations, the costs of collector arrays, mounting, piping, storage, etc. are becoming more well established, and standard engineering cost estimation can be used.

Historically, incentives have been used to make an investment in an active solar energy system more attractive by reducing the effective capital cost. For the commercial owner, incentives of energy and tax credits, depreciation and loan-interest-expense tax deductions have been beneficial. For the residential owner, energy tax credits and loan-interest-expense tax deductions have been of major importance. The effect of large tax credits has been dramatic, having the effect of reducing the capital outlay and effective cost of the system by as much as 40 to 50%. The tax credits subsidized the production of energy by solar means and have facilitated the development of the solar industry. However, solar energy and investment tax credits have been largely phased out.

22.3   Active Solar Cooling Technologies

In this section, the performance of solar-driven absorption, Rankine, and desiccant systems is discussed. In the past 10 years, there have been many projects to advance the state of the art of active solar cooling. The NSF Research Applied to National Needs program, the Solar Demonstration program, the NASA 404 program, and the Solar Research and Development program have developed many systems, making many advances in technology. Operational problems have often prevented installed systems from meeting their design expectations as regards percentage of the cooling load that will be met by the solar system and the amount of energy savings. Hardware and control problems can seriously affect performance.

For a solar cooling system that cannot generate electric power, the maximum energy savings are determined by the energy required to meet the space-conditioning load by conventional means. Considerable advance in fossil-fired heating and cooling equipment is expected between now and the year 2000. The development of pulse combustion gas-fired furnaces has increased heating efficiency to 90% or greater. The performance of conventional heat pumps and vapor compression air conditioners will be improved by the use of more efficient motors and optimized heat exchanger areas. The projected seasonal COP of commercial cooling systems is assumed to be 4.0. For residential cooling systems, the seasonal COP is assumed to be 2.75 (Hass and Hall 1982).

22.3.1   Active Solar Absorption Cooling Systems
A thorough discussion of research in active solar absorption cooling systems is included in chapter 20, Absorption Systems and Components. A comprehensive assessment of the energy-savings potential of active solar

cooling systems using current absorption cooling technology was performed to establish future cost goals and to define research needs (Hughes et al. 1981; Choi and Morehouse 1982). Commercial systems (25 tons of cooling) in a 7-zone light commercial office building and residential systems (3 tons) for a single-story residence were analyzed with computer simulations (Choi and Morehouse 1982) in four cities representative of different climatic regions: Miami, warm and humid; Phoenix, hot and dry; Fort Worth, warm and humid with mild winters; and Washington, DC, warm summers and cold winters.

The energy-savings potential of advanced absorption cooling systems has been estimated from detailed computer simulations by projecting electrical and thermal performance and cost improvements (Warren 1982a, 1982c). The thermal and electrical characteristics of absorption solar-driven cooling systems have been evaluated for different concepts, including both single and double-effect absorption systems. A solar-driven absorption cooling system is considered as representative of the range of technical options that may be significant in the future.

The performance of conventional and advanced solar-driven absorption chillers as a function of boiler firing temperature is shown in figure 22.1. Today's lithium-bromide single-effect absorption chillers have a maximum thermal performance, $COP_{chiller}$, of 0.72, while the COP of double-effect chillers can approach 1.15. Advanced ammonia/water absorption chillers presently under development, if successful, will operate at higher temperatures with improved performance: the "2R" regenerative chiller is projected to have a COP of 1.25 at 280°F (140°C) and the "1R" regenerative chiller is projected to have a COP of 1.55 at 280°F. Efficient collection of solar energy at these elevated temperatures requires evacuated-tube collectors and high-temperature fluid or pressurized water storage. These requirements may add to the cost of the solar collection system.

22.3.1.1 Commercial Absorption Cooling System

A solar-driven commercial absorption system with a backup electric chiller is shown in figure 22.2. The baseline 25-ton absorption chiller operates with a COP of 0.7 at a firing temperature of 90°C (195°F) and has an electrically operated backup chiller. At this low thermal efficiency, natural gas firing of the absorption chiller is not cost effective compared to an electric vapor-compression chiller unless relative prices of natural gas and electricity strongly favor gas usage. On the other hand, for a large solar-

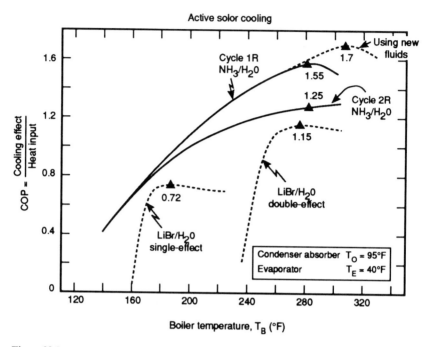

Figure 22.1
Performance of solar-driven absorption chiller. Source: Wahlig 1982, private
communication, unpublished.

fraction system, the added capital cost of the electric vapor-compression
chiller is difficult to justify. For single-effect absorption chillers sized for
low annual solar fraction, the solar absorption chiller would be sized to
match the collector array, and a separate backup vapor compression chiller would be used. For advanced chillers or single-effect chillers sized for
large (>75%) annual solar fraction, the solar driven absorption chiller
would be sized for full design load, and natural gas backup would be
provided.

Advanced absorption chillers proposed or currently under development
are projected to operate with a COP as high as 1.55 at an operating
temperature of 140°C (280°F). For the baseline systems, the collector loop
consists of about 232 m² (2500 ft²) of evacuated-tube collectors. For the
advanced systems providing the same solar fraction, the collector loop
consists of about 93 m² (1000 ft²) of advanced integrated CPC evacuated-

ACTIVE SOLAR COOLING AND HEATING SYSTEM

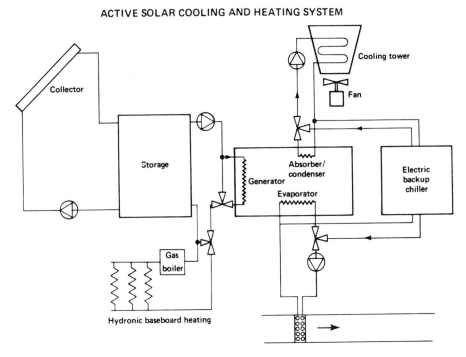

Figure 22.2
A solar-driven absorption cooling system with backup electric chillers. Source: Warren 1982a.

tube collectors or equivalent high-performance linear trough collector. Backup can be provided with a conventional electric chiller or with a gas boiler firing the absorption chiller. For advanced systems, the gas backup boiler can be integrated into the absorption chiller.

22.3.1.2   Estimates of Energy Performance
Estimates of performance of the baseline system have been obtained from the detailed simulation (Choi and Morehouse 1982). The performance of the improved systems has been estimated, assuming a reduction in the parasitic power consumption for the chiller by about 20% and for heat rejection by about a factor of two. The energy savings of the baseline and improved solar-driven absorption cooling systems are shown in table 22.3. The first-year energy-cost savings based on 1982 NEPP-4 energy prices are shown in the table. Also shown are the annual energy cost savings

Table 22.3
Energy savings in Phoenix for a commercial-scale solar-driven absorption cooling system

System type	Baseline	Improved	Advanced	Advanced
Backup energy source	Electric	Electric	Electric	Gas
Conventional electric energy	91.4 GJ$_e$	91.4 GJ$_e$	91.4 GJ$_e$	91.4 GJ$_e$
Solar electric energy required	52.3 GJ$_e$	40.0 GJ$_e$	34.4 GJ$_e$	23.7 GJ$_e$
Net electric energy savings	39.1 GJ$_e$	51.4 GJ$_e$	57.0 GJ$_e$	67.8 GJ$_e$
	37.2 MBtu$_e$	48.7 MBtu$_e$	54.1 MBtu$_e$	64.3 MBtu$_e$
Solar natural gas required				30.6 GJ$_g$
Net natural gas energy savings				-30.6 GJ$_g$
				-29.0 MBtu$_e$
1982 first-year annual fuel cost savings (gas \$5.00/MBtu, electricity \$20.11/MBtu)	\$748	\$979	\$1,088	\$1,148
Year 2000 10-yr. levelized annual fuel cost savings (gas \$12.56/MBtu, electricity \$35.78/MBtu)	\$1,331	\$1,742	\$1,936	\$1,936
Incremental system cost (from table 22.4)	\$136,600	\$50,700	\$35,600	\$22,400
Approximate year 2000 payback	> 100 yrs	29 yrs	18 yrs	12 yrs
Performance indices				
Collector area	232 m^2	232 m^2	93 m^2	93 m^2
COP$_{chiller}$	0.71	0.71	1.55	1.55
COP$_{system}$	0.16	0.16	0.41	0.41
ECOP$_{system}$	7.5	10.9	10.9	13.5

Source: Warren 1982a.

based on the 10-year levelized energy costs for a system built in the year 2000. Assuming an 18% discount rate, a 2% fuel-escalation rate, and year 2000 energy prices (NEPP-4) from table 22.1 of \$8.97/MBtu for natural gas and \$25.56/MBtu for electricity, the 10-year levelized cost is a factor (from table 22.2) of 1.40 times the year 2000 energy costs.

The performance of the systems with advanced collectors and chillers has been estimated, assuming: (1) the thermal COP of the absorption chiller is increased from 0.7 to about 1.55, and the operating temperature is increased from 90°C (195°F) to 140°C (280°F), based on projections of advanced chiller performance to be achieved by research in progress; (2) the collector area is reduced to give the same solar fraction for cooling, considering both the increased chiller COP and the use of the integrated compound parabolic-reflector evacuated-tube collector that has been pro-

Table 22.4
Hypothetical component and systems costs (1981 dollars) for commercial-scale solar-driven absorption cooling systems

System type	Baseline	Improved	Advanced	Advanced
Backup energy source	Electric	Electric	Electric	Gas
Collector area m^2	232 m^2	232 m^2	93 m^2	93 m^2
Collector area ft^2	2500 ft^2	2500 ft^2	1000 ft^2	1000 ft^2
Chiller COP	0.7	0.7	1.55	1.55
Collectors (installed)	$88,500	$22,400	$9,000	$9,000
Storage	$8,200	$8,200	$8,200	$8,200
Solar-driven chiller	$24,500	$12,300	$12,300	$12,300
Cooling tower	$5,800	$5,800	$4,300	$4,300
Backup chiller	$13,200	$13,200	$13,200	
Controls	$10,600	$3,000	$3,000	$3,000
Misc. pumps, piping, etc.	$9,400	$9,400	$9,200	$9,200
Total system cost	$160,200	$74,300	$59,200	$46,000
Conventional cost	$23,600	$23,600	$23,600	$23,600
Incremental system cost	$136,600	$50,700	$35,600	$22,400

Source: Warren 1982a.

posed and tested; and (3) the backup cooling is provided by an integrated gas boiler that operates at a COP of 1.55 (Warren 1982a).

Table 22.3 shows the projected improvement in electric energy savings for a 25-ton solar-driven absorption cooling system, as the electrical efficiency for heat rejection, chiller operation, and collector loop operation are improved. The main benefit of the advanced systems is a reduction in collector area and cost and in the heat rejection required by the chiller. Integrating the gas backup into the chiller decreases the total system cost. The effect of advanced system performance on electric energy savings is also shown in table 22.3.

The cost of today's systems has been estimated by engineering analysis (SAI 1983). It is clear that the cost of active solar cooling systems must be reduced if they are to be competitive in the market. Assumptions must be made about the cost reductions for future systems. The costs that can be reached are a matter of speculation. The hypothetical costs of present, improved, and advanced solar-driven absorption cooling systems in the year 2000, expressed in constant 1981 dollars, are shown in table 22.4.

Present-day 20-ton electric-driven air-cooled vapor compression chillers have an installed cost of about $490/ton. If the cost of advanced solar-driven absorption chillers with integral gas backup can be reduced to that of today's chillers, the ultimate cost of the advanced absorption chillers would be reduced from about \$820/ton to \$490/ton. Today's solar collectors cost about \$380/m^2 (\$35/ft^2). The costs in table 22.4 project a reduction in solar collector cost to about \$100/m^2 (\$9/ft^2). There is considerable controversy as to whether such low costs could be reached.

Based on this analysis, an active solar-driven high-performance absorption heat pump in a cooling-only mode with an integrated gas-fired heat pump backup should achieve costs that give approximately a 12-year payback for a system installed in the year 2000, as shown in table 22.3. Of course, the payback period depends on the actual fuel costs and fuel-escalation rates at the time the system is built.

It should be pointed out that the solar equipment (collectors, storage tanks, pumps, and piping) used for supplying heat to absorption chillers can also provide space heating during seasons when cooling is not needed. The capital cost of those components should therefore be equally prorated between cooling and heating functions, the proportions being dependent on the relative loads. Assuming 50% of the costs of collection, storage, and miscellaneous items is allocated to the heating operation, incremental system costs for cooling only in table 22.4 become, for the four cases, \$83,550, \$30,700, \$9,200, and \$9,200, respectively. The corresponding payback periods (table 22.3) only for cooling become 63 years, 18 years, 4.7 years, and 4.7 years. These figures show that the advanced systems can have an attractive potential when combined with solar space heating.

22.3.2 Active Solar Rankine Cooling Systems

The application of Rankine heat-engine-driven systems for solar cooling and heating has been under active development in the DOE program. There have been basically three approaches: solar driven heat pumps; solar driven chillers with electric power generation capability; and advanced systems such as fossil-boost schemes to raise the temperature of the Rankine working fluid in order to achieve higher thermodynamic efficiencies (Curran 1975). These technologies are discussed in detail in chapter 19, Mechanical Systems and Components.

22.3.2.1 Rankine-Powered Heat Pump

In the Rankine-driven active solar cooling system shown in figure 22.3 about 232 m^2 (2500 ft^2) of parabolic trough concentrating collectors, or an equivalent area of evacuated-tube collectors, supply heat to a boiler that

EVAPORATIVELY COOLED RANKINE COOLING
SYSTEM WITH POWER GENERATION

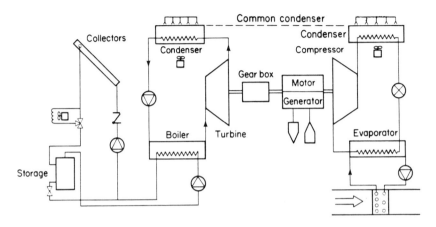

Figure 22.3
Evaporatively cooled commercial Rankine system with power generation. Source: Warren 1982c.

generates high temperature 150°C (300°F) vapor to drive a turbine. The turbine produces shaft power to drive a vapor compression air conditioner or heat pump. In the heat pump configuration, insulated and pressurized thermal storage is used, and backup is provided by a gas-fired boiler that drives the turbine. In the power-generating configuration, the shaft power can be used to generate electric power, or an electric motor can be used for backup.

In the heating mode the heat available to the space includes both the heat pumped from the ambient environment and the waste heat from the condenser of the Rankine engine. Rankine-powered heat pumps are efficient heating devices. One must evaluate the improved performance and additional costs compared to those of an active direct solar space heating system.

In the cooling mode, the cooling effect comes from the evaporator coil of the heat pump. The waste heat from both the heat engine and heat pump condensers must be discharged to the environment. Because of the low efficiency of the Rankine cycle, the rejected heat can be 2 or 3 times that from an electrically driven heat pump.

One can define the chiller thermal performance, $COP_{chiller}$, for a Rankine chiller system as the cooling effect delivered, divided by the energy supplied to the boiler that generates vapor for the turbine. For the one NSDN Rankine site in Phoenix, operating at about 200°F, $COP_{chiller}$ averaged 0.24 during five months of 1981. This ratio is consistent with a Rankine cycle efficiency of 0.08 (mechanical/thermal), a mechanical efficiency of 0.85, and a vapor compression chiller COP of 3.5 (thermal/mechanical). If the Rankine cycle is operated at a temperature of 300°F and a cycle efficiency about 0.14, a mechanical efficiency of 0.85, and the compressor performance ratio (cooling effect to shaft power) of 8.0, the overall cooling thermal performance is about 0.7, which is comparable with today's absorption chillers.

The economics of solar Rankine-driven heat pumps depends on the ratio of annual heating load to cooling load, the availability of sunshine, and the ratio of cost of delivered thermal energy to the cost of electric energy to drive a conventional heat pump. A simplified analysis (Khalifa et al. 1983) indicates the minimum efficiency of the Rankine heat engine necessary to compete with conventional heat pump, as a function of the ratio of the cost of thermal energy to electric energy. The cost of thermal energy is either the cost of heat provided by gas or oil with the appropriate combustion efficiency, or it is the annualized cost of providing solar energy at the appropriate temperature to drive the heat engine (cooling) or heat the spaces (heating).

The required Rankine power loop thermal efficiency to achieve parity with primary energy cost for electrically and thermally driven heat pumps in heating and cooling for various ratios of heating to cooling has been calculated by Khalifa and Melikian (1982) and is shown in figure 22.4. For cooling only, $Q_H/Q_C = 0$, the minimum Rankine efficiency must be 0.25 when the fuel cost ratio is 4. Present turbines, fired at a temperature of about 300°F, have a Rankine power loop efficiency of about 0.14. At that efficiency, a cooling-only system would become cost-effective only if the cost of electricity became greater than 7 times the cost of thermal energy. The greater the heating load, the smaller is the Rankine power loop efficiency required. For a heating-only system, Rankine efficiency of zero corresponds to an electric-driven heat pump. It is cost-effective when the ratio of electric to thermal energy costs are the same as the COP of the heat pump. The application of heating significantly improves the perfor-

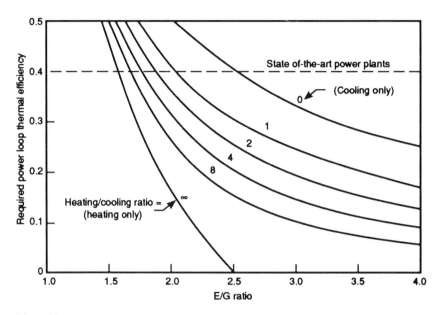

Figure 22.4
Required power loop thermal efficiency for thermally driven heat pumps. Source: Khalifa and Melikian 1982.

mance and reduces the required Rankine efficiency to compete with conventional systems.

The required power loop thermal efficiency depends on the performance (ECOP) of the compressor. Figure 22.5 shows the required power loop thermal efficiency for a thermally driven heat pump system compared with a system comprising an electric air conditioner for cooling and a gas furnace for heating; the heating load is assumed double the cooling load, $Q_H/Q_C = 2$. The required power loop efficiency decreases with improved cooling ECOP of the heat pump. A comparison with an electric heat pump (EHP) (heating and cooling) is also shown on figure 22.5; the required power loop efficiency is independent of the heat pump cooling ECOP, assuming that the same heat pump is used for both systems. To compete, the power loop efficiency of the thermally driven heat pump must be much higher than in the case of the electric-gas combination, unless the cost of electricity is more than about five times the cost of thermal energy.

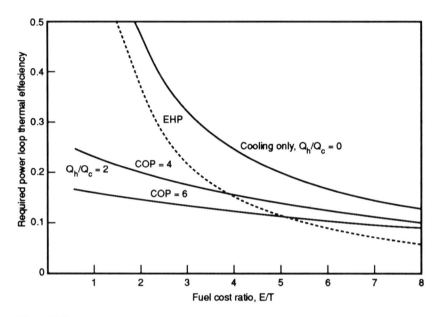

Figure 22.5
Required power loop thermal efficiency for different heating/cooling load ratios for a
thermally driven heat pump system competing with gas heating and electric
air-conditioning, as a function of the electric to thermal energy cost ratio. Adapted from
Khalifa and Melikian 1982.

As has been discussed previously, the annualized cost of useful energy
delivered by a solar energy system depends on the cost of the hardware to
provide the energy, the annual useful solar energy delivered by the system,
and the economics of the capital investment that is used to establish the
annualized cost of owning and operating the system. Clearly, for the solar
system to be cost-effective, the annualized cost of the system per unit of
delivered energy must be less than the levelized cost of delivering the same
energy by use of fossil fuel. The NEPP-4 energy costs for the year 2000
indicate a fuel-cost ratio of about 4. This ratio will place an upper limit on
the capital cost of the solar delivery system. In addition, for a given ratio
of electric to thermal energy cost and a specific heating to cooling load
ratio, there is a required thermal efficiency of the Rankine power loop that
must be met.

The thermal performance of the Rankine systems can be improved by
advanced turbine design and by increasing the turbine operating tempera-

ture. Electrical performance can be improved by reduced parasitic mechanical and electrical losses. The mechanical efficiency can be improved somewhat by direct coupling of the compressor to the turbine and improving the efficiency of the compressor.

22.3.2.2 Fossil-Boost Systems

To overcome the poor thermal efficiency of power generation at typical solar collector temperatures, the performance of Rankine expanders (turbo compressor) can be improved by using solar energy to generate steam at low ($< 150°C$) temperature and by increasing the operating temperature to $1100°F$ ($580°C$) using a fossil-boost scheme (Curran 1981b; Lior and Koai 1984a, 1984b). A base-case system using evacuated-tube collectors with a water-cooled power-cycle condenser analyzed in Phoenix showed an overall Rankine cycle efficiency of 13.5% and an overall thermal COP of 0.8. It was able to meet 73% of the total summer cooling load. Feasible reduction in the condenser fan-power and the use of a commercially available high COP chiller would increase the overall thermal COP for the system from 0.85 to 1.35.

22.3.2.3 Rankine Chiller with Power Generation

For a system that is used primarily for cooling, much better annual utilization of heat-engine-driven systems can be obtained if excess collected solar energy is used to generate electric power. Electric power can be sold to the grid if generated in excess of that needed within the building. A Rankine turbine system with electric power generation capability and minimal thermal storage capacity connected with the utility can provide better utilization of the solar resource (Gossler et al. 1980; Aasen 1982, 1983; Curran 1981a; Biancardi et al. 1982). Generation of electric power with an induction motor-generator is an inherently safe and inexpensive method of interfacing to the utility grid.

For a representative second-generation high-temperature ($300°F$) system in Phoenix with Rankine power loop thermal efficiency of 0.14, an input of 1032 GJ_{th} produces 148 GJ of mechanical shaft output. After subtracting mechanical loses, electrical parasitic requirements, and generator losses, a net electrical output of 97 GJ_e is obtained. This combination gives an efficiency of 0.094 (electric/thermal). However, when a collection efficiency of 0.54 is included, the overall system conversion efficiency is only (5%) (electric/solar). Nevertheless, the addition of power-generation capabilities significantly improves the overall energy savings for the systems.

22.3.2.4   Boil-in-Collector Systems

A promising system from a thermal performance consideration is the boil-in-collector system in which working fluid is boiled in the collector and used to drive the turbine directly (Hedstrom 1983). The boil-in-collector system can benefit from reduced installation costs and higher energy savings resulting from higher engine efficiency due to elimination of heat exchanger temperature differences, lower parasitic power losses due to elimination of solar loop pump, and lower engine cost due to elimination of two heat exchangers (preheater and boiler). However, the trade-off may result in a loss in system modularity and potentially higher maintenance costs.

22.3.2.5   Cost and Performance Projections

Performance estimates of advanced systems have been obtained by using the results of the detailed simulation analysis of a Rankine heat pump and a Rankine cooling system with power generation (Choi and Morehouse 1982). The collector loop consists of 232 m^2 (2500 ft^2) of parabolic trough concentrating collectors or an equivalent area of evacuated-tube collectors. Heat rejection to prevent boiling of the storage liquid, an expansion tank, and insulated piping are required. The collector loop fluid is 50% glycol and water. A pump circulates the fluid in the collector loop. A Rankine turbine drives a 25-ton vapor compression centrifugal chiller with a COP of 4.4. The unit includes a shaft-mounted motor/generator that may be connected to the utility grid. Separate (or common) evaporatively cooled condensers reject heat from the Rankine and chiller condensers to the environment.

The net electric power produced by a Rankine cooling system (with power generation) that meets the annual cooling load in Phoenix, for different collector types and areas, is shown in figure 22.6. The systems with flat-plate, evacuated-tube, and the Brookhaven National Laboratory low-cost polymer collectors were operated at a Rankine generator temperature of (90°C) 195°F. It should be noted, however, that the prototype BNL polymer collectors could not withstand the required operating temperatures in testing at Colorado State University (Lof, Westhoff, and Karaki 1984).

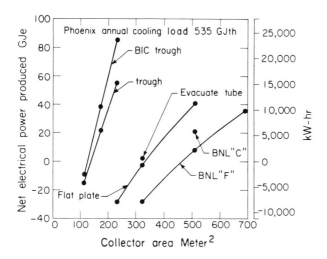

Figure 22.6
Net electric power produced by a commercial Rankine cooling system with power generation in Phoenix as a function of different collector areas and types. Source: Warren 1982c.

Table 22.5 summarizes the projected energy savings, estimated incremental system cost, and approximate payback period based on the SAI simulation analysis for the high-temperature 150°C (300°F) Rankine turbine systems operated as a thermally fired heat pump with air-cooled condenser, as an evaporatively cooled chiller with power generation, and as an evaporatively cooled chiller with power generation driven directly by boil-in-collectors. The breakdown of present and hypothetical system costs is given in table 22.6.

22.3.2.6   Advanced System Development
Active solar Rankine cooling will be a useful technology option only if research and development is pursued to maximize the performance and minimize the cost of these systems. Hardware development effort is needed to determine the extent of potential cost reduction, to identify reliability problems and solutions, and to develop advanced controls. Fluids research is needed to find the stable combination of fluids that could give better performance and economics. Mechanical and electrical losses for the turbo-compressor can be reduced to a minimum by high-speed direct coupling of turbine and generator. For commercial application, the per-

Table 22.5
Comparison of energy savings, year 2000 cost goals, and system costs for high-temperature (300°F) Rankine turbine systems in Phoenix

System type	Baseline	Improved	Advanced	Advanced
Backup energy source	Gas	Gas	Electric	Electric
Heat rejection	Air-cooled heat pump	Air-cooled heat pump	Evap.-cooled power gen.	Evap. BIC
Collector area	232 m^2	186 m^2	232 m^2	232 m^2
Collector area	2500 ft^2	2000 ft^2	2500 ft^2	2500 ft^2
Conventional electric energy	91.4 GJ$_e$	91.4 GJ$_e$	91.4 GJ$_e$	91.4 GJ$_e$
Solar electric energy required	30.8 GJ$_e$	29.9 GJ$_e$	−54.5 GJ$_e$	−85.0 GJ$_e$
Net electric energy savings	60.6 GJ$_e$ 57.5 MBtu$_e$	61.6 GJ$_e$ 58.4 MBtu$_e$	145.8 GJ$_e$ 138.4 MBtu$_e$	176.4 GJ$_e$ 167.4 MBtu$_e$
Solar gas energy required	148.8 GJ$_g$	71.3 GJ$_g$	0.0 GJ$_g$	0.0 GJ$_g$
Net gas energy savings	−148.8 GJ$_g$ −141.2 MBtu$_g$	−71.3 GJ$_g$ −67.6 MBtu$_g$	0.0 GJ$_g$	0.0 GJ$_g$
1982 first-year annual fuel cost savings (gas \$5.00/MBtu, electricity \$20.11/MBtu)	\$450	\$836	\$2,783	\$3,336
Year 2000 10-yr. levelized annual fuel cost savings (gas \$12.56, electricity \$35.87)	\$280	\$1,241	\$4,952	\$5,990
Estimated incremental system cost	\$237,200	\$76,100	\$76,700	\$76,700
Approximate year 2000 payback	—	61 yrs	16 yrs	13 yrs

Source: Warren 1982c.

formance of Rankine expanders (turbo compressor) can be improved by increasing the operating temperature from 300°F to 450°F or higher using high-performance collectors or fossil-boost schemes.

Advanced Rankine systems may become cost-effective in the future, but much research and development is required before active solar cooling technology can compete in the marketplace. Rankine technology has two major obstacles. First, it involves high-speed rotating machinery with the inherent concern over reliability. Second, high temperatures are required to achieve significant conversion efficiencies from thermal energy to shaft power. Primarily for these reasons, it is not likely that Rankine

Table 22.6
Hypothetical component and systems costs (1981 dollars) for a commercial-scale
solar-driven Rankine cooling systems

System type	Present Heat pump	Improved Heat pump	Advanced Power gen.	Advanced Power gen.
Backup energy source	Gas	Gas	Electric	Electric
Collector area m²	232 m²	186 m²	232 m²	232 m²
Collector area ft²	2500 ft²	2000 ft²	2500 ft²	2500 ft²
Collectors (installed)	$100,000	$18,000	$22,500	$22,500
Collectors, piping, etc.	$9,200	$9,200	$9,200	$9,200
Storage	$26,000	$14,200	$6,200	$6,200
Rankine turbine	$99,600	$39,600	$39,600	$39,600
Cooling tower	$6,900	$46,900	$6,900	$6,900
Backup chiller	$5,400	$5,400		
Controls	$13,700	$6,400	$6,400	$6,400
Heating system			$9,500	$9,500
Total system cost	$260,800	$98,700	$100,300	$100,300
Conventional cost	$23,600	$23,600	$23,600	$23,600
Incremental system cost	$237,200	$76,100	$76,700	$76,700

Source: Warren 1982c.

technology will find major application in residential and commercial space-conditioning.

22.3.3 Active Solar Desiccant Cooling Systems

Active solar desiccant cooling systems can be divided into four types: open-cycle solid desiccant systems; open-cycle solid desiccant hybrid systems; open-cycle liquid desiccant systems; and closed-cycle solid desiccant systems (SERI 1982). Each of these technologies has been under active development as discussed in detail in chapter 21, Desiccant Systems and Components.

22.3.3.1 Open-Cycle Solid Desiccant Systems

Desiccant materials regenerated by use of conventional fuels have been widely used for space-conditioning environments and processes in which very low humidity is required. Desiccants are attractive for use with solar energy because they can be regenerated at temperatures that can be attained by flat-plate air and liquid solar collectors. There are several different configurations of an open-cycle desiccant chiller: recirculation, ventilation, and vapor compression hybrid. In the recirculation mode of operation, air from the building is recirculated through dehumidifies and cooling side of the cycle, while from the outside is heated and used to regenerate the desiccant. The recirculation configuration is preferable when the outside air humidity is high and the ventilation configuration cannot operate efficiently. In the ventilation configuration, outside air is drawn into the cooling side of the system, and all of the building air is exhausted through the regeneration side. In the vapor-compression hybrid configuration, heat from the vapor-compression condenser is used to regenerate the desiccant material, assisted by thermal energy from either solar energy or fossil fuels.

Preliminary evaluation of the performance and economics of desiccant cooling systems was undertaken by Shelpuk et al. (1979). Based on annual thermal simulations and life-cycle cost analysis of both solar and conventional residential heating and cooling systems, it was concluded that solar heating/desiccant cooling systems could be nearly cost-competitive with conventional residential heating and cooling on a life-cycle basis. Solar desiccant coolers operate best as dehumidifiers, and combined heating and cooling systems have the best economics in climates with nearly balanced heating and cooling loads.

Desiccant cooling systems have four principal components: desiccant dehumidifier, heat exchanger, evaporative coolers, and solar or auxiliary heat source. Computer simulations show that modification of desiccant properties can produce only limited increases in COP. The real key to better performance is improved design of the dehumidifier wheel. Systems employing a spiral-wound parallel-passage desiccant dehumidifier made from a plastic-film base with fine silica-gel particles adhering to the surface have been built and tested at SERI (Schlepp & Schultz, 1984). The results show performance predicted by the analytical model, and achievement of the COP of 1.2 required for a cost-effective solar desiccant cooling system appears possible. Additional computer simulation was used to predict the performance of a desiccant cooling system containing the new dehumidifier. There is a trade-off between the spacing of the channels in the dehumidifier (and the amount of desiccant in the wheel) and the work required to pump the air through the dehumidifier. Small variations in channel spacing and dimensional charges due to temperature differences will have to he corrected for this design to become practical.

A simple model with balanced air flows, 90% effective evaporative coolers, and 95% effective heat exchangers, with a wheel one meter in diameter,

0.20 meter deep, with a 0.7 mm channel width, rotating at 15 revolutions per hour, and a regeneration temperature of 80°C predicts a system thermal COP of 1.07 and an electrical ECOP of 7.9 (EER = 27 Btu/Wh). Based on this analysis and laboratory test data, a 10 kW (3 ton) residential cooling system has been under development. In an experimental evaluation of advanced desiccant cooling systems (Huskey, 1982), the COP of advanced desiccant cooling systems in a test bed has exceeded 1.0 at ARI conditions.

Today's state of the art assumes sensible heat-exchanger effectiveness and electric power consumption as demonstrated by an experimental system. The system is very sensitive to the effectiveness of the regenerator heat exchange wheel and to air stream seal leakage. The latter must be kept as low as possible. A crucial element in the performance of a desiccant system is the parasitic electric power required to move the air through the system and to turn the rotary heat exchangers and desiccant wheels.

A recent field test of a prototype, residential desiccant cooling unit at Colorado State University, Fort Collins (Löf, Cler, and Brisbane 1987), has shown that an average cooling thermal COP of 1.0 can be achieved under mild weather conditions. At a cooling output of 7kW with supply hot water at 60 to 70°C for desiccant regeneration, the average electrical COP is about 5 to 6. If delivering rated output, 10.5 kW cooling, the electrical COP would be about 9.0. During this test, solar collection efficiency was 40 to 50%. At storage temperatures of 55° to 70°C, and with evacuated-tube collector, the average cooling delivered per unit of solar radiation incident on collector exceeded 0.4 during four 10-day periods and reached an average of 0.51 during one 10-day period. This was the highest solar cooling performance factor of any cooling system tested at CSU.

In another study (Hass and Hall 1982), it was concluded that advanced solar/gas and gas-only desiccant systems can be economically competitive with advanced electric heat pumps and solar/electric systems in the 1990s in regions typified by Philadelphia and Dallas/Fort Worth if they operate at COP levels greater than 1.0.

22.3.3.2   Open-Cycle Solid-Desiccant Vapor-Compression Hybrid

In hybrid desiccant cooling system, a desiccant dehumidifier provides latent cooling, an indirect evaporative cooler provides sensible cooling, and

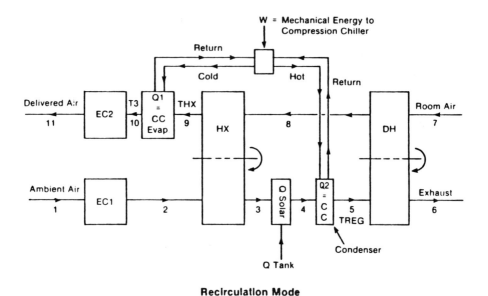

Recirculation Mode

Figure 22.7
Schematic of the hybrid desiccant/vapor-compression chiller in the recirculation mode.
Source: SERI 1982.

a vapor compression chiller delivers any additional sensible cooling needed (Sheridan and Mitchell 1982). The purpose of hybrid systems is to reduce electricity use and the amount of cooling required of the vapor compression unit. A schematic of the hybrid system is shown in figure 22.7. The rejected condenser heat from the vapor-compression chiller can be used as part of the thermal source for regeneration of the desiccant. In addition, fuel-fired backup or solar heat can also be used for regeneration.

Figure 22.8 shows the effective chiller thermal performance, COP_a, and chiller electrical performance, ECOP, for fossil- and solar-fired systems, at various levels of performance. The solid curve represents the points where the primary energy use of the desiccant system just equals that of the vapor compression machine with an ECOP of 2.5. To save energy, the system should have a COP_a and an ECOP that lie above and to the right of this curve (Jurinak, Mitchell, and Beckman 1984), indicating better electrical and thermal performance than a conventional chiller.

Table 22.7 shows the performance of different systems in Phoenix and Miami based on TRNSYS simulation. (Howe, Beckman, and Mitchell

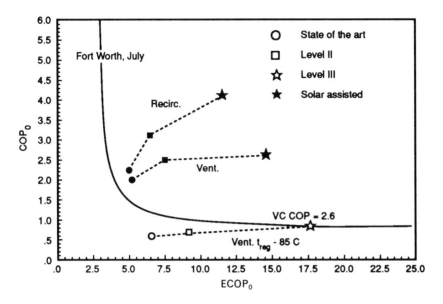

Figure 22.8
Solar-assisted thermal performance, COP_a, and electrical performance, ECOP, for different desiccant cooling cycles in Fort Worth in July. Adapted from Jurinak, Mitchel, and Beckman 1984.

1983a, 1983b). The building is a commercial structure with 900 m² (9700 ft²) of floor space and with a variable air volume system. The desiccant-condenser system uses the waste heat from the chiller condenser in combination with auxiliary heating from natural gas (as needed) to regenerate the desiccant. On the supply air side, the outside dehumidified air is indirectly evaporatively cooled, then cooled further in the chiller evaporator coil. The solar system has 200 m² (2150 ft²) of collector area.

The hybrid systems show significant reduction in electric energy use and first-year energy cost savings. Hybrid system in Phoenix shows a greater percentage reduction in energy use than in Miami because of the dryer climate that provides a better heat sink for the indirect evaporative cooler. It is clear from the table that the desiccant dehumidifier in combination with a vapor-compression chiller has energy savings both in Miami and Phoenix that are comparable to that achieved with a solar-regenerated desiccant system. The use of solar energy to supply the thermal regeneration energy has no distinct advantage over using condenser and auxiliary heat. Combining solar with condenser regeneration increases the energy

Table 22.7
Performance of desiccant systems in a 900 m^2 (9700 ft^2) building, electricity $20.11/MBtu, gas $5.39/MBtu, 1982 dollars

System type	(1) Vapor compression	(2) Desiccant condenser	(3) Desiccant solar condenser	(4) Desiccant solar
Phoenix				
Solar area (ft^2)	0	0	2150 ft^2	2150 ft^2
Electricity use (GJ$_e$)				
Vapor compression	673 GJ$_e$	200 GJ$_e$	169 GJ$_e$	173 GJ$_e$
Fan	108 GJ$_e$	166 GJ$_e$	166 GJ$_e$	166 GJ$_e$
Gas use (GJ$_g$)	—	2 GJ$_g$	21 GJ$_e$	0 GJ$_e$
First-year energy cost	$14,900	$7,460	$6,495	$6,457
First-year cost savings	—	$7,440	$8,405	$8,443
5-yr. cost goal	$0	$37,200	$42,000	$42,200
Miami				
Solar area (ft^2)	0	0	2150 ft^2	2150 ft^2
Electricity use (GJ$_e$)				
Vapor compression	900 GJ$_e$	378 GJ$_e$	356 GJ$_e$	403 GJ$_e$
Fan	112 GJ$_e$	176 GJ$_e$	176 GJ$_e$	176 GJ$_e$
Gas use (GJ$_g$)	—	57 GJ$_g$	345 GJ$_g$	15 GJ$_g$
First-year cost	$19,300	$10,870	$11,930	$11,135
First-year cost savings	—	$8,430	$7,370	$8,160
5-yr. cost goal	$0	$42,150	$36,850	$40,820

Adapted from Howe, Beckman, and Mitchell 1983a, 1983b.

savings slightly; however, the additional savings probably would not justify the solar addition.

As part of the Active Program Research Requirements Assessment, desiccant technologies were evaluated (SERI 1982). Table 22.8 summarizes the system incremental cost, first-year annual savings, and simple payback period for the several base-case systems representative of the best of today's technology (Scholten and Morehouse 1983). The hybrid open-cycle solid desiccant systems are the most attractive, having a projected payback of 14 years or less. Two other technologies that are being pursued include closed-cycle solid desiccant systems and open-cycle liquid desiccant systems.

22.3.3.3 Closed-Cycle Solid Desiccant Systems

A limited evaluation of the performance of a closed-cycle solid desiccant system developed by Zeopower Company was also performed (SERI

Table 22.8
Energy savings, year 2000 first-year energy cost, incremental system cost, and simple payback for light commercial (C) and residential (R) solar desiccant systems

Desiccant cooling systems			Energy saved MBtu/yr.		NEPP-4 $/MBtu		Year 2000 1982$/yr. savings	Incremental system cost (1982$)	Simple payback (yrs.)
System type	Location	C/R	Electric	Gas	Electric	Gas			
Open cycle solid	Phoenix	R	35.2	−20.7	$17.90	$8.67	$451	$13,000	28.8
Open-cycle solid vapor-compression hybrid	Phoenix	R	31.1	0	$17.90	$8.67	$557	$7,500	13.5
Open-cycle solid vapor-compression hybrid	Wash. DC	C	81.1	0	$27.00	$9.79	$2,190	$23,000	10.5
Closed-cycle solid	Wash. DC	R	−3.1	22.2	$25.80	$10.37	$150	$93,400	621.7
Open-cycle liquid	Wash. DC	R	12.6	0	$25.80	$10.37	$325	$16,600	51.1
Open-cycle liquid	Wash. DC	C	76.1	0	$27.00	$9.79	$2,055	$78,200	38.1

Adapted from Scholten and Morehouse 1983.

1982), based on information in the final report (Tchernev 1981). As shown in table 22.8, this system saves only a small amount of energy, and with its large first cost, it is not economically attractive at this time.

22.3.3.4 Open-Cycle Liquid-Desiccant Systems

Open-cycle liquid-desiccant systems depend on the absorption of moisture from air by a liquid, typically aqueous solutions of lithium chloride, calcium chloride, and lithium bromide, or certain glycol solutions. The dry air can be evaporatively cooled, and the solution must then be regenerated by vaporizing the water from solution. Regeneration can be done in a packed bed supplied with solar-heated air, or on a rooftop regenerator where the solution is sprayed or trickled over a south facing roof surface. Liquid-desiccant solar cooling systems have been studied since the 1950s. Open-cycle LiCl liquid desiccant (absorbent) systems with packed-bed regenerators have been investigated at Colorado State University (Lenz et al. 1982). Rooftop regenerator systems have been evaluated at the University of South Carolina (Robison 1983). The performance of glazed

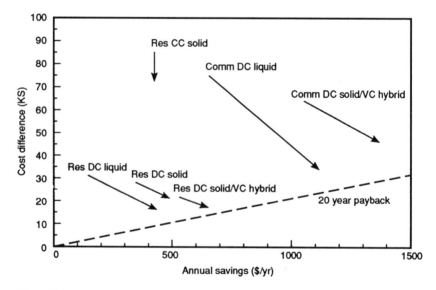

Figure 22.9
Cost and savings improvement projection for different desiccant technologies: open cycle (OC), closed cycle (CC), residential system (Res), and commercial system (Comm). The dashed line indicates a 20-year simple payback. Adapted from Shelpuck 1979.

open-flow liquid-desiccant solar collectors has also been evaluated (McCormick, Brown, and Tucker 1983).

Recent evaluation of the performance of a liquid-desiccant system with rooftop regenerators, shown in table 22.8 (SERI 1982), has been based on the performance results (Robison 1983). The system was based on 100 ft^2 of site-built collector/regenerator, a 900-gallon uninsulated steel tank, and a calcium chloride brine solution. Air-conditioning was accomplished with a spray dehumidifier and an evaporative cooler with a backup packaged air conditioner. The overall system thermal COP was 0.6 from insolation at collectors to average daily cooling output. The parasitic power was estimated at an ECOP of 27.

Improvement of open- and closed-cycle systems could yield improved performance. Figure 22.9 shows the projected cost differences and annual savings for commercial and residential systems (SERI 1982). The dashed line indicates a simple payback period of 20 years. Areas of possible improvement include better dehumidifiers and less expensive but efficient air-to-air heat exchange devices, low-cost collectors, or rooftop regenerators.

22.4   Comparison of Different Technologies

22.4.1   Performance Figures of Merit

The overall performance of active solar cooling systems depends on the performance of the collection, storage, chiller, and heat rejection subsystems. The collection efficiency depends on the average operating temperature of the system, the optical and thermal characteristics of the collector, and the effectiveness of the collector control system. The storage efficiency depends on the matching of the load demand to the available solar resource and the thermal losses through conduction, boiling, or heat rejection.

The operating thermal performance of an absorption chiller system depends on the design COP, the operating controls and conditions, and the matching of the load to the available resource. The operating thermal performance of a Rankine chiller system depends on the efficiency of the Rankine engine, the mechanical losses in the drive train, and the efficiency of the compression chiller. The operating electrical performance depends on the pumping power required to transport heat to the generator or boiler and to the evaporator and the internal solution pumps, and on the pumping and fan power required for the heat rejection condenser.

The solar cooling performance factor, SCPF, is a measure of the effectiveness of converting incident solar energy at the collector to useful delivered cooling effect. The system electrical performance, $ECOP_{system}$, is a measure of the electric energy required to operate all parts of the solar system and is equal to the solar cooling effect delivered, divided by the total electric energy required to run the solar system.

Table 22.9 shows the performance figures of merit for different absorption, Rankine, desiccant, and photovoltaic cooling systems based on computer simulations of baseline and proposed advanced systems. The measured performance of two operating systems is also shown. Collection efficiencies range from 0.31 to 0.54. The storage efficiencies vary from 0.36 to 0.60, except for systems with electric power generation capability, which can have storage efficiencies of 0.90 and make much more effective use of collected solar energy.

For single-effect absorption systems, the chiller COP is typically 0.65. However, for systems that do not have an auxiliary heat source on the generator side, observed seasonal performance is rarely that high. Absorption systems fired by solar alone operate at off-design conditions more of the time and have poorer performance. Advanced systems are proposed with a chiller COP as high as 1.55.

For Rankine systems, the effective chiller COP (0.24) depends on the Rankine engine efficiency (0.08 at 95°C), the drivetrain efficiency (0.85), and the compression chiller performance (3.5 thermal/mechanical). If the Rankine power loop efficiency can be increased to 0.14 by going to higher temperatures (about 150°C) and if the chiller electrical coefficient of performance can be increased to 5.0, then the effective chiller COP can reach 0.7. Only by increasing the operating temperature significantly with fossil-fuel-boost schemes can the effective chiller COP for Rankine systems be increased to levels approaching 1.0.

The purpose of thermally driven cooling systems is to use solar energy to displace the electric energy to run a conventional vapor-compression air conditioner or chiller. Therefore, the electrical performance of the chiller, $ECOP_{chiller}$, and of the system as a whole, $ECOP_{system}$, plays a crucial role in determining the overall electric energy savings. As shown in table 22.9, the projected chiller ECOP ranges from 6.4 to 15.5, and the system ECOP ranges from 7.5 to 13.5. These figures can be compared with to-

Table 22.9
Performance comparison of light commercial (C) and residential (R) active solar cooling systems

System type	Location	C/R	Collector area ft²	Collection efficiency	Storage efficiency	Chiller COP	Chiller ECOP	SCPF	System ECOP
Absorption cooling systems									
Baseline electric backup	Phoenix	C	2500	.39	.6	.7	8.9	.16	10.9
Improved electric backup	Phoenix	C	2500	.39	.6	.7	13.2	.16	13.5
Advanced electric backup	Phoenix	C	1000	.32	.3	1.55	15.5	.41	
Absorption									
(El Torro Library)	So. Cal.	C[a]	1427	.31	.51	.43	6.4	.10	
Baseline gas backup	Phoenix	R	430	.4	.58	.65		.15	7.5
Advanced gas backup	Phoenix	R	160	.45	.58	1.55	12	.40	9.9
Rankine cooling systems									
Mod-2 heat pump	Phoenix	C	2500	.54	.43	.64[b]	12.4	.15	9.7
Mod-3 heat pump	Phoenix	C	2000	.54	.36	1[b]	13.4	.19	10.4
Power generation									
(Salt River 1982)	Phoenix	C[a]	8208	.44	.9	.24[b]	7.84	.07	10.7
Desiccant cooling systems									
Open cycle	Phoenix	R				1.1	8.2		
Open-cycle/ vapor-compression hybrid	Phoenix	R				1.5	8.8		
Open cycle	Ft. Collins	R[a]						0.5	
Photo voltaic cooling									
Current				.073[c]		2.0		.15	
Advanced				.073[c]		4.0		.30	

Adapted from Scholten and Morehouse 1983.
a. Indicates measured performance.
b. Based on thermal supply to Rankine engine cycle.
c. Conversion efficiency.

day's conventional chiller ECOP of 2.3 to 4.4. Reduction of parasitic power is a major research goal for future systems. Depending on the cooling load requirements, system sizing, and operating characteristics, Rankine systems that have power-generation capability can generate net electric power, primarily during periods of high insolation and low cooling demand.

The observed performance of two installations in the National Solar Data Network, an absorption system at El Torro Library in California and a Rankine system at Salt River in Phoenix, are shown in table 22.9. The figures are based on results from the 1980, 1981, and 1982 cooling seasons (NSDN 1981; Wetzel and Pakkala 1982; Logee and Kendall 1983). The absorption chiller thermal performance, COP_{chill}, in three other NSDN sites ranged from 0.22 to 0.45 in the summer of 1982, and 0.43 to 0.60 in the summer of 1981. These values are all below the thermal COP of 0.7 at the steady-state design operating point. Problems that led to degradation of performance included rapid cycling between heating and cooling because of control problems, cycling of the machine because of high condenser temperatures in humid environments, improper staging of auxiliary chillers, and the inability of one absorption chiller to hold a vacuum.

Of the five NSDN cooling sites reported in the summer of 1982, the El Torro Library, with 1427 ft^2 of evacuated-tube collectors and a 25-ton solar absorption chiller, had the highest solar cooling performance factor of .10 with good solar collection efficiency (31%), reasonable storage and transport losses, and a fairly good chiller. The next best system was the Rankine turbine/vapor compression chiller in Phoenix, Arizona, which had a SCPF of 0.07 with 8,208 ft^2/of collectors (efficiency 44%) driving a Rankine engine and two 25-ton chillers. Storage and transport losses were relatively low, the Rankine engine efficiency was 7%, and the vapor compression chiller had an ECOP of about 4.0. The overall electric COP of the system was 10.7.

In the NSDN sites, the highest system electrical coefficient of performance was 6.4 in 1982 and 7.0 in 1981 for the Rankine system in Phoenix. At the other sites, it ranged from 1.0 to 2.3 in 1982. These low ratios show that the operating electric energy for many solar systems was as large as that for a typical vapor compression air conditioner!

Closely controlled testing of absorption cooling systems in buildings at Colorado State University show a measured monthly average COP of 0.7

with a high daily average COP of 0.84 and a peak COP of 0.9 (Karaki, Vattano, and Brothers 1984). The system operated for three months with a solar cooling performance factor of 0.31 to 0.33. These results indicate that when properly designed, engineered, installed, and operated, a solar cooling system can work near its potential. The real challenge is to develop superior cooling systems that will use solar energy to produce cooling in a cost-effective manner.

The performance of thermally driven solar cooling systems can be compared with the performance of a system of photovoltaic electric power driving a conventional vapor compression chiller. Assuming unencapsulated solar cells with an efficiency of 0.12, a cell packing density of 0.76, glazing losses of 0.08, and inverter losses of 0.15, the efficiency of the solar array is 0.073. For 1,000 watts of solar power incident at the collector, 73 watts of electrical power is available to run the chiller. Assuming a vapor-compression chiller with an $ECOP_{VC}$ of 2.0, the cooling output would be 146 watts for a solar cooling performance factor of 0.146, which is comparable with the best single-effect absorption chillers and with advanced Rankine systems. If a vapor-compression chiller with a COP of 4.0 is assumed, an impressive SCPF of approximately 0.3 would be realized. A photovoltaic system requires no additional electric energy to operate and has few moving parts. Clearly, the photovoltaic compression cooling option has significant promise and represents a challenge to thermally drive cooling processes.

The simple payback period is the ratio of the installed cost to the value of energy saved in the first year of operation. The value of energy saved depends on assumptions regarding future escalation in energy costs. Table 22.10 shows the location, energy saved, first-year energy savings in the year 2000, estimated incremental system costs, and simple payback for different system types, based on the results of the Active Program Research Requirements assessment (Scholten and Morehouse 1983). Since the focus of the study was the evaluation of cooling system performance, systems with significant heating loads were not included in the analysis.

Under the economic conditions assumed in the analysis, simple payback periods of 20 to 25 years are projected for absorption systems, even of advanced types. A commercial Rankine system with power generation has a projected simple payback of 24 years. For desiccant systems, only the open-cycle solid desiccant/vapor-compression hybrid seems attractive on the basis of the present state of technology.

Table 22.10
Comparison of energy savings, year 2000 First-year energy cost, incremental system cost, and simple payback for light commercial (C) and residential (R) active solar cooling systems

System type	Location	C/R	Energy saved MBtu/yr.		NEPP-4 energy cost 1982$/MBtu		Year 2000 first-yr. 1982$/yr. savings	Incremental system cost (1982$)	Simple payback (yrs.)
			Electric	Gas	Electric	Gas			
Absorption cooling systems									
Open cycle	Wash., DC	R	−.3	50.2	$25.80	$10.37	$513	$19,700	38.4
Evap. cooled gas backup	Phoenix	R	27.8	1.3	$17.90	$8.67	$509	$13,100	25.7
Water-cooled gas backup	Phoenix	C	61.6	−29.0	$19.30	$8.63	$939	$20,400	21.7
Water-cooled electric backup	Phoenix	C	53.4	.0	$19.30	$8.63	$1,031	$23,800	23.1
Rankine cooling systems									
300°F evap. cooled with power generation	Phoenix	C	138.3	.0	$19.30	$8.63	$2,669	$65,000	24.4
300°F evap. cooled with power generation	Phoenix	R	15.6	.0	$17.90	$8.67	$279	$11,300	40.5
Desiccant cooling systems									
Open-cycle solid	Phoenix	R	35.2	−20.7	$17.90	$8.67	$451	$13,000	28.8
Open-cycle solid/vapor-compression hybrid	Phoenix	R	31.1	.0	$17.90	$8.67	$557	$7,500	13.5
Open-cycle solid/vapor-compression hybrid	Wash., DC	C	81.1	.0	$27.00	$9.79	$2,190	$23,000	10.5
Open cycle liquid	Wash., DC	C	76.1	.0	$27.00	$9.79	$2,055	$78,200	38.1

Adapted from Scholten and Morehouse 1983.

22.4.2   Cost of Delivered Energy

The cost of delivered energy has been used in comparisons of different solar heating and cooling technologies by estimating the annual energy delivered by the system and by allocating the capital cost for the system at a yearly rate. In this case, no assumptions need to be made about the cost of competing fuel, but realistic assumptions must be made about the cost of the solar systems. Löf (1983) has examined the unit cost of delivered solar heat for domestic water alone and for combined water and space heating. Figure 22.10 shows the unit cost of delivered solar heat in $/GJ as a function of solar fraction for domestic hot water systems and for combined space heating and domestic hot water systems in Denver. These costs may be compared with electricity at $0.06/kWh ($16.7/GJ) and $0.10/kWh ($27.8/GJ); natural gas at $0.20/m^3 and 60% efficiency ($8.9/GJ), and heating oil at $0.30/liter ($13.9/GJ). In general, the space heating

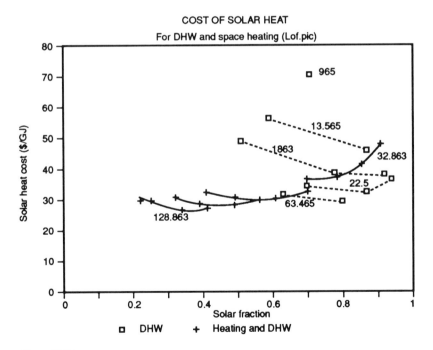

Figure 22.10
Solar energy expansion diagram showing the unit cost of delivered solar energy for domestic hot water and space heating systems in Denver. Adapted from Lof 1983.

systems have a lower delivered unit energy cost at smaller solar fractions. At these costs, solar energy for space and water heating in Denver is competitive with electric resistance heat for supply of limited load fractions.

In a study of solar cooling costs (Bankston 1986), absorption and desiccant systems have been compared. Solar energy expansion diagrams relating the cost of delivered heat and cooling to the annual solar fraction achieved by the system have been developed. The cost of delivered energy includes the annualized capital cost of the solar system using standard economic assumptions of a 20-year system life and a capital annualization factor of 0.0764 corresponding to a simple payback period of 13.1 years.

Figure 22.11(a) shows the unit cost of delivered solar heat in \$/GJ as a function of solar fraction. Single-effect, 25-ton and 5-ton absorption system and advanced absorption and desiccant systems are compared. The delivered solar energy cost depends on both the ability of the system to collect and use solar energy and on assumptions regarding the capital costs of these systems. The current average cost of delivered thermal energy from natural gas is about \$7.5/GJ (\$27/MWh or \$0.75/therm). Even the advanced solar systems have difficulty competing with conventional heat.

The cost of delivered cooling includes the annual average performance of the energy conversion device and is shown in figure 22.11(b). Because of the assumed higher thermal performance of advanced absorption and desiccant systems, the cost of cooling delivered by these systems is lower than that from single-effect absorption machines. The current national average cost of electricity is about \$0.085/kWh or \$24/GJ (\$85/MWh). Assuming an electrical COP of 4.0, the cost of delivered conventional cooling would be about \$6/GJ (\$20/MWh) plus the annualized capital cost of the chiller of about \$3/GJ (\$10/MWh). The advanced systems are not competitive with conventional cooling at today's prices but could possibly compete at higher electricity costs. Computing the unit cost of delivered solar heat or cooling over a range of solar fractions provides a method that allows comparison of widely differing technologies on an equal basis.

22.4.3 Improving Cost Effectiveness

The cost, performance, and reliability of all active solar technologies must be improved if solar is to play a major role in the marketplace. The cost of collecting solar energy must be reduced, and an advanced cooling technol-

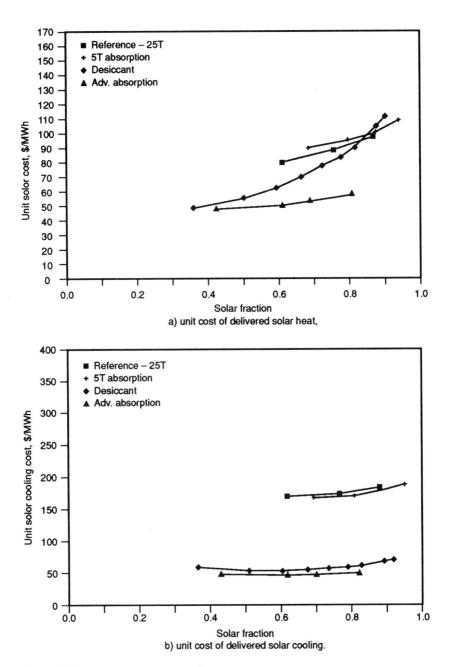

Figure 22.11
Solar energy expansion diagrams showing the unit cost of delivered solar heat and solar cooling for cooling technologies in Miami: (a) unit cost of delivered solar heat, (b) unit cost of delivered solar cooling. Adapted from Bankston 1986.

ogy must be developed if solar is to make a contribution to space cooling. Although today's active solar cooling systems generally produce only small energy savings and are quite expensive, the energy savings can be increased and the costs can be reduced. Four steps are required: (1) thermal performance improvement of collectors, chillers, and heat rejection; (2) electric performance improvement to reduce parasitic electric power consumption; (3) decrease of system component costs; and (4) system integration to simplify the system and reduce duplication of function.

The system thermal performance can be improved by increasing the absorption or Rankine chiller thermal performance, $COP_{chiller}$, thereby reducing the solar heat input, collector area, and heat rejection requirements for meeting a given fraction of the cooling load. A crucial issue in the development of cost-effective cooling systems is improving the operating electrical performance by the reduction of parasitic mechanical and electric energy consumption.

To improve the thermal performance of chillers, research is required. Thermodynamic cycles of absorption chillers for a variety of refrigerant-absorbent pairs must be developed and tested. Prototypes of advanced absorption chillers with improved thermal COP up to 1.55 must be built and refined on the basis of laboratory and field tests. The natural-gas backup should be integrated into the chiller design to reduce the number of system components by eliminating the backup chiller and to operate at maximum chiller efficiency in the gas-fired model. Advanced Rankine cycle air-conditioner systems must be designed, built, and tested to determine the extent of potential cost reduction, to identify reliability problems and solutions, and to develop advanced controls. Advanced solar desiccant and hybrid desiccant systems should be developed.

The electric operating energy of the chiller consists of energy to run pumps in the collector-storage loop and in the generator, solution, and chilled-water circuits in the absorption chiller; energy for pumps in the boiler and chilled-water loops in the Rankine system; and fan energy to move air through equipment in desiccant systems. Electric energy consumption can be reduced by proper design and sizing of pumps, piping, and heat exchangers in the collector loop and in the chiller.

Absorption chiller and Rankine turbocompressor costs can possibly be reduced by mass production and, packaging techniques. There is a need to achieve integration and standardization of components in active solar cooling systems to eliminate duplication of function and to minimize the number and size of system parts. Efficient preassembly of equipment can minimize the number of pumps and valves and reduce piping runs to a minimum. Use of direct-fire backup of an absorption chiller in both the cooling and heating modes will increase fuel efficiency and reduce total system energy consumption. System control costs can be reduced by application of integrated microcomputer control. Careful integration of the control of different system functions can minimize the number of sensors required, maximize reliability, and improve system performance.

The utilization of collected solar energy can be maximized if the system is designed for supply of space heating and hot water as well as cooling The domestic hot water load is relatively constant throughout the year. The space heating load and the summer cooling load cover the winter and summer. For Rankine systems that develop shaft power, the addition of electric power generation will improve solar utilization. Thermally fired heat pumps have the potential to operate with high efficiencies in the heating mode and provide cooling at a small additional cost.

Development and improvement of low-cost, high-performance, compound parabolic concentrating (CPC) evacuated-tube solar collectors and low-cost, high-performance, linear tracking parabolic trough concentrating solar collectors are needed. Lightweight but rigid materials that can be used for the parabolic reflector in concentrating collectors should be developed. Advanced evacuated-tube collectors with internal parabolic concentrating reflectors will have improved performance characteristics, particularly at the high temperatures needed for some of the active solar cooling applications (Winston and O'Gallagher 1981). At operating temperatures of 300°F, low-cost parabolic trough collectors have also been developed (Gee 1983).

The installed cost of high-performance, compound parabolic, evacuated-tube or linear-trough concentrating collectors should be reduced to about $9 to $10 /ft^2 if the value of solar heat collected is to be competitive with natural gas. The current costs of manufacturing, distribution, and installation of residential water heating and space conditioning systems using flat-plate and evacuated tube collectors are high. Based on an examination of the cost-reduction opportunities and of the range of system costs that may be anticipated by a mature solar industry, one study (Jacobsen and Ackerman 1981) predicts a significant reduction in the cost of flat-plate and evacuated-tube collectors as production increases (figure 22.12).

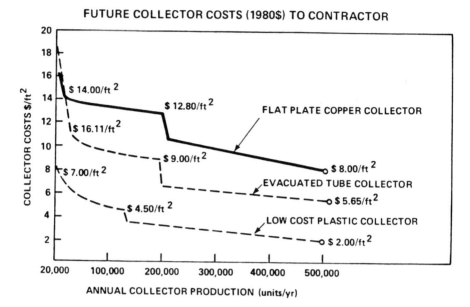

Figure 22.12
Profected collector costs through distributor markup (in 1980 dollars) as a function of annual collector production. Adapted from Haas and Jacobsen 1980.

The research on solar cooling systems has developed technology that, when operated by a fossil-fuel source, may compete seriously with solar-driven systems. Solar cooling research has therefore helped to develop its own competition. Efficient cooling technologies will help reduce our dependence on fossil fuel. The opportunity for solar energy to displace fossil fuel will ultimately depend on the cost of the solar systems that collect thermal energy for use as a heat source.

References

Aasen, R. K. 1982. *Second-generation solar Rankine cycle air conditioning system study.* Honeywell Inc. final report.

———. 1983. *Final report: Advanced solar heating and cooling system design and development (404) program.* Honeywell Inc.

Bankston, C. 1986. *Comparative systems analysis final report.* DOE DE-AC03–84SF 12217.

Biancardi, F. R., et al. 1982. Development and test of solar Rankine cycle heating and cooling systems. *ASHRAE Transactions* 88, part 1, paper HO-82–8 no. 1.

Choi, M., and J. Morehouse. 1982. *Analysis of commercial and residential solar absorption and Rankine cooling systems.* Science Applications, Inc., draft final report, contract DE-AC03–81SF11573.

Curran, H. M. 1975. *Assessment of solar-powered cooling of buildings.* Hittman Associates, Inc., final report COO/2622–1.

Curran, H. M. 1977. Coefficient of performance of solar-powered space cooling systems. *Solar Energy* 19(1977): 601.

———. 1981a. *On-site solar thermal electric power generation.* Hittman report.

———. 1981b. Solar/fossil Rankine cooling. *Proc. 16th IECEC,* Atlanta, GA, August 1981.

Gee, R. C. 1983. *Low-temperature IPH parabolic troughs: Design variations and cost-reduction potential.* SERI/TR-253–1662.

Gossler, A. A., et al. 1980. *Solar space and hot water heating systems with power generation: Performance and economic report.* Honeywell Inc. report.

Hass, S. A., and A. S. Jacobsen. 1980. *Evaluation of residential and commercial solar/gas heating and cooling technologies.* Gas Research Institute report GRI-79/0150, vol. 1.

Hass, S. A., and K. D. Hall. 1982. *Competitive assessment of desiccant solar/gas systems for single family residences.* Booz, Allen & Hamilton Inc. final report for Gas Research Institute, no. GRI-81/0063.

Hedstrom, J. C. 1983. *Performance of a solar air conditioning system utilizing boiling collectors.* LANL report LA-9724.

Howe, R. R., W. A. Beckman, and J. W. Mitchell. 1983a. Factors influencing the performance of commercial hybrid desiccant air conditioning systems. *Proc. 18th IECEC,* pp. 1940–45.

———. 1983b. Commercial applications for solar hybrid desiccant systems. *Proc. 1983 Annual Meeting of AS/ISES.*

Hughes, P. J., et al. 1981. Status of several solar systems with respect to cost and performance goals. SAI draft final report.

Huskey, B., et al. 1982. *Advanced solar/gas desiccant cooling system.* Exxon final report for Gas Research Institute, no. GRI-81/0064.

Jacobsen, A. S., and P. D. Ackerman. 1981. Cost reduction projections for active solar systems. *Proc. AS/ISES Conference,* Philadelphia, pp. 1291–95.

Jurinak, J. J., J. W. Mitchell, and W. A. Beckman. 1984. Open cycle desiccant air conditioning as an alternative to vapor compression cooling in residential applications. *J. Solar Energy Eng.*

Karaki S., P. Vattano, and P. Brothers. 1984. *Performance evaluation of a solar heating and cooling system consisting of an evacuated-tube heat pipe collector and air cooled lithium bromide chiller.* Report SAN-30569–32, Colorado State University, Fort Collins.

Khalifa, H. E., et al. 1983. *Limitations and potential improvements of heat engine driven heat pumps.* UTRC project no. 151358.

Khalifa, H. E., and G. Melikian. 1982. *Solar/gas Rankine/Rankine-cycle heat pump assessment.* UTRC report R82–955621.

Leboeuf, C. 1980. *Standard assumptions and methods for solar heating and cooling system analyses.* SERI/TR-351–402.

Lenz, T. G., 1982. *Investigation of the removal of non-condensable gases in an open cycle lithium chloride absorption air conditioner.* Final report, Colorado State University.

Lior N. 1977. Solar energy and the steam Rankine cycle for driving and assisting heat pumps in heating and cooling modes. *Energy Converion* 16(1977):

Lior, N., H. Yeh, and I. Zinnes. 1980. Solar cooling via vapor compression cycles. *Proc. AS/ISES 1980 Annual Meeting*, Phoenix, AZ, pp. 210–13.

Lior N., and K. Koai, 1984a. Solar-powered/fuel assisted Rankine cycle power and cooling system: Sensitivity analysis. *J. Solar Energy Eng.* 106(1984):

————. 1984b. Solar powered/fuel-assisted Rankine cycle power and cooling system: Simulation method and seasonal performance. *J. Solar Energy Eng.* 106 (1984): 142–52. A briefer version also published as ASME paper no. WA/Sol-8, 1982.

Logee, T., and P. Kendall. 1983. *Comparative report: Performance of active solar space cooling systems, 1982 cooling season.* U.S. DOE National Solar Data Program, SOLAR/0005-83/40.

Löf, G. O. G. 1983. Solar heating economics—a simple and rational basis for replacing fuel with solar energy. *Proc. ISES Solar World Congress*, Perth, Australia, p. 256.

Löf, G. O. G., M. A. Westhoff, and S. Karaki. 1984. *Performance evaluation of an active solar cooling system utilizing low cost plastic collectors and an evaporatively-cooled absorption chiller.* Report SAN-30569–30, Colorado State University, Fort Collins.

Löf, G. O. G., G. Cler, and T. Brisbane. 1988. Performance of a solar dessicant cooling system. *J. Solar Energy Eng.* 110(1988): 165.

McCormick, P. O., S. R. Brown, and S. P. Tucker. 1983. *Performance of a glazed open flow liquid desiccant solar collector for both summer cooling and winter heating.* Final report, Lockheed, LMSC-HREC TR D867353.

NEPP-IV (National Energy Policy Plan). 1983. *Energy projections to the year 2010.* DPE/PE-0029/2, draft.

NSDW (National Solar Data Network). 1981. *Comparative report: Performance of active solar space cooling systems, 1980 cooling season.* SOLAR/0005-81/40.

OTA (Office of Technology Assessment). 1983. *US natural gas availability, conventional gas supply through the year 2000.* Technical memorandum, September 1983.

Pogany, D., D. S. Ward, and G. O. G. Löf. 1975. Economics of solar heating and cooling systems. Paper presented at the Annual Meeting of the International Solar Energy Society, Los Angeles, CA, 1975 Solar Energy Congress and Exposition.

Robison, H. I. 1983. *Open-cycle chemical heat pump and energy storage system.* Final report, University of South Carolina.

SAI (Science Applications, Inc.). 1983. *Active solar component cost catalog, version 3.* Prepared by JRB Associates/Science Applications, Inc.

Schlepp, D. R., and K. J. Schultz. 1984. High Performance Solar Desiccant Cooling Systems: Performance Evaluation and Research Recommendations, SERI/TR-252-2497, Sept.

Scholten, W. B., and J. H. Morehouse. 1983. *Active program research requirements.* Final report, Science Applications, Inc., DOE/SF/115 73T1 (vol. 1).

SERI (Solar Energy Research Institute). 1982. *APRR: Initial review and system level ranking—desiccant systems.* SERI APRR submission, 14 May 1982.

Shelpuk, B. C., et al. 1979. *Simulation and economic analyses of desiccant cooling systems.* SERI report SERI/TR-34–090.

Sheridan, J. C., and J. W. Mitchell. Hybrid solar desiccant cooling systems. Proc. of the 1982 Annual Meeting of AS/ISES, 585–590.

Tchernev, D. 1981. *The development of a low cost integrated zeolite collector: Final report for the period September 25, 1978—September 24, 1980*. The Zeopower Company, DOE/SAN/ 2117-1.

Warren, M. L., and M. A. Wahlig. 1983. Methodology to determine cost and performance goals for active solar cooling systems. LBL-12343. *J. Solar Energy Eng.* 105(1983): 217–23.

Warren M. L., and M. A. Wahlig. 1982. Cost and performance goals for active solar absorption cooling systems. Paper presented at the ASME Solar Energy Technical Conference, Albuquerque, NM, April 1982, LBL-13050. *J. Solar Energy Eng.*

Warren, M. L., and H. S. Liers. 1983. Research goals for active solar cooling systems. Paper presented at the ASES Meeting, Minneapolis, MN, June 1983.

Warren, M. L. 1982a. *Active Program Research Requirements for water and air cooled closed cycle commercial absorption systems*. LBID-651.

———. 1982b. *Active Program Research Requirements for residential solar absorption and Rankine cooling systems*. LBID-652.

———. 1982c. *Active Program Research Requirements for commercial solar Rankine cooling systems*. LBID-650.

Wetzel, P., and P. Pakkala. 1982. *Comparative Report: performance of active solar space cooling systems, 1981 cooling season*. U.S. DOE National Solar Data Program, SOLAR/0023-82/40.

Winston, R., and J. O'Gallagher. 1981. Engineering development studies for intergrated evacuated CPC arrays. *Proc. Active System Heating and Cooling Contractor's Review Meeting*, Washington, DC, September 1981.

Contributors

William A. Beckman

William A. Beckman is the Ouweneel-Bascom Professor of Mechanical Engineering, director of the Solar Energy Laboratory, and former chairman of the mechanical engineering department at the University of Wisconsin-Madison. He has been involved in solar energy research since 1963. He has worked for the Chrysler Corporation, the Jet Propulsion Laboratory in Pasadena California, and the CSIRO Division of Mechanical Engineering in Melbourne, Australia. His research is in the systems analysis area, which has led to the development of the TRNSYS simulation program and the F-Chart solar system design method. He received his BS, MS, and PhD degrees from the University of Michigan.

Dwayne S. Breger

Dwayne S. Breger is currently a senior research associate in the department of mechanical engineering at the University of Massachusetts at Amherst. Between 1982 and 1988 he was a U.S. participant in the International Energy Agency, Task on Central Solar Heating Plants with Seasonal Storage. Since 1988, Breger has led an effort at the University of Massachusetts to develop a first national CSHPSS project. His research also has involved seasonal storage of winter ice for cooling applications. He received a BS degree in engineering from Swarthmore College and an M.S. in technology and policy from the Massachusetts Institue of Technology.

Henry M. Curran

Henry M. Curran received his bachelor's degree from the University of British Columbia, his master's and doctoral degrees from the University of Texas, all in mechanical engineering. His professional career includes service as a university professor and administrator, and as a consulting engineer. His areas of expertise include solar systems, refrigeration, turbomachinery, water desalting, and controls. He is the author or coauthor of over fifty papers and reports. In the American Society of Mechanical Engineers he is a Life Fellow, and has served as chairman of the Solar Energy Division and as an associate editor of the *Journal of Solar Energy Engineering*. He is a registered professional engineer in Texas and Florida.

John A. Duffie

John A. Duffie has been involved with many aspects of solar energy research since 1954. He is now emeritus professor of chemical engineering at the University of Wisconsin-Madison, was a founder (with Farrington Daniels) of the Solar Energy Laboratory there, and was its director from 1954 to 1988. He has also been affiliated with the University of Queensland, CSIRO (Australia), and the University of Birmingham, UK. His group has developed TRNSYS, the solar process simulation program, FCHART, the solar heating design program, utilizability methods for thermal design, and has done wide-ranging studies of solar process systems, their design, and economics. He and his colleagues are authors of several books, including *Solar Energy Thermal Processes, Solar Heating Design*, and *Solar Engineering of Thermal Processes*. He holds BChE and MChE degrees from Rensselaer Polytechnic Institute and a PhD from the University of Wisconsin.

Sharon M. Embrey

Sharon (Sherry) M. Embrey is the assistant department head of the support department at Vitro Corporation, a large systems integration and engineering firm. From 1979 to 1984, Sherry was the group supervisor of the energy technology project. The group analyzed active, passive, and hybrid solar installations for over 170 commercial and residential sites in the DOE National Solar Data Network. Performance reports and related papers were presented at such conferences as ASME, ASHRAE, and AS/ISES. The group performed similiar contracts with several other companies. Sherry has a BS in mechanical engineering from the University of Rhode Island and has completed all the course work for her masters in mechanical engineering.

Susumu Karaki

Susumu Karaki is professor emeritus of civil engineering at Colorado State University, specializing in hydraulics, fluid mechanics, and solar energy applications. From 1977 to 1989, he was the director of the Solar Energy Applications Laboratory at Colorado State. Karaki's honors include a NASA Fellowship awarded in 1969, the Gleddon Fellowship at the Univ. of Western Australia in 1981, and the 1984 Abbot Award of the American Solar Energy Society. He served as vice president of the International Solar Energy Society, and is a member of several other professional societies. He received a doctorate in civil engineering from Stanford University in 1968.

Sanford A. Klein

Professor Klein received his PhD degree in chemical engineering at the University of Wisconsin-Madison in 1976. He is currently a professor in mechanical engineering at the University of Wisconsin. His research work has dealt with analysis and design of solar energy systems and simulation methodology. He is the author or co-author of more than fifty technical papers in this field that have been published in *Solar Energy, Transactions of the American Society of Mechanical Engineers, International Journal of Heat and Mass Transfer, Transactions of American Society of Heating, Refrigeration and Air Conditioning Engineers*, and elsewhere. He is co-author (with W. A. Beckman and J. A. Duffie) of *Solar Heating Design* (Wiley Interscience, 1977). He is a member of American Society of Mechanical Engineers, American Society of Heating, Refrigeration and Air Conditioning Engineers, and International Solar Energy Society. He was a visiting scientist at the National Bureau of Standards in 1983.

In 1983, the Solar Energy Laboratory, in which Professor Klein is a principal investigator, received the first Achievement Through Action Award of the International Solar Energy Society.

Noam Lior

Noam Lior is professor and graduate group chairman of mechanical engineering and applied mechanics at the University of Pennsylvania. He has been intensively involved in active, passive, and solar thermal energy research since 1973. Some of the specific projects include heat transfer and fluid mechanics of solar collectors and salt-gradient solar ponds, heat transfer in phase change thermal storage, development of a novel solar-powered/fuel-assisted hybrid Rankine cycle, open-cycle OTEC, solar heating and cooling, solar retrofit, and applications of solar energy in the urban environment. He designed and brought into operation the first solar-heated home in Philadelphia (a retrofit), and served as director of the American Solar Energy Society, chairman of its engineering division, and associate editor of the ASME *Journal of Solar Energy Engineering*. In the solar energy field he was consultant to ERDA, DOE, Argonne National Laboratory, and the Solar Energy Research Institute. He has published over eighty papers and is the editor of *Measurements and Control in Water Desalination* (Elsevier, 1986).

George Löf

George Löf has been involved with active solar energy research since 1942. Dr. Löf has been professionally affiliated with the University of Colorado, University of Denver, University of Wisconsin, and the Solaron Corporation. He established and directed the Solar Energy Applications Laboratory at Colorado State University, where he is now Professor Emeritus.

He directed one of the nation's principal research programs in applying solar energy to the heating and cooling of buildings; pioneered and developed solar air heating technology; developed and marketed commercial systems for solar space heating; authored 200 research papers and articles in this field.

Dr. Löf co-authored books on *Technology in American Water Development, The Economies of Water Use in the Beet Sugar Industry, Water Demand for Steam Electric Generation, Manual on Solar Distillation of Saline Water, Solar Heating and Cooling of Buildings—Design of Systems; Sizing, Installation, and Operation of Systems.*

John W. Mitchell

Professor Mitchell received his BS (1956), MS (1957), Engineer (1959), and PhD (1963) from Stanford University. He has been on the faculty of the mechanical engineering department of the University of Wisconsin since 1962 and was the chairman of the department from 1983 to 1987. In 1981, he was a visiting scientist in the mechanical engineering division of the Commonwealth Scientific and Industrial Research Organization— Australia. In 1987, he was a visiting researcher at Ecole des Mines, Paris, and University of Liege, Belgium.

His areas of interest are thermodynamics and heat transfer applied to solar and building energy systems. He has conducted research and published about 130 articles on topics such as building energy use, energy management systems, and solar cooling systems. His recent studies are on optimal control strategies for HVAC systems. In 1983, he authored the book *Energy Engineering.* Professor Mitchell is a member of ASME, ASHRAE, and Sigma Xi, and is a registered professional engineer of the state of Wisconsin. He has consulted for heat exchanger companies on heat exchanger designs, Alaska and New Zealand on energy use, and several companies on energy utilization.

In 1983, the Solar Energy Laboratory, in which Professor Mitchell is a principal investigator, received the first Achievement Through Action Award of the International Solar Energy Society. He received the Amoco Award for Distinguished Teaching in 1976 and the Pi Tau Sigma Distinguished Teaching Award of the Department of Mechanical Engineering in 1967, 1970, 1973, 1974, and 1983.

Jeffrey H. Morehouse

Jeff Morehouse is an associate professor of mechanical engineering at the University of South Carolina. From 1978 to 1984 Morehouse was a division manager with Science Applications International Corporation in Washington, D.C., where he managed major analytical and support contracts for the DOE Active Solar Program involving system simulations and economic analyses. Prior to that, Morehouse taught at Texas A & M University for two years, spent four years in the Navy as a nuclear power instructor, and worked for two years in the IBM office products plant in Austin, Texas. He received a PhD from Auburn University, an MS from Rice University, and a BS from Rutgers University, all in mechanical engineering.

Jefferson Shingleton

Jefferson Shingleton is a private consulting engineer providing industrial, commercial, government, and private clients with a range of renewable energy system services. Since 1978 Mr. Shingleton has specialized in providing research and development services to renewable energy system manufacturers, suppliers, and fabrication and installation contractors in the areas of product development; structural system design and integration; site analysis and field investigations; system fabrication and construction; and structural analysis, testing, and certification. He recieved a BS in civil engineering from the West Virginia University.

Charles Smith

Charles Smith has been involved with active solar energy since 1973. He has been a research associate in the civil engineering department at Colorado State University since that time; his areas of specialization include solar energy for heating, cooling, and agriculture; heat transfer; fluid mechanics; control systems; and electrical and mechanical equipment. He received his MS in mechanical engineering in 1968 from the University of Wisconsin, and his BS in 1966 from Michigan Technological University, also in mechanical engineering. He is a member of the American Society of Mechanical Engineers, the International Solar Energy Society, and the American Society of Heating, Refrigerating and Air-Conditioning Engineers. He has presented over twenty papers in eight countries.

Elmer R. Streed

Elmer R. Streed has been involved with active/passive solar energy research since 1960. Mr. Streed has been professionally affiliated with the building research division, National Bureau of Standards (research engineer) (1974–1980): Lockheed Palo Alto Research Laboratories (staff scientist). (1960–1964 and 1970–1973); Ames Research Center, NASA (Chief, Vehicle Systems Design Branch) (1964–1970). His major accomplishments in solar energy research include the development and evaluation of standard test methods for the thermal performance and durability of solar collectors. He contributed to the development of data requirements, performance criteria, and standards for components and systems for the National Solar Heating and Cooling Demonstration Program. Mr. Streed served as chairman of ASTM E21:10 Subcommittee on Solar Energy Heating and Cooling Applications from 1975 to 1980.

At LMSC he initiated the design and evaluation of honeycomb and selective surface absorbers for solar collectors, and developed apparatus to measure the thermophysial properties of solar absorber and thermal radiation surfaces. At NASA he conducted applied research for the development of materials and devices for spacecraft thermal-control and on-board space power, Mr. Streed served as general and technical Chairman, AIAA Thermophysics Conference, San Franciseo, 1969.

His other publications include contributions to the *Durability of Building Materials and Components*, ASTM, 1980; *Thermophysics of spacecraft and Planetary Bodies*, Academic Press, 1967; *Thermophysics and Temperature Control of Spacecraft and Entry Vehicles*, Academic Press, 1966.

Michael A. Wahlig

Michael Wahlig is a senior scientist at the Lawrence Berkeley Laboratory. From 1976 to 1986 he was leader of the solar energy program within the applied science division at LBL; he currently serves as leader of the building energy systems program, which includes solar buildings research. He has been principal investigator for solar cooling research activities at LBL since 1975, concentrating primarily on absorption cooling technology. He also provided technical program support to the U.S. Department of Energy Solar Cooling Program from 1979 to 1984. He holds a PhD in physics from the Massachusetts Institute of Technology.

Mashuri L. Warren

Mashuri L. Warren, P. E. is Product Manager for ASI Controls, a developers of unitary and application-specific direct digital controls for the heating, ventilating, and air conditioning industry. Previously he was a staff scientist in the building energy system program in the applied science division of Lawrence Berkeley Laboratory. Since 1980 he has headed the systems simulation and analysis group for the active solar cooling project at LBL and has actively participated in the DOE Active Program Research Assessment. From 1978 to 1980 Warren was responsible for the solar controls test facility at LBL. Before joining the LBL staff, Warren taught at California State University, Hayward, and at San Francisco State University. He received a PhD in physics from the University of California, Berkeley.

C. Byron Winn

C. Byron Winn has been involved with active and passive solar energy research since 1973. Professor Winn has been professionally affiliated with Colorado State University, Stanford University, the University of Colorado, Santa Clara University, the University of Texas, the University of Newcastle, Lockheed Missiles & Space Co., Martin-Marietta, Office National D'Etudes et de Recherches Aerospatiales, Solar Control Corporation, Western Solar Corporation, Solar Environmental Engineering Co., ECO-ERA.

His major accomplishments in solar energy research include the development of controllers and control strategies, software development and testing (SIMSHAC, SOLCOST, etc.), design and construction of a reconfigurable passive solar research facility (REPEAT), and starting of three solar companies. Other publications include (with Hunn), "Active System Design and Sizing Methods," chapter 10, Solar Design Workbook, SERI, May 1980; (with F. Dubin and B. Hunn), "Design Development," chapter 14, Solar Design Wookbook, SERI, May 1980; (with J. Hayes). *Progress in Passive Solar Energy Systems*, American Solar Energy Society, 1982; and "Controls in Solar Energy Systems," *Advances in Solar Energy*, ASES. 1982.

Ronald M. Wolosewicz

Ronald M. Wolosewicz is a principal engineer with the Rockwell Graphic Systems Division of Rockwell International in Westmont, Illinois. Prior to joining Rockwell, he was associated with Argonne National Laboratory. While at ANL, he was active in the thermal energy storage area and supervised the reliability activities associated with DOE's Solar Data Project. He participated in the ASHRAE task force that wrote the standard for electrically charged thermal energy storage devices, was a major contributor to ANL's Reliability and Design Guide-lines for Solar Domestic Hot Water Systems, the Reliability and Design Guide-lines for Combined Space-heating and Domestic Hot Water, wrote the rock-bed storage chapter for ANL's Design and Installation Manual for Thermal Energy Storage Systems. He has published extensively in refereed engineering journals. He is a registered professional engineer in Illinois, was a Fulbright-Hayes post-doctoral fellow at the Polish Academy of Sciences in Warsaw, elected to Pi Tau Sigma, active in the design education subcommittee of ASME's design division, chairman of ASME-Triodyne Safety Award Selection Committee and listed in *Who's Who in the Midwest*. He received a PhD in mechanical engineering from Northwestern University.

Index